W0245957

Progress in Mathematics
Volume 115

Integrable Systems

The Verdier Memorial Conference

Actes du Colloque International de Luminy

Olivier Babelon
Pierre Cartier
Yvette Kosmann-Schwarzbach
Editors

Springer Science+Business Media, LLC

Olivier Babelon
CNRS-Université de Paris VI
L.P.T.H.E. Tour 16
4, place Jussieu
75252 Paris Cedex 05
France

Pierre Cartier
Ecole Normale Supérieure de Paris
15 rue d'Ulm
75005 Paris Cedex
France

Yvette Kosmann-Schwarzbach
Centre de Mathématiques
Ecole Polytechnique
91128 Palaiseau
France

Library of Congress Cataloging-in Publication Data

Integrable systems : in memory of Jean-Louis Verdier : actes du colloque international de Luminy / Olivier Babelon, Pierre Cartier, Yvette Kosmann-Schwarzbach, editors.
p. cm. -- (Progress in mathematics : v. 115)
Includes bibliographical references and index.

1. Geometry, Differential--Congresses. 2. Hamiltonian systems--Congresses. I. Verdier, Jean-Louis. II. Babelon, Olivier, 1951-. III. Cartier, P. (Pierre) IV. Kosmann-Schwarzbach, Yvette, 1941-. V. Series: Progress in mathematics ; vol. 115.
QA641.I55 1993 93-40972
514'.74--dc20 CIP

Printed on acid-free paper

Birkhäuser 

Typeset by the Authors in TEX and AMSTEX

ISBN 978-1-4612-6703-4 ISBN 978-1-4612-0315-5 (eBook)
DOI 10.1007/978-1-4612-0315-5

9 8 7 6 5 4 3 2 1

Jean-Louis Verdier (1935–1989)
Photograph by Joëlle Pichaud

Contents

Preface ix

Introduction

Hommage à Jean-Louis Verdier: au jardin des systèmes intégrables
D. Bennequin 1

Part I. Algebro-Geometric Methods and τ-Functions

Compactified Jacobians of Tangential Covers
A. Treibich 39

Heisenberg Action and Verlinde Formulas
B. van Geemen and E. Previato 61

Hyperelliptic Curves that Generate
Constant Mean Curvature Tori in $\mathbb{R}^3$
N. M. Ercolani, H. Knörrer, and E. Trubowitz 81

Modular Forms as τ-Functions for Certain Integrable
Reductions of the Yang-Mills Equations
L. A. Takhtajan 115

The τ-Functions of the $\mathfrak{g}$AKNS *Equations*
G. Wilson 131

On Segal-Wilson's Definition of the τ-Function
and Hierarchies AKNS-D *and* mcKP
L. A. Dickey 147

The Boundary of Isospectral Manifolds, Bäcklund Transforms
and Regularization
P. van Moerbeke 163

Part II. Hamiltonian Methods

The Geometry of the Full Kostant-Toda Lattice
N. M. Ercolani, H. Flaschka, and S. Singer 181

Deformations of a Hamiltonian Action of a Compact Lie Group
V. Guillemin 227

Linear-Quadratic Metrics "Approximate" any Nondegenerate, Integrable Riemannian Metric on the 2-Sphere and the 2-Torus
A. T. Fomenko 235

Canonical Forms for Bihamiltonian Systems
P. J. Olver 239

Bihamiltonian Manifolds and Sato's Equations
P. Casati, F. Magri, and M. Pedroni 251

Part III. Solvable Lattice Models

Generalized Chiral Potts Models and Minimal Cyclic Representations of $U_q(\widehat{\mathfrak{gl}}(n, C))$
E. Date 275

Infinite Discrete Symmetry Group for the Yang-Baxter Equations and their Higher Dimensional Generalizations
M. Bellon, J.-M. Maillard, and C. Viallet 277

Part IV. Topological Field Theory

Integrable Systems and Classification of 2-Dimensional Topological Field Theories
B. Dubrovin 313

List of Participants 361

Index 365

Preface

This book constitutes the proceedings of the International Conference on Integrable Systems in memory of J.-L. Verdier. It was held on July 1–5, 1991 at the Centre International de Recherches Mathématiques (C.I.R.M.) at Luminy, near Marseille (France). This collection of articles, covering many aspects of the theory of integrable Hamiltonian systems, both finite- and infinite-dimensional, with an emphasis on the algebro-geometric methods, is published here as a tribute to Verdier who had planned this conference before his death in 1989 and whose active involvement with this topic brought integrable systems to the fore as a subject for active research in France.

The death of Verdier and his wife on August 25, 1989, in a car accident near their country house, was a shock to all of us who were acquainted with them, and was very deeply felt in the mathematics community. We knew of no better way to honor Verdier's memory than to proceed with both the School on Integrable Systems at the C.I.M.P.A. (Centre International de Mathématiques Pures et Appliquées in Nice), and the Conference on the same theme that was to follow it, as he himself had planned them. D. Bennequin, P. Cartier and A. Chenciner agreed to join O. Babelon and Y. Kosmann-Schwarzbach to form a new organizing committee, chaired by P. Cartier. The final list of speakers at the Conference was very close to the original list of invitations discussed with Verdier himself, and the invited participants included ten students chosen from among those who had attended the C.I.M.P.A. School, as originally planned by Verdier.

The refereed articles in this volume represent the advances in the field of complete integrability that were reported at the Luminy conference. In many cases, articles have been updated for publication. In two instances, where the results had been previously published, only summaries with references appear here. The articles represent very diverse methods and report very diverse results. This is a reflection of the complexity and richness of the field of research that is referred to as the theory of completely integrable systems.

In his preliminary text, D. Bennequin takes us into "the garden of integrable systems." He surveys the evolution of the subject, from Abel onwards, explaining the connections with the classical theory of elliptic functions, describing how algebraic curves and the infinite Grassmannian came to play a prominent role in the theory. He then analyzes the important contributions that Verdier made in this area.

The first part of this book contains articles that make essential use of Riemann surfaces and their theta functions in order to construct classes of solutions of integrable systems, and articles dealing with the tau-functions that generalize the classical theta functions. The first three papers in this part exemplify the algebro-geometric methods, while the next four deal more specifically with the tau-functions in their various guises.

A. Treibich, who was a close collaborator of Verdier, studies the family of elliptic solitons of the Kadomtsev–Petviashvili hierarchy associated with a projective curve, showing that if Γ is a tangential cover of an elliptic curve E, then its compactified Jacobian covers a symmetric power of E.

In an article written with B. van Geemen, E. Previato, who also collaborated with Verdier, describes recent work on the space of higher-order nonabelian theta functions over a Riemann surface of genus at least two, whose dimension is given by the Verlinde numbers, arising in the fusion rules of conformal field theory, and some connections of nonabelian theta functions with the Schottky problem.

The lecture of H. Knörrer, describing his work with N. Ercolani and E. Trubowitz, deals with the immersed submanifolds of $\mathbb{R}^3$ with constant mean curvature. The link with integrable systems lies in the fact that such immersions can be found by solving the elliptic sinh–Gordon equation, quasi-periodic solutions of which can be constructed by means of the Riemann theta function of hyperelliptic curves.

In the following articles the tau-functions of various integrable systems play a prominent part.

L. Takhtajan's paper shows that classical modular forms generate a tau-function for several integrable reductions of the self-dual Yang–Mills equations, and therefore general classes of solutions of many equations, in $0+1$, $1+1$, and $2+1$ dimensions.

G. Wilson reviews the generalized Ablowitz–Kaup–Newell–Segur equations associated with a simple Lie algebra, $\mathfrak{g}$, namely the evolution equations equivalent to the zero-curvature equation for a connection obtained from a "bare" connection by the dressing action of the associated loop group. He shows that, when $\mathfrak{g}$ is simply laced, solutions of the $\mathfrak{g}$AKNS equations can be obtained in terms of tau-functions, which he defines by means of the canonical trivialization of the fibration of the central extension of the loop group over the "big cell" in the loop group.

By redefining the tau-function of Segal and Wilson, L. Dickey is able to determine a tau-function for the hierarchies generated by matrix first-order differential operators which generalize the AKNS equations and for the multi-component KP hierarchies.

P. van Moerbeke studies the blowing-up of a solution of the Korteweg–de Vries equation with respect to a complexified time variable, and, more generally, the compactification of isospectral manifolds of differential operators. He then poses and answers analagous questions for the isospectral families of periodic Jacobi matrices.

In the second part of the book, the main emphasis is on the Hamiltonian formalism, and for the last two papers in this part, on the bihamiltonian formalism, while the first one is an illustration of the interplay of the Hamiltonian and the algebro-geometric methods in the study of integrability.

The joint work of N. M. Ercolani, H. Flaschka and S. Singer was presented in Flaschka's lecture on the geometry of the Kostant–Toda lattice, whose Lax matrix has entries below, on and just above the diagonal, the latter being all equal to 1. Using ideas from complex algebraic geometry, they prove the complete integrability of this system, and they determine its constants of motion, which are rational (rather than polynomial) functions on the phase-space.

Studying the vertex set of a Hamiltonian action of a compact commutative Lie group on a compact symplectic manifold, i.e., the image under the moment map of the set of fixed points, V. Guillemin proves a local rigidity theorem for Hamiltonian actions.

In his short contribution, A.T. Fomenko states a conjecture regarding the determination of all the integrable geodesic flows on two-dimensional compact manifolds.

P.J. Olver uses his classification of bihamiltonian structures based on the double Darboux theorem of Turiel in order to draw a list of canonical forms for bihamiltonian systems, and he obtains criteria for both the local and global integrability of such systems.

In the lecture of F. Magri, written in collaboration with P. Casati and M. Pedroni, the theory of soliton equations is presented from the Hamiltonian point of view, the hierarchies of bihamiltonian equations being generated by the Casimir functions of a pencil of Poisson brackets. It is shown that Sato's operator corresponds to the differential of a Casimir function expressed in terms of pseudodifferential operators, and Sato's equations to the vanishing of Poisson brackets on one-forms.

The third part of the book contains two papers that deal with the theory of two-dimensional solvable lattice models in which the quantum Yang–Baxter equation plays a fundamental role.

E. Date's contribution is a summary of his work on the finite-dimensional cyclic representations of the quantum groups at roots of unity and their relation with the two-dimensional lattice models associated with higher genus algebraic curves.

In his lecture, J.-M. Maillard presented his joint work with M. Bellon and C. M. Viallet, showing that the quantum Yang–Baxter equation admits a symmetry group which acts by birational projective transformations on the algebraic varieties which parametrize the solutions of the equation, and he discussed the generalization of these results to the tetrahedron and hyper-simplicial equations that are the analogues of the QYBE for lattices of dimension 3 or more.

In the concluding article, B. Dubrovin shows the interrelations of many aspects of the integrability of the hierarchies of Hamiltonian systems with topological field theory (TFT). He defines the limiting or "averaging" process of a hierarchy and of the tau-function which yields a system of partial equations whose coefficients induce a Frobenius structure on the invariant manifolds of the hierarchy, and hence a solution of the Witten–Dijkgraff–E. Verlinde–H. Verlinde (WDVV) equations from 2-dimensional TFT. The bihamiltonian structure of the integrable hierarchy is used in the calculation of higher genus corrections, and the WDVV equations are shown to specify the periods of the Abelian differentials of Riemann surfaces as functions on moduli spaces of these surfaces, so that both the bihamiltonian approach and the algebro-geometric methods enter the theory.

The Conference was funded by the C.I.R.M. and by the Société Mathématique de France. Additional support was provided by the French Ministry of Foreign Affairs, and grants were offered by the University of Paris VII, the U.F.R. de Mathématiques et Informatique of the University of Paris VII and the C.N.R.S. research unit U.R.A. 212 (Théories Géométriques). Financial support also came from the Fédération Française des Sociétés d'Assurances and is hereby gratefully acknowledged. We thank in particular our friend and colleague, Prof. M. Flato, who helped us with the fundraising even though he was not free to participate in the Conference.

It is a pleasure to thank the C.I.R.M. and its director, G. Lachaud, for a well-organized and very pleasant conference in the beautiful surroundings of Luminy, and the C.I.M.P.A. and its managing director, J.-M. Lemaire, for providing financial support for the students of the School on Integrable Systems who participated in the Conference.

We thank the editors of the series Progress in Mathematics for offering to publish this volume, and the staff of Birkhäuser Publishing Company for their help and efficient work.

The long-term planning and organizational work were performed with the secretarial assistance of Madame Claudine Roussel, from the University of Paris VII, who joined us at Luminy to handle some of the day-to-day needs of the conference. She also assisted us in the editing of this volume and the forthcoming volume of the courses given at the C.I.M.P.A. school. In the name of all the participants in the conference, we thank her for her expert and invariably cheerful collaboration.

Paris, February 1993
O. Babelon
P. Cartier
Y. Kosmann-Schwarzbach

INTRODUCTION

Hommage à Jean-Louis Verdier : au jardin des systèmes intégrables

Daniel Bennequin

Quel chemin mystérieux mène des théories cohomologiques et des catégories dérivées aux équations non linéaires et aux solitons elliptiques ? Jean-Louis Verdier a été sensible à la même beauté formelle en explorant la dualité de Poincaré et les systèmes complètement intégrables. Un des premiers en France il a noué des contacts avec les physiciens, dans son séminaire à l'Ecole Normale, autour des équations de Yang-Mills, du problème de Riemann-Hilbert, des matrices de transfert, de l'équation de Hill...

A partir de 1977, à côté d'études sur la correspondance de McKay, la transformation de Fourier géométrique et la monodromie modérée, il a publié sur les équations différentielles algébriques, les instantons, les algèbres de Lie affines et les systèmes hamiltoniens, les matrices S, les modèles σ, les groupes quantiques... Ses plus récentes contributions originales concernent les applications harmoniques de la sphère de dimension 2 dans la sphère de dimension 4 et les solutions doublement périodiques de KdV et KP.

Sans doute Jean-Louis Verdier (avec d'autres) se réappropriait-il ainsi le passé classique en remontant aux origines de la géométrie algébrique, à l'inspiration d'Abel, Jacobi et Riemann, mais en même temps, dans le même mouvement, il pouvait donner libre cours à sa volonté de classifier, de géométriser, d'algébriser dans la ligne montrée par Hilbert, Weil ou Grothendieck.

A-t-il vu les étranges liens qui unissent les topos et les cordes? Il devinait sûrement les structures à l'œuvre dans ces nouvelles théories des invariants, nées de rapprochement avec la théorie des particules, et qui sont des motifs secrets entre les grandes colonnes de l'algèbre.

Parmi tous les livres de mathématiques, Jean-Louis Verdier préférait les œuvres d'Abel. Dès le début, les fonctions abéliennes furent attachées à la solution des problèmes de Mécanique (voir les leçons de Jacobi sur la Dynamique) et leur présence reste un attrait des systèmes intégrables. Au départ de l'histoire que je veux raconter est un problème de Mécanique. Cette histoire n'est pas destinée aux spécialistes (à part peut être quelques remarques), ils la connaissent bien, je voudrais juste qu'elle donne un peu le goût des fruits qui attiraient Jean-Louis.

En 1971 F. Calogero découvrit un système de mécanique quantique complètement intégrable: celui de n points matériels $x_1, x_2, \ldots, x_n$, sur une droite soumis au potentiel

$$U(x_1, x_2, \ldots, x_n) = -2 \sum_{i \neq j} (x_i - x_j)^{-2} \tag{1}$$

C'est à dire que Calogero sut calculer jusqu'au bout son spectre d'énergie.

Quelques temps après, en 1975, J. Moser démontra l'intégrabilité complète du système de mécanique classique correspondant, celui dont le hamiltonien est

$$H_1(x_1, \ldots, p_1, \ldots) = \frac{1}{2} \sum_i p_i^2 - 2 \sum_{i \neq j} (x_i - x_j)^{-2} \tag{2}$$

C'est à dire que Moser sut trouver $n-1$ fonctions $H_2, \cdots, H_n$, des x et des p qui avec H_1 forment une famille de n intégrales premières du mouvement, indépendantes, en involution.

(La dynamique des observables classiques est régie par l'équation $\dot{f} = \{H_1, f\}$ où $\{f, g\} = \sum_i \frac{\partial f}{\partial p_i} \frac{\partial g}{\partial x_i} - \frac{\partial f}{\partial x_i} \frac{\partial g}{\partial p_i}$ est le crochet de Poisson; l'involutivité des H signifie que $\forall i, j,\ \{H_i, H_j\} = 0$ et leur indépendance est l'indépendance linéaire des champs de vecteurs hamiltoniens

$$\left(\frac{\partial H_i}{\partial p_1}, \ldots, \frac{\partial H_i}{\partial p_n}, -\frac{\partial H_i}{\partial x_1}, \ldots, -\frac{\partial H_i}{\partial x_n} \right)$$

sur un ouvert dense de l'espace des phases, de coordonnées $x_1, \ldots, x_n, p_1, \ldots, p_n$.)

En fait, le système de Calogero-Moser se laisse mettre sous la forme de Lax:

$$\frac{dL}{ds} = [M, L] \tag{3}$$

où M et L sont des matrices carrées $n \times n$; les coefficients diagonaux de L sont les impulsions $p_1, p_2, \ldots, p_n$, et en dehors de la diagonale $(L)_{ij} = 2/(x_i - x_j)$. Alors $H_k = \frac{1}{k+1} Tr(L^{k+1})$.

(Si l'on souhaite comprendre mieux les raisons de l'intégrabilité, on doit se tourner du côté des algèbres de Lie semi-simples et lire les articles de M. A. Olshanetsky et A. M. Perelomov en 1976. Par exemple, dans une Lettere al Nuovo Cimento (10 Luglio 1976) ils expliquent comment le système (2) se ramène à une partie du flot géodésique sur l'espace symétrique $SL_n(\mathbb{C})/SU_n$, et comment il se déforme jusqu'au jeu de billard dans une chambre de Weyl de sl_n. Depuis, bien des systèmes intégrables ont été reconnus sur les orbites coadjointes des groupes de Lie (Adler, Kostant, Symes, van Moerbeke...). Sur le procédé, avec des systèmes intégrables de dimension finie, mais à partir d'algèbres de Lie de dimensions infinies, citons l'exposé de J.-L. Verdier au Séminaire Bourbaki (n° 566, 1980-81) sur les travaux de M. Adler et P. van Moerbeke; on y trouvera le traitement moderne de référence de la toupie de Lagrange à partir d'une orbite coadjointe d'une algèbre de Kac-Moody.)

Dans leur (première) lettre de 76, Olshanetsky et Perelomov résolvent le problème de Cauchy posé par H_1: si la condition initiale est $(x_1(0), \ldots; p_1(0), \ldots)$ l'ensemble des $x_i(s)$ est le spectre de la matrice $X(s)$ où $(X)_{ij} = x_i(0)\delta_{ij} + s \cdot (L)_{ij}(0)$.
(*Remarque:* si $\forall i$, $x_i(0) \in \mathbb{R}$, $p_i(0) \in \sqrt{-1}\mathbb{R}$ et $x_1(0) > \cdots > x_n(0)$ l'évolution en temps imaginaire pur $s = \sqrt{-1}y$, $y \in \mathbb{R}$, garde les x_i réels et évite les collisions.)

En 1977, H. Airault, H. McKean et J. Moser, ainsi que les frères D.V. et G.V. Choudnovsky, découvrent la relation étonnante de (1) avec un autre système complètement intégrable (de dimension infinie cette fois): l'équation d'évolution non linéaire KdV (du nom de ses inventeurs, Korteweg et de Vries en 1895):

$$4\frac{\partial u}{\partial t} = 6u\frac{\partial u}{\partial x} + \frac{\partial^3 u}{\partial x^3} \tag{4}$$

où le potentiel $u(x)$ évolue avec le temps t. Toute solution de KdV rationnelle en x, pour toutes les valeurs de t (et nulle à l'infini) est de la forme:

$$u(x,t) = -2\sum_{i=1}^{n}(x - x_i(t))^{-2} \tag{5}$$

où les pôles $x_1(t), \ldots, x_n(t)$ doivent satisfaire aux équations

$$\forall\, i, \qquad \sum_{j\neq i}(x_i - x_j)^{-3} = 0 \tag{6}$$

et se déplacent suivant t selon les équations

$$\forall\, i, \qquad \dot{x}_i = 3\sum_{j\neq i}(x_i - x_j)^{-2}. \tag{7}$$

Or cela revient à dire que le mouvement des pôles de la solution est exactement celui que réclame le flot hamiltonien de $H_2' = -\frac{3}{4}H_2 = -\frac{1}{4}TrL^3$ en restriction au lieu critique de H_1. (En effet

$$H_2' = -\frac{1}{4}\sum_i p_i^3 + \sum_i \sum_{j\neq i}(2p_i + p_j)(x_i - x_j)^{-2};$$

le lieu critique de H_1 est défini par $\forall i,\ \ p_i = 0$ et $\frac{\partial U}{\partial x_i} = 0$; $\dot{x}_i = \frac{\partial H_2'}{\partial p_i} = -\frac{3}{4}p_i^2 + 3\sum_{j\neq i}(x_i - x_j)^{-2}$ donne (7) si $p_i = 0$ et $\dot{p}_i = -\frac{\partial H_2'}{\partial x_i} = -12\sum_{j\neq i}(p_i + p_j)(x_i - x_j)^{-3}$ préserve $\{\forall i,\ p_i = 0\}$.)

Réciproquement, si des fonctions $x_1, \ldots, x_n$ de t évoluent suivant le gradient symplectique de $-\frac{3}{4}H_2$ en restant dans l'ensemble où le gradient symplectique de H_1 s'annule (c'est à dire qu'elles vérifient (6) et (7)), la formule (5) donne une solution (rationnelle en x) de KdV (4).

Airault, McKean et Moser remarquent aussi que l'ensemble défini par (6) est non vide si et seulement si n est un nombre entier triangulaire: $n = \frac{m(m+1)}{2}, m \in \mathbb{N}^{\times}$.

(Les articles originaux sont:
H. Airault, H. McKean, J. Moser; *Rational and elliptic solutions of the KdV equation*, Comm. on pure and appl. Math., vol 30, pp 95-148 (auquel on se réferera par A.McK.M. dans la suite), et
D.V. Choudnovsky, G.V. Choudnovsky; *Pole expansion of non linear partial differential equations*, Il Nuovo Cimento, vol 40B, pp 339–353 (cité D.V. & G.V.Ch).)

Dès 1975 il avait semblé naturel à nos auteurs (Calogero, Moser, Olshanetsky, Perelomov) de généraliser l'étude à d'autres potentiels. Moser avait déjà remplacé les $(x_i - x_j)^{-2}$ de (1) par $\sin^{-2}(x_i - x_j)$ et Calogero, dans sa Lettere al Nuovo Cimento (vol 13) du 12 Luglio 1975, avait introduit des fonctions elliptiques $\wp(x_i - x_j)$. $\wp$ est la fonction de Weierstrass d'un réseau $\Lambda = 2\omega\mathbb{Z} + 2\omega'\mathbb{Z}$ de $\mathbb{C}$:

$$\wp_\Lambda(x) = x^{-2} + \sum_{w\in\Lambda^{\times}} ((x+w)^{-2} - w^{-2}).$$

C'est une fonction méromorphe et doublement périodique de la variable complexe x; on doit la considérer comme une fonction sur la courbe elliptique $X = \mathbb{C}/\Lambda$. La fonction $\sin^{-2}(x) - \frac{1}{3}$ est une dégénerescence (normalisée) de $\wp$ lorsqu'une des périodes tend vers l'infini, la courbe X devenant un tore multiplicatif $\mathbb{G}_m = \mathbb{C}^{\times}$, et la fonction rationnelle x^{-2} apparaît lorsque les deux périodes indépendantes sont devenues infinies, la courbe est alors la droite affine $\mathbb{G}_a = \mathbb{C}$.

(Comme il est expliqué dans le livre d'André Weil, *Elliptic functions according to Eisenstein and Kronecker*, en partant de x^{-2} Eisenstein avait su développer avant Weierstrass une théorie élémentaire des fonctions circulaires et des fonctions elliptiques; dans ses notations

$$\begin{aligned} \epsilon_2 &= \sum_{\mu \in \mathbb{Z}} (x+\mu)^{-2} = \pi^2 \sin^{-2} \pi x \\ E_2 &= x^{-2} + \sum\nolimits'_e (x+w)^{-2} = \wp(x) + e_2 \end{aligned}$$

où $e_2 = \sum'_e w^{-2}$ n'est pas un invariant modulaire.)

Avec Moser et Calogero remplaçons le terme $(x_i - x_j)^{-2}$ du potentiel U de (1) par $\sin^{-2}(x_i - x_j)$ ou par $\wp(x_i - x_j)$;

$$H(x_1, \ldots, p_1, \cdots) = \frac{1}{2} \sum_i p_i^2 - 2 \sum_i \sum_{j \neq i} \wp(x_i - x_j) \tag{8}$$

et l'on trouvera encore un système complètement intégrable.

Le cas trigonométrique $U = -2 \sum \sum \sin^{-2}(x_i - x_j)$ correspond à un problème quantique qui avait également été intégré auparavant: l'équation de Sutherland (1972). Dans ce cas Moser écrit une paire de Lax (L, M) (comme en (3), la matrice L a les p_i sur la diagonale et les coefficients $\cot(x_i - x_j)$ hors de la diagonale; l'équation (3) résulte alors du théorème d'addition de la cotangente $\cot(a+b) = (\cot a \cot b - 1)/(\cot a + \cot b)$).

Pour les systèmes (8), Calogero (en 1975) découvre une écriture sous la forme de Lax. Son point de départ: une solution (α, β) de l'équation fonctionnelle

$$\alpha'(x)\alpha(y) - \alpha(x)\alpha'(y) = \alpha(x+y)(\beta(x) - \beta(y)), \tag{9}$$

avec $\beta(-x) = \beta(x)$, donne une solution (L, M) de (3)

$$(L)_{ij} = \delta_{ij} p_i + (1 - \delta_{ij}) \alpha(x_i - x_j) \tag{10}$$

$$(M)_{ij} = \delta_{ij} \sum_{k \neq i} \beta(x_i - x_k) + (1 - \delta_{ij}) \alpha'(x_i - x_j) \tag{11}$$

qui équivaut aux équations de Hamilton dérivant de l'énergie potentielle

$$U(x_1, \cdots, x_n) = \sum_i \sum_{j > i} \alpha(x_i - x_j) \alpha(x_j - x_i). \tag{12}$$

Donc si $\alpha(x)\alpha(-x)$ ne diffère de $-4\wp(x)$ que par une constante on retrouve les équations de (8); Calogero propose comme α les fonctions elliptiques 2dn/sn et 2cn/sn (alors $\beta = -2\wp$).

(La fonction de Jacobi $v = \mathrm{sn}_k(x)$ est l'inverse de l'intégrale de Legendre $x = \int_0^v \frac{du}{\sqrt{(1-u^2)(1-k^2u^2)}}$ et les fonctions cn_k, dn_k, vérifient $\mathrm{cn}_k^2 + \mathrm{sn}_k^2 = 1$,

$\mathrm{dn}_k^2 + k^2\mathrm{sn}_k^2 = 1$. Soit Λ le réseau engendré par la demi-période $2K = 2\int_0^1 \frac{du}{\sqrt{(1-u^2)(1-k^2u^2)}}$ et la période $2iK' = 2\int_0^{\frac{1}{k}} \frac{du}{\sqrt{(1-u^2)(1-k^2u^2)}}$; notons X le quotient $\mathbb{C}/\Lambda$; on dit que k est le (un) module de X et l'on a

$$\wp_\Lambda = \frac{1}{\mathrm{sn}_k^2} - \frac{1}{3} - \frac{k^2}{3}.$$

L'équation (9) appartient à la théorie des fonctions elliptiques: sa solution la plus simple est $\alpha(x) = a/\mathrm{sn}(bx)$, $\beta(x) = -ab/\mathrm{sn}^2(bx)$ et cela traduit la formule d'addition

$$\mathrm{sn}(x+y) = \frac{\mathrm{sn}x\mathrm{cn}y\mathrm{dn}y + \mathrm{cn}x\mathrm{dn}x\mathrm{sn}y}{1 - k^2\mathrm{sn}^2x\mathrm{sn}^2y},$$

compte tenu de l'équation différentielle $\mathrm{sn}'x = \mathrm{cn}x\ \mathrm{dn}x$. Pour les solutions dn/sn et cn/sn, on utilise les deux autres équations différentielles (d'Euler): $\mathrm{cn}'x = -\mathrm{sn}x\ \mathrm{dn}x$ et $\mathrm{dn}'x = -k^2\mathrm{sn}x\ \mathrm{cn}x$.

Dans la Note sur la théorie des fonctions elliptiques (1894), Ch. Hermite montre que la fonction $\varphi(x)$ qui inverse une intégrale de Jacobi $g(v) = \int_0^v \frac{du}{\sqrt{1-2\delta u^2+\epsilon u^4}}$ satisfait à la relation

$$\varphi(x+y)(\varphi(x)\varphi'(y) - \varphi'(x)\varphi(y)) = \varphi^2(x) - \varphi^2(y);$$

ce qui revient à dire que $\alpha(x) = 1/\varphi(x)$ vérifie l'équation (9) (et $\beta(x) = -\alpha^2(x)$). Dans ce cas $v = \alpha(x)$ inverse l'intégrale $x = \int_v^\infty \frac{du}{\sqrt{1-2\delta u^2+\epsilon u^4}}$. Dans leur article de 1976 aux Inventiones, Olshanetsky et Perelomov démontrent une réciproque: si α, fonction méromorphe et *impaire* de x, vérifie (9) alors $1/\alpha$ est l'inverse d'une intégrale de Jacobi (éventuellement dégénérée).

Ensuite, Olshanetsky et Perelomov ont donné une raison d'algèbre linéaire à tout cela et ont généralisé (8) en bâtissant des potentiels sur des systèmes de racines d'algèbre de Lie semi-simples (ici les $(x_i - x_j)$ forment le système A_{n-1}). Dans les cas trigonométriques et elliptiques (réels) la dynamique se tient dans une chambre du groupe de Weyl affine. (Voir Inventiones Mathematicae, vol. 37, pp. 109–120 (1976).) Le système trigonométrique s'identifie à une partie du flot géodésique d'un espace symétrique et s'intègre explicitement, cf. la Lettere al Nuovo Cimento, vol. 17, 1976 d'Olshanetsky et Perelomov. Enfin, en 1977, Perelomov démontre que dans le cas elliptique aussi les hamiltoniens $H_k = \frac{1}{k+1}TrL^{k+1}$ sont indépendants et en involution (mais il ne donne pas de moyen pour intégrer le système (8)).

Or, l'équation KdV stationnaire $6uu_x + u_{xxx} = 0$ ne possède pas que la solution particulière $-2x^{-2}$, elle en a d'autres aussi remarquables:

$-2\sinh^{-2}(x) - \frac{10}{3}$ (ou $-2\sin^{-2}(x) + \frac{2}{3}$) et $-2\wp_\Lambda(x)$ pour tout réseau Λ de $\mathbb{C}$.

La fonction $v = \wp_\Lambda(x)$ inverse l'intégrale elliptique de Weierstrass $x = \int_v^\infty \frac{du}{\sqrt{4u^3 - g_2 u - g_3}}$, où $g_2 = 60 \sum_{\Lambda^\times} w^{-4}$ et $g_3 = 140 \sum_{\Lambda^\times} w^{-6}$, donc $\wp'^2 = 4\wp^3 - g_2\wp - g_3$ et en dérivant deux fois $\wp''' = 12\wp\wp'$.

(*Remarque:* l'homothétie $x' = \epsilon x$ change le réseau Λ en $\Lambda' = \epsilon\Lambda$ et la fonction $\wp_\Lambda$ en $\wp_{\Lambda'} = \epsilon^{-2}\wp_\Lambda$. Soient u_1, u_2, u_3 les valeurs de $\wp_\Lambda$ aux $\frac{1}{2}$-périodes, i.e. les racines de $4u^3 - g_2 u - g_3 = 0$; pour chacune des douze valeurs $\sqrt{u_j - u_k}$ de ϵ on détermine une valeur de k (égale au rapport $\sqrt{u_l - u_k}/\sqrt{u_j - u_k}$) telle que $\Lambda' = 2K(k)\mathbb{Z} + 2iK'(k)\mathbb{Z}$. Ainsi chaque réseau "possède" une fonction $\wp$ et six triplets de fonctions $\mathrm{sn}_k, \mathrm{cn}_k, dn_k$. Voir le cours d'Analyse de Jordan, chapitre VII.)

D'après Airault, McKean et Moser (1977), toute solution de KdV (non nécessairement stationnaire) qui possède deux périodes indépendantes en $x \in \mathbb{C}$ (pour toute valeur de t) s'écrit

$$u(x,t) = -2\sum_1^n \wp(x - x_i(t)) + c, \tag{13}$$

où l'entier n, les fonctions $x_i(t)$ et la constante c sont uniquement déterminés par u. Ses pôles $x_i(t)$ sont deux à deux distincts pour au moins une valeur du temps t. De plus on a

$$\forall i,\ \forall t,\ \ \sum_{j\neq i} \wp'(x_i - x_j) = 0 \tag{14}$$

et la dynamique des pôles est donnée par

$$\forall i,\ \ \dot{x}_i = 3\sum_{j\neq i} \wp(x_i - x_j) - \frac{3c}{2} \tag{15}$$

ce qui s'interprète encore comme les équations de Hamilton de H_2 sur le lieu critique de H. (Précisément $-\frac{3}{4}H_2$ guide $x_i(t)$ à une fonction affine de t prés.) Et, comme dans le cas rationnel, la réciproque est vraie: si les $x_i(t)$ satisfont à (14) et (15), alors (13) est solution de KdV. Cependant l'ensemble défini par (14) est bien plus difficile à cerner que ne l'était l'ensemble (6) et l'on pénètre dans une forêt de problèmes nouveaux.

Avec A. Treibich-Kohn et J.-L. Verdier, nous appellerons les solutions de la forme (13) telles que $c = 0$ des *solitons elliptiques de KdV* (de degré n).

Remarque: si la fonction $u(x,t)$ est solution de KdV, la fonction $v(x,t) = u(x+\frac{3}{2}ct, t) + c$ est également solution de KdV pour toute constante c.

Bien sûr il y avait des exemples; les plus classiques avaient pour conditions initiales les potentiels de Lamé $u(x) = -m(m+1)\wp(x)$ $(m \in \mathbb{N})$. Dans ces exemples, au temps 0 tous les pôles confluent, et $n = \frac{m(m+1)}{2}$. (Plus souvent on considérait seulement $u(x) = -m(m+1)\wp(x+K\tau)$ avec $\Lambda = 2K(\mathbb{Z}+\tau\mathbb{Z})$ et τ imaginaire pur, car u est alors fonction réelle périodique continue de la variable x, ou bien l'écriture équivalente $-m(m+1)k^2\mathrm{sn}_k^2(x)+c$, car $\mathrm{sn}_k(x+iK') = 1/k\mathrm{sn}_k(x)$ et $\wp(x+K\tau) = k^2\mathrm{sn}_k^2(x) - \frac{1+k^2}{3}$ si $\tau = \frac{iK'}{K}$.)

Et puis, en 1974, Dubrovin et Novikov avaient répertorié tous les exemples (réels) de degré n égal à 3; Airault, McKean et Moser montrent que dans ce cas, pour une courbe elliptique $X = \mathbb{C}/\Lambda$ donnée, l'ensemble des solitons elliptiques est constitué d'une variété de dimension 2 contenant le point de Lamé $-6\wp_\Lambda$, et d'un nombre fini de courbes isolées (plus précisément une surface fibrée sur X de fibre une autre courbe elliptique et 9 courbes exceptionnelles isogènes à X).

En 1967, Gardner, Greene, Kruskal, Miura avaient découvert une méthode pour résoudre le problème de Cauchy de KdV (par diffusion inverse). Cette méthode a d'abord été appliquée aux potentiels à décroissance rapide (Lax, Zakharov, Faddeev), mais entre 1974 et 1976, Novikov, Dubrovin, Adler, Lax, Matveev, Its, McKean, van Moerbeke, Kac ont su l'adapter aux potentiels périodiques "à nombre fini de zones d'instabilité". A cette catégorie appartiennent nos exemples elliptiques. (L'algèbre sous jacente à la méthode fut clairement dégagée par Gelfand et Dickey (1975).)

La théorie spectrale de l'équation de Hill

$$\psi'' + u\psi = \lambda\psi$$

entre en jeu. En effet, lorsque u suit l'évolution imposée par l'équation de Korteweg-de Vries (4), le spectre de l'équation de Hill ne change pas.

La recherche des fonctions propres ψ périodiques pour u elliptique (surtout quand $u = -m(m+1)\wp$) semble être un thème assez ancien; par exemple A. McK.M. citent un travail de Guerritore en 1909, et l'analyse de Ince en 1940 sur les fonctions de Lamé. La période moderne fut inaugurée par Akhiezer (1961) et Hochstadt (1965); et puis ce fut l'explosion, en 1974. (Comme références, il y a le recueil de G. Wilson, *Integrable systems*, Cambridge University Press 1981, et le Séminaire de l'Ecole Normale Supérieure en 1977 animé par A. Douady, J. H. Hubbard et J.-L. Verdier.)

Après l'article A. McK.M. cela permettait de calculer effectivement les solutions du système dynamique (14), (15), pour certaines conditions initiales $(x_1, \ldots, x_n)$, à l'aide des fonctions Thêta de courbes hyperelliptiques:

$$\mu^2 = -(\lambda - \lambda_0) \cdots (\lambda - \lambda_{2m}), \quad (n = \frac{m(m+1)}{2}).$$

(On dispose alors de coordonnées actions-angles partielles.)

Les progrès suivants reviennent à I. M. Krichever. En 1978 (Functional Analysis, vol. 12, pp 76-78), il élargit l'étude aux solutions rationnelles en x d'une équation d'évolution non linéaire portant sur les fonctions u de deux variables x, y, l'équation KP (pour Kadomtsev et Petviashvili), et trouve ainsi une interprétation de la dynamique de H_1 lui même (celui de la formule (2) du début).

Puis en 1980 paraît l'article, *Elliptic solutions of the Kadomtsev-Petviashvili equation and integrable system of particle* (Functional Analysis, vol. 14, pp 282–290). D'abord Krichever y construit une famille de matrices (L, M) dépendant d'un paramètre a sur la courbe X, d'où une "courbe spectrale" Γ, revêtement ramifié de X, d'équation $\det(L(a) - \lambda) = 0$. Ensuite il applique la belle théorie qu'il avait développée les années précédentes afin de résoudre les équations KP et KdV pour les potentiels quasi-périodiques. Résultat: ce sont les fonctions Thêta de Γ qui vont permettre d'intégrer explicitement tout le système (8); si bien qu'avec les variables d'actions $H_1, \ldots, H_n$, on disposera de n variables d'angles: ce seront les coordonnées sur la jacobienne de la courbe spectrale. (Toutefois, ces résultats sont subordonnés à des hypothèses de généricité qui assurent la lissité de Γ et qui font, par exemple, que les solitons elliptiques de KdV, où la variable y n'apparaît pas, ne sont pas considérés explicitement dans l'article de 1980.)

Pour comprendre les paires de Lax construites par Krichever, revenons à l'équation fonctionnelle (9) étudiée par Calogero, et posons, quel que soit $a \in X = \mathbb{C}/\Lambda$,

$$\alpha(x, a) = -2\frac{\sigma(x-a)}{\sigma(a)\sigma(x)} e^{x\zeta(a)}, \quad \text{et} \quad \beta(x) = -2\wp(x),$$

où la fonction σ est celle qu'avait définie Weierstrass

$$\sigma_\Lambda(x) = x \prod_{w \in \Lambda^\times} \left(1 - \frac{x}{w}\right) e^{\frac{x}{w} + \frac{1}{2}\frac{x^2}{w^2}},$$

et la fonction ζ égale $\frac{\sigma'}{\sigma}$, si bien que $\wp = -\zeta' = -\frac{d^2}{dx^2} \log \sigma$. Sur le réseau $\Lambda = 2\omega\mathbb{Z} + 2\omega'\mathbb{Z}$ on trouve

$$\sigma(u + 2\omega) = -\sigma(u)e^{2\eta(u+\omega)} \quad \text{et} \quad \sigma(u + 2\omega') = -\sigma(u)e^{2\eta'(u+\omega')}.$$

Krichever constate que (α, β) est solution de (9) et que

$$\alpha(x)\alpha(-x) = 4(\wp(a) - \wp(x)),$$

donc les formules (10) et (11) ramènent la dynamique de Calogero-Moser (8) à une forme de Lax (3) (pour toute valeur de a). Lorsque a vaut ω, ω' ou $(\omega+\omega')$ on obtient des solutions impaires équivalentes aux $2\mathrm{sn}^{-1}, 2\mathrm{cn}\,\mathrm{sn}^{-1}$, $2\mathrm{dn}\,\mathrm{sn}^{-1}$ de Calogero.

(Dans l'ancienne littérature la fonction $\varphi(x, a) = \frac{1}{2}\alpha(x, -a)$ s'appelle une fonction elliptique de deuxième espèce; c'est la plus jolie fonction propre du premier potentiel de Lamé $u = -2\wp$; on a

$$\frac{d^2\varphi}{dx^2} - 2\wp(x)\varphi = \wp(a)\varphi\,.$$

Signalons que dès 1877 Hermite avait exprimé les fonctions de Lamé (généralisées), solutions de l'équation $\frac{d^2\varphi}{dx^2} - m(m+1)\wp(x)\varphi = \lambda\varphi$, comme sommes de produits de $n = \frac{m(m+1)}{2}$ fonctions $\varphi(x, a_i)$ et qu'il avait montré comment les $\wp(a_i)$ s'obtiennent en résolvant une équation polynomiale de degré n. Cf. Hermite, Oeuvres, tome 3, pp. 266 à 418, *Sur quelques applications des fonctions elliptiques*, ou le tome 2 du traité de Halphen.)

Et surtout Krichever en 1980 démontre que le sujet s'inscrit dans le cadre général de ses constructions de 1976 et 1977. Maintenant des fonctions abéliennes de tous les genres fleurissent partout dans le domaine des solitons elliptiques.

C'est le moment de donner un aperçu de la théorie de Krichever. En même temps je préciserai le contenu de son travail de 1980.

L'équation de Korteweg-de Vries avait deux origines.

D'abord elle a été inventée pour engendrer l'onde de translation de Russel (1834), la vague solitaire en eau peu profonde, baptisée soliton: $u(x, t) = 2a^2 \cosh^{-2}(ax + 4a^3 t)$. Elle a su générer aussi des n-solitons, systèmes de n vagues pointues qui retrouvent leurs formes après l'interaction (ceux ci ont été observés expérimentalement bien avant que Hirota ne découvre leur formule en 1971; les solitons trigonométriques font partie de cette famille).

D'autre part, autour de 1925, l'équation KdV est apparue tout naturellement dans l'examen des invariants d'équations différentielles: plusieurs articles de J. L. Burchnall, T. W. Chaundy et H. E. Baker parurent à Londres entre 1922 et 1931 et puis furent oubliés jusqu'à ce que la théorie soit retrouvée et enrichie (indépendamment) par Krichever en 1976. Ensuite de nombreux auteurs sont intervenus (Drinfeld, Mumford, Gelfand, Dickey,...) Une excellente référence est le texte de J.-L. Verdier, *Equations différentielles algébriques*, Séminaire Bourbaki 1977-78 n° 512, LN

710 pp 105-134. La question est la suivante: étant donné un opérateur d'ordre l, en dimension 1, dont les coefficients sont fonctions de x, $Q_l = a_0(x)\left(\frac{\partial}{\partial x}\right)^l + \cdots + a_l(x)$; quels sont les opérateurs qui commutent avec lui? Plus généralement: déterminer les sous-algèbres commutantes de l'algèbre des opérateurs différentiels! (Suivant les hypothèses précises sur les coefficients, analytiques, méromorphes, formels, différentiels,... le problème est différent; avec des coefficients polynomiaux la question est étudiée par Amitsur (1958) et Dixmier (1968). Ici je parlerai surtout de coefficients méromorphes.)

Dans ce contexte KdV apparait vite: si $Q_2 = \frac{\partial^2}{\partial x^2} + u(x)$ (dit "opérateur de Liouville"), KdV stationnaire est la condition pour qu'il existe un Q_3 tel que $[Q_3, Q_2] = 0$. Et pour avoir KdV avec le temps t (comme en (4)) il suffit de chercher Q_3 tel que $[\frac{\partial}{\partial t} - Q_3, Q_2] = 0$, où les coefficients sont fonctions de x et t. Pour un nombre entier m quelconque, en écrivant

$$[\frac{\partial}{\partial t_{2m+1}} - Q_{2m+1}, Q_2] = 0,$$

on forme les équations de la hierarchie KdV; ce sont d'autres équations d'évolution non linéaires et toutes ces équations sont compatibles: c'est à dire qu'elles définissent des flots permutables dans l'espace des potentiels u. En prenant comme point de départ Q_l au lieu de Q_2, on obtient les hierarchies $\mathrm{KdV}_{(l)}$. (Par exemple, $\mathrm{KdV}_{(3)}$ est la hierarchie de Boussinesq.) Ces flots sont hamiltoniens pour plusieurs structures symplectiques naturelles sur l'espace des u (ou des Q_l) comme l'ont montré Zakharov-Faddeev et Gelfand-Dickey; voir aussi Drinfeld et Sokolov, Doklady, vol. 23 (1981).

Les équations de KP mettent plus de démocratie entre les variables: on s'interesse à des fonctions $u(t_1, t_2, t_3, \ldots)$ et l'on demande

$$[\partial_m - Q_m, \partial_n - Q_n] = 0$$

quels que soient n et m. La première équation de la hierarchie KP est:

$$3\frac{\partial^2 u}{\partial y^2} = \frac{\partial}{\partial x}\left(4\frac{\partial u}{\partial t} - \frac{\partial^3 u}{\partial x^3} - 6\frac{\partial u}{\partial x}\right). \tag{16}$$

On l'obtient en faisant $m = 3$, $n = 2$, $t_1 = x$, $t_2 = y$ et $t_3 = t$. Auparavant (pour les besoins des plasmas?) B. B. Kadomtsev et V. I. Petviashvili (1970) avaient découvert cette équation en imposant une perturbation transverse, dans la direction des y, du soliton de KdV; ce qui donne

$$u(x, y, t) = \frac{(a-b)^2}{2}\cosh^{-2}(\frac{1}{2}((a-b)x + (a^2 - b^2)y + (a^3 - b^3)t + c))).$$

Même si ça n'est pas complètement clair du point de vue physique, le passage de KdV à KP représente un précieux gain de symétrie mathématique (par exemple, en oubliant les variables dont le numéro est divisible par l on retrouve les $\mathrm{KdV}_{(l)}$, qui sont souvent notées KP_l).

En 1978 et 1980, I.M. Krichever considère les solutions de (16) qui sont fonctions méromorphes des trois variables complexes x, y, t et qui peuvent se mettre sous la forme suivante

$$u(x,y,t) = -2\sum_1^n \wp(x - x_i(y,t)). \tag{17}$$

Nous les nommerons *solitons elliptiques de KP* (et l'entier n sera le degré du soliton).

(Dans l'appendice A3 de leur article sur les *"variétés de Krichever"*, A. Treibich et J.-L. Verdier démontrent que toute solution méromorphe de KP doublement périodique sur le réseau Λ de $\mathbb{C}$ se ramène à la forme (17), c'est à dire à un soliton elliptique de KP. Plus précisément, $u(x,y,t)$ doit être de la forme $-2\sum \wp(x - x_i(y,t)) + a(t)y + b(t)$ et l'on peut faire disparaitre la fonction affine de y par un changement de variables de la forme $(x,y,t) \to (x + c(t)y + d(t), y + e(t), t)$.)

Si aux temps $(y,t) = (0,0)$ les x_i sont distincts, la condition nécessaire et suffisante pour que (17) définisse une solution de KP (16) (c'est à dire un soliton elliptique) est que les x_i vérifient

$$\forall i, \quad \frac{\partial^2 x_i}{\partial y^2} = 4\sum_{j\neq i} \wp'(x_i - x_j) \tag{18}$$

$$\forall i, \quad 4\frac{\partial x_i}{\partial t} = -3\left(\frac{\partial x_i}{\partial y}\right)^2 + 12\sum_{j\neq i} \wp(x_i - x_j). \tag{19}$$

(La nécessité se voit facilement en substituant dans (16), en utilisant $\wp''' = 12\wp\wp'$ et en identifiant les termes polaires.)

Ces équations traduisent des évolutions hamiltoniennes dans l'espace de phases des $(x_1, \ldots, x_n, p_1, \ldots, p_n)$: c'est H (cf (8)) qui régit l'évolution suivant y car si l'on pose

$$\frac{\partial x_i}{\partial y} = \frac{\partial H}{\partial p_i} = p_i,$$

on trouve bien

$$\frac{\partial p_i}{\partial y} = \frac{\partial^2 x_i}{\partial y^2} = -\frac{\partial H}{\partial x_i}.$$

D'autre part, considérons l'une des matrices $L(a)$, $a \in X$, de Krichever (page 9) et posons $H_2'(a) = -\frac{1}{4}Tr(L(a)^3)$; on a

$$\begin{aligned} H_2'(a) &= -\frac{1}{4}\sum_i p_i^3 - \frac{3}{4}\sum_i p_i \sum_{j\neq i} \alpha(x_i - x_j, a)\alpha(x_j - x_i, a) \\ &\quad -\frac{1}{4}\sum_i \sum_{j\neq i} \sum_{k\neq i,j} \alpha(x_i - x_j, a)\alpha(x_j - x_k, a)\alpha(x_k - x_i, a). \end{aligned}$$

On sait déja que $\alpha(x,a)\alpha(-x,a) = 4(\wp(a) - \wp(x))$. Par ailleurs, quels que soient $x, y, z \in \mathbb{C}$, tels que $x + y + z = 0$, on a la jolie formule:

$$\begin{aligned} \wp'(a)\sigma^3(a)\sigma(x)\sigma(y)\sigma(z) &= \sigma(a-x)\sigma(a-y)\sigma(a-z) \\ &\quad -\sigma(a+x)\sigma(a+y)\sigma(a+z) \end{aligned}$$

donc

$$\alpha(x,a)\alpha(y,a)\alpha(z,a) + \alpha(-x,a)\alpha(-y,a)\alpha(-z,a) = 8\wp'(a)$$

si bien que

$$\begin{aligned} H_2'(a) &= -\frac{1}{4}\sum_i p_i^3 + 3\sum_i p_i \sum_{j\neq i} \wp(x_i - x_j) \\ &\quad -3(n-1)\wp(a)\sum_i p_i - n(n-1)(n-2)\wp'(a). \end{aligned}$$

Les premières équations de Hamilton s'écrivent, pour i entre 1 et n,

$$\frac{dx_i}{dt} = \frac{\partial H_2'}{\partial p_i} = -\frac{3}{4}p_i^2 + 3\sum_{j\neq i} \wp(x_i - x_j) - 3(n-1)\wp(a).$$

Par conséquent, toujours en posant $\frac{\partial x_i}{\partial y} = p_i$, si $x_i(y,t)$ vérifie (19), la fonction $x_i(t) - 3(n-1)\wp(a)t$ évolue selon H_2'. Pour ce même flot de H_2' le deuxième groupe d'équations de Hamilton donne

$$\frac{dp_i}{dt} = -\frac{\partial H_2'}{\partial x_i} = -3p_i\sum_{j\neq i} \wp'(x_i - x_j) - 3\sum_{j\neq i} p_j\wp'(x_i - x_j);$$

mais, justement, c'est compatible avec (18) car (19) implique

$$\begin{aligned} 4\frac{\partial^2 x_i}{\partial y \partial t} &= -6\frac{\partial^2 x_i}{\partial y^2}\left(\frac{\partial x_i}{\partial y}\right) + 12\sum_{j\neq i} \wp'(x_i - x_j)\left(\frac{\partial x_i}{\partial y} - \frac{\partial x_j}{\partial y}\right) \\ &= -24\frac{\partial x_i}{\partial y}\sum_{j\neq i} \wp'(x_i - x_j) + 12\sum_{j\neq i} \wp'(x_i - x_j)\left(\frac{\partial x_i}{\partial y} - \frac{\partial x_j}{\partial y}\right) \\ &= 4\left(-3p_i\sum_{j\neq i} \wp'(x_i - x_j) - 3\sum_{j\neq i} p_j\wp'(x_i - x_j)\right). \end{aligned}$$

(Notons que l'annulation du crochet de Poisson $\{H, H_2'\}$ équivaut à la forme suivante des formules d'addition de la fonction $\wp$:

$$\sum_i \sum_{j\neq i} \sum_{k\neq i,j} \wp(x_i - x_j)\wp'(x_i - x_k) = 0.)$$

Finalement: le point de coordonnées $(x_i - 3(n-1)\wp(a)t, p_i = \frac{\partial x_i}{\partial y})$ évolue selon H_1 pour y et selon H_2' pour t. Autrement dit, en convenant que $H_0 = \sum_1^n p_i$, le point de coordonnées (x_i, p_i) évolue selon H_1 pour y et selon la fonction $H_2'(a) + 3(n-1)\wp(a)H_0$ pour t.

Une découverte essentielle de l'école russe (Novikov, Dubrovin, Krichever...) dans les années 70 est qu'on peut associer des solutions de KP à toute courbe algébrique sur $\mathbb{C}$ et qu'inversement, il est possible de reconnaître sur les propriétés dynamiques d'une solution si celle ci provient d'une courbe. Dès lors on peut parler de "solutions géométriques" (ou "algébriques") de KP; c'est le sujet de la célèbre "construction de Krichever".

Supposons d'abord qu'on a affaire à une courbe projective lisse Γ; choisissons un point p_0 sur Γ et plongeons Γ dans sa jacobienne via l'application d'Abel: la jacobienne est le quotient du dual des formes holomorphes par le réseau des périodes $(\omega \to \oint_\gamma \omega)$, $Jac(\Gamma) = (H^{1,0}(\Gamma))^*/H_1(\Gamma, \mathbb{Z})$; l'application d'Abel $A : \Gamma \to Jac(\Gamma)$ est l'intégrale définie, $A(p)(\omega) = \int_{p_0}^p \omega$. En privilégiant une base canonique $A_1, \cdots, A_g, B_1, \cdots, B_g$ de l'homologie $H_1(\Gamma, \mathbb{Z})$ $(A_i \cdot A_j = 0, A_i \cdot B_j = \delta_{i,j}, B_i \cdot B_j = 0)$, on récupère une base $\omega_1, \cdots, \omega_g$ de l'espace vectoriel $H^{1,0}$ des formes holomorphes (en demandant $\oint_{A_i} \omega_j = \delta_{ij}$), et l'on forme une "matrice des périodes" Ω suivant Riemann: $(\Omega)_{ij} = \oint_{B_i} \omega_j$. Les relations de Riemann disent que Ω est symétrique $({}^t\Omega = \Omega)$ et que sa partie imaginaire est définie positive $(Im(\Omega) >> 0)$. La jacobienne s'identifie au quotient concret $\mathbb{C}^g/(\mathbb{Z}^g + \Omega\mathbb{Z}^g)$ et la fonction Thêta de Riemann est la fonction suivante de $\vec{z} \in \mathbb{C}^g$:

$$\theta(\vec{z}; \Omega) = \sum_{\vec{n}\in\mathbb{Z}^g} \exp(2i\pi\vec{n}\cdot\vec{z} + i\pi\vec{n}\Omega\vec{n}).$$

A présent choisissons une coordonnée analytique z près de p_0, centrée en p_0; considérons le vecteur vitesse $\vec{u}$ de l'application A, son accélération $\vec{v} = \frac{d^2A}{dz^2}$ et son vecteur du troisième ordre $\vec{w} = \frac{1}{2}\frac{d^3A}{dz^3}$, le théorème s'énonce ainsi:

Il existe une constante $c \in \mathbb{C}$ (que nous expliquerons un peu plus loin) telle que, pour tout $\vec{e} \in \mathbb{C}^g$ la fonction de trois variables

$$u(x, y, t) = 2\frac{\partial^2}{\partial x^2} \log \theta(x\vec{u} + y\vec{v} + t\vec{w} + \vec{e}) + c \tag{20}$$

soit solution de KP. De plus, si Γ est hyperelliptique, que p_0 est un point de Weierstrass et que la coordonnée z est antisymétrique (vis à vis de l'involution hyperelliptique), le vecteur $\vec{v}$ s'annule et $u(x,t)$ est solution de KdV.

(La formule (20) porte souvent les noms de Dubrovin et Novikov et/ou de Its et Matveev, qui l'ont écrite entre 1974 et 1975 pour l'équation de Korteweg-de Vries, dans le cas général elle est due à Krichever, 1977.)

Encore plus remarquable est la réciproque conjecturée par S.P. Novikov et démontrée en 1986 par T. Shiota (après des résultats de Mulase); c'est une réponse à une question de Riemann étudiée par Schottky: quelles sont les matrices carrées $g \times g$, Ω, symétriques et de partie imaginaire définie positive qui sont des matrices de périodes de courbes algébriques? Le théorème de Shiota dit que si le diviseur $\Theta = \{\vec{z} \in \mathbb{C}^g \mid \theta(\vec{z},\Omega) = 0\}$ est irréductible et si l'on peut trouver $\vec{u},\vec{v},\vec{w}$ dans $\mathbb{C}^g$ et c dans $\mathbb{C}$ tels que pour tout $\vec{e} \in \mathbb{C}^g$, la formule (20) définisse une solution de (16), alors Ω vient d'une courbe projective lisse Γ de genre g. (D'après Torelli, Γ est déterminée par Ω.)

Toute la théorie des fonctions elliptiques, double périodicité (Painlevé), théorèmes d'addition (transférence), etc, peut se déduire de l'équation de KdV stationnaire $\wp''' = 12\wp\wp'$; de même on peut dire que toute la théorie des fonctions abéliennes est contenue dans KP, il reste à décrypter.

Le meilleur cadre pour comprendre ces résultats et pour les étendre aux courbes singulières, afin de couvrir aussi les solutions rationnelles et les solitons, est celui qu'a présenté M. Sato en hiver 80-81: la "grassmannienne infinie". Elle provient de la troisième source de systèmes intégrables: après la Mécanique classique, et les E.D.P. non linéaires, la Mécanique statistique, et plus précisément le modèle d'Ising. Il va apparaître (entre autres choses) des fonctions spéciales généralisant les fonctions Thêta.

Une référence toute indiquée est un exposé de Jean-Louis au Séminaire Bourbaki: *Les représentations des algèbres de Lie affines, applications à quelques problèmes de physique* (d'après Date, Jimbo, Kashiwara, Miwa); on peut aussi se reporter à l'article original de M. Sato et Y. Sato, *Soliton equations as dynamical systems on Infinite Dimensional Grassmann Manifold*, Lecture Notes in Num. Appl. Anal. 5 (1982), ou l'article de Date, Jimbo, Kashiwara, Miwa, *Transformation group for soliton equations*, World Science Publ., 1983, ou encore à un texte que Jean-Louis affectionait particulièrement: *Loop groups and equations of KdV type* de G. Segal et G. Wilson, publications de l'I. H. E. S, 1985.

Soient E_- l'espace vectoriel (sur $\mathbb{C}$) de base dénombrable $e_0, e_{-1}, e_{-2}, \cdots$ et E_+ celui dont la base est $e_1, e_2, e_3, \ldots$; notons E^- (resp. E^+) la complétion formelle de E_- (resp. E_+), c'est à dire l'espace des suites $(a_0, a_{-1}, a_{-2}, \ldots)$ (resp. $(a_1, a_2, a_3, \ldots)$); E désigne la somme directe $E_- \oplus$

E^+. La grassmannienne de Sato Gr_E est l'ensemble des sous-espaces vectoriels de E dont la projection sur E_- est Fredholm (c'est à dire de noyau et conoyau de dimensions finies; l'indice étant la différence entre la dimension du noyau et celle du conoyau); elle décrit les sous-espaces de E comparables à E_-. Segal et Wilson considèrent une version plus adaptée au travail de la topologie: ils remplacent E_- par l'espace de Hilbert H_- et E^+ par l'espace de Hilbert H_+; un point de la "grassmannienne compacte" Gr_H sera un sous espace de H dont la projection sur H_- est Fredholm et dont la projection sur H_+ est compacte. (*Attention:* dans Segal et Wilson H_+ désigne notre H_- et H_- notre H_+).

Des points spéciaux de Gr_E (ou Gr_H) sont associés à chaque suite décroissante Y d'entiers positifs presque tous nuls $f_1 \geq \cdots \geq f_n \geq \cdots$ et à chaque entier relatif $m \in \mathbb{Z}$: l'espace $E_{Y,m}$ a pour base $e_{i_m}, e_{i_{m-1}}, e_{i_{m-2}}, \ldots$ où $i_{m-l} + l - m = f_{l+1}$; si bien que $i_{-k} = -k$ lorsque k est assez grand et que $E_{Y,m}$ est d'indice m. Notons $E^{Y,m}$ le sous-espace de E constitué par les suites (a_n) $n \in \mathbb{Z}$ telles que $a_n = 0$ si $e_n \in E_{Y,m}$; une partition en "cellules de Schubert" $C_{Y,m}$ (de dimensions et codimensions infinies) de Gr_E s'obtient en rassemblant les sous-espaces de E qui se projettent bijectivement sur $E_{Y,m}$ parallèlement à $E^{Y,m}$.

Chaque élément F de Gr_E (ou de Gr_H) donne lieu à une solution de KP: Soit $F \in Gr_E$ d'indice $m \in \mathbb{Z}$; notons E_m (resp. E^m) l'espace $E_{(0),m}$ (resp. $E^{(0),m}$) et $f = (f_-, f_+)$ un monomorphisme de E_m dans $E = E_m \oplus E^m$ d'image F tel que $f_-(e_{-k}) = e_{-k}$ pour les grandes valeurs de k. Interprétons tout point de E comme une série de Laurent formelle $\sum_{n>>-\infty} a_n z^n$, introduisons une infinité de variables "auxiliaires" $t_1, t_2, t_3, \cdots$ et considérons (avec Sato) "l'opérateur d'évolution universelle" $h(t)$ (sur E) de multiplication par $\exp(\sum t_n z^{-n}) = \sum_{k\geq 0} p_k(t_1, \cdots, t_k) z^{-k}$. (Les $p_k(t)$ sont les polynômes de Schur, $p_1 = t_1$, $p_2 = \frac{1}{2}t_1^2 + t_2$, $p_3 = \frac{1}{6}t_1^3 + t_1 t_2 + t_3, \cdots$)

Dans la décomposition $E = E_m \oplus E^m$, la matrice de $h(t)$ est de la forme $\begin{pmatrix} a & b \\ 0 & d \end{pmatrix}$. Soit u_0 une autre variable auxiliaire; on définit une fonction des variables t et de u_0 par la formule:

$$\tau_F(t_1, t_2, \ldots; u_0) = u_0^m \det(af_- + bf_+). \tag{21}$$

(Les choix arbitraires font que τ_F n'est définie qu'à une constante multiplicative près.)

Remarque: la régularisation du déterminant de (21) dépend du type d'espace F; ça peut être un déterminant de Fredholm (version Gr_H) mais un meilleur choix peut être celui de Von Koch.

Lorsque $F \in Gr_E$ se trouve être le graphe d'une application linéaire

$f_+ : E_- \to E^+$, ce qui est le cas le plus simple d'indice nul, on a aussi:

$$\tau_F(t_1, t_2, \ldots) = \det(1 + a^{-1}bf_-).$$

Et si F appartient à Gr_H, $a^{-1}bf_-$ est nucléaire (i.e. à trace) et on a affaire à un déterminant de Fredholm.

Le résultat est qu'en posant $\check{\tau}(t_1, t_2, \ldots) = \tau(-t_1, -t_2, \ldots)$ on trouve une solution de la hiérarchie KP:

$$u = 2\left(\frac{\partial}{\partial t_1}\right)^2 \log \check{\tau}. \tag{22}$$

(*Attention*: $\check{\tau}$ est la fonction appelée τ par S.W.; les t_n de Sato étant les $-t_n$ de Segal et Wilson. Et le z de S.W. est notre $1/z$. Il est vrai qu'en mettant τ à la place de $\check{\tau}$ dans (22) on obtient aussi une solution de la hiérarchie KP, cette solution n'est autre que $\check{u}$. Cependant la relation des équations avec la géométrie des courbes nous invite à choisir la formule (22).)

Par exemple si F est l'espace $E_Y = E_{Y,0}$ ($Y = (f_1, \cdots, f_n)$), τ_Y est la fonction de Schur (du tableau d'Young Y):

$$\tau_Y(t) = \begin{vmatrix} p_{f_1}(t) & p_{f_1+1}(t) & \cdots & p_{f_1+n-1}(t) \\ p_{f_2-1}(t) & p_{f_2}(t) & \cdots & p_{f_2+n-2}(t) \\ \vdots & & & \\ p_{f_n-n-1}(t) & p_{f_n-n}(t) & \cdots & p_{f_n}(t) \end{vmatrix} \tag{23}$$

et la solution u_Y de KP qui lui correspond est une fonction rationnelle.

D'après H. Weyl τ_Y est le caractère d'une représentation irréductible de $GL_N(\mathbb{C})$; en définissant les variables spectrales z_i par les identités (de Newton) $nt_n = \sum_1^N z_i^n$ et en posant $l_k = n - k + f_{n-k}$: $\tau_Y(t) = \det(z_i^{l_j - 1}) / \det(z_i^{j-1})$. Quand l'indice de F n'est plus égal à zéro, on doit ajouter une variable u_0; alors avec $F = E_{Y,m}$ on aura $\tau_{Y,m} = u_0^m \tau_Y(t)$ et cela donnera tous les caractères irréductibles des groupes unitaires U_N.

Retenons encore l'élégante formule de Kontsevitch: Si $F \subset E$ a pour base des séries $f_i(z) = z^{-i} + O(z^{-i})$ (et toujours si $t_n = \frac{1}{n}\sum_1^N z_i^n$) on a $\tau_F(t) = \det(f_i(z_j)) / \det(z_j^{-i})$.

Remarque: la formule (21) est le sens "normal" d'une expression naturelle en théorie quantique des champs; la formule

$$\det(a + bf_-) = \det\left(\left(h(t)\begin{pmatrix} 1 & 0 \\ f_- & 1 \end{pmatrix}\right)_{--}\right)$$

correspond à un coefficient de la représentation spin(∞) considérée en théorie des fermions libres: $< e_- g(t) e_F >$ où e_- et e_F sont des états de Fock associés à E_- et F (deux vides possibles) et $g(t)$ une quantification de la rotation $h(t)$ (voir l'exposé cité de J.-L. Verdier sur les travaux de Date, Jimbo, Kashiwara, Miwa). C'est la bosonisation, au fort parfum de cohomologie: on peut voir l'algèbre S des polynômes en l'infinité de variables t comme une algèbre de cohomologie (limite) de la grassmannienne et l'application qui associe τ à F plonge (la composante d'indice nul de) Gr_E dans une complétion convenable de S en envoyant les centres des cellules de Schubert E_Y sur les caractères τ_Y. C'est aussi la formule de Giambelli (cf. Griffith et Harris) qui exprime le (co)cycle de Schubert en fonction des classe de Chern (ici les t_i).

Pour retrouver la solution de KP associée à $(\Gamma, p, z, \vec{e})$, cf. (20) on choisit un diviseur (q) de degré g étranger à p_0, disons $(q) = q_1 + q_2 + \cdots + q_g$, dans Γ et on prend comme espace F celui des séries de Laurent $\sum_{n>>-\infty} a_n z^n$ qui convergent dans un voisinage épointé de 0 et qui s'étendent sur Γ en des fonctions méromorphes uniformes avec un diviseur des pôles majoré par (q) (i.e. avec au plus un pôle pour chaque q_i). Le théorème de Riemann-Roch dit que F est d'indice 0.

Afin de relier τ à θ (et de dire le rapport entre $\vec{e}$ et (q)) on développe l'application d'Abel:

$$\forall k, \quad 1 \leq k \leq g, \quad \omega_k(z) = \sum_{n \geq 1} n a_{kn} z^{n-1} dz$$

(ce sont les formes holomorphes de base) et on choisit des formes sans A_i-périodes $(1 \leq i \leq g)$ avec pôles seulement en p_0:

$$\forall n, \quad n \geq 1, \quad \beta_n(z) = -n z^{-n-1} dz - \sum_{m \geq 1} q_{nm} z^{m-1} dz$$

(on a $\oint_{B_k} \beta_n = -2\pi i a_{kn}$ par réciprocité). Notons $Q(t)$ la forme quadratique $\sum_n \sum_m q_{nm} t_n t_m$; choisissons des chemins γ_i de p_0 aux q_i et notons $l(t)$ la forme linéaire

$$\sum_{i=1}^{g} \text{p.f.} \int_{\gamma_i} \sum_n t_n \beta_n - \frac{1}{2} \sum_n \sum_{l+k=n} t_n \frac{q_{lk}}{lk}$$

(ici, pour un chemin γ de p_0 à un point de coordonnée z on considère la partie finie p.f. $\int_\gamma \beta_n = z^{-n} - \sum q_{nm} \frac{z^m}{m}$ et on l'étend analytiquement à $q \in \Gamma$ quelconque). Enfin, introduisons le vecteur $A(q)$ dans $\mathbb{C}^g$ de composantes $A_k(q) = \sum_{i=1}^{g} \int_{\gamma_i} \omega_k$ qui relève l'image d'Abel, $A(q) = \sum_1^g A(q_i)$.

Soient Δ le vecteur de Riemann (un certain vecteur fixe associé à p_0 dans $\mathbb{C}^g$) et A la matrice de format $g \times \infty$ de coefficients $A_{kl} = l a_{kl}$, alors (avec une constante C indéterminée à priori) on a

$$\tau_F(t_1, t_2, \cdots) = Ce^{l(t)+\frac{1}{2}Q(t)}\theta(-\Delta + A(q) - A \cdot t) \tag{24}$$

si bien que le vecteur $\vec{e}$ de la solution géométrique de KP (formule (20)) est égal à $A(q) - \Delta$.

Dans le potentiel u (de (20) ou (22)) le terme linéaire l disparait (ainsi que la constante C) et de Q il ne reste que la constante q_{11} (notée c dans (20)). La forme quadratique Q, également introduite par Riemann, est une version infinie de la matrice des périodes.

Remarque: La même construction marche pour les diviseurs (q) de tout degré et même pour les diviseurs non-effectifs, c'est à dire que (q) peut être n'importe quelle combinaison à coefficients dans $\mathbb{Z}$ de points de Γ distincts de p_0, seulement l'espace F n'est plus d'indice 0; on s'y ramène en jouant avec z en p_0: si le degré de (q) est $(g - m)$, $m \in \mathbb{Z}$, on considère l'espace F' des séries $f(z) = \sum_{n>>-\infty} a_n z^n$ convergentes autour de 0, telles que $z^{-m} f(z)$ s'étende à Γ avec un diviseur des pôles majoré par (q). D'ailleurs la formule (24) garde un sens quel que soit (q) et donne toujours une solution de KP. (Souvent dans (24) on divise par $\theta(A(q) - \Delta)$ pour faire $\tau(0) = 1$, mais cela demande une hypothèse supplémentaire sur (q) et de toute façon, à proprement parler, τ n'a été définie qu'à la constante multiplicative près.) Il faut faire attention que dans Segal-Wilson la donnée géométrique est un peu différente: (q) y est remplacé par un fibré en droite trivialisé près de p_0 et cela influe sur le terme $\exp(l(t))$.

En particulier, pour une courbe elliptique $\mathbb{C}/\Lambda$ (marquée par l'origine 0) et $(q) = \emptyset$, on a

$$\tau(x, 0, \cdots) = \sigma(-x) = C \exp(-\frac{e_2}{2} x^2) \theta_{11}(-x).$$

Si $\Lambda = \mathbb{Z} + 2\omega'\mathbb{Z}$, on pose $q = \exp(4\pi i\omega')$, alors

$$\tau(x) = e^{-\frac{e_2}{2}x^2} \left(2\frac{i\omega'}{\pi}\right) \prod_n (1 - q^n)^3 \sum_n e^{-2i\pi nx} q^{\frac{n^2+n}{2}} (-1)^n$$

vérifie $\tau(0) = 1$ et donne bien $u(x) = -2\wp(x)$. (Dans ce cas la forme holomorphe ω_1 est dz, donc $a_{11} = 1$ et $a_{1n} = 0$ dès que $n \geq 2$; τ ne dépend des t_n, $n \geq 2$, que par l'intermédiaire de Q et de l; la forme β_1 est égale à $d(\zeta(z) - 2\eta z) = -E_2 dz$ et $q_{11} = -e_2$. Dès qu'on fait intervenir un diviseur $q_1 \in \Gamma$ il apparait un facteur $\exp(x\zeta(q_1))$ dans $\tau(x, 0, \cdots)$ en accord avec la tradition.)

Dans le travail entrepris par Treibich et Verdier il intervient d'autres données géométriques que les courbes lisses; or un avantage de la présentation grassmannienne est la généralisation facile aux courbes singulières

(Mumford, Segal, Wilson).

La donnée géométrique est une courbe Γ complète (i.e. projective) et intègre (i.e. réduite et irréductible), c'est à dire une surface de Riemann compacte connexe avec un nombre fini de points singuliers (cf. le livre de J. P. Serre: *Groupes algébriques et corps de classes*).

En plus on fixe un point p_0 de la partie lisse Γ^0 et une coordonnée analytique z centrée en p_0; ce qui joue le rôle du diviseur (q) est un faisceau algébrique cohérent $\mathcal{L}$ génériquement de rang 1 sans torsion muni d'une trivialisation φ (des sections locales) au voisinage de p_0. L'espace F est alors celui des séries de Laurent $\sum_{n>>-\infty} a_n z^n$, vues comme sections (méromorphes) de $\mathcal{L}$ près de p_0 à l'aide de φ, qui s'étendent en sections globales au dessus de Γ. Si $\chi(\mathcal{L}) = \dim H^0(\Gamma, \mathcal{L}) - \dim H^1(\Gamma, \mathcal{L})$ est la caractéristique d'Euler de $\mathcal{L}$, l'indice de F est $\chi(\mathcal{L}) - 1$. La formule (21) dans le cas générique d'indice 0, ou une formule analogue dans le cas général, définit encore une fonction τ et la formule (22) donne encore une solution de KP; le choix de φ n'influe pas sur le potentiel u.

Les classes d'isomorphismes de $\mathcal{L}$ possibles à χ fixé font une variété projective (donc compacte) $W(\Gamma)$. Les (classes de) fibrés holomorphes de rang 1 appartiennent à $W(\Gamma)$; ils donnent tout lorsque Γ est lisse et presque tout lorsque Γ est tracée sur une surface complexe lisse.

Même quand Γ est singulière on peut définir sa jacobienne (généralisée) $Jac(\Gamma)$ (Rosenlicht, Serre, Grothendieck...): c'est l'ensemble des classes de fibrés de rang 1 de degré 0. Le produit tensoriel en fait un groupe abélien qui agit naturellement sur les modules $W(\Gamma)$; la partie W^0 de W sur laquelle Jac agit librement est l'ensemble des faisceaux qui ne sont pas images directes de faisceaux (sans torsion, etc...) sur une courbe moins singulière, les faisceaux "maximaux". Notons que l'espace F ne change pas par image directe (cf. S.W.).

Dans les cas qu'on verra les singulariés de Γ sont celles d'une courbe plane, alors $W(\Gamma)$ est intègre (en particulier irréductible) et W^0 coincide avec l'ensemble des fibrés en droite (du bon degré). (Réf. Altman, Kleiman et Iarrobino.)

En général $Jac(\Gamma)$ n'est pas compacte: si $\Gamma' \to \Gamma$ est la désingularisation minimale de Γ, une courbe lisse connexe de genre g, on obtient $Jac(\Gamma)$ comme extension de groupes abéliens; c'est l'extension de $Jac(\Gamma')$ (une variété abélienne $\mathbb{C}^g/\Lambda'$) par un groupe $U \times T$ où U est unipotent (analytiquement un $\mathbb{C}^r$) et T un tore (analytiquement $(\mathbb{C}^\times)^s$). Cette extension n'est triviale (i.e. un produit) que si $\Gamma' = \Gamma$ ou $\Gamma' = \mathbb{P}^1$. La dimension de $Jac(\Gamma)$ est le genre arithmétique $\pi = g + \delta$ de Γ, et $Jac(\Gamma)$ peut aussi se voir comme le quotient d'un espace vectoriel, le dual des 1-formes régulières sur Γ (définies par des conditions de résidus aux singularités, cf. le livre de Serre) par le sous-groupe discret $H_1(\Gamma^0; \mathbb{Z})$. Une application d'Abel A plonge la partie lisse Γ^0 dans la jacobienne de Γ.

Par exemple les n-solitons de Hirota proviennent des courbes Γ rationnelles ($g = 0$) avec n points doubles ordinaires ($\delta = n$). Donnons nous $2n$ points distincts de $\mathbb{C}^\times = \mathbb{P}^1 \setminus (\{0\} \cup \{\infty\})$ et répartissons les en couples, soient $(p_1, q_1), \cdots, (p_n, q_n)$, puis choisissons n constantes non nulles $(\lambda_1, \cdots, \lambda_n)$ (par exemple $(1, \cdots, 1)$); un espace F (qui donnera un n-soliton) est celui des polynômes f de z et z^{-1} de degré $\leq n$ en z^{-1}, soumis aux n contraintes $f(p_i) = \lambda_i f(q_i)$. La courbe Γ s'obtient à partir de $\mathbb{P}^1$ en collant p_i à q_i ($\forall i,\ 1 \leq i \leq n$), la coordonnée locale en 0 est z; la jacobienne est $(\mathbb{C}^\times)^n$, elle s'identifie à l'orbite des flots de la hierarchie KP. Pour une solution de KdV il faut imposer $p_i = -q_i, \forall i$ (cf.S.W.).

Toujours avec une courbe Γ rationnelle, mais possédant cette fois une singularité irreductible, on trouve le bosquet des solutions rationnelles de KP. La désingularisation Γ' est la droite projective $\mathbb{P}^1$ et la singularité x_0 est un point cuspidal (le plus simple est le point de rebroussement ordinaire sur la courbe plane $y^2 = x^3$); on choisit une coordonnée rationnelle z sur Γ (une bijection avec $\mathbb{P}^1$) qui vaut ∞ en x_0 et 0 en p_0. L'anneau des fonctions sur Γ régulières hors de p_0 est représenté par une sous algèbre R des polynômes de Laurent en z. On suppose que la trivialisation φ du faisceau $\mathcal{L}$ est donnée sur toute la partie lisse $\Gamma^0 = \mathbb{C}$; comme $\mathcal{L}$ est décrit par ses sections sur $\Gamma \setminus \{p_0\}$ on peut l'identifier par le sous-espace F de l'espace E des séries de Laurent. Cet espace F doit être un module sur l'anneau R. Les points F de la grassmannienne qu'on attrape ainsi sont ceux qui sont coincés entre un $z^{-l}E_-$ et un z^kE_- (i.e. $\exists\, l, k \in \mathbb{N}, z^{-l}E_- \subset F \subset z^kE_-$ où E_- est l'espace des polynômes de z^{-1}); c'est la petite grassmannienne Gr_f de Sato qui paramètre les sous espaces vectoriels de $E_f = E_- \oplus E_+$, Fredholm au dessus de E_-; elle correspond au sous-ensemble Gr_0 de Gr_H chez Segal et Wilson, les espaces $E_{Y,m}$ aux centres des cellules de Schubert sont dedans.

Un cas spécialement intéressant pour nous est celui où $z^{-2}F \subset F$ et où F est d'indice 0, ce cas est traité dans les paragraphes 7 et 8 de S.W.: alors Γ est isomorphe à la courbe plane $y^2 = x^{2m+1}$ complétée par un point lisse p_0 à l'infini; l'équation paramétrique est $x = z^{-2},\ y = z^{-1-2m}$ (le nombre de croisements δ en x_0 est égal à m; c'est un invariant analytique de Γ). L'anneau local (algébrique) R en x_0 est l'algèbre des polynômes en z^{-2} et z^{-1-2m}; le sous-espace F de E, pour être un R-module doit satisfaire à $z^{-2}F \subset F$ et $z^{-1-2m}F \subset F$ et pour l'indice 0 on demande que $z^{-m}E_- \subset F \subset z^mE_-$.

Tous ces points F de Gr_E vont donner des solutions de la hiérarchie KdV. Le plus simple est F_m qui est engendré, comme espace vectoriel sur $\mathbb{C}$, par les monômes

$$z^m, z^{m-2}, z^{m-4}, \ldots, z^{2-m}, z^{-m}, z^{-m-1}, z^{-m-2}, z^{-m-3}, \ldots;$$

c'est un E_Y, correspondant au diagramme d'Young $f_1 = m, f_2 = m -$

$1,\dots,f_m = 1$. Le potentiel u_m fourni par les formules (23) et (22) n'est autre que $-m(m+1)x^{-2}$ (en faisant $t_1 = x, t_2 = 0, t_3 = 0,\dots$), c'est la dégénerescence rationnelle du potentiel de Lamé. Soit U l'ensemble des polynômes de degré $2m-1$ en z^{-1} de la forme $1 + a_1 z^{-1} + a_2 z^{-2} + \dots + a_{2m-1} z^{1-2m}$; c'est un groupe unipotent pour la multiplication usuelle de polynômes tronquée à l'ordre $-2m$ (cf. Serre p. 100). La formule $\exp\left(\sum_1^{2m-1} t_k z^{-k}\right) \equiv \sum a_l z^{-l}$ modulo z^{-2m} établit un isomorphisme de groupes entre $\mathbb{C}^{2m-1}$ et U. La jacobienne de Γ s'identifie au sous-groupe C_m de U, image de l'espace des coordonnées impaires $(t_1, 0, t_3, 0, \cdots, t_{2m-1})$; on a $a_1 = t_1, a_2 = \frac{1}{2}t_1^2, a_3 = \frac{1}{6}t_1^3 + t_3$, etc. Le sous-espace F associé à un polynôme $p(z^{-1})$ dans C_m est engendré par $z^m p(z^{-1}), z^{m-2}p(z^{-1}),\dots, z^{2-m}p(z^{-1}), z^{-m}, z^{-m-1},\dots$ Les m premiers flots de KdV agissent transitivement sur C_m en translatant les variables t_{2k-1}. (La série $z^m \exp(t_1 z^{-1} + \dots + t_{2m-1}z^{1-2m})$ appartient à la complétion hilbertienne de F, et F est l'unique sous-espace de E d'indice 0 tel que $z^{-2}F \subset F$, $z^{-m}E_- \subset F \subset z^m E_-$ contenant $z^m p(z^{-1})$.)

L'ensemble $P_E = Gr_0^{(2)}$ (cf. S.W.) des F d'indice 0 de Gr_E tels que $z^{-2}F \subset F$ se laisse aussi décrire comme l'orbite de E_- dans Gr_E sous l'action du semi-groupe $GL_2[z^{-1}, z]$ des polynômes de Laurent à valeurs dans $GL_2(\mathbb{C})$: à une série de Laurent $f(z) = f_1(z^2) + z^{-1}f_2(z^2)$, la matrice $M(z) = \begin{pmatrix} a(z) & b(z) \\ c(z) & d(z) \end{pmatrix}$ associe la série $(Mf)(z) = a(z^2)f_1(z^2) + b(z^2)f_2(z^2) + z^{-1}(c(z^2)f_1(z^2) + d(z^2)f_2(z^2))$.

L'espace homogène P_E admet donc une décomposition cellulaire très simple: pour chaque degré $m \in \mathbb{N}$, une seule cellule C_m de dimension m (centrée sur $u_m = -m(m+1)x^{-2}$). Et on obtient comme cela toutes les solutions rationnelles de KdV qui tendent vers 0 (A.McK.M. 77).

Cette décomposition avait été reconnue par Airault, Mc Kean, et Moser en 1977. Posons $n = \frac{1}{2}m(m+1)$. Dans le produit symétrique $S_n = \mathbb{C}^n/\mathbb{S}_n$ considérons le sous-ensemble L_n^0 des n-uples de points distincts $(x_1,\dots,x_n)$ vérifiant les équations (6) $\forall i, \sum_{j\neq i}(x_i - x_j)^{-3} = 0$ et notons L_n l'adhérence de L_n^0 dans S_n; Airault, McKean, et Moser montraient que les flots de KdV agissent sur L_n via la formule (5): $u(x; t_1, t_3, \cdots) = -2\sum_{i=1}^n (x - x_i(t_1, t_3, \cdots))^{-2}$, que pour $k \geq m$ le champ ∂_{2k+1} est identiquement nul et que l'action des m premiers flots est libre et transitive. Donc L_n est une cellule de dimension m.

Remarque: près de ∞, $u(x) = -2(nx^{-2} + 2s_1 x^{-3} + 3s_2 x^{-4} + \cdots)$ où $s_k = x_1^k + \cdots + x_n^k$; en faisant appel à la relation de récurrence de Gelfand-Dickey $\partial_{2k+1} = P_3 \partial^{-1} \partial_{2k-1}$ où $P_3 = \frac{1}{4}\partial^3 + \frac{1}{2}u\partial + \frac{1}{2}\partial o u$ (et $\partial = \partial/\partial x = \partial_1$), on trouve que le terme dominant à l'infini de $\partial_{2k+1}u$ est égal à $4x^{-3-2k} \cdot \frac{3}{2} \cdot \frac{5}{3} \cdots \frac{2k-1}{k-1} \cdot \frac{2k+1}{k} \cdot n(n-1)(n-3)(n-6)\cdots(n - \frac{1}{2}k(k+1))$. C'est de là que sort la contrainte de rationalité $n = \frac{1}{2}m(m+1)$. Et il faut faire ce détour par KdV pour voir que les L_n sont vides lorsque n n'est pas triangulaire (cf. A. McK.M. pp. 110 et 132).

Nous passons alors près de quelques coins d'ombres. Il semble difficile de décrire directement sur les x_i le lieu L_n, c'est à dire d'examiner les confluences permises des pôles de u (elles ne peuvent se faire que par paquets triangulaires); on aimerait également comparer les flots KdV ∂_{2k+1} et les flots hamiltoniens $H_{2k} = \frac{1}{2k+1}Tr(L^{2k+1})$ qui se déduisent de la matrice L de Moser (3) (pour $k = 3$, M. Adler a calculé que ces champs sont proportionnels), on voudrait aussi résoudre les problèmes de Cauchy en t_5, t_7, etc. comme l'ont fait Olshanetsky et Perelomov pour $t = t_3$.

Puis nous nous engageons dans un fourré plus épais: que sait-on de l'espace M_E des solitons trigonométriques de KdV (les n-solitons de Hirota qui seraient des solutions périodiques en x de la hierarchie KdV)? A-t-on une décomposition de M_E en tores dont certains renferment les potentiels $-m(m+1)(\sin^{-2} x - \frac{1}{3})$? (L'ensemble L_n^0 est remplacé par l'ensemble des n-uples de points distincts $(z_1, \cdots, z_n)$ de $\mathbb{C}^\times = \mathbb{C} \setminus \{0\}$ vérifiant

$$\forall i, \quad \sum_{j\neq i} z_i z_j (z_i + z_j)/(z_i - z_j)^3 = 0,$$

A.McK.M. en disent peu de chose.) Pourtant dans ce fourré on aperçoit des fruits alléchants, une déformation trigonométrique du projectif infini (qu'on peut voir cyclique ou torique; il suffit de penser à la manière dont les opérateurs vertex engendrent les solitons par les matrices élémentaires pour respirer l'odeur de la K-théorie et des espaces de lacets).

Les matrices L de Calogero, Moser, et Krichever restent mystérieuses, quel rapport avec les formes de Lax des équations KP et KdV? Nous allons voir comment Krichever a levé un coin du voile et cela va nous permettre d'accéder à la question initiale de Burchnall, Chaundy et Krichever sur les sous-algèbres commutatives d'opérateurs différentiels en une variable:

La théorie KP établit une correspondance entre deux sortes d'objets, d'une part les ensembles $(\Gamma, p_0, z, \mathcal{L}, \varphi)$ constitués par une courbe (complète, intègre), un point lisse dessus, une coordonnée centrée en ce point (à l'ordre fini), un faisceau de rang 1 (cohérent, sans torsion, maximal, tel que $h^0 = h^1 = 0$) et une trivialisation locale, et d'autre part les algèbres abéliennes R possèdant au moins deux opérateurs $\partial^k + \cdots$ et $\partial^l + \cdots$ (où $\partial = \partial/\partial x$) d'ordres k et l premiers entre eux. Les éléments de R sont ceux qui commutent avec un (unique) opérateur pseudodifférentiel infini

$$L' = \partial + \sum_{i\geq 1} u_i(t)\partial^{-i} \tag{25}$$

Comment obtenir L'? Dans l'espace vectoriel F attaché aux données géométriques (et convenablement complété du côté de z^{-1}) on retient l'unique élément $\psi(z;t_1,t_2,\cdots)$ de la forme

$$\psi = \exp\left(\sum_{n\geq 1} t_n z^{-n}\right)\cdot\left(1+\sum_{m\geq 1} a_m(t) z^m\right); \tag{26}$$

c'est la fonction de Baker-Akhiezer. (Dans les cas géométriques lisses, ψ est l'unique fonction uniforme sur Γ avec le développement asymptotique en p_0 donné par l'exponentielle et avec des pôles majorés par le diviseur (q).) Alors on pose

$$S(t) = 1 + \sum_{m\geq 1} a_m(t)\partial^{-m} \tag{27}$$

et

$$L'(t) = S(t)\partial S^{-1}(t)$$

On a $L'\psi = z^{-1}\psi$ et si B_n est la partie polynomiale en ∂ de $(L')^n$, on a $\partial_n\psi = B_n\psi$ et $\partial_n L' = [B_n, L']$. La hierarchie KP est équivalente aux équations de Zakharov-Shabat:

$$\partial_n B_m - \partial_m B_n = [B_n, B_m] \tag{28}$$

L'algèbre R est l'ensemble des combinaisons linéaires finies $\sum_{n\geq 0} c_n B_n$ telles que $(\sum c_n\partial_n)L' = 0$. Le potentiel u de $B_2 = \partial^2 + u$ est solution de KP. Lorsque $(L')^2$ est égal à l'opérateur différentiel B_2, l'algèbre R est le commutant de B_2, la fonction ψ ne dépend que des variables impaires $t_1, t_3, \cdots$, et elle est fonction propre de tous les opérateurs B_n. Le potentiel u est solution de KdV. Et dans les cas géométriques la courbe Γ est hyperelliptique et la coordonnée z est impaire.

(Pour les sous-algèbres abéliennes contenant au moins un opérateur $\partial^k + \cdots$, Krichever a montré qu'il faut considérer des faisceaux de rang $r \geq 1$ sur Γ et a trouvé des "solutions de rang r" de KP. Comme le demandait Jean-Louis: que dire des sous algèbres abéliennes qui ne contiennent aucun $\partial^k + \cdots$?)

Les fonctions ψ et τ se déduisent l'une de l'autre par une célèbre formule de Sato et Krichever:

$$\psi(z;t_1,t_2,\cdots) = \frac{e^{\sum t_n z^{-n}}\check{\tau}(t_1 - z, t_2 - \frac{z^2}{2}, \cdots)}{\check{\tau}(t_1, t_2, \cdots)} \tag{29}$$

(*Attention*: rappelons que $\check{\tau}(t_1,t_2,\ldots) = \tau(-t_1,-t_2,\ldots)$. Le choix de τ est "imposé" par la théorie quantique des champs, celui de ψ par la théorie des équations différentielles; ce qui explique les signes dans la formule (29). Si l'on change $\check{\tau}$ en τ et t_n en $-t_n$ dans le membre de droite on trouve la "fonction conjuguée" $\psi^*(z;t_1,t_2,\ldots)$ qui n'est pas dans F (en général), mais qui est associée à l'adjoint formel de l'opérateur L'. La fonction $\check{\psi}^*$ engendre une "solution duale" de la hierarchie KP, qui répond à la dualité de Serre sur les courbes algébriques.)

La formule (29) est l'expression différentielle des équations de Plücker de la grassmannienne. Elle a donc une portée très générale, elle concerne toute solution de la hierarchie KP. Quand les données sont géométriques et lisses on la doit à Matveev et Its pour KdV et à Krichever pour KP.

C'est par ce chemin que Krichever arrive aux solitons elliptiques et fait le lien entre les matrices L et M des systèmes hamiltoniens classiques ((3) et (8)) et les opérateurs L' et B_n de KP ((25) et (28)).

Etant donnée une courbe elliptique $X = \mathbb{C}/\Lambda$, pour tout point a de X nous avons les matrices de taille $n \times n$, $L(a)$ et $M(a)$ définies par (10) et (11), avec les fonctions α et β de la page 9; elles dépendent de $(x_1, \ldots, x_n)$ et $(p_1, \ldots, p_n)$, elles sont doublement périodiques en a et se transforment par conjugaison simple $Ad(g(a, w))$ lorsqu'on déplace x_i d'une période $w \in \Lambda$. Soient λ une variable complexe supplémentaire et Γ la courbe d'équation $\det(L(a) - \lambda \text{ Id}) = 0$ dans le plan de coordonnées (a, λ); on complète Γ par n points lisses au dessus de l'origine $a = 0$ de X et on la marque par le point p_0 où $-\frac{\lambda}{2} + \frac{1}{a}$ présente un pôle simple. Lorsque les (x_i, p_i) sont guidés par les hamiltoniens $H_k = \frac{1}{k+1} Tr(L(a))^{k+1}$, elle ne bouge pas; c'est la "courbe spectrale" de L. Les fonctions H_k dépendent de a mais nous avons vu (page 14) que les gradients symplectiques de H_0, H_1 et $H_2'' = -\frac{3}{4}H_2 + 3(n-1)\wp(a)H_0$ n'en dépendent pas. De plus tous les crochets de Poisson $\{H_k(a), H_{k'}(a')\}$ sont nuls. A présent laissons évoluer $u(x, y, t) = -2\sum_1^n \wp(x - x_i(y, t))$ selon les champs de vecteurs, $\partial_2 = \partial_y$, $\partial_3 = \partial_t$ de KP, formules (18) et (19); les points x_i et leurs dérivées $p_i = \frac{\partial x_i}{\partial y}$ évoluent alors suivant H_1 pour y et suivant H_2'' pour t. Pour tout point $p = (a, \lambda)$ de Γ choisissons un vecteur propre $\vec{v}(y, t; p)$ de $L(a; y, t)$ associé à la valeur propre λ et notons $(v_1, \cdots, v_n)$ ses composantes; enfin considérons la fonction suivante sur Γ:

$$\psi(p; x, y, t) = e^{-\frac{\lambda}{2}x + \frac{\lambda^2}{4}y - \frac{\lambda^3}{8}t} \sum_{i=1}^{n} v_i(y, t; p)\alpha(x - x_i(y, t), a);$$

l'observation de Krichever est que ψ est la fonction de Baker-Akhiezer provenant d'une donnée géométrique $(\Gamma, p_0, z; (q))$ (pour une normalisation convenable en (y, t) de $\vec{v}$).

La courbe Γ est de genre n, z est une coordonnée de l'ordre de a/n au voisinage de p_0 et le diviseur (q) est de degré $(n-1)$; le tout (y compris Γ) dépend des conditions initiales $x_i(0,0), p_i(0,0)$ et doit être générique. Le développement asymptotique de ψ au voisinage de p_0 s'écrit:

$$\psi(p;x,y,t) = e^{xz^{-1}+yz^{-2}+tz^{-3}}\left(\frac{1}{z}+\sum_{m\geq 0} a_m(x,y,t)z^m\right).$$

Si bien que, par l'intermédiaire des fonctions α (qui sont des fonctions de Baker-Akhiezer pour la première équation de Lamé, $n=1$), la matrice L représente le générateur L' et M représente $B_2 = \partial^2 + u$. (On peut de même identifier des matrices représentant tous les B_n; Treibich, Verdier (89).) La fonction θ de Γ s'exprime comme un produit de fonctions $\sigma(x-x_i)$ de $\mathbb{C}/\Lambda$, tout au moins en restriction aux trois premiers flots $\partial_1, \partial_2, \partial_3$, de KP dans la jacobienne de Γ.

Malgré tout, le tableau des solitons elliptiques restait incomplet au début des années 80 et l'on savait encore trop peu sur la spécialisation à KdV.

Le problème que Jean-Louis Verdier et Armando Treibich-Kohn se sont posé était de trouver *toutes* les solutions elliptiques de KP et de KdV. Et ils ont assez largement rempli leur programme dans les deux articles:

Solitons elliptiques, in Grothendieck Festschrift, volume 3, Progress in Math. 88, Birkhauser, 1990, pp 437–480, et *Variétés de Krichever des solitons elliptiques*, à paraitre dans les proceedings de la conférence Indo-Francaise de Géométrie de Bombay (1989), édités par A. Beauville et S. Ramanan. (Ces textes sont rassemblés, avec des suppléments, dans la thèse de A. Treibich-Kohn soutenue à Rennes le 8 Janvier 1991; à cela il faut ajouter à présent l'article paru en Novembre 1992 au Duke Math. Journal: *Revêtements exceptionnels et sommes de 4 nombres triangulaires*, vol. 68, No.2, pp.217 à 236.)

Afin de se libérer des hypothèses de généricités ils ont placé les idées de Krichever sous un éclairage de géométrie algébrique et en ont tiré une jolie récolte de structure et d'arithmétique.

Tout d'abord ils ont utilisé un peu de dualité pour comprendre la relation entre la courbe elliptique (X) et les courbes spectrales (Γ) qui interviennent dans les constructions de Krichever: soient Γ une courbe algébrique sur $\mathbb{C}$ projective et intègre (c'est à dire une surface de Riemann compacte connexe avec un nombre fini de points singuliers), p_0 un point lisse de Γ et π un morphisme fini de degré n (c'est à dire un revêtement ramifié à n feuillets) de Γ sur une courbe elliptique X qui envoie p_0 sur l'élément neutre $q_0 = 0$ de X.

Définition 1. On dit que π est un *revêtement tangentiel* si l'application d'image réciproque π^* entre fibrés en droites de degré 0 envoie X (identifiée à sa jacobienne) sur une courbe tangente à l'image des points lisses Γ^0 de

Γ par l'application d'Abel A dans $Jac(\Gamma)$.

(Il s'agit de la jacobienne généralisée et $A(p)$ est le fibré associé au diviseur $p - p_0$.) On dira que π est *minimal* s'il ne peut se factoriser par un morphisme birationnel $\Gamma \to \Gamma'$ suivi d'un revêtement tangentiel $\pi' : \Gamma' \to X$, (et primitif s'il ne peut se factoriser par un revêtement tangentiel $\Gamma \to X'$ suivi d'une isogénie $X' \to X$).

Le premier résultat important décrit à isomorphisme près l'ensemble des revêtements tangentiels d'une courbe X; pour le formuler on introduit la surface $S = S(X)$:

Soit $\Delta \to X$ l'espace des modules de fibrés de rang 1 (et de degré 0) sur X munis d'une connexion holomorphe (fibrés plats); Δ est aussi l'unique extension de groupes algébrique non scindable de X par la droite affine $G_a = \mathbb{C}$ (surface de Serre); la *surface* S est la complétion projective fibre à fibre de Δ (Δ se plonge canoniquement comme sous-fibré affine d'un fibré vectoriel de rang 2 sur X et S est le fibré en droites projectives associé). C'est une surface réglée dont les génératrices sont les fibres $S_x, x \in X$ de $\pi_S : S \to X$; elle contient une unique courbe elliptique C_0 d'auto-intersection nulle (la section à l'infini; p_0 désignera le point $C_0 \cap S_{q_0}$. Les automorphismes se S au dessus de X sont les éléments du groupe affine de dimension 1 qui agit naturellement sur S en fixant C_0.

Pour tout entier n on appellera S_n le système linéaire $|nC_0 + S_0|$ (i.e. l'ensemble des diviseurs effectifs de S linéairement équivalents à $nC_0 + S_0$; ce système n'a pas de composante fixe et son unique point fixe est p_0; son élément générique est lisse de genre n.

En passant par une caractérisation cohomologique de revêtements tangentiels, Treibich et Verdier démontrent:

Théorème 1. *La condition nécessaire et suffisante pour qu'un morphisme de degré n, $\pi : (\Gamma, p) \to (X, q_0)$ soit tangentiel est qu'il se factorise par un morphisme birationnel sur une courbe de S : $j : \Gamma \to \Gamma' = j(\Gamma) \subset S(X)$ tel que $j(p) = p_0$. Alors Γ' appartient à S_n, et j est un plongement de Γ dans S si et seulement si π est minimal. Inversement toute courbe réduite et irréductible du système S_n passe par p_0, est lisse en p_0 et sa projection sur X est un revêtement tangentiel minimal. Si deux applications j_1 et j_2 relèvent le même revêtement il existe un unique automorphime μ de S au dessus de X tel que $j_2 = \mu \circ j_1$.*

De là on voit que tout revêtement tangentiel domine un minimal et surtout on met la main sur un véritable espace des modules de revêtements tangentiels minimaux de degré n de X: dans (l'espace projectif) S_n le complémentaire $V_n(X)$ de l'ensemble des diviseurs réductibles possède une structure naturelle d'espace affine de dimension n. L'ensemble des classes d'équivalence de revêtements tangentiels (minimaux) s'identifie à l'espace affine de dimension $n-1$, $V_n(X)/\mathrm{Aut}(S, X)$.

Remarques. 1) Tout morphisme de groupes algébrique $X \to Jac(\Gamma)$ est induit par un revêtement ramifié $\Gamma \to X$, qui n'est pas forcément tangentiel (cf. A. Treibich, CRAS, 1988).

2) En remplaçant X par une courbe Y quelconque on pourrait de même chercher les morphismes plats $\pi : \Gamma \to Y$ tels que $\pi^* : Jac(Y) \to Jac(\Gamma)$ envoie Y^0 tangentiellement à Γ^0 en 0, mais A. Treibich (Duke M. J. 1988) montre que Y doit être de genre arithmétique 1 sauf si deg(π)=2 et que Y et Γ sont des courbes hyperelliptiques.

3) Plus près des arguments de Krichever est la notion de polynôme tangentiel de Treibich (89); la courbe spectrale correspond à $j(\Gamma)$ par équivalence birationnelle entre $\mathbb{P}^1 \times X$ et $S(X)$. De cette façon A. Treibich a développé des algorithmes pour faire des calculs explicites de revêtements tangentiels.

Puisque tout élément Γ de $V_n(X)$ est une courbe à singularités planaires, l'ensemble des fibrés de rang 1 est dense dans la variété projective $W(\Gamma)$; c'est l'ensemble noté tout à l'heure $W^0(\Gamma)$ où $Jac(\Gamma)$ agit librement. Nous noterons $W_n(X)$ le schéma projectif réunion des $W(\Gamma)$ lorsque Γ décrit $V_n(X)$ (et nous nous restreindrons aux faisceaux de caractéristique d'Euler zéro).

On fixera un réseau Λ de $\mathbb{C}$ tel que X soit isomorphe à $\mathbb{C}/\Lambda$ et on notera a la coordonnée de $\mathbb{C}$; sur S il existe alors une unique fonction méromorphe κ, infinie sur $C_0 + S_0$ seulement, anti-invariante (pour l'involution ι qui remonte $a \to -a$) et telle que $\kappa + \frac{1}{a}$ possède un pôle simple sur C_0 en p_0; on choisira comme coordonnée locale z sur Γ en p_0 (quel que soit $\Gamma \in V_n(X)$) la fonction $z = (\kappa + a^{-1})^{-1}$.

Soit $Sol_n(X)$ l'ensemble des solitons elliptiques de KP sur X (cf.(17), i.e. les $u(x,y,t) = -2\sum_1^n \wp(x - x_i(y,t)))$; A. Treibich et J.-L. Verdier démontrent que la construction de Krichever (éventuellement généralisée) via la fonction τ (et la coordonnée z) définit une application sol_n de $W_n(X)$ dans $Sol_n(X)$, et le principal résultat de leur article sur les "variétés de Krichever" s'énonce

Théorème 2. *sol_n est une bijection.*

On peut préciser beaucoup en adoptant le point de vue hamiltonien (cf.(8)) comme l'avaient fait Calogero, Moser et Krichever à l'origine: notons $Sym^n(X)$ le quotient $X^n/\mathbb{S}_n$ (du produit cartésien X^n de X par elle même n fois par le groupe symétrique) et $TSym^n(X)$ son fibré tangent; ci_n désignera l'application de $Sol_n(X)$ dans $TSym^n(X)$ qui à $u(x,y,t)$ associe les conditions initiales

$$(x_1(0,0),\ldots;\frac{\partial x_1}{\partial y}(0,0),\ldots).$$

Par ailleurs, quels que soient $\Gamma \in V_n(X)$ et $\mathcal{L} \in W(\Gamma)$, l'orbite $X \cdot \mathcal{L}$ de $\mathcal{L}$ sous l'action de $X \subset Jac(\Gamma)$ rencontre le diviseur Θ_Γ canonique de $W(\Gamma)$ (l'ensemble des faisceaux de caractéristique 0 tels que $h^0 > 0$) en n points (avec multiplicités) et cela définit une application $i_n : W_n(X) \to Sym^n(X)$. La composée $W_n \overset{sol_n}{\to} Sol_n \overset{ci_n}{\to} TSym^n \overset{\pi_n}{\to} Sym^n$ coincide avec i_n.

Au dessus des n-uples de points de X deux à deux *distincts* l'application ci_n est une bijection (cela caractérise les solutions "en position générale"); mais en général ci_n n'est ni surjective ni injective. (Il est important de remarquer que l'application i_n est partout bien définie sur chaque $W(\Gamma)$, c'est à dire que $X \cdot \mathcal{L}$ n'est contenue dans Θ_Γ pour aucun faisceau $\mathcal{L}$. A ce point intervient la formule de Fay-Segal-Wilson qui dit que $\tau_F(t_1, 0, 0, \cdots)$ n'est jamais identiquement nulle; en fait l'ordre d'annulation de cette fonction en $t_1 = 0$ est égal à la codimension de la strate de la grassmannienne à laquelle F appartient.)

Soient Δ_n la grande diagonale de Sym^n ($\exists\ i, j, x_i = x_j$); au dessus de son complémentaire, sur $T(Sym^n - \Delta_n)$, il y a une forme symplectique naturelle définie par coordonnées de Darboux, $\omega = \sum_1^n dx_i \wedge dp_i$, et cette forme s'étend à $TSym^n X$ en une forme à pôle logarithmique le long de Δ_n. Il résulte d'un théorème de Deligne (sur une variété complète, toute forme à singularité logarithmique est fermée) que ci_n envoit chaque $W(\Gamma)$ sur une sous-variété lagrangienne de $TSym^n(X)$ (c'est l'image qui s'appelle variété de Krichever). Le champ ∂_y de KP sur $W_n(X)$ s'en va ainsi sur le gradient symplectique de H_1 dans $TSym^n$; c'est lui qui relève i_n de W_n à $TSym^n$.

A. Treibich-Kohn et J.-L. Verdier montrent aussi comment des matrices de Calogero-Moser-Krichever arrivent par image directe de la fonction méromorphe $\kappa_{|\Gamma}$ et pourquoi les hamiltoniens qui s'en déduisent sur $TSym^n(X)$ s'identifient aux flots ∂_m de KP (en principe ce sont des quantités calculables).

Venons en au point plus délicat: comment déterminer dans $Sol_n(X)$ les solitons elliptiques de KdV? Il faut d'abord suivre l'effet des involutions:

Définition 2. Disons qu'un revêtement tangentiel $\pi : \Gamma \to X$ est *symétrique* s'il existe une involution ι de Γ fixant p_0 au dessus de l'involution canonique ($a \to -a$) de X et qu'il est *hyperelliptique* si le quotient de Γ par ι est la droite projective.

Nous dirons *hypertangentiel* en raccourci de tangentiel hyperelliptique; il faut faire attention qu'un revêtement hypertangentiel minimal parmi les hypertangentiels peut ne pas être minimal parmi les tangentiels.

Sur la surface $S(X)$ on a une involution canonique ι qui relève celle de X; elle a huit points fixes. Notons $\omega_0 = q_0, \omega_1, \omega_2, \omega_3$, les 4 points d'ordre 2 de X, $s_0 = p_0, s_1, s_2, s_3$, les 4 points de C_0 qui se trouvent au dessus des ω_i, et encore r_0, r_1, r_2, r_3, les 4 points d'ordre 2 du groupe Δ, avec $\pi_S(r_i) = \omega_i$; les huit points s, r sont les points fixes de ι.

Si $\pi : (\Gamma, p) \to (X, 0)$ est tangentiel symétrique, il existe une unique application $j : \Gamma \to S(X)$ relevant π invariante par ι (envoyant p en p_0); dans $S(X)$ la courbe $\Gamma' = j(\Gamma)$ est lisse en $s_0 = p_0$, évite s_1, s_2, s_3, et a pour multiplicité des nombres μ_i en r_i. On a nécessairement $\mu_0 - n$ impair et $\mu_i - n$ pair si $i \geq 1$; un tel quadruplet $(\mu_0, \mu_1, \mu_2, \mu_3)$ s'appelle un "type adapté" (à n) de revêtement (tangentiel symétrique, de degré n).

Lorsque de plus π est hyperelliptique on doit avoir $\mu_0^2 + \mu_1^2 + \mu_2^2 + \mu_3^2 \leq 2n+1$, donc le nombre de types adaptés possibles pour un degré donné est fini. Par ailleurs on sait que le genre arithmétique de $\Gamma' \in V_n(X)$ est égal à n, mais si $\pi : \Gamma \to X$ est hypertangentiel de type μ son genre arithmétique m devra satisfaire à l'inégalité $2m < \mu_0 + \mu_1 + \mu_2 + \mu_3$, d'où $\frac{1}{2}m(m+1) \leq n$. Et l'on voit que Γ' peut être singulière; en fait, mis à part des cas triviaux, les courbes Γ' (tangentielles minimales) qui vont intervenir pour KdV sont toujours singulières. Quand Γ est minimale au sens hyperelliptique on a $m = \frac{1}{2}(\mu_0 + \mu_1 + \mu_2 + \mu_3 - 1)$.

Soit $S^\perp$ la surface S après qu'on ait éclaté ses huit points r et s; l'involution ι se relève à $S^\perp$ et le quotient $S^\perp/\iota$ est la désingularisation minimale $\tilde{S}$ de S/ι. Tout revêtement hypertangentiel $\Gamma \to X$ se factorise par un unique morphisme $j^\perp : \Gamma \to S^\perp$, et l'image $j^\perp(\Gamma)/\iota$ dans $\tilde{S}$ est une courbe rationnelle. Les courbes $j^\perp(\Gamma)$ donnent les hypertangentiels minimaux; elles sont lisses en dehors des quatre droites exeptionnelles au dessus des points r. En étudiant de près les courbes rationnelles de certains systèmes linéaires de $\tilde{S}$, Treibich et Verdier établissent:

Théorème 3. *Pour tout $n > 0$ il n'existe qu'un nombre fini de classes d'isomorphismes de revêtements hypertangentiels minimaux de degré n de X, tous de genre arithmétique inférieur ou égal au nombre $\gamma(n) = \sup\{m | n \geq \frac{1}{2}m(m+1)\}$.*

Les revêtements avec $\sum_0^3 \mu_i^2 = 2n + 1$ correspondent à des droites exeptionnelles de $\tilde{S}$ (des droites d'auto-intersection -1); il y en a *exactement un* pour chaque type possible. Un travail arithmétique de J. Oesterlé montre que le nombre de types adaptés possibles est croissant en $O(n \log\log n)$. On peut aussi estimer le nombre des revêtements de genre arithmétique égal à $\gamma(n)$; on trouve qu'il en existe un unique si et seulement si n est triangulaire ($n = \frac{1}{2}m(m+1), m = \gamma(n)$); son type est fixe: $\mu_0 = 3\left[\frac{m}{2}\right] - m + 1, \mu_i = m - \left[\frac{m}{2}\right], i > 0$, et il correspond au potentiel de Lamé $-2n\wp(x)$.

Maintenant on peut dire ce qui se passe pour les solitons elliptiques de KdV: Rappelons que Airault, McKean et Moser ont prouvé que les pôles de $u(x,t) = -2\sum_1^n \wp(x - x_i(t))$, solution de KdV (équation (4)), sont astreints à demeurer dans le sous-ensemble L_n de $Sym^n(X)$ défini par les n équations (14) ($\forall i, \quad \sum_{j\neq i}\wp'(x_i - x_j) = 0$), et aussi qu'ils ne passent pas toute leur vie dans la diagonale Δ_n (c'est à dire que deux pôles de u finissent toujours par se séparer).

Notons $H_n(X)$ le sous-ensemble des éléments hypertangentiels de $V_n(X)$; c'est une réunion de composantes de l'ensemble $SV_n(X)$ des points fixes de ι dans V_n. (SV_n correspond aux symétriques.) Soit $WH_n(X)$ la réunion des $W(\Gamma)$ concernés; la fonction τ permet d'envoyer WH_n dans $SolH_n$, ensemble des solitons elliptiques de KdV. Puisque, pour tout i entre 1 et n, on a $p_i = \frac{\partial x_i}{\partial y} = 0$, la restriction de $ci_n sol_n$ de W_n dans $TSym^n$ à WH_n va entièrement dans $L_n \subset Sym^n \subset TSym^n$, partie de la section nulle, et elle coincide avec la restriction de $i_n : W_n \to Sym^n$ à WH_n.

Théorème 4. *De $WH_n(X)$ dans $L_n(X)$, i_n ($= ci_n sol_n$) est un revêtement ramifié* bijectif. *Il n'y a qu'un nombre fini de composantes connexes dans $L_n(X)$. Si Γ appartient à $H_n(X)$, l'image de $W(\Gamma)$ (compactification de la jacobienne de Γ) est une composante irréductible de $L_n(X)$ et toute composante est ainsi couverte.*

Armando Treibich-Kohn et Jean-Louis Verdier sortent de cette analyse des "nouveaux solitons elliptiques"; leurs conditions initiales sont de la forme

$$u(x) = -2\sum_{i=0}^{3}\frac{1}{2}a_i(a_i+1)\wp(x-\omega_i) \tag{30}$$

où les a_i sont des nombres entiers positifs, et ils ne sont pas tous dans la famille de Lamé. En fait, tous les quadruplets (a_0, a_1, a_2, a_3) fournissent des potentiels de revêtements exceptionnels. Si bien que pour tout n il y a des solitons elliptiques de degré n et pas seulement lorsque n est triangulaire; par exemple

$$u(x) = -6\wp(x) - 2\wp(x-\omega_3)$$

est dans $SolH_4$. Que dire des non-exceptionnels?

(Dans l'article au Duke Math. Journal en 92 ou dans la note au C. R. A. S de 90 également intitulée *Revêtements tangentiels et sommes de 4 nombres triangulaires*, on trouve les résultats détaillés sur les a_i et les μ_i qui leur correspondent.)

C'est la question de l'équilibre des températures à l'intérieur d'un ellipsoïde qui avait amené Lamé (vers 1840) aux équations

$$\psi''(x) - m(m+1)\wp(x)\psi(x) = \lambda\psi(x).$$

(Il les écrivait avec les notations de Jacobi $\psi'' - m(m+1)k^2\mathrm{sn}_k^2(x)\psi = \mu\psi$ après translation en x d'une $\frac{1}{2}$-période et dilatation de x.

En posant

$$y = \mathrm{sn}_k^2(x) \quad \text{et} \quad P = y(y+1)(1-k^2y)$$

on trouve la forme algébrique

$$4P\frac{d^2\psi}{dy^2} + 2P'\frac{d\psi}{dy} - k^2m(m+1)y\psi = \mu\psi,$$

qui provient de l'écriture du laplacien de $\mathbb{R}^3$ en coordonnées elliptiques.)

La question du refroidissement à l'intérieur d'une sphère dans l'espace à trois dimensions donne les cas très particuliers, intégrables avec les fonctions élémentaires, de l'équation de Bessel:

$$\psi''(x) - m(m+1)x^{-2}\psi(x) = \lambda\psi(x); \quad m \in \mathbb{N}.$$

On reconnait les potentiels rationnels des pages 21 à 22, c'est à dire le point de départ de Calogero. Le potentiel $-\frac{m(m+1)}{x^2}$ (avec $x = \frac{\sqrt{2M}}{\hbar}r$) est aussi "l'énergie centrifuge" pour le facteur radial de la fonction d'onde d'une particule de moment cinétique m (et de masse M) dans un champ central symétrique en mécanique quantique.

(Il est amusant de faire apparaitre les dégénerescences circulaires du type soliton $-m(m+1)\sin^{-2}(x)$: soient φ et θ les angles d'Euler, θ la latitude comptée à partir du pôle nord et φ la longitude; le laplacien de la sphère S^2 de $\mathbb{R}^3$ s'écrit

$$\Delta_S V = \frac{\partial^2 V}{\partial\theta^2} + \cot\theta\frac{\partial V}{\partial\theta} + \frac{1}{\sin^2\theta}\frac{\partial^2 V}{\partial\varphi^2}.$$

Si bien que la fonction d'onde d'un spineur de Neveu-Schwarz sur S^2 de "moment" $m+\frac{1}{2}$ autour de l'axe des z est de la forme $V(\theta)\exp(i(m+\frac{1}{2})\varphi)$ où $V(\theta)$ satisfait à une équation

$$V''(\theta) + \cot\theta V'(\theta) - (m+\frac{1}{2})^2\sin^{-2}(\theta)V(\theta) = \lambda V(\theta),$$

et en posant $V(\theta) = \psi(\theta)\cdot(\sin\theta)^{-\frac{1}{2}}$ on retrouve

$$\psi''(\theta) - m(m+1)\sin^{-2}(\theta)\psi(\theta) = (\lambda - \frac{1}{4})\psi(\theta).$$

En coordonnées $z = \rho e^{i\varphi}$, $\rho = \cot(\theta)$, cela donne une équation de Gegenbauer $(1+\rho^2)(V'' - \rho V' + (m+\frac{1}{2})^2V) = \lambda V$.)

Lamé, puis Liouville et Heine, s'intéressaient surtout aux solutions uniformes sur la courbe elliptique (X), alors que la fonction ψ de Baker-Akhieser, formule (26), elle, est uniforme sur la courbe spectrale (Γ). Mais ψ se rattache aussi à de très anciennes formules.

En donnant la solution par les fonctions elliptiques du problème de la rotation d'un corps solide autour d'un point fixe dans le cas où il n'y a point de forces accélératrices, Jacobi (avant 1850) devait résoudre la première équation de Lamé ($m = 1$). Et l'on peut lire dans les leçons de Weierstrass la solution

$$\varphi(x,a) = e^{-x\zeta(a)} \frac{\sigma(x+a)}{\sigma(x)\sigma(a)}$$

qui a des multiplicateurs en x (voir page 9) mais qui est doublement périodique en a; la valeur propre est $\lambda = \wp(a)$ et l'autre solution est $\varphi(x,-a)$. Cette expression de φ est bien le premier cas de la formule (29) de Its, Matveev, Krichever, et Sato, car σ est la fonction τ de X (voir page 19).

Plus tard Hermite a montré que l'intégration du pendule sphérique (mouvement d'un point pesant sur une sphère) dépend de la deuxième équation de Lamé ($m = 2$), et Halphen a écrit les fonctions propres comme produits de deux fonctions φ associées à des paramètres a_1, a_2; on peut y voir la formule (29) d'une fonction ψ associée à un revêtement ramifié à deux feuillets de X. Pour $m \in \mathbb{N}^\times$ quelconque Hermite et Halphen ont analysé les fonctions ψ.

Cependant, à ma connaissance, il n'apparait pas dans ces travaux centenaires de fonctions θ (ou τ) de courbes hyperelliptiques. Elles vont être révélées par la théorie de Floquet: Prenons les potentiels de Lamé sous leurs formes réelles $u_m(x) = -m(m+1)k^2\mathrm{sn}_k^2(x)$, autorisons $m \in \mathbb{R}$, supposons k réel entre 0 et 1 et demandons nous pour quelles valeurs de m et de λ l'équation $\psi'' + u_m\psi = -\lambda\psi$ possède une ou plusieurs solutions de période $4K$ (i.e. le double d'une période de $\mathrm{sn}_k^2(x)$, la période réelle de $\mathrm{sn}_k(x)$). Autrement dit, cherchons le spectre joint de l'opérateur de Sturm-Liouville $-B : \psi \to -\psi'' - u\psi$ et de la translation $\psi(x) \to \psi(x+4K)$ (le "spectre périodique"). Et pour cela nous devons considérer tout le spectre L^2 de $-B$ (problème de Hill); les "zones d'instabilité" sont les intervalles du complémentaire du spectre (continu) de $-B$.

Après l'étude partielle de Guerritore (1909) le travail marquant est celui de Ince (Proc.Roy.Soc.Edinburgh; 1940): il y est démontré que pour chaque valeur entière de m on a $m+1$ intervalles d'instabilité $]-\infty, \lambda_0[,]\lambda_1, \lambda_2[, \dots,$ $]\lambda_{2m-1}, \lambda_{2m}[$ (pour chaque λ_i, $0 \leq i \leq 2m$, il y a une solution périodique unique), et qu'au contraire lorsque m appartient à $\mathbb{R}\backslash\mathbb{N}^\times$ le nombre de zones d'instabilité est infini (le spectre périodique est entièrement de multiplicité 1).

(Les fonctions périodiques satisfaisant à $\psi'' + u_m\psi + \lambda\psi = 0$ s'appellent fonctions de Lamé. Lorsque m est entier, on parle de polynômes de Lamé car ce sont des polynômes de degré m en les $\mathrm{sn}_k, \mathrm{cn}_k, dn_k$; ceux qu'étudiait Lamé précisément. Ces fonctions sont analogues aux fonctions de Mathieu, ou "fonctions du cylindre elliptique", qui se présentent à la "limite thermodynamique" $m \to \infty$, $k \to 0$, $m(m+1)k^2 \to l \in \mathbb{R}$, $u(x) = -l\sin^2(x)$; cf. le livre de Whittaker et Watson, *Modern Analysis*.)

Mais pour les courbes hyperelliptiques il faut attendre les années 70 et les jardiniers Novikov, Dubrovin, Adler, Lax, McKean, van Moerbeke, Airault, Moser, Matveev, Its, Krichever,... La courbe Γ est le revêtement de $\mathbb{P}^1(\mathbb{C})$ ramifié en $\lambda_0, \lambda_1, \cdots, \lambda_{2m}$ et ∞.

Ainsi le travail de Krichever sur les solitons elliptiques de KP et celui de Treibich et Verdier sur les solitons elliptiques de KdV apparaissent comme des cas de "réduction des intégrales abéliennes aux fonctions elliptiques" qui auraient ravi les grands amateurs de ce sujet classique, de Jacobi à Poincaré en passant par Hermite et Picard.

Les équations du second ordre de Treibich et Verdier (celles dont la formule (30) donne les potentiels) ont aujourd'hui 111 ans; Darboux les avait découvertes à propos d'équations des ondes en dimension $1+1$ (cf. *Sur une équation linéaire*, C.R.A.S. 1882, T. XCIV no. 25, pp. 1645–1648). Bien sûr il n'y avait alors ni KdV, ni le souci du nombre de zones d'instabilité, ni les revêtements tangentiels.

Je me rappelle la joie de Jean-Louis Verdier quand il a présenté ses nouveaux solitons à la R. C. P. 25. Ce qui les rattachait à la tradition lui plaisait bien, mais ce qui les reliait à la géométrie algébrique lui plaisait encore plus. Son style l'entraînait vers la perspective des structures.

Quel intérêt y a-t-il à décrire l'ensemble de toutes les solutions elliptiques de l'équation de Korteweg-de Vries?

Eh bien! Pour l'entrevoir il suffit de se reporter à l'ensemble des solutions rationnelles et à sa structure en cellules (pp.21 à 22); on s'est déja interrogé sur ce que pouvait être la déformation circulaire M_E (cf. p.23), un éventail infini de cellules toriques où l'on voit des dessins de nombres triangulaires. (On y trouvera sûrement les sources $-m(m+1)\sin^{-2}(x) - m'(m'+1)\cos^{-2}(x)$, est-ce qu'il y en aura d'autres? La lettre M évoque une théorie multiplicative.)

Ici, avec la variété des courbes elliptiques, on s'attend à découvrir une structure bien plus riche encore. L'ensemble *SolEll* de tous les solitons elliptiques de KdV est une réunion infinie de groupes algébriques commutatifs (le théorème 4 de Treibich-Verdier précise comment, si l'on se concentre sur les faisceaux maximaux); il est important de comprendre de quelle façon les blocs s'attachent les uns aux autres dans la grassmannienne infinie. On ne sait presque rien (sauf des finitudes) sur les "cellules" "non-exceptionnelles", mais déjà l'arrangement des exceptionnelles (qui sont "les plus grosses") a des couleurs éblouissantes, puisqu'il dépend de la décomposition des entiers (n) en sommes de 4 nombres triangulaires ($\frac{a_i(a_i+1)}{2}$) et de la décomposition des entiers impairs $(2n+1)$ en sommes de 4 carrés (μ_i^2) (Oesterlé, Treibich, Verdier). A cet endroit on rencontre une autre application (la plus étonnante) des fonctions elliptiques, à la théorie des nombres, car Jacobi a montré que les fonctions Thêta comptent les décompositions des entiers en carrés et en triangles (*Fundamenta Nova*,

1829). On ne pourra pas manquer de voir apparaître à nouveau les formes modulaires quand on s'occupera de la variation du réseau Λ. On devine caché le spectre de la cohomologie elliptique.

Peut-on imaginer des espaces mieux motivés que $P_E = SolRat$, $M_E = SolTrig$ et $SolEll$? Un problème est de décrire sur ces ensembles des structures naturelles d'espace de lacets infinis! Si on y arrive: de quelles cohomologies s'agit-il? Pourquoi les trouve-t-on dans cette région du monde?

Notre promenade s'arrête là. Je garde le souvenir d'allées bien taillées et puis d'un vaste parc avant d'entrer dans les bois. Ensuite j'ai marché plus longtemps que prévu, sans prendre le temps de m'arrêter; entre les feuilles et les ronces j'ai vu des ornements néolithiques (les premières figures de nombres triangulaires) et une lumière platonicienne (le silence du fermion libre), un temple de Kyoto où brille tout l'or du vide parfait et des chateaux d'architecture classique (les statues d'Abel, Jacobi, Hermite, Riemann,...). Je pensais aux excursions de Jean-Louis Verdier (la dualité, les catégories,...), et maintenant j'ai l'impression de respirer l'air des cimes enneigées, des montagnes en Inde où s'adosse le jardin.

Dans ce jardin on peut contempler les arbres vénérables d'une haute futaie, et sentir dans le sol se tordre de futures racines.

Le texte de l'histoire développe un exposé du 19 octobre 1989, à Paris 7, lors de la journée à la mémoire de Jean-Louis Verdier. J'aurais eu du mal à réussir sans les notes d'Alain Chenciner. L'aide la plus précieuse m'est venue de J.-L. Verdier lui-même (exposé à la R. C. P. 25 en 1988) et de A. Treibich-Kohn (sa thèse et puis ses commentaires). Enfin jamais le texte n'aurait vu le jour sans le soutien et les conseils de Yvette Kosmann-Schwarzbach et d'Olivier Babelon.

L'apport de Jean-Louis Verdier aux systèmes intégrables ne s'est pas limité aux travaux que je viens de mentionner: il y eut la description de l'ensemble des applications harmoniques de S^2 dans S^4 en relation avec les twisteurs de Penrose, puis l'étude des solitons elliptiques de l'équation de Boussinesq, avec Emma Previato. Et aussi un effort d'éclaircissement sur bien des sujets; par exemple son exposé sur les travaux de Drinfeld au Séminaire Bourbaki (no. 685, 1986–87) *Groupes quantiques* est une des meilleures références sur le passage des groupes de Lie-Poisson aux groupes quantiques. Dans les discussions avec d'autres chercheurs Jean-Louis a su exercer son infuence et communiquer son enthousiasme; par exemple avec Jimbo, Miwa, Ueno pour leur grand travail sur l'isomonodromie, ou avec les jeunes spécialistes de mécanique statistique Jaekel et Maillard, un rapprochement fécond et une invitation à la géométrie algébrique.

Enfin, Jean-Louis n'était pas avare de questions enrichissantes; l'une de celles qui lui tenaient le plus à cœur était la suivante: comment marier les solutions géométriques, attachées aux courbes algébriques (que nous avons vues dans l'exposé) avec les diffusons qui viennent du scattering-inverse (solutions L^2 de KdV par exemple)? Grâce aux opérateurs vertex qui sont des transformations de Bäcklund de KdV ou KP on peut implanter un n-soliton sur toute solution; comment faire de même avec les solutions quasi-périodiques? Ou avec toute solution sans coefficient de réflexion?

Beaucoup ont été très choqués par la disparition de Jean-Louis et de sa femme Yvonne. Sans doute le dramatique accident, l'arrêt brutal, y sont pour quelque chose. Mais surtout on sentait partout la place à part de Jean-Louis, faite de confiance non-aveugle, une chaleur réconfortante qui tout d'un coup manque.

Université de Strasbourg
IRMA
Laboratoire de Mathématiques
7, rue René Descartes
67084 – Strasbourg Cedex, France

PART I
Algebro-Geometric Methods and tau-functions

Compactified Jacobians of Tangential Covers

Armando Treibich

0. Introduction

0.1. The main purpose of this paper is to complete several geometric constructions, developed in [T-V, II] for the study of elliptic solitons associated with a given elliptic curve E. In fact we show that the compactified Jacobian of any tangential cover of degree n over E covers $E^{(n)}$, the nth symmetric product of E. It then follows that the *theta* divisor of that Jacobian is ample and naturally equipped with a *theta* function. Last but not least we prove a Torelli theorem for tangential covers within the frame of degree $n!$ coverings of $E^{(n)}$.

0.2. Let us recall that the moduli space of tangential covers of arithmetic genus and degree equal to n over E is isomorphic to $\mathbb{C}^{n-1}$. Moreover, its generic point corresponds to a smooth tangential cover whose compactified Jacobian is isomorphic to a principally polarized Jacobian variety. Nevertheless, for any $n \geq 2$, there exist families of singular tangential covers of arithmetic genus n to which our results apply, generalizing the classical properties of the Jacobian variety.

0.3. This paper is organized as follows: We start by stating very briefly the main properties of the Jacobians of a general integral projective curve over $\mathbb{C}$ (Section 1) and, more specifically, of a smooth curve (Section 2). The general background of the Krichever dictionary and the KP hierarchy is explained in Section 3, with special attention to vanishing-order formulas for the *tau* function, τ.

Starting from Section 4 we restrict ourselves to tangential covers $\pi\colon \Gamma \longrightarrow E$ as well as to elliptic KP solitons of degree $n \geq 1$. After recalling their basic properties we explicitly relate the compactified Jacobian of Γ, denoted by $W(\Gamma)$, to the family of elliptic KP solitons associated with Γ through the Krichever dictionary. The basic tool is a Calogero–Moser–Krichever-type matrix defined in terms of the initial condition of the elliptic KP solitons. We finally show that $W(\Gamma)$ is a ramified covering of $E^{(n)}$, $I\colon W(\Gamma) \longrightarrow E^{(n)}$, and we calculate the image of particular subvarieties of $W(\Gamma)$.

A more detailed study of the finite morphism, $I\colon W(\Gamma) \longrightarrow E^{(n)}$, carried out in Section 5, shows that the *theta* divisor Θ_Γ of $W(\Gamma)$ comes from $E^{(n)}$ and it is easily seen to be ample and admitting of a *theta* function. Furthermore, every tangential cover of genus and degree equal to n, $\pi\colon \Gamma \longrightarrow E$, defines a ramified covering of degree $n!$, $I\colon W(\Gamma) \longrightarrow E^{(n)}$, and conversely, given the morphism I, one can recover the curve Γ and the projection $\pi: \Gamma \longrightarrow E$.

1. Integral Projective Curves and Their Jacobians

1.1. Let Γ be an integral projective curve over $\mathbb{C}$, i.e., a compact Riemann surface with a finite number of singularities. Fix a point $p \in \Gamma^0$ of the dense open subset of smooth points of Γ. Canonically associated with the pointed curve (Γ, p) there is a Jacobian variety $\operatorname{Jac}\Gamma$ and the Abel embedding $Ab: \Gamma^0 \longrightarrow \operatorname{Jac}\Gamma$ which maps a point $p' \in \Gamma^0$ onto the divisor class of $p' - p$. The manifold $\operatorname{Jac}\Gamma$ is defined as the moduli space of degree 0 line-bundles over Γ. It is a connected commutative algebraic group over $\mathbb{C}$ whose dimension i s the arithmetic genus $g = h^1(\Gamma, \mathcal{O}_\Gamma)$. Let Γ_h denote the analytical complex space associated to Γ, and let $\mathcal{O}_{\Gamma_h}$ be its structural sheaf. The exponential sequence

$$0 \longrightarrow \mathbb{Z} \longrightarrow \mathcal{O}_{\Gamma_h} \xrightarrow{\exp} \mathcal{O}^*_{\Gamma_h} \longrightarrow 0$$

gives rise to the following exact cohomology sequence,

$$0 \longrightarrow H^1(\Gamma_h, \mathbb{Z}) \longrightarrow H^1(\Gamma_h, \mathcal{O}_{\Gamma_h}) \xrightarrow{\exp} H^1(\Gamma_h, \mathcal{O}^*_{\Gamma_h}) \longrightarrow H^2(\Gamma_h, \mathbb{Z}).$$

Moreover, it is known that $H^1(\Gamma_h, \mathbb{Z})$ has a finite number of generators and, by J.-P. Serre's (GAGA) results, that $H^1(\Gamma_h, \mathcal{O}_{\Gamma_h})$ and $H^1(\Gamma_h, \mathcal{O}^*_{\Gamma_h})$ are isomorphic to $H^1(\Gamma, \mathcal{O}_\Gamma)$ and to the Picard group $\operatorname{Pic}(\Gamma)$, respectively. It follows that $\operatorname{Jac}\Gamma$ is isomorphic to the quotient-variety $\frac{H^1(\Gamma, \mathcal{O}_\Gamma)}{H^1(\Gamma_h, \mathbb{Z})}$, and that $H^1(\Gamma, \mathcal{O}_\Gamma)$ is isomorphic to $T_{Jac\,\Gamma, 0}$, the tangent space at the origin of $\operatorname{Jac}\Gamma$. Whenever Γ is a smooth curve of genus $g = h^1(\Gamma, \mathcal{O}_\Gamma)$, Γ_h is topologically homeomorphic to a two-dimensional real sphere with g handles. In that case, $H^1(\Gamma_h, \mathbb{Z})$ $(\simeq \mathbb{Z}^{2g})$ is a lattice of $H^1(\Gamma, \mathcal{O}_\Gamma)$ and their quotient, $\operatorname{Jac}\Gamma$, is an Abelian variety.

1.2. The moduli space of all torsion-free, rank 1, coherent sheaves over Γ of degree $g-1$, denoted by $W(\Gamma)$, exists and will henceforth be called the *compactified Jacobian* of Γ. It is a projective variety on which $\operatorname{Jac}\Gamma$ acts via the tensor product. Whenever Γ is a smooth curve, $W(\Gamma)$ is isomorphic to $\operatorname{Jac}\Gamma$, but, in general, its structure encodes all the information about the possible desingularizations of Γ and it can be very complicated. Nevertheless, if Γ is integral and can be embedded in a smooth algebraic surface, $W(\Gamma)$ is reduced and irreducible, and $\operatorname{Jac}^{g-1}\Gamma$ is a dense open subset of $W(\Gamma)$. In other words, $W(\Gamma)$ appears in that case as a natural compactification of $\operatorname{Jac}^{g-1}\Gamma \simeq \operatorname{Jac}\Gamma$ ([A-I-K]).

1.3. Let $j: \overline{\Gamma} \longrightarrow \Gamma$ be a birational morphism between two curves, i.e., j is a partial desingularization of Γ, and denote by N_j the quotient-sheaf defined by the canonical injection $\mathcal{O}_\Gamma^* \longrightarrow j_*(\mathcal{O}_{\overline{\Gamma}}^*)$. It is a sheaf of Abelian groups with finite support over Γ and the cohomology sequence of

$$0 \longrightarrow \mathcal{O}_\Gamma^* \longrightarrow j_*(\mathcal{O}_{\overline{\Gamma}}^*) \longrightarrow N_j \longrightarrow 0$$

gives rise (by 1.1) to the following exact sequence,

$$0 \longrightarrow H^0(\Gamma, N_j) \longrightarrow \operatorname{Jac}\Gamma \xrightarrow{j^*} \operatorname{Jac}\overline{\Gamma} \longrightarrow 0. \tag{1.3.1}$$

In other words, the natural homomorphism $j^*\colon \operatorname{Jac}\Gamma \longrightarrow \operatorname{Jac}\overline{\Gamma}$, which associates with any line-bundle on Γ of degree 0 its inverse image by $j\colon \overline{\Gamma} \longrightarrow \Gamma$, is surjective and its kernel is a linear algebraic group isomorphic to $H^0(\Gamma, N_j)$. The extension (1.3.1) of $\operatorname{Jac}\overline{\Gamma}$ by $H^0(\Gamma, N_j)$ is trivial if and only if $\operatorname{Jac}\overline{\Gamma} = 0$, i.e. $\overline{\Gamma} \simeq \mathbb{P}^1(\mathbb{C})$, or $\operatorname{Jac}\Gamma \simeq \operatorname{Jac}\overline{\Gamma}$, i.e., $\Gamma \simeq \overline{\Gamma}$.

On the other hand for any sheaf $\mathcal{F}$ which belongs to $W(\overline{\Gamma})$, the direct image $j_*(\mathcal{F})$ is in $W(\Gamma)$. The corresponding map defines, by the projection formula, an equivariant embedding $j_*\colon W(\overline{\Gamma}) \longrightarrow W(\Gamma)$ with respect to the canonical actions of $\operatorname{Jac}\overline{\Gamma}$ and $\operatorname{Jac}\Gamma$. Furthermore, the *theta* divisors Θ_Γ and $\Theta_{\overline{\Gamma}}$ have as supports the subsets $\{\xi \in W(\Gamma),\ h^0(\xi) > 0\}$ and $\{\overline{\xi} \in W(\overline{\Gamma}),\ h^0(\overline{\xi} > 0\}$, respectively, and satisfy the relation $(j_*)^*(\Theta_\Gamma) = \Theta_{\overline{\Gamma}}$.

2. Smooth Projective Curves and Their Jacobians

2.1. Throughout this section let Γ be a smooth projective curve of genus $g \geq 1$. Tensoring $\operatorname{Jac}\Gamma$ by any sheaf in $W(\Gamma)$ defines an isomorphism of both Jacobians, and $(W(\Gamma), \Theta_\Gamma)$ becomes a principally polarized Abelian variety.

We will now present some of the main classical properties of the polarized variety $(W(\Gamma), \Theta_\Gamma)$ and its relation with Γ.

As usual, we shall start by recalling the existence of a holomorphic (*theta*) function, θ_Γ, over the universal covering of $W(\Gamma)$ which vanishes to order 1 along the ample divisor, $\Theta_\Gamma = \{\xi \in W(\Gamma),\ h^0(\xi) > 0\}$. A careful study of the complete linear systems $|m\Theta_\Gamma|$ $(m \geq 1)$ leads to the following results:

2.1.1. for any $m \geq 1$, $\dim|m\Theta_\Gamma| = m^g - 1$;

2.1.2. for any $m \geq 2$, the associated map $i_{m\Theta_\Gamma}\colon W(\Gamma) \longrightarrow \mathbb{P}^{m^g-1}$ is everywhere well defined;

2.1.3. for any $m \geq 3$, $i_{m\Theta_\Gamma}$ is an embedding, i.e., $m\Theta_\Gamma$ is a very ample divisor.

Let ω_Γ be the canonical sheaf of Γ with $\deg\omega_\Gamma = 2g-2$, and let $\tau\colon W(\Gamma) \longrightarrow W(\Gamma)$ be the involution $\xi \longmapsto \omega_\Gamma \otimes \xi^{-1}$. By the Riemann–Roch theorem, the divisor Θ_Γ is τ-invariant, and has 2^{2g} fixed points called θ-characteristics. The group-variety $\operatorname{Jac}\Gamma$ also has a canonical involution,

$\mathcal{L} \longmapsto \mathcal{L}^{-1}$, and tensoring by any θ-characteristic defines an equivariant isomorphism between $\operatorname{Jac}\Gamma$ and $W(\Gamma)$. The quotient of $W(\Gamma)$ by τ, called the *Kummer variety* of $(W(\Gamma), \Theta_\Gamma)$ is isomorphic to the image $i_{2\Theta_\Gamma}(W(\Gamma)) \subset \mathbb{P}^{2^g-1}$. Moreover, the morphism $i_{2\Theta_\Gamma}$ factors into the projection $W(\Gamma) \longrightarrow W(\Gamma)/\sim\tau$ followed by the embedding of $W(\Gamma)/\sim\tau$ in $\mathbb{P}^{2^g-1}$.

2.2. By choosing a point p of Γ we can identify $\operatorname{Jac}\Gamma$ with $W(\Gamma)$, $\mathcal{L} \longmapsto \mathcal{L}((g-1)p)$, and define another *Abel embedding* $A_p\colon \Gamma \longrightarrow W(\Gamma)$, $p' \longmapsto \mathcal{O}_\Gamma((g-2)p+p')$. Furthermore, for any $d < g$ there is a natural extension of A_p to the whole dth symmetric power of Γ,

$$A_p\colon \Gamma^{(d)} \longrightarrow W(\Gamma), \ \sum_{1\le i\le d} p_i \longmapsto \mathcal{O}_\Gamma\Big((g-d-1)p + \sum_{1\le i\le d} p_i\Big)$$

such that $A_p(\Gamma^{(d)})$ is an integral normal and Cohen–Macaulay subvariety birational to $\Gamma^{(d)}$ (cf., [A-C-G-H]). Remarkably, for $d = g-1$, $A_p\colon \Gamma^{(g-1)} \longrightarrow W(\Gamma)$ does not depend upon the point p chosen and $A_p(\Gamma^{(g-1)}) = \Theta_\Gamma$.

2.3. For any d-codimensional subvariety C of $W(\Gamma)$, where $0 \le d \le g$ we denote by $[C] \in H^{2d}(W(\Gamma), \mathbb{Z})$ the cohomology class associated to C and $[C]^m \in H^{2dm}(W(\Gamma), \mathbb{Z})$ where $m \ge 1$ the mth cup-power of $[C]$. For example, $[\Theta_\Gamma]^m = m![A_p(\Gamma^{(g-m)})]$ if $m < g$, and $[\Theta_\Gamma]^g = g![\text{point}] = g!$. Conversely, the Matsusaka criterion asserts that a principally polarized Abelian variety of dimension g, (W, θ), containing a smooth curve C such that $[\Theta]^{g-1} = (g-1)![C]$, is isomorphic to $(W(C), \Theta_C)$, the polarized Jacobian of C.

2.4. Take another smooth curve Γ' of genus g. The polarized Jacobians $(W(\Gamma), \Theta_\Gamma)$ and $(W(\Gamma'), \Theta_{\Gamma'})$ are isomorphic if and only if $\Gamma' \simeq \Gamma$. In other words, if $\Gamma \not\simeq \Gamma'$ there may exist an isomorphism $f\colon W(\Gamma') \xrightarrow{\sim} W(\Gamma)$, but the divisor $f^*(\Theta_\Gamma)$ will not be equal or cohomologous to $\Theta_{\Gamma'}$. This result, known as the *Torelli theorem*, may be made more precise: the curve Γ can be effectively recovered from the data $(W(\Gamma), \Theta_\Gamma)$ by looking at the image of Θ_Γ in $\mathbb{P}^{g-1}$ under the Gauss map (cf., [A-C-G-H]).

3. The Krichever Dictionary and the Wave (ψ) and Tau (τ) Functions

3.1. For the rest of the section, let Γ be a projective and integral curve and let λ be a local coordinate of Γ at a smooth point p. For any sheaf ξ in the compactified Jacobian of Γ, choosing a trivialization φ of ξ around p enables us to identify the meromorphic sections of ξ which are holomorphic over $\Gamma - \{p\}$ with a Laurent series in the variable λ. The ring $\bigoplus_{1\le m} H^0(\xi(mp))$ corresponds under the latter identification to a subspace of the Hilbert space $L^2(S^1)$, $S^1 = \{|\lambda| = 1\}$, whose closure w belongs to the Grassmannian $Gr = Gr(L^2(S^1))$ (cf., [S-W] §6). The so-called Krichever dictionary or correspondence associates with the algebraic geometry data $\mathcal{D} = (\Gamma, p, \lambda, \xi, \varphi)$ the point $w \in Gr$. On the other hand, any $w \in Gr$ possesses a *wave* (or Baker) function,

$$\psi_w(\lambda, (x, \mathbf{t})) = \exp(x\lambda^{-1} + t_2\lambda^{-2} + t_3\lambda^{-3} + \cdots)\left(1 + \sum_{j\geq 1} a_j(x, \mathbf{t})\lambda^j\right),$$

and a pseudo-differential operator

$$Q_w(x, \mathbf{t}) = \frac{\partial}{\partial x} + \sum_{j\geq 1} a_j(x, \mathbf{t})\left(\frac{\partial}{\partial x}\right)^{-j}$$

with meromorphic coefficients in the variables $(x, \mathbf{t}) = (x, (t_2, t_3, \ldots)) \in \mathbb{C} \times \mathbb{C}^\infty$, which satisfy the following evolution equations:

$$(3.1.1)_m \qquad \frac{\partial}{\partial t_m}\psi_w = P_m\psi_w, \quad m \geq 1;$$

where P_m is the differential operator part of $(Q_w)^m$, i.e., $P_m = [(Q_w)^m]_+$. We finally obtain the compatibility equations of the system of evolution equations $\{(3.1.1)_m\}$, namely for any integers $m, \ell \geq 2$:

$$(3.1.2)_{m,\ell} \qquad \left[\frac{\partial}{\partial t_m} - P_m, \frac{\partial}{\partial t_\ell} - P_\ell\right] = 0.$$

For example a straightforward calculation shows that

$$P_2 = \frac{\partial^2}{\partial x^2} + u \quad \left(u(x, \mathbf{t}) = -2\frac{\partial}{\partial x}a_1(x, \mathbf{t})\right);$$

$$P_3 = \frac{\partial^3}{\partial x^3} + \frac{3}{2}u\frac{\partial}{\partial x} + v \quad \left(v(x, \mathbf{t}) = -3\frac{\partial}{\partial x}a_1 + 3\frac{\partial^2}{\partial x^2}a_1 + 3\frac{\partial}{\partial x}a_2\right),$$

and from the compatibility equation $(3.1.2)_{2,3}$ we conclude that the function $u(x, \mathbf{t})$ above is a solution of the Kadomtsev–Petviashvili (KP) equation.

3.2. Every line bundle $\mathcal{F} \in \operatorname{Jac}\Gamma$ is isomorphic to the restriction to $\Gamma \times \{\mathcal{F}\}$ of the Poincaré line bundle over $\Gamma \times \operatorname{Jac}\Gamma$. The latter being trivial along $\{p\} \times \operatorname{Jac}\Gamma$, the sheaf $\mathcal{F}$ inherits a natural trivialization $\varphi_{\mathcal{F}}$, around p and acts on every data $\mathcal{D}$ as follows,

$$\mathcal{F} : \mathcal{D} = (\Gamma, p, \lambda, \xi, \varphi) \longmapsto (\Gamma, p, \lambda, \xi \otimes \mathcal{F}, \varphi \otimes \varphi_{\mathcal{F}}).$$

The Krichever correspondence $\mathcal{D} \longmapsto w \in Gr$ extends to the whole $\operatorname{Jac}\Gamma$-orbit of $\mathcal{D}$ and gives a natural embedding in the Grassmannian Gr of $\xi \operatorname{Jac}\Gamma$, the $\operatorname{Jac}\Gamma$-orbit of ξ in $W(\Gamma)$.

For any $n \geq 1$ and $t \in \mathbb{C}$, $\exp(t\lambda^{-n})$ acts on Gr by multiplication, $w \longmapsto \exp(t\lambda^{-n})w$, and generates the so-called nth KP flow. Let us now consider the nth derivative of the Abel map at 0 (see 1.1), $\frac{\partial^n}{\partial^n\lambda}Ab(\lambda)_{|\lambda=0}$, as a tangent vector at the origin of $\operatorname{Jac}\Gamma$, and write

$$U_n = -\frac{1}{n}\frac{\partial^n}{\partial^n\lambda}Ab(\lambda)_{|\lambda=0} \in T_{Jac\,\Gamma,0}.$$

A straightforward verification shows that the restriction of the nth KP flow to the orbit $\xi \operatorname{Jac}\Gamma$, $\xi \in W(\Gamma)$ naturally embedded in Gr (see 3.2) is generated by U_n. In other words, the map $\xi \operatorname{Jac}\Gamma \longrightarrow Gr$ is equivariant with respect to the actions of $\exp(t\lambda^{-n})$ on Gr and $\exp(tU_n) \in \operatorname{Jac}\Gamma$ on $\xi \operatorname{Jac}\Gamma$.

3.3. There exists a third important object attached to any $w \in Gr$, the *tau*-function $\tau_w(x, \mathbf{t})$, which is related to and clarifies the relations between ψ_w and Q_w. We now state its main properties:

3.3.a. $\tau_w(x, \mathbf{t})$ is defined and holomorphic over $\mathbb{C} \times \mathbb{C}^\infty$ viewed as the tangent space at w to the whole KP hierarchy flow.

3.3.b. Whenever w corresponds to a geometrico-algebraic data $\mathcal{D} = (\Gamma, p, \lambda, \xi, \varphi)$, the inverse image of τ_w by the embedding $\xi \operatorname{Jac}\Gamma \longrightarrow Gr$ is a holomorphic $H^1(\Gamma_h, \mathbb{Z})$-multiplicative function over $\xi T_{Jac\,\Gamma,0} = \xi H^1(\Gamma, \mathcal{O}_\Gamma)$ (see 1.1), whose zero divisor is well defined on $\xi \operatorname{Jac}\Gamma$ and equal to $(\xi \operatorname{Jac}\Gamma) \cap \Theta_\Gamma$. The restrictions of τ_w to any orbit $\xi \operatorname{Jac}\Gamma$ where $\xi \in W(\Gamma)$ generalize the usual Riemann *theta* function of the Jacobian variety.

3.3.c. The wave (or Baker) function ψ_w is equal, up to an exponential factor, to a quotient of suitable translates of τ_w. We can then easily deduce that the coefficients of the KP hierarchy differential operators $\{P_m, m \geq 2\}$, relating ψ_w and Q_w (see 3.1.1), can be expressed in terms of τ_w and its derivatives. For example, $P_2 = \frac{\partial^2}{\partial x^2} - 2\frac{\partial^2}{\partial x^2}\log\tau_w$ (see [S-W]), and it follows (3.1) that $-2\frac{\partial^2}{\partial x^2}\log\tau_w$ is a solution of the KP equation.

3.3.d. Putting together 3.3.b and c we get a large family of KP solutions parametrized by all data $\{\mathcal{D} = (\Gamma, p, \lambda, \xi, \varphi)\}$. In fact, we can drop the trivialization φ everywhere, since changing it amounts to multiplying the corresponding point $w \in Gr$ by some constant function, $c_0+c_1\lambda+c_2\lambda^2+\cdots$, $c_0 \neq 0$, and τ_w by a simple exponential, whereas the KP solution $u_w = -2\frac{\partial^2}{\partial x^2}\log\tau_w$ does not change.

3.4. The one-parameter subgroup of $\operatorname{Jac}\Gamma$ generated by the tangent vector $U_1 = \frac{\partial}{\partial\lambda}Ab(\lambda)|_{\lambda=0}$, is independent of the local coordinate λ and will be denoted by $G(\Gamma, p)$. A remarkable result of [S-W] §8 asserts that the orbit of any $\xi \in W(\Gamma)$ under the action of the first KP flow is equal to $\xi G(\Gamma,p)$ and is not contained in the *theta* divisor Θ_Γ. Furthermore Segal and Wilson calculate the multiplicity of intersection at ξ of Θ_Γ and $\xi G(\Gamma,p)$, denoted henceforth $\text{mult}_\xi(\Theta_\Gamma, \xi G(\Gamma,p))$, as the vanishing order at $x = 0$ of $\tau_w(x,0,0,\ldots)$. They prove in fact that $\text{mult}_\xi(\Theta_\Gamma, \xi G(\Gamma,p))$ is equal to the codimension of the stratum of Gr containing w. There is another formula for that multiplicity, proved only for *smooth* curves (cf., [F]) but very simple to calculate in many cases.

Proposition 3.5 (Fay's multiplicity formula). *Let Γ be a smooth curve of genus $g \geq 1$, ξ a sheaf in $W(\Gamma) = \text{Pic}^{g-1}\Gamma$ and $d = h^0(\Gamma,\xi)$. Taking $\{m_i,\ i = 1,\ldots,d\}$ and $\{n_i,\ i = 1,\ldots,d\}$ as the gap sequences of integers, defined by*

$$\begin{cases} h^0(\xi(-m_i p)) = d - i + 1 = h^1(\xi(n_i p)), \\ h^0(\xi(-(m_i+1)p)) = d - i = h^1(\xi((n_i+1)p)), \quad i = 1,\ldots,d, \end{cases}$$

we get the formula

$$\text{mult}_\xi(\Theta_\Gamma, \xi G(\Gamma,p)) = d + \sum_{1\leq i\leq d} (m_i + n_i).$$

3.6. The general Segal and Wilson multiplicity formula is a bit more involved. Let $w \in Gr$ correspond to the data $(\Gamma, p, \lambda, \xi, \varphi)$, $\xi \in W(\Gamma)$, and define the strictly increasing sequence of integers $\mathcal{S} = (s_i)_{i\in\mathbb{N}}$ by the condition $s \in \mathcal{S}$ if and only if $h^0(\xi(sp)) > h^0(\xi((s-1)p))$. The virtual cardinal of $\mathcal{S}$ is equal to $\chi(\xi) - 1 = h^0(\xi) - h^1(\xi) - 1 = -1$, i.e., for any $i >> 1$, $i - s_i = -1$ and [S-W] §8.6 asserts that:

Proposition 3.7 (Segal–Wilson multiplicity formula). *The vanishing order at $x = 0$ of $\tau_w(x,0,0,\ldots)$, the restriction of τ_w to the first* KP *flow orbit of w, is equal to*

$$\text{mult}_\xi(\Theta_\Gamma, \xi G(\Gamma,p)) = \sum_{0\leq i} (i - s_i + 1).$$

4. Elliptic Solitons Versus Tangential Covers and Polynomials

4.1. Let us now consider a lattice Λ of $\mathbb{C}$ and denote by E the elliptic curve $\mathbb{C}/\Lambda$, by $q \in E$ its origin and by z the natural local coordinate of E at q. The Weierstrass function $\mathfrak{p}$ associated with (E,z) is the unique section $\mathfrak{p}$ of $\mathcal{O}_E(2q)$ such that $\mathfrak{p}(z) - z^{-2}$ vanishes at q $(z=0)$. Given any meromorphic solution of KP, $u(x,y,t)$, that is Λ-periodic in the variable x, there exist (cf., [T-V, II] A3) an integer $n \geq 0$, functions $a(t)$ and $b(t)$, and an analytic map $(y,t) \longmapsto \sum_{1\le i\le n} x_i(y,t)$ from $\mathbb{C}^2$ to $E^{(n)}$, the nth symmetric power of E such that

$$u(x,y,t) = \sum_{1\le i\le n} \wp(x - x_i(y,t)) + b(t)y + a(t).$$

Furthermore, up to a simple reparametrization of $u(x,y,t)$ in the variables y and t, we can assume that $u(x,y,t) = \sum_{1\le i\le n} \wp(x-x_i(y,t))$ (and $x_i(0,0) \neq x_j(0,0)$ if $i \neq j$). These x-Λ-periodic KP solutions, the so-called elliptic solitons of degree n over E, define a completely integrable Hamiltonian system of n particles in $\mathbb{C}$ with a first Hamiltonian function,

$$H = \frac{1}{2} \sum_{1\le i\le n} p_i^2 + 2\sum_{i\neq j} \wp(x_i - x_j).$$

The set of functions $\{x_i(y,t),\ i = 1,\dots,n\}$ is uniquely determined by the following system of partial differential equations,

$$\begin{cases} \dfrac{\partial^2}{\partial y^2} x_i = 4 \displaystyle\sum_{\substack{1\le j\le n\\ j\neq i}} \wp'(x_i - x_j) & i = 1,\dots,n, \\ \dfrac{\partial}{\partial t} x_i = \dfrac{3}{4}\left(\dfrac{\partial}{\partial y} x_i\right)^2 - 3 \displaystyle\sum_{\substack{1\le j\le n\\ j\neq i}} \wp(x_i - x_j) & i = 1,\dots,n, \end{cases}$$

where $\wp'$ is the derivative of $\wp$.

Finally, it can be proved that the elliptic soliton $u(x,y,t)$ above is simply parametrized by the *initial condition* point $(\Sigma x_i,\ \frac{\partial}{\partial y}\Sigma x_i)(0,0)$ of $T_{E^{(n)}}$, the tangent-fiber space of $E^{(n)}$. In other words, for any point $(\Sigma\,\alpha_i,\ (p_i)) \in T_{E^{(n)}}$ such that $\alpha_i \neq \alpha_j$ if $i \neq j$, there exists a unique elliptic soliton

$$u = 2\sum_{1\le i\le n} \wp(x - x_i(y,t))$$

with initial condition point equal to $(\Sigma\,\alpha_i, (p_i))$.

4.3. Let $\pi\colon (\Gamma,p) \longrightarrow (E,q)$ be a finite pointed morphism of an integral curve Γ onto the elliptic curve E, and let $p \in \Gamma^0$ be a smooth point such that $\pi(p) = q$. Associated to π we have the Abel map, $Ab\colon \Gamma^0 \longrightarrow \operatorname{Jac}\Gamma$, $p' \longmapsto \mathcal{O}_\Gamma(p'-p)$, as well as the non-trivial homomorphism $\pi^*\colon E \longrightarrow \operatorname{Jac}\Gamma$, $\alpha \longmapsto \pi^*(\mathcal{O}_E(\alpha - q))$. The origin of $\operatorname{Jac}\Gamma$ belongs to both images ($Ab(p) = \pi^*(q) = \mathcal{O}_\Gamma$) and it is quite natural to look for tangency (see also (4.5)).

Definition 4.4. The finite pointed morphism $\pi\colon (\Gamma,p) \longrightarrow (E,q)$ is a *tangential cover* if and only if $\pi^*(E)$ and $Ab(\Gamma^0)$ are tangent at the origin of $\operatorname{Jac}\Gamma$.

Tangential covers are relevant to the elliptic soliton theory because "they stand on the other side of the mirror." In fact, their compactified Jacobians represent the algebraic geometry counterpart of elliptic solitons, as shown in (4.5) below:

Theorem 4.5 ([T-V, I & II]). *If (Γ, p) is the source of a tangential cover of degree $n \geq 1$, there exists a local coordinate of Γ at p, λ, such that for any sheaf ξ in $W(\Gamma)$, the* KP *solution associated with the data $(\Gamma, p, \lambda, \xi)$ (see (3.1)) is an elliptic soliton of degree n. Furthermore, the wave function $\psi(x, \mathbf{t})$ and the whole* KP *hierarchy of differential operators $\{P_m(x, \mathbf{t}),\ m \geq 2\}$ attached to the data $(\Gamma, p, \lambda, \xi)$ are Λ-periodic in the variable x. Conversely any elliptic soliton of degree $n \geq 1$ corresponds to some data $(\Gamma, p, \lambda, \xi)$ where (Γ, p) is the source of a tangential cover of degree n.*

4.6. An important feature of this new approach to elliptic solitons can be described in one word, effectiveness. For example, we will show how to construct, on a down-to-earth basis (4.14), all tangential covers, and then how to calculate algebraically the initial conditions of the corresponding elliptic solitons (4.18).

Let us start however with a non-effective tangency criterion.

Theorem 4.7 ([K], [T]). *A finite pointed morphism $\pi\colon (\Gamma, p) \longrightarrow (E, q)$ (see (4.1.3)) is a tangential cover if and only if there exists a meromorphic function k on Γ, called the tangential function for π, satisfying the following conditions:*

4.7.a. *k is a section of the sheaf $\mathcal{O}_\Gamma(\pi^*(q))$;*

4.7.b. *$k + \pi^*(z^{-1})$ has a simple pole at p;*

4.7.c. *$k + \pi^*(z^{-1})$ is defined at any point of $\pi^{-1}(q) - \{p\}$.*

Definition 4.8. Let

$$P(T) = T^n + \sum_{1 \leq j \leq n} \alpha_j T^{n-j}$$

be a monic polynomial of degree $n \geq 1$ with coefficients in the function field of E. We say that $P(T)$ is a *tangential polynomial* if and only if its coefficients are holomorphic over $E - \{q\}$ and it satisfies the condition that

(4.9) the coefficients of $zP(T - z^{-1})$ are holomorphic at q.

Remark 4.10. The tangency criterion 4.7 is of no help for the construction or for the study of the tangential covers unless coupled with the notion of tangential polynomials (4.11 & 4.14).

Theorem 4.11. *Let* $\pi: (\Gamma, p) \longrightarrow (E, q)$ *be a tangential cover of degree* n. *Then the tangential function of* π *is unique up to an additive constant and its characteristic polynomial, with respect to* π, *is a tangential polynomial of degree* n. *Conversely, any tangential polynomial corresponds to a unique tangential cover, equipped with the choice of a tangential function.*

4.12. The main tool in 4.11 is a particular ruled surface $S \longrightarrow E$ which factors any tangential cover $\Gamma \longrightarrow E$. Recall that a ruled surface over E is an algebraic surface R together with a surjective morphism $\pi_R: R \longrightarrow E$ whose fibers are isomorphic to $\mathbb{P}^1$. Let R_q denote the fiber of π_R over the point $q \in E$, and take a point r of R_q. The strict transform of R_q by the blow-up of R at r is an exceptional divisor of the first kind, in other words, a smooth rational curve of self-intersection -1. We can therefore contract it, getting another ruled surface $\mathrm{elm}_r(R) \longrightarrow E$, and a birational map R- - - > $\mathrm{elm}_r(R)$ over E, called the *elementary transformation with center* r.

4.13. Let us consider now the variable T as a rational function on $\mathbb{P}^1 = \mathbb{C} \cup \{\infty\}$ having a simple pole at ∞. We choose $\{T^{-1}, z\}$ as local coordinates of the trivial ruled surface $\mathbb{P}^1 \times E$ at $p_0 = (\infty, q)$, and denote by p_1 the *infinitely near* point corresponding to the tangent direction -1 at p_0. By blowing up p_0, then p_1 and contracting the strict transform of the fibers passing through p_0 and p_1, we construct a rational map from $\mathbb{P}^1 \times E \longrightarrow E$ to a new ruled surface $S \longrightarrow E$. That map, denoted by $\mathrm{elm}_{p_1,p_0}: \mathbb{P}^1 \times E \longrightarrow S$, is an isomorphism outside the fiber $\mathbb{P}^1 \times \{q\}$. Furthermore, the strict transform of $\{\infty\} \times E$, denoted by C_0, is the unique section of $S \longrightarrow E$ having zero self-intersection. The latter property characterizes the couple $(S \longrightarrow E, C_0)$ up to isomorphism.

Let S_q be the fiber of S over q and let $p_s = S_q \cap C_0$. The set of tangential polynomials of degree n over (E, z), denoted by $\mathcal{P}(n, E, z)$, is an n-dimensional affine space constructed by formal integration of the elements of $\mathcal{P}(n-1, E, z)$ ([T]). Theorem 4.14 now enables one to go backwards, from a tangential polynomial to a tangential cover endowed with a tangential function.

Theorem 4.14 ([T]). *For any* $P(T) \in \mathcal{P}(n, E, z)$ *the strict transform, by* elm_{p_1,p_0}, *of its divisor of zeroes* $\{P = 0\} \subset \mathbb{P}^1 \times E$ *is a divisor of* S, *called* $\Gamma(P)$, *satisfying the following properties:*

4.14.1. $\Gamma(P)$ *is an integral curve linearly equivalent to* $nC_0 + S_q$ *passing through the point* $p_s \in S$;

4.14.2. *the natural pointed morphism* $\pi_P: (\Gamma(P), p_s) \longrightarrow (E, q)$ *is a tangential cover of degree and arithmetic genus equal to* n;

4.14.3. *the strict transform of the projection* $\mathbb{P}^1 \times E \longrightarrow \mathbb{P}^1$, *restricted to* $\Gamma(P)$, *is a tangential function of* π_P, *and its characteristic polynomial is* *equal to* $P(T)$.

Furthermore, the map $P \longmapsto \Gamma(P)$ *defines an affine isomorphism between* $\mathcal{P}(n, E, z)$ *and* $I|nC_0+S_q|$, *the (dense open) subset of integral divisors in the complete linear system* $|nC_0 + S_q|$.

Remark 4.15. Start from any tangential cover of degree $n \geq 1$ equipped with a tangential function k and let $P(T) = P_k(T) \in \mathcal{P}(n, E, z)$ be its characteristic polynomial.

4.15.1. Associated with $P(T)$ there is another tangential cover $\pi_P: (\Gamma(P), p_s) \longrightarrow (E, q)$ and a birational morphism $j: \Gamma \longrightarrow \Gamma(P)$ which sends p to p_s and factors π as $\pi = \pi_P \circ j$. We conclude that the affine space $\mathcal{P}(n, E, z)$ (cf., 4.13) parametrizes all tangential covers of degree and arithmetic genus n, equipped with a tangential function.

4.15.2. Any other tangential function for π is equal to k + cst where cst $\in \mathbb{C}$ and its characteristic polynomial is equal to $P(T\text{-cst})$. It follows that we can uniquely normalize the choice of k by requiring that its trace, $\mathrm{Tr}_\pi(k)$, vanishes at q or, equivalently, that $P(T) = T^n + \alpha_2 T^{n-2} + \cdots + \alpha_n$. In other words the affine sub-space

$$\left\{P(T) \in \mathcal{P}(n, E, z),\ \frac{\partial^{n-1}}{\partial T^{n-1}} P_{|T=0} = 0\right\}$$

parametrizes all tangential covers of degree and arithmetic genus n.

4.16. Let

$$u(x, y, t) = 2 \sum_{1 \leq i \leq n} \wp(x - x_i(y, t))$$

be a KP elliptic soliton of degree n satisfying the generic condition, $x_i(0,0) \neq x_j(0,0)$ if $i \neq j$. Such a soliton clearly corresponds biunivocally to the initial condition point

$$\{(\alpha_i, p_i)\} = \left\{\left(x_i(0,0),\ \frac{\partial}{\partial y} x_i(0,0)\right)\right\},$$

as well as to some spectral data $(\Gamma, p, \lambda, \xi)$ (cf., 4.5). We get a natural map $\{(\alpha_i, p_i)\} \longmapsto (\Gamma, p, \lambda, \xi)$ and its inverse which will be formulated quite explicitly in (4.18) and (4.20).

Let $\sigma(z)$ denote the classical *sigma* function, i.e., Λ-multiplicative in $z \in \mathbb{C}$, such that $\frac{\partial^2}{\partial z^2}\sigma(z) = -\wp(z)$, and for any $\{(\alpha_i, p_i)\} \in (E \times \mathbb{C})^{(n)}$, where $\alpha_i \neq \alpha_j$ if $i \neq j$, define the so-called Calogero–Moser–Krichever matrix $A(\{(\alpha_i, p_i)\}, z)$ as follows,

$$\text{(4.16.1)} \qquad A_{\ell j}(\{(\alpha_i, p_i)\}, z) = \begin{cases} \frac{1}{2} p_\ell & \text{if } \ell = j, \\ \frac{\sigma(z+(\alpha_\ell - \alpha_j))\sigma(z+\alpha_j)\sigma(\alpha_\ell)}{\sigma(z)\sigma(z+\alpha_\ell)\sigma(\alpha_\ell - \alpha_j)\sigma(\alpha_j)} & \text{if } \ell \neq j. \end{cases}$$

Observe that as a function, $A_{\ell j}$ is meromorphic in $(\{(\alpha_i, p_i)\}, z)$ and Λ-periodic with respect to $z \in \mathbb{C}$. We will consider the matrix A above as a homomorphism of vector bundles $A: \mathcal{E} \longrightarrow \mathcal{E}(q)$ for

$$\mathcal{E} = \bigoplus_{1 \le i \le n} \mathcal{O}_E((-\alpha_i) - q).$$

4.17. Let ξ belong to $W(\Gamma)$, the compactified Jacobian of Γ, and let $k \in H^0(\Gamma, \mathcal{O}_\Gamma(\pi^*(q)))$ be a tangential function for the tangential cover $\pi: (\Gamma, p) \longrightarrow (E, q)$ and denote by λ the local coordinate of Γ at p such that $\lambda^{-1} = k + \pi^*(z^{-1})$. The multiplication by k defines sheaf homomorphisms $k: \xi \longrightarrow \xi(\pi^*(q))$ and $\pi_*(k): \pi_*(\xi) \longrightarrow \pi_*(\xi(\pi^*(q))) \simeq \pi_*(\xi)(q)$ such that

Theorem 4.18. *For every ξ in a dense open subset of $W(\Gamma)$, the initial condition point $\{(\alpha_i, p_i)\}$ of the elliptic soliton associated with $(\Gamma, p, \lambda, \xi)$ satisfies the following conditions:*

4.18.1. $\alpha_i \neq \alpha_j \neq q$ *if* $i \neq j$;

4.18.2. $\pi_*(\xi)$ *is isomorphic to the vector bundle* $\bigoplus_{1 \le i \le n} \mathcal{O}_E((-\alpha_i) - q)$;

4.18.3. $\pi_*(k): \pi_*(\xi) \longrightarrow \pi_*(\xi)(q)$ *is equal to the* C-M-K *matrix* $A(\{(\alpha_i, p_i)\}, z)$.

4.19. Conversely, $(\Gamma, \xi) \in W(n, E)$ is uniquely and effectively determined by the knowledge of the vector bundle $\mathcal{E} = \pi_*(\xi)$ and the homomorphism $A = \pi_*(k): \mathcal{E} \longrightarrow \mathcal{E}(q)$.

Theorem 4.20 ([T-V, II]). *The characteristic polynomial of A: $\mathcal{E} \longrightarrow \mathcal{E}(q)$, $P(T) = \det(TId - A)$, is tangential of degree n and Γ is isomorphic to the integral divisor $\Gamma(P)$ of S defined by 4.14. Furthermore, the multiplication by $\pi^*(A)$ and the tangential function k send $\pi^*(\mathcal{E})$ into $\pi^*(\mathcal{E}(q))$ and the sheaf $\xi \in W(\Gamma)$ we are looking for is isomorphic to the cokernel of $k - \pi^*(A)$ modulo torsion.*

4.21. Consider now a general pointed finite morphism $\pi: (C, r) \longrightarrow (E, q)$ such that C is an integral curve and $r \in C$ a smooth point. The one parameter subgroup of $\operatorname{Jac} C$ generated by the tangent to C at r (see 3.4), denoted by $G(C, r)$, acts on $W(C)$ and any $G(C, r)$-orbit $\xi G(C, r)$, where $\xi \in W(C)$, intersects properly the *theta* divisor Θ_C. On the other hand we know that any $\pi^*(E)$-orbit, $\xi\pi^*(E)$, intersects Θ_C with multiplicity equal to $\deg(\pi)$ ([T-V, II; App. 2]). In particular, for any tangential cover $\pi: (\Gamma, p) \longrightarrow (E, q)$ it is clear that $G(\Gamma, p) = \pi^*(E)$ and for any $\xi \in W(\Gamma)$, the inverse image of Θ_Γ by the orbit morphism $\operatorname{Orb}_\xi: E \longrightarrow W(\Gamma)$, $\alpha \longmapsto \xi(\pi^*(\alpha - q))$, is an effective divisor of degree n. In fact we obtain a morphism $I: W(\Gamma) \longrightarrow E^{(n)}$, $\xi \longmapsto I(\xi) = \operatorname{Orb}_\xi^*(\Theta_\Gamma)$

whose basic properties will be developed in Section 5. Let us only recall here that $I\colon W(\Gamma) \longrightarrow E^{(n)}$ factors through the morphism $AI\colon W(\Gamma) \longrightarrow T_{E^{(n)}}$, which sends ξ to the *initial condition* of the elliptic soliton associated to $(\Gamma, p, \lambda, \xi)$, followed by the natural projection $T_{E^{(n)}} \longrightarrow E^{(n)}$. It is also known that the morphism $AI\colon W(\Gamma) \longrightarrow AI(W(\Gamma))$ has degree one, and the projection $AI(W(\Gamma)) \longrightarrow E^{(n)}$ is a finite morphism ([T-V, II]).

Finally we carry out a straightforward calculation of the canonical sheaf ω_Γ, as well as of particular values of I (see 4.23, 25) needed for our Torelli assertion (5.14).

4.22. Let $S \longrightarrow E$ be the particular ruled surface defined earlier, let C_0 be its unique section of self-intersection zero, S_q the fiber over $q \in E$, and ω_S the canonical sheaf of S. Fix a rational function $\underline{k}$ of S, having $C_0 + S_q$ as divisor of poles (see [T]). We know that any tangential cover of degree and arithmetic genus n, $\pi\colon (\Gamma, p) \longrightarrow (E, q)$, $(\Gamma \in I|nC_0 + S_q|)$, is canonically equipped with an embedding $\iota\colon \Gamma \longrightarrow S$ and a tangential function k, equal to $\iota^*(\underline{k})$, the restriction of $\underline{k}$ to Γ. Let us denote by $P(T) \in \mathcal{P}(n, E, z)$ its characteristic polynomial, and by $P^{(j)}(T) = \frac{\partial^j}{\partial T^j} P(T)$ $(j = 1, \ldots, n)$ the jth derivative of $P(T)$. The adjunction formula tells us that ω_Γ, the canonical sheaf of Γ is locally free and isomorphic to $\iota^*(\omega_S(\Gamma))$. Moreover

Proposition 4.23. *For every $\Gamma \in I|nC_0 + S_q|$, the canonical sheaf ω_Γ is isomorphic to $\mathcal{O}_\Gamma((n-2)p + \pi^*(q))$ and $\{P^{(j)}(k)P^{(1)}(k)^{-1}dz, j = 1, \ldots, n\}$ is a basis of $H^0(\Gamma, \omega_\Gamma)$.*

Proof. The canonical sheaf of S is equal to $\omega_S = \mathcal{O}_S(-2C_0)$ and replacing in the adjunction formula gives $\omega_\Gamma = \mathcal{O}_\Gamma(\pi^*(q) + (n-2)p)$. On the other hand one can easily verify that $P^{(n)}(\underline{k})P^{(1)}(\underline{k})^{-1}d\underline{k} \wedge dz$ is a holomorphic section of $\omega_S(\Gamma - (n-2)C_0 - S_q)$ and that for any $j = 2, \ldots, n-1$, $P^{(j)}(\underline{k})P^{(1)}(\underline{k})^{-1}d\underline{k} \wedge dz$ is a holomorphic section of $\omega_S(\Gamma - (j-2)C_0)$. Taking the Poincaré residue over Γ of the latter $n-1$ meromorphic differential forms leads us to the following set of linearly independent *holomorphic* differentials:

$$w_n = P^{(n)}(k)P^{(1)}(k)^{-1}dz \in H^0(\Gamma, \omega_\Gamma((n-2)p - \pi^*(q))),$$
$$w_j = P^{(j)}(k)P^{(1)}(k)^{-1}dz \in H^0(\Gamma, \omega_\Gamma(-(j-2)p)), \quad j = 2, \ldots, n-1,$$

which we complete into a basis of $H^0(\omega_\Gamma)$ by adding $w_1 = dz$. □

4.24. The morphism $A_p\colon (\Gamma^0)^{(d)} \longrightarrow W(\Gamma)$,

$$\sum_{1 \leq j \leq d} p_i \longmapsto \mathcal{O}_\Gamma\Big((n-d-1)p + \sum_{1 \leq j \leq d} p_i\Big),$$

where $d \in \mathbb{N}^+$, is an embedding for $d = 1, \ldots, n-1$, and the closure in $W(\Gamma)$ of each image ($d \leq n$) is an integral subvariety of dimension d. A straightforward application of the Segal–Wilson formula (3.7), plus 4.23 and Serre's duality, enables us to evaluate the morphism $I: W(\Gamma) \longrightarrow E^{(n)}$ over the latter subvarieties:

Corollary 4.25. *For every $d = 1, \ldots, n-1$, the image $A_p((\Gamma^0)^{(d)})$ is contained in $I^{-1}((n-d)qE^{(d)})$. Furthermore, if $d = 1$ the map A_p can be extended to an embedding of Γ into $W(\Gamma)$ (cf., [A-K]) and $I \circ A_p(p_1) = (n-1)q + (-\pi(p_1))$ for any point $p_1 \in \Gamma$.*

Proof. For any element

$$\sum_{1 \leq j \leq d} p_j \in (\Gamma^0)^{(d)} \quad (1 \leq d \leq n-1)$$

and for $\ell = n - d - 1$, the locally free sheaves $\mathcal{L} = A_p(\Sigma p_j)$ and $\mathcal{L}(-\ell p)$ correspond to effective divisors and $H^0(\mathcal{L}(-\ell p))$ is a non-zero subspace of $H^0(\mathcal{L})$, i.e., $h^0(\mathcal{L}) \geq h^0(\mathcal{L}(-\ell p)) > 0$. It follows that $\text{mult}_{\mathcal{L}}(\Theta_\Gamma, \mathcal{L}\pi^*(E)) \geq n - d$ (see 3.5). Hence the sheaf $\mathcal{L}$ belongs to $I^{-1}((n-d)qE^{(d)})$.

For $d = 1$, on the other hand, $\mathcal{L} = \mathcal{O}_\Gamma((n-2)p + p_1)$, and its dual is isomorphic to $\mathcal{L}^\nu \simeq \omega_\Gamma \otimes \mathcal{L}^{-1} \simeq \omega_\Gamma(-(n-2)p - p_1) \simeq \mathcal{O}_\Gamma(\pi^*(q) - p_1)$ (4.23). Moreover, by the Riemann–Roch theorem and Serre's duality, $H^0(\mathcal{L}(\pi^*(q-\alpha))) \simeq H^0(\mathcal{L}^\nu(\pi^*(\alpha - q))) = H^0(\mathcal{O}_\Gamma(\pi^*(\alpha) - p_1))$, which is non-zero if we choose $\alpha = \pi(p_1)$. It easily follows that $I(\mathcal{O}_\Gamma((n-2)p+p_1)) = (n-1)q + (-\pi(p_1))$. □

5. Compactified Jacobians as Ramified Coverings of $E^{(n)}$

5.1. The moduli space $V(n, E)$ of all tangential covers of degree and arithmetic genus $g = n$ (equipped with a tangential function) is effectively constructed via $\mathcal{P}(n, E, z)$, the affine space of all tangential polynomials of degree n, and isomorphic to it. The same approach leads to a parametrization of $V(n, E)$ by $I|nC_0 + S_q|$, the set of all integral divisors of the ruled surface S linearly equivalent to $nC_0 + S_q$. It is also shown, by Bertini arguments, that the generic tangential cover $\Gamma \in V(n, E)$ is smooth ([T]). There exist however ("big") families of singular curves in $V(n, E)$. For example, choose any point r of S not lying in C_0 and an integer m such that $0 < m(m-1) \leq 2n-2$, then the subset of curves $\Gamma \in V(n, E)$ having multiplicity m at r makes up an affine subspace of positive dimension. As it will turn out later in this section, the corresponding compactified Jacobians are ramified coverings of degree $n!$ of $E^{(n)}$, and their polarized structures become easier to handle. Consequently, we will manage to generalize some classical results about polarized Jacobians of smooth curves to all curves $\Gamma \in V(n, E)$ (see Section 2).

5.2. The family of integral curves over $V(n,E)$, $\Gamma(n,E)=\{\{\Gamma\},\Gamma\in V(n,E)\}$ is naturally equipped with a point or section $p(n,E)\colon V(n,E)\longrightarrow\Gamma(n,E)$, a projection $\pi(n,E)\colon\Gamma(n,E)\longrightarrow E$, and a meromorphic function $k(n,E)$ whose restrictions to the fiber over any $\Gamma\in V(n,E)$ permit recovering, respectively, the smooth point $p\in\Gamma$, the tangential cover $\pi\colon(\Gamma,p)\longrightarrow(E,q)$, and the tangential function k canonically associated to Γ. Since $V(n,E)$ is parametrized by $I|nC_0+S_q|$ (5.1), $\Gamma(n,E)$ is known to be a projective and flat variety over $V(n,E)$, and we can apply the results of [A-K] & [A-I-K] to $W(n,E)=\{W(\Gamma),\Gamma\in V(n,E)\}$, the relative compactified Jacobian of $\Gamma(n,E)$. We now collect the most relevant properties, for our purposes, of the general fiber $W(\Gamma)$, $\Gamma\in V(n,E)$:

5.2.1. $W(\Gamma)$ is a locally complete intersection, Cohen–Macaulay, integral and projective variety of dimension n over $\mathbb{C}$;

5.2.2. The $\operatorname{Jac}\Gamma$-orbit of the sheaf $\mathcal{O}_\Gamma((n-1)p)\in W(\Gamma)$ is a dense open subset of $W(\Gamma)$, isomorphic to $\operatorname{Jac}\Gamma$, whose complement is made up of the compactified Jacobians of all partial desingularizations of Γ;

5.2.3. The support of the *theta* divisor Θ_Γ, $\{\xi\in W(\Gamma),h^0(\xi)>0\}$, does not contain any $\pi^*(E)$-orbits.

5.3. Consider the group homomorphism $\alpha\longmapsto\mathcal{O}_{\Gamma(n,E)}(\pi^*(n,E)(\alpha-q))$ of E into the relative generalized Jacobian of $\Gamma(n,E)$ and the corresponding action, via tensor product, of E on $W(n,E)$. Denote by m (resp., π_2): $E\times W(n,E)\longrightarrow W(n,E)$, the morphism $(\alpha,(\Gamma,\xi))\longmapsto(\Gamma,\xi(\pi^*(\alpha-q)))$ (resp., the projection $(\alpha,(\Gamma,\xi))\longmapsto(\Gamma,\xi)$) and by $\Theta(n,E)=\{\Theta_\Gamma,\Gamma\in V(n,E)\}$ the relative flat divisor of $W(n,E)$ over $V(n,E)$. The inverse image $m^*(\Theta(n,E))$ is a relative divisor of $E\times W(n,E)$ with respect to π_2 and the map $I(n,E)$: $W(n,E)\longrightarrow E^{(n)}$, $(\Gamma,\xi)\longmapsto\{\alpha,(\alpha,\xi)\in m^*(\Theta(n,E))\}$, is well defined over all of $W(n,E)$ (cf., [A-C-G-H] ch. IV, p. 165), globalizes the morphism $I\colon W(\Gamma)\longrightarrow E^{(n)}$ defined in Section 4. The seemingly unrelated lemmas (5.4) and (5.5) deal with basic technical results needed for the study of I as well as of the polarized variety $(W(\Gamma),\Theta_\Gamma)$.

Lemma 5.4. *For any tangential cover $\pi\colon(\Gamma,p)\longrightarrow(E,q)$ of degree $n,\Gamma\in V(n,E)$, and any $\xi\in W(\Gamma)$, the direct image sheaf $\pi_*(\xi)$, is isomorphic to a unique direct sum of indecomposable vector bundles, $\pi_*(\xi)\simeq\bigoplus_{j\in J}V_j$, having the following properties:*

5.4.1. $\operatorname{rank}(\pi_*(\xi))=\Sigma r_j=n$ $(r_j=\operatorname{rank}(V_j))$;

5.4.2. $\deg V_j=0$ *for any* $j\in J$;

5.4.3. *let*

$$I(\xi)=\sum_{1\leq i\leq n}\alpha_i\in E^{(n)}$$

and, for any V_j, let $\beta_j\in E$ be uniquely defined by the condition $h^0(V_j(\beta_j-$

$q)) > 0$. *Then* $\{\alpha_i\} = \{\beta_j\}$.

Proof. 5.4.1. π is a flat and finite morphism of degree n and therefore $\pi_*(\xi)$ is locally free of rank n.

5.4.2. For any $\alpha \in E$, the Riemann–Roch theorem and the projection formula warrant the following identifies

$$\begin{cases} 0 = \chi(\xi) = \chi(\pi_*(\xi)) = \deg \pi_*(\xi) = \sum_{j\in J} \deg V_j, \\ \pi_*(\xi(\pi^*(\alpha - q))) \simeq \pi_*(\xi)(\alpha - q) \simeq \bigoplus_{j\in J} V_j(\alpha - q), \\ h^0(\xi(\pi^*(\alpha - q))) = \sum_{j\in J} h^0(E, V_j(\alpha - q)). \end{cases}$$

For a generic $\alpha \in E$ the sheaf $\xi(\pi^*(\alpha - q))$ does not belong to Θ_Γ (see 5.2.3), meaning that

$$0 = h^0(\xi(\pi^*(\alpha - q))) = \sum_j h^0(V_j(\alpha - q)) = h^0(V_j(\alpha - q)) \text{ for all } j \in J.$$

On the other hand, any indecomposable vector bundle over E without non-zero sections must have degree ≤ 0 ([A]). In other words, we have proved the inequalities $\sum_j \deg V_j = 0$ and $h^0(V_j(\alpha - q)) = 0 \geq \deg V_j$, from which we finally deduce that $\deg V_j = 0$ for any $j \in J$.

5.4.3. Let us prove that $\mathrm{supp}(I(\xi)) = \{\alpha_i\}$ is contained in $\{\beta_j\}$ and vice-versa. An element $\alpha \in E$ belongs to $\{\alpha_i\}$ if and only if $0 < h^0(\xi(\pi^*(\alpha - q))) = \sum_j h^0(V_j(\alpha - q))$, and that inequality is equivalent to $0 < h^0(V_j(\alpha - q))$ for at least one direct term V_j, proving that α should also belong to $\{\beta_i\}$. □

Lemma 5.5. *The image of the embedding* $E^{(n-1)} \longrightarrow E^{(n)}$,

$$\sum_{2\leq i\leq n} \alpha_i \longmapsto q + \sum_{2\leq i\leq n} \alpha_i,$$

denoted by $qE^{(n-1)}$, *is an integral ample divisor of* $E^{(n)}$ *whose nth cup-power is equal to* $[qE^{(n-1)}]^n = 1$.

Proof. The natural projection $r\colon E^n \longrightarrow E^{(n)}$ is a finite morphism of degree $n!$ and $(E^n, r^*(qE^{(n-1)}))$ is a principally polarized Abelian variety. It follows that $r^*(qE^{(n-1)})$ is an ample divisor and $[r^*(qE^{(n-1)})]^n = n!$. Analogously, applying Serre's or Nakai's criterion and the projection formula to the finite projection r, we deduce that $qE^{(n-1)}$ is an ample divisor and that

$$n! = [r^*(qE^{(n-1)})]^n = \deg r[qE^{(n-1)}]^n = n![qE^{(n-1)}]^n. \qquad \square$$

Theorem 5.6. *For any $\Gamma \in V(n,E)$, the associated morphism $I: W(\Gamma) \longrightarrow E^{(n)}$ (see 4.21) is flat and finite.*

Proof. Let us fix a point $\sum_{1\leq i\leq n} \alpha_i$ of $E^{(n)}$ and study the fiber $I^{-1}\left(\sum_i \alpha_i\right)$. For any $\xi \in I^{-1}\left(\sum_i \alpha_i\right)$ the direct image $\pi_*(\xi)$ is isomorphic to a direct sum, $\bigoplus_{j\in J} V_j$, of indecomposable vector bundles, and for any $j \in J$ there exists a unique $\beta_j \in \{\alpha_i\}$ such that $h^0(V_j(\beta_j - q)) > 0$ (5.4.3). Conversely, Atiyah's classification [A] implies that V_j ($j \in J$) is uniquely determined by the knowledge of the data $(\beta_j, rk\, V_j)$, and we conclude, using 5.4, that the set of vector bundles $\left\{\pi_*(\xi),\ \xi \in I^{-1}\left(\sum_i \alpha_i\right)\right\}$ has a finite number of elements, up to isomorphism. The latter property gives us the final argument in proving the finiteness of I. The spectral theory of ([T-V, II] App. 1 & §4), shows indeed how one can recover the sheaf $\xi \in W(\Gamma)$ from a knowledge of $\pi_*(\xi)$ and of the homomorphism $\pi_*(k): \pi_*(\xi) \longrightarrow \pi_*(\xi)(q)$. The latter spectral theory defines an embedding of the complete sub-variety $I^{-1}\left(\sum_i \alpha_i\right)$ into the finite-dimensional affine space $\bigoplus_V \operatorname{Hom}(V, V(q))$, where V belongs to the finite set $\left\{\pi_*(\xi),\right.$ $\left.\xi \in I^{-1}\left(\sum_i \alpha_i\right)\right\}$. In particular the fiber $I^{-1}\left(\sum_i \alpha_i\right)$ must be finite and the morphism I, being proper and quasi-finite, turns out to be finite.

One can now prove, from general facts, that I or any other finite surjective morphism having a Cohen–Macaulay source (5.2.1) and a smooth image must be flat (cf., [E-G-A] §14.4.4, §15.4.2 or [H] III, Ex. 9.3a, 10.9). □

Remark 5.7.

5.7.1. Let $\Delta^{(n)} \subset E^{(n)}$ denote the *generalized diagonal*, image of the embedding $E \times E^{(n-2)} \longrightarrow E^{(n)}$,

$$\left(\alpha, \sum_{3\leq i\leq n} \alpha_i\right) \longmapsto 2\alpha + \sum_{3\leq i\leq n} \alpha_i,$$

and let $T_{E^{(n)}} \longrightarrow E^{(n)}$ be the tangent bundle of $E^{(n)}$. There exist a birational morphism, $AI: W(n,E) \longrightarrow T_{E^{(n)}}$, and a rational surjective map, $(H_i)_{1\leq i\leq n}: T_{E^{(n)}} \longrightarrow \mathbb{C}^n \simeq V(n,E)$, that are well defined over the dense open subset $T_{E^{(n)}-\Delta^{(n)}}$, such that the morphism $I(n,E): W(n,E) \longrightarrow E^{(n)}$ (resp., $W(n,E) \longrightarrow V(n,E)$) factors through AI followed by the canonical projection $T_{E^{(n)}} \longrightarrow E^{(n)}$ (resp., $(H_i)_{1\leq i\leq n}$: $T_{E^{(n)}} \longrightarrow \mathbb{C}^n \simeq V(n,E)$).

5.7.2. For any $\Gamma \in V(n,E)$, the closure of $(H_i)^{-1}(\Gamma)$ in $T_{E^{(n)}}$, equal to $AI(W(\Gamma))$, is a birational and proper model of $W(\Gamma)$.

5.7.3. Up to compactifying $T_{E^{(n)}}$ and blowing it up along appropriate subvarieties of $T_{\Delta^{(n)}}$, we can suppose that (H_i) is the restriction of a proper morphism $(\overline{H}_i): \overline{T} \longrightarrow \mathbb{P}^n$, while $\overline{T}$ still projects onto $E^{(n)}$. In the latter case, for any $\Gamma \in V(n,E) \simeq \mathbb{C}^n \subset \mathbb{P}^n$ and any $\sum_i \alpha_i$ in the dense open subset $E^{(n)} - \Delta^{(n)}$, the fibers of the corresponding projections over Γ and $\sum_i \alpha_i$ intersect each other in a 0-cycle whose degree is well defined and does not depend upon the Γ that was chosen. On the other hand, the support of this 0-cycle is always contained in $T_{E^{(n)}-\Delta^{(n)}}$, hence its degree is equal to the degree of the finite morphism $I: W(\Gamma) \longrightarrow E^{(n)}$.

Theorem 5.8. *For any $\Gamma \in V(n, E)$, the degree of the associated finite morphism $I: W(\Gamma) \longrightarrow E^{(n)}$ is equal to $n!$ and the theta divisor Θ_Γ is ample and equal to $I^*(qE^{(n-1)})$.*

Proof. It easily follows from (5.4.3) that a sheaf ξ belongs to Θ_Γ if and only if $I(\xi)$ belongs to $qE^{(n-1)}$, which amounts to saying that the divisors Θ_Γ and $I^*(qE^{(n-1)})$ have same support. On the other hand $I: W(\Gamma) \longrightarrow E^{(n)}$ is equivariant with respect to the natural actions of E on $W(\Gamma)$ ($\xi \longmapsto \xi(\pi^*(\alpha - q))$), and on $E^{(n)}$ ($\Sigma\alpha_i \longmapsto \Sigma(\alpha_i + \alpha)$) and its discriminant divisor, disc I, is E-invariant.

We conclude that the ample (5.5) non-E-invariant divisor $qE^{(n-1)}$ must properly intersect disc I, hence $I^*(qE^{(n-1)})$ is an ample reduced divisor equal to Θ_Γ.

We still need to check that $\deg I = n!$. The latter degree is independent of the curve $\Gamma \in V(n, E)$ and equal to $[\Theta_\Gamma]^n = I_*(I^*([qE^{(n-1)}]^n)) = \deg I [qE^{(n-1)}]^n = \deg I$, which we know (Section 2) to be $[\Theta_\Gamma]^n = n!$ whenever Γ is smooth. □

5.9. Let $\Lambda \subset \mathbb{C}$ be a lattice such that $(E, q) \simeq (\mathbb{C}/\Lambda, 0)$ and denote by $\theta_E(z)$ the usual *theta* (holomorphic, Λ-multiplicative, vanishing over Λ) function considered as a section of the line bundle $\mathcal{O}_E(q)$. Analogously, the principally polarized Abelian variety (E^n, Θ_{E^n}) is obtained as the quotient of $\mathbb{C}^n = \bigoplus_{1\le i\le n} e_i\mathbb{C}$ by the lattice $\Lambda^n = \bigoplus_{1\le i\le n} e_i\Lambda$, coupled with the *theta* function $\theta_{E^n}(z_1, \ldots, z_n) = \prod_i \theta_E(z_i)$, considered as a section of $\mathcal{O}_{E^n}(\Theta_{E^n})$. Moreover, Θ_{E^n} and θ_{E^n} are invariant under permutations of the coordinates and get identified respectively by descent theory to the divisor $qE^{(n-1)}$ of $E^{(n)}$ and a non-trivial section of $\mathcal{O}_{E^n}(qE^{(n-1)})$. A straightforward application of 5.8 gives

Corollary 5.10. *The inverse image of θ_{E^n} by the morphism $I: W(\Gamma) \longrightarrow E^{(n)}$ ($\Gamma \in V(n, E)$) is a theta function for the polarized variety $(W(\Gamma), \Theta_\Gamma)$. In other words, $I^*(\prod_i \theta_E(z_i))$ is a section of $\mathcal{O}_{W(\Gamma)}(\Theta_\Gamma)$ vanishing along $I^*(qE^{(n-1)}) = \Theta_\Gamma$.*

5.11. Associated to $E^{(n)}$ and any point $\sum_{1\le\ell\le j} \beta_\ell$ of $E^{(j)}$ ($j = 1, \ldots, n-1$) there are an *addition* map

$$s: E^{(n)} \longrightarrow E, \quad \sum_{1\le i\le n} \alpha_i \longmapsto \alpha_1 + \cdots \alpha_n$$

and the sub-variety

$$\left(\sum_{\ell}\beta_\ell\right)E^{(n-j)}=\left\{\Sigma\beta_\ell+\Sigma',\Sigma'\in E^{(n-j)}\right\}$$

of $E^{(n)}$. We also associate to any $\Gamma\in V(n,E)$, besides π and I, an embedding $A_p\colon\Gamma\longrightarrow W(\Gamma)$, defined at each smooth point p_1 by the equality $A_p(p_1)=\mathcal{O}_\Gamma((n-2)p+p_1)$. We shall now state some useful properties of these morphisms.

Lemma 5.12.

5.12.1. *The tangential cover* $\pi\colon\Gamma\longrightarrow E$ *factors as* $\pi=-s\circ I\circ A_p$;

5.12.2. *The* j*th cup-power of* $[qE^{(n-1)}]$ *is equal to* $[jqE^{(n-j)}]$.

Proof. 5.12.1. This is just a rephrasing of the formula, obtained in the proof of Corollary 4.25,

$$I(\mathcal{O}_\Gamma((n-2)p+p_1))=(n-1)q+(-\pi(p_1)).$$

5.12.2. For all subsets $\{\beta_\ell,\ell=1,\dots,j\}\in E$ of j distinct points, the equalities

$$\bigcap_{1\le\ell\le j}(\beta_\ell E^{(n-1)})=\left(\sum_{1\le\ell\le j}\beta_\ell\right)E^{(n-j)}$$

hold and imply, at the cup-product level, the existence of n positive integers $\{b_j,\ j=1,\dots,n\}$ such that $[qE^{(n-1)}]^j=b_1\dots b_j[jqE^{(n-j)}]$. But we already knew that $[qE^{(n-1)}]^n=[nq]=1$ (5.5). Hence $\prod_1^n b_j=1$ and $[qE^{(n-1)}]^j=[jqE^{(n-j)}]$ for any $j=1,\dots,n$. □

We easily deduce from 5.6 and 5.12 the following result:

Proposition 5.13. *For any* $\Gamma\in V(n,E)$ *and* $j=1,\dots,n-1$, $\mathcal{W}_j=I^*(jqE^{(n-j)})$ *is a pure* $(n-j)$*-dimensional cycle of* $W(\Gamma)$ *such that*

1) $\text{support}(\mathcal{W}_j)\subset\text{support}(\mathcal{W}_{j-1})$;
2) $j![\mathcal{W}_j]=[\Theta_\Gamma]^j$;
3) $A_p(\Gamma)$ *is an irreducible component of* $\mathcal{W}_{n-1}$.

Corollary 5.14. *For any* $\Gamma\in V(n,E)$ *the associated finite morphism,* $I\colon W(\Gamma)\longrightarrow E^{(n)}$, *equipped with the action of* $\operatorname{Jac}\Gamma$ *on* $W(\Gamma)$, *determines effectively the isomorphism class of the tangential cover* $\pi\colon(\Gamma,p)\longrightarrow(E,q)$.

Proof. The restriction of the finite flat morphism I to $(A_p(\Gamma),A_p(p))$ is a tangential cover of degree n over $((n-1)qE,nq)$, and the E-orbit of

$A_p(p) = \mathcal{O}_\Gamma((n-1)p)$ is tangent to $A_p(\Gamma)$ at $A_p(p)$. Conversely, let C be any irreducible component of $\mathcal{W}_{n-1} = I^*((n-1)qE)$ such that

a) the E-orbit of r is tangent to C at r;

b) the natural projection $\pi_C: (C,r) \longrightarrow (E,q)$ ($\pi_C = s \circ I|_C$) is a tangential cover of degree and arithmetic genus equal to n;

c) there exists an isomorphism $\varphi: W(C) \longrightarrow W(\Gamma)$ over $E^{(n)}$, equivariant with respect to the corresponding actions of $\operatorname{Jac} C$ and $\operatorname{Jac}\Gamma$.

Let us write

$$\pi_{\hat{C}}: (\hat{C}, r) \longrightarrow (C,r) \xrightarrow{\pi_C} (E,q)$$

and

$$\pi_{\hat{\Gamma}}: (\hat{\Gamma}, p) \longrightarrow (\Gamma,p) \xrightarrow{\pi} (E,q),$$

the tangential covers obtained after smoothing π_C and π_Γ ([T-V, I] §2.10). The corresponding polarized compactified Jacobians naturally embed in $(W(C), \Theta_C)$ and $W(\Gamma, \Theta_\Gamma)$, where they are uniquely characterized as the Jac-orbits of minimal dimension. It follows that our Jac-invariant isomorphism φ restricts to an isomorphism between $(W(\hat{C}), \Theta_{\hat{C}})$ and $(W(\hat{\Gamma}, \Theta_{\hat{\Gamma}})$. Hence, by Torelli (2.4) plus the well-known result $\operatorname{Aut}(\hat{\Gamma}) \cong \operatorname{Aut}(W(\hat{\Gamma}, \Theta_{\hat{\Gamma}}))$ mod(± 1), if Γ is nonhyperelliptic, it reduces to an isomorphism between

$$A_r: (\hat{C}, r) \longrightarrow W(\hat{C}) \quad \text{and} \quad A_p: (\hat{\Gamma}, p) \longrightarrow W(\hat{\Gamma}).$$

The identification $A_r \simeq A_p$, coupled with the natural projections onto E (cf., [T] 1.3), finally give us an isomorphism between $\pi_{\hat{C}}$ and $\pi_{\hat{\Gamma}}$. On the other hand it follows from (4.11) that any tangential cover of degree n uniquely factors through a unique tangential cover of degree equal to the arithmetic genus n. Hence π is isomorphic to π_C. □

REFERENCES

[A] M. Atiyah, *Vector bundles over an elliptic curve*, Proc. London Math. Soc. (3) **7** (1957), 414–452.

[A-C-G-H] E. Arbarello, M. Cornalba, P.A. Griffiths and J. Harris, *Geometry of Algebraic Curves I*, (1985), Springer-Verlag, New York.

[A-I-K] A. Altman, A. Iarrobino and S. Kleiman, *Irreducibility of the compactified jacobian*, Real and Complex Singularities, Ed.: P. Holm, Proc. Nordic Summer School NAVF (Oslo) (1977), Sijthoff and Noordhoff, Amsterdam.

[A-K] A. Altman and S. Kleiman, *Compactifying the Picard scheme*, Adv. in Math. **35** (1980), 50–112.

[E-G-A IV] A. Grothendieck and J. Dieudonné, *Eléments de Géométrie algébrique* IV (3), Publ. Math. I.H.E.S. **28** (1966).

[F] J.D. Fay, *On the even-order vanishing of Jacobian theta functions*, Duke Math. J. **51** (1984), 109–132.

[G-H] P. Griffiths and J. Harris, *Principles of Algebraic Geometry* (1978), Wiley Interscience.

[H] R. Hartshorne, *Algebraic Geometry*, Grad. Texts in Math. **52** (1977), Springer-Verlag, New York.

[K] I.M. Krichever, *Elliptic solutions of the K-P equation and integrable systems of particles*, Funct. Anal. **14** (1980), no. 4, 45–54 (Russian), 282–290 (English).

[S-W] G. Segal and G. Wilson, *Loop groups and equations of KdV type*, Publ. Math. I.H.E.S. **61** (1985), 5–65.

[T] A. Treibich, *Tangential Polynomials and Elliptic Solitons*, Duke Math. J. **59** (1989), 611-627.

[T-V, I] A. Treibich and J.-L. Verdier, *Solitons Elliptiques*, Ed.: P. Cartier et al., Grothendieck Festschrift, Progress in Math. **88** (1990), Birkhäuser, Boston.

[T-V, II] A. Treibich and J.-L. Verdier, *Variétés de Kritchever des Solitons Elliptiques de KP*, Eds.: A. Beauville and S. Ramanan, Proceedings of the Franco-Indian Colloquium, Bombay (1989).

URA au CNRS 751
Université de Lille I
59655 Villeneuve d'Ascq
France

Heisenberg Action and Verlinde Formulas

Bert van Geemen and Emma Previato*

In this paper we review some of the recent work on 'nonabelian theta functions'. We discuss various links between abelian and nonabelian theta functions as well as links with the Schottky problem and open questions.

We let $\mathcal{M}_0(C)$ ($\mathcal{M}_p(C)$, resp.) denote the space of S-equivalence classes (cf. [S]) of semistable bundles of rank 2 and determinant equal to $\mathcal{O}_C$ ($\mathcal{O}_C(p)$, resp.) over a fixed Riemann surface C of genus ≥ 2, p a point of C. Both spaces have Picard group isomorphic to $\mathbf{Z}$; if we call $\mathcal{L}_0(\mathcal{L}_p)$ the ample generator, the spaces $H^0(\mathcal{M}_\epsilon, \mathcal{L}_\epsilon^k)$ ($\epsilon = 0$ or p) consist of what is regarded as the kth order nonabelian theta functions and their dimensions $N_{k,\epsilon}$ are predicted by the Verlinde formulas, cf. Section 1 (we call them the Verlinde spaces below, for short); a geometric derivation of the formulas is now available [Sz], [BeSz].

Nonabelian theta functions can be restricted via the natural map j : Jac $C \to \mathcal{M}_0$ to (the image of) the Jacobian. Beauville [B1] proved that $j^* : H^0(\mathcal{M}_0, \mathcal{L}_0) \to H^0(J, 2\Theta) = V$ is an isomorphism (the notation is introduced in (2.2) below) so that forms of degree k in V can be viewed as k^{th} order nonabelian theta functions via the multiplication map $S^kV \to H^0(\mathcal{M}_0, \mathcal{L}_0^{\otimes k})$. Beauville showed [B2] that when $k = 2$ this map is an isomorphism if and only if C has no vanishing (classical) thetanulls, and the present authors showed that if C has no vanishing thetanulls the map is surjective for $k = 4$.

In Sections 2–4 below, we review the geometry of these maps in light of the Heisenberg action; in Sections 5–7 we collect further directions, examples, and questions which stem from our geometric overview. Most results, with the exception of a few remarks, have appeared elsewhere and we give references to the best of our knowledge; nevertheless, we think it worthwhile further to unify the two viewpoints, Schottky and nonabelian thetas: in particular we put the methods of [B2] in the framework of [vGvdG] with changing moduli.

The second author wishes to thank the organizers of the Luminy workshop where the work [vGP] was presented; the workshop, to honor the memory of J.-L. Verdier, was an exposure of great depth to some of his many mathematical interests.

* Research partially supported by NSF Grant DMS-9105221

1. The Verlinde formulas

E. and H. Verlinde (cf. [V]) defined a fusion algebra on representations of the Kac Moody algebra $\hat{su}(2)$ in order to get a conformal field theory for any Riemann surface (the fusion rules prescribe how to add a handle). As a result, they obtained the following numbers:

$$N_{0,k} = \sum_{j=1}^{k+1} \left(\frac{k+2}{2\sin^2 \frac{j\pi}{k+2}} \right)^{g-1} \qquad (\text{"even case"})$$

$$N_{1,k} = \sum_{j=1}^{2k+1} (-1)^{j+1} \left(\frac{k+1}{\sin^2 \frac{j\pi}{2k+2}} \right)^{g-1} \qquad (\text{"odd case"}).$$

Note that the terms in $N_{1,k}$ are, up to sign, the same as those in $N_{0,2k}$; this too has a representation–theoretic explanation and for that reason the $N_{1,k}$'s are also called twisted Verlinde numbers. As stated in the introduction:

(1.1) **Theorem** [Sz,BeSz]. $N_{\epsilon,k} = \dim H^0(\mathcal{M}_\epsilon, \mathcal{L}_\epsilon^k)$.

At first sight it may not be clear that the $N_{\epsilon,k}$ are integers; however, as J.–B. Zuber explained to the second author at this workshop, they had arisen as integers in the Verlinde setup. They are made up, using the fusion rules, of the integers N_{jlm}:

$$V_j \text{ "}\otimes\text{" } V_l = \oplus_{m=0}^{k/2} N_{jlm} V_m,$$

the V_j being highest weight representations of level k for $\hat{su}(2)$, and the "$\otimes$" being a truncated tensor product defined by the fusion rules. After further decomposing into eigenspaces for one operator $V_j = \oplus V_{jn}$, and using the Kac-Weyl character formula and the Weyl denominator formula, the numbers

$$S_{jl} = \sqrt{\frac{2}{k+2}} \sin \frac{\pi(2j+1)(2l+1)}{4(k+2)}$$

are obtained as entries of a matrix that represents the action of (a suitable element of order 2 of) the modular group on

$$\chi_j(q) = q^{c_j} \sum_n (\dim\ V_{jn}) q^n$$

(c_j a suitable constant related to the central charge)

and Verlinde's work provides the link between S_{jl} and N_{jlm} (cf. [ABI], [CIZ]).

More directly, D. Zagier [Z] calculated the $N_{1,k}$ in a different manner, suggested by Bott [Bo] and obtained an expression for them that involves the residues of $(z-z^{-1})(z^k-z^{-k})/(z-z^{-1}-2)$ at g points of the Riemann sphere; through this work, A. Szenes succeeded in writing the odd-case Verlinde numbers as an Atiyah–Bott fixed-point formula [Sz]; Zagier also obtained several other expressions for the Verlinde numbers and for various generating functions and gave a different proof of the odd case, based on the topology of $\mathcal{M}_1$. For the even case, a Hecke correspondence was used [BeSz], see (3.5) below.

(1.2) Some examples of Verlinde numbers:

$N_{0,0}=1$ $\quad$ $N_{1,0}=1$

$N_{0,1}=2^g$ $\quad$ $N_{1,1}=2^{g-1}(2^g-1)$

$N_{0,2}=2^{g-1}+2^{2g-1}=2^{g-1}(2^g+1)$ $\quad$ $N_{1,2}=3^{g-1}2^{2g-1}-2^{2g-1}+3^{g-1}$

$$N_{0,3}=2(5+\sqrt{5})^{g-1}+2(5-\sqrt{5})^{g-1}$$

$$\begin{aligned}N_{0,4}&=3^{g-1}2^{2g-1}+2^{2g-1}+3^{g-1}\\&=(3^g+1)/2+(2^{2g}-1)(3^{g-1}+1)/2\end{aligned}$$

2. Hyperplanes

In this section we study the space $H^0(\mathcal{M}_0,\mathcal{L}_0)$. A nonzero element of this space defines a hyperplane, or a linear section of $\phi_{\mathcal{L}}(\mathcal{M}_0)$, with $\phi_{\mathcal{L}}$ the natural map:

$$\phi_{\mathcal{L}}:\mathcal{M}_0\longrightarrow \mathbf{P}H^0(\mathcal{M}_0,\mathcal{L}_0).$$

We recall the more general situation of rank n bundles studied in [BNR]. As traditional, $\mathcal{U}(n,d)$ will denote the moduli space of semistable bundles of rank n and degree d, and $\mathcal{SU}(n,L)$ those with fixed determinant $L\in \mathrm{Pic}^d(C)$. Let $\mathcal{L}$ be the ample generator of $\mathrm{Pic}\mathcal{SU}(n,\mathcal{O})$ and let $\Theta_C\subset Pic^{g-1}(C)$ be the theta divisor.

(2.1) **Theorem** [BNR]. *The (rational) map which associates to any point e of $\mathcal{SU}(n,\mathcal{O})$ the divisor $\Delta_E=\{\xi\in \mathrm{Pic}^{g-1}C:\Gamma(E\otimes\xi)\neq 0\}$, where E is a semistable bundle in the class of e, induces an isomorphism of $\Gamma(\mathrm{Pic}^{g-1}C,\mathcal{O}(n\Theta_C))$ with $\Gamma(\mathcal{SU}(n,\mathcal{O}),\mathcal{L})^*$. In particular:*

$$\dim\Gamma(\mathcal{SU}(n,\mathcal{O}),\mathcal{L})=n^g.$$

This result shows that a ('nonabelian') Verlinde space and an ('abelian') space of theta functions (more precisely, its dual) coincide. Before sketching the proof of the theorem, we recall the natural dualities in the $n = 2$ case and a theorem of Beauville clarifying the geometry related to $\mathcal{L}_0$.

(2.2) Let $\kappa \in Pic^{g-1}(C)$ be a theta characteristic, i.e. a line bundle on C with $\kappa^{\otimes 2} \cong \Omega^1_C$. We say κ is even resp. odd if $h^0(C, \kappa)$ is an even resp. odd number. Each theta characteristic defines a symmetric divisor $\Theta_\kappa := \Theta_C - \kappa$ in $J := \operatorname{Jac} C = Pic^0(C)$, and all effective symmetric divisors giving the principal polarization are obtained in this way. For any integer k, the divisors $2k\Theta_\kappa$ on J are linearly equivalent; we simply write $2k\Theta$ for this equivalence class.

(2.3) The Mumford group $\mathcal{G}$ is the group of automorphisms of $\mathcal{O}(2\Theta)$ which lift the action of an element of J_2(= the two–torsion subgroup of J) on $V := H^0(J, \mathcal{O}(2\Theta))$. A similar Heisenberg group, with $\mathbf{Z}/2\mathbf{Z}$ replaced by $\mathbf{R}$ plays a role in quantum mechanics; the analogue of the representation on V is called the Schrödinger representation [C]. We fix a *theta structure*, namely an isomorphism between $\mathcal{G}$ and the Heisenberg group:

$$H := H_g := \mathbf{C}^* \times (\mathbf{Z}/2)^g \times \operatorname{Hom}((\mathbf{Z}/2)^g, \mathbf{C}^*)$$

with multiplication $(s, \alpha, \alpha^*)(t, \beta, \beta^*) = (st\beta^*(\alpha), \alpha + \beta, \alpha^*\beta^*)$. Using the theta structure, the action on V can be described as follows: there is a basis $\{X_\sigma\}$ $(\sigma \in (\mathbf{Z}/2\mathbf{Z})^g)$ for which $(t, \alpha, \alpha^*)X_\sigma = t\alpha^*(\alpha + \sigma)X_{\sigma+\alpha}$.

The choice of a symplectic homology basis for C with corresponding period matrix τ, determines a theta structure and the basis $\{X_\sigma\}$ of V, the space of second-order theta functions, is then given by the $X_\sigma = \vartheta\left[\begin{smallmatrix}\sigma\\0\end{smallmatrix}\right](2z, 2\tau)$, $\sigma \in (\mathbf{Z}/2\mathbf{Z})^g$ (cf. [B1, 2.6]).

(2.4) The action of the Heisenberg group on V induces an action on $V \otimes V$, which will be a direct sum of non-isomorphic 1 dimensional representations, each inducing a duality $V \to V^*$. The representations in $\operatorname{Sym}^2 V$ correspond to even theta characteristics and the ones in $\wedge^2 V$ to odd theta characteristics. We denote the corresponding duality by ξ_κ.

(2.5) A rather direct connection between the abelian and nonabelian theta functions is given by the map:

$$j : J \longrightarrow \mathcal{M}_0, \qquad L \mapsto L \oplus L^{-1},$$

note that j factors over $J/\pm 1$, the Kummer variety of J, which in fact injects into $\mathcal{M}_0$. These facts were put together by Beauville to refine Theorem 2.1 in the $n = 2$ case:

(2.6) **Theorem** [B1]. *(i) The isomorphism* $\rho : H^0(\mathcal{M}_0, \mathcal{L}_0) \to H^0(J, 2\Theta_C)^*$ *of (2.1) is induced by the morphism*

$$\delta : \mathcal{M}_0 \to |2\Theta_C| \qquad E \mapsto \Delta_E := \{\xi \in Pic^{g-1}C :\ h^0(E \otimes \xi) > 0\}.$$

(ii) We have $j^*\mathcal{L}_0 \cong \mathcal{O}(2\Theta)$ *and the map* $j^* : H^0(\mathcal{M}_0, \mathcal{L}_0) \to H^0(J, \mathcal{O}(2\Theta))$ *is an isomorphism.*
(iii) For any theta characteristic $\kappa \in \mathrm{Pic}^{g-1}(C)$ *we have an isomorphism* $|2\Theta_C| \cong_\kappa |2\Theta|$, $D \mapsto D - \kappa$, *and the diagram*

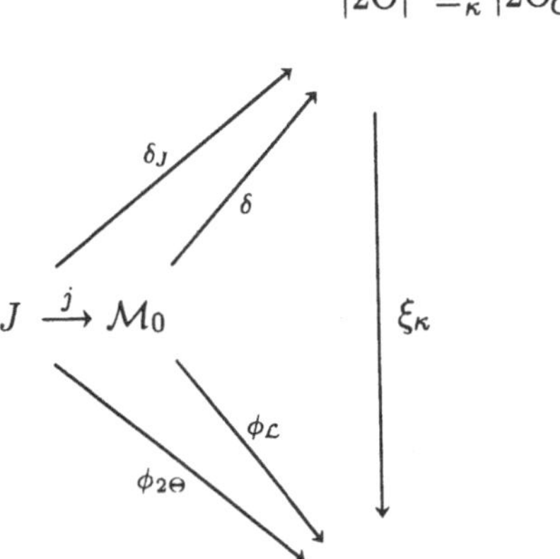

$$\mathbf{P}V \cong |2\Theta|^*$$

is commutative, where $\delta_J: L \to T_L^*\Theta \oplus T_{L^{-1}}^*\Theta$.

(2.7) The proof of the theorems (2.1), (2.6) and further applications depend on the Prym geometry. We analyze certain covers of a fixed curve; this is best done for the general (possibly ramified) n-sheeted cover $\pi: \tilde{C} \to C$ of Riemann surfaces, as occurs in the spectral-curve situation of [BNR]. Let Nm: Jac $\tilde{C} \to$ Jac C be the norm homomorphism, which can be identified with the transpose of π^*, and suppose that π^*: Jac $C \to$ Jac $\tilde{C}$ is *injective.* We cut some corners and describe the geometric situation: let $\tilde{\Theta}$ be a theta divisor of Jac $\tilde{C}$; Ker $Nm = P$ and π^* Jac $C = N$ can be thought, in a way, as being orthogonal inside Jac $\tilde{C} = A$. There is a canonical isomorphism $|\Theta_P|^* \to |\Theta_N|$ which completes the diagram

$$\begin{array}{ccc} P & \longrightarrow & |\Theta_P|^* \\ \downarrow{\scriptstyle \phi} & & \\ |\Theta_N| & & \end{array}$$

where the rational map $P \to |\Theta_N|$ is defined by $p \mapsto T_p^*(\tilde{\Theta})\Big|_N$. In [BNR], in order to avoid choices of theta divisors, this is applied to $N' = \pi^*\,\mathrm{Pic}^{g-1}C$, $A' = \mathrm{Pic}^{h-1}\tilde{C}$ and $P' = Nm^{-1}(D)$ where $h =$ genus $\tilde{C}$, $g =$ genus C, $Nm : \mathrm{Pic}^{h-g}\tilde{C} \to \mathrm{Pic}^{h-g}C$ and $D = \det(\pi_*\mathcal{O}_{\tilde{C}})^{-1}$. Then there is a natural isogeny $\mu: N' \times P' \to A$ and there exists a natural line bundle τ on P' so that $\mu_{P'}^*\tau \otimes \mu_{N'}^* n\Theta_C = \mu^*\Theta_{\tilde{C}}$ ($\mu_{P'}, \mu_{N'}$ are the projections). Thus we have a natural element $\mu^*\Theta_{\tilde{C}} \in H^0(P', \tau) \otimes H^0(\mathrm{Pic}^{g-1}C, \mathcal{O}(n\Theta_C)) = \mathrm{Hom}(H^0(P', \tau)^*, H^0(\mathrm{Pic}^{g-1}C, \mathcal{O}(n\Theta_C)))$ which is an isomorphism:

(2.8) **Theorem** [BNR]. *The spaces $H^0(\mathrm{Pic}^{g-1}C, \mathcal{O}(n\Theta_C))$ and $H^0(P', \tau)^*$ are naturally isomorphic, and the isomorphism is compatible with the rational maps of P' given by $p \mapsto T_p^*\Theta_{\tilde{C}} \cap \pi^* \mathrm{Pic}^{g-1}C$ and by the linear system of τ, respectively.*

A suitable choice of $\pi: \tilde{C} \to C$ will ensure [BNR] that $\pi_*\mathrm{Pic}^m\tilde{C}$ ($m = d - n(n-1)(g-1)$) (rather, π_* of the bundles whose image is semistable) is dense in $\mathcal{U}(n,d)$. From this it follows that dim $H^0(\mathcal{U}(n, n(g-1)), \mathcal{O}(\Theta_{\mathcal{U}})) = 1$, with $\Theta_{\mathcal{U}} := \{E \in \mathcal{U}(n, n(g-1)) : h^0(E) > 0\}$ the natural divisor on $\mathcal{U}(n, n(g-1))$.

Similarly one can arrange for π_*P' to be dense in $\mathcal{SU}(n, \mathcal{O})$; one more ingredient, namely the irreducibility of the action of Mumford's theta group on $H^0(P', \tau)$ plus the equivariance of $H^0(\mathcal{SU}(n), \mathcal{L}) \to H^0(P', \tau)$ under the action of the n-torsion of Jac C, shows that dim $H^0(\mathcal{SU}(n), \mathcal{L}) = n^g$.

(2.9) From now on we shall only be concerned with the case in which the rank is $n = 2$. Our notation, as set up in the introduction, will simply be

$$\mathcal{M}_0 := \mathcal{SU}(2, \mathcal{O}), \quad \mathcal{M}_1 := \mathcal{M}_p := \mathcal{SU}(2, \mathcal{O}(p)), \quad (p \in C).$$

Now we consider an unramified double cover $\pi_x: \tilde{C} \to C$ (in particular π_x^* is *not* injective); this corresponds to a point of order 2, $x \in J_2$ and the attendant Prym, $P_x \subset \mathrm{Jac}\tilde{C}$, is the identity component of the kernel of the norm map. The Prym has dimension $g-1$ and has a natural principal polarization Ξ_x. Now we have a map

$$\phi_x : P_x \longrightarrow \mathcal{M}_0, \qquad \eta \mapsto \pi_{x*}(\eta) \otimes \xi \tag{2.10}$$

(used in both [B2] and [vGP], where $\xi^2 = x$, with the only drawback that the map depends on the choice of ξ). This gives a $(g-1)$ dimensional subvariety of $\mathcal{M}_0$ and (in analogy to Theorem 2.6) an injection: $H^0(P_x, \mathcal{O}(2\Xi_x))^* \hookrightarrow H^0(Pic^{g-1}C, \mathcal{O}(2\Theta_C))$. This configuration is exploited in two somewhat different directions in [B2] and [vGP], cf. Section 3 and Section 4 below, resp.

3. Quadrics and the odd determinant case

(3.1) To study $H^0(\mathcal{M}_0, \mathcal{L}_0^2)$ we use the following diagram which relates a 'nonabelian' multiplication map m_2 with an 'abelian' map n_2:

$$\begin{array}{ccc} S^2V & \xrightarrow{m_2} & H^0(\mathcal{M}_0, \mathcal{L}_0^{\otimes 2}) \\ & \searrow^{n_2} & \downarrow j^* \\ & & H^0(J, \mathcal{O}(4\Theta))_+ \end{array} \tag{3.2}$$

(recall $V := H^0(J, \mathcal{O}(2\Theta)) \cong_{j^*} H^0(\mathcal{M}_0, \mathcal{L}_0))$

It turns out that $\dim S^2V = 2^{g-1}(2^g+1) = \dim H^0(J, \mathcal{O}(4\Theta))_+ = N_2$, the second Verlinde number; here the subscript + stands for the subspace of even theta functions. This might suggest that all maps are isomorphisms, but the situation is actually a little more delicate.

Beauville [B2] exhibited bases for $H^0(\mathcal{M}_0, \mathcal{L}_0^{\otimes 2})$ and $H^0(\mathcal{M}_p, \mathcal{L}_p)$. These spaces have dimension equal to the number of even (resp. odd) theta characteristics, as per (1.2). In both cases, the construction is suggested by the fact that the dualizing sheaf to $\mathcal{M}_\epsilon(\epsilon = 0, 1)$ on one hand has at $E \in \mathcal{M}_\epsilon$ the fiber $\wedge^{max} H^0(\mathcal{E}nd_0 E \otimes K_C)$ where $\mathcal{E}nd_0$ is the sheaf of trace-zero endomorphisms, on the other hand is the square of $\mathcal{L}_0^{\otimes 2}$ (resp. $\mathcal{L}_1$). The theta characteristics give sections as follows:

(3.3) **Theorem** [B2]. *Let κ be an even (resp. odd) theta characteristic and let D_κ be the reduced subvariety of $\mathcal{M}_0$ (resp. $\mathcal{M}_1$) given by*

$$D_\kappa := \{E \in \mathcal{M}_\epsilon : \ h^0(\mathcal{E}nd_0 E \otimes \kappa) > 0\}.$$

(i) There is a section $d_\kappa \in H^0(\mathcal{M}_0, \mathcal{L}_0^{\otimes 2})$ (resp. $H^0(\mathcal{M}_1, \mathcal{L}_1)$) whose divisor is D_κ and these sections give bases of these spaces.
(ii) Let $\vartheta_\kappa \in H^0(J, \mathcal{O}(\Theta_\kappa))$ be a section defining Θ_κ. Then a basis for $H^0(J, \mathcal{O}(4\Theta))_+$ is given by the $[2]^*\vartheta_\kappa$ with κ even, where $[2] : J \to J$, $x \mapsto 2x$ (similarly $H^0(J, \mathcal{O}(4\Theta))_-$ is spanned by the $[2]^*\vartheta_\kappa$ with κ odd).
(iii) For any even κ there is a $c_\kappa \in \mathbf{C}$ such that:

$$j^* d_\kappa = c_\kappa \cdot ([2]^*\vartheta_\kappa) \quad \text{and} \quad c_\kappa = 0 \Leftrightarrow h^0(C, \kappa) > 0.$$

In particular, j^* is an isomorphism iff $h^0(C, \kappa) = 0$ for all κ, i.e. J has no vanishing theta nulls.
(iv) The map m_2 is an isomorphism iff J has no vanishing thetanulls (in fact $m_2^* d_\kappa$ lies in the representation space corresponding to κ, and is zero iff $h^0(C, \kappa) > 0$).

(3.4) The results of (3.3) combined with the diagram (3.2) show that n_2 is an isomorphism iff J has no vanishing thetanulls; this can easily be proven directly by classical theta function theory.

The method of proof involves comparing Ξ_x and the pull-back of $\mathcal{L}_\epsilon$ via the maps $\phi_x : P_x \to \mathcal{M}_\epsilon$ of (2.10), which send a line bundle M over the covering $\pi_x : \tilde{C} \to C$ to the vector bundle $\pi_{x*}(M \otimes \tilde{\epsilon})$, with $\tilde{\epsilon} \in Pic(\tilde{C})$ such that $Nm\,\tilde{\epsilon} = x \otimes \epsilon$; here by abuse of notation $\epsilon = \mathcal{O}_C$, resp. $\mathcal{O}_C(p)$. In other words, J does not give enough theta divisors to calculate with, but without making recourse to the larger Pryms of 2.7, the various P_x do.

(3.5) **The odd determinant case.** For the odd case, the Hecke correspondence is used; it plays a fundamental role in [BeSz] as well, in that it allows one to lift geometric statements on $\mathcal{M}_p := \mathcal{M}_1$ to a $\mathbf{P}^1$ bundle $\mathcal{P}$ over it and deduce corresponding statements for $\mathcal{M}_0$: consider the diagram

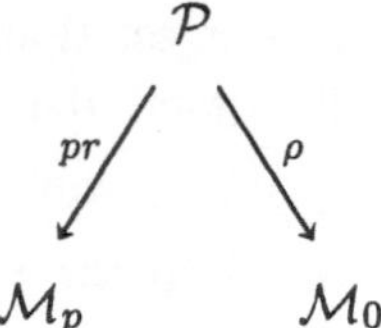

where $p \in C$ is a point, $\mathcal{P} := \mathbf{P}(U_p)$, where $U_p = U\Big|_{\mathcal{M}_1 \times p}$ and U is the universal rank 2 bundle over $\mathcal{M}_p \times C$ such that det $U_p = \mathcal{L}_1$. A point of $\mathcal{P}$ consists of a bundle E in $\mathcal{M}_p$ and a nonzero homomorphism to the skyscraper sheaf supported at $p, u\colon E \to \mathbf{C}_p$ (up to homothety). The fibre of ρ at a point F of $\mathcal{M}_0$ is the line $\mathbf{P}(Ext^1_{\mathcal{O}_C}(\mathbf{C}_p, F))$.

(3.6) [B2, 3.3] If $l \in J$ does not have order 2 (so $l \not\cong l^{-1}$), then one can define a bundle $F_l \in \mathcal{M}_1$ which fits in an exact sequence $0 \to l \oplus l^{-1} \to F_l \to \mathbf{C}_p \to 0$; the map $l \mapsto F_l$ can be extended to a morphism $j_p\colon \hat{J} \to \mathcal{M}_p$, which factors through the involution to define a morphism of $\hat{K}$ to $\mathcal{M}_p$. The circumflexes indicate blow-up at the points of order 2.

(3.7) [B2, 3.4, 3.5]. If the fibre of pr is mapped to $\mathcal{M}_0$ by sending $(E,u) \mapsto$ ker u, the image of the composite $\mathbf{P}(U_p)\Big|_E \to \mathcal{M}_0 \overset{\phi_{\mathcal{L}}}{\to} \mathbf{P}(V)$ is a line, and thus a point in $\wedge^2 V$; let this define $\varphi_p\colon \mathcal{M}_p \to \mathbf{P}(\Lambda^2 V)$; the composite $\varphi_p \circ j_p$ is the Gauss map associated to the tangent vector to C at p. The diagram

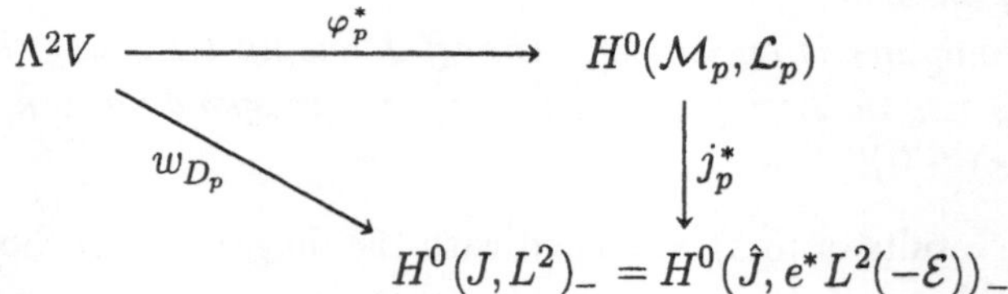

(where $e\colon \hat{J} \to J$ is the blow–up of J at J_2 and $\mathcal{E}$ the exceptional divisor) is commutative; to define w_{D_p}, one applies the vector field D_p on J, gotten from the tangent vector to $C \subset J$ at p, to the Wahl homomorphism $(s,t) \mapsto t^{\otimes 2} d(\frac{s}{t})$

$$w\colon \Lambda^2 H^0(J,L) \to H^0(J, L^2 \otimes \Omega^1_J).$$

Finally, if κ is an odd theta characteristic, $j_p^* d_\kappa = [2]^* \vartheta_\kappa$ if $h^0(\kappa(-p)) = 0$, while $j_p^* d_\kappa = 0$ if $h^0(\kappa(-p)) > 0$. In particular, φ_p^* and w_{D_p} have the same kernel and are isomorphisms iff, for all odd theta characteristics κ of C, $h^0(\kappa(-p)) = 0$.

4. Quartics

The spaces S^kV can be explicitly decomposed into irreducible pieces for the H-action; when k is even the action factors over the abelian quotient $H/H' \cong \mathbf{C}^* \times (\mathbf{Z}/2)^g \times \mathrm{Hom}((\mathbf{Z}/2)^g, \mathbf{C}^*)$; denote by $X(H)$ the group of characters of $(\mathbf{Z}/2)^g \times \mathrm{Hom}((\mathbf{Z}/2)^g, \mathbf{C}^*)$ and by 0 the trivial character; then

$$S^4V = \oplus_{\chi \in X(H)} S^4_\chi V, \qquad \dim S^4_0 V = d(g), \quad \dim S^4_\chi V = d(g-1),$$

with $d(g) := (2^g+1)(2^{g-1}+1)/3$, cf. [vG].

A nonzero $F \in S^4V$ defines a quartic hypersurface $Z(F) \subset \mathbf{P}(V)$, the projective space which contains $\phi_{\mathcal{L}}(\mathcal{M}_0)$ and the images of all the Pryms $\phi_{\mathcal{L}}(\phi_x(P_x)) \subset \phi_{\mathcal{L}}(\mathcal{M}_0)$ (cf. (2.10)). In fact, the map $K := \phi_{\mathcal{L}} \circ \phi_x$ can be identified with the natural map $P_x \to \mathbf{P}H^0(P_x, \mathcal{O}(2\Xi_x))$ and the diagram:

$$\text{(4.1)} \qquad \begin{array}{ccc} P_x & \xrightarrow{K} & \mathbf{P}H^0(P_x, \mathcal{O}(2\Xi_x)) \\ {\scriptstyle \phi_x}\downarrow & & \cap \\ \mathcal{M}_0 & \xrightarrow{\phi_{\mathcal{L}}} & \mathbf{P}V \end{array}$$

identifies $\mathbf{P}H^0(P_x, \mathcal{O}(2\Xi_x)) \subset \mathbf{P}V$ with an eigenspace $\mathbf{P}V_x$ of the action of x (more precisely, of a corresponding element in the Heisenberg group). In particular, $\phi_x^*\mathcal{L}_0 = \mathcal{O}(2\Xi_x)$.

Using the ϕ_x and j, we restrict sections of $\mathcal{L}_0^{\otimes 4}$ to the union of all Pryms and J (that is, to their images in $\mathcal{M}_0$). It will be convenient to write $P_0 := J$ and to agree that $0 \in J_2$ corresponds to the trivial character $0 \in X(H)$.

The following diagram exhibits the maps we will be interested in; note that we write $m = \oplus m_\chi$, $n = \oplus_\chi n_\chi$ for the multiplication maps originating from $S^4V = \oplus_\chi S^4_\chi V$.

$$\begin{array}{ccc} S^4_\chi V & \xrightarrow{m_\chi} & H^0(\mathcal{M}_0, \mathcal{L}_0^{\otimes 4}) \\ & {\scriptstyle n_\chi}\searrow & \downarrow {\scriptstyle res} \\ & & \oplus_{x \in J_2} H^0(P_x, \mathcal{O}(8\Xi_x))_+ \end{array}$$

The diagram again relates multiplication maps for nonabelian theta functions (the m_χ) with those of abelian theta functions (the n_χ). Since the Heisenberg group acts on both S^4V and on $H^0(\mathcal{M}_0, \mathcal{L}_0^{\otimes 4})$ and m is equivariant, we have that $\ker m = \oplus_\chi \ker m_\chi$. The diagram implies that $\dim\ker m_\chi \le \dim\ker n_\chi$, and thus we obtain the following lower bound for the fourth Verlinde number

$$\begin{aligned} N_4 &= \dim H^0(\mathcal{M}_0, \mathcal{L}_0^{\otimes 4}) \\ &\ge \dim S^4V - \dim\ker m \\ &\ge \textstyle\sum_\chi (\dim S^4_\chi V - \dim\ker n_\chi) \end{aligned}$$

The main result of [vGP], explained in the remainder of this section, is that for a generic curve one has $\dim S^4_\chi V - \dim\ker n_\chi = e(g)$ if $\chi = 0$ and $= e(g-1)$ if $\chi \neq 0$ with $e(g) := (3^g+1)/2$. The lower bound for N_4 thus obtained actually coincides with the value of N_4 predicted by Verlinde, and again the non-abelian theta functions are 'identified' with certain abelian theta functions:

(4.2) **Theorem.** *Let C be curve with no vanishing thetanulls. Then the following hold:*
(i) The multiplication map $m : S^4V \to H^0(\mathcal{M}_0, \mathcal{L}^{\otimes 4})$ is surjective.
(ii) ker m_χ= ker n_χ for all χ, thus also ker m=ker n.
(iii) The map res : $H^0(\mathcal{M}_0, \mathcal{L}^{\otimes 4}) \to \oplus_{x\in J_2} H^0(P_x, \mathcal{O}(8\Xi_x))$ is injective.

The map n_0 was already studied in [vG], for the other n_χ we need the results of [vGP]. It is a curious phenomenon that quartics in the kernel of the n_χ's must satisfy a certain condition on their singular locus:

(4.3) **Theorem.** (a weaker version of Theorem 1 in [vG], less technical to state). *Let $F \in S^4_0V$. Then F vanishes on all Pryms (i.e. $K(P_x) \subset Z(F)$ for all $x \in J_2$) if and only if $K(J) := \phi_{\mathcal{L}}(j(P_0)) \subset Sing Z(F)$.*

(4.4) **Theorem** (Theorem 1 in [vGP]). *Let $F \in S^4_\chi V$, $\chi \neq 0$, and let $x \in J_2$ correspond to χ. Then F vanishes on all Pryms if and only if $K(P_x) \subset Sing Z(F)$.*

The proofs of these theorems are discussed in (4.6) below; they imply, with χ corresponding to x:

$$\ker(n_\chi) = \{F \in S^4_\chi V : \ K(P_x) \subset Sing Z(F)\}.$$

(4.5) Therefore $F \in \ker(n_\chi)$ iff $\frac{\partial F}{\partial X_\sigma}$ is a (cubic) equation for $K(P_x)$ for all $\sigma \in (Z/2Z)^g$ (in view of the Heisenberg action, it suffices in fact that $\frac{\partial F}{\partial X_0}$ is a cubic equation). There are interesting relations between the cubics, quartics and the action of H. We have an isomorphism

$$\frac{1}{4}\frac{\partial}{\partial X_0} : S^4_0V \longrightarrow S^3_0V, \quad \text{with } S^3_0V := \{G \in S^3V : \ (1,0,\alpha^*)G = G, \ \forall \alpha^*\}.$$

and homomorphisms

$$M(\chi) : S^3_0V \to S^4_\chi V, \quad G \mapsto \sum_\sigma \alpha^*(\sigma) X_{\sigma+\alpha}((1,\sigma,0)G),$$

where $\chi = (\alpha, \alpha^*) \in X(H)$, $\sigma \in (\mathbf{Z}^g/2\mathbf{Z})^g$; $M(0)$ is the inverse of $\frac{1}{4}\frac{\partial}{\partial X_0}$. Moreover, if χ corresponds to x then the restrictions of $M(0)F$ and $M(\chi)F$ to V_x, the eigenspace of x in V, differ by a non-zero multiplicative constant.

Using the isomorphism $\frac{1}{4}\frac{\partial}{\partial X_0} : S^4_0V \to S^3_0V$, we see that $\ker n_0 = \ker(S^3_0V \to H^0(J, \mathcal{O}(6\Theta))_+)$, and the fact that the multiplication map

$S^3V \to H^0(J, \mathcal{O}(6\Theta))_+$ is surjective for any C with no vanishing thetanull then implies that dim ker $n_0 = d(g) - e(g)$, as desired.

In case $\chi \neq 0$ one proceeds in a rather similar way to derive dim ker $n_\chi = d(g-1) - e(g-1)$, replacing $J = P_0$ by P_x with x corresponding to χ; for the details, see [vGP].

(4.6) The proofs of the theorems (4.3) and (4.4) involve a comparison between the multiplication maps (restricted to subspaces of)

$$S^4V_x \to H^0(P_x, \mathcal{O}(8\Xi_x))$$

for the various x. The main point is that, w.r.t. suitable bases, the entries of the matrices defining these maps are theta constants which become equal in virtue of the Schotty-Jung (SJ) (when comparing $\chi = 0$ with a nonzero x) and the Donagi (D) relations (when comparing two orthogonal x) of (4.7) below. It is interesting to note that these relations can be proven in a natural way using rank 2 bundles.

Let x, y be nonzero in J_2 and x, y be orthogonal with respect to the Weil pairing; we denote by a bar the coset of, say, y in $(x)^\perp / < x > \cong (P_x)_2$. The action of $< x >^\perp / < x >$ on $\mathbf{P}V_x$ coincides with the action of H_{g-1}, defined as in (2.3) and the same holds for the action of $< x >^\perp \cap < y >^\perp / < x, y >$ on $\mathbf{P}V_{x,y}$, an eigenspace of $\bar{y}$ in $\mathbf{P}V_x$.

Finally for a principally polarized abelian variety (ppav) A, and a nonzero $x \in A_2$, if we let $S_x = \{z \in A : \ 2z = x\}$ then the image in $\mathbf{P}V$ of the (A_2 principal homogeneous) space S_x lies in the union of the two eigenspaces of x, and equals the union of the two $K_A \cap \mathbf{P}V_x$ if A is indecomposable. In particular, $K(J)$ and $\mathbf{P}V_x$ meet in a finite number of points; however, (4.8) below permits to compare the vanishing of certain (quartic) polynomials on $K(J)$ and P_x.

(4.7) **Theorem** *The following relations hold (here we use the auxiliary notation $K := \phi_{\mathcal{L}} j$ and $K_x := \phi_{\mathcal{L}} \phi_x$ for the maps from J and the P_x to $\mathbf{P}V$):*

(SJ) $$K(J) \cap \mathbf{P}V_x = K_x((P_x)_2)$$

(D) $$K_x(S_{\bar{y}}) \cap \mathbf{P}V_{x,y} = K_y(S_{\bar{x}}) \cap \mathbf{P}V_{x,y}$$

Sketch of Proof: (SJ) is a consequence of (4.1) and of H-equivariance; indeed, if $z_x^{\otimes} \cong x$, on one hand $j(z_x) = z_x^{-1} \oplus z_x \in K \cap \mathbf{P}V_x$, on the other $j(z_x) = (\mathcal{O}_C \oplus x) \otimes z_x^{-1} = \phi_x(\mathcal{O}_{\tilde{C}})$ and $\mathcal{O}_{\tilde{C}} \in (P_x)_2$. As remarked, $K \cap \mathbf{P}V_x$ is a homogeneous space under the action of $H_{g-1} = < x >^\perp / < x >$ and so is $(P_x)_2$, equivariantly with respect to the maps. (D) is proved similarly in [vGP], Proposition 2, by explicitly constructing an irreducible representation of $\pi_1(C)$ in $SU(2)$, namely a stable rank 2 bundle over C, with trivial determinant, and showing that it equals both $\phi_x(p_x)$ and $\phi_y(p_y)$ with $p_x \in P_x$ and $p_x^{\otimes 2} = \overline{y}, p_y^{\otimes 2} = \overline{x}$; again equivariance gives the conclusion.

The advertised comparison between multiplication maps is given by explicit matrices: a choice of theta structure gives a period matrix, hence as in [vG] explicit bases for the spaces involved. For a suitable choice of bases, Riemann's formula gives (cf. [I], IV.1, [vG] Prop. 4–6, [vGP], §2 Lemma):

(4.8) **Corollary.**
(i)(SJ) For any non-zero $x \in J_2$ corresponding to χ, the restriction map gives an isomorphism $S^4_\chi V \cong S^4_0 V_x$, and the multiplication maps $S^4_\chi V \to H^0(J, \mathcal{O}(8\Theta))$ and $S^4_0 V_x \to H^0(P_x, \mathcal{O}(8\Xi_x))$ are defined (when restricted to their image) by the same matrices up to a (nonzero) constant.
(ii)(D) Let x, y be orthogonal elements of $J_2\backslash\{0\}$, corresponding to characters χ, μ, and let $\overline{\chi}, \overline{\mu}$ be the characters of H_{g-1} induced on $(P_x)_2$, $(P_y)_2$. Then under the isomorphism given by the restriction:

$$S^4_{\overline{\chi}} V_x \cong S^4_0 V_{x,y} \cong S^4_{\overline{\chi}} V_y,$$

the multiplication maps $m_{4,x,\overline{\mu}}$ and $m_{4,y,\overline{\chi}}$ differ by a nonzero multiplicative constant.

Sketch of proof of (4.3), (4.4). We indicate the argument used to obtain (4.3). Let $F \in S^4_0 V$ and define $G := M(0)^{-1}F$; $F_\chi := M(\chi)G \in S^4_\chi V$ (note $F_0 = F$).

If $K(J) \subset Sing(Z(F))$, then $G = \frac{1}{4}\frac{\partial}{\partial X_0}F$ is a cubic equation for $K(J)$, and, for all $\alpha \in (\mathbf{Z}/2\mathbf{Z})^g$, $G_\alpha := (1, \alpha, 0)G$ is also an equation for $K(J)$ since $K(J)$ is invariant under the action of H. Therefore each $F_\chi \in S^4_\chi V$ is a quartic equation for $K(J)$. Using (4.8)(ii), we see that F_χ also vanishes on $K(P_x) \subset \mathbf{P}V_x$ (with x corresponding to χ). Since F and F_χ have the same restriction to $\mathbf{P}V_x$ (see (4.5)), F also vanishes on $K(P_x)$. This shows that for all non-zero x F vanishes on $K(P_x)$ and since by assumption $K(J) \subset Z(F)$, F vanishes on all Pryms, i.e. $F \in \ker n_0$.

Conversely, assume that $F = F_0 \in \ker n_0$, i.e. F_0 vanishes on all Pryms, in particular F_0 vanishes on $K(J)$. We will first show that all other F_χ's, $\chi \neq 0$, also vanish on $K(J)$. For this we restrict F_χ to $\mathbf{P}V_x$, with x corresponding to χ. Since F and F_χ coincide, up to scalar multiple, and since F vanishes by assumption on all Pryms, in particular on $P_x \in \mathbf{P}V_x$, we see that F_χ vanishes on P_x. Using (4.8)(ii) again, we see that for any non-trivial χ, F_χ is also an equation for $K(J)$. To finish the argument, we take suitable linear combinations of the various F_χ, and we find that

$K(J) \subset Z(X_\alpha \frac{\partial}{\partial X_\alpha} F)$. Since X_α is not zero on $K(J)$, each $\frac{\partial}{\partial X_\alpha} F$ is an equation for $K(J)$, so we get $K(J) \subset Sing(Z(F))$.

The same principles, with the role of (SJ) being played by (D) give the proof of (4.4); for details see [vGP].

(4.9) **Corollary.** *For a curve with no vanishing thetanulls, a quartic vanishes on $\phi(\mathcal{M}_0)$ iff it vanishes on $\phi_x(P_x)$ for all Pryms of C.*

The Corollary uses the fact that the inequality obtained for N_4 is an equality [Bel].

As a curiosity, we cite other instances which relate quartic hypersurfaces of $\mathbf{P}V$ to the geometry of the curve:

(4.10) The Kummer variety K can be defined by quartics. For $g = 3$, the image of $\mathcal{M}_0$ is a quartic hypersurface (cf. [NR]).

5. Heat Equation

(5.1) Rank 1. For a period matrix τ belonging to the Siegel upper-half space $\mathbf{H}_g$, the Riemann theta function ϑ is a solution to the heat equations [M2, Chapter I]:

$$\left(\nabla_{\frac{\partial}{\partial \tau_{kl}}} \vartheta\right)(z,\tau) = 0, \quad \text{with } \nabla_{\frac{\partial}{\partial \tau_{kl}}} := \frac{\partial}{\partial \tau_{kl}} - 2\pi i(1+\delta_{kl})\frac{\partial^2}{\partial z_k \partial z_l}.$$

The heat equations and the functional equation expressing the fact that $\vartheta(\tau, z)$ is a global section of $H^0(X_\tau, \mathcal{O}(\Theta_\tau))$ determine the function $\vartheta : \mathbf{C}^g \times \mathbf{H}_g \to \mathbf{C}$ up to multiplication by a constant. The line bundle over $\mathbf{H}_g$ whose fibers are the $H^0(X_\tau, \mathcal{O}(\Theta_\tau))$ thus has a natural connection for which ϑ is flat. The covariant differentiation of the connection is given by the $\nabla_{\frac{\partial}{\partial \tau_{kl}}}$'s. Note that a holomorphic section $\tau \mapsto f(\tau)\vartheta(\tau, z)$ of the bundle is flat exactly when $\frac{\partial}{\partial \tau_{kl}} f = 0$ for all k, l, i.e. when f is a constant.

More generally, the theta functions which form the basis $\{\vartheta[{r \atop 0}](nz, n\tau)\}_{r \in (\mathbf{Z}/n\mathbf{Z})^g}$ of $H^0(X_\tau, \mathcal{O}(n\Theta_\tau))$ satisfy a (slightly modified) heat equation. This heat equation thus determines a connection on the bundle over $\mathbf{H}_g$ whose fibers are the $H^0(X_\tau, \mathcal{O}(n\Theta_\tau))$ and the elements of the basis given define a basis of flat sections.

More intrinsically, the heat equation can be interpreted as a calculation of a first order infinitesimal deformation of a pair (A, Θ), a ppav and a divisor giving the principal polarization. In particular, given an abelian scheme over some parameter space B and a relatively ample line bundle $\mathcal{L}$ over it, the family of effective divisors on the fibres which are defined by the sections of the line bundles induced by $\mathcal{L}$ (that is, the projectivization of the bundles considered above) is endowed with a canonical flat connection, the holonomy being the monodromy on the set of theta structures (cf. [W] for this interpretation). In particular, the monodromy is isomorphic to a subgroup of $\Gamma_g/\Gamma_g(2n,4n)$ when $\mathcal{L}_b \cong \mathcal{O}(2n\Theta_b)$ $(b \in B)$, with $\Gamma_g := Sp(2g,\mathbf{Z})$ and $\Gamma(2n,4n)$ Igusa's theta group of level n.

(5.2) Higher rank. The results of (5.1) were generalized by Hitchin [H2], cf. also [ADPW]; he constructs, by a heat equation, a projectively flat connection on the global sections of $\mathcal{L}_1$ over $\mathcal{M}_1$, and more generally one has projective flat connections on the bundle with fibers $H^0(\mathcal{M}_\epsilon, \mathcal{L}_\epsilon^{\otimes k})$ over any parameter space of curves. However, the corresponding holonomy has not been worked out. When working with the universal curve over the moduli space $M_g(n)$ of curves with a level n structure ($n \geq 3$), the holonomy group $Q_{(k,\epsilon,n)}$ of the bundle over $M_g(n)$ with fibers $\mathbf{P}H^0(\mathcal{M}_\epsilon, \mathcal{L}_\epsilon^{\otimes k})$ will be a quotient of $\pi_1(M_g(n))$, which in turn is a subgroup of the Teichmüller group T_g (= mapping class group). There should be a representation $\rho_{k,\epsilon}$ of T_g on $\mathbf{P}H^0(\mathcal{M}_\epsilon, \mathcal{L}_\epsilon^{\otimes k})$ such that $Q_{(k,\epsilon,n)} = \rho_{k,\epsilon}(\pi_1(M_g(n)))$. There is a natural surjective homomorphism (given by the action on the integral homology of a curve) $T_g \to \Gamma_g$, but for large k it is improbable that $\rho_{k,\epsilon}$ factors over Γ_g.

In case $k = 1,2$ however, there exists a relation between the abelian and nonabelian theta functions, in that the restriction maps j^*:

$$H^0(\mathcal{M}_0, \mathcal{L}_0) \longrightarrow H^0(J, \mathcal{O}(2\Theta)) \quad \text{and} \quad H^0(\mathcal{M}_0, \mathcal{L}_0^{\otimes 2}) \longrightarrow H^0(J, \mathcal{O}(4\Theta))_+$$

are both isomorphisms as recalled in Sections 2, 3, the second one when the curve is generic. Since these maps should be compatible with the flat connections on both sides, that would determine the holonomy on the nonabelian theta functions for $k = 1,2$. For $k = 4$ the situation might be especially interesting: the (generically) injective restriction map (cf. (4.2)):

$$H^0(\mathcal{M}_0, \mathcal{L}_0^{\otimes 4}) \longrightarrow \oplus_{x \in J_2} H^0(P_x, \mathcal{O}(8\Xi_x))$$

and the natural 'abelian' connection on the right–hand side (for both a g dimensional and $g-1$ dimensional abelian varieties!) should determine the connection on the left–hand side; it may be that the holonomy group in this case is related to Brylinski's dihedral levels [Br].

6. Examples and Questions

In this section we discuss in some special cases the geometry of the maps

$$\delta : \mathcal{M}_0 \longrightarrow \mathbf{P}H^0(\mathcal{M}_0, \mathcal{L}_0) \quad \text{and} \quad \varphi_p : \mathcal{M}_p \longrightarrow \mathbf{P}\wedge^2 V \cong \mathbf{P}H^0(\mathcal{M}_p, \mathcal{L}_1),$$

as well as questions on the geometry which arises in $\mathbf{P}V$ via δ.

Hyperelliptic case

(6.1) For all $g \geq 2$, the map δ factors over $\mathcal{M}_0/\iota^*$ with ι the hyperelliptic involution (and only if $g = 2$ does ι^* act trivially). In [DR], $\mathcal{M}_0/\iota$ is given as the subvariety of a homogeneous space, namely the grassmannian Gr of maximal isotropic subspaces of $\mathbf{C}^{2g+2}$ (with the standard quadratic form). It turns out that the maps δ for the various hyperelliptic curves 'patch' together and extend to a map: $\tilde{\delta} : Gr \to \mathbf{P}V$. Moreover, this map is equivariant for the action of the Spin group of $SO(2g+2)$, acting via a half-spin representation on $\mathbf{P}V$ (see [vG]).

The odd case is also treated in [DR]: $\mathcal{M}_1$ can be given as the scheme of zeros of a section of a homogeneous bundle on a grassmannian; set-theoretically, the totality of $(g-2)$-dimensional planes lying in the intersection of two quadrics in $\mathbf{P}^{2g+1}$; both [L1] and [Sz] use this essentially for the calculation of the Verlinde numbers.

(6.2) For $g = 2$ it was shown in [NR1] that $\mathcal{M}_0 \cong Gr \cong \mathbf{P}^3$; in fact $\delta = \tilde{\delta}$ is now an isomorphism; the Kummer surface $K(J)$ is defined by a quartic polynomial and now the $K(P_x) \cong \mathbf{P}V_x \cong \mathbf{P}^1$ are $2 \cdot 15$ lines. Since $\mathcal{L}$ corresponds to $\mathcal{O}(1)$ one has, for $g = 2$: $N_{0,k} = \dim H^0(\mathbf{P}^3, \mathcal{O}(k)) = \binom{k+3}{3}$.

$\mathcal{M}_1$ is not isomorphic to it but it is described in say, [B2, 3.15]: for $p \in C$ not a Weierstrass point, the morphism $\varphi_p: \mathcal{M}_p \to \mathbf{P}(\Lambda^2 V) \cong \mathbf{P}^5$ is an embedding; its image is the intersection of the Grassmannian Gr $(2, V)$ and another quadric; $\hat{K}$ is the trace on $\mathcal{M}_p$ of a third quadric (notation as in Section 3).

(6.3) For $g = 3$, the example [B1, 3.5] continues: the morphism δ has degree 2 onto a smooth quadric $Q \cong Gr$ (this isomorphism is related to the triality for $SO(8)$, which is of type D_4) in $\mathbf{P}V \cong \mathbf{P}^7$, this quadric is in fact an element of $V \otimes V$ which spans the representation of H corresponding to the unique theta characteristic on C with $h^0(C, \kappa) = 2$.

The canonical divisor is $\mathcal{L}_0^{-4} = \delta^*(K_Q(2))$, where K_Q is the canonical divisor of Q; since δ is defined by $\mathcal{L}_0$, its branch locus is the trace on Q of a hypersurface H of degree 4; the Kummer is the singular locus of $Q \cap H$, hence is defined set theoretically by Q, H and the $\binom{8}{2}$ minors of the matrix of partials.

The geometry of the case $g = 3$, C non hyperelliptic

(6.4) In this case δ defines an isomorphism of $\mathcal{M}_0$ to a quartic hypersurface Y of $\mathbf{P}^7$ ([NR2]). The singular locus of Y is the Kummer variety, which is thus defined by 8 cubics. The quartic surfaces $K(P_x)$ are obtained by intersecting Y with $\mathbf{P}V_x \cong \mathbf{P}^3$. The bundle $\mathcal{L}_0$ corresponds to $\mathcal{O}_Y(1)$, thus $H^0(\mathcal{M}_0, \mathcal{L}_0^{\otimes k}) \cong H^0(\mathbf{P}^7, \mathcal{O}(k))/H^0(\mathbf{P}^7, \mathcal{O}(k-4))$ and thus, for $g = 3$, $N_{0,k} = \binom{k+7}{7} - \binom{k+3}{7}$.

Questions

(6.5) By virtue of the Heisenberg group action, there is a natural map

$$Th : \mathbf{H}_g \longrightarrow \mathbf{P}V, \qquad \tau \mapsto (\ldots, \vartheta[{\sigma \atop 0}](\tau, 0), \ldots),$$

which factors over $\mathcal{A}_g(2,4) := \mathbf{H}_g/\Gamma_g(2,4)$. In case a ppav A is the Jacobian of a curve, one knows that the intersection of $Th(\mathbf{H}_g)$ with $K(A)$, the image of A in $\mathbf{P}V$, contains a surface (this is related to Prym varieties of 2:1 covers of C ramified in 2 points, cf. [vGvdG]), but for general A it seems likely the the intersection consists just of points, in fact the set $K(A_2)$ (note that, with 0_A the identity element of A one has $K(0_A) = Th(\tau)$, with τ a suitable period matrix of A).

It would be interesting to know more about the intersection of $Th(\mathbf{H}_g)$ and $\delta(\mathcal{M}_0)$ as well as to have a geometric interpretation for it. In case $g = 3$, both spaces are of codimension 1 in $\mathbf{P}^7$, of degree 16 and 4 respectively, so the intersection will be 5 dimensional. For $g = 4$, they have codimension 5 resp. 6 in $\mathbf{P}^{15}$, so the intersection should be at least 4 dimensional.

(6.6) A related but probably easier question is to find, for a Jacobian J, the intersection of the tangent space $T_{K(0_J)}$ to $Th(\mathbf{H}_g)$ at the point $K(0_J) \in Th(\mathbf{H}_g)$ with $\delta(\mathcal{M}_0)$.

The intersection of $T_{K(0_J)}$ with $K(J)$ is well understood. In fact, identifying $\mathbf{P}V \cong |2\Theta|^*$, the hyperplanes in $\mathbf{P}V$ passing through $K(0_J)$ are those $D \in |2\Theta|$ with $0_J \in D$, and the hyperplanes which contain $T_{K(0_J)}$ correspond to the divisors wich have multiplicity at least 4 in 0_J (here one uses the heat equations, cf. [vGvdG]); we denote that space by:

$$|2\Theta|_{00} := \{D \in |2\Theta| : \mathit{mult}_0(D) \geq 4\}.$$

Writing $C - C := \{x - y \in J : x, y \in C\}$, which is a surface in J, a well known result of Welters, cf. [BD], is (except for $g = 4$, when there can be two additional points):

$$\cap_{D \in |2\Theta|_{00}} D = C - C, \quad \text{so} \quad T_{K(0_J)} \cap K(J) = K(C - C).$$

It would be nice to have a similar explicit description of $T_{K(0_J)} \cap \delta(\mathcal{M}_0)$. One might first want to consider the intersection of $T_{K(0_J)}$ with the tangent cone to $\delta(\mathcal{M}_0)$ at the (singular) point $\delta(\mathcal{O} \oplus \mathcal{O})$, however we do not know a suitable description of this cone.

(6.7) The geometry of the divisors from $|2\Theta|_{00}$ (and also from $|2\Theta|$ itself) is not yet very well understood. The theory of rank two bundles provides an interesting way to produce elements in these spaces via the map $\delta : E \mapsto \Delta_E \in |2\Theta|$, cf. (2.6)(i). The recent paper of Laszlo [L2] provides a good starting point, in particular he proves an analogue of the Riemann-Kempf vanishing theorem for the divisor $\Theta_{\mathcal{U}}$ (defined after (2.8)) as well as the following result [L2, V.2]:

$$mult_L(\Delta_E) \geq h^0(E \otimes L),$$

(here we change the notation of (2.6) to have $\det E = \Omega^1_C$ and $\Delta_E := \{x \in J : h^0(E \otimes x) > 0\}$; one can get things right again by replacing E by $E \otimes \kappa^{-1}$ for a theta characteristic κ). A nice example is provided by the following 'canonical' rank two bundle E with $\det E = \Omega^1_C$ on a non-hyperelliptic curve of genus 3. Its dual E^* is the rank two bundle which fits in the exact sequence:

$$0 \longrightarrow E^* \longrightarrow H^0(C, \Omega^1_C) \otimes_{\mathbb{C}} \mathcal{O}_C \xrightarrow{\sigma} \Omega^1_C \longrightarrow 0,$$

where σ is the "tautological" multiplication map. The results of Laszlo ([L2, IV.9]) show that

$$\Delta_E = C - C$$

the unique element in $|2\Theta|_{00}$ for $g = 3$.

7. Spectral Curves

The notion of a spectral curve was first introduced, in the late sixties, in the theory of integrable systems. We saw it playing a surprising and crucial role in the proofs of Section 2. It would be interesting to look at the relationship in even greater detail and perhaps find applications to integrable systems. In this section we indicate some 'historical' motivation and questions.

The fibre of the tangent bundle to $\mathcal{U}(n,d)$ at a stable point E is given by $H^1(\mathcal{E}nd E)$, thus the fibre of the cotangent bundle by $\mathrm{Hom}(E, E \otimes K_C)$. Hitchin proved that for an open set of stable bundles the spectral curve of the homomorphism $E \to E \otimes K_C$ has genus $n^2(g-1)+1$ and the spectral invariants are in involution with respect to the natural symplectic structure on the cotangent bundle [H1]. Elements of $H^0(C, \mathcal{E}nd_0 E \otimes \kappa)$ enter the definition of a basis of sections for two of the Verlinde spaces, as we saw in Section 3. The spectral curves that correspond to them in the manner of [H1] should be linked to points of order 4 in J, at least in the case of even κ where $H^0(\mathcal{M}_0, \mathcal{L}_0^{\otimes 2}) \to H^0(J, \mathcal{O}(4\Theta))$ (by restriction). A correspondence between these and Donagi's curves should be found, which he constructs in his proof of (D) as fiber products of two étale double covers of C, cf. [D]. The corresponding rank 4 bundles over C and their image in $|4\Theta|$ according to [BNR] should give an interpretation of (D) in $|4\Theta|$, in the spirit of the Schottky relations, and come from a classical theta formula. The rank 4 bundles that appear in v([B2], 2.3), namely $E \otimes G$ where $E \in \mathcal{M}_0$ is fixed and

$$\Delta_\kappa(E) = \{G \in \mathcal{M}_0 | h^0(C, E \otimes \kappa \otimes G) \geq 1\}$$

will be among the ones looked for.

A geometric description of the images of $\phi_p(\mathcal{M}_1)$ has not been pursued, nor have the images of K in $\mathbf{P}(\Lambda^2 V)$, both for fixed p and their union as p runs on C, or their analog if p is replaced by a degree 1 line bundle; the equations for the image are certainly linkable (explicitly) with equations for the curve in canonical space, in view of the Gauss map interpretation (3.7) for $\varphi_p \circ j_p$. As noticed in [B2], these maps give the trisecants which characterize the Schottky locus. The geometry of these maps should be particularly explicit in the hyperelliptic case in view of possible spin equivariance. To close the circle, in the hyperelliptic case the 'homogeneous' model of (6.1) enters explicitly the description of Neumann's integrable system.

REFERENCES

[ABI] D. Altschüler, M. Bauer and C. Itzykson, The branching rules of conformal embeddings, *Comm. Math. Phys.* **132** (1990), 349–364.

[ADPW] S. Axelrod, S. Della Pietra and E. Witten, Geometric quantization of Chern–Simons gauge theory, *J. Differential Geom.* **33** (1991), 787–902.

[B1] A. Beauville, Fibrés de rang 2 sur une courbe, fibré déterminant et fonctions thêta, *Bull. Soc. Math. France* **116** (1988), 431–448.

[B2] A. Beauville, Fibrés de rang 2 sur une courbe, fibré déterminant et fonctions thêta, II, *Bull. Soc. Math. France* **119** (1991), 259–291.

[BD] A. Beauville and O. Debarre, Sur les fonctions thêta du second ordre, *Arithmetic of complex manifolds*, 27–39, Springer-Verlag, Berlin 1989.

[BNR] A. Beauville, M.S. Narasimhan and S. Ramanan, Spectral curves and the generalized theta divisor, *J. Reine Angew. Math.* **398** (1989), 169–179.

[Be1] A. Bertram, A partial verification of the Verlinde formulae for vector bundles of rank 2, Preprint 1991.

[Be2] A. Bertram, Moduli of rank 2 vector bundles, theta divisors, and the geometry of curves in projective space, Preprint 1991.

[BeSz] A. Bertram and A. Szenes, Hilbert polynomials of moduli spaces of rank 2 vector bundles II, Preprint 1991.

[Bo] R. Bott, Stable bundles revisited, *J. Differential Geom. Supplement* **1** (1991), 1–18.

[Br] J.–L. Brylinski, Propriétés de ramification à l'infini du groupe modulaire de Teichmüller, *Ann. Sci. École Norm. Sup.* **12** (1979) 295–333.

[CIZ] A. Cappelli, C. Itzykson and J.-B. Zuber, The A-D-E classification of minimal and $A_1^{(1)}$ conformal invariant theories, *Comm. Math. Phys.* **113** (1987), 1–26.

[C] P. Cartier, Quantum mechanical commutation relations and theta functions, *Proc. Sympos. Pure Math.* **9**, eds. A. Borel and G. Mostow, pp. 361–383.

[DR] U.V. Desale and S. Ramanan, Classification of vector bundles of rank 2 on hyperelliptic curves, *Invent. Math.* **38** (1976), 161–185.

[D] R. Donagi, Non-Jacobians in the Schottky loci, *Ann. of Math.* **126** (1987), 193–217.

[DN] J.-M. Drezet and M.S. Narasimhan, Groupe de Picard des variétés de modules de fibrés semi-stables sur les courbes algébriques, *Invent. Math.* **97** (1989), 53–94.

[vG] B. van Geemen, Schottky-Jung relations and vector bundles on hyperelliptic curves, *Math. Ann.* **281** (1988), 431–449.

[vGvdG] B. van Geemen and G. van der Geer, Kummer varieties and the moduli spaces of abelian varieties, *Amer. J. Math.* **108** (1986), 615–642.

[vGP] B. van Geemen and E. Previato, Prym varieties and the Verlinde formula, MSRI Preprint 04829-91.

[H1] N. Hitchin, Stable bundles and integrable systems, *Duke Math. J.* **54** (1987), 91–114.

[H2] N. Hitchin, Flat connections and geometric quantization, *Comm. Math. Phys.* **131** (1990), 347–380.

[I] J.-I. Igusa, Theta functions, Springer-Verlag, Berlin 1972.

[L1] Y. Laszlo, Dimension de l'espace des sections du diviseur thêta généralisé, *Bull. Soc. Math. France* **119** (1991), 293–306.

[L2] Y. Laszlo, Un théorème de Riemann pour les diviseurs thêta sur les espaces des modules de fibrés stables sur une courbe, *Duke Math. J.* **64** (1991), 333–347.

[M1] D. Mumford, Prym varieties I, in *Contributions to Analysis*, 325–350, Acad. Pren. New York, 1974.

[M2] D. Mumford, Tata lectures on theta I, Birkhäuser, Boston 1983.

[NR1] M.S. Narasimhan and S. Ramanan, Moduli of vector bundles on a compact Riemann surface, *Ann. of Math.* **89** (1969), 19–51.

[NR2] M.S. Narasimhan and S. Ramanan, 2Θ-linear systems on abelian varieties, in *Vector bundles on algebraic varieties*, p. 415–427, Oxford University Press, 1987.

[S] C.S. Seshadri, Space of unitary vector bundles on a compact Riemann surface, *Ann. of Math.* **85** (1967), 303–336.

[Sz] A. Szenes, Hilbert polynomials of moduli spaces of rank 2 vector bundles I, Preprint 1991.

[V] E. Verlinde, Fusion rules and modular transformations in 2d conformal field theory, *Nuclear Phys. B* **300** (1988), 360–376.

[W] G. Welters, Polarized abelian varieties and the heat equation, *Compositio Math.* **49** (1983), 173–194.

[Z] D. Zagier, The cohomology ring of the moduli space of rank 2 vector bundles, in preparation.

Department of Mathematics
University of Utrecht
3508 TA
Utrecht, The Netherlands

Department of Mathematics
Boston University
Boston, MA 02215, USA

Hyperelliptic Curves that Generate Constant Mean Curvature Tori in $\mathbb{R}^3$

N. M. Ercolani,* H. Knörrer and E. Trubowitz

This paper is dedicated,
with many fond recollections, to Jean-Louis Verdier

Introduction

Let u be a solution of the elliptic-sinh Gordon equation

$$u_{w\bar{w}} + \sinh u = 0 \tag{1}$$

on the simply connected domain $\Omega \subset \mathbb{C}$. There is an algorithm that associates an immersion F of Ω in $\mathbb{R}^3$ to u with constant mean curvature $\frac{1}{2}$ (see e.g. [3]). To implement ite first solves

$$\psi_w = -\frac{1}{2}\begin{pmatrix} u_w & i \\ i & -u_w \end{pmatrix}\psi \tag{2}$$

$$\psi_{\bar{w}} = \frac{1}{2.i}\begin{pmatrix} 0 & e^{-u} \\ e^u & 0 \end{pmatrix}\psi \tag{3}$$

for

$$\psi = \begin{pmatrix} \psi_1(w) \\ \psi_2(w) \end{pmatrix}$$

on Ω. Equation (1) is the consistency condition for (2) and (3). The immersion F is obtained by integrating

$$dF_1 = \frac{1}{d}\left(e^u\bar{\psi}_1^2 + \psi_2^2\right)dw + \frac{1}{d}\left(e^u\psi_1^2 + \bar{\psi}_2^2\right)d\bar{w} \tag{4}$$

$$dF_2 = \frac{2i}{d}e^{u/2}\bar{\psi}_1\psi_2 dw + \frac{2(-i)}{d}e^{u/2}\psi_1\bar{\psi}_2 d\bar{w} \tag{5}$$

$$dF_3 = \frac{(-i)}{d}\left(e^u\bar{\psi}_1^2 - \psi_2^2\right)dw + \frac{i}{d}\left(e^u\psi_1^2 - \bar{\psi}_2^2\right)d\bar{w} \tag{6}$$

where

$$d = \left(|\psi_1|^2 e^{u/2} + |\psi_2|^2 e^{-u/2}\right)$$

---

* Nick Ercolani wishes to thank the Forschunginstitut für Mathematik at Zürich for its hospitality while this work was in progress. He also acknowledges support from the NSF (DMS-9001897) as well as the Arizona Center for the Mathematical Sciences, sponsored by AFOSR Contract FY8671-900589.

is independent of z. The forms on the right hand side of (4-6) are closed by (2-3). It follows at once from (4-6) that

$$\langle F_w, F_{\bar{w}} \rangle = 2e^u \tag{7}$$

$$\langle F_w, F_w \rangle = \langle F_{\bar{w}}, F_{\bar{w}} \rangle = 0 \tag{8}$$

and in particular that the first fundamental form

$$\langle dF, dF \rangle = 4e^u dw\, d\bar{w}$$

We see that $Re\, w$, $Im\, w$ are isothermal coordinates on the image of F. Here

$$\langle v, w \rangle = v_1 w_1 + v_2 w_2 + v_3 w_3 \,.$$

Let

$$N(w) = \frac{1}{d}\left(i(\psi_1\psi_2 - \bar{\psi}_1\bar{\psi}_2), -\left(e^{u/2}|\psi_1|^2 - e^{-u/2}|\psi_2|^2\right)\right), -(\psi_1\psi_2 + \psi_1\psi_2)\Big)$$

be the unit normal vector to the image of F at $F(z)$. Then, the second fundamental form

$$-\langle dF, dN \rangle = dw\, dw + 2e^u dw\, d\bar{w} + d\bar{w}\, d\bar{w} \tag{10}$$

so that the mean curvature is $\frac{1}{2}$.

There is also an algorithm that associates quasi-periodic solutions of (1) on $\mathbb{R}^2$ to hyperelliptic curves

$$X : y^2 = x \prod_{i=1}^{2g} (x - e_i) \tag{11}$$

where the branch points are distinct and satisfy

$$e_{i+g} = \frac{1}{\bar{e}_i}, \; i = 1, \ldots, g \tag{12}$$

Let $A_1, \ldots, A_g$ be the cycles on the hyperelliptic curve obtained by lifting the paths in the x-plane

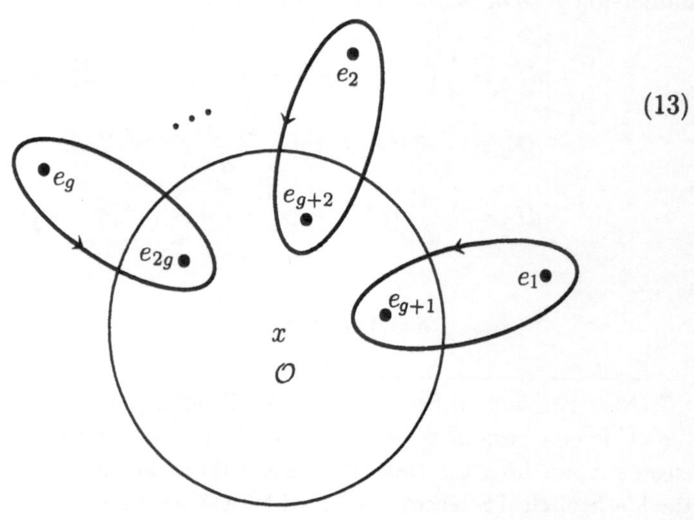

(13)

and let $B_1, \ldots, B_g$ be their duals with respect to the intersection form. Let Ω_0 and Ω_∞ be the unique meromorphic differentials satisfying

$$\int_{A_i} \Omega_0 = \int_{A_i} \Omega_\infty = 0, \quad i = 1, \ldots, g$$

$$\Omega_0 = \frac{dx}{x^{3/2}} + \mathcal{O}\left(\frac{1}{x^{1/2}}\right) dx \quad \text{for } x \text{ near } 0$$

$$\Omega_\infty = -x^{3/2} d\frac{1}{x} + \mathcal{O}\left(x^{1/2}\right) d\frac{1}{x} \quad \text{for } x \text{ near } \infty \tag{15}$$

and holomorphic everywhere else.

The map σ

$$(x,y) \longrightarrow \left(\frac{1}{\bar{x}}, \frac{\left(\prod_{i=1}^{2g} e_i\right)^{1/2} \bar{y}}{\bar{x}^{g+1}} \right)$$

is an antiholomorphic involution of X. Clearly,

$$\sigma^* \Omega_0 = \bar{\Omega}_\infty .$$

It follows that the vectors

$$\nu(0) = \left(\int_{B_i} \Omega_0, i = 1, \ldots, g \right)$$

$$\nu(\infty) = \left(\int_{B_i} \Omega_\infty, \, i = 1, \ldots, g \right) \tag{16}$$

are complex conjugate. For $\zeta \in \mathbb{C}^g$ set

$$u(\zeta) = 2 \log \frac{\theta\left(\zeta + \left(\frac{1}{2}, \ldots, \frac{1}{2}\right)\right)}{\theta(\zeta)} \tag{17}$$

where θ is the Riemann theta function of X for the chosen homology basis. Then, after scaling w,

$$u\left(\zeta + w\,\nu(0) + \bar{w}\,\nu(\infty)\right))$$

is a real, quasi-periodic solution of (1), for every $\zeta \in \mathbb{R}^g$. In particular, every hyperelliptic curve (11) and (12) generates a $(g-2)$ parameter family of constant mean curvature immersions of $\mathbb{C}$ in $\mathbb{R}^3$.

Pinkall-Sterling [6] and then Bobenko [2] showed that every constant mean curvature immersion of $\mathbb{C}$ in $\mathbb{R}^3$ that is periodic with respect to a lattice in $\mathbb{C}$, in other words an immersed torus, is generated by a curve (11) and (12). Observe that non isomorphic curves yield different tori. Bobenko [2] determined necessary and sufficient conditions for a curve (11 and 12) to generate a constant mean curvature torus. They are

(a)] Ω_∞ has a root p, lying over a point p' on the unit circle in the x plane.

(b) Let

$$\gamma' = \{tp' \mid t \geq 1\}$$

and γ its lift to X starting at p. Then, the span of the vectors

$$\left(\int_\gamma \Omega_0, \int_{B_1} \Omega_0, \ldots, \int_{B_g} \Omega_0\right)$$

$$\left(\int_\gamma \Omega_\infty, \int_{B_1} \Omega_\infty, \ldots, \int_{B_g} \Omega_\infty\right)$$

in $\mathbb{C}^{g+1}$ must contain two linearly independent rational vectors.

It is known that there are no curves satisfying (a) and (b) for $g = 1$.

Wente's [7] original example of an immersed constant mean curvature torus corresponds to the case $g = 2$ (see [2§13]) and the examples exhibited in [8] correspond to $g = 3$. It is not known whether there are curves (11) and (12) of every genus larger than three satisfying (a) and (b). We shall show however

Theorem 1. *For each even $g \geq 2$ there are infinitely many curves (11) and (12) satisfying (a) and (b).*

Each such curve generates a $(g-2)$-parameter family of constant mean curvature immersions of a 2-torus in $\mathbb{R}^3$. All the curves constructed in this paper have the additional property that the set of branch points is invariant under $e_i \longrightarrow \frac{1}{e_i}$.

A condition similar to (a) and (b) above arises in Hitchin's investigation of harmonic maps of a two-torus to S^3. In sections 9-12 of [5] the existence of curves of genus ≤ 3 fulfilling Hitchin's conditions is established.

Preliminaries

For each $2g$-tuple $\lambda = (\lambda_1, \ldots, \lambda_{2g})$ of complex numbers different from ± 2 put

$$R_\lambda(z) := \prod_{i=1}^{2g} (z - \lambda_i)$$

and let $C_\lambda^{(\nu)}$ be the hyperelliptic curve associated to the equation

$$y^2 = (z + (-1)^\nu 2)\, R_\lambda(z) \quad \nu = 1, 2\,.$$

Whenever the sets $\{\lambda_1, \lambda_2\}, \{\lambda_3, \lambda_4\}, \ldots, \{\lambda_{2g-1}, \lambda_{2g}\}$ are mutually disjoint one can choose homotopy classes $h_1, \ldots, h_g$ of nonintersecting loops in $\mathbb{C}\backslash\{\lambda_1, \ldots, \lambda_{2g}, +2, -2\}$ such that h_i has winding number one around λ_{2i-1} and λ_{2i} and winding number zero around all the other points λ_j and also around the point 2 (see the figure below). Lifting these loops by the projections $\pi^{(\nu)} : (y, z) \mapsto z$ to the curves $C_\lambda^{(\nu)}$ determines (up to sign) nonintersecting homology classes $a_1^{(\nu)}, \ldots, a_g^{(\nu)} \in H_1(C_\lambda^{(\nu)}, \mathbb{Z})$. This set can be completed in a unique way to a normalized basis $a_1^{(\nu)}, \ldots, a_g^{(\nu)}, b_1^{(\nu)}, \ldots, b_g^{(\nu)}$ of $H_1(C_\lambda^{(\nu)}, \mathbb{Z})$. Then there is a unique differential form $\Omega^{(\nu)}$ of the second kind on $C_\lambda^{(\nu)}$ that is holomorphic outside infinity, fulfills

$$\int_{a_i^{(\nu)}} \Omega^{(\nu)} = 0 \quad \text{for } i = 1, \ldots, g$$

and

$$\Omega^{(\nu)} = -z^{3/2} d\left(\frac{1}{z}\right) + \mathcal{O}(z^{1/2}) d\left(\frac{1}{z}\right) \qquad \text{for } z \to \infty\,.$$

Here, holomorphic differential means a regular section of the canonical sheaf. If $C_\lambda^{(\nu)}$ is singular at p then $\Omega^{(\nu)}$ is a Rosenlicht differential at p. After we fix a choice of a square root of z near ∞ outside the real axis, $\Omega^{(\nu)}$ is uniquely determined.

Observe that $\Omega^{(\nu)}$ is left fixed by the hyperelliptic involution $(y, z) \mapsto (-y, z)$, so we may write

$$\Omega^{(\nu)} = \frac{dz}{\sqrt{(z + (-1)^\nu 2)\, R_\lambda(z)}} \cdot \prod_{i=1}^{g} (z - \alpha_i^{(\nu)})\,.$$

Since each h_i has winding number zero around the point $+2$ there is a homotopy class $\tilde{c}$ of a path in $\mathbb{C} \smallsetminus \{\lambda_1, \ldots, \lambda_{2g}\}$ starting and ending at $+2$ that does not intersect any of the h_i and has winding number 1 around all the points $\lambda_i, i = 1, \ldots, 2g$, and the point -2. Up to orientation $\tilde{c}$ determines a path c in $C_\lambda^{(2)}$ connecting the two points lying over $+2$.

Let M_g be the space of sequences $\lambda = (\lambda_1, \ldots, \lambda_{2g})$ as above, together with all the choices described. We denote points of M_g by s, and the associated data by $\lambda(s) = (\lambda_1(s), \ldots, \lambda_{2g}(s))$, $R_s(z), C_s^{(\nu)}, h_i(s), a_i^{(\nu)}(s), b_i^{(\nu)}(s)$, $\Omega^{(\nu)}(s)$, $\alpha_i^{(\nu)}(s), \tilde{c}(s), c(s)$. At each point s of M_g where the roots $\alpha_1^{(\nu)}(s), \ldots, \alpha_g^{(\nu)}(s)$ of $\Omega^\nu(s)$ are pointwise distinct, $\nu = 1, 2$, the branch points $\lambda_1, \ldots, \lambda_{2g}$ are local coordinates on M_g.

On the open subset of M_g where $\lambda_i(s), i = 1, \dots, 2g$ are pairwise distinct we define

$$\begin{aligned} Z(s) &:= \alpha_1^{(1)}(s) - 2 \\ I_1(s) &:= \left(\int_{b_1^{(1)}(s)} \Omega^{(1)}(s), \dots, \int_{b_g^{(1)}(s)} \Omega^{(1)}(s) \right) \\ I_2(s) &:= \left(\int_{c(s)} \Omega^{(2)}(s); \int_{b_1^{(2)}(s)} \Omega^{(2)}(s), \dots, \int_{b_g^{(2)}(s)} \Omega^{(2)}(s) \right) \end{aligned}$$

To see that I_1 and I_2 extend to analytic maps on all of M_g introduce the holomorphic differentials $\omega_j^{(\nu)}(s)$ on $C_s^{(\nu)}$ characterised by

$$\int_{a_i^{(\nu)}(s)} \omega_j^{(\nu)}(s) = 2\pi i \delta_{ij}\,.$$

By reciprocity ([4, p. 67])

$$\omega_i^{(\nu)}(s) = -\left(\frac{1}{4} \int_{b_i^{(\nu)}(s)} \Omega^{(\nu)}(s) \right) \sqrt{z}\, d\left(\frac{1}{z}\right) + \mathcal{O}\left(\sqrt{\frac{1}{z}}\right) d\left(\frac{1}{z}\right) \quad \text{for } z \to \infty$$

whenever $\lambda_{2i-1}(s) \neq \lambda_{2i}(s)$. The leading term of the expansion of $\omega_i^{(\nu)}(s)$ at infinity can thus be used to define I_ν everywhere.

In M_g we distinguish the real subset $M_{g,\mathbb{R}}$ consisting of all $s \in M_g$ for which

(i) $\lambda_{2i-1}(s) = \bar{\lambda}_{2i}(s)$.

(ii) $h_i(s)$ is invariant under complex conjugation, and meets the real axis transversally in only two points that both lie in the interval $(-2, 2)$. The lift $a_i^{(\nu)}(s)$ is chosen such that over the point where $h_i(s)$ meets the real axis with positive orientation the y-component of $a_i^{(\nu)}(s)$ has positive imaginary part when $\nu = 1$ and is purely real and positive when $\nu = 2$.

(iii) $\tilde{c}(s)$ is invariant under complex conjugation and $c(s)$ starts at the point with $y > 0$.

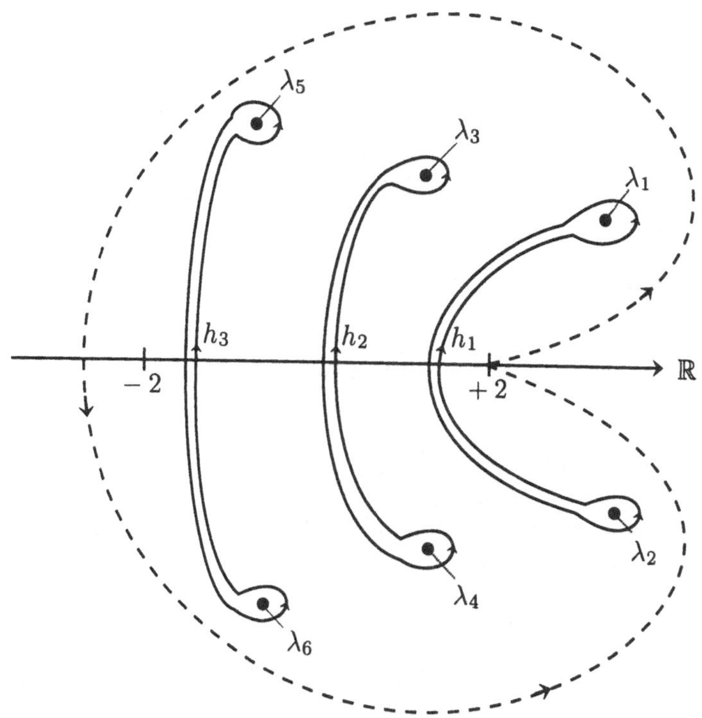

The relation of points of M_g and $M_{g,\mathbb{R}}$ to constant mean curvature tori on $\mathbb{R}^3$ is the following. For $s \in M_g$ let $X(s)$ be the fibered product of $C^{(1)}(s)$ and $C^{(2)}(s)$ with respect to the maps $\pi^{(1)}$ and $\pi^{(2)}$ to $\mathbb{P}^1$. Denote by τ_ν the canonical projections of $X(s)$ to $C^{(\nu)}(s)$.

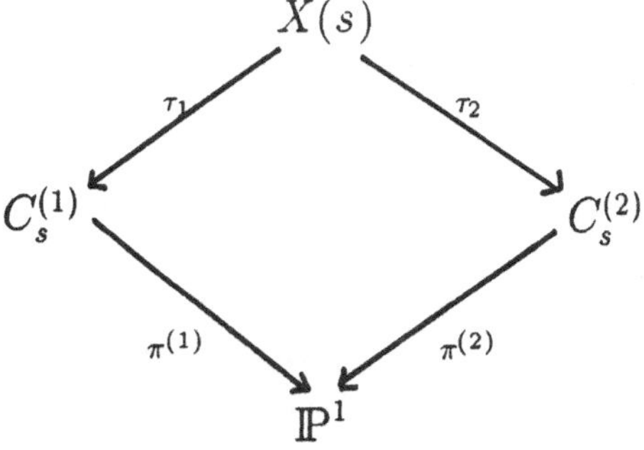

Then, $X(s)$ is the hyperelliptic curve associated to the equation

$$y^2 = x \prod_{\substack{e\in\mathbb{C} \\ e+\frac{1}{e}\in\{\lambda_1,\dots,\lambda_{2g}\}}} (x-e)$$

and the maps τ_1, τ_2 are given by

$$\tau_\nu : (y,x) \longmapsto \left(\frac{x+(-1)^\nu}{x^{g+1}} y, x + \frac{1}{x} \right) .$$

They are the quotient maps by the involutions

$$i_\nu : (y,x) \longmapsto \left((-1)^\nu \frac{y}{x^{2g+1}}, \frac{1}{x} \right)$$

on $X(s)$ and ramify at the points lying over $x = (-1)^\nu$ on $X(s)$. Therefore, $\tau_2^*(\Omega^{(2)})$ always vanishes at the points of $X(s)$ lying above $x = 1$ of $X(s)$, and $\tau_1^*(\Omega^{(1)})$ vanishes at these points if and only if $\Omega^{(1)}$ has $+2$ as a root, so certainly whenever $Z(s) = 0$.

Furthermore

$$\Omega_0 := \frac{1}{2}\left(\tau_1^*(\Omega^{(1)}) - \tau_2^*(\Omega^{(2)})\right)$$
$$\Omega_\infty := \frac{1}{2}\left(\tau_1^*(\Omega^{(1)}) + \tau_2^*(\Omega^{(2)})\right)$$

are differentials on $X(s)$ that are holomorphic outside 0 and ∞ and are normalized to look like

$$\Omega_0 = \frac{dx}{x^{3/2}} + O\left(\frac{1}{x^{1/2}}\right) dx \qquad \text{for } x \text{ near } 0$$
$$\Omega_\infty = -x^{3/2} d\left(\frac{1}{x}\right) + O(x^{1/2}) d(\frac{1}{x}) \quad \text{for } x \text{ near } \infty .$$

Clearly Ω_∞ vanishes at $x = 1$ whenever $Z(s) = 0$.

If $s \in M_{g,\mathbf{R}}$, then the set E of all $e \in \mathbb{C}$ for which $e + \frac{1}{e} \in \{\lambda_1, \dots, \lambda_{2g}\}$ is invariant under the antiholomorphic involution $e \mapsto \bar{e}$ and $e \mapsto \frac{1}{\bar{e}}$. In particular, the hyperelliptic curve $X(s)$, $s \in M_{g,\mathbf{R}}$, satisfies (11) and (12).

In addition, the lift of each of the loops h_i under the map $x \mapsto x + \frac{1}{x}$ consists of two loops h_i', h_i'' each of which is invariant under the involution $x \mapsto \frac{1}{\bar{x}}$ and encircles precisely one pair $e, \frac{1}{\bar{e}}$ of branch points of the hyperelliptic cover $\pi : X(s) \longrightarrow \mathbb{C}$, $(y,x) \mapsto x$. (Observe that the image of the unit circle under the map $x \mapsto x + \frac{1}{x}$ is the interval $[-2,2]$!)

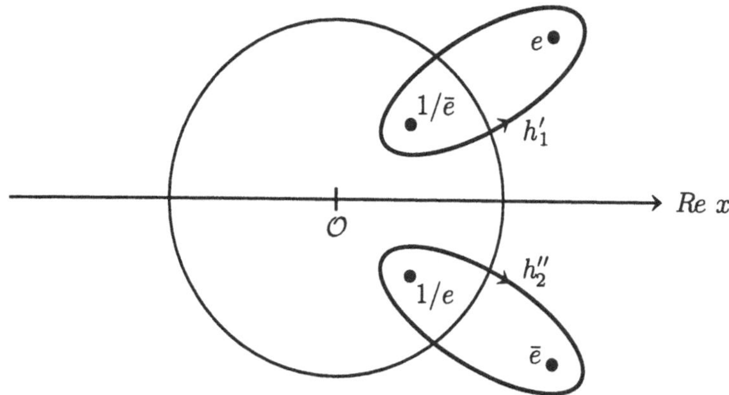

Lifts of these loops to $X(s)$ determine homology classes A_i', A_i'' on $X(s)$ with $(\tau_\nu)_*(A_i') = (-1)^\nu(\tau_\nu)_*(A_i'') = a_i^{(\nu)}$ for $i = 1, \ldots, g$. Clearly,

$$\int_{A_i'} \Omega_0 = \int_{A_i''} \Omega_0 = \int_{A_i'} \Omega_\infty = \int_{A_i''} \Omega_\infty = 0 \text{ for } i = 1, \ldots, g\,.$$

In particular, Ω_0 and Ω_∞ are differentials of the second kind. We complete the A_i', A_i'' to a canonical homology basis

$$A_1', \ldots, A_g', A_1'', \ldots, A_g'', B_1', \ldots, B_g', B_1'', \ldots, B_g'' \text{ of } H_1\left(X(s), \mathbb{Z}\right)\,.$$

Finally, let γ be the path on $X(s)$ joining the two points over $x = 1$ such that $\tau_2(\gamma) = c(s)$ and $\pi(\gamma)$ has winding number one around all the points of E.

As explained in the introduction, Bobenko [2] has shown that the curve $X(s), s \in M_{g,\mathbf{R}}$, corresponds to a $(2g-2)$-parameter family of constant mean curvature tori in $\mathbb{R}^3$ if

(a) Ω_∞ vanishes at the points over $x = 1$.

(b) The two vectors

$$\left(\int_\gamma \Omega_0; \int_{B_1'} \Omega_0, \ldots, \int_{B_g'} \Omega_0, \int_{B_1''} \Omega_0, \ldots, \int_{B_g''} \Omega_0\right)$$

$$\left(\int_\gamma \Omega_\infty; \int_{B_1'} \Omega_\infty, \ldots, \int_{B_g'} \Omega_\infty, \int_{B_g''} \Omega_\infty\right)$$

span a plane in $\mathbb{C}^{2g+1}$ that contains two linearly independent rational vectors.

As was said before, (a) is ensured by $Z(s) = 0$. For (b) we add and subtract the two vectors above to get

$$\left(\int_\gamma \tau_1^*\left(\Omega^{(1)}\right); \int_{B_1'} \tau_1^*\left(\Omega^{(1)}\right), \ldots, \int_{B_g''} \tau_1^*\left(\Omega^{(1)}\right)\right)$$

$$\left(\int_\gamma \tau_2^*\left(\Omega^{(2)}\right); \int_{B_1'} \tau_2^*\left(\Omega^{(2)}\right), \ldots, \int_{B_g''} \tau_2^*\left(\Omega^{(2)}\right)\right).$$

Now observe that $\tau_1(\gamma)$ is homologous to $a_1^{(1)}(s) + \ldots + a_g^{(1)}(s)$; thus

$$\int_\gamma \tau_1^*(\Omega^{(1)}) = 0.$$

On the other hand

$$\int_\gamma \tau_2^*(\Omega^{(2)}) = \int_{c(s)} \Omega^{(2)}(s),$$

$$\int_{B_i'} \tau_\nu^*(\Omega^{(\nu)}) = (-1)^\nu \int_{B_i''} \tau_\nu^*(\Omega^{(\nu)}) = \int_{b_i^{(\nu)}(s)} \Omega^{(\nu)}(s).$$

Therefore, after adding the B_i''-components of the vectors above to the B_i'-components we get the vectors

$$\left(0; 0, \ldots, 0; -\int_{b_1^{(1)}(s)} \Omega^{(1)}(s), \ldots, -\int_{b_g^{(1)}(s)} \Omega^{(1)}(s)\right)$$

$$\left(\int_{c(s)} \Omega^{(2)}(s); 2\int_{b_1^{(2)}} \Omega^{(2)}(s), \ldots, 2\int_{b_g^{(2)}(s)} \Omega^{(2)}(s);\right.$$

$$\left.\int_{b_1^{(2)}(s)} \Omega^{(2)}(s), \ldots, \int_{b_g^{(2)}(s)} \Omega^{(2)}(s)\right).$$

The plane spanned by these two vectors contains two linearly independent rational vectors if and only if $I_1(s)$ and $I_2(s)$ are multiples of rational vectors,i.e., if they represent rational points in the projective space $\mathbb{P}^{g-1}$ resp. $\mathbb{P}^g$.

Thus, the proof of the theorem stated in the introduction and in particular the existence of families of constant mean curvature tori in $\mathbb{R}^3$ follows from

Proposition 2 *For each $g \geq 1$ there exists $s_0 \in M_{g,\mathbb{R}}$ such that $\lambda_i(s_0) \neq \lambda_j(s_0)$ for $i \neq j$ and*
(i)

$$Z(s_0) = 0$$

(ii)

The derivative of the map $F : M_g \longrightarrow \mathbb{C} \times \mathbb{P}^{g-1} \times \mathbb{P}^g$,

$$s \mapsto (Z(s), [I_1(s)], [I_2(s)])$$

is invertible at s_0.

It is clear that (i) and (ii) yield the existence of infinitely many points s with the desired property arbitrarily close to s_0. Since all the maps involved are real analytic, it even shows that the set of such points is dense on the component of $\{s \in M_{g,\mathbb{R}} | Z(s) = 0\}$ containing s_0.

We prove the statement of Proposition 2 by induction on g. In the induction step we will need a technical condition which we formulate now.

If for $\nu = 1, 2$ the numbers $\alpha_j^{(\nu)}(s), j = 1, \ldots, g$ are pairwise different then $\frac{\Omega^{(\nu)}(s)}{z - \alpha_j^{(\nu)}(s)}, j = 1, \ldots, g$ is a basis of holomorphic differentials on $C_s^{(\nu)}$. Therefore the matrix

$$\left(\int_{a_i^{(\nu)}(s)} \frac{\Omega^{(\nu)}(s)}{z - \alpha_j^{(\nu)}(s)} \right)_{i,j=1,\ldots,g}$$

is invertible. In particular the system of equations

$$\frac{3}{2} \int_{a_i^{(\nu)}(s)} z\Omega^{(\nu)}(s) + \sum\nolimits_{j=1}^{g} \gamma_j^{(\nu)}(s) \frac{\Omega^{(\nu)}(s)}{z - \alpha_j^{(\nu)}(s)} = 0 \quad i = 1, \ldots, g$$

has a unique solution $\gamma_1^{(\nu)}(s), \ldots, \gamma_g^{(\nu)}(s)$. We put

$$\tilde{\Omega}^{(\nu)}(s) := \frac{3}{2} z\Omega^{(\nu)}(s) + \sum_{j=1}^{g} \frac{\gamma_j^{(\nu)}(s)}{z - \alpha_j^{(\nu)}(s)} \Omega^{(\nu)}(s) .$$

By construction, $\tilde{\Omega}^{(\nu)}(s)$ is a differential with a pole of order 4 at ∞ and vanishing a-periods. Put

$$J_1(s) := \left(\int_{b_1^{(1)}(s)} \tilde{\Omega}^{(1)}(s), \ldots, \int_{b_g^{(1)}(s)} \tilde{\Omega}^{91)}(s) \right)$$

$$J_2(s) := \left(\int_{c(s)} \tilde{\Omega}^{(2)}(s), \int_{b_1^{(2)}(s)} \tilde{\Omega}^{(2)}(s), \ldots, \int_{b_g^{(2)}(s)} \tilde{\Omega}^{(2)}(s) \right)$$

Clearly the derivative of F is invertible at s if and only if the vectors

$$(0, I_1(s), 0), (0, 0, I_2(s)), \left(\frac{\partial Z}{\partial \lambda_j}(s), \frac{\partial I_1}{\partial \lambda_j}(s), \frac{\partial I_2}{\partial \lambda_j}(s) \right), j = 1, \ldots, 2g$$

form a basis of $\mathbb{C}\times\mathbb{C}^g\times\mathbb{C}^{g+1}$. In this case there are unique numbers $\eta^{(\nu)}(s)$ such that

$$\left(\gamma_1^{(1)}(s),-J_1(s),-J_2(s)\right)-\eta^{(1)}(s)\,(0,I_1(s),0)-\eta^{(2)}(s)\,(0,0,I_2(s))$$

lies in span $\left\{\left(\frac{\partial Z}{\partial\lambda_j}(s),\frac{\partial I_1}{\partial\lambda_j}(s),\frac{\partial I_2}{\partial\lambda_j}(s)\right),j=1,\dots,2g\right\}$. Finally put

$$D_\nu(s):=-(-1)^\nu+\frac{1}{2}\sum_{j=1}^{2g}\lambda_j(s)-\sum\nolimits_{j=1}^{g}\alpha_j^{(\nu)}(s).$$

The statement we actually prove by induction on g and which is stronger than Proposition 2 is

Proposition 3 *For each $m\geq 1$ and $g\geq 0$ with $g\leq m$ there exists $s_0\in M_{g,\mathbb{R}}$ such that $\lambda_i(s_0)\neq\lambda_j(s_0)$ for $i\neq j$ and*

(i) $Z(s_0)=0$

(ii) The vectors $(0,I_1(s_0),0)\,,(0,0,I_2(s_0))$ and $\left(\frac{\partial Z}{\partial\lambda_j}(s_0),\frac{\partial I_1}{\partial\lambda_j}(s_0),\frac{\partial I_2}{\partial\lambda_j}(s_0)\right)$ form a basis of $\mathbb{C}\times\mathbb{C}^g\times\mathbb{C}^{g+1}$

(iii) For $\nu=1,2$ the numbers $\alpha_i^{(\nu)}(s_0)$ are pairwise different

(iv) If $T(s_0)$ denotes the map

$$x\mapsto-(D_1(s_0)-D_2(s_0))\,\frac{3x+4\,(D_1(s_0)-D_2(s_0))}{4x+5\,(D_1(s_0)-D_2(s_0))}$$

and $T^n(s_0)$ denotes the n-th iterate of this map then

$$5\,(D_1(s_0)-D_2(s_0))+4T^n(s_0)\left(\eta^{(1)}(s_0)-\eta^{(2)}(s_0)\right)\neq 0\quad\text{for } 0\leq n\leq m-g.$$

Observe that (i) and (ii) above reformulate Proposition 2. Part (iii) is used in the course of the induction to ensure the existence of $\eta^{(1)}(s_0)$ and $\eta^{(2)}(s_0)$, which in turn, together with (iv), will be used to prove that dF is invertible at some point of $M_{g+1,\mathbb{R}}$.

Also, observe that the fractional linear transformation $T(s_0)$ has $x=-(D_1(s_0)-D_2(s_0))$ as a fixed point, and this is the only fixed point of $T(s_0)$. In the group of fractional linear transformations it is therefore conjugate to a translation $x\mapsto x+$ const. In particular, there is at most one $n>0$ such that

$$5\,(D_1(s_0)-D_2(s_0))+4T^n\left(\eta^{(1)}(s_0)-\eta^{(2)}(s_0)\right)=0.$$

For the proof of Proposition 3 we fix a number m and perform induction on g. In Section 3 we carry out the induction step, and in Section 4 we treat the case $g=1$.

Induction Step

For any point $s \in M_{g,\mathbb{R}}$ with $\lambda_i(s) \neq \lambda_j(s)$ for $i \neq j$, any $\mu \in (-2,2)$ and any $t \in \mathbb{R}$ denote by $s'(s,\mu,t)$ the point of $M_{g+1,\mathbb{R}}$ characterized by

$$\begin{aligned} \lambda_i(s') &= \lambda_i(s) \qquad \text{for } i = 1,\ldots,2g \\ \lambda_{2g+1}(s') &= \mu + it, \\ \lambda_{2g+2}(s') &= \mu - it \end{aligned}$$

for $i = 1,\ldots,g$ the loop $h_i(s')$ is homotopic to $h_i(s)$ in $\mathbb{C}\setminus\{\lambda_1(s),\ldots,\lambda_{2g}(s), +2,-2\}$, and it can be represented by a conjugation-invariant path that meets the real axis only along the interval $(-2,\mu)$ and $h_{g+1}(s')$ is close to the segment joining $\mu + it$ and $\mu - it$.

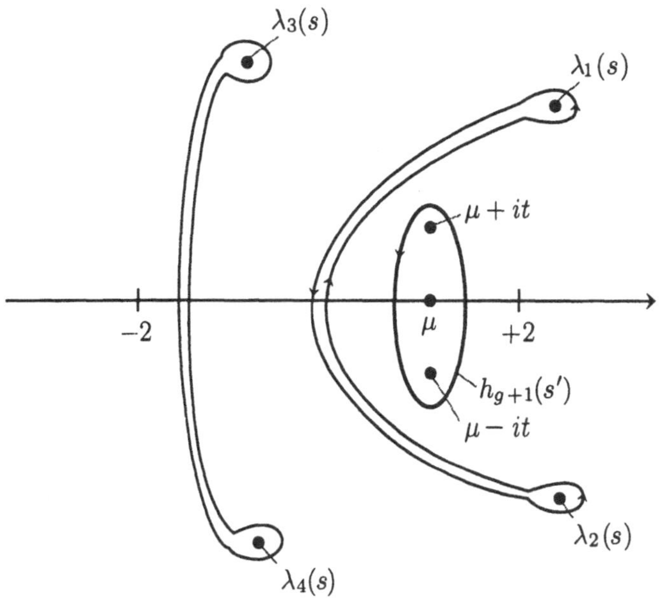

Now fix a point $s_0 \in M_{g,\mathbb{R}}$ that fulfills all the conditions of Proposition 2. We are going to show that for generic $\mu \in (-2,2)$ there is a point in $M_{g+1,\mathbb{R}}$ arbitrarily close to $s'(s_0,\mu,0)$ that also satisfies the conditions of Proposition 2.

To make the notation more transparent, we write $R_0(z)$ for $R_{s_0}(z)$, s_0' for $s'(s_0,\mu,0)$, and denote the maps I_ν, J_ν from M_g to $\mathbb{C}^{g-1+\nu}$ resp. from M_{g+1} to $\mathbb{C}^{g+\nu}$ by $I_\nu^{(g)}, J_\nu^{(g)}$ resp. $I_\nu^{(g+1)}, J_\nu^{(g+1)}$. Similarly we attach a superscript g to the map $Z : M_g \longrightarrow \mathbb{C}$.

Lemma 4

(i) For all $\mu \in (-2,2)$ one has (with suitable ordering) $\alpha_j^{(\nu)}(s_0') = \alpha_j^{(\nu)}(s_0)$ for $j = 1,\ldots,2g$, $\alpha_{g+1}^{(\nu)}(s_9') = \mu$ and therefore $Z^{(g+1)}(s_0') = Z^{(g)}(s_0)$,

$D_\nu(s_0') = D_\nu(s_0)$. *Furthermore,*

$$\frac{\partial Z^{(g+1)}}{\partial \lambda_j}(s')\Big|_{s'=s_0'} \neq 0 \quad \textit{for some} j = 1, \ldots, 2g$$

$$\frac{\partial Z^{(g+1)}}{\partial t}(s')\Big|_{s'=s_0'}, \quad \frac{\partial I_\nu^{(g+1)}}{\partial t}(s')\Big|_{s'=s_0'} = 0 \qquad \text{for } \nu = 1, 2 \,.$$

(ii) For generic $\mu \in (-2,2)$ (here, "generic" means "for all $\mu \in (-2,2)$ outside the zero set of an analytic function") the vectors

$$\left(0; I^{(g+1)}(s_0'), 0\right), \left(0; 0; I_2^{(g+1)}(s_0')\right),$$

$$\frac{\partial}{\partial \lambda_j}\left(Z^{(g+1)}(s'), I_1^{(g+1)}(s'), I_2^{(g+1)}(s')\right)\Big|_{s'=s_0'}, j = 1, \ldots, 2g$$

$$\frac{\partial}{\partial \mu}\left(Z^{(g+1)}(s'), I_1^{(g+1)}(s'), I_2^{(g+1)}(s')\right)\Big|_{s'=s_0'}$$

$$\frac{\partial^2}{\partial t^2}\left(Z^{(g+1)}(s'), I_1^{(g+1)}(s'), I_2^{(g+1)}(s')\right)\Big|_{s'=s_0'}$$

form a basis of $\mathbb{C} \times \mathbb{C}^{g+1} \times \mathbb{C}^{g+2}$.

(iii) If $\left(\gamma_1^{(1)}(s_0'); -J_1^{(g+1)}(s_0'); -J_2^{(g+1)}(s_0')\right) =$

$$\eta_1^{(1)}(s_0')\left(0, I_1^{(g+1)}(s_0'); 0\right) + \eta^{(2)}(s_0')\left(0; 0; I_2^{(g+1)}(s_0')\right)$$

modulo the space spanned by $\frac{\partial}{\partial \lambda_j}(Z^{(g+1)}, I_1^{(g+1)}, I_2^{(g+1)})\Big|_{s'=s_0'}$ $j = 1, \ldots, 2g,$

$\frac{\partial}{\partial \mu}(Z^{(g+1)}, I_1^{(g+1)}, I_2^{(g+1)})\Big|_{s'=s_0'},$

$\frac{\partial^2}{\partial t^2}(Z^{(g+1)}, I_1^{(g+1)}, I_2^{(g+1)})\Big|_{s'=s_0'}$, *then for generic μ*

$$5\,(D_1(s_0') - D_2(s_0')) + 4T^n(s_0')\,(\eta_1(s_0') - \eta_2(s_0')) \neq 0$$

for $0 \leq n \leq m - g - 1$.

Before we prove Lemma 4, we show how it can be used to perform the induction step.

Choose $\mu \in (-2,2)$ such that (ii) and (iii) hold. By part (i) of Lemma 4 the point s_0'n lies in $N := \{s' \in M_{g+1,\mathbb{R}} \mid Z^{(g+1)}(s') = 0\}$, N is smooth at s_0' and the tangent hyperplane of N at s_0' contains the t-direction. Therefore, there exists a smooth curve

$$\mathbb{R} \ni \tau \mapsto s'(\tau) = (s(\tau), \mu(\tau), \tau) \in M_{g+1,\mathbb{R}}$$

lying in N and beginning at s_0' with tangent at s_0' in the t-direction, that is,

$$s'(0) = s_0', \quad s'(\tau) \in N \text{ for all } \tau,$$
$$\frac{d}{d\tau}s(\tau)\Big|_{\tau=0} = 0, \quad \frac{d}{d\tau}\mu(\tau)\Big|_{\tau=0} = 0.$$

Clearly, for small τ, $s'(\tau)$ fulfills conditions (iii) of Proposition 3.

Since

$$\frac{\partial}{\partial t}\left(Z^{(g+1)}(s'(\tau)), I_1^{(g+1)}(s'(\tau)), I_2^{(g}(s'(\tau))\right)$$
$$= \tau \cdot \frac{\partial^2}{\partial t^2}\left(Z^{(g+1)}(s'), I_1^{(g+1)}(s'), I_2^{(g+1)}(s')\right)\Big|_{s'=s_0'} + 0(\tau^2)$$

for sufficiently small $\tau \neq 0$ the vectors

$$\left(0; I_1^{(g+1)}(s'(\tau)); 0\right), \left(0; 0; I_2^{(g+1)}(s'(\tau))\right)$$
$$\frac{\partial}{\partial \lambda_i}\left(Z^{(g+1)}(s'(\tau)), I_1^{(g+1)}(s'(\tau)), I_2^{(g+1)}(s'(\tau))\right) \quad i = 1, \dots, 2g$$
$$\frac{\partial}{\partial \mu}\left(Z^{(g+1)}(s'(\tau)), I_1^{(g+1)}(s'(\tau)), I_2^{(g+1)}(s'(\tau))\right)$$
$$\frac{\partial}{\partial t}\left(Z^{(g+1)}(s'(\tau)), I_1^{(g+1)}(s'(\tau)), I_2^{(g+1)}(s'(\tau))\right)$$

are a basis of $\mathbb{C} \times \mathbb{C}^{g+1} \times \mathbb{C}^{g+2}$, so that part (ii) of Proposition 2 holds. Finally, $T(s'(\tau))$ is close to $T(s'0)$, and $D_\nu(s'(\tau))$ (resp. $\eta^{(\nu)}(s'(\tau))$) is close to $D_\nu(s_0')$ (resp. $\eta^{(\nu)}(s_0')$), so part (iv) of the Proposition is also fulfilled for all sufficiently small τ. Therefore the induction step follows from Lemma 4.

Proof of Lemma 4. Since the differentials $\Omega^{(\nu)}(s')$ are characterized by

$$\int_{a_i^{(\nu)}(s')} \Omega^{(\nu)}(s') = 0 \quad \text{for } i = 1, \dots, g+1 \tag{19}$$

and their behaviour at infinity, we can, for s' near s_0', write

$$\Omega^{(\nu)}(s')$$
$$= \prod_{j=1}^{g+1}\left(z - \alpha_j^{(\nu)}(s')\right)\frac{dz}{\sqrt{(z+(-1)^\nu 2)(z-(\mu+it))(z-(\mu-it))R_s(z)}}$$

with analytic functions $\alpha_j^{(\nu)}(s')$ satisfying

$$\begin{cases} \alpha_j^{(\nu)}(s'(s,\mu,0)) = \alpha_j^{(\nu)}(s) & \text{for } j = 1, \dots, g \\ \alpha_{g+1}^{(\nu)}(s'(s,\mu,0)) = \mu. \end{cases} \tag{20}$$

This yields the first half of part (i) of Lemma 4.

To complete the proof of part (i) we have to consider the t-derivatives. For convenience of notation we define for any function $f(s'(s,\mu,t))$.

$$\dot{f} = \frac{\partial}{\partial t} f(s')\Big|_{s'=s'_0} \quad \ddot{f} = \frac{\partial}{\partial t^2} f(s')\Big|_{s'=s'_0} .$$

Observe that

$$\dot{\Omega}^{(\nu)} = \left(-\sum_{j=1}^{g} \frac{\dot{\alpha}_j^{(\nu)}}{z - \alpha_j^{(\nu)}(s_0)} - \frac{\dot{\alpha}_{g+1}}{z-\mu} \right) \cdot \prod_{j=1}^{g} \left(-\alpha_j^{(\nu)}(s_0) \right) \frac{dz}{\sqrt{(z+(-1)^\nu 2)\, R_0(z)}} \tag{21}$$

Taking the t-derivative of (19) for $i = g+1$, we get

$$res_{z=\mu} \dot{\Omega}^{(\nu)} = 0 .$$

This implies that $\dot{\alpha}_{g+1} = 0$, unless $\mu = \alpha_j^{(\nu)}(s_0)$ for some $j = 1, \dots, g$. In either case (21) shows that $\dot{\Omega}^{(\nu)}$ lifts to a holomorphic differential on the normalization $C_{s_0}^{(\nu)}$ of $C_{s'_0}^{(\nu)}$. From (21) we see that

$$\int_{\alpha_j^{(\nu)}(s'_0)} \dot{\Omega}^{(\nu)} = 0 \quad \text{for } i = 1, \dots, g ,$$

so this holomorphic differential on $C_{s_0}^{(\nu)}$ has vanishing a-periods. This shows that

$$\dot{\Omega}^{(\nu)} = 0 , \tag{22}$$

Since $b_1^{(\nu)}(s'), \dots, b_g^{(\nu)}(s'), c(s')$ can, for s' close to s'_0 be represented by loops in $\mathbb{P}^1$ that are independent of s', this implies tht

$$\frac{\partial}{\partial t} \left(\int_{b_i^{(\nu)}(s')} \Omega^{\nu}(s') \right) \Big|_{s'=s'_0} = 0 \qquad \text{for } i = 1, \dots, g$$

$$\frac{\partial}{\partial t} \left(\int_{C(s')} \Omega^{(2)}(s') \right) \Big|_{s'=s'_0} 0$$

Thus, all but the last components of $\dot{I}_{\nu}^{(g+1)}$ are zero.

To deal with the last two components of these maps we use the holomorphic differentials $\omega^{(\nu)}(s')$ on $C_{s'}^{(\nu)}$ characterized by

$$\begin{aligned} &\int_{\alpha_i^{(\nu)}(s')} \omega^{(\nu)}(s') = 0 \qquad \text{for } i = 1, \dots, g \\ &\int_{\alpha_{g+1}^{(\nu)}(s')} \omega^{(\nu)}(s') = 2\pi i \end{aligned} \tag{23}$$

Write

$$\omega^{(\nu)}(s') = L^{(\nu)}(s') \cdot \prod_{j=1}^{g} \left(z - \beta_j^{(\nu)}(s')\right) \cdot \frac{dz}{\left((z + (-1)^\nu 2)\, R_{s'}(z)\right)^{1/2}} \,.$$

By the reciprocity formula

$$\int_{b_{g+1}^{(\nu)}(s')} \Omega^{(\nu)}(s') = 4L^{(\nu)}(s') \,. \tag{24}$$

As before one sees that

$$\dot{\omega}^{(\nu)} = 0 \tag{25}$$

and so one concludes from (24) that also the last components of the t-derivatives of $I_1^{(g+1)}$ and $I_2^{(g+1)}$ vanish at s_0'. This finishes proof of part (i) of Lemma 4.

To prove parts (ii) and (iii) of Lemma 4 we have to show that the determinant of the matrix formed by the vectors in (ii) (resp. the quantities $5\,(D_1(s_0') - D_2(s_0')) + 4T^n(s_0')\left(\eta^{(1)}(s_0') - \eta^{(2)}(s_0')\right)$, $0 \le n \le m - g - 1$), are not identically zero as functions of μ. For this purpose, we consider the analytic continuations of these functions as μ moves in the complex plane in a little half-circle with non-negative imaginary part around the point $+2$ and then to $+\infty$ along the real axis.

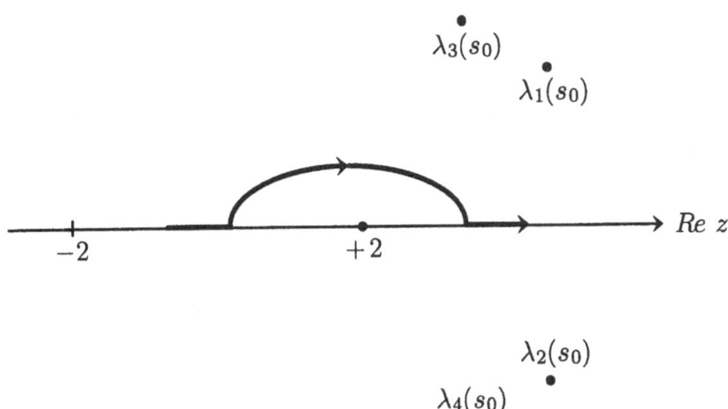

Observe that for s near s_0, μ in the region described above and all sufficiently small t, the construction of $s'(s,\mu,t)$ still makes sense and defines an analytic map to M_{g+1}. We now determine the asymptotics of the relevant quantities as $\mu \longrightarrow \infty$ along the path described above. To complete the proof of Lemma 4 we require.

Lemma 5

(a) $I_\nu^{(g+1)}(s_0') = \left(I_\nu^{(g)}(s_0), 4\mu^{1/2}\left(1 - \frac{D_\nu(s_0)}{\mu}\right) + O(\mu^{-3/2})\right)$

(b) For $j = 1, \ldots, 2g$

$$\frac{\partial Z^{(g+1)}}{\partial \lambda_j}(s_0') = \frac{\partial Z^{(g)}}{\partial \lambda_j}(s_0)$$

and $\frac{\partial I_\nu^{(g+1)}}{\partial \lambda_j}(s_0') = \left(\frac{\partial I_\nu^{(g)}}{\partial \lambda_j}(s_0); O(\mu^{-1/2})\right)$.

(c) $\frac{\partial Z^{(g+1)}}{\partial \mu}(s_0') = 0,$

$$\frac{\partial I_\nu^{(g+1)}}{\partial \mu}(s_0') = \left(0, \ldots, 0; 2\mu^{-1/2}\left(1 + \frac{D_\nu(s_0)}{\mu}\right) + O(\mu^{-5/2})\right)$$

(d) $\ddot{Z} = \frac{\gamma_1^{(1)}(s_0)}{\mu^3} + O\left(\frac{1}{\mu^4}\right)$

$$\ddot{I}_\nu^{(g+1)} = \left(\frac{1}{\mu^2}\left(-\frac{1}{2} + \frac{D_\nu(s_0)}{\mu}\right) I_\nu^{(g)}(s_0) - \frac{1}{\mu^3} J_\nu^{(g)}(s_0) + O\left(\frac{1}{\mu^4}\right)\right);$$

$$\frac{3}{2\mu^{3/2}}\left(1 + \frac{3D_\nu(s_0)}{\mu}\right) O(\mu^{-7/2})$$

(e) $\gamma_1^{(1)}(s_0') = \gamma_1^{(1)}(s_0)$

$$J_\nu^{(g+1)}(s_0') = \left(J_\nu^{(g)}(s_0); 2\mu^{3/2}\left(1 + 3\frac{D_\nu(s_0)}{\mu}\right) + O(\mu^{-1/2})\right)$$

We first prove Lemma 5 and then continue the proof of Lemma 4.

Proof of Lemma 5. To simplify notation we suppress the sub- or superscript ν whenever convenient. Also, we write $a_i^{(\nu)}$ (resp. $b_i^{(\nu)}$) for $a_i^{(\nu)}$ (resp. $b_i^{(\nu)}(s_0)$)

(a) From (20) and the construction of $s'(s,\mu,t)$ it is obvious that

$$\int_{b_i^{(\nu)}(s'(s,\mu,0))} \Omega^{(\nu)}\left(s'(s,\mu,0)\right) = \int_{b_i^{(\nu)}(s)} \Omega^{(\nu)}(s) \qquad \text{for } i = 1, \ldots, g$$

$$\int_{c(s'(s,\mu,0))} \Omega^{(2)}\left(s'(s,\mu,0)\right) = \int_{c(s)} \Omega^{(2)}(s) \tag{26}$$

This proves (a) for all but the last components of the vectors involved. Observe that for $t \neq 0$ the last components of $I^{(g+1)}$ are $\int_{b_{g+1}(s'(s,\mu,t))} \Omega\left(s'(s,\mu,t)\right)$ where the integration contour $b_{g+1}\left(s'(s,\mu,t)\right)$ is as in the figure that follows.

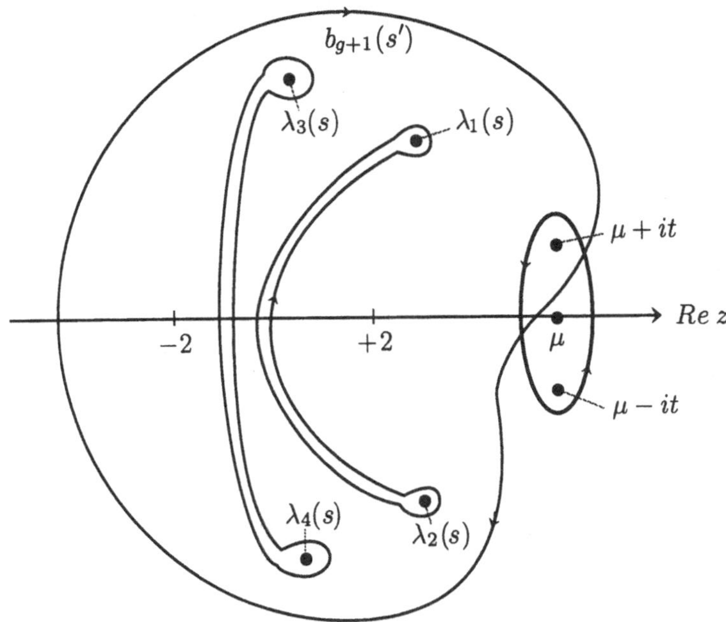

This function extends continuously to $t = 0$, so the last component of $I^{(g+1)}(s'(s, \mu, 0))$ is

$$-\int_{|z|=\mu} \Omega(s)\,. \tag{27}$$

Since

$$\frac{1}{\sqrt{(z + (-1)^\nu 2)\, R_s(z)}} \prod_{j=1}^{g} \left(z - \alpha_j^{(\nu)}(s)\right) = \frac{1}{z^{1/2}} + \frac{D_\nu(s)}{z^{3/2}} + O\left(\frac{1}{z^{5/2}}\right) \tag{28}$$

this immediately gives the desired estimate.

(b) By definition $Z^{(g+1)}(s') = \alpha_1^{(1)}(s') - 2$, so the first statement is immediate from (20). The statement about all but the last component of $\frac{\partial I^{(g+1)}}{\partial \lambda_j}(s_0')$ follows directly from (26). As $\frac{\partial \Omega}{\partial \lambda_j}(s, z) = O\left(\frac{1}{z^{3/2}}\right) dz$, the estimate for the last component is obtained from (27).

(c) All statements but the one about the last component of $\frac{\partial I}{\partial \mu}(s_0')$ follow again from (20) resp. (26). The result about the last component is a consequence of (a).

(d) A direct computation using the fact (see (22)) that $\dot{\Omega} = 0$ gives

$$\ddot{\Omega} = \left(-\sum_{j=1}^{g} \frac{\ddot{\alpha}_j}{z - \alpha_j} - \frac{\ddot{\alpha}_{g+1}}{z - \mu} - \frac{1}{(z - \mu)^2}\right) \Omega(s_0')\,. \tag{29}$$

By (19), $res_{z=\mu}\ddot{\Omega} = 0$, so

$$\ddot{\alpha}_{g+1}\left(\frac{\Omega(s_0)}{dz}\right)\Big|_{z=\mu} = -\left(\frac{d}{dz}\left(\frac{\Omega(s_0)}{dz}\right)\right)\Big|_{z=\mu}.$$

Using (28) this gives

$$\ddot{\alpha}_{g+1} = \frac{1}{2\mu}\left(1+\frac{2D}{\mu}\right)+O\left(\frac{1}{\mu^3}\right). \tag{30}$$

The homology classes $a_i(s_0'), i = 1,\ldots,g$, can be represented by loops whose π_ν-images lie in a compact, μ-independent region K of the z-plane. In K we have the estimates

$$\frac{1}{z-\mu} = -\frac{1}{\mu} - \frac{z}{\mu^2} + O\left(\frac{1}{\mu^3}\right)$$
$$\frac{1}{(z-\mu)^2} = \frac{1}{\mu^2} + 2\frac{z}{\mu^3} + O\left(\frac{1}{\mu^4}\right).$$

If we insert this and (30) into (29) we get that in K

$$\ddot{\Omega} = -\left(\frac{3}{2}\frac{z}{\mu^3} + \sum_{j=1}^{g}\frac{\ddot{\alpha}_j}{z-\alpha_j}\right)\Omega(s_0) - \frac{1}{\mu^2}\left(\frac{1}{2}-\frac{D}{\mu}\right)\Omega(s_0) \tag{31}$$
$$+O\left(\frac{1}{\mu^4}\right)\Omega(s_0). \tag{32}$$

From (19) we know that $\int_{a_i(s_0')}\ddot{\Omega} = 0$ for $i = 1,\ldots,g$. If we insert (31) into this and use the fact that $\int_{a_i}\Omega(s_0) = 0$, we get

$$\frac{3}{2}\int_{a_i} z\Omega(s_0) + \mu^3\sum_{j=1}^{g}\ddot{\alpha}_j\int_{a_i}\frac{\Omega(s_0)}{z-\alpha_j} = O\left(\frac{1}{\mu}\right) \quad \text{for } i = 1,\ldots,g.$$

If we compare the last expression with the definition of γ_j and $\tilde{\Omega}$ in Section 2 we get

$$\ddot{\alpha}_j = \frac{1}{\mu^3}\gamma_j + O\left(\frac{1}{\mu^4}\right) \tag{33}$$
$$\ddot{\Omega} = -\frac{1}{\mu^3}\tilde{\Omega}(s_0) - \frac{1}{\mu^2}\left(\frac{1}{2}-\frac{D}{\mu}\right)\Omega(s_0) + O\left(\frac{1}{\mu^4}\right). \tag{34}$$

Since, by definition, $\ddot{Z} = \ddot{\alpha}_1^{(1)}$ we immediately get the statemet about $\ddot{Z}$. Furthermore, observe that $c(s') + a_{g+1}^{(2)}(s')$, $b_i^{(\nu)}(s') + a_{g+1}^{(\nu)}(s')$, $i = 1,\ldots,g$, are homologous to paths whose projections lie in K (see the following figure) so we also get the statement about all but the last component of $\ddot{I}^{(g+1)}$.

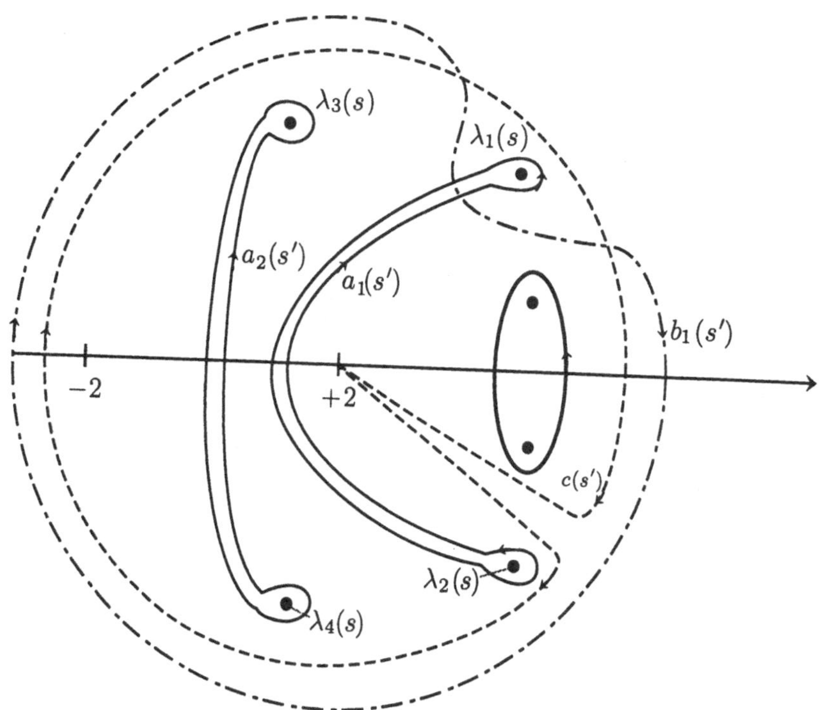

To deal with the last component of $\ddot{I}^{(g+1)}$ we again use the reciprocity formula (24). Observe that for $z \in K$ and t bounded

$$\omega^{(\nu)}\left(s'(s,\mu,t)\right) = L^{(\nu)}\left(s'(s,\mu,t)\right)\left(\frac{1}{\sqrt{\mu^2+t^2}} + O\left(\frac{1}{\mu^2}\right)\right)$$
$$\left(\frac{dz}{\sqrt{(z+(-1)^\nu 2)\,R_s(z)}}\prod_{j=1}^{g}\left(z-\beta_j\left(s'(s,\mu,t)\right)\right)\right) \tag{35}$$

where the O-estimate is uniform in t and its derivatives. Since by definition $\int_{a_i^{(\nu)}(s')}\omega^{(\nu)}(s') = 0$ for $i = 1, \ldots, g$, we get

$$\int_{a_i^{(\nu)}(s')}\frac{dz}{\sqrt{(z+(-1)^\nu 2)\,R_s(z)}}\prod_{j=1}^{g}\left(z-\beta_j^{(\nu)}(s')\right) = O\left(\frac{1}{\mu}\right).$$

The integrand above defines a differential form on $C_s^{(\nu)}$ which is holomorphic outside ∞, looks like $\frac{dz}{\sqrt{z}} + O\left(\frac{1}{z^{3/2}}\right)dz$ at infinity and whose

a-periods are $O\left(\frac{1}{\mu}\right)$. Therefore, subtracting a suitable holomorphic differential on $C_s^{(\nu)}$

$$\frac{dz}{\sqrt{(z+(-1)^\nu 2)\, R_s(z)}} \prod_{j=1}^{g} (z-\beta_j(s')) = \Omega^{(\nu)}(s) + O\left(\frac{1}{\mu}\right),$$

on K. In particular

$$\beta_j(s_0') = \alpha_j(s_0) + O\left(\frac{1}{\mu}\right). \tag{36}$$

Furthermore, using $res_{z=\mu}\omega(s_0') = 1$, we get

$$L \cdot \prod_{j=1}^{g} \frac{\mu - \beta_j}{\sqrt{(\mu+(-1)^\nu 2)\, R_0(\mu)}} = 1,$$

hence by (36) and (28)

$$L \cdot \left(\frac{1}{\mu^{1/2}} + \frac{D}{\mu^{3/2}} + O\left(\frac{1}{\mu^{5/2}}\right)\right) = 1$$

(Note that this, together with (24) once again yields the last part of (a)).

Therefore by (35)

$$\omega(s_0') = \frac{1}{\sqrt{\mu}}\Omega(s_0) + O\left(\frac{1}{\mu^{3/2}}\right) \qquad \text{on } K. \tag{37}$$

Next, using $\dot{\omega} = 0$ one has

$$\ddot{\omega} = \left(\frac{\ddot{L}}{L} - \sum_{j=1}^{g} \frac{\ddot{\beta}_j}{z-\beta_j(s_0')}\right)\omega(s_0') - \frac{\omega(s_0')}{(z-\mu)^2}. \tag{38}$$

As $\int_{a_i(s_0')}\omega = \int_{a_i(s_0')}\ddot{\omega} = 0$ for $i = 1, \ldots, g$ we get from (38)

$$\sum_{j=1}^{g} \ddot{\beta}_j \int_{a_i(s_0')} \frac{\omega(s_0')}{z-\beta_j(s_0')} = -\int_{a_i(s_0')} \frac{\omega(s_0')}{(z-\mu)^2}.$$

Using (37) and the expansion of $(z-\mu)^{-2}$ in K one obtains

$$\begin{aligned}\int_{a_i(s_0')} \frac{\omega(s_0')}{(z-\mu)^2} &= \frac{2}{\mu^3}\int_{a_i(s_0')} \left(z + O\left(\frac{1}{\mu}\right)\right)\omega(s_0') \\ &= \frac{2}{\mu^{7/2}}\int_{a_i(s_0')} z\Omega(s_0) + O\left(\frac{1}{\mu^{9/2}}\right).\end{aligned}$$

Clearly

$$\sum_{j=1}^{g} \ddot{\beta}_j \sqrt{\mu} \int_{a_i(s_0')} \frac{\omega(s_0')}{z-\beta_j(s_0')} = 0\left(\frac{1}{\mu^3}\right) \tag{39}$$

since $\int_{a_i(s_0')} z\Omega(s_0)$ is independent of μ.

By (36) and (37) the matrix $\left(\sqrt{\mu}\int_{a_i(s_0')}\frac{\omega(s_0')}{z-\beta_j(s_0')}\right)_{i,j=1,\dots,g}$ differs from the invertible, μ-independent matrix $\left(\int_{a_i}\frac{\Omega}{z-\alpha_j}\right)_{i,j}$ by $O\left(\frac{1}{\mu}\right)$. It now follows from (39) that

$$\ddot{\beta}_j = O(\mu^{-3})\,. \tag{40}$$

Applying the fact that $res_{z=\mu}\ddot{\omega} = 0$, and $res_{z=\mu}\omega(s_0') = 1$ to (38) we get

$$\begin{aligned}
\frac{\ddot{L}}{L(s_0')} - \sum_{j=1}^{g} \frac{\ddot{\beta}_j}{\mu-\beta_j(s_0')} &= res_{z=\mu}\frac{\omega(s_0')}{(z-\mu)^2} \\
&= \frac{1}{2}\frac{d^2}{dz^2}\left(\frac{L(s_0')}{\sqrt{z+(-1)^{\nu}R_0(z)}}\prod_{j=1}^{g}(z-\beta_j(s_0'))\right)\Bigg|_{z=\mu} \\
&= \frac{L(s_0')}{2}\left(\frac{3}{4}\frac{1}{\mu^{5/2}} + \frac{15}{4}\frac{D(s_0)}{\mu^{7/2}} + O\left(\frac{1}{\mu^{9/2}}\right)\right)
\end{aligned}$$

by (36) and (28). We know that (see the expression preceding (37))

$$L(s_0') = \mu^{1/2}\left(1 - D(s_0)\mu^{-1}\right) + O(\mu^{-3/2})\,.$$

Putting this together with (36) we get

$$\begin{aligned}
\ddot{L} &= \frac{3}{8}\mu^{-3/2}\left(1-D(s_0)\mu^{-1}\right)^2\left(1+5D(s_0)\mu^{-1}\right) + O(\mu^{-7/2}) \\
&= \frac{3}{8\mu^{3/2}}\left(1+\frac{3D(s_0)}{\mu}\right) + O(\mu^{-7/2}) \quad .
\end{aligned}$$

With reciprocity (24) this gives the desired result

(e) is proven the same way as (a).

We now continue the proof of Lemma 4

To show (ii) it is enough to prove that the determinant of the matrix whose rows are the vectors

$$
\begin{aligned}
u_1(\mu) &:= \left(0, I_1^{(g+1)}(s_0'), 0\right)\\
u_2(\mu) &:= \left(0, 0, I_2^{(g+1)}(s_0')\right)\\
v_j(\mu) &:= \left(\frac{\partial Z^{(g+1)}}{\partial \lambda_j}(s_0'), \frac{\partial I_1^{(g+1)}}{\partial \lambda_j}(s_0'), \frac{\partial I_2^{(g+1)}}{\partial \lambda_j}(s_0')\right), j = 1, \dots, 2g\\
w_1(\mu) &:= \left(0, \frac{\partial I_1^{(g+1)}}{\partial \mu}(s_0'), \frac{\partial I_2^{(g+1)}}{\partial \mu}(s_0')\right)\\
w_2(\mu) &:= \left(\ddot{Z}^{(g+1)}, \ddot{I}_1^{(g+1)}, \ddot{I}_2^{(g+1)}\right)
\end{aligned}
$$

is not identically zero as function of $\mu \in (-2, 2)$. This function is analytic, so it suffices to prove this for its analytic continuation as $\mu \longrightarrow +\infty$ in the manner described above.

Observe that by Lemma 5

$$
\begin{aligned}
&w_2(\mu) =\\
&\frac{1}{\mu^3}\left(\gamma_1^{(1)}(s_0) + O\left(\frac{1}{\mu}\right); -J_1^{(g)}(s_0) + O\left(\frac{1}{\mu}\right), 0; -J_2^{(g)}(s_0) + O\left(\frac{1}{\mu}\right), 0\right)\\
&+\frac{1}{\mu^2}\left(0; \left(-\frac{1}{2} + \frac{D_1(s_0)}{\mu}\right) I_1^{(g)}(s_0), 0; \left(-\frac{1}{2} + \frac{D_2(s_0)}{\mu}\right) I_2^{(g)}(s_0), 0\right)\\
&+\frac{3}{2\mu^{3/2}}\left(0; 0, 1 + \frac{3D_1(s_0)}{\mu} + O(\mu^{-2}); 0, 1 + \frac{3D_2(s_0)}{\mu} + O(\mu^{-2})\right) .
\end{aligned}
$$

By the induction hypothesis the vectors $\left(0, I_1^{(g)}(s_0), 0\right)$, $\left(0, 0, I_2^{(g)}(s_0)\right)$, $\left(\frac{\partial Z^{(g)}}{\partial \lambda_j}(s_0), \frac{\partial I_1^{(g)}}{\partial \lambda_j}(s_0), \frac{\partial I_2^{(g)}}{\partial \lambda_j}(s_0)\right)$ form a basis of $\mathbb{C} \times \mathbb{C}^g \times \mathbb{C}^{g+1}$ and by definition of $\eta^{(\nu)}(s_0)$ there exists $\zeta_j \in \mathbb{C}$ such that

$$
\begin{aligned}
\left(\gamma_1^{(1)}(s_0), -J_1^{(g)}(s_0), -J_2^{(g)}(s_0)\right) &= \eta^{(1)}(s_0)\left(0, I_1^{(g)}(s_0), 0\right)\\
&+ \eta^{(2)}(s_0)\left(0, 0, I_2(g)(s_0)\right) \qquad (41)\\
&+ \sum_{j=1}^{2g} \zeta_j \left(\frac{\partial Z^{(g)}}{\partial \lambda_j}(s_0), \frac{\partial I_1^{(g)}}{\partial \lambda_j}(s_0), \frac{\partial I_2^{(g)}}{\partial \lambda_j}(s_0)\right) .
\end{aligned}
$$

Therefore, there are $\tilde{\zeta}_1, \dots, \tilde{\zeta}_{2g}, \tilde{\eta}_1, \tilde{\eta}_2$ with

$$
\tilde{\zeta}_j = \zeta_j + O\left(\frac{1}{\mu}\right), \tilde{\eta}_\nu = \eta^{(\nu)}(s_0) + O\left(\frac{1}{\mu}\right)
$$

such that

$$\left(\gamma_1^{(1)}(s_0)+O\left(\frac{1}{\mu}\right);-J_1^{(g)}(s_0)+O\left(\frac{1}{\mu}\right),0;-J_2^{(g)}(s_0)+O\left(\frac{1}{\mu}\right),0\right)$$
$$-\tilde{\eta}_1u_1(\mu)-\tilde{\eta}_2u_2(\mu)-\sum_{j=1}^{2g}\tilde{\zeta}_jv_j(\mu)$$
$$=\left(0;0,-4\eta^{(1)}(s_0)\mu^{1/2}+O(\mu^{-1/2});0,-4\eta^{(2)}(s_0)\mu^1/2+O(\mu^{-1/2})\right).$$

Similarly

$$\left(0;\left(-\frac{1}{2}+\frac{D_1(s_0)}{\mu}\right)I_1^{(g)}(s_0),0;\left(-\frac{1}{2}+\frac{D_2(s_0)}{\mu}\right)I_2^{(g)}(s_0),0\right)$$
$$-\left(-\frac{1}{2}+\frac{D_1(s_0)}{\mu}\right)u_1(\mu)-\left(-\frac{1}{2}+\frac{D_2(s_0)}{\mu}\right)u_2(\mu)$$
$$=\left(0;0,\mu^{1/2}\left(2-\frac{6D_1(s_0)}{\mu}\right)+O\left(\mu^{-3/2}\right);0,\mu^{1/2}\left(2-\frac{6D_2(s_0)}{\mu}\right)+O(\mu^{-3/2})\right).$$

Therefore,

$$\tilde{w}_2(\mu):=w_2(\mu)-\frac{7}{4\mu}w_1(\mu)-\frac{1}{\mu^3}\sum_{j=1}^{2g}\tilde{\zeta}_j\nu_j(\mu)$$
$$-\sum_{\nu=1}^{2}\left(\frac{1}{\mu^3}\tilde{\eta}_\nu+\frac{1}{\mu^2}\left(-\frac{1}{2}+\frac{D_\nu(s_0)}{\mu}\right)\right)u_\nu(\mu) \qquad (42)$$

equals

$$\left(0;0,-\left(5D_1(s_0)+4\eta_1^{(1)}(s_0)\right)\mu^{-5/2}+O(\mu^{-7/2});\right.$$
$$\left.0,-\left(5D_2(s_0)+4\eta^{(2)}(s_0)\right)\mu^{-5/2}+O(\mu^{-7/2}\right).$$

It follows from part (ii) in the induction hypothesis that the determinant above is non-zero whenever $w_1(\mu)$ and $\tilde{w}_2(\mu)$ are linearly independent. For $\mu\longrightarrow\infty$ this is certainly the case whenever

$$5D_1(s_0)+4\eta^{(1)}(s_0)\neq 5D_2(s_0)+4\eta^{(2)}(s_0).$$

This is a consequence of part (iv) of the induction hypothesis.

Finally to prove part (iii) of Lemma 4 we compute the asymptotics of $\eta^{(\nu)}(s_0')$. Since $\gamma_1^{(1)}(s_0')=\gamma^{(1)}(s_0)$ by part (e) of Lemma 5, (41) and part (e) of Lemma 5 imply that

$$\left(\gamma_1^{(1)}(s_0'), -J_1^{(g+1)}(s_0'), -J_2^{(g+1)}(s_0')\right)$$

$$- \sum^2 \zeta_j v_j(\mu) - \eta^{(1)}(s_0)u_1(\mu) - \eta^{(2)}(s_0)u_2(\mu)$$

$$= \Big(0; 0, -2\mu^{3/2} - \left(6D_1(s_0) + 4\eta^{(1)}(s_0)\right)\mu^{1/2} + O(\mu^{-1/2}); 0, -2\mu^{3/2}$$

$$- \left(6D_2(s_0) + 4\eta^{(2)}(s_0)\right)\mu^{1/2} + O(\mu^{-1/2})\Big) \quad = A(\mu)w_1(\mu) + B(\mu)\tilde{w}_2(\mu)$$

with

$$A(\mu) = -\mu^2 + O(\mu)$$

$$B(\mu) = \frac{4\left(D_1(s_0) - D_2(s_0)\right) + 4\left(\eta^{(1)}(s_0) - \eta^{(2)}(s_0)\right)}{5\left(D_1(s_0) - D_2(s_0)\right) + 4\left(\eta^{(1)}(s_0) - \eta^{(2)}(s_0)\right)}\mu^3 + O(\mu^2).$$

By definition of $\eta^{(\nu)}(s_0')$ we get, using (42)

$$\eta^{(\nu)}(s_0') = \eta^{(\nu)}(s_0) + \frac{1}{2}\frac{B(\mu)}{\mu^2} - \frac{1}{\mu^3}B(\mu)\left(\eta^{(\nu)}(s_0) + D_\nu(s_0)\right) + O\left(\frac{1}{\mu}\right)$$

hence

$$\eta^{(1)}(s_0') - \eta^{(2)}(s_0') = T(s_0)\left(\eta^{(1)}(s_0) - \eta^{(2)}(s_0)\right) + O\left(\frac{1}{\mu}\right)$$

where, as we have defined in Proposition 3

$$T(s_0): x \longmapsto x - \frac{4\left(D_1(s_0) - D_2(s_0)\right) + 4x}{5\left(D_1(s_0) - D_2(s_0)\right) + 4x}\left(\left(D_1(s_0) - D_2(s_0)\right) + x\right)$$

$$= -\left(D_1(s_0) - D_2(s_0)\right)\frac{3x + 4\left(D_1(s_0) - D_2(s_0)\right)}{4x + 5\left(D_1(s_0) - D_2(s_0)\right)}$$

Since $T(s_0') = T(s_0)$ by part (i) of Lemma 4 this shows that

$$5\left(D_1(s_0') - D_2(s_0')\right) + 4T^n(s_0')\left(\eta^{(1)}(s_0') - \eta^{(2)}(s_0')\right) \neq 0$$

for $n \leq m - g - 1$.

This concludes the proof of Lemma 4 and thus the induction step in the proof of Proposition 3.

It may be helpful to observe that $T(s_0)$ has to be a fractional linear transformation. After all, $A(\mu)$ and $B(\mu)$ are determined by the matrix equation

$$\begin{pmatrix} 2\mu^{-1/2}\left(1 + \frac{D_1(s_0)}{\mu}\right) & -\mu^{-5/2}\left(5D_1(s_0) + 4\eta^(1)(s_0)\right) \\ 2\mu^{-1/2}\left(1 + \frac{D_2(s_0)}{\mu}\right) & -\mu^{-5/2}\left(5D_2(s_0) + 4\eta^(2)(s_0)\right) \end{pmatrix}\begin{pmatrix} A \\ B \end{pmatrix}$$

$$= \begin{pmatrix} -2\mu^{3/2} - \left(6D_1(s_0) + 4\eta^{(1)}(s_0)\right)\mu^{1/2} \\ -2\mu^{3/2} - \left(6D_2(s_0) + \eta^{(2)}(s_0)\right)\mu^{1/2} \end{pmatrix}.$$

Considering the case $D_1(s_0) = D_2(s_0) = 0$ one sees that the coefficient of $\eta^{(1)}(s_0) - \eta^{(2)}(s_0)$ in the numerator and denominator must be the same, and hence $T(s_0)$ is a fractional linear transformation. Also it is obvious that $x = -(D_1(s_0) - D_2(s_0))$ is a fixed point. Putting $\eta^{(\nu)} = -D_\nu$ in the matrix equation above one sees directly that the highest order term of B is zero; therefore $x = -(D_1(s_0) - D_2(s_0))$ is even a fixed point of multiplicity 2.

Beginning the Induction; the case $g = 1$

In the case $g = 1$ the curves $C_s^{(1)}$ and $C_s^{(2)}$ are of course elliptic. The tori they generate are those first discovered by Wente 7, the connection to elliptic functions was discovered by Abresch [1], and written in the present form by Bobenko [2]. We now show that there is $s_0 \in M_{1,\mathbf{R}}$ satisfying Proposition 3.

To simplify the notation we delete the subscript 1 from $a_1^{(\nu)}(s), b_1^{(\nu)}(s)$, $\alpha_1^{(r)}(s), \gamma_1^{(r)}(s)$, and we make the change of variables $\zeta = z - 2$.

For $r > 0, \theta \in (0, \pi)$ we define the point $s = s(r,\theta) \in M_{1,\mathbf{R}}$ by

$$\lambda_1(s) = 2 + r\,e^{i\theta}\,,\; \lambda_2(s) = 2 + r\,e^{-i\theta}\,.$$

If $f(s)$ is a function of s we often write $f(r,\theta)$ for $f\left(s(r,\theta)\right)$.

The hyperelliptic curve $C_{s(r,\theta)}^{(1)}$ is

$$y^2 = \zeta(\zeta^2 - 2r\zeta\cos\theta + r^2)$$

and

$$\Omega^{(1)}(r,\theta) = \frac{\zeta - Z(r,\theta)}{y}d\zeta\,.$$

As in [2] Section 13 one sees that

$$\int_{a^{(1)}(r,\theta)} \frac{\zeta d\zeta}{y} = 2\sqrt{2r}\int_\theta^\pi \frac{\cos t\,dt}{\sqrt{\cos\theta - \cos t}}.$$

and that there is a unique $\theta_0 \in \left(0, \frac{\pi}{2}\right)$ for which the latter integral vanishes. Thus

$$Z(r,\theta) = 0 \Leftrightarrow \theta = \theta_0\,.$$

To further simplify notation write $a^{(1)}$ for $a^{(1)}(1,\theta_0)$ and similarly with $b^{(1)}, \gamma^{(1)}, \Omega^{(1)}$. Because the curves $C_{s(r,\theta_0)}^{(1)}, r > 0$ are all isomorphic, the quantities in Proposition 3 that are defined using the curves $C_s^{(1)}$ can be computed in terms of θ_0 and the elliptic integral

$$P := \frac{1}{2\pi i}\int_{a^{(1)}} \frac{d\zeta}{y}\quad.$$

Observe that this number is a period of the holomorphic one form on $C^{(1)}_{s(1,\theta_0)}$, so different from zero.

Lemma 6

(i) $\frac{\partial Z}{\partial r}(r,\theta_0) = 0,$

$\frac{\partial Z}{\partial \theta}(r,\theta)\Big|_{\theta=\theta_0} = \frac{-r}{2\sin\theta_0},$

$\gamma^{(1)}(r,\theta_0) = \frac{r^2}{2}$

(ii) $\int_{b^{(1)}(r,\theta_0)} \Omega^{(1)}(r,\theta_0) = \frac{4}{P}\sqrt{r}$

$\frac{\partial}{\partial\theta}\int_{b^{(1)}(r,\theta)} \Omega^{(1)}(r,\theta)\Big|_{\theta=\theta_0} = \frac{2}{P}\sqrt{r}\frac{\cos\theta_0}{\sin\theta_0}$

$\int_{b^{(1)}(r,\theta_0)} \tilde{\Omega}^{(1)}(r,\theta_0) = \frac{4}{P}(3+2r\cos\theta_0)\sqrt{r}.$

Proof. Since $Z(r,\theta_0)$ vanishes identically the first statement of (i) is obvious. For the first statement of (ii) observe that

$$\int_{b^{(1)}(r,\theta_0)} \Omega^{(1)}(r,\theta_0) = \sqrt{r}\int_{b^{(1)}} \Omega^{(1)}$$

and that by reciprocity ([4, p. 67])

$$\int_{b^{(1)}} \Omega^{(1)} = \frac{4}{P} \quad .$$

Next

$$\frac{\partial}{\partial\theta}\Omega^{(1)}(1,\theta)\Big|_{\theta=\theta_0} = \left(\frac{-\partial Z}{\partial\theta}(1,\theta)\right)\Big|_{\theta=\theta_0} \cdot \frac{d\zeta}{y} - \sin\theta_0\frac{\zeta^3 d\zeta}{y^3} \tag{43}$$

and

$$d\left(\frac{-\zeta^2\cos\theta_0+\zeta}{y}\right) = -\sin^2\theta_0\frac{\zeta^3 d\theta}{y^3} - \frac{1}{2}\cos\theta_0\frac{\zeta d\zeta}{y} + \frac{1}{2}\frac{d\zeta}{y} \quad . \tag{44}$$

Since $\int_{a^{(1)}}\frac{\partial}{\partial\theta}\Omega^{(1)}(1,\theta)\Big|_{\theta=\theta_0}$ identity (43) yields

$$\begin{aligned}\frac{\partial Z}{\partial\theta}(1,\theta)\Big|_{\theta=\theta_0}\cdot\int_{a^{(1)}}\frac{d\zeta}{y} &= -\sin\theta_0\int_{a^{(1)}}\frac{\zeta^3 d\zeta}{y^3}\\ &= \frac{1}{2}\frac{\cos\theta_0}{\sin\theta_0}\int_{a^{(1)}}\Omega^{(1)} - \frac{1}{2\sin\theta_0}\int_{a^{(1)}}\frac{d\zeta}{y}.\end{aligned}$$

Therefore

$$\frac{\partial Z}{\partial\theta}(1,\theta)\Big|_{\theta=\theta_0} = -\frac{1}{2\sin\theta_0} \quad .$$

As $Z(r,\theta) = rZ(1,\theta)$ we get the second part of (i).

Next by (43) and (44)

$$\int_{b^{(1)}} \frac{\partial}{\partial\theta}\Omega^{(1)}(1,\theta)\Big|_{\theta=\theta_0}$$
$$= \frac{1}{2\sin\theta_0}\int_{b^{(1)}}\frac{d\zeta}{y} - \sin\theta_0\int_{b^{(1)}}\frac{\zeta^3 d\zeta}{y^3} = \frac{1}{2}\frac{\cos\theta_0}{\sin\theta_0}\int_{b^{(1)}}\Omega^{(1)}.$$

This gives the second statement of (ii).

By definition

$$\begin{aligned}\tilde{\Omega}^{(1)}(r,\theta_0) &= \frac{3}{2}(\zeta+2)\Omega^{(1)}(r,\theta_0) + \gamma^{(1)}(r,\theta_0)\frac{\Omega^{(1)}(r,\theta_0)}{\zeta - Z(r,\theta_0)}\\ &= \frac{3}{2}(\zeta+2)\frac{\zeta d\zeta}{y} + \gamma^{(1)}(r,\theta_0)\frac{d\zeta}{y}\quad,\end{aligned}$$

so

$$\tilde{\Omega}^{(1)}(r,\theta_0) - dy = (3 + 2r\cos\theta_0)\frac{\zeta d\zeta}{y} + \left(\gamma^{(1)}(r,\theta_0) - \frac{r^2}{2}\right)\frac{d\zeta}{y} \tag{45}$$

Also by definition $\int_{a^{(1)}(r,\theta_0)}\tilde{\Omega}^{(1)}(r,\theta_0) = 0$. Therefore, $\gamma^{(1)}(r,\theta_0) = \frac{r^2}{2}$ and the last statement of (ii) also follows immediately.

Next we compute the asymptotics as $r \to 0$ of the quantities in Proposition 3 associated to the curves

$$C^{(2)}_{s(r,\theta)} : y^2 4V_S G + 4)(\zeta^2 - 2r\zeta\cos\theta + r^2)\,.$$

Lemma 7

(i) $\int_{b^{(2)}(r,\theta)}\Omega^{(2)}(r,\theta) = 8 + r\cos\theta + O(r^2)$

(ii) $\alpha^{(2)}(r,\theta) = 2 + r\cos\theta + O(r^2)$

(iii) Denote by $\tilde{c}(r,\theta)$ the lift of a path starting at $\zeta = 0$, going straight to $re^{i\theta}$, in a circle with positive orientation around $\zeta = re^{i\theta}$ and straight back to $\zeta = O$ such that $y > O$ at the starting point of $\tilde{c}(r,\theta)$. Then

$$\int_{\tilde{c}(r,\theta)}\Omega^{(2)}(r,\theta) = -r + O(r^2)$$

(iv) $\int_{b^{(2)}(r,\theta)}\tilde{\Omega}^{(2)}(r,\theta) = -8 + O(r)$

(v) $\int_{\tilde{c}(r,\theta)}\tilde{\Omega}^{(2)}(r,\theta) = -3r + O(r^2)$

Proof. (i) Put

$$Q(r,\theta) := \frac{1}{2\pi i}\int_{a^{(2)}(r,\theta)} \frac{d\zeta}{y} .$$

This is an analytic function of r and θ. Clearly

$$Q(0,\theta) = res_{\zeta=0}\frac{d\zeta}{\zeta\sqrt{\zeta+4}} = \frac{1}{2}$$

$$\frac{\partial}{\partial r}Q(r,\theta) = res_{\zeta=0}\left(\frac{\partial}{\partial \tau}\frac{d\zeta}{y}\bigg|_{r=0}\right) = res_{\zeta=0}\frac{\cos\theta d\zeta}{\zeta^2\sqrt{\zeta+4}} = -\frac{1}{16}\cos\theta .$$

Therefore,

$$Q(r,\theta) = \frac{1}{2}\left(1-\frac{r}{8}\cos\theta\right) + O(r^2) \tag{46}$$

Using the reciprocity formula

$$\int_{b^{(2)}(r,\theta)} \Omega^{(2)}(r,\theta) = \frac{4}{Q(r,\theta)}$$

part (i) of the Lemma immediately follows from (46).

(ii) in a similar way as (46) one shows that

$$\frac{1}{2\pi i}\int_{a^{(2)}(r,\theta)} \frac{\zeta d\zeta}{y} = \frac{r}{2}\cos\theta + O(r^2) \tag{47}$$

Since the integral of

$$\Omega^{(2)}(r,\theta) = \frac{\zeta d\zeta}{y} - \left(\alpha^{(2)}(r,\theta)-2\right)\frac{d\zeta}{y}$$

over $a^{(2)}(r,\theta)$ is identically zero, the claim follows from (46) and (47).

(iii) The integral under consideration is

$$2\int_0^{re^{i\theta}} \frac{\zeta-\alpha}{\sqrt{(\zeta+4)}\cdot\sqrt{\zeta^2-2r\zeta\cos\theta+r^2}}d\zeta \tag{48}$$

where we write

$$\alpha := \alpha^{(2)}(r,\theta) - 2 .$$

We expand

$$\frac{2}{\sqrt{\zeta+4}} = 1 + O(\zeta) .$$

Since by scaling

$$\int_0^{re^{i\theta}} \frac{\zeta^k d\zeta}{\sqrt{\zeta^2-2r\zeta\cos\theta+r^2}} = r^k\int_0^{e^{i\theta}} \frac{\zeta^k d\zeta}{\sqrt{\zeta^2-2\zeta\cos\theta+1}} \tag{49}$$

the integral (48) is

$$\int_0^{re^{i\theta}} \frac{(\zeta-\alpha)d\zeta}{\sqrt{\zeta^2-2r\zeta\cos\theta+r^2}}+O(r^2)\,.$$

Using the expansion of α from part (ii) of the Lemma we get the desired result.

(iv) By construction the integral of the differential form

$$\tilde{\Omega}^{(2)}(r,\theta)=\frac{3}{2}\frac{\zeta(\zeta-\alpha)}{y}d\zeta+3\Omega^{(}(r,\theta)+\gamma^{(2)}(r,\theta)\frac{d\zeta}{y}$$

over $a^{(2)}(r,\theta)$ vanishes identically. Using (46), (47) the fact that

$$\int_{a^{(2)}(r,\theta)}\zeta^2\frac{d\zeta}{y}=O(r^2)$$

and the expansion of α from part (ii) of the Lemma, one deduces that

$$\gamma^{(2)}(r,\theta)=O(r^2) \tag{50}$$

The cycle $b^{(2)}(r,\theta)$ can be represented by the lift under $\pi^{(2)}$ of a path as indicated in the figure below

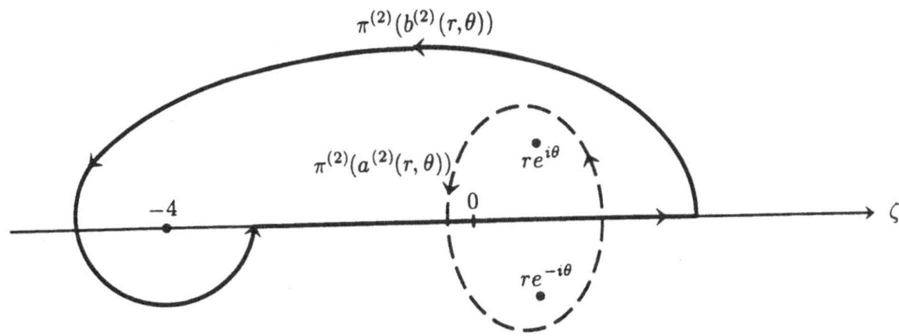

Since $a^{(2)}(r,\theta)$ and $b^{(2)}(r,\theta)$ have intersection number one, the lift is such that $y<0$ on the piece of curve lying over the interval $(-4,0)$. Therefore by (49) and the expansion of α stated in part (ii) of the Lemma

$$\lim_{r\to 0}\int_{b^{(2)}(r,\theta)}\left(\tilde{\Omega}^{(2)}(r,\theta)-3\Omega^{(2)}(r,\theta)\right)=2\int_{-4}^{0}\frac{3}{2}\zeta^2\frac{d\zeta}{\zeta\sqrt{\zeta+4}}=-32\quad .$$

This, together with (i), gives part (iv) of Lemma 7.

(v) As in the proof of part (iii) one sees that

$$\int_{\tilde{c}(r,\theta)} \left(\tilde{\Omega}^{(2)}(r,\theta) - 3\Omega^{(2)}(r,\theta)\right)$$
$$= \int_0^{re^{i\theta}} \left(1 - \frac{\zeta}{8}\right) \frac{\frac{3}{2}(\zeta^2 - \alpha\zeta) + \gamma^{(2)}(r,\theta)}{\sqrt{\zeta^2 - 2r\zeta\cos\theta + r^2}} d\zeta + O(r^3).$$

Using (49), (50) and the expansion of α we see that this integral is $O(r^2)$. Together with (i) this gives the desired result.

We are now ready to prove Proposition 3 in the case $g = 1$.

As observed above $Z(r,\theta) = 0$ if and only if $\theta = \theta_0$. So part (i) of the Proposition is fulfilled whenever we choose $\theta = \theta_0$. Part (iii) of the Proposition is trivial in the case $g = 1$. To prove part (ii) we have to show that for some $r > 0$ the vectors $(0; I_1(r,\theta_0); 0)$, $(0; 0; I_2(r,\theta_0))$, $\frac{\partial}{\partial r}(Z(r,\theta_0), I_1(r,\theta_0), I_2(r,\theta_0))$, $\frac{\partial}{\partial\theta}(Z(r,\theta), I_1(r,\theta), I_2(\mu,\theta)))\big|_{\theta=\theta_0}$ are linearly independent. For part (iv) we have to determine $\eta^{(1)}(r,\theta_0), \eta^{(2)}(r,\theta_0)$ by writing $\left(\gamma_1^{(1)}(r,\theta_0), -J_1(r,\theta_0), -J_2(r,\theta_0)\right)$ as a linear combination of these vectors and study the images of $\eta^{(1)}(r,\theta_0) - \eta^{(2)}(r,\theta_0)$ under iterates of the map $T(r,\theta_0)$.

Since $\tilde{c}(r,\theta)$ is homologuous to $c(r,\theta) + b^{(2)}(r,\theta) + a^{(2)}(r,\theta)$ (see figure)

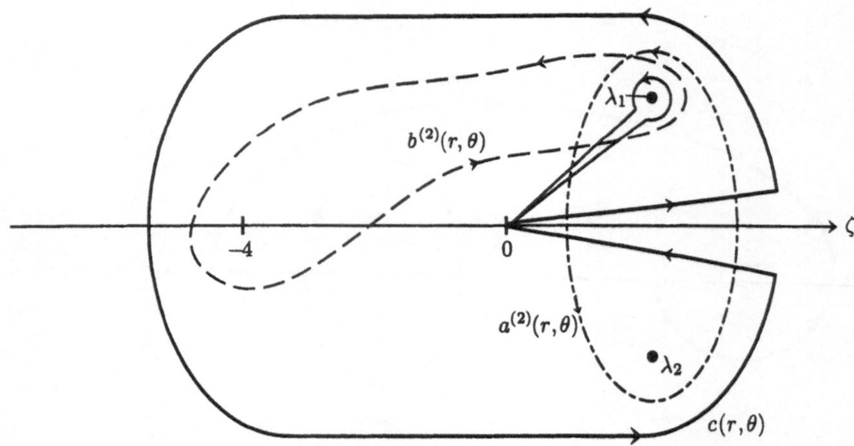

we may replace I_2 (resp. J_2) in the vectors above by

$$\hat{I}_2(r,\theta) := \left(\int_{b^{(2)}(r,\theta)} \Omega^{(2)}(r,\theta), \int_{\tilde{c}(r,\theta)} \Omega^{(2)}(r,\theta)\right) \quad \text{resp.}$$
$$\hat{J}_2(r,\theta) := \left(\int_{b^{(2)}(r,\theta)} \tilde{\Omega}^{(2)}(r,\theta), \int_{\tilde{c}(r,\theta)} \tilde{\Omega}^{(2)}(r,\theta)\right).$$

We put

$$\begin{aligned}
R_1(r) &:= \Big(0 \quad ; \quad I_1(r,\theta_0) \quad ; \quad 0\Big)\\
R_2(r) &:= \Big(0 \quad ; \quad 0 \quad ; \quad \hat{I}_2(r,\theta_0)\Big)\\
R_3(r) &:= \frac{\partial}{\partial r}\Big(Z(r,\theta_0) \quad ; \quad I_1(r,\theta_0) \quad ; \quad I_2(r,\theta_0)\Big)\\
R_4(r) &:= \frac{\partial}{\partial \theta}\Big(Z(r,\theta) \quad ; \quad I_1(r,\theta) \quad ; \quad I_2(r,\theta)\Big)\Big|_{\theta=\theta_0}\\
R_5(r) &:= \Big(\gamma^{(1)}(r,\theta_0) \quad ; \quad -J_1(r,\theta_0) \quad ; \quad -\hat{J}_2(r,\theta_0)\Big).
\end{aligned}$$

By Lemma 6 and 7

$$\begin{aligned}
R_1(r) &:= \Big(0 \quad ; \quad \tfrac{4}{P}r^{1/2} \quad ; \quad 0 \quad ; \quad 0\Big)\\
R_2(r) &:= \Big(0 \quad ; \quad 0 \quad ; \quad 8+O(r) \quad ; \quad -r+O(r^2)\Big)\\
R_3(r) &:= \Big(0 \quad ; \quad \tfrac{2}{P}r^{-1/2} \quad ; \quad O(1) \quad ; \quad -1+O(r)\Big)\\
R_4(r) &:= \Big(\tfrac{-r}{2\sin\theta_0} \quad ; \quad O(r^{1/2}) \quad ; \quad O(r) \quad ; \quad O(r)\Big)\\
R_5(r) &:= \Big(\tfrac{r^2}{2} \quad ; \quad -\tfrac{12}{P}r^{1/2}+O(r^{3/2}) \quad ; \quad 8+O(r) \quad ; \quad 3r+O(r^2)\Big) \quad .
\end{aligned}$$

Consider the 5×4 matrix whose rows are $R_1(r), \ldots, R_5(r)$. If we multiply the third row of this matrix r, the fourth row $r \sin\theta_0$, the first column by $2/r^2$, the second by $\frac{1}{2}Pr^{-1/2}$ and the last column by r^{-1} one obtains the matrix

$$\begin{pmatrix} 0 & 2 & 0 & 0 \\ 0 & 0 & 8 & -1 \\ 0 & 1 & 0 & -1 \\ -1 & 0 & 0 & 0 \\ 1 & -6 & 8 & 3 \end{pmatrix} + O(r) \tag{51}$$

One immediately sees that the first four rows of the leading term of the matrix (51) are linearly independent and that

$$(-1,1,-4,-1,-1)\begin{pmatrix} 0 & 2 & 0 & 0 \\ 0 & 0 & 8 & -1 \\ 0 & 1 & 0 & -1 \\ -1 & 0 & 0 & 0 \\ 1 & -6 & 8 & 3 \end{pmatrix} = 0 \quad .$$

Therefore for sufficiently small $r > 0$ the vectors $R_1(r), \ldots, R_4(r)$ are linearly independent, and $\lim_{r\to 0} \eta^{(1)}(r,\theta_0) = -1$, $\lim_{r\to 0} \eta^{(2)}(r,\theta_0) = 1$. Thus part (ii) of Proposition (ii) is proven, for all sufficiently small $r > 0$ and for part (iv) it suffices to show that for $0 \leq n \leq m-1$

$$\lim_{r\to 0} 5\,(D_1(r,\theta_0) - D_2(r,\theta_0)) + 4T^n(r,\theta_0)\left(\eta^{(1)}r,\theta_0) - \eta^{(2)}(r,\theta_0)\right) \neq 0\,.$$

But

$$\lim_{r\to 0}(D_1(r,\theta_0) - D_2(r,\theta_0)) = 2\,,$$

so $T := \lim_{r\to 0} T(r,\theta_0)$ is the map

$$x \longmapsto -2\frac{3x+8}{4x+10}\quad,$$

which has $-2 = \lim_{r\to 0}\left(\eta^{(1)}(r,\theta_0) - \eta^{(2)}(r,\theta_0)\right)$ as a fixed point.

With this the proof of Proposition 3 for the case $g = 1$ is complete.

REFERENCES

[1] U. Abresch, "Constant mean curvature tori in terms of elliptic functions", *J. reine u. angew. Math.*, **394** (1987), 169–192 .

[2] A. Bobenko, "All constant mean curvature tori in $\mathbb{R}^3, S^3, \mathbb{H}^3$ in terms of theta-functions", *Math. Ann.*,**290** (1991), 209–245.

[3] A. Bobenko, "Constant mean curvature surfaces and integrable equations", *Russ. Math. Surveys*, **46** (1991), 1–45.

[4] H. Farkas, I. Kra, "Riemann Surfaces". Graduate Texts in Mathematics, **71**, Springer Verlag, 1980.

[5] N. Hitchin, "Harmonic maps from a 2-torus to the 3-sphere", *J. Diff. Geom.*, **31** (1990), 627–710.

[6] U. Pinkall, I. Sterling, "On the classification of constant mean curvature tori", *Ann. Math.*, **130** (1989), 407–451.

[7] H. Wente, "Counterexample to a conjecture of Hopf", *Pacific J. Math.*, **121** (1986), 193–246.

[8] H. Wente, "Constant Mean Curvature Immersions of Enneper Type", *Mem. AMS* **100** (1992).

N.M. Ercolani
The University of Arizona
Tucson, AZ 85721
USA

H. Knörrer
ETH-Zentrum
CH 8092 Zürich
Switzerland

E. Trubowitz
ETH-Zentrum
CH 8092 Zürich
Switzerland

Modular Forms as τ-Functions for Certain Integrable Reductions of the Yang-Mills Equations

Leon A. Takhtajan

Abstract

We study properties of integrable reductions of the Self-Dual Yang-Mills Equations discovered by S. Chakravarty, M. Ablowitz and P. Clarkson [Phys. Rev. Lett., **65**, 1085–1087, 1990]. We show how classical modular forms and functions appear as their general solutions. We discuss possible consequences of this novel phenomenon in integrable systems which indicates deep connections between integrable equations, group representations, modular forms and moduli spaces.

1. Introduction

The famous self-dual Yang-Mills (SDYM) equations, first discovered in particle physics, play a significant role in mathematics, with applications from mathematical physics to geometry (see, e.g., lectures [1]). They are equations for the self-dual unitary connection A in a Hermitian vector bundle E associated with the principal G-bundle over a Riemannian four-manifold M. In terms of the curvature form $F = dA + A \wedge A$, the SDYM equations are written as

$$F = *F, \tag{1}$$

where $*$ denotes the Hodge star-operator.

System (1) is an elliptic system of non-linear first-order PDE's. Namely (leaving aside the global properties), assume that E is a trivial vector bundle over the Euclidean space $\mathbb{R}^4$ (with fibers isomorphic as linear spaces to Lie algebra g of a gauge Lie group G) so that in standard coordinates x_μ, $\mu = 0, 1, 2, 3$, on $M = \mathbb{R}^4$ the connection A and curvature F are given by $A = \sum_{\mu=0}^{3} A_\mu dx_\mu$, $F = \sum_{\mu,\nu=0}^{3} F_{\mu\nu} dx_\mu \wedge dx_\nu$. Then equations (1) can be written as the system

$$F_{01} = F_{23},\ F_{02} = F_{31},\ F_{03} = F_{12}, \tag{2}$$

which has the following remarkable properties.

1. The system (2) is an infinite dimensional covariant gradient flow with respect to the famous Chern-Simons action, i.e.

$$D_0 A_i = \frac{\delta S_{CS}}{\delta A_i}(A), \quad i = 1, 2, 3. \tag{3}$$

Here D_0 is a covariant derivative in the time direction ($t = x_0$), the Chern-Simons functional is defined by the following integral

$$S_{CS}(A) = \int_Y \mathrm{Tr}(A \wedge dA + \frac{2}{3} A \wedge A \wedge A),$$

over the time slice $Y \subset M$ (in our case $Y = \mathbb{R}^3$) and Tr denotes the invariant bilinear form on g.

2. The SDYM equations, as was discovered by R.Ward [2] and A.Belavin and V.Zakharov [3], can be written as a compatibility condition for an auxiliary linear system. In the null-coordinates (complex coordinates on $\mathbb{R}^4$) $u = t + \sqrt{-1}x_3$, $\bar{u} = t - \sqrt{-1}x_3$, $v = x_1 + \sqrt{-1}x_2$, $\bar{v} = x_1 - \sqrt{-1}x_2$; $A = A_u du + A_{\bar{u}} d\bar{u} + A_v dv + A_{\bar{v}} d\bar{v}$, this system has the form

$$\begin{aligned}(\partial_u + \zeta\partial_{\bar{v}})\Psi &= (A_u + \zeta A_{\bar{v}})\Psi, \\ (\partial_v - \zeta\partial_{\bar{u}})\Psi &= (A_v - \zeta A_{\bar{u}})\Psi.\end{aligned}$$

Its compatibility condition (for all values of spectral parameter ζ) yields the equations

$$F_{uv} = F_{\bar{u}\bar{v}} = 0, \; F_{u\bar{u}} + F_{v\bar{v}} = 0,$$

which are equivalent to the SDYM system (2) in null-coordinates.

3. The SDYM system is a "universal system" for classical integrable "solitonic" equations in the sense that all these equations are its reductions (it was conjectured by R. Ward [4] and established by many researchers; see important papers by L. Mason and G. Sparling [5] and by M.bAblowitz, S. Chakravarty and P. Clarkson [6], as well as references there). Such reductions can be achieved by utilizing the choice of the gauge Lie algebra g, the gauge freedom of SDYM equations and by reducing the dependence on space-time variables. Lax pairs for these equations can be obtained in the same way from the "master Lax pair" – a compatibility condition for the SDYM system. and provide a powerful tool for their solution. In addition, Hamiltonian, or gradient flow properties of these equations follow from the "universal" property (3) of the SDYM equations.

Particular examples of such reductions include all classical tops from Euler to Kovalevskaya (0 + 1-dimensions), the famous KdV, Nonlinear Schrödinger, Sine-Gordon, Toda lattice and N-wave equations (1 + 1-dimensions), KP and D-S equations (2 + 1-dimensions) (see, e.g., [7]). These systems were studied extensively by different methods: complex analysis – Riemann-Hilbert and $\bar{\partial}$-problems; algebraic geometry – formalism of Baker-Akhiezer functions; algebraic analysis – universal Grassmannians and τ-functions (see, i.e., [7, 8] and references there). For typical boundary conditions, like rapidly decreasing, periodic or self-similar cases, τ-functions, which provide universal formulas for general solutions, are expressed in terms of Fredholm determinants, Riemann theta functions or Painlevé transcendents (and their generalizations) respectively. For instance, the KdV solution can be written as a second derivative of the logarithm of the corresponding τ-function.

The examples above by no means exhaust all SDYM reductions (if this was the case there will be no need in this lecture!) and even the simplest $0+1$ case yields a system of ODE's with unusual properties of solutions and a new type of τ-function. The first example of such a reduction was discovered by M. Ablowitz, S. Chakravarty and P. Clarkson [6] and will be described in the next section.

Analysing this example, we can formulate the paradigm of our approach to SDYM reductions developed in [9, 10]. It states that SDYM systems with finite-dimensional gauge groups (say, a simple Lie group G) lead to "standard" integrable reductions described above, whereas reductions with infinite dimensional gauge groups (say, Diff(G)) lead to novel integrable equations with new τ-functions and analytic properties of solutions. Here we illustrate this paradigm with the simplest case $G = SU(2)$.

2. Basic Example: from Lagrange to Halphen

The simplest SDYM reduction (see, e.g., [6]) can be described as follows. Fix the gauge by setting $A_t = 0$ and require that remaining Yang-Mills potentials depend only on the variable t, i.e. $A_i = A_i(t)$, $i = 1, 2, 3$. Then (2) reduces to the following system of ODE's

$$\begin{aligned}\frac{dA_1}{dt} &= [A_2, A_3],\\ \frac{dA_2}{dt} &= [A_3, A_1],\\ \frac{dA_3}{dt} &= [A_1, A_2],\end{aligned}$$

which appears in the theory of static non-abelian monopoles and is called Nahm's system [11].

After the choice $G = SU(2)$ and the assumption that

$$A_i = \omega_i(t) t_i \in su(2),\ i = 1, 2, 3,$$

where t_1, t_2, t_3 are $su(2)$-generators (i.e. they satisfy commutation relations $[t_i, t_j] = \epsilon_{ijk} t_k$), Nahm's system can be reduced to:

$$\frac{d\omega_1}{dt} = \omega_2\omega_3,$$

$$\begin{aligned}\frac{d\omega_2}{dt} &= \omega_1\omega_3,\\ \frac{d\omega_3}{dt} &= \omega_1\omega_2.\end{aligned}$$

This system of equations goes back to Lagrange and we will call it the Lagrange system.

Now, following [6], let us present another SDYM reduction. Consider the Nahm system and choose the (infinite dimensional) group $SDiff(G)$ of volume preserving diffeomorphisms of $G = SU(2)$ to be a gauge group instead of $SU(2)$. In other words, assume that Yang-Mills potentials $A_i(t)$ belong to a Lie algebra $sdiff(G)$ of all "divergence-free" vector fields on $SU(2)$. To specialize this choice further, denote by X_i, $i = 1,2,3$, the standard basis of the left-invariant vector fields on G (i.e. X_i are $su(2)$-generators in the regular representation of G). Let $O : SU(2) \mapsto SO(3)$ be the covering defined by $O(g) = (O_{ij}(g))_{i,j=1}^3 \in SO(3)$, $g\sigma_i g^{-1} = \sum_{j=1}^3 O_{ij}(g)\sigma_j$, $g \in SU(2)$, where σ_i are standard Pauli matrices. Now assume the special form for the Yang-Mills potentials, proposed in [6]:

$$A_i(t) = \sum_{j=1}^{3} \omega_j(t) O_{ji}(g) X_j \in sdiff(G).$$

Using the basic properties of the operators X_i: $X_i(g) = gt_i$, $X_i(g^{-1}) = -t_i g^{-1}$ (where $t_i = \sigma_i/2\sqrt{-1}$), Nahm's system can be reduced to the following system for ω_i's

$$\begin{aligned}\frac{d\omega_1}{dt} &= \omega_2\omega_3 - \omega_1(\omega_2 + \omega_3),\\ \frac{d\omega_2}{dt} &= \omega_1\omega_3 - \omega_2(\omega_1 + \omega_3),\\ \frac{d\omega_3}{dt} &= \omega_1\omega_2 - \omega_3(\omega_1 + \omega_2).\end{aligned}$$

This system arises in general relativity as a self-dual reduction of the Bianchi IX cosmological model [12]. (Recall that the theory of gravity can be considered as the Yang-Mills theory with a diffeomorphism's group as a gauge group). Actually, the origin of this system of ODE's goes back more than 100 years ago: it was solved by Halphen in 1881 [13, 14] (see next section) and we call it the Halphen system.

Lagrange's and Halphen's systems have the following properties which are the consequence of the "master" properties **1. - 2.** of the SDYM system.

1. They are gradient flows with the same gradient function $f = \omega_1\omega_2\omega_3$ but with respect to the different metrics:

$$\frac{d\omega_i}{dt} = \frac{\partial f}{\partial \omega_i},$$

for the Lagrange system (standard metric on $\mathbb{R}^3$) and

$$\frac{d\omega_i}{dt} = \sum_{j=1}^{3} h_{ij} \frac{\partial f}{\partial \omega_j}, \quad i = 1, 2, 3,$$

for the Halphen system, where $h_{ij} = 2\delta_{ij} - 1$ is another (pseudo-) Riemannian metric on $\mathbb{R}^3$.

2. They admit Lax representation, i.e. they are equivalent to the Lax equation

$$\frac{dL}{dt} = [L, M],$$

where "Lax operators" L and M have the following form:

$$\begin{aligned} L(\zeta) &= \omega_1 t_1 - \sqrt{-1}\omega_2 t_2 - 2\sqrt{-1}\zeta\omega_3 t_3 + \zeta^2(\omega_1 t_1 + \sqrt{-1}\omega_2 t_2), \\ M(\zeta) &= \zeta^{-1}(\omega_1 t_1 - \sqrt{-1}\omega_2 t_2) - \zeta(\omega_1 t_1 + \sqrt{-1}\omega_2 t_2), \end{aligned}$$

for the Lagrange system and

$$\begin{aligned} L(g,\zeta) &= \frac{d\zeta}{d\tilde{\zeta}}(\omega_1 X_1 - \sqrt{-1}\omega_2 X_2 - 2\sqrt{-1}\tilde{\zeta}\omega_3 X_3 \\ &\quad + \tilde{\zeta}^2(\omega_1 X_1 + \sqrt{-1}\omega_2 X_2)), \\ M(g,\zeta) &= \frac{1}{\zeta}\frac{d\zeta}{d\tilde{\zeta}}((\omega_1 X_1 - \sqrt{-1}\omega_2 X_2) - \tilde{\zeta}^2(\omega_1 X_1 + \sqrt{-1}\omega_2 X_2)) \end{aligned}$$

for the Halphen system. Here

$$\tilde{\zeta} = g^t(\zeta) = \frac{\alpha\zeta - \bar{\beta}}{\beta\zeta + \bar{\alpha}}, \quad g = \begin{pmatrix} \alpha & \beta \\ -\bar{\beta} & \bar{\alpha} \end{pmatrix} \in SU(2),$$

is a fractional linear action of $SU(2)$ on $\mathbf{P}^1 = \mathbb{C} \cup \{\infty\}$ and

$$\frac{d\zeta}{d\tilde{\zeta}} = \left(\frac{d\tilde{\zeta}}{d\zeta}\right)^{-1} = \left(\frac{dg^t(\zeta)}{d\zeta}\right)^{-1}.$$

Remark 1. The Lax pair for the Lagrange system is of a standard form (for $t_i = \sigma_i/2\sqrt{-1}$ Lax operators L and M are 2×2 matrices) and algebro-geometrical methods can be used for its solution. The Lax pair for the Halphen system is of a different nature. Although we are still dealing with a system of ODE's, the Lax operators L and M are now first-order differential operators with respect to the "auxiliary" variable g taking values in the $SU(2)$ group manifold. Lax pairs of this type seem to be new in the theory of integrable systems. It would be extremely interesting to develop a theory of such Lax pairs which, for the Halphen case, will yield the solution presented below.

Remark 2. It is possible to obtain Halphen's Lax pair from that for Lagrange system. Namely, let T be a three-dimensional representation of $SU(2)$ realized in the vector space of polynomials $f(\zeta)$ of degree ≤ 2:

$$T(g)f(\zeta) = \frac{d\zeta}{dg^t(\zeta)} f(g^t(\zeta)).$$

Replacing generators t_i in Lagrange's $L(\zeta)$ by left-invariant differential operators X_i (i.e. replacing two-dimensional representation of $su(2)$ by the regular representation) we obtain

$$L_{Halphen}(g,\zeta) = T(g)L_{Lagrange}(\zeta),$$

where $T(g)$ acts on Lagrange's $L(\zeta)$ considered as a polynomial of degree 2 in ζ. Analogously

$$M_{Halphen}(g,\zeta) = \frac{1}{\zeta}T(g)(\zeta M_{Lagrange}(\zeta)).$$

3. Elliptic Functions and Modular Forms

Case A. Lagrange system and elliptic functions

Lagrange system has two obvious quadratic integrals $I_1 = \omega_1^2 - \omega_2^2$, $I_2 = \omega_1^2 - \omega_3^2$, that confine the flow to the intersection of two quadrics – the locus of elliptic curve C. It also arises, via algebraic equation $\det(L(\zeta) - \mu) = 4\mu^2 + I_1\zeta^4 - 2(I_1 - 2I_2)\zeta^2 + I_1 = 0$, as a genus 1 spectral curve with the modulus τ defined by $\kappa(\tau) = (\sqrt{I_2} - \sqrt{I_2 - I_1})/\sqrt{I_1}$. Either solving Lagrange system directly "à la Jacobi" (using differential equations for elliptic functions) or specializing general formalism of Baker-Akhiezer functions, it is easy to show that

$$\begin{aligned} \omega_1(t) = \sqrt{I_1}\,\mathrm{sn}(\sqrt{I_2}t;\ \frac{\tau}{2}),\ \omega_2(t) &= \sqrt{-I_1}\,\mathrm{cn}(\sqrt{I_2}t;\ \frac{\tau}{2}), \\ \omega_3(t) &= -\sqrt{-I_2}\,\mathrm{dn}(\sqrt{I_2}t;\ \frac{\tau}{2}), \end{aligned} \tag{4}$$

where we have used standard notations for Jacobi elliptic functions.

The Lagrange system has all the properties of integrable top-like ODE's: its solutions are single-valued meromorphic functions of a complex "time" t with simple poles as movable singularities. In addition (genus 1 case), they are doubly periodic in the complex t-plane. We say (using Adler-van Moerbeke terminology [15]) that Lagrange system is algebraically integrable and, therefore, satisfies the classical Painlevé property.

Case B. Halphen system and modular functions and forms

The Halphen system has the following remarkable properties which differ dramatically from those for the Lagrange system.

i) It has no quadratic, or, more generally, polynomial integrals, so that it is not algebraically integrable! (However, it was shown by S. Chakravarty [16] that it possesses "multi-valued integrals").

ii) Transformation property: if the set $\omega_i(t)$ is Halphen's solution, then a new set

$$\tilde{\omega}_i(t) = \frac{1}{(ct+d)^2}\omega_i(\frac{at+b}{ct+d}) + \frac{c}{ct+d}, \quad \begin{pmatrix} a & b \\ c & d \end{pmatrix} \in PSL(2,\mathbf{C}),$$

is also a solution [13].

iii) Elementary symmetric functions of ω_i's satisfy a system of ODE's which reduces to a third-order nonlinear ODE

$$y''' = 2y'y'' - 3(y')^2 \tag{5}$$

for $y = -2(\omega_1 + \omega_2 + \omega_3)$, where prime stands for d/dt. This particular ODE was studied by Chazy [17, 18] in 1910 and we will call (5) the Chazy equation.

iiii) The Halphen system can be integrated in terms of elliptic integrals of "variable" modulus t [13, 14, 17, 18, 19] which are related to the elliptic modular function $\kappa(t)$. The latter serves as a classical example in the general theory of modular functions and forms.

Recall (see, e.g., [20, 21]) that a modular form of the integer weight k for the Fuchsian group Γ is a meromorphic function f on the upper half-plane $H = \{t \in \mathbf{C} \mid \mathrm{Im}\, t > 0\}$ (it is meromorphic near the cusps of Γ with respect to the local uniformizer) with the property

$$f(\frac{at+b}{ct+d}) = (ct+d)^k f(t), \text{ for all } \gamma = \begin{pmatrix} a & b \\ c & d \end{pmatrix} \in \Gamma.$$

Let $\Gamma = PSL(2,\mathbf{Z})$ be the modular group and $\Gamma(2)$ be its principal congruence subgroup of level 2 defined by $\gamma \equiv I \pmod 2$, where I is 2×2 unit matrix. Corresponding Riemann surfaces H/Γ and $H/\Gamma(2)$ have genus 0, a cusp at ∞, and (after a suitable normalization) elliptic fixed points of orders $2,3$ at $\sqrt{-1}$ and $\sqrt[3]{-1}$ for the former case and two more cusps at $0,1$ for the latter case. Denote by J and λ holomorphic mappings which establish isomorphisms $H/\Gamma \cong \mathbf{P}^1$ and $H/\Gamma(2) \cong \mathbf{P}^1$. These "Hauptmodules" (in the classical terminology of Felix Klein), are modular functions for Γ and $\Gamma(2)$ respectively, have a single simple pole at ∞ and are normalized by $J(\sqrt{-1}) = \lambda(1) = 1$.

Theorem 1. *Transformation property* **ii)** *and the formulas*

$$\omega_1 = -\frac{1}{2}\frac{d}{dt}\log\frac{\lambda'}{\lambda}, \quad \omega_2 = -\frac{1}{2}\frac{d}{dt}\log\frac{\lambda'}{\lambda-1}, \quad \omega_3 = -\frac{1}{2}\frac{d}{dt}\log\frac{\lambda'}{\lambda(\lambda-1)} \tag{6}$$

provide the general solution of the Halphen system.

Proof. Using (6) it is easy to see that the Halphen system is equivalent to the following third-order nonlinear ODE

$$\frac{\lambda'''}{\lambda'} - \frac{3}{2}(\frac{\lambda''}{\lambda'})^2 = -\frac{1}{2}(\frac{1}{\lambda^2} + \frac{1}{(\lambda-1)^2} - \frac{1}{\lambda(\lambda-1)})(\lambda')^2.$$

This ODE is nothing but the famous Schwartz equation for the elliptic modular function $\kappa = 1 - 1/\lambda$ (see, e.g., [22])!

Therefore Halphen's solutions possess the property of a natural movable boundary, i.e. they exist, are holomorphic and single-valued inside a certain circle D on the complex t-plane. Its boundary ∂D is movable in the sense that it depends on initial conditions (because of **ii**) and contains a dense set of essential singularities.

This property could be interpreted as a "non-Euclidean" (or hyperbolic) version of the Painlevé property; the latter should be considered as inherent in Euclidean geometry.

The meromorphic function λ' is a modular form of weight 2 for $\Gamma(2)$ and can be used for constructing the basis of the three-dimensional linear space of regular (i.e. without poles) forms of weight 2:

$$E_1 = \frac{\lambda'}{\lambda},\ E_2 = \frac{\lambda'}{\lambda-1}, E_3 = \frac{\lambda'}{\lambda(\lambda-1)}$$

Up to a normalization these are holomorphic Eisenstein series of weight 2 for $\Gamma(2)$ and Theorem 1 reads

$$\omega_i = -\frac{1}{2}\frac{d}{dt}\log E_i,$$

$i = 1,2,3$; so that modular forms E_i play the role of τ-functions for the Halphen system!

Next, consider the Chazy equation.

Theorem 2. *Transformation property*

$$y(t) \mapsto \tilde{y}(t) = \frac{1}{(ct+d)^2}y(\frac{at+b}{ct+d}) - \frac{6c}{ct+d}, \quad \begin{pmatrix} a & b \\ c & d \end{pmatrix} \in PSL(2,\mathbb{C}),$$

and the formula

$$y = \frac{1}{2}\frac{d}{dt}\log\frac{(J')^6}{J^4(J-1)^3} \tag{7}$$

provide the general solution for the Chazy equation.

We will discuss its proof later.

Remark 3. Comparing Theorems 1 and 2, we can see that the Chazy solution y can be represented by two "Hauptmodules": λ for the congruence subgroup $\Gamma(2)$ and J for the modular group Γ, which suggests a differential relation between these two functions. Indeed, we have the following

Lemma 1.

$$\frac{(J')^6}{J^4(J-1)^3} = 432\frac{(\lambda')^6}{\lambda^4(\lambda-1)^4}. \tag{8}$$

Proof. Use the classical relation between the Hauptmodules J and λ (see, e.g., [22]):

$$J = \frac{4}{27}\frac{(\lambda^2-\lambda+1)^3}{\lambda^2(\lambda-1)^2}.$$

The function $(J')^6$, which appears in Lemma 1, is a modular form of weight 12 for Γ and has a pole of order 6 at the cusp ∞. Recall that 12 is a minimal weight (for the modular group Γ) admitting cusp forms (i.e. modular forms that vanish at ∞). There is only one (up to a scalar factor) cusp form of weight 12 given by the classical discriminant function

$$(2\pi)^{-12}\Delta = q\prod_{n=1}^{\infty}(1-q^n)^{24},$$

where $q = \exp(2\pi\sqrt{-1}t)$ is the local uniformizer at ∞.

We have the following.

Lemma 2.

$$(2\pi)^{-12}\Delta = 1728\frac{(J')^6}{J^4(J-1)^3}. \tag{9}$$

Proof. Inside the fundamental domain of H/Γ function J takes every value once (with proper count at the elliptic fixed points) so that the right hand side of (9) is a regular modular form of weight 12 for Γ. Since J has a simple pole at ∞ (with a residue 1/1728) the right hand side of (9) vanishes at ∞, i.e. is a cusp form and hence is proportional to Δ.

Using (9), we can rewrite the Chazy solution as

$$y = \frac{1}{2}\frac{d}{dt}\log\Delta, \tag{10}$$

so that modular discriminant Δ plays the role of τ-function for the Chazy equation!

Remark 4. Using elementary facts from classical theory of modular forms (see, e.g., [21]) formula (10) can be rewritten as

$$y = \pi\sqrt{-1}E_2,$$

where now E_2 is a normalized Eisenstein series of weight 2 for the modular group Γ (not a modular form due to a "correction term" in its transformation property).

The proof of Theorem 2 is based on another simple result – a certain specialization of that of Rankin [23].

Lemma 3. *Let f be a modular form of weight 2 for a Fuchsian group Γ with a multiplier system χ, i.e.*

$$\frac{d\gamma(t)}{dt} f(\gamma(t)) = \chi(\gamma) f(t) \quad \text{for all} \quad \gamma \in \Gamma.$$

Set $y = 3f'/f$, $Y_1 = y' - y^2/6$ and define, for $n > 1$,

$$Y_n = Y'_{n-1} - \frac{n}{3} y Y_{n-1}.$$

Then the functions Y_n, $n \geq 1$, are modular forms of weight $2n+2$ for the group Γ with trivial multiplier system.

Proof. Induction.

Remark 5. Functions Y_n from the lemma also appeared in P.Clarson's approach [24] for generalizing Chazy's original method using hypergeometric equations.

Lemma 3 provides a very simple proof of Theorem 2. Namely, consider the function $f = J'/J^{2/3}(J-1)^{1/2}$. It is a regular (use arguments of Lemma 2) modular form of weight 2 for Γ with a certain multiplier system χ (because of fractional powers). By Lemma 3 the function

$$Y_3 = y''' - 2y'y'' + 3(y')^2 - 4(y' - \frac{1}{6}y^2)^2$$

is a holomorphic modular form of weight 8 for Γ. Since $y(t) \to c$ as $t \to \sqrt{-1}\infty$, $Y_1 \to -c^2/6$, $Y_3 \to -c^4/9$, we have $Y_3 + 4Y_1^2 = O(q)$ as $q \to 0$, so that $Y_3 + 4Y_1^2$ is a cusp form of weight 8 for the modular group Γ. But there are no cusp forms of this weight which implies $Y_3 + 4Y_1^2 = 0$ – a Chazy equation!

Remark 6. One can generalize this approach for triangular Fuchsian groups of the type $(\pi/m, \pi/n, 0)$ (recall that Γ has the type $(\pi/2, \pi/3, 0)$ and $\Gamma(2)$ – $(0,0,0)$) by applying Lemma 3 to the appropriate function f of the form

$$f = \frac{J'}{J^{\alpha_1}(J-1)^{\alpha_2}},$$

where J is the corresponding Hauptmodule and exponents α_1 and α_2 are determined by the condition that f is regular whenever $J = 0$ or $J = 1$. Then, using the Riemann-Roch theorem (in a manner similar to the proof of Theorem 2) it is possible to represent the Chazy solution in terms of certain other Hauptmodules. This explains the origin of two representations for y discussed earlier and provides new ones. For instance, the group $(\pi/3, \pi/3, 0)$ yields a representation

$$y = \frac{1}{2} \frac{d}{dt} \log \frac{(J')^6}{J^4(J-1)^4},$$

whereas for $\Gamma(2)$ one gets another one:

$$y = \frac{1}{2}\frac{d}{dt}\log\frac{(\lambda')^6}{\lambda^5(\lambda-1)^5}.$$

Viewed as a representations of the same Chazy solution these formulas should be considered as "infinitesimal versions" of algebraic relations between different Hauptmodules (c.f. Lemma 1); for instance, the latter representation is related to the modular equation in its simplest form – an algebraic equation for $J(t)$ and $J(2t)$.

Remark 7. Let Γ_n be a triangular group of the type $(\pi/2, \pi/3, \pi/n)$, $n > 6$, and let J_n be the corresponding Hauptmodule. Specializing Lemma 3 for the following modular form of weight 8: $Y_3 + 4n^2Y_1^2/(n^2-36)$ and using the arguments of Theorem 2, we conclude that

$$y = \frac{1}{2}\frac{d}{dt}\log\frac{(J_n')^6}{J_n^4(J_n-1)^3}$$

satisfies the generalized Chazy equation

$$y''' - 2y'y'' + 3(y')^2 + \frac{144}{n^2-36}(y' - \frac{1}{6}y^2)^2 = 0.$$

This equation also has the property of a movable natural boundary, i.e. its solutions are single-valued, meromorphic (with one pole in the fundamental domain of Γ_n) inside a certain circle in the complex t-plane depending on inital conditions.

Remark 8. The present approach can be generalized to other higher order non-linear ODE's and can be described as follows. Consider the infinitely generated graded polynomial ring $R = \mathbf{C}[Y_1, Y_2, ...]$ with generators Y_n having degree $2n+2$. Using a suitable Fuchsian group Γ, function f from Lemma 3 and Riemann-Roch type arguments from the proof of Theorem 2 we can get explicit solutions for a certain ODE's of the form $P(Y_1, Y_2, ...) = 0$, where $P \in R$ is a homogeneous polynomial (in other words, $P(Y_1, Y_2, ...)$ is a modular form for Γ of a weight equals to the total degree of P). The polynomial $Y_3 + 4Y_1^2$, which yields the Chazy equation, provides the simplest example.

Remark 9. Generalized Chazy equation makes sense for values $n \leq 6$ as well. Case $n < 6$ corresponds to finite subgroups of $SU(2)$ (groups of regular solids) with Hauptmodules (and solutions) given by elementary functions [25]. For the case $n = 6$ the generalized Chazy equation reduces to $y' - 1/6y^2 = 0$ and also has an elementary solution. We can say that the case $n < 6$ corresponds to the Riemann geometry of $\mathbf{P}^1$ (positive constant curvature), case $n = 6$ – to the Euclidean geometry on $\mathbf{C}$ (zero curvature) and case $n > 6$ – to the non-Euclidean (hyperbolic) geometry on H (constant negative curvature).

Remark 10. Signatures $(2, 3, n)$ also appear in Dynkin diagrams for Lie algebras E_{n+3}, $n = 3, 4, 5$. It is tempting to speculate that there exists a deep connection between generalized Chazy equations and such Lie algebras for $n < 6$, affine Lie algebra $\hat{E}_8$ for $n = 6$, and general Kac-Moody algebras for $n > 6$. In particular, limiting case $n = \infty$ yields the Kac-Moody algebra E_∞ which should, somehow, be related with the Chazy equation! The Kac-Moody algebra A_∞ plays an important role in integrable systems (see e.g., [26]), so once again one can speculate that Lie algebra E_∞ (whatever it means) presumably plays an important role as well.

4. Conclusion

Here I will briefly indicate important though not yet completely understood and solved problems (some of them are currently under investigation).

1. The fact that modular forms in general, and Δ in particular, could appear as a τ-function is not so suprising as it may seem. Indeed, recall that hyperbolic plane H can be considered as the Teichmüller space of Riemann surfaces of genus 1 (elliptic curves), modular group Γ – as the Teichmüller modular group and the quotient H/Γ – as the moduli space, so that the modular form Δ can be interpreted as a holomorphic cross-section of a certain line bundle over this moduli space. Since τ-function appears as a cross-section of a certain line bundle over the universal Grassmannian and moduli spaces of Riemann surfaces and vector bundles can be imbedded into this Grassmannian via the Krichever map, modular forms can be treated as τ-functions.

The quotient $H/\Gamma(2)$ is also a moduli space of elliptic curves with additional restrictions on the 2-division points, so that modular forms E_i are also subjects of this interpretation. To make this picture clear, one should understand how these modular forms arise from the corresponding Lax pair, i.e. develop the analog of Baker-Akhiezer functions formalism for such Lax pairs.

2. The moduli parameter τ, which was the integral of motion for the Lagrange system, plays the role of "physical time" t for Halphen and Chazy equations. The latter situation is typical for models of Topological Conformal Field Theory, where two-point correlation functions are related to dynamical systems on moduli spaces (see [27]). These systems can be considered as a far-reaching geometrical generalization of Whitham's modulation theory (see, e.g. [28]). It would be instructive to relate Halphen and Chazy equations with that from [27].

3. It is possible to obtain multi-dimensional generalizations (i.e. to include spatial variables x_μ) of the Halphen system by another SDYM reductions – this was done in [9]. The most important open problem here is to develop the analog of the Inverse Scattering Transform for the corresponding Lax operator. It should be extremely interesting, since even in the Halphen's "stationary in x" case, corresponding τ-function is given by modular forms.

4. In principle, it is possible to generalize the Lagrange (and Halphen) system for an arbitrary simple Lie group. These generalizations could lead to new τ-functions given by certain holomorphic sections of line bundles over moduli spaces (say, Siegel modular forms).

However, we are at the very beginning of our understanding of the exciting properties of this new class of integrable systems obtained by SDYM reductions with infinite dimensional gauge groups of which the Halphen system is the simplest example.

Acknowledgments. This paper represents an extended version of the lecture given at the memorial *Colloque Verdier*, Luminy, Summer 1991 and it is my pleasure to thank the organizers for their hospitality.

I am grateful to Mark Ablowitz and Sarbarish Chakravarty, who introduced me to the fascinating field of SDYM reductions and explained their beautiful result; to Alexander Its for extremely fruitful discussions of differential equations for modular forms and to Willy Hereman for computer calculations of certain relevant hypergeometric equations.

REFERENCES

[1] M.F. Atiyah, *Classical Geometry of Yang-Mills Fields*, Fermi Lectures, Scuola Normale Pisa, 1980.

[2] R. Ward, *On self-dual gauge fields*, Phys. Lett., **61A**, 81-82, 1977.

[3] A.A. Belavin, V.E. Zakharov, *Yang-Mills equations as inverse scattering problem*, Phys. Lett., **73B**, 53-57, 1978.

[4] R. Ward, *Integrable and solvable systems, and relations among them*, Phil. Trans. Roy. Soc. London A, **315**, 451-457, 1985.

[5] J. Mason, G. Sparling, *Nonlinear Schrödinger and Korteweg -de Vries equations are reductions of self-dual Yang-Mills equations*, Phys. Lett., **137 A**, 29-33, 1989.

[6] S. Chakravarty, M.J. Ablowitz and P.A. Clarkson, *Reductions of self-dual Yang-Mills equations and classical systems*, Phys. Rev. Lett., **65**, 1085-1087, 1990.

[7] M.J. Ablowitz, P.A. Clarkson, *Solitons, Nonlinear Evolution Equations and Inverse Scattering*, London Math. Soc. Lecture Notes Series **149**, Cambridge Univ. Press, 1991.

[8] G. Segal, G. Wilson, *Loop groups and equations of KdV type*, Inst. Hautes Études Sci. Publ. Math., **61** 5-65, 1985.

[9] S. Chakravarty, M.J. Ablowitz and L. Takhtajan, *Self-dual Yang-Mills equation and new special functions in integrable systems*, PAM preprint **108**, University of Colorado, Boulder, 1991.

[10] M.J. Ablowitz, S. Chakravarty and L. Takhtajan, *Integrable systems, self-dual Yang-Mills equations and modular forms*, PAM preprint **113**, University of Colorado, Boulder, 1991.

[11] W. Nahm, *The algebraic geometry of multi-monopoles*, in *"Group Theoretical Methods in Physics"*, Eds. M. Serdagorlu, E.Inonu, Lect. Notes in Physics, **180**, 456-466, 1982, Springer Verlag.

[12] G.W. Gibbons, C.N. Pope, *The positive action conjecture and asymptotically Euclidean metrics in quantum gravity*, Commun. Math. Phys., **66**, 267-290, 1979.

[13] G.-H. Halphen, *Sur un système d'équations différentielles*, C.R. Acad. Sc. Paris, **92**, 1001-1003, 1881.

[14] G.-H. Halphen, *Sur certains systèmes d'équations différentielles*, C.R. Acad. Sc. Paris, **92**, 1004-1007, 1881.

[15] M. Adler, P. van Moerbeke, *The algebraic integrability of geodesic flow on $SO(4)$*, Invent. Math., **67**, 297-326, 1982.

[16] S. Chakravarty, *private communication.*

[17] J. Chazy, *Sur les équations différentielles dont l'intégrale générale possède une coupure essentielle mobile*, C.R. Acad. Sc. Paris, **150**, 456-458, 1910.

[18] J. Chazy, *Sur les équations différentielles du trousième et d'ordre supeérieur dont l'intégrale générale à ses points critiques fixés*, Acta Math., **34**, 317-385, 1911.

[19] F.J. Bureau, *Sur des systèmes différentiels non linéaires du troisième ordre et les équations différentielles non linéaires associées*, Acad. Roy. Belg. Bull. Cl. Sc. (5), **73**, 335-353, 1987.

[20] S. Lang, *Elliptic Functions*, Addison-Wesley, 1973.

[21] N. Koblitz, *Introduction to Elliptic Curves and Modular Forms*, Springer-Verlag, 1984.

[22] L. Ford, *Automorphic functions*, Chelsea Pub. Co., 1951.

[23] R.A. Rankin, *The construction of automorphic forms from the derivatives of a given form*, J. Indian Math. Soc., **20**, 103-116, 1956.

[24] P.A. Clarkson, *SERC postdoctoral fellowship report* B/RF/6935, 1986.

[25] F. Klein, *Vorlesungen über das ikosaeder und die auflösung der gleichungen vom fünften grade*, Leipzig, 1884.

[26] J.-L. Verdier, *Algèbres de Lie, systèmes Hamiltoniens, courbes algébriques*, Séminaire Bourbaki, 33-e année, **566**, 1-10, 1980/1981.

[27] B. Dubrovin, *Differential geometry of moduli spaces and its applications to soliton equations and to topological conformal field theory*, Preprints di Matematica **117**, Scuola Normale, Pisa, 1991.

[28] B. Dubrovin, S. Novikov, *Hydrodynamics of weakly deformed soliton lattices, differential geometry and Hamiltonian theory*, Russian Math. Surveys, **44**:6, 35-101, 1989.

Department of Mathematics
State University of New York at Stony Brook
Stony Brook, NY 11794-3651
USA

The τ-Functions of the $\mathfrak{g}$AKNS Equations

George Wilson

1. Introduction

The AKNS equations (see [1])

$$\left.\begin{aligned} iq_t &= -\tfrac{1}{2}q_{xx} + q^2 r \\ ir_t &= \tfrac{1}{2}r_{xx} - qr^2 \end{aligned}\right\} \tag{1.1}$$

are generally agreed to be one of the most basic examples of an integrable evolutionary system. They may be regarded as a complexified version of the physically important nonlinear Schrödinger equation(s)

$$iq_t = -\frac{1}{2}q_{xx} \pm q^2\bar{q} \tag{1.2}$$

to which they reduce if we impose one of the conditions $q = \pm\bar{r}$. The equations (1.1) are the simplest (in some sense) integrable system associated with the Lie algebra $\mathfrak{sl}(2, \mathbb{C})$; as such they have natural generalizations in which $\mathfrak{sl}(2, \mathbb{C})$ is replaced by some other simple Lie algebra $\mathfrak{g}$. I shall refer to these as the $\mathfrak{g}$AKNS equations. In the case when $\mathfrak{g}$ is $\mathfrak{sl}(n, \mathbb{C})$ the $\mathfrak{g}$AKNS equations are exactly the equations referred to as the AKNS-D equations in [3]; for a general simple Lie algebra $\mathfrak{g}$ it seems that they were first introduced in [9]. The equations (1.2), in which x and t must be considered to be real, are associated with the real subalgebras $\mathfrak{su}(2)$ and $\mathfrak{su}(1,1)$ of $\mathfrak{sl}(2, \mathbb{C})$, and will not be discussed explicitly here.

The main purpose of this paper is to understand the formulae

$$q = 2i(\tau_+/\tau), \quad r = -2i(\tau_-/\tau) \tag{1.3}$$

expressing solutions of (1.1) in terms of three τ-functions. This substitution would be the starting point for the study of (1.1) by "Hirota's direct method" (cf. [6], where (1.2) is treated by that method); but it is encountered by most authors who try to solve these equations. In the theory of the algebro-geometric solutions of (1.1), for example, the τ-functions in (1.3) are expressed in terms of the theta function of a Riemann surface (cf. [8]). We shall place ourselves in a slightly wider context, namely, the class of solutions obtained from the trivial one $q = r \equiv 0$ by the so-called "dressing" action of the loop group $\mathrm{LSL}(2, \mathbb{C})$. In that context too the τ-functions have an independent group-theoretical definition and have no *a priori* connexion with the system (1.1). In the spirit of my papers [10,11] (which discussed a different example, the modified KdV equation), I want to show how the formulae (1.3) arise naturally out of the definition of the dressing action. I also want to analyse, from as elementary a point of view as possible, why it is that these formulae have a natural generalization to the $\mathfrak{g}$AKNS equations only in the case when $\mathfrak{g}$ is simply laced (that is, of type A, D or E). Perhaps the most interesting feature of the paper is that much of the proof of (1.3) is rather different from the proof in [10,11] of a similar looking formula for the very closely related mKdV equation. I think this illustrates the general principle that one should not expect a uniform explanation of all the details of soliton theory.

The paper is set out as follows. In §2 I review the definition of the $\mathfrak{g}$AKNS equations and of the special class of solutions obtained by the action of the loop group L on the trivial solution. In §3 I introduce the basic holomorphic function σ on the universal central extension of L, and formulate a trivial looking lemma about it (Lemma 3.7). Then in §4 I define (in terms of σ) the appropriate τ-functions, and show (modulo Lemma 3.7) that if $\mathfrak{g}$ is simply laced then a formula generalizing (1.3) gives solutions of the $\mathfrak{g}$AKNS equations. Finally, §5 gives a proof of the fundamental lemma (3.7).

The article makes very little claim to originality: all the ideas explained here are standard, or at least implicit, in the literature. The formula (1.3) itself is proved, in a context similar to ours, in [2]. The main difference between the proof in [2] and the one given here is that the authors of [2] emphasize the role played by the homogeneous realization of the basic representation of the loop algebra $L\mathfrak{sl}(2)$; indeed, as V. Kac has pointed out to me, it is easy to prove our basic lemma (3.7) (in the simply laced case) if one assumes the representation theory known, and that is essentially what is done in [2]. Here I have tried to prove (1.3) using as little machinery as possible, in particular, I have made no use of representation theory. Also, although there are fundamental reasons (cf. [10,11]) why there cannot be a formula like (1.3) for the solutions of the $\mathfrak{g}$AKNS equations if $\mathfrak{g}$ is not simply laced, I have tried to avoid using that hypothesis for as long as possible. The result of this effort is the most curious thing in the paper: it is as follows. The dependent variables in the $\mathfrak{g}$AKNS equations are naturally indexed by the roots of $\mathfrak{g}$; recall that in general the roots are of two lengths, called *long* and *short*. It turns out that a formula of type (1.3) always holds for the $\mathfrak{g}$AKNS variables indexed by the long roots. The simply laced algebras $\mathfrak{g}$ are exactly those for which the roots all have the same length (we may consider them all to be long); so it is only in that case that we get a formula for all the unknown functions in our equations.

Notation. Here is some notation that will be used throughout the paper. We denote by $\mathfrak{g}$ a (finite dimensional) simple complex Lie algebra; $L\mathfrak{g} \equiv C^\infty(S^1;\mathfrak{g})$ will be the corresponding loop algebra, consisting of all smooth $\mathfrak{g}$-valued functions on the unit circle $S^1 \subset \mathbb{C}$. Where necessary we fix also a Cartan subalgebra $\mathfrak{h} \subset \mathfrak{g}$. We shall denote by $L_+\mathfrak{g}$ (resp. $L_-\mathfrak{g}$) the subalgebras of $L\mathfrak{g}$ consisting of all loops with a Fourier expansion of the form $\sum_i X_i z^i$ with $i \geq 0$ (resp. with $i \leq 0$); and by $M_+\mathfrak{g}$ (resp. $M_-\mathfrak{g}$) the smaller subalgebras consisting of all loops $\sum_i X_i z^i$ with $i > 0$ (resp. with $i < 0$). We therefore have (as linear spaces)

$$L\mathfrak{g} = M_-\mathfrak{g} \oplus L_+\mathfrak{g} = M_-\mathfrak{g} \oplus \mathfrak{g} \oplus M_+\mathfrak{g} = L_-\mathfrak{g} \oplus M_+\mathfrak{g},$$

where $\mathfrak{g}$ is identified with the subalgebra of $L\mathfrak{g}$ consisting of the constant loops. If G is the simply connected Lie group with Lie algebra $\mathfrak{g}$, we shall denote by L the loop group $C^\infty(S^1;G)$; the subgroups of L corresponding to the above subalgebras of $L\mathfrak{g}$ will be denoted by L_+, L_-, M_+ and M_-. The above decompositions of $L\mathfrak{g}$ do not correspond to global decompositions of the group L, but we have a dense open subset

$$M_-L_+ = M_-GM_+ = L_-M_+ \subset L$$

consisting of all loops γ that can be factorized in the form $\gamma = \gamma_-\gamma_+$ with $\gamma_- \in M_-$, $\gamma_+ \in L_+$. We shall refer to this subset of L as the *big cell.*

2. The $\mathfrak{g}$AKNS Equations

As above, we let $\mathfrak{g}$ be any simple (complex) Lie algebra and $\mathfrak{h} \subset \mathfrak{g}$ a Cartan subalgebra. We fix also a regular element $\Lambda \in \mathfrak{h}$. The equations of the AKNS hierarchy associated with $(\mathfrak{g}, \Lambda)$ are indexed by pairs (M, r), where $M \in \mathfrak{h}$ and r is a non-negative integer. We fix such a pair (M, r); then, by definition, the solutions of the corresponding AKNS equation will be in one-to-one correspondence with one-parameter families of flat connexions

$$\mathcal{A} = \{\partial/\partial x - U(x,t;z),\ \partial/\partial t - V(x,t;z)\}$$

(over some region of (x,t)-space) such that the potentials U and V depend in a certain way on the "spectral" parameter z. The certain way amounts to requiring that $\mathcal{A}$ should be obtained by "dressing" the bare (that is, independent of x and t) connexion

$$\mathcal{A}_0 = \{\partial/\partial x - \Lambda z,\ \partial/\partial t - Mz^r\}.$$

Note that $\mathcal{A}_0$ is indeed flat, since Λ and M commute. In general, dressing a connexion means acting on it by a gauge transformation that does not change the form of its singularities as a function of the spectral parameter; thus in our case, the potentials U and V in $\mathcal{A}$ will be required to be polynomials in z, with leading terms Λz and Mz^r, respectively.

To describe the equations, we consider first gauge transformations of $\mathcal{A}_0$ by elements of the group of *formal* expressions of the form

$$\phi_-(x,t;z) = \exp\left(\sum_1^\infty \chi_i(x,t)z^{-i}\right).$$

Clearly, such a ϕ_- transforms $\mathcal{A}_0$ into a family of formal connexions $\mathcal{A}$ with components $\partial/\partial x - U$ and $\partial/\partial t - V$, where U and V have the form

$$U = \Lambda z + \sum_{-\infty}^{0} u_i(x,t)z^i, \quad V = Mz^r + \sum_{-\infty}^{r-1} v_i(x,t)z^i$$

for some $\mathfrak{g}$-valued functions u_i and v_i. Of course, $\mathcal{A}$ is still flat, so we have the zero-curvature equation

$$[\partial/\partial x - U,\ \partial/\partial t - V] = 0, \tag{2.1}$$

or, equivalently,

$$U_t = V_x - [U,V].$$

As explained above, we want to consider only ϕ_- such that the terms in $\mathcal{A}$ involving negative powers of z are absent, so that U and V are polynomials in z; they therefore take the form

$$U = \Lambda z + u(x,t), \quad V = Mz^r + \sum_{i=0}^{r-1} v_i(x,t)z^i. \tag{2.2}$$

From the equation $\partial/\partial x - U = \phi_-^{-1}(\partial/\partial x - \Lambda z)\phi_-$ we find that

$$u(x,t) = [\Lambda, \chi_1(x,t)] \tag{2.3}$$

so that u is a function with values in the subspace $\operatorname{Im}\operatorname{ad}(\Lambda) \subset \mathfrak{g}$. Thus if we choose for each root α of $(\mathfrak{g},\mathfrak{h})$ a non-zero vector e_α in the corresponding root space, so that $\mathfrak{g}$ decomposes

$$\mathfrak{g} = \bigoplus_\alpha \mathbb{C}e_\alpha \oplus \mathfrak{h},$$

then u can be written in the form $u(x,t) = \sum_\alpha u_\alpha(x,t)e_\alpha$ where the u_α are scalar-valued functions. It is characteristic of soliton theory that the functions v_i in (2.2) are given by universal (that is, independent of ϕ_-) local expressions in the u_α, so that (2.1) reduces to an evolutionary system for the functions $u_\alpha(x,t)$. More precisely, we have the following.

Proposition 2.4. *Let $\mathcal{A} = \{\partial/\partial x - U,\ \partial/\partial t - V\}$ be obtained as above by formally dressing the bare connexion $\mathcal{A}_0$, and suppose the potentials U and V have the polynomial form* (2.2). *Then the zero-curvature equation* (2.1) *is equivalent to a system of evolution equations*

$$\partial u_\alpha / \partial t = f_\alpha(u_\beta)$$

where the f_α are certain universal (that is, depending only on M and r) polynomials in the u_β and their x-derivatives $u_\beta^{(j)}$. Furthermore, the f_α are homogeneous of degree $r+1$ if we give $u_\beta^{(j)}$ degree $j+1$.

We call this system the $\mathfrak{g}$AKNS equation associated with the data $(\mathfrak{g}, \Lambda; M, r)$. I omit the proof of (2.4), since it has been given many times in the literature (perhaps the best treatment would be obtained by extracting the relevant material from [4]). The basic system (1.1) is the special case when $\mathfrak{g}$ is $\mathfrak{sl}(2,\mathbb{C})$, $\Lambda = M = \operatorname{diag}(i,-i)$ and $r = 2$, so that the potential U has the form

$$U = \begin{bmatrix} i & 0 \\ 0 & -i \end{bmatrix} z + \begin{bmatrix} 0 & q \\ r & 0 \end{bmatrix},$$

and the right-hand side of (1.1) is homogeneous of degree 3.

We now restrict our attention to the special class of solutions of the $\mathfrak{g}$AKNS equation obtained by choosing the formal group element ϕ_- above to be a genuine function of z, defined for sufficiently large values of z. For definiteness we shall suppose that $\phi_-(x,t;z)$ is a function of x and t with values in the group M_- defined at the end of §1. The problem of finding such ϕ_- so that the transformed connexion $\mathcal{A} = \phi_-^{-1}\mathcal{A}_0\phi_-$ has the polynomial form (2.2) is solved as follows (see [12]). Consider also the group L_+ introduced in §1; and suppose that in addition to ϕ_- we can find a function $\phi_+(x,t;z)$ with values in L_+ such that

$$\mathcal{A} = \phi_-^{-1}\mathcal{A}_0\phi_- = \phi_+\mathcal{A}_0\phi_+^{-1}. \tag{2.5}$$

We may think of L_+ as the group of all smooth functions from the unit circle S^1 to G that extend to holomorphic G-valued functions on the disk $\{z : |z| < 1\}$; similarly M_- is the group of all smooth functions $S^1 \to G$ that extend holomorphically to the disk $\{z : |z| > 1\}$ and take the value

1 at infinity. Thus the first expression for $\mathcal{A}$ in (2.5) shows that it has no singularities outside S^1, except for the poles at infinity; and the second expression shows that it has no singularities inside S^1. Hence the potentials U and V have the desired form (2.2). If now we set $\phi = \phi_-\phi_+$, then we have $\phi^{-1}\mathcal{A}_0\phi = \mathcal{A}_0$, so that ϕ must have the form

$$(2.6) \qquad \phi(x,t;z) = \exp(x\Lambda z + tMz^r)g(z)\exp(-x\Lambda z - tMz^r)$$

where g belongs to the loop group $L \equiv C^\infty(S^1;G)$. Conversely, if g belongs to the big cell $M_-L_+ \subset L$, and we define ϕ by (2.6), then (at least for a dense open set of values of x and t) ϕ can be factorized $\phi = \phi_-\phi_+$ as above, and we can define a connexion $\mathcal{A}$ by (2.5). Combining these observations with (2.4) and (2.3), we obtain the following.

Proposition 2.7. *Let g belong to the big cell of the loop group L, let $\phi(x,t;z)$ be defined by* (2.6), *and let $\phi = \phi_-\phi_+$ be its factorization with $\phi_- \in M_-$, $\phi_+ \in L_+$. Suppose*

$$\phi_-(x,t;z) = \exp\left(\sum_1^\infty \chi_i(x,t)z^{-i}\right).$$

Then the $\mathfrak{g}$-valued function $u(x,t) = [\Lambda, \chi_1(x,t)]$ satisfies the $\mathfrak{g}$AKNS *equation indexed by the pair (M,r) occurring in* (2.6).

Some explanation is needed, since unfortunately (see [5]) an element of M_- does not necessarily have the form $\exp\left(\sum_1^\infty \chi_i z^{-i}\right)$ for any *convergent* series $\sum_1^\infty \chi_i z^{-i}$. Probably the simplest point of view is the following: choose any matrix representation of G, so that $\phi_- \in M_-$ can be expanded in a Fourier series $\phi_- = 1 + \sum_1^\infty \phi_i z^{-i}$ with the ϕ_i matrices; then the χ_i are the coefficients of the formal series $\log \phi_-$. They belong canonically to the Lie algebra $\mathfrak{g}$, because they can also be described in terms of the successive derivatives at infinity of ϕ_- with respect to z^{-1}.

3. The Central Extension $\hat{L}$ and the Function σ

The function σ that forms the subject of this section can be thought of in several ways: one of them is as follows. Consider the Grassmannian-like homogeneous space L/L_+. The image in L/L_+ of the complement of the big cell in L is a divisor in L/L_+; it therefore corresponds to a holomorphic line bundle $\mathcal{L}$ over L/L_+ together with a section of this bundle. If $\hat{L}$ denotes the group of automorphisms of $\mathcal{L}$, then the pullback of $\mathcal{L}$ to $\hat{L}$ is canonically trivial, and the section therefore becomes a function on $\hat{L}$. The group $\hat{L}$ turns out to be a central extension of L by $\mathbb{C}^\times$. I refer to [7] for a rigorous discussion along these lines. Here we shall merely give a simple characterization of σ, assuming certain properties of the group $\hat{L}$ which we make plausible by considering the corresponding Lie algebra $\hat{L}\mathfrak{g}$.

Let $\langle\ ,\ \rangle$ denote the (unique) invariant symmetric bilinear form on $\mathfrak{g}$ such that if α is a *long* root and h_α the corresponding coroot, then $\langle h_\alpha, h_\alpha\rangle = 2$. Then $\hat{L}\mathfrak{g}$ is the central extension of $L\mathfrak{g}$ by $\mathbb{C}$ defined by the cocycle

$$\omega(X(z), Y(z)) = \operatorname{res}_{z=0}\langle dX(z)/dz, Y(z)\rangle. \tag{3.1}$$

Thus as a linear space, we have $\hat{L}\mathfrak{g} = L\mathfrak{g} \oplus \mathbb{C}$, and the bracket is given by

$$[(X,\lambda),(Y,\mu)] = ([X,Y], \omega(X,Y)).$$

Corresponding to the extension $0 \to \mathbb{C} \to \hat{L}\mathfrak{g} \to L\mathfrak{g} \to 0$ of Lie algebras there is a central extension

$$1 \to \mathbb{C}^\times \to \hat{L} \to L \to 1 \tag{3.2}$$

of groups, with $\hat{L}$ simply connected (that depends on the form $\langle\ ,\ \rangle$ being normalized in the above way: cf. [7], ch. 4 and 6). We denote by

$$s : L\mathfrak{g} \to \hat{L}\mathfrak{g} \equiv L\mathfrak{g} \oplus \mathbb{C}$$

the splitting $s(X) = (X, 0)$ of the above exact sequence of Lie algebras. Of course, s is not a Lie algebra homomorphism, so does not induce a splitting of (3.2); indeed, the fibration (3.2) is not even topologically trivial, which is the main reason why it is not easy to construct $\hat{L}$. However, let us consider the subalgebra $L_+\mathfrak{g}$ of $L\mathfrak{g}$. The cocycle (3.1) vanishes on $L_+\mathfrak{g}$, so the restriction of s to $L_+\mathfrak{g}$ is a Lie algebra homomorphism. Since the group L_+ is simply connected, we expect there to be a corresponding splitting $s : L_+ \to \hat{L}$ of groups, giving a canonical holomorphic trivialization of the part of the fibration (3.2) lying over L_+. We admit this fact, which, strictly speaking, would need to be proved using some explicit construction of $\hat{L}$. Similar remarks apply to the subgroup L_- of L, and, *a fortiori*, to the smaller subgroups M_- and M_+. More generally, suppose γ belongs to the big cell in L, that is, that $\gamma = \gamma_-\gamma_+$ with $\gamma_- \in M_-$, $\gamma_+ \in L_+$. Then if we set

$$s(\gamma) = s(\gamma_-)s(\gamma_+)$$

we get a canonical trivialization of the fibration (3.2) over the big cell in L. We shall call the dense open subset of $\hat{L}$ that lies over the big cell of L

the big cell of $\hat{L}$. Thus any element $\hat{\gamma}$ in the big cell of $\hat{L}$ can be written uniquely in the form $\hat{\gamma} = \lambda s(\gamma)$, where $\lambda \in \mathbb{C}^\times$ belongs to the centre of $\hat{L}$ and $\gamma \in L$. We shall call λ the *central component* of $\hat{\gamma}$. The function σ can now be characterized as follows: *σ is the unique holomorphic function on $\hat{L}$ that assigns to an element of the big cell of $\hat{L}$ its central component.* This characterization of σ will suffice for our purposes: but we should note that it is unsatisfactory because it is not clear from what we have said that σ is a holomorphic function on the whole of $\hat{L}$, rather than just on the big cell. To see that we should have to define it in a different way, for example in the way sketched at the beginning of this section, or as a matrix coefficient in the basic representation of $\hat{L}$ (cf. [7], ch. 11).

We record a few simple properties of the function σ. The first two follow immediately from the definition above.

Lemma 3.3. *If $\lambda \in \mathbb{C}^\times$ belongs to the centre of $\hat{L}$, then*

$$\sigma(\lambda\hat{\gamma}) = \lambda\sigma(\hat{\gamma}) \quad \textit{for any } \hat{\gamma} \in \hat{L}.$$

Lemma 3.4. *The function σ is invariant with respect to left multiplication by elements of the subgroup $s(L_-)$ of $\hat{L}$, and with respect to right multiplication by elements of $s(L_+)$.*

The next lemma is the origin of the formula (1.3) that we are studying.

Lemma 3.5. *Let $\phi = \phi_-\phi_+$ belong to the big cell of L, and let $\hat{\phi}$ be any element of $\hat{L}$ lying over ϕ. Then for any $\hat{\psi} \in \hat{L}$ we have*

$$\sigma(\hat{\psi}s(\phi_-)) = \sigma(\hat{\psi}\hat{\phi})/\sigma(\hat{\phi}).$$

Proof. By (3.3), the right-hand side does not depend on the choice of $\hat{\phi}$, since the possible choices differ only by a factor in the centre of $\hat{L}$. Thus we may choose $\hat{\phi} = s(\phi)$, so that $\sigma(\hat{\phi}) = 1$. But then, using (3.4), we have

$$\sigma(\hat{\psi}\hat{\phi}) = \sigma(\hat{\psi}s(\phi_-)s(\phi_+)) = \sigma(\hat{\psi}s(\phi_-)),$$

as stated.

To end this section, we formulate the main lemma of the paper. Let $\gamma \in M_-$, so that (at least formally, see the end of §2) we can write

$$\gamma(z) = \exp\left(\sum_1^\infty \gamma_i z^{-i}\right), \quad \gamma_i \in \mathfrak{g}. \tag{3.6}$$

We decompose γ_1 in the form $\gamma_1 = X + \sum_\alpha \gamma_1^\alpha e_\alpha$, where $X \in \mathfrak{h}$, and e_α is our chosen basis for the root space of α. Thus for each root α we have a map $A_\alpha : M_- \to \mathbb{C}$ defined by

$$A_\alpha(\gamma) = \gamma_1^\alpha.$$

On the other hand, if $h_\alpha \in \mathfrak{h}$ is the coroot corresponding to α, then we have a homomorphism ψ_α from $\mathbb{C}^\times$ to H defined by $\psi_\alpha(z) = \exp((\log z)h_\alpha)$. Here of course H denotes the Cartan subgroup of G corresponding to $\mathfrak{h}$. Restricting to $S^1 \subset \mathbb{C}^\times$, we may regard ψ_α as a loop in H, hence as an element of L. Then there is a one-dimensional space of elements $\hat{\psi}_\alpha$ of $\hat{L}$ lying over ψ_α. Fixing one of them, we define a map $B_\alpha : M_- \to \mathbb{C}$ by

$$B_\alpha(\gamma) = \sigma(\hat{\psi}_\alpha s(\gamma)).$$

Lemma 3.7. *Suppose that α is a long root (in particular, if $\mathfrak{g}$ is simply laced, then α can be any root). Then for a suitable choice of $\hat{\psi}_\alpha$ we have $A_\alpha(\gamma) = B_\alpha(\gamma)$ for all $\gamma \in M_-$.*

4. τ-Functions and the $\mathfrak{g}$AKNS Equations

It is probably not wise to attempt to define the notion of a τ-function; in practice it means a function obtained by restricting one of the functions like σ to a suitable many-parameter subspace of $\hat{L}$, usually some coset of part of a Heisenberg subgroup of $\hat{L}$. Here we define just the τ-functions required to solve the $\mathfrak{g}$AKNS equation defined in §2, corresponding to the choices $\Lambda \in \mathfrak{h}$, $(M, r) \in \mathfrak{h} \times \mathbb{N}$. We fix from now on an element g in the big cell of L, and an element $\hat{g}$ of $\hat{L}$ lying over it. Then we define a function τ of two variables by

$$\tau(x,t) = \sigma(s[\exp(x\Lambda z + tMz^r)]\hat{g}). \tag{4.1}$$

Note that the expression $\exp(x\Lambda z + tMz^r)$ belongs to the subgroup L_+ of L, hence to the big cell, over which the splitting s is defined. We shall need also a τ-function corresponding to each root α of $\mathfrak{g}$. As in §3, let $\psi_\alpha \in LH \subset L$ be the homomorphism defined by α, and $\hat{\psi}_\alpha$ the lifting of it to $\hat{L}$ to be specified in §5 below. Then we define

$$\tau_\alpha(x,t) = \sigma(\hat{\psi}_\alpha s[\exp(x\Lambda z + tMz^r)]\hat{g}). \tag{4.2}$$

The following is the main result of this paper.

Theorem 4.3. *Fix (as above) any g in the big cell of L, and let τ and τ_α be defined by* (4.1) *and* (4.2). *For each root α, set*

$$u_\alpha(x,t) = \alpha(\Lambda)\tau_\alpha(x,t)/\tau(x,t).$$

Suppose that the Lie algebra $\mathfrak{g}$ is simply laced. Then the $\mathfrak{g}$-valued function $u(x,t) = \sum_\alpha u_\alpha(x,t)e_\alpha$ coincides with the solution to the $\mathfrak{g}$AKNS equation described in Proposition 2.7.

In the case when $\mathfrak{g}$ is $\mathfrak{sl}(2,\mathbb{C})$ and $\Lambda = \text{diag}(i,-i)$, this gives the formula (1.3); there τ_+ and τ_- denote the τ-functions (4.2) corresponding to the (unique) positive and negative roots of $\mathfrak{g}$.

Returning to Theorem 4.3 in the general case, notice that u does not depend on the choice of lifting $\hat{g}$ of g, because, by (3.3), a different choice would simply multiply all the τ-functions by the same constant, which would cancel out in the formula for u_α. On the other hand, u does depend on the choice of root vectors e_α, and τ_α depends on the choice of lifting $\hat{\psi}_\alpha$ of ψ_α. In §5 we shall see how to make these choices compatibly, so as to obtain the Theorem.

To prove (4.3) (modulo Lemma 3.7) we have only to combine some of the statements of Sections 2 and 3. In the notation of §3, the solution to the $\mathfrak{g}$AKNS equation in Proposition 2.7 is given by

$$u_\alpha = \alpha(\Lambda)A_\alpha(\phi_-).$$

Using (3.7) and (3.5), we get

$$A_\alpha(\phi_-) = B_\alpha(\phi_-) = \sigma(\hat{\psi}_\alpha s(\phi_-)) = \sigma(\hat{\psi}_\alpha\hat{\phi})/\sigma(\hat{\phi}); \tag{4.4}$$

and as the element $\hat{\phi}$ of $\hat{L}$ lying over ϕ we may choose

$$\hat{\phi} = s[\exp(x\Lambda z + tMz^r)]\hat{g}s[\exp(-x\Lambda z - tMz^r)].$$

The right-hand factor $s[\exp(-x\Lambda z - tMz^r)]$ belongs to $s(L_+)$, hence by (3.4) it can be omitted without changing the last expression in (4.4). That proves the Theorem.

5. Proof of Lemma 3.7

There are two steps in the proof of (3.7): the first is to reduce the problem to the case when γ has the form $\gamma(z) = \exp(\lambda z^{-1}e_\alpha)$, the second is to prove the lemma in that case. Both steps use the hypothesis that α is a long root. We observe first:

Lemma 5.1. *The map $A_\alpha : M_- \to \mathbb{C}$ is a homomorphism.*

Proof. This follows from the fact that the map $\gamma \mapsto \gamma_1$ is a homomorphism from M_- to $\mathfrak{g}$ (where γ_1 is as in (3.6)). Indeed, set $k = z^{-1}$, so that

$$\gamma_1 = \left(\frac{d\gamma}{dk}\gamma^{-1}\right)\bigg|_{k=0}.$$

Then we have

$$\frac{d(\gamma\eta)}{dk}(\gamma\eta)^{-1} = \frac{d\gamma}{dk}\gamma^{-1} + \gamma\left(\frac{d\eta}{dk}\eta^{-1}\right)\gamma^{-1},$$

and $\gamma(k)$ is the identity when $k = 0$; hence $(\gamma\eta)_1 = \gamma_1 + \eta_1$, as stated.

Now let M_0 be the kernel of the homomorphism A_α, and let M_α be the one dimensional subgroup $\{\exp(\lambda z^{-1}e_\alpha)\colon \lambda \in \mathbb{C}\}$ of M_-.

Lemma 5.2. *Each $\gamma \in M_-$ has a unique decomposition $\gamma = \gamma_0\gamma_\alpha$ with $\gamma_0 \in M_0$, $\gamma_\alpha \in M_\alpha$.*

Proof. It is obvious that $M_0 \cap M_\alpha = \{1\}$, hence the uniqueness. Given $\gamma \in M_-$, let $\lambda = A_\alpha(\gamma)$; then we have

$$\gamma = [\gamma\, \exp(-\lambda z^{-1}e_\alpha)]\exp(\lambda z^{-1}e_\alpha).$$

By (5.1), the factor inside the square brackets belongs to M_0.

It follows from (5.1) and (5.2) that $A_\alpha(\gamma) = A_\alpha(\gamma_\alpha)$. We now show that the same is true for the map B_α. We have

$$B_\alpha(\gamma) = \sigma(\hat{\psi}_\alpha s(\gamma)) = \sigma(\hat{\psi}_\alpha s(\gamma_0)s(\gamma_\alpha)) = \sigma(\hat{\psi}_\alpha s(\gamma_0)\hat{\psi}_\alpha^{-1}\hat{\psi}_\alpha s(\gamma_\alpha)).$$

Because of (3.4), the desired result $B_\alpha(\gamma) = B_\alpha(\gamma_\alpha)$ follows from the next lemma.

Lemma 5.3. *For any $\gamma_0 \in M_0$, we have $\hat{\psi}_\alpha s(\gamma_0)\hat{\psi}_\alpha^{-1} \in s(L_-)$.*

Proof. Note first that since the centre of $\hat{L}$ acts trivially by conjugation, $\hat{\psi}_\alpha s(\gamma_0)\hat{\psi}_\alpha^{-1}$ does not depend on the choice of $\hat{\psi}_\alpha$ lying over $\psi_\alpha \in L$; we shall denote it loosely by $\mathrm{Ad}\,\psi_\alpha(s(\gamma_0))$. Although the exponential map from $M_-\mathfrak{g}$ to M_- is not in general surjective, it is certainly true that the image of $M_-\mathfrak{g}$ contains a neighbourhood of the identity in M_-, hence generates M_-. So it is enough to prove the lemma in the case when γ_0 has the form $\gamma_0 = \exp q(z)$, where $q(z) = \sum_1^\infty q_i z^{-i} \in M_-\mathfrak{g}$, and the coefficient of

e_α in the root space decomposition of q_1 is zero. Since the adjoint action of a group commutes with the exponential map, it is enough to show that $\mathrm{Ad}\,\psi_\alpha s(q)$ belongs to the subalgebra $s(L_-\mathfrak{g}) \equiv L_-\mathfrak{g} \oplus 0$ of $\hat{L}\mathfrak{g} = L\mathfrak{g} \oplus \mathbb{C}$. But we have the formula (see [7], p. 44)

$$\mathrm{Ad}\,\psi_\alpha(q,0) = (\psi_\alpha q \psi_\alpha^{-1}, \mathrm{res}_{z=0}\langle \psi_\alpha^{-1}\psi_\alpha', q\rangle),$$

where $\psi_\alpha' \equiv d\psi_\alpha/dz$. Since $\psi_\alpha^{-1}\psi_\alpha' = z^{-1}h_\alpha$ and q involves only negative powers of z, the central term in this expression is indeed zero. It remains to see that the Fourier expansion of $\mathrm{Ad}\,\psi_\alpha(q) \in L\mathfrak{g}$ involves no positive powers of z. Now, $\mathrm{Ad}\,\psi_\alpha$ acts trivially on $\mathfrak{h}$, and if β is any root of $\mathfrak{g}$, the action of $\mathrm{Ad}\,\psi_\alpha$ on the corresponding root space is given by

$$\begin{aligned}\mathrm{Ad}\,\psi_\alpha(e_\beta) &= \mathrm{Ad}\,\exp((\log z)h_\alpha)e_\beta \\ &= \exp\,\mathrm{ad}((\log z)h_\alpha)e_\beta \\ &= \exp((\log z)\beta(h_\alpha))e_\beta \\ &= z^{\beta(h_\alpha)}e_\beta.\end{aligned}$$

Hence the lemma will be true provided that $\beta(h_\alpha) < 2$ for all roots $\beta \neq \alpha$. But from the elementary theory of root systems, we have

$$\beta(h_\alpha) = 2(\alpha,\beta)/(\alpha,\alpha)$$

where $(\ ,\ )$ is a Weyl group invariant inner product on $\mathfrak{h}^*$. So the condition we want holds exactly when α is a long root.

Obviously, $A_\alpha(\exp(\lambda z^{-1}e_\alpha)) = \lambda$; so we have reduced the proof of (3.7) to the following lemma.

Lemma 5.4. *For a suitable choice of $\hat{\psi}_\alpha \in \hat{L}$ covering ψ_α, we have*

$$\sigma(\hat{\psi}_\alpha s(\exp(\lambda z^{-1}e_\alpha))) = \lambda \quad \textit{for any } \lambda \in \mathbb{C}.$$

The main step in the proof of (5.4) is to describe the "suitable choice" of $\hat{\psi}_\alpha$. For that we shall use the two embeddings i_α and j_α of $\mathfrak{sl}(2)$ into $L\mathfrak{g}$ defined as follows. Let $f_\alpha \in \mathfrak{g}$ be the root vector for $-\alpha$ such that $[e_\alpha, f_\alpha] = h_\alpha$ is the coroot corresponding to α; and let

$$e = \begin{bmatrix} 0 & 1 \\ 0 & 0 \end{bmatrix}, \quad h = \begin{bmatrix} 1 & 0 \\ 0 & -1 \end{bmatrix}, \quad f = \begin{bmatrix} 0 & 0 \\ 1 & 0 \end{bmatrix}$$

be the standard basis for $\mathfrak{sl}(2)$. Then $i_\alpha : \mathfrak{sl}(2) \to \mathfrak{g} \subset L\mathfrak{g}$ is the usual embedding associated with the root α, defined by

$$i_\alpha(e) = e_\alpha, \quad i_\alpha(f) = f_\alpha, \quad i_\alpha(h) = h_\alpha;$$

and $j_\alpha : \mathfrak{sl}(2) \to L\mathfrak{g}$ is defined by

$$j_\alpha(e) = zf_\alpha, \quad j_\alpha(f) = z^{-1}e_\alpha, \quad j_\alpha(h) = -h_\alpha.$$

We denote the induced maps of groups from SL(2) to L by the same symbols i_α and j_α. Let

$$w = \begin{bmatrix} 0 & 1 \\ -1 & 0 \end{bmatrix}$$

be the usual representative for the non-trivial element of the Weyl group of SL(2).

Lemma 5.5. *In L we have $\psi_\alpha = i_\alpha(w)j_\alpha(w)$.*

The proof comes down to a calculation in LSL(2), and is based on the formula

$$w = \begin{bmatrix} 1 & 0 \\ -1 & 1 \end{bmatrix} \begin{bmatrix} 1 & 1 \\ 0 & 1 \end{bmatrix} \begin{bmatrix} 1 & 0 \\ -1 & 1 \end{bmatrix} = \exp(-f)\exp(e)\exp(-f).$$

So if j is the embedding of $\mathfrak{sl}(2)$ in $L\mathfrak{sl}(2)$ defined by

$$j(e) = zf, \quad j(f) = z^{-1}e, \quad j(h) = -h,$$

then we have

$$j(w) = \begin{bmatrix} 1 & -z^{-1} \\ 0 & 1 \end{bmatrix} \begin{bmatrix} 1 & 0 \\ z & 1 \end{bmatrix} \begin{bmatrix} 1 & -z^{-1} \\ 0 & 1 \end{bmatrix} = \begin{bmatrix} 0 & -z^{-1} \\ z & 0 \end{bmatrix}$$

so that

$$wj(w) = \begin{bmatrix} z & 0 \\ 0 & z^{-1} \end{bmatrix} = \exp((\log z)h).$$

Applying the map i_α to this, we obtain the lemma.

We can now define liftings $\hat{i}_\alpha$ and $\hat{j}_\alpha$ of i_α and j_α to embeddings of $\mathfrak{sl}(2)$ into $\hat{L}\mathfrak{g}$, as follows. First, $\hat{i}_\alpha$ is just the composition si_α of i_α with the splitting $s : L\mathfrak{g} \to \hat{L}\mathfrak{g}$. This is indeed a homomorphism, since i_α takes values in $\mathfrak{g} \subset L\mathfrak{g}$, and the cocycle (3.1) vanishes on $\mathfrak{g}$. To obtain $\hat{j}_\alpha$ we lift the images of e and f using s; that then determines $\hat{j}_\alpha(h)$. For simplicity, for the rest of the paper we shall use the same notation for an element X of $L\mathfrak{g}$ and for the corresponding element $s(X) = X \oplus 0$ of $\hat{L}\mathfrak{g}$; then we have $\hat{j}_\alpha(e) = zf_\alpha$, $\hat{j}_\alpha(f) = z^{-1}e_\alpha$, and

$$\hat{j}_\alpha(h) = -h_\alpha + \langle e_\alpha, f_\alpha \rangle. \tag{5.6}$$

We denote the maps of groups from SL(2) to $\hat{L}$ determined by $\hat{\imath}_\alpha$ and $\hat{\jmath}_\alpha$ by the same symbols; then the choice of lifting $\hat{\psi}_\alpha$ to be used in (5.4) is

$$\hat{\psi}_\alpha = \hat{\imath}_\alpha(w)\hat{\jmath}_\alpha(w).$$

Now, $\hat{\imath}_\alpha(w)$ belongs to the subgroup $s(G) \subset s(L_-)$ of $\hat{L}$; hence by (3.4), Lemma 5.4 reduces to the assertion that

$$\sigma(\hat{\jmath}_\alpha(w)\exp(\lambda z^{-1}e_\alpha)) = \lambda \quad \text{for any } \lambda \in \mathbb{C}. \tag{5.7}$$

But since $\exp(\lambda z^{-1}e_\alpha) = \hat{\jmath}_\alpha \exp(\lambda f)$, the proof of this is just a calculation in the group $\hat{\jmath}_\alpha\mathrm{SL}(2)$. In SL(2) we have (for $\lambda \neq 0$)

$$w\exp(\lambda f) = \exp(-\lambda^{-1}f)\exp((\log\lambda)h)\exp(\lambda^{-1}e),$$

so applying the map $\hat{\jmath}_\alpha$, we find

$$\hat{\jmath}_\alpha(w)\exp(\lambda z^{-1}e_\alpha) = \exp(-\lambda^{-1}z^{-1}e_\alpha)\exp((\log\lambda)\hat{\jmath}_\alpha(h))\exp(\lambda^{-1}zf_\alpha).$$

Since the first and last factors here belong to $s(L_-)$ and $s(L_+)$ (respectively), we get

$$\sigma(\hat{\jmath}_\alpha(w)\exp(\lambda z^{-1}e_\alpha)) = \sigma(\exp((\log\lambda)\hat{\jmath}_\alpha(h))). \tag{5.8}$$

But since α is a long root, we have

$$2 = \langle h_\alpha, h_\alpha\rangle = \langle h_\alpha, [e_\alpha, f_\alpha]\rangle = \langle f_\alpha, [h_\alpha, e_\alpha]\rangle = \langle f_\alpha, 2e_\alpha\rangle,$$

so that $\langle e_\alpha, f_\alpha\rangle = 1$. Hence from (5.6) we get

$$\exp((\log\lambda)\hat{\jmath}_\alpha(h)) = \exp(-(\log\lambda)h_\alpha) \times \exp(\log\lambda) \in H \times \mathbb{C}^\times.$$

The central component of this is $\exp(\log\lambda) = \lambda$. In view of (5.7) and (5.8), that completes the proof.

REFERENCES

1. M.J. Ablowitz, D.J. Kaup, A.C. Newell and H. Segur, *The inverse scattering transform: Fourier analysis for nonlinear problems*, Studies in Appl. Math. **53** (1974), 249–315.
2. M.J. Bergveld and A.P.E. ten Kroode, *τ-functions and zero curvature equations of Toda-AKNS type*, J. Math. Phys. **29** (1988), 1308–1320.

3. L.A. Dickey, *On Segal–Wilson's definition of the τ-function and hierarchies AKNS-D and mcKP*, this volume.
4. V.G. Drinfel'd and V.V. Sokolov, *Lie algebras and equations of Korteweg–de Vries type*, Itogi Nauki i Tekhniki, ser. Sovremennye Problemy Matematiki, **24** (1984), 81–180; J. Sov. Math. **30** (1985), 1975–2036.
5. R. Goodman and N.R. Wallach, Erratum to the paper *Structure and ... of the circle*, J. Reine Angew. Math. **347** (1984), 220.
6. R. Hirota, *Exact envelope-soliton solutions of a nonlinear wave equation*, J. Math. Phys. **14** (1973), 805–809.
7. A. Pressley and G. Segal, *Loop Groups*, Clarendon Press, Oxford, 1986.
8. E. Previato, *Hyperelliptic quasi-periodic and soliton solutions of the nonlinear Schrödinger equation*, Duke Math. J. **52** (1985), 329–377.
9. G. Wilson, *The modified Lax and two-dimensional Toda lattice equations associated with simple Lie algebras*, Ergod. Th. and Dynam. Sys. **1** (1981), 361–380.
10. G. Wilson, *Habillage et fonctions τ*, C.R. Acad. Sc. Paris, t. 299, Série I (1984), 587–590.
11. G. Wilson, *Infinite-dimensional Lie groups and algebraic geometry in soliton theory*, Phil. Trans. R. Soc. London A **315** (1985), 393–404.
12. V.E. Zakharov and A.B. Shabat, *Integration of the nonlinear equations of mathematical physics by the inverse scattering method*, II, Funct. Anal. Appl. **13**:3 (1979), 13–22 (Russian), 166–174 (English).

Mathematics Department
Imperial College
London SW7 2BZ
England

On Segal-Wilson's Definition of the τ-Function and Hierarchies AKNS-D and mcKP

L.A. Dickey

1. Segal and Wilson in their well-known article [1] developed Sato's ideas (see e.g. [8]) and devised a construction for solutions to KP hierarchy of integrable equations based on Grassmannians in Hilbert spaces. To any element of a Grassmannian a Baker function of a solution can be assigned which, in turn, enables one to find the solution itself. Segal and Wilson also suggested a construction of τ-functions based on Grassmannians, and a τ-function is connected with the Baker function by the famous Sato formula

$$\hat{w}(t_1, t_2, ...) = \frac{\tau(t_1 - 1/z, t_2 - 1/2z^2, t_3 - 1/3z^3, ...)}{\tau(t_1, t_2, t_3, ...)}. \tag{1.1}$$

This formula determines the significance of τ-functions in the theory of integrable systems.

We present here, first of all, another form of a definition of the τ-function; although it is equivalent to the original Segal-Wilson definition, it is, we believe, easier to handle. However, our main goal is to define the τ-function for hierarchies generated by matrix first order differential operators (AKNS-D hierarchies) and closely connected with them multi-component KP hierarchies (mcKP). The problem of how to define the τ-function for these hierarhies in Segal-Wilson's terms was posed in our paper [2]. We have managed to solve it only after we modified the definition of the τ-function since the new definition fits very well the construction of known examples of solutions (soliton solutions). For these examples it is easy to find a function that can be called a τ-function, and this leads to a general definition.

2. KP hierarchy, soliton solutions and Grassmannians

2.1. Recall a few well-known facts about the KP hierarchy. It is generated by a pseudo-differential operator

$$L = \partial + u_1\partial^{-1} + u_2\partial^{-2} + ..., \quad \partial = d/dx.$$

Let $B_m = (L^m)_+$ where $+$ symbolizes retaining only positive powers of ∂. Let t_k be parameters (time variables). Then the KP hierarchy is the set of equations

$$\partial_k L = [B_m, L], \quad \partial_k = \partial/\partial t_k. \tag{2.1.1}$$

All the equations commute. The variable t_1 can be identified with x.

If we represent L as

$$L = \hat{w}\partial\hat{w}^{-1}$$

where $\hat{w}(\partial) = \sum_0^\infty w_i\partial^{-i}$, $w_0 = 1$ then the *Baker function* is

$$w(t,z) = \hat{w}(\partial)\exp(\sum_1^\infty t_s z^s) = \hat{w}(z)\exp(\sum_1^\infty t_s z^s).$$

We shall also use the notations $\xi(t,z) = \sum_1^\infty t_s z^s, g(t,z) = \exp\xi(t,z)$. This function satisfies the equations

$$Lw = zw, \text{ and } \partial_k w = B_k w.$$

2.2. Recall the notion of infinite-dimensional Grassmannians used by Segal and Wilson. They considered both a very general and abstract notion and a more concrete one; here we shall use the latter.

Let H be the Hilbert space of functions on the unit circle $|z| = 1$: $H = \{\sum_{\infty}^{\infty} v_i z^i\}$ (we do not discuss convergence etc). Let $H_+ = \{\sum_0^\infty v_i z^i\}$ and $H_- = \{\sum_{-\infty}^{-1} v_i z^i\}$, p_+ and p_- denoting projections on H_+ and H_-.

An element of the Grassmannian $W \in \text{Gr}$ is a subspace $W \subset H$ such that when p_+ is restricted to W, $p_+|_W$ is a Fredholm operator, i.e. has a finite-dimensional kernel and cokernel, and $p_-|_W$ is a compact operator or, better, a trace class operator (in our example it will be even finite-dimensional). We take a narrower class of W, namely, we assume that ind $p_+|_W$ = dim ker $p_+|_W-$dim coker $p_+|_W = 0$. Furthermore, we confine ourselves to a generic (the so-called "transversal") case where $p_+|_W$ is a one-to-one correspondence.

Example. Let W consist of series $v(z) = \sum_{-N}^\infty v_i z^i$ satisfying equations

$$v(\alpha_i) - \tilde{a}_i v(\beta_i) = 0,\ i = 1,...,N,\ \tilde{a}_i = a_i(\beta_i/a_i)^N$$

where α_i, β_i and a_i are some complex numbers. Comparing W with H_+ we see that there are N extra terms and N additional equations allowing us to recover $v(z)$ uniquely if p_+v is known (in a generic case).

Let an element $W \in \text{Gr}$ be given. Then there is a unique element of W having the form $w(t,z) = \sum_0^\infty w_s(t)z^{-s}\exp\xi(t,z) \in W$, $w_0 = 1$, where t is a parameter. It can be proven (see [1] or [3]) that this is a Baker function of the KP hierarchy. To emphasize its dependence on the choice of W it can be denoted by $w_W(t,z)$.

Let us turn to our example. According to definitions, the Baker function is a function $w(t,z)=\sum_0^N w_s z^{-s}\exp\xi(t,z)$ satisfying the equations

$$\sum_0^N w_s[\alpha_i^{(N-s)}\exp\xi(t,\alpha_i)-a_i\beta_i^{(N-s)}\exp\xi(t,\beta_i)]=0$$

or $\sum_0^N w_s\partial^{N-s}y_i(t)=0,\ i=1,...,N,\ w_0=1$, where $y_i(t)=\exp\xi(t,\alpha_i)-a_i\exp\xi(t,\beta_i)$. Let us write this equation in matrix form:

$$(w_N,...,w_0)\cdot\begin{pmatrix} y_1 & \cdots & y_N \\ \cdots & \cdots & \cdots \\ y_1^{(N-1)} & \cdots & y_N^{(N-1)} \\ y_1^{(N)} & \cdots & y_N^{(N)} \end{pmatrix}=0.$$

Then $w_N,...,w_0$ must be proportional to signed minors of this $(n+1)\times n$ matrix. Therefore, $\sum_0^N w_s z^{N-s}=z^N\hat{w}(t,z)$ is the determinant

$$\begin{vmatrix} y_1 & \cdots & y_N & 1 \\ \cdots & \cdots & \cdots & \cdots \\ y_1^{(N)} & \cdots & y_N^{(N)} & z^N \end{vmatrix}$$

which must be divided by $w_0=\Delta$. Thus,

$$\hat{w}(t,z)=\frac{z^{-N}}{\Delta}\begin{vmatrix} y_1 & \cdots & y_N & 1 \\ \cdots & \cdots & \cdots & \cdots \\ y_1^{(N)} & \cdots & y_N^{(N)} & z^N \end{vmatrix}. \tag{2.2.1}$$

2.3. Let us transform Eq.(2.2.1) by making zeros appear in the last column.

$$\hat{w}(t,z)=\frac{z^{-N}}{\Delta}\begin{vmatrix} y_1-z^{-1}y_1' & \cdots & y_N-z^{-1}y_N' & 0 \\ \cdot\cdot & \cdot\cdot & \cdot\cdot & \cdot\cdot \\ y_1^{(N-1)}-z^{-1}y_1^{(N)} & \cdots & y_N^{(N-1)}-z^{-1}y_N^{(N)} & 0 \\ y_1^{(N)} & \cdots & y_N^{(N)} & z^N \end{vmatrix}.$$

We have

$$y_i^{(j)}-z^{-1}y_i^{(j+1)}=(1-\frac{\alpha_i}{z})\alpha_i^j\exp\xi(t,\alpha_i)-a_i(1-\frac{\beta_i}{z})\beta_i^j\exp\xi(t,\beta_i).$$

Using $(1-\alpha_i/z)=\exp\ln(1-\alpha_i/z)=\exp(-\sum_1^\infty\alpha_i^s/sz^s)$ we can write this difference as $y_i^{(j)}(t_1-1/z,t_2-1/2z^2,t_3-1/3z^3,...)$. Now

$$\hat{w}(t,z)=\frac{\Delta(t_1-1/z,t_2-1/2z^2,t_3-1/3z^3,...)}{\Delta(t_1,t_2,t_3,...)}.$$

If we set $\tau(t) = \Delta(t)$ we will get precisely Eq(1.1).

2.4. We have seen that the τ-function in our example is the determinant of the matrix

$$Y = \begin{pmatrix} y_1 & \dots & y_N \\ \dots & \dots & \dots \\ y_1^{(N-1)} & \dots & y_N^{(N-1)} \end{pmatrix}. \tag{2.4.1}$$

The general idea of Segal and Wilson is that a τ-function is the determinant of some mapping in the Grassmannian. They suggest the following mapping. If $W \in \mathrm{Gr}$ is an element determining the Baker function $w_W(t,z)$, then they consider the combined mapping

$$H_+ \overset{(p_+|_W)^{-1}}{\longrightarrow} W \overset{\beta g^{-1}}{\longrightarrow} Wg^{-1} \overset{p_+}{\to} H_+ \overset{\mathbf{g}}{\to} H_+$$

where $\mathbf{g}$ is the transformation in H multiplying any element $v(z)$ by g, i.e.

$$\tau_W(t) \equiv \tau_W(g) = \det[\mathbf{g} \circ p_+ \circ \mathbf{g}^{-1} \circ (p_+|_W)^{-1}]. \tag{2.4.2}$$

It can be proven (see [1] or [3]) that this $\tau_W(t)$ and the above defined $w_W(t,z)$ are connected by Eq(1.1).

2.5. We have seen in our example that the τ-function is the determinant of some matrix Y. The same idea that a τ-function must be a determinant of some important mapping in the Grassmannian inspired us to look for a mapping with the matrix Y which can be formulated in terms of the Grassmannian. This resulted in the following:

Definition. Let W be a generic element of the Grassmannian (in the above sense). Let $A : H_+ \to H_-$ be a mapping with the graph W, i.e. $A = p_-(p_+|_W)^{-1}$. Let $l_W : H \to H_-$ denote projection parallel to W, i.e. $l_W = p_- - Ap_+$. Consider the mapping $l_W \circ \mathbf{g} : H_- \to H_-$. Then

$$\tau_W(t) \equiv \tau_W(g) = \det(l_W \circ \mathbf{g}). \tag{2.5.1}$$

Remark. If we consider the same mapping on the H_- extended by constants: $\bar{H}_- = \{\sum_{-\infty}^{0} v_i z^i\}$, then the Baker function $\hat{w}_W$ can be defined as an element of the (one-dimensional) kernel of the mapping $l_W \circ \mathbf{g} : \bar{H}_- \to H_-$ (this element must be normalized). And the definition of the τ-function means that this is in a sense a measure of deviation of this mapping from the projector $\bar{H}_- \to H_-$. Both functions have closely connected definitions.

2.6. Proposition. *Definitions (2.4.2) and (2.5.1) of the τ-function are equivalent.*

Proof. The main difficulty in this proof is in the fact that we must compare two mappings in different spaces, H_+ in the first case and H_- in the second. Let $g^{-1}(t,z) = \sum_0^\infty Q_s z^s$, $Q_0 = 1$ be the expansion in z (the coefficients Q_s are the so-called Schur polynomials in $-t$).

We start with the second mapping. It sends the base elements to

$$z^{-k} \mapsto (gz^{-k})_- - A(gz^{-k})_+, \ k = 1, 2, ...$$

The addition to any line of a linear combination of the previous lines does not affect the determinant. Therefore, the following mapping must have the same determinant:

$$z^{-k} \mapsto (gz^{-k}\sum_0^{k-1} Q_s z^s)_- - A(gz^{-k}\sum_0^{k-1} Q_s z^s)_+$$

$$= (gz^{-k}(\sum_0^{k-1} Q_s z^s - g^{-1}))_- + z^{-k} - A(gz^{-k}\sum_0^{k-1} Q_s z^s)_+$$

$$= z^{-k} - A(g\sum_0^{k-1} Q_s z^{-k+s})_+. \tag{2.6.1}$$

The definition (2.4.2) is connected with the mapping

$$x \in H_+ \mapsto g((x + Ax)\cdot g^{-1})_+ = x + g(Ax \cdot g^{-1})_+ \in H_+. \tag{2.6.2}$$

The mapping is a sum of the identity mapping and a contracting (more precisely, trace class) operator: $Tx = g(Ax \cdot g^{-1})_+$. If Ax is expanded in the basis of H_-, i.e. in z^{-k}, then the image of T will be contained in the span of

$$f_k = g(z^{-k}g^{-1})_+ \in H_+.$$

For example, in the case of soliton solutions it is enough to take a finite number of f_k, $k = 1, ..., N$, and T is a finite dimensional mapping.

Now it is clear that $\det(I + T)$ can be calculated by restricting $I + T$ to the span of f_k. We have

$$f_k \mapsto g(Af_k \cdot g^{-1})_+ + f_k.$$

If $Af_k \in H_-$ is expanded in the basis $Af_k = \sum_{j=1}^\infty a_{kj} z^{-j}$, then the above mapping is $f_k \mapsto f_k + \sum_{j=1}^\infty a_{kj} f_j$, i.e. the determinant we are looking for is $\det(a_{ij} + \delta_{ij})$. Now, the following mapping has the same determinant,

$$H_- \to H_- : \ z^{-k} \mapsto z^{-k} + Af_k = z^{-k} + Ag(z^{-k}g^{-1})_+$$

(observe that here we have passed to mappings in H_-). It remains to transform:

$$z^{-k} + Ag(z^{-k}g^{-1})_+ = z^{-k} + Ag(g^{-1} - \sum_0^{k-1} Q_s z^s)z^{-k}$$

$$= z^{-k} + A(1 - \sum_{0}^{k-1} Q_s z^s g) z^{-k} = z^{-k} - A \sum_{0}^{k-1} (Q_s z^{-k+s} g)_+$$

which coincides with (2.6.1). □

2.7. As an example of how easy is it to work with this new definition, let us prove

Proposition. *The Baker function $\hat{w}_W(t,z)$ and the τ-function are connected by Eq.(1.1).*

Proof. The plan of the proof will be the same as in [1]; first we prove

Lemma. *For two different transformation* $\mathbf{g}$ *and* $\mathbf{g}_1$ *we have*

$$\tau_W(gg_1) = \tau_W(g)\tau_{Wg^{-1}}(g_1).$$

Then we apply this formula for $g_1(t,z) = 1 - z/\zeta$ where ζ is a parameter.

Proof of the lemma. $\mathrm{T}_W(g)$ will denote the restriction of the operator $l_W \circ g$ to H_-; thus $\tau_W(g) = \det \mathrm{T}_W(g)$. Let $y \in H_-$. Then we can decompose yg_1 (in accordance with $H = H_- \oplus \mathbf{g}^{-1}W$) as $yg_1 = wg^{-1} + y_1$ where $w \in W$ and $y_1 \in H_-$. This means that $y_1 = \mathrm{T}_{Wg^{-1}}$. Multiply the equality just obtained by g; we have $yg_1g = w + y_1g$. Again we decompose $y_1g = w_1 + y_2$ where $w_1 \in W$ and $y_2 \in H_-$ which means $y_2 = \mathrm{T}_W(g)y_1$. Now $yg_1g = w + w_1 + \mathrm{T}_W(g)\mathrm{T}_{Wg^{-1}}(g_1)$ which means that

$$\mathrm{T}_W(g_1g) = T_W(g)T_{Wg^{-1}}(g_1).$$

Thus we have obtained the required formula not only for the determinants but even for the mappings themselves, which also implies the correctness for determinants. And this formula follows directly from the definitions. □

Now let $g_1 = 1 - z/\zeta$, $|\zeta| > 1$. Then $gg_1 = \exp \sum_1^\infty t_s z^s \exp \ln(1 - z/\zeta) = \exp \sum_1^\infty (t_s - 1/s\zeta^s) z^s$, i.e. multiplication of g by g_1 corresponds to the translation of arguments $(t_1, t_2, ...) \mapsto (t_1 - 1/\zeta, t_2 - 1/2\zeta^2, ...)$. It remains for us to prove that $\tau_{Wg^{-1}}(g_1)$ is equal to $\hat{w}_W(t,\zeta)$.

To this end we calculate the mapping $\mathrm{T}_{Wg^{-1}}(g_1)$ on the basis. For z^{-k}, $k \geq 2$ we have that $z^{-k}(1 - z/\zeta) \in H_-$. Thus

$$\mathrm{T}_{Wg^{-1}}(g_1)z^{-k} = z^{-k} - \frac{z^{-k+1}}{\zeta}, \quad k \geq 2.$$

For $k = 1$ we have $z^{-1}(1 - z/\zeta) = z^{-1} - 1/\zeta$. In order to obtain the projection to H_- parallel to Wg^{-1}, recall that $\hat{w}_W(t,z) \in Wg^{-1}$ and

$\hat{w}_W - 1 \in H_-$. We write:

$$z^{-1}g_1 = z^{-1} - \frac{1}{\zeta}\hat{w}_W(z) + \frac{1}{\zeta}(\hat{w}_W(z) - 1)$$

and $T_{W_{g^{-1}}}(g_1)z^{-1} = z^{-1} + (1/\zeta)\sum_1^\infty w_i z^{-i}$ where w_i are the coefficients of the expansion of the Baker function $\hat{w}$.

The matrix of this mapping is

$$\begin{pmatrix} 1+w_1/\zeta & w_2/\zeta & w_3/\zeta & \dots \\ -1/\zeta & 1 & 0 & \dots \\ 0 & -1/\zeta & 1 & \dots \\ \dots & \dots & \dots & \dots \end{pmatrix}$$

The determinant can easily be calculated and it is equal to $1 + w_1/\zeta + w_2/\zeta^2 + w_3/\zeta^3 + \dots = \hat{w}_W(t, \zeta)$, as required. □

3. AKNS-D hierarchies

3.1. Let

$$L = \partial + U - zA, \quad A = diag(a_1, ..., a_n) = const, \ U = (u_{\alpha\beta}), \ u_{\alpha\alpha} = 0$$

with distinct a_α. *Resolvents* are series $R = \sum_{i_0}^\infty R_j z^{-j}$ that commute with L, i.e. $[L, R] = 0$. Their elements are differential polynomials in elements $u_{\alpha\beta}$ of the matrix U. Resolvents form an n-dimensional commutative algebra over the field of constant diagonal series $C(z) = \sum_{i_1}^\infty C_i z^{-1}$. A basis is formed of n resolvents, $R_\alpha = \sum_0^\infty R_{j\alpha} z^{-j}$, where $R_{0\alpha} = E_\alpha$, and E_α is the matrix with the only non-vanishing element 1, at the $\alpha\alpha$ place, and all elements of matrices R_j, $j > 0$, are differential polynomials without constants. (For the proof of existence and uniqueness, see [3], Ch.9). The basic resolvents satisfy the relation $R_\alpha R_\beta = \delta_{\alpha\beta} R_\alpha$. Let

$$B_{k\alpha} = (z^k R_\alpha)_+ = \sum_{j=0}^k R_{j\alpha} z^{k-j}.$$

The subscript + means taking non-negative powers of z. The hierarchy is the set of equations

$$\partial_{k\alpha} L = [B_{k\alpha}, L]$$

where $\partial_{k\alpha}$ means $\partial/\partial t_{k\alpha}$, and $t_{k\alpha}$, $k = 0, 1, 2, ...$, $\alpha = 1, ..., n$, is a set of variables, called the "times".

Remark (see [3] or [5]). Variables x and $t_{k\alpha}$ are not independent, in fact

$$\partial = \sum_\alpha a_\alpha \partial_{1\alpha}.$$

3.2. *Formal Baker function.* Let us represent L in a "dressing" form

$$L = w\partial w^{-1}, \ \ R_\alpha = wE_\alpha w^{-1},$$

where

$$w = \hat{w}(z)\exp(\sum_{k=0}^{\infty}\sum_{\alpha=1}^{n} z^k E_\alpha t_{k\alpha}), \ \hat{w}(z) = \sum_0^\infty w_j z^{-j}, \ w_0 = I.$$

The equations of the hierarchy are equivalent to

$$\partial_{k\alpha} w = B_{k\alpha} w, \text{ or } \partial_{k\alpha}\hat{w} = -(z^k R_\alpha)_-\hat{w}.$$

The equation of the dressing can also be written as $L(w) = 0$ where $L(w)$ means the action of the operator L on the function w in contrast to the notation Lw which means the product of two operators, of the first and of the zero-order, respectively.

The function w is called the wave Baker function.

3.3. *Definition of the Grassmannian.* Let H be $L^2(S, \mathbf{C}^n)$, a space of series $v(z) = \sum_{-\infty}^{\infty} v_k z^k$ where $v_k \in \mathbf{C}^n$, $|z| = 1$; let H_+ and H_- be the spaces of truncated series $H_+ = \{\sum_0^\infty\}$, $H_- = \{\sum_{-\infty}^{-1}\}$; $H = H_+ \oplus H_-$. Let $p_\pm$ be the natural projections of H onto these subspaces. We shall think of vectors as vector-rows.

The Grassmannian Gr is the set of all subspaces $W \subset H$ possessing properties i) $p_+|_W$ is a one-to-one correspondence (in the generic case), and ii) $zW \subset W$.

(Following Wilson [4] we can prove that this Grassmannian is isomorphic to the Grassmannian $\mathrm{Gr}^{(n)}$ of scalar functions; we shall not use this fact now.)

We say that a matrix function belongs to W if all its rows do. Let $g(t,z) = \exp\xi = \exp\sum_{k=0}^{\infty}\sum_{\alpha=1}^{n} z^k E_\alpha t_{k\alpha}$. We consider a transformation $\mathbf{g}^{-1}$ of the space H

$$\mathbf{g}^{-1} : H \to H, \ v \mapsto vg^{-1}.$$

If $W \in \mathrm{Gr}$ then for almost all t the subspace $Wg^{-1} \in \mathrm{Gr}$.

3.4. *Baker function.* For a $W \in \mathrm{Gr}$, $w_W(t,z)$ is a Baker function if i) for any $t = (t_1, t_2, ...)$, $w_W \in W$ as a function of z, ii) $p_+(w_W g^{-1}) = 1$. Together the conditions mean that $w_W(t,z)$ is the only element of W of the form

$$w_W(t,z) = (1 + \sum_1^\infty w_i z^{-i})g(t,z) \equiv \hat{w}_W(t,z)g(t,z).$$

3.5. Proposition. *The Segal-Wilson (S-W) Baker function $w_W(t,z)$ is a Baker function in the e sense of Section 3.2.*

Proof. Let $\partial = \sum_1^n a_\alpha \partial_{1\alpha}$. It must be proved that i) $L(w) = 0$ for some $L = \partial + U - zA$, where $w = w_W(t,z)$ and ii) $\partial_{k\alpha} w = B_{k\alpha} w$, $B_{k\alpha} = (z^k R_\alpha)_+$, and $R_\alpha = w E_\alpha w^{-1}$. From the definition of w_W it follows that $w' w^{-1} = Az - U + (w'w^{-1})_-$ where $U = -(w'w^{-1})_0$. Let $(w'w^{-1})_- = Q$. One must prove that $Q = 0$. We have

$$w' + (U - zA)w = Qw.$$

The left-hand side is an element of W. Then $(w' + (U - zA)w)g^{-1} = Qwg^{-1}$ is an element of Wg^{-1}. The right-hand side is $O(z^{-1})$. This means that $p_+(Qwg^{-1}) = 0$. We recall that p_+ is one-to-one mapping of Wg^{-1} onto H_+. Therefore $w' + (U - zA)w = 0$, i.e. $w'w^{-1} = zA - U$, so i) is proven. Further, for R_α and $B_{k\alpha}$ defined as above, we have

$$(\partial_{k\alpha} - B_{k\alpha})w = (\partial_{k\alpha}\hat{w} + (z^k R_\alpha)_- \hat{w}) \cdot g, \quad (\hat{w} = wg^{-1}).$$

The left-hand side belongs to W. Hence $(\partial_{k\alpha}\hat{w} + (z^k R_\alpha)_- \hat{w})$ is an element of Wg^{-1} whose p_+- projection vanishes (since this is $O(z^{-1})$). Thus

$$\partial_{k\alpha}\hat{w} + (z^k R_\alpha)_- \hat{w} = 0$$

which is equivalent to ii. □

3.6. *An example of elements of the Grassmannian.* Let $m_i, i = 1, \ldots, Nn$, be points inside the unit circle, $|m_i| < 1$. Here N is a natural number ("the soliton number"). Let η_i be Nn vector-columns in $\mathbb{C}^n$ which span this space. Let W be the set of all elements of H (vector-rows) of the form $v(z) = \sum_{-N}^{\infty} v_k z^k$ satisfying the relations $v(m_i) \cdot \eta_i = 0$. This W is for almost all sets $\{m_i, \eta_i\}$ an element of the Grassmannian. The corresponding solutions are solitons.

3.7. We can find an explicit expression for the Baker function in exactly the same way that we did for the KP hierarchy. The result will be the following.

Let $\xi_\beta(t,z) = \sum_{s=1}^{\infty} t_{s\beta} z^s$ and $y_{\beta i} = \exp \xi_\beta(t, m_i) \cdot \eta_{i\beta}$. The $\alpha\beta$th element of the matrix $\hat{w}(z)$ is given by the determinant of the $(Nn+1)$th order,

$$\hat{w}_{\alpha\beta} = \frac{z^{-N}}{\Delta}\left|\begin{array}{cccccc|c}
y_{11} & \dots & \dots & \dots & y_{1,Nn} & & 0 \\
\dots & \dots & \dots & \dots & \dots & & 1 \\
y_{n1} & \dots & \dots & \dots & y_{n,Nn} & & 0 \\
- & - & - & - & - & - & - \\
\partial_{11} y_{11} & \dots & \dots & \dots & \partial_{11} y_{1,Nn} & & 0 \\
\dots & \dots & \dots & \dots & \dots & & z \\
\partial_{1n} y_{n1} & \dots & \dots & \dots & \partial_{1n} y_{n,Nn} & & 0 \\
- & - & - & - & - & - & - \\
\dots & \dots & \dots & \dots & \dots & \dots & \dots \\
- & - & - & - & - & - & - \\
\partial_{11}^{N-1} y_{11} & \dots & \dots & \dots & \partial_{11}^{N-1} y_{1,Nn} & & 0 \\
\dots & \dots & \dots & \dots & \dots & & z^{N-1} \\
\partial_{1n}^{N-1} y_{n1} & \dots & \dots & \dots & \partial_{1n}^{N-1} y_{n,Nn} & & 0 \\
- & - & - & - & - & - & - \\
\partial_{1\alpha}^{N} y_{\alpha 1} & \dots & \dots & \dots & \partial_{1\alpha}^{N} y_{\alpha,Nn} & & z^N \delta_{\alpha\beta}
\end{array}\right|.$$

We have here N blocks of rows containing n rows each, and one separate (the last) row. Non-zero elements of the last column are at the βth place in each block, and also the last element (when $\alpha \neq \beta$). Δ is the minor of the first Nn rows and columns.

3.8. *The τ-function for this example.* We proceed in the same way as for the KP hierarchy, i.e. making zeros appear in the last column. If $\alpha = \beta$ then we can annul all elements of this column except the last one. If $\alpha \neq \beta$ then the last element of the last column is zero, and it is possible to annul all elements of the last column except that of the order z^{N-1}.

An easy calculation shows that

$$\partial_{1\beta}^{j} y_{\beta i} - \frac{1}{z}\partial_{1\beta}^{j+1} y_{\beta i} = \partial_{1\beta}^{j} \exp \sum_{s=1}^{\infty} (t_{s\beta} - \frac{1}{sz^s}) m_i^s \cdot \eta_{i\beta}.$$

Therefore, for diagonal elements we obtain

$$\hat{w}_{\alpha\alpha}(t,z) = \frac{\Delta(\dots, t_{s\alpha} - 1/sz^s, \dots)}{\Delta(t)}; \qquad (3.8.1)$$

the term $1/sz^s$ being only subtracted from the time variables with the second subscript α.

For non-diagonal elements we have

$$\hat{w}_{\alpha\beta}(t,z) = z^{-1} \frac{\Delta_{\alpha\beta}(\dots, t_{s\beta} - 1/sz^s, \dots)}{\Delta(t)}, \qquad (3.8.2)$$

where $\Delta_{\alpha\beta}$ is the algebraic adjunct of the element z^{N-1} of the last column. This can also be described as the minor Δ where the βth row of the last block is replaced by the αth row of the next block (which is not involved in Δ), and the sign must be changed.

Now it is quite natural to consider as the τ-function the following matrix

$$\tau_{\alpha\beta}(t) = \begin{cases} \Delta_{\alpha\beta}(t), & \text{if } \alpha \neq \beta \\ \Delta(t), & \text{if } \alpha = \beta \end{cases} .$$

3.9. *The general definition of the τ-function for the AKNS-D hierarchy.* Just as in the case of KP, the general definition can be obtained from the analysis of the above example.

Let W be an element of the Grassmannian. We introduce the same operator l as before, the projection $H \to H_-$ parallel to W. Then

$$\tau_W(t) \equiv \tau_W(g) = \det \mathrm{T}_w(g), \text{ where } \mathrm{T}_W(g) = l \circ \mathbf{g} : H_- \to H_-. \tag{3.9.1}$$

Let $R_{\alpha\beta} : H_- \to H$ be an operator given by its action on basic elements

$$R_{\alpha\beta} z^{-k} \mathbf{e}_p = \begin{cases} -\mathbf{e}_\alpha, & \text{if } k = 1, p = \beta \\ z^{-k}\mathbf{e}_p & \text{otherwise} \end{cases} .$$

Set

$$\tau_{W\alpha\beta}(t) \equiv \tau_{W\alpha\beta}(g) = \det \mathrm{T}_{W\alpha\beta}(g), \tag{3.9.2}$$

where $\mathrm{T}_{W\alpha\beta}(g) = l \circ \mathbf{g} \circ R_{\alpha\beta} : H_- \to H_-$.

3.10. Proposition. *Elements of the Baker function can be expressed in terms of the τ-function as*

$$\hat{w}_{W\alpha\alpha}(t, z) = \frac{\tau_W(\ldots, t_{s\alpha} - 1/sz^s, \ldots)}{\tau_W(t)}, \tag{3.10.1}$$

and

$$\hat{w}_{W\alpha\beta}(t, z) = z^{-1} \frac{\tau_{W\alpha\beta}(\ldots, t_{s\beta} - 1/sz^s, \ldots)}{\tau_W(t)}, \ \alpha \neq \beta. \tag{3.10.2}$$

Proof. First we prove a lemma which follows directly from the definitions:

Lemma.

$$\mathrm{T}_W^{-1}(g)\mathrm{T}_W(gg_1) = \mathrm{T}_{Wg^{-1}}(g_1). \tag{3.10.3}$$

(The proof does not differ from that in the case of KP, see Section 2.7). Now, let $g_1 = \mathrm{diag}\ (1, \ldots, 1 - z/\zeta, 1, \ldots, 1)$ where $1 - z/\zeta$ is at the αth place. Then gg_1 is the same as g where $t_{s\alpha}$ are replaced by $t_{s\alpha} - 1/s\zeta^s$ but

$t_{s\beta}$ where $\beta \neq \alpha$ remain unchanged. If we take the determinant of both sides of Eq.(3.10.3) then the left-hand side will coincide with the right-hand side of Eq.(3.10.1). The mapping in the right-hand side of Eq.(3.10.3) preserves all basic elements $z_k\mathbf{e}_p$ with $p \neq \alpha$, sends elements $z^{-k}\mathbf{e}_\alpha$ with $k \geq 2$ to $z^{-k}\mathbf{e}_\alpha - z^{-k+1}/\zeta\mathbf{e}_\alpha \in H_-$, and it remains to find its action on $z^{-1}\mathbf{e}_\alpha$. We have

$$\mathbf{g}_1 z^{-1}\mathbf{e}_\alpha = z^{-1}\mathbf{e}_\alpha - \frac{1}{\zeta}\mathbf{e}_\alpha = z^{-1}\mathbf{e}_\alpha - \frac{1}{\zeta}(\mathbf{e}_\alpha - \hat{w}_{W\alpha}) - \frac{1}{\zeta}\hat{w}_{W\alpha},$$

where $w_{W\alpha}$ is the αth row of the Baker function. The operator l annihilates the last term belonging to W and does not change the rest of the terms belonging to H_-.

Now, the determinant of the mapping $\mathrm{T}_{Wg^{-1}}(g_1)$ can be calculated in the space H_- modulo the subspace spanned by the basic elements $z^{-k}\mathbf{e}_\beta$, with $\beta \neq \alpha$, and is equal to

$$\begin{vmatrix} 1 + w_{(W\alpha\alpha)1}/\zeta & w_{(W\alpha\alpha)2}/\zeta & w_{(W\alpha\alpha)3}/\zeta & \cdots \\ -1/\zeta & 1 & 0 & \cdots \\ 0 & -1/\zeta & 1 & \cdots \\ \cdots & \cdots & \cdots & \cdots \end{vmatrix} = \hat{w}_{W\alpha\alpha}(t,\zeta),$$

which proves Eq.(3.10.1).

In order to calculate the non-diagonal elements, let us multiply Eq.(3.10.3) to the right by the operator $R_{\alpha\beta}$. The determinant of the left-hand side is equal to the right-hand side of Eq.(3.10.2). The mapping in the right-hand side preserves all the basic elements $z_k\mathbf{e}_p$, with $p \neq \beta$, sends $z^{-k}\mathbf{e}_\beta$ with $k \geq 2$ to $z^{-k}\mathbf{e}_\beta - z^{-k+1}/\zeta\mathbf{e}_\beta \in H_-$ and it remains to find its action on $z^{-1}\mathbf{e}_\beta$. The operator $R_{\alpha\beta}$ sends this element to $-\mathbf{e}_\alpha$, the operator $\mathbf{g}_1$ does not do anything. We have $-\mathbf{e}_\alpha = (-\mathbf{e}_\alpha + \hat{w}_{W\alpha}) - \hat{w}_{W\alpha}$. The projector l annihilates the last term belonging to W and does not change the other terms. The proof is completed by the calculation of the determinant:

$$\begin{vmatrix} w_{(W\alpha\beta)1} & w_{(W\alpha\beta)2} & w_{(W\alpha\beta)3} & \cdots \\ -1/\zeta & 1 & 0 & \cdots \\ 0 & -1/\zeta & 1 & \cdots \\ \cdots & \cdots & \cdots & \cdots \end{vmatrix} = \hat{w}_{W\alpha\beta}(t,\zeta) \cdot \zeta$$

as required. □

3.11. In [5] we dealt with another, algebraic-geometrical, example of elements of the Grassmannian, found the corresponding Baker and τ-functions (in terms of θ-functions) and proved formulas (3.10.1) and (3.10.2).

4. Multi-component KP

4.1. Very small modifications are needed to transfer this theory to the multi-component KP hierarchy.

Let

$$L = A\partial + u_0 + u_1\partial^{-1} + \cdots$$

where u_i are $n \times n$ matrices, $A = \text{diag}(a_1, ..., a_n)$, and a_i are distinct non-zero constants. Diagonal elements of u_0 are assumed to be zero. Let $R_\alpha = \sum_{j=0}^{\infty} R_{j\alpha}\partial^{-j}$, $\alpha = 1, ..., n$, where $R_{0\alpha} = E_\alpha$ such that

$$[L, R_\alpha] = 0.$$

It can be shown that such matrices exist and their elements are differential polynomials in elements of u_i, and

$$R_\alpha R_\beta = \delta_{\alpha\beta} R_\alpha, \quad \sum R_\alpha = I.$$

Let $B_{k\alpha} = (L^k R_\alpha)_+$. The mcKP hierarchy (multicomponent KP) is

$$\partial_{k\alpha} L = [B_{k\alpha}, L],$$

which implies $\partial_{k\alpha} R_\beta = [B_{k\alpha}, R_\beta]$.

Variables x and $t_{k\alpha}$ are not independent,

$$\partial = \sum_\alpha a_\alpha^{-1} \partial_{1\alpha}.$$

4.2. *Baker function.* Let

$$L = \hat{w} A \partial \hat{w}^{-1}, \; \hat{w} = \sum_0^\infty w_i \partial^{-i}, w_0 = I.$$

Then $R_\alpha = \hat{w} E_\alpha \hat{w}^{-1}$. Set

$$w = \hat{w} \exp(\sum_{k=1}^{\infty} \sum_{\alpha=1}^{n} z^k E_\alpha t_{k\alpha}).$$

This Baker function satisfies the equations

$$Lw = zw, \quad \text{and} \quad \partial_{k\alpha} w = B_{k\alpha} w.$$

The latter equation is equivalent to

$$\partial_{k\alpha} \hat{w} = -(L^k R_\alpha)_- \hat{w}.$$

4.3. *Grassmannian.* The same definition of the Grassmannian as in the case of AKNS-D (Sect. 3.3) is retained here with a single distinction: we drop the requirement ii, $zW \subset W$. Thus, the AKNS-D Grassmannian is a subset of the mcKP Grassmannian.

Example. Let W be the set of all elements of H of the form $v(z) = \sum_{-N}^{\infty} v_k z^k$ satisfying the relations

$$v(m_i^{(1)}) \cdot \eta_i^{(1)} + v(m_i^{(2)}) \cdot \eta_i^{(2)} = 0, \; i = 1, ..., Nn,$$

i.e. we have here, compared to AKNS-D, a doubled set of points and vectors.

4.4. *The Baker function of an element of the Grassmannian.* The definition is the same as for the AKNS-D hierarchy, Sect. 3.4.

Proposition. *The Baker function of an element W of the Grassmannian, $w_W(t,z)$, is a Baker function in the sense of Sect. 4.2 of some operator L.*

Proof. Let $\partial = \sum_1^n a_\alpha^{-1} \partial_{1\alpha}$. First we prove that w_W satisfies a differential equation of the form $\partial_{k\alpha} w = B_{k\alpha} w$ where $B_{k\alpha}$ is a polynomial in ∂ with matrix coefficients. Evidently,

$$\partial_{k\alpha} w = (E_\alpha z^k + O(z^{k-1})) \exp \xi(t,z),$$

$$\partial^q w = (A^{-q} z^q + O(z^{q-1})) \exp \xi(t,z).$$

Therefore,

$$\partial_{k\alpha} w - A^k E_\alpha \partial^k = O(z^{k-1}) \exp \xi(t,z).$$

We can proceed by subtracting terms of the form $a_k \partial^q$ and reducing the order of the remainder. At the end we get $\partial_{k\alpha} w - B_{k\alpha} w = O(z^{-1}) \exp \xi(t,z)$. Now, $(\partial_{k\alpha} w - B_{k\alpha} w) \cdot g^{-1} = O(z^{-1})$. This implies that the p_+-projection of this expression vanishes. On the other hand, it belongs to Wg^{-1}. We assumed that $Wg^{-1} \in \mathrm{Gr}$. This, in particular, means that $p_+|_W$ is a one-to-one correspondence, and the vanishing of the projection yields the vanishing of the expression itself. Thus,

$$\partial_{k\alpha} w - B_{k\alpha} w = 0.$$

Put $L = \hat{w} A \partial \hat{w}^{-1} = A\partial + u_0 + u_1 \partial^{-1} + ...$, $R_\alpha = \hat{w} E_\alpha \hat{w}^{-1}$. Then for the above constructed $B_{k\alpha}$ we have $B_{k\alpha} w = \partial_{k\alpha} w = (\partial_{k\alpha} \hat{w}) g + \hat{w} E_\alpha z^k g$, i.e. $[B_{k\alpha} \circ \hat{w} - (\partial_{k\alpha} \hat{w}) - \hat{w} E_\alpha z^k] g = 0$ which implies $B_{k\alpha} - (\partial_{k\alpha} \hat{w}) \circ \hat{w}^{-1} - L^k R_\alpha = 0$. Taking the positive part, we obtain $B_{k\alpha} = (L^k R_\alpha)_+$. Taking the negative part we have $\partial_{k\alpha} \hat{w} = -(L^k R_\alpha)_- \hat{w}$. This is an equation which is equivalent to the mcKP hierarchy. □

4.5. *Definition of the τ-function.* We retain the same definition of the τ-function, Eqs.(3.9.1) and (3.9.2), as for the AKNS-D hierarchy.

Proposition. *The Baker and the τ-function are connected by the same relations (3.10.1) and (3.10.2) as those of the AKNS-D hierarchy.*

Proof. All the considerations in Sect. 3.10 hold since we never used there the invariance of W under multiplication by z. □

Relations (3.10.1) and (3.10.2) were written without proof by Takasaki and Ueno [6]. There is an apparent misprint in their formulas: they omitted z^{-1} in (3.10.2).

We have already observed (see [2]) that a Baker function for AKNS-D is, at the same time, a Baker function for mcKP (not vice versa). However, if a solution for mcKP is constructed with the help of this function, then L reduces to $L = A\partial + u_0$.

A τ-function for the AKNS hierarchy ($n = 2$) from the point of view of the theory of representations was defined by Bergvelt and ten Kroode [7], who did not consider equalities of the type (3.10.1)-(3.10.2).

REFERENCES

[1] G. Segal and G. Wilson, *Loop groups and equations of KdV-type*, Publ. Math. I.H.E.S, **63** (1985), 1–64.

[2] L.A. Dickey, *Another example of the τ-function*, Proceedings of the CRM Workshop on Hamiltonian Systems, Transformation Groups and Spectral Transform Methods, 39–44, Montréal, October 1989.

[3] L.A. Dickey, *Soliton equations and Hamiltonian systems*, World Scientific, 1991.

[4] G. Wilson, *Infinite-dimensional Lie groups and algebraic geometry in soliton theory,* Phil. Trans. Royal Soc. London A315, 1985, 393–404.

[5] L.A. Dickey, *On the τ-function of matrix hierarchies of integrable equations*, Journal Math. Physics, bf 32 (1991), 2996–3002.

[6] K. Ueno and K. Takasaki, "Toda lattice hierarchy", in: *Advanced Studies in Pure Mathematics* **4** (1984), 1–95.

[7] M.J. Bergvelt, and A.P.E. ten Kroode, *τ functions and zero curvature equations of Toda-AKNS type*, J.Math.Phys., **29**:6 (1988), 1308–1320.

[8] E. Date, M. Jimbo, M. Kashiwara, and T. Miwa, in Jimbo and Miwa (ed), *Non-linear Integrable Systems — Classical Theory and Quantum Theory*, Proc. R.I.M.S. Symposium, Singapore, 1983.

Department of Mathematics

University of Oklahoma

Norman, OK 73019

USA

The Boundary of Isospectral Manifolds, Bäcklund Transforms and Regularization

Pierre van Moerbeke*

A la mémoire de Jean-Louis Verdier

Let the potential

$$q(t_1, t_3, t_5, \ldots) = 2\frac{d^2}{dx^2}\log\tau(t_1, t_3, \ldots), \qquad t = (t_1, t_3, t_5, \ldots) \in \mathbb{C}^\infty$$

satisfy the Korteweg-de Vries hierarchy, and in particular the KdV equation

$$\frac{\partial q}{\partial t} + 3q\frac{\partial q}{\partial x} + \frac{1}{2}\frac{\partial^3 q}{\partial x^3} = 0, \qquad (x = t_1, t = t_3).$$

Then if q blows up at a time $t^* \in \mathbb{C}^\infty$, it behaves as

(0.1)

$$q(t_1^* + x, t_3^*, \ldots) = -\frac{j(j-1)}{x^2} + \text{ higher order terms } (x \sim 0), j = 2, 3, \ldots;$$

this can be verified by direct calculation. Apply now the Bäcklund-Darboux transformation to the second-order differential operator P, with wave function Ψ (eigenfunction behaving as $\exp\sum_1^\infty t_i z^i$ for large values $z \in \mathbb{C}$ of the spectral parameter)

$$P\Psi \equiv (\frac{d^2}{dx^2} + q)\Psi = z\Psi;$$

define, for large but fixed $z_1 \in \mathbb{C}$ and corresponding Ψ_1, the operator A and A^T

$$A = \Psi_1\frac{d}{dx}\Psi_1^{-1} = \frac{d}{dx} - v \text{ and } A^T = -\frac{d}{dx} - v$$

with $v = \Psi_1'/\Psi_1$. Then $P - z_1$ admits the following decomposition

$$P = \frac{d^2}{dx^2} + q = \frac{d^2}{dx^2} - \frac{\Psi_1''}{\Psi_1} + z_1 = -A^T A + z_1$$

with $q = -\Psi_1''/\Psi_1 + z_1 = -v' - v^2 + z_1$. The new linear operator $\tilde{P}$ is now defined by conjugation

$$\tilde{P} = APA^{-1} = -AA^T + z_1 = \frac{d^2}{dx^2} + v' - v^2 + z_1 = \frac{d^2}{dx^2} + \tilde{q}.$$

---

*The support of a National Science Foundation grant DMS - 9203246 is gratefully acknowledged.

It induces the following transformation on q

$$q \mapsto \tilde{q} = q + 2v' = 2\frac{d^2}{dx^2}\log\tau + 2\frac{d^2}{dx^2}\log\Psi_1 = 2\frac{d^2}{dx^2}\log\tau\Psi_1,$$

and thus a transformation on τ

$$\tau \mapsto \tilde{\tau} \equiv \tau\Psi_1; \tag{0.2}$$

since $\tilde{P}(A\Psi) = APA^{-1}A\Psi = AP\Psi = Az\Psi = zA\Psi$, we also have the map

$$\Psi \mapsto \tilde{\Psi} \equiv A\Psi = -\frac{\Psi_1'}{\Psi_1}\Psi + \Psi'. \tag{0.3}$$

That is to say *the new function $\tilde{\tau}$ is obtained by multiplying τ with Ψ_1* and the *new wave function $\tilde{\Psi}$* is *a linear combination of the old Ψ and Ψ'.*

Moreover if q behaves as (0.1), then since the function v is a solution of the Ricatti equation $v^2 + v' + q - z_1 = 0$, it behaves as

$$v = \frac{\alpha}{x} + \ldots \qquad \text{with } \alpha^2 - \alpha - j(j-1) = 0. \tag{0.4}$$

Choosing the *positive root* $\alpha = j$, we see that

$$\begin{aligned} \tilde{q} = q + 2v' &= -\frac{j(j-1)}{x^2} - 2\frac{j}{x^2} + \text{ higher order terms} \\ &= \frac{-j(j+1)}{x^2} + \ldots; \end{aligned} \tag{0.5}$$

that is the (integer) leading term $j(j-1)$ of $-q$ is increased to $(j+1)j$. Picking the *negative root* $\alpha = -j+1$ yields

$$\begin{aligned} \tilde{q} = q + 2v' &= -\frac{j(j-1)}{x^2} - \frac{2(-j+1)}{x^2} + \ldots \\ &= -\frac{(j-1)(j-2)}{x^2} + \ldots, \end{aligned} \tag{0.6}$$

thus decreasing the leading term of $-q$ to $(j-1)(j-2)$. That is to say, in the latter case, the Bäcklund transformation has the effect of lowering the leading term of $-q$. In particular, if $q = -2/x^2 + \ldots$, then the Bäcklund transformed potential $\tilde{q}$ is finite. It is now this simple idea which extends to general differential operators.

Let differential operators

$$P \equiv D^p + q_2(x, t_2, \ldots)D^{p-2} + \ldots + q_p(x, t_2, \ldots), \; D \equiv \frac{d}{dx} \tag{0.7}$$

flow according to the isospectral equations[1]

$$\frac{\partial P}{\partial t_k} = [P_+^{k/p}, P], \qquad k = 1, 2, \ldots \qquad (t_1 \equiv x); \tag{0.8}$$

[1] $(\sum_{i:-\infty}^{\infty} b_i D^i)_\pm = \sum_{\substack{i \geq 0 \\ i < 0}} b_i D^i$

such P's form a (generically infinite-dimensional) isospectral manifold $\mathcal{M}$ of differential operators. I now adress the following questions [A-vM2]:

(i) What is the behavior of P near its blow up locus $t = t^*$; that is, how do the functions $q_i(t)$ blow up near t^*, and what does it depend on ?

(ii) How does one regularize P near t^* ? In geometrical language, how does one compactify the isospectral manifold $\mathcal{M}$?

The blowing-up of the solution of the KdV equation is a phenomenon, which is usually not seen in real time, but which occurs in complex time $t \in \mathbb{C}^\infty$. For instance the scattering or periodic solutions of KdV (among them the finite band potentials) remain finite for real time, but blow up *in the complex* (see for instance [B]). This should be compared to the Euler top (expressed in body coordinates), whose solution is given by an elliptic function of time which is finite in the real, but which blows up for a certain complex time.

These results, to some extent, also put the Painlevé analysis for integrable partial differential equations, initiated by the applied community, chiefly Tabor, Weiss and co-workers, on a mathematical footing. The point of this lecture is to show that a special class of Bäcklund-Darboux transformations provides precisely the tools necessary to study limits of isospectral differential operators; a special class, because general Bäcklund transforms turn out to be more complicated (see § 3 in [vM2]).

Questions similar to (i) and (ii) can now be posed for isospectral difference operators and in particular for the isospectral family of periodic Jacobi matrices [A-H-vM]. According to [vM-Mu] and [vM1], for each fixed values of the spectrum, the manifold of isospectral periodic Jacobi matrices parametrizes the affine part of a hyperelliptic Jacobian; this affine part is the whole Jacobian, but a number of specific translates of the theta-divisor, where some entries of the matrices cease to be finite. In this situation, the questions above take on a particularly nice form: can the whole Jacobian be parametrized by matrices or more precisely can the Jacobi matrices be expressed in a new basis such that their limits exist near the theta-translates mentioned above ? ([A-H-vM]).

1. Deformations of pseudo-differential operators and the Grassmannian

It is more convenient to pose questions (i) and (ii) for a family of pseudo-differential operators

$$L = D + \sum_{j:-1}^{-\infty} a_j(t) D^j \qquad t = (x, t_2, t_3, \ldots) \tag{1.1}$$

with holomorphic coefficients in x, depending on $t_2, t_3, \ldots$ and flowing according to

$$\frac{\partial L}{\partial t_k} = [(L^k)_+, L] \qquad k = 1, 2, \ldots. \tag{1.2}$$

A precise answer to questions (i) and (ii) can be given, thanks to Sato's celebrated discovery [S] that[2]

$$L = SDS^{-1} \text{ with } S = \sum_{n=0}^{\infty} \frac{p_n(-\tilde{\partial})\tau(t)}{\tau(t)} D^{-n}; \tag{1.3}$$

that is the solution L of (1.2) is expressible in terms of a single function $\tau(t_1, t_2, \ldots)$, solution of the KP hierarchy. It is also well-known [DJKM] that the wave functions $\Psi(t, z)$ and $\Psi^*(t, z)$, solution of

$$L\Psi = z\Psi \text{ and } \frac{\partial \Psi}{\partial t_n} = (L^n)_+ \Psi, \tag{1.4}$$

and

$$L^\top \Psi^* = z\Psi^* \text{ and } \frac{\partial \Psi^*}{\partial t_n} = -(L^\top)^n_+ \Psi^*,$$

can be represented in terms of τ (for large $z \in \mathbb{C}$) as follows

$$\Psi(t, z) = S e^{\sum_1^\infty t_j z^j} = e^{\sum_1^\infty t_j z^j} \frac{\tau(t - [z^{-1}])}{\tau(t)} \equiv e^{\sum t_j z^j} \psi(t, z), \tag{1.5}$$

and

$$\Psi^*(t, z) = (S^\top)^{-1} e^{-\sum_1^\infty t_i z^i} = e^{-\sum_1^\infty t_i z^i} \frac{\tau(t + [z^{-1}])}{\tau(t)}$$

$$\text{with } [s] \equiv (s, \frac{s^2}{2}, \frac{s^3}{3}, \ldots).$$

To the wave function Ψ one associates a plane, generated by all its partial derivatives with regard to $t_1, t_2, \ldots$ at $t = 0$, viewed as functions of z (see [S], [S-W]), which in view of (1.4) can always be expressed in terms of partial derivatives with regard to $x = t_1$:

$$W^0 = \text{span}\{\Psi(t, z) \mid_{t=0}, \frac{\partial}{\partial x}\Psi(t, z) \mid_{t=0}, \frac{\partial^2}{\partial x^2}\Psi(t, z) \mid_{t=0}, \ldots\};$$

at a generic time $t \in \mathbb{C}^\infty$, one shows

$$W^t \equiv e^{-\sum_1^\infty t_i z^i} W^0 = \text{span}\{\psi(t, z), \nabla\psi(t, z), \nabla^2\psi(t, z), \ldots\} \tag{1.6}$$

where $\nabla = \frac{\partial}{\partial x} + z$. Let Gr be the infinite-dimensional Grassmannian consisting of such planes and their limits.

---

[2] $e^{\sum_1^\infty t_i z^i} = \sum_0^\infty p_n(t) z^n$, $p_n(-\hat{\partial}) = p_n(-\frac{\partial}{\partial t_1}, -\frac{1}{2}\frac{\partial}{\partial t_2}, -\frac{1}{3}\frac{\partial}{\partial t_3}, \ldots)$, where $p_n(t)$ are the Schur polynomials.

For generic $t \in \mathbb{C}^\infty$ (as above), it is readily seen from (1.6) and the form of ∇ that W^t contains functions

$$\nabla^k \psi(t,z) = z^k\big(1 + O(z^{-1})\big)$$

of all orders $k = 0, 1, 2, 3, \ldots$, whereas for special t's, the plane W^t may have a basis $\varphi_0, \varphi_1, \ldots$ of finite order, that is

$$\varphi_i(z) = z^{s_i}\big(1 + O(z^{-1})\big)$$

$$s_0 < s_1 < s_2 < \ldots \text{ and } s_i = i \text{ for large } i.$$

Thus to each W^t one associates a finite sequence (see [S], [S-W], and [P-S])

$$\nu(W^t) \equiv (\nu_0, \nu_1, \nu_2, \ldots) \text{ where } \nu_i = i - s_i,$$

which in turn defines a Young diagram. Statement (1.6) shows that $\nu(W^t) = 0$ for generic t, and the manifold Gr has a cellular decomposition into Birkhoff strata, all parametrized by Young diagrams.

2. Bäcklund transformations acting on the Grassmannian

For W^t and $W_1^t \in Gr$, let Ψ and Ψ_1 be the corresponding wave functions and τ and τ_1 the corresponding τ-functions. Then if $\tau(t), \tau_1(t) \neq 0$, the *following three statements are equivalent*[3] [A-vM2]:

(2.1) (i) $zW_1^t \subset W^t$

(ii) $z\Psi_1(t,z) = \frac{\partial}{\partial x}\Psi(t,z) - \alpha\Psi(t,z)$ for some function $\alpha = \alpha(t)$

(iii) $\{\tau(t - [z^{-1}]), \tau_1(t)\} + z(\tau(t - [z^{-1}])\tau_1(t) - \tau_1(t - [z^{-1}])\tau(t)) = 0.$

If any of these statements hold, then $\alpha = (\log(\tau_1/\tau))' = \tau_1'/\tau_1 - \tau'/\tau$. If τ corresponds to the plane W in Gr, then the new τ-functions (vertex operators)

$$\tau_1(t) \equiv X(t,z_1)\tau(t) \equiv e^{\Sigma t_i z_1^i}\tau(t - [z_1^{-1}]) = \Psi(t,z_1)\tau(t) \tag{2.2}$$

and

$$\tilde{\tau}_1(t) \equiv \tilde{X}(t,z_1)\tau(t) \equiv e^{-\Sigma t_i z_1^i}\tau(t + [z_1^{-1}]) = \Psi^*(t,z_1)\tau(t)$$

correspond respectively to linear spaces W_1 and $\tilde{W}_1 \in$ Gr satisfying

$$zW_1^t \subset W^t \text{ and } zW^t \subset \tilde{W}_1. \tag{2.3}$$

The associated wave functions

(2.4)

$$\begin{aligned}\Psi_1(t,z) &= -\frac{z_1}{z}\Psi(t - [z_1^{-1}], z) = z^{-1}A_{\Psi(t,z_1)}\Psi(t,z)\\ \tilde{\Psi}_1(t,z) &= -\frac{z}{z_1}\Psi(t + [z_1^{-1}], z) \equiv zA^{-1}_{\Psi(t,z_1)}\Psi(t,z),\end{aligned}$$

[3] $\{f, g\} = f'g - fg'$, where $' = \partial/\partial x$.

are expressed in terms of the Bäcklund-Darboux transformation

$$A_{\Psi(t,z_1)}\Psi(t,z) = \frac{\{\Psi(t,z),\Psi(t,z_1)\}}{\Psi(t,z_1)} \tag{2.5}$$

and its "inverse" $A^{-1}_{\Psi(t,z_1)}$; although the latter does not really exist, it is defined by means of formula (2.4).

Then compounding Bäcklund transformations

$$\Psi_k(t,z) \equiv z^{-1}A_{\Psi_{k-1}(t,z_k)}\Psi_{k-1}(t,z)$$

for fixed but arbitrary $z_1,\ldots,z_k$ near $z=\infty$, we have that

$$\begin{aligned}\Psi_k(t,z) &= z^{-k}A_{\Psi_{k-1}(t,z_k)}\cdots A_{\Psi_1(t,z_2)}A_{\Psi(t,z_1)}\Psi(t,z)\\ &= z^{-k}\frac{\text{Wronskian}[\Psi(t,z_1),\ldots,\Psi(t,z_k),\Psi(t,z)]}{\text{Wronskian}[\Psi(t,z_1),\ldots,\Psi(t,z_k)]}\end{aligned}$$

is a wave function associated to a plane W_k related to the original plane W by the inclusion

$$z^kW^t_k \subset W^t.$$

Similarly, compounding inverse Bäcklund transformations

$$\tilde\Psi_k(t,z) \equiv zA^{-1}_{\tilde\Psi_{k-1}(t,z_k)}\tilde\Psi_{k-1}(t,z), \qquad \tilde\Psi_0 \equiv \Psi$$

leads to the wave function

$$\tilde\Psi_k(t,z) = z^{-k}A^{-1}_{\tilde\Psi_{k-1}(t,z_k)}\cdots A^{-1}_{\tilde\Psi_0(t,z_1)}\Psi(t,z) \tag{2.6}$$

associated to the plane $\tilde W_k$ related to the original plane W by the inclusion

$$z^kW^t \subset \tilde W^t_k. \tag{2.7}$$

The main point in verifying these statements is to check that the functions τ and τ_1 (respectively τ and $\tilde\tau_1$) satisfy the bilinear differential equation (2.1(iii)); this amounts to the (differential) Fay identity which in terms of Ψ and Ψ^* assumes the form

$$\Psi^*(t,\lambda)\Psi(t,\mu) = \frac{1}{\mu-\lambda}\frac{\partial}{\partial x}\left(e^{\sum_1^\infty t_i(\mu^i-\lambda^i)}\frac{\tau(t+[\lambda^{-1}]-[\mu^{-1}])}{\tau(t)}\right).$$

The details of the proof can be found in [A-vM2] and [vM2].

3. Young diagrams, order of vanishing of τ and regularization

Given a Young diagram $\nu = (\nu_0 \geq \nu_1 \geq \ldots \geq \nu_n \geq 0)$, the corresponding Schur polynomial is defined as (see footnote 2 for definition of p_k)

$$F_\nu(t) \equiv \det(p_{\nu_i-i+j})_{0\leq i\leq j\leq n}, \qquad (t=(t_1,t_2,\ldots)).$$

A *first ingredient* due to Sato [S] is that a τ-function admits a Fourier expansion in terms of Schur polynomials

$$\tau(t^* + t) = \sum_\nu \xi_\nu(W^{t^*})F_\nu(t), \text{ over all Young diagrams } \nu$$

where W^{t^*} is the plane associated with t^* and[4] with Fourier coefficients

$$(3.1) \qquad \xi_\nu(W^{t^*}) = \det \operatorname{proj}(W^{t^*} \to H_\nu),$$

satisfying Plücker relations. The latter implies that near a point t^*, where the Young diagram $\nu^* \equiv \nu(W^{t^*}) \neq 0$, we have $\xi_\nu(W^{t^*}) = 0$ for all ν such that $\nu \not\geq \nu^*$; therefore near the point t^*, $\tau(t^* + t)$ has the following Fourier series:

$$\tau(t^* + t) = \xi_{\nu^*}(W^{t^*})F_{\nu^*}(t) + \sum_{\substack{\nu \geq \nu^* \\ |\nu| > |\nu^*|}} \xi_\nu(W^{t^*})F_\nu(t), \text{ with } \xi_{\nu^*}(W^{t^*}) \neq 0.$$

A *second ingredient* is that the polynomial $F(t - [s])$ admits the following Taylor series in s about[5] $s = 0$

$$\begin{aligned} F_\nu(t - [s]) &= F_\nu(t) + \ldots + s^k p_k(-\tilde{\partial})F_\nu(t) + \ldots + s^{\nu_0} p_{\nu_0}(-\tilde{\partial})F_\nu(t) \\ &= F_\nu(t) + \ldots + s^k F_{\nu\backslash(k)}(t) + \ldots + s^{\nu_0} F_{\nu\backslash\text{first row}}(t), \end{aligned}$$

where $F_{\nu\backslash(k)}$ is the skew-Schur polynomial associated with the Young diagram $\nu\backslash(k)$ (i.e. Young diagram ν with a row of k boxes removed in the upper left hand corner; see MacDonald):

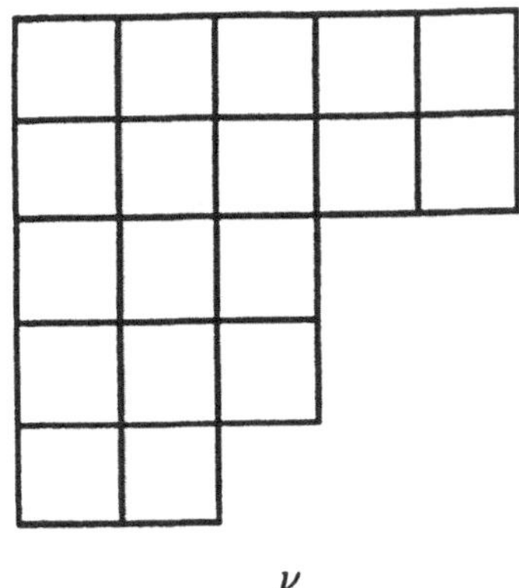

ν

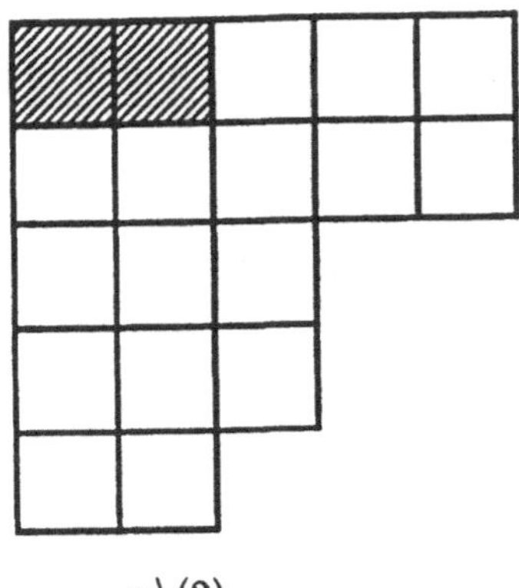

$\nu\backslash(2)$

These two facts now lead to the following three theorems.

[4] H_ν is the plane

$$H_\nu = \{z^{i-\nu_i}, i = 0, 1, 2, \ldots\}$$

[5] Given a Young diagram μ_i st. $\nu_i \geq \mu_i$, define the skew-Schur polynomial

$$F_{\nu\backslash\mu}(t) = \det(p_{\nu_i - \mu_j - i + j})_{0 \leq i,j \leq n}$$

Theorem 1. *[A-vM2] At a point t^* where $\tau(t^*)=0$, let the plane $W^{t^*}\in$ (Stratum $\nu_0\geq\nu_1\geq\ldots$). Let $W_1,W_2,\ldots$ be planes obtained by inverse Bäcklund transformations (2.4) such that*

$$z^{\nu_0}W^t\subset z^{\nu_0-1}W_1^t\subset z^{\nu_0-2}W_2^t\subset\ldots\subset W_{\nu_0}^t,$$

and $\tau_1,\tau_2,\ldots$ the corresponding τ-functions; then

$$\text{Young diagram }(W_k^{t^*})=\text{Young diagram }W^{t^*}\backslash\,(\text{first } k \text{ columns}),$$

and

$$\tau(t^*)=\tau_1(t^*)=\ldots=\tau_{\nu_0-1}(t^*)=0,\ \text{and } \tau_{\nu_0}(t^*)\neq 0.$$

More specifically, for $\bar{x}=(x,0,0,\ldots)$ small [6],

$$\tau_k(t^*+\bar{x})=a_kx^{|\nu|-\sum_1^k\hat{\nu}_i}+\ldots \text{(this is the KP-analogue of (0.5))}$$

The successive Bäcklund transformations constitute a ladder enabling one to "climb" out of the singularity by knocking off each time the left most column of the Young diagram. Bäcklund transforming repeatedly gives the τ-function a softer and softer zero, according to a well-defined pattern, until it ultimately does not vanish any more.

In the next theorem we indicate how certain differential polynomials $p_k(-\tilde{\partial})$, applied to τ behave in the t_1-direction near the stratum [A-vM2]. The exact exponent in the estimate below was announced by Laumon [La] in a footnote of [S-W].

Theorem 2. *At a point t^* where $\tau(t^*)=0$, let the plane $W^{t^*}\in$ (Stratum $\nu_0\geq\nu_1\geq\ldots$). Then we have the following estimates ($\bar{x}=(x,0,0,\ldots)$) :*

$$\begin{aligned} p_k(-\tilde{\partial})\tau(t^*+\bar{x}) &= c_k^\nu x^{|\nu|-k}+\ldots && \text{for } 0\leq k\leq\nu_0,\ \text{with } c_k^\nu\neq 0,\\ &= c_k^\nu x^{|\nu|-\nu_0}+\ldots && \text{for } k\geq\nu_0, \end{aligned}$$

where the c_k^ν $(0\leq k\leq\nu_0)$ are numbers determined by the partition ν by means of the following polynomial identity

$$P(z)=\sum_{k=0}^{\nu_0}\frac{c_k^\nu}{c_0^\nu}z(z-1)\ldots(z-\nu_0+k+1)=\prod_{i=0}^{\nu_0-1}(z-(\nu_0+\hat{\nu}_i-i-1))$$

or, more explicitly [7]

$$c_k^\nu=(-1)^k\det\Big(\frac{1}{(\nu_i-k\delta_{j0}-i+j)!}\Big)_{0\leq i,j\leq\hat{\nu}_0}=(-1)^k\frac{f^{\nu/(k)}}{(|\nu|-k)!}\neq 0.$$

[6] the dual Young diagram $\hat{\nu}$ of ν is the Young diagram obtained by flipping the Young diagram around its diagonal; define

$$|\nu|\equiv\sum_1^{\hat{\nu}_0}\nu_i=\sum_1^{\nu_0}\hat{\nu}_i.$$

[7] $f^{\nu|\mu}$ denotes the number of standard tableaux of the skew Young diagram $\nu-\mu$.

Theorem 1 and 2 describe how the function $\tau(t)$, its derivatives and Bäcklund transforms behave near its point t^* of vanishing; of course this depends on the (stratum $\nu_0 \geq \nu_1 \geq \ldots$) containing W^{t^*}. The next theorem tells us then that $\Psi(t,z)$ multiplied with an appropriate factor (independent of z) tends to a finite limit, when $t \to t^*$ in the t_1-direction and that this limit is - up to a multiplicative factor - a *new wave function* evaluated at $t = t^*$. In fact this new wave function yields a new frame with regard to which all the constituents of the limiting plane W^{t^*} can be expressed. Theorem 3 tells us how to express the element of W^{t^*} which blows up maximally (i.e., behaving as z^{ν_0} for $z \nearrow \infty$) as a limit (for $t \to t^*$) of elements in W^t; for a full statement and theorem 3, see [A-vM2]. The reader is reminded (1.3) that ψ is defined as $\Psi = e^{\Sigma t_j z^j}\psi$.

Theorem 3. *Consider a point t^* where $\tau(t^*) = 0$, with $W^{t^*} \in$ (stratum $\nu_0 \geq \nu_1 \geq \ldots$); then the following limit exists and equals*

$$\lim_{x \to 0} \psi(t^* + \bar{x}, z) \frac{\tau(t^* + \bar{x})}{p_{\nu_0}(-\tilde{\partial})\tau(t^* + \bar{x})} = z^{-\nu_0}\tilde{\psi}_{\nu_0}(t^*, z) \in W^{t^*} \tag{3.2}$$

where $\tilde{\Psi}_{\nu_0}(t,z)$ is obtained by ν_0 "inverse" Bäcklund transforms associated with arbitrary parameters $z_1, \ldots, z_{\nu_0}$ near $z = \infty$; therefore the associated plane $\tilde{W}^t_{\nu_0}$ satisfies (see (2.6) and (2.7))

$$z^{\nu_0} W^t \subset \tilde{W}^t_{\nu_0}.$$

Corollary. *$\tilde{\Psi}_{\nu_0}(t,z)$ evaluated at $t = t^*$ is independent of the parameters $z_1, \ldots, z_{\nu_0}$ figuring in the Bäcklund transformations.*

This paradox follows at once from Theorem 3, because the right hand side of (3.2), which seemingly depends on the parameters $z_1, \ldots, z_{\nu_0}$, equals the left hand side of (3.2) which does not. For instance, in the case of a simple zero t^* of $\tau(t)$, the independence, mentioned in the corollary follows from the Fay identity, expressed in the following form

$$\frac{z}{z - z_1} \frac{\tau(t - [z^{-1}] + [z_1^{-1}])}{\tau(t + [z_1^{-1}])} - \frac{z}{z - z_2} \frac{\tau(t - [z^{-1}] + [z_2^{-1}])}{\tau(t + [z_2^{-1}])} \tag{3.3}$$

$$= \frac{z(z_1 - z_2)}{(z - z_1)(z - z_2)} \frac{\tau(t)\ \tau(t + [z_1^{-1}] + [z_2^{-1}] - [z^{-1}])}{\tau(t + [z_1^{-1}])\tau(t + [z_2^{-1}])}.$$

Up to a multiplicative factor, the left hand side of (3.3) equals

$$A^{-1}_{\Psi(t,z_1)}\Psi(t,z) - A^{-1}_{\Psi(t,z_2)}\Psi(t,z)$$

which, in view of its right hand side, vanishes along the locus $\{\tau(t^*) = 0\}$.

4. Compactifying isospectral manifolds of Jacobi matrices

If N-periodic Jacobi matrices $(a_N \equiv a_0,\ b_N \equiv b_0)$

$$(4.1)\qquad L(z) = \begin{pmatrix} b_1 & a_1 & & & z \\ 1 & b_2 & a_2 & & \\ & & & & \\ & & & & a_{N-1} \\ a_N z^{-1} & & & 1 & b_N \end{pmatrix}, \sum_1^N b_i = 0, \prod_1^N a_i \neq 0,$$

flow according to the Toda hierarchy[8]

$$(4.2)\qquad \frac{\partial L}{\partial t_j} = [L, (L^j)_+], \qquad j = 1, 2, \ldots, N-1,$$

then the hyperelliptic curve (of genus $g = N-1$)

$$X : \left\{ \begin{array}{ll} (z,\lambda)|0 = \det(\lambda I - L(z)) & \equiv R(\lambda) - (z + Az^{-1}) = 0 \\ & \equiv \lambda^N + I_2\lambda^{N-2} + \ldots + (-1)^N I_N - (z + Az^{-1}) \end{array} \right\}$$

is isospectral, i.e. the I_j and $A \equiv \prod_1^N a_i$ are constants of motion of (4.2); the eigenvector f of L (see [vM-Mu] and [vM])

$$f(z,\lambda)L(z) = \lambda f(z,\lambda) \qquad f = (f_0 = 1, f_1, \ldots, f_{N-1})$$

yields meromorphic functions f_k on the isospectral curve X (with two points P and Q covering ∞) such that for some divisor $D \geq 0$ of order g (common to all f_k):[9]

$$(4.4)\qquad (f_k) \geq -kP + kQ - D \text{ and } (z) = -NP + NQ.$$

The set of matrices $L(z)$ with fixed spectral curve X or what is the same, the set obtained by letting $L(z)$ flow according to the vector fields (4.2), parametrizes an affine part $\mathcal{A}_X$ of the hyperelliptic Jacobian J_X (complex torus), obtained by removing from J_X a number of divisors, one for each dot in the (extended) Dynkin diagram going with the Toda equations; the case discussed here is the Kac-Moody Lie algebra $a_N^{(1)}$. In precise terms, if $\Theta_0 \equiv \{P + \sum_1^{g-1} x_i,\ x_i \text{ generic} \in X\} \subset J_X$ denotes the theta-divisor, and $\Theta_r = \Theta_0 + r(Q-P)$ translates of Θ_0 on J_X (by means of $Q - P \equiv \int_P^Q \omega \in J_X$), then

$$\mathcal{A}_X = J_X \backslash (\Theta_0 \cup \Theta_1 \cup \ldots \cup \Theta_{N-1})$$

is parametrized by the *isospectral* set[10] of N-periodic Jacobi matrices. If one approaches the Θ_i's, several entries will blow up, while others will tend

[8] $(\,)_+$ means the (strictly) upper triangular part of the matrix $(\,)$.

[9] the divisor (f) of a meromorphic function f is the set of its zeros with $+$ signs and its poles with $-$ signs (codimension 1 subvarieties)

[10] maintaining all the coefficients of the curve X defined in (4.3): A and all the coefficients of $R(\lambda)$.

to zero. Can the whole Jacobian J_X rather than the affine part $\mathcal{A}_X$ be parametrized by (isospectral) matrices ? That is, when the t_1-trajectories (solution of $\frac{\partial L}{\partial t_1} = [L, L_+]$) hit a point t^* of the theta-divisor Θ_0 or any of the translates, can the matrix $L(z,t)$ be conjugated

$$L_{\text{new}}^{\top} = B^{-1} L^{\top} B$$

by means of a matrix B of polynomial entries in the a's and b's such that $\lim_{t\to t^*} L_{\text{new}}$ exists and, if so, what is this limit ? This is the analogue of questions (i) and (ii) posed in the introduction for differential operators. The answer to these questions was given in [A-H-vM].

Let me work out the simplest case, where $t^* \in \Theta_1, t^* \notin \Theta_i$ $(i \neq 1)$, then guessing the matrix B for which $\lim_{t\to t^*} L_{\text{new}}$ exists is quite easy, in view of the Painlevé analysis for the Toda lattice. Indeed one shows that (see [A-vM1])
(4.5)

$$\begin{array}{lll} a_1 = -\frac{1}{(t_1-t_1^*)^2} + \dots & b_1 = -\frac{1}{t_1-t_1^*} + \dots & b_2 = \frac{1}{t_1-t_1^*} + \dots \\ a_0 = \alpha(t_1 - t_1^*) + \dots & a_2 = \beta(t_1 - t_1^*) + \dots, & \alpha\beta \neq 0 \end{array}$$

with all the other entries a_i and b_i bounded near t^* and a_i bounded away from 0. Then conjugating $L^{\top}$ by the matrix

$$B = \left(\begin{array}{cc|c} 0 & 1 & \\ 1 & -b_1 & O \\ \hline \multicolumn{2}{c|}{O} & I \end{array}\right)$$

leads to $(a_N = a_0)$

$$B^{-1}L^{\top}B = \left(\begin{array}{cc|cccccc} I_1^{(1)} & -I_2^{(1)} & 1 & 0 & & \dots & 0 & a_N b_1 z^{-1} \\ 1 & 0 & 0 & 0 & & \dots & 0 & a_N z^{-1} \\ \hline a_2 & -a_2 b_1 & b_3 & 1 & 0 & \dots & & \\ 0 & 0 & a_3 & b_4 & 1 & & O & \\ 0 & 0 & 0 & a_4 & b_5 & & & \\ \vdots & \vdots & & & & & & \\ & & & O & & & & 1 \\ 0 & z & & & & & a_{N-1} & b_N \end{array}\right)$$

where $I_1^{(1)}$ and $I_2^{(1)}$, defined by

$$\det\left(\lambda I - \begin{pmatrix} b_1 & a_1 \\ 1 & b_2 \end{pmatrix}\right) = \lambda^2 - (b_1 + b_2)\lambda + b_1 b_2 - a_1 = \lambda^2 - I_1^{(1)}\lambda + I_2^{(1)}$$

are invariant polynomials of $s\ell(2, \mathbb{C})$.

Then letting $t_1 \to t_1^*$, the matrix above tends to a finite limit, as is seen from the leading behaviors (4.5)

$$\lim_{t_1\to t_1^*} B^{-1}L^\top B = \left(\begin{array}{cc|cccccc} I_1^{(1)} & -I_2^{(1)} & 1 & 0 & & \ldots & 0 & -\alpha z^{-1} \\ 1 & 0 & 0 & 0 & & \ldots & 0 & 0 \\ \hline 0 & \beta & b_3 & 1 & 0 & & & \\ 0 & 0 & a_3 & b_4 & 1 & & O & \\ 0 & 0 & 0 & a_4 & b_5 & & & \\ & & & & & \ddots & & \\ \vdots & \vdots & & & & \ddots & & \\ & & & & & \ddots & & \\ 0 & 0 & & O & & & & 1 \\ 0 & z & & & & & a_{N-1} & b_N \end{array}\right),$$

where the a_i and b_i are all evaluated at $t = t^*$.

When t^* belongs to the intersection of several theta-translates, the limit of L_{new} when $t \to t^*$ is described as follows: when

$$t^* \in \Theta_1 \cap \Theta_2 \cap \ldots \cap \Theta_s, \qquad t^* \notin \text{ other } \Theta_i \;\; (1 \le s \le N-3)$$

then in the limiting $N\times N$ matrix the upper-left $(s+1)\times(s+1)$ block (which blew up in the matrix $L^\top$) gets replaced by its associated "companion matrix" B_1, the rest of the matrix being nearly unchanged, more specifically

$$\lim_{t\to t^*} L_{\text{new}} = \left(\begin{array}{c|c} B_1 & A_2 \\ \hline A_1 & B_2 \end{array}\right)$$

where the A_i and B_i are the matrices below, all evaluated at $t = t^*$:

$$B_1 = \begin{pmatrix} I_1^{(s)} & -I_2^{(s)} & & & (-1)^s I_{s+1}^{(s)} \\ 1 & 0 & & & \\ & & & O & \\ & \ddots & & \ddots & \vdots \\ & O & & & \\ & & & 1 & 0 \end{pmatrix},$$

$$B_2 = \begin{pmatrix} b_{s+2} & 1 & & & \\ a_{s+2} & b_{s+3} & 1 & & O \\ & & & & \\ & \ddots & \ddots & \ddots & \\ O & & a_{N-2} & b_{N-1} & 1 \\ & & & a_{N-1} & b_N \end{pmatrix},$$

$$A_1 = \begin{pmatrix} & (-1)^s a_{s+1} I_s^{(s-1)} \\ & 0 \\ O & \vdots \\ & 0 \\ & z \end{pmatrix},$$

and

$$A_2 = \begin{pmatrix} 1 & 0 & \dots & 0 & (-1)^{s+1} z^{-1} \sum_{\ell=1}^{s} I_\ell^{(s-1)} (\frac{\partial}{\partial t_1})^{s-\ell} a_N \\ & & O & & \end{pmatrix}.$$

The expressions $I_\ell^{(k)}$ appearing in the matrices above are the invariant polynomials of $\mathrm{sl}(s+1, \mathbb{C})$ defined by

$$\det \begin{pmatrix} \lambda - b_1 & a_1 & & 0 \\ 1 & \lambda - b_2 & a_2 & \\ & & & \\ & & & a_s \\ 0 & & 1 & \lambda - b_{k+1} \end{pmatrix}$$

$$= \lambda^{k+1} - I_1^{(k)} \lambda^k + I_2^{(k)} \lambda^{k-1} + \dots + (-1)^{k+1} I_{k+1}^{(k)}.$$

The Birkhoff strata employed in the operator case get here replaced by Birkhoff strata for a flag constructed from the sequence of meromorphic functions. Details can be found in [A-H-vM].

REFERENCES

[A-vM1] Adler, M., van Moerbeke, P., Kowalewski's asymptotic method, Kac-Moody Lie algebras and regularization, Commun. Math. Phys. 83, 83-106 (1982) and The Toda lattice, Dynkin diagrams, singularities and Abelian varieties, Invent. Math. 103, 223-278 (1991).

[A-vM2] Adler, M., van Moerbeke, P., Birkhoff strata, Bäcklund transformations and limits of isospectral operators, Adv. in Math. (1993).

[A-H-vM] Adler, M., Haine, L., van Moerbeke, P., Limit matrices for the Toda flow and periodic flags for loop groups, Math. Ann. (1993).

[B] Birnir, Bj., Singularities of the complex KdV flows, Comm. pure & Applied Math. 29, 283-305 (1986).

[DJKM] Date, E., Jimbo, M., Kashiwara, M. and Miwa, T., Transformation groups for soliton equations, Proceedings of RIMS Symposium on Non linear Integrable Systems-Classical and Quantum theory-Kyoto (May 1981), ed. by M. Jimbo & T. Miwa (World Scientific Co. Singapore, 1983) (pp. 39-119).

[Eh-Kn] Ehlers, F., Knörrer, H., An algebro-geometric interpretation of the Bäcklund transformation for the KdV equation, Comm. Math. Helv. 57, 1-10 (1982).

[Fl] Flaschka, H., The Toda lattice in the complex domain, in Algebraic analysis, Acad. Press 1, 1988, 141-154.

[Fl-Ha] Flaschka, H., Haine, L., Variétés de drapeaux et réseaux de Toda, Math. Z. 208 (1991) 545-556.

[K] Kac, V., Infinite-dimensional Lie algebras, 3rd edition, Cambridge Univ. Press.

[La] Laumon, footnote to [S-W].

[McD] Mac Donald, I., Symmetric functions and Hall polynomials, Oxford University Press, 1979.

[P-S] Pressley, A., Segal, G., Loop groups, Clarendon Press, Oxford, 1986.

[S] Sato, M., Soliton equations as dynamical systems on infinite dimensional Grassmann manifolds, RIMS Kokyuroku 439 (1981) 30-46.

[S] Sato, M., Soliton equations and the universal Grassmann manifold (by Noumi, in Japanese), Math. Lect. Notes n^{0} 18 (Sophia University 1984).

[S-S] Sato, M. and Sato, Y., Soliton equations as dynamical systems on infinite dimensional Grassmann manifolds, Lect. Notes in Num. Appl. Anal. 5, 259-271 (1982).

[S-W] Segal, G., Wilson, G., Loop groups and equations of KdV type, Publ. Math. I.H.E.S. 61, 5-65 (1985).

[Ta] Tabor, M., Weiss, J., Carnevale, G., The Painlevé property for partial differential equations, J. Math. Phys. 24, 522-526 (1983).

[vM1] van Moerbeke, P., The isospectral deformations of discrete Laplacians, Springer Verlag Lecture Notes, 755 (1979) 313-370 and The spectrum of Jacobi matrices, Invent. Math. 37, 45-81 (1976).

[vM2] van Moerbeke, P., Integrable foundations of string theory, CIMPA lectures on Integrable Systems (June 1991), World Scientific, to appear.

[vM-Mu] van Moerbeke, P. and Mumford, D., The spectrum of difference operators and algebraic curves, Acta Math. 143 (1979) 93-154.

Department of Mathematics
University of Louvain
1348 Louvain-la-Neuve, Belgium
and
Brandeis University
Waltham, MA 02254

PART II
Hamiltonian Methods

The Geometry of the Full Kostant-Toda Lattice

N. M. Ercolani, H. Flaschka, and S. Singer

1. Introduction

The equations for the (finite, nonperiodic) Toda lattice can, as is well known, be written in Lax form,

$$\dot{X}(t) = [X(t), \mathbf{\Pi} X(t)]. \tag{1}$$

Here X is a symmetric tridiagonal matrix and $\mathbf{\Pi}$ is the projection onto the skew-symmetric summand in the decomposition of X into skew-symmetric plus upper triangular. The eigenvalues of X are constants of motion of (1), and the Toda lattice turns out to be a completely integrable Hamiltonian system.

Deift, Li, Nanda and Tomei [DLNT] proved that the Toda equations (1) remain completely integrable even when X is allowed to be an arbitrary (sufficiently "generic") symmetric $n \times n$ matrix. They constructed the constants of motion (on the order of $n^2/4$ are needed), the associated angle variables, and described the typical level set of the constants of motion. Our aim in this paper is to interpret some of the techniques and results in [DLNT] in the language of Lie groups and homogeneous spaces.

We will consider the Lax equation (1) not for symmetric matrices, but for Hessenberg matrices

$$\begin{pmatrix} * & 1 & 0 & \cdots & 0 \\ * & * & 1 & \ddots & \vdots \\ * & \ddots & \ddots & \ddots & 0 \\ \vdots & \ddots & * & * & 1 \\ * & \cdots & * & * & * \end{pmatrix}. \tag{2}$$

An explanation of this choice is given in Subsection 2.4; the point, in brief, is that such matrices admit a unique transformation to a certain normal form, and this feature is crucial to our approach. Furthermore, because we use ideas of complex algebraic geometry, we take the entries of (2) to be *complex*.

Denote the Lie algebra of lower-triangular matrices by $\mathcal{B}_-$, and let ϵ stand for the nilpotent matrix that remains when all entries "$*$" in (2) are set to zero. Our phase space will be $\epsilon + \mathcal{B}_-$. The projection $\mathbf{\Pi}X$ in (1) now refers to the strictly lower triangular part of X. In his study [Ko2] of the Toda lattice, Kostant used *tridiagonal* matrices of the form (2); these can be conjugated to tridiagonal symmetric matrices by a unique diagonal matrix. There is no analogous simple transformation of elements of $\epsilon + \mathcal{B}_-$ to the symmetric form, and one might expect the properties of our modification of the DLNT Toda lattice to bear little relation to those of the symmetric version. In fact, the two systems have a common origin, and thus many results in [DLNT] carry over almost without change [S1]. The two versions of the Toda lattice are compared in the appendix.

The Lax equation (1), with $X \in \epsilon + \mathcal{B}_-$, will be called the *full Kostant-Toda lattice* to distinguish it from the DLNT version. In this Introduction, we will describe our results about this system. We keep the discussion on an informal and intuitive level, and include several ideas that have helped us to think about the full Kostant-Toda lattice, even though we have not always been able to use them to deduce hard results.

First, we must understand why the full Kostant-Toda lattice should be completely integrable. One needs a Poisson bracket on $\epsilon + \mathcal{B}_-$, of course, but we do not require its explicit form just yet. Powerful evidence for integrability comes from the so-called *factorization method* for solving (1) with initial condition $X(0) = X_0$. Factor $\exp tX_0$ as $n(t)b(t)$, with $n(t)$ lower unipotent and $b(t)$ upper triangular. Then

$$X(t) \stackrel{\text{def}}{=} n(t)^{-1}X_0 n(t) \tag{3}$$

solves (1) [RSTS, R]. (When the factorization cannot be done for some t_0, the solution $X(t)$ has a pole at $t = t_0$.) Since this is an almost explicit solution (only the exponential of tX_0 is not effectively computable), one expects the equations to possess the requisite number of constants of motion in involution. The problem is to find them.

As in [DLNT], the constants of motion can be taken to be the *Ritz values* of the matrix X.

Definition 1.1 *For $k = 0, \ldots, [(n-1)/2]$, denote by $(X - \lambda\,\mathrm{Id})_{(k)}$ the result of removing the first k rows and last k columns from $(X - \lambda\,\mathrm{Id})$ (we call this the* k-chop of X*), and let λ_{rk}, $r = 1, \ldots, n-2k$, denote the roots of*

$$\widetilde{Q}_k(X,\lambda) \stackrel{\text{def}}{=} \det(X - \lambda\,\mathrm{Id})_{(k)} = E_{0k}\lambda^{n-2k} + \ldots + E_{n-2k,k}. \tag{4}$$

The λ_{rk} turn out to be constants of motion. Set

$$Q_k(X,\lambda) \stackrel{\text{def}}{=} \frac{\det(X - \lambda\,\mathrm{Id})_{(k)}}{E_{0k}} = \lambda^{n-2k} + I_{1k}\lambda^{n-2k-1} + \ldots + I_{n-2k,k}. \tag{5}$$

The I_{rk} are constants of motion equivalent to the λ_{rk}; we call them the k-chop integrals. *(The functions $I_{1k} = \sum_r \lambda_{rk}$ are Casimirs on $\epsilon + \mathcal{B}_-$.)*

Note that $Q_0(X,\lambda)$ is the characteristic polynomial of X, and that the I_{r0} are its coefficients.

Here we see one of the many unusual features of the full Toda lattice: the constants of motion are *rational*, rather than polynomial, functions on phase space. They are extremely unwieldy. For instance, the integral I_{21} for $\mathcal{SL}(4,\mathrm{C})$ is (see Section 7)

$$\frac{X_{21}X_{32}X_{43} - X_{21}X_{33}X_{42} - X_{22}X_{31}X_{43} + X_{42}X_{31}}{X_{41}} + X_{22}X_{33} - X_{32}.$$

In a nutshell, our paper exploits the following remarkable fact (stated here for $\mathcal{SL}(4,\mathrm{C})$): when the set of matrices $X \in \epsilon + \mathcal{B}_-$ with fixed spectrum $\{\lambda_1, \lambda_2, \lambda_3, \lambda_4\}$ is realized as subvariety of the flag manifold $SL(4,\mathrm{C})/$ upper triangular matrices (see Definition 1.3 below), I_{21} assumes the simple and symmetric form

$$-\frac{\lambda_1\lambda_2\lambda_3\pi_4\pi_4^* + \lambda_1\lambda_2\lambda_4\pi_3\pi_3^* + \lambda_1\lambda_3\lambda_4\pi_2\pi_2^* + \lambda_2\lambda_3\lambda_4\pi_1\pi_1^*}{\lambda_4\pi_4\pi_4^* + \lambda_3\pi_3\pi_3^* + \lambda_2\pi_2\pi_2^* + \lambda_1\pi_1\pi_1^*}, \tag{6}$$

where π_i, π_i^* are certain homogeneous coordinates on P^3 and $(\mathrm{P}^3)^*$. We appeal to representation theory to generalize this formula, and to explain why the chopped integrals should have this and other useful properties. Then we use the very special form (6) to build the level set of all the constants of motion.

DLNT base their study of the I_{rk} on an invariance property that will also be crucial for us.

Definition 1.2 *Let $\mathbf{P}_k$ be the parabolic subgroup of $SL(n,C)$ whose entries below the diagonal in the first k columns and to the left of the diagonal in the last k rows are zero. In particular, $\mathbf{P}_0 = SL(n,C)$.*

If E_{rk} is a coefficient of $\widetilde{Q}_k$ (cf. (4)), then (as in [DLNT])

$$E_{rk}(p^{-1}Yp) = \chi_k(p)E_{rk}(Y), \quad p \in \mathbf{P}_k,\ Y \in \mathcal{G}. \tag{7}$$

where χ_k is a certain character of $\mathbf{P}_k$ (see Example 2.1). The ratios $I_{rk} = E_{rk}/E_{0k}$ of (5) are *invariant* under the action of $\mathbf{P}_k$.

Our interpretation of these ideas is motivated by *invariant theory* and *symplectic reduction*[1].

One general result from invariant theory is easy to prove (see Proposition A.1):

[1]The phase space $\epsilon + \mathcal{B}_-$ is (almost) never a symplectic manifold, because the Poisson bracket can be degenerate. We should really use *Poisson reduction*. However, since reduction is used as motivation rather than working tool, we ignore such technicalities.

Proposition 1.1 *If a rational function F, defined on a dense open subset of $\mathcal{G}$ and on a dense open subset of $\epsilon + \mathcal{B}_-$, is invariant under conjugation by the group $\mathbf{B}_+$ of upper triangular matrices,*

$$F(b^{-1}Yb) = F(Y), \quad b \in \mathbf{B}_+, \; Y \in \mathcal{G},$$

then its restriction to $\epsilon + \mathcal{B}_-$ is a constant of motion for the full Kostant-Toda lattice.

Such functions are plentiful. Let $\mathcal{CAS}$ be the ring of invariant polynomials, let J be the ideal generated by $\mathcal{CAS}$, and let J^+ be the set of $f \in J$ with zero constant term. The ring S is a product $J^+S \otimes H$, where H is a $\mathbf{G}$-module complement of J^+S, graded by degree (the harmonic polynomials, for instance) [Ko1, Dix]. The representation of $SL(n,\mathbf{C})$ on H, in turn, decomposes into an infinite direct sum of irreducible finite-dimensional representations. If F^χ is the highest weight vector of a summand with highest weight χ, then

$$F^\chi(b^{-1}Xb) = \chi(b)F^\chi(X), \quad b \in \mathbf{B}_+.$$

We call such functions *semi-invariants*. The ratio of two highest weight vectors for two copies of this representation is then *invariant* under $\mathbf{B}_+$, and will by Proposition 1.1 produce a constant of motion of the full Kostant-Toda lattice.

A complete picture of all the constants of motion would follow from a thorough understanding of the Poisson algebra generated by the highest weight vectors in the decomposition of a suitable H. Calculations (using MACSYMA) of representations of groups of small rank (e.g., C_2, B_3, C_3) show that there are distinct families of involutive constants of motion for the full Kostant-Toda lattice. The set I_{rk} in (5) is only one of at least two possible even for $SL(4,\mathbf{C})$ (see the Conclusion); it may be the most interesting for geometers. We believe, but have not proved, that the full Kostant-Toda lattice (at least for the classical algebras) is integrable in the non-commutative sense.

Reduction is in some sense a converse to invariant theory. Instead of building up invariants along a chain of parabolic subgroups $\mathbf{P}_k$, one uses integrals at each level of the chain to reduce to smaller symplectic manifolds. We describe the procedure for the $\mathcal{SL}(n,\mathbf{C})$ lattice; adaptations to Lie algebras of types B_n, C_n are sketched in the body of the paper, and the problems with D_n and G_2 are explained in the conclusion.

At first, one has the constants of motion I_{r0}, the Chevalley invariants, or equivalently the eigenvalues $\{\lambda_{r0}\}$ of X. That these are integrals follows from the Adler-Kostant-Symes Theorem just as in the tridiagonal case. The angles conjugate to the $\{\lambda_{r0}\}_{r=2}^{r=n}$ are given by the resolvent prescription [Mo, DLNT] that works for tridiagonal Toda: $\theta_{r0} = \log \gamma_{r0}/\gamma_{10}$, where

$$\gamma_{r0} = \lim_{\lambda \to \lambda_{r0}} \det(X - \lambda \, \mathrm{Id}) \, \langle (X-\lambda)^{-1} e_n, e_n \rangle. \tag{8}$$

To carry out symplectic reduction, one must first find good coordinates on the level set of fixed λ_{r0} $(r = 1, \dots, n)$ (the level set of the momentum map for the Chevalley invariants of X). Our approach relies on an embedding of this isospectral submanifold (of the phase space $\epsilon + \mathcal{B}_-$) into the flag manifold $\mathbf{Flag}_n = SL(n, \mathrm{C})/\mathbf{B}_+$.

Proposition 1.2 ([Ko3]) *There is a unique lower unipotent L that conjugates $X \in \epsilon + \mathcal{B}_-$ to a* companion matrix C_{Λ_0},

$$X = L C_{\Lambda_0} L^{-1}, \tag{9}$$

$$C_{\Lambda_0} = \begin{pmatrix} 0 & 1 & 0 & \dots & 0 \\ 0 & 0 & 1 & \dots & 0 \\ \vdots & \ddots & \ddots & \ddots & \vdots \\ 0 & \dots & \ddots & \ddots & 1 \\ s_n & \dots & \dots & s_2 & 0 \end{pmatrix}.$$

(The characteristic polynomial of C_{Λ_0} is $\lambda^n - s_2\lambda^{n-2} - \dots - s_n$.)

We cannot overemphasize the importance of this result. Because L is unique, an isospectral set in $\epsilon + \mathcal{B}_-$ can be mapped isomorphically to the flag manifold, and our picture of the full Kostant-Toda lattice depends crucially on this embedding. Notice also that the entries of C_{Λ_0} are polynomials in the entries of X; this feature makes it possible to do MACSYMA calculations, and indeed some of our basic formulas (such as (6)) were discovered in that way.

Definition 1.3 *Let $(\epsilon + \mathcal{B}_-)_{\Lambda_0}$ be the set of elements of $\epsilon + \mathcal{B}_-$ with fixed eigenvalues* $\Lambda_0 = \mathrm{diag}\,(\lambda_1, \dots, \lambda_n)$. *Define*

$$\begin{aligned} K_{\Lambda_0} : (\epsilon + \mathcal{B}_-)_{\Lambda_0} &\to \mathbf{Flag}_n \\ X &\mapsto L^{-1} \bmod \mathbf{B}_+. \end{aligned}$$

Proposition 1.3 *Under the map K_{Λ_0}, the Toda flow becomes*

$$X(t) \mapsto (\exp C_{\Lambda_0} t)\, L^{-1} \bmod \mathbf{B}_+.$$

More generally, the "basic" Toda flows of the Hamiltonians Trace X^j *generate the linear algebraic torus action on $\mathbf{Flag}_n$ of the centralizer of C_{Λ_0} in $SL(n, C)$.*

The second half of reduction requires a choice of "angles" on the level set of the momentum map. We use (9) to rewrite (8) as

$$\gamma_{r0} = \lim_{\lambda \to \lambda_{r0}} \det(X - \lambda\,\mathrm{Id})\,\langle (C_{\Lambda_0} - \lambda)^{-1} e_n, L^t e_n \rangle. \tag{10}$$

From this one can deduce that the γ_{r0} do not depend on the full flag L^{-1} mod $\mathbf{B}_+$, but only on the partial flag L^{-1} mod $\mathbf{P}_1$ (cf. Definition 1.2), which is represented by an element of the Heisenberg group

$$L_0 = \begin{pmatrix} 1 & 0 & 0 & \dots & \dots & 0 \\ a_1 & 1 & 0 & \dots & \dots & 0 \\ a_2 & 0 & 1 & \ddots & \ddots & \vdots \\ \vdots & \vdots & \ddots & \ddots & \ddots & \vdots \\ a_{n-2} & 0 & \dots & 0 & 1 & 0 \\ c & b_{n-2} & \dots & b_2 & b_1 & 1 \end{pmatrix}. \tag{11}$$

This group is an open dense submanifold of $SL(n,\mathbf{C})/\mathbf{P}_1$.

To get the reduced phase space, one induces a Poisson structure on a level set of $\{\theta_{r0}\}$ in $\epsilon + \mathcal{B}_-$. Fixing the angles singles out a certain subvariety Θ_0 of matrices of the form (11), and the reduced phase space fibers over Θ_0. A dense open subset of the fiber over a point (11) is coordinatized by the $(n-2)\times(n-2)$ matrices in which the subdiagonal zeros in (11) are replaced by nonzero entries $*$:

$$\begin{pmatrix} 1 & 0 & 0 & \dots & \dots & 0 \\ 0 & 1 & 0 & \dots & \dots & 0 \\ 0 & * & 1 & \ddots & \ddots & \vdots \\ \vdots & \vdots & \ddots & \ddots & \ddots & \vdots \\ 0 & * & \dots & * & 1 & 0 \\ 0 & 0 & \dots & 0 & 0 & 1 \end{pmatrix}$$

The fiber can be identified with the smaller flag manifold

$$\mathbf{Flag}_{n-2} = SL(n-2,\mathbf{C})/\mathbf{B}_+^{(n-2)};$$

here $\mathbf{B}_+^{(n-2)}$ is the subgroup of upper triangular matrices in $SL(n-2,\mathbf{C})$.

A point in $\mathbf{Flag}_n$ represents a chain of subspaces

$$\{0\} \subset V_1 \subset \dots \subset V_k \subset \dots \subset V_{n-1} \subset \mathbf{C}^n,$$

and in particular, $X \in (\epsilon + \mathcal{B}_-)_{\Lambda_0}$ is identified (via $X \mapsto L^{-1}$ mod $\mathbf{B}_+$ in Definition 1.3) with the flag generated by the columns of L^{-1}. The angles θ_{r0}, we assert, depend only on the *partial flag* $V_1 \subset V_{n-1}$.

We now sketch what is, for us, the crucial next step—reduction by the momentum map for the 1-chop integrals I_{r1} from Definition 1.1. Here reduction makes contact with invariant theory: the equations $I_{r1} = c_{r1}, r = 1, \dots, n-2$, for the level set of the 1-chop integrals, which are very messy in the affine coordinates on $\epsilon + \mathcal{B}_-$, again depend only on the partial flag $V_1 \subset V_{n-1}$. In fact, they are quadratic (bilinear) equations in the entries of (11) (cf. (6)). They define a unique point (11) in the base Θ_0 of our fibered reduced phase space. The Hamiltonian flows generated by the I_{r1} move in the fiber $\mathbf{Flag}_{n-2}$, where they are coordinatized by a new set of angles, θ_{r1}.

Note that we are now essentially back at the starting point. The level set of the eigenvalues λ_{r0} of X was embedded into $\mathbf{Flag}_n$; after the first reduction, the level set of the λ_{r1} has been embedded in $\mathbf{Flag}_{n-2}$. The analogy goes further. The polynomial Q_1 (cf. (5)) is the characteristic polynomial of elements of $\mathcal{GL}(n-2, \mathrm{C})$ associated to points in $\mathbf{Flag}_{n-2}$ by a variant of Kostant's map K_{Λ_0} in Definition 1.3; its coefficients I_{r1} are "reduced" Chevalley invariants, and its roots λ_{r1} play the role of the spectrum, λ_{r0}, at the start.

When we next look at the angles θ_{r1} and the 2-chop integrals I_{r2}, we find that in each fiber $\mathbf{Flag}_{n-2}$ they involve only the partial flag $V_2 \subset V_{n-2}$: the pattern repeats.

The reduction procedure would leave us, eventually, with a one-point reduced space. We do not actually carry out the reduction; rather, we use it as motivation in a study of the geometry of the level set of all the integrals I_{rk} and of their Hamiltonian flows on the level set. Let $\mathcal{LSV}$ ("level set variety") denote the closure, in $\epsilon + \mathcal{B}_-$, of this level set. We will make assumptions that force a dense open subset of $\mathcal{LSV}$ to lie in a generic symplectic leaf of the Poisson structure on $\epsilon + \mathcal{B}_-$; its boundary, however, contains non-generic, lower-dimensional leaves.

As one might expect from the reduction discussion, $\mathcal{LSV}$ is a tower of bundles of level sets of the successive momentum maps for the k-chop integrals I_{rk}. We now summarize the picture as developed in Sections 4-6.

First fix the Chevalley (or "0-chop") integrals I_{r0} of X and embed $(\epsilon + \mathcal{B}_-)_{\Lambda_0}$ (Definition 1.3) in the flag manifold $\mathbf{Flag}_n$ by the Kostant map in Definition 1.3. The Hamiltonian flows become the action of a certain Cartan subgroup $\mathbf{H}$ on the flag manifold $\mathbf{Flag}_n$, as in Proposition 1.3.

The 1-chop integrals I_{r1}, as functions on $\mathbf{Flag}_n$, depend only on the partial flag $V_1 \subset V_{n-1}$. $\mathbf{Flag}_n$ is a fiber bundle

$$\begin{array}{ccc} \mathbf{P}_1/\mathbf{B}_+ & \longrightarrow & SL(n,\mathrm{C})/\mathbf{B}_+ \\ & & \downarrow \\ & & SL(n,\mathrm{C})/\mathbf{P}_1. \end{array} \tag{12}$$

The level set of the I_{r1} in $\mathbf{Flag}_n$ is a bundle over a certain variety $\mathcal{F}_1$ cut out by the equations $I_{r1} = c_{r1}, r = 1, \dots, n-2$, in the partial flag variety $SL(n,\mathrm{C})/\mathbf{P}_1$. Because the I_{r1} are constant along the I_{r0} flows,

$\mathcal{F}_1$ is invariant under $\mathbf{H}$. Using the theory of toric varieties to calculate the degree of $\mathcal{F}_1$, we show that it is in fact a single generic orbit of the torus $\mathbf{H}$ on $SL(n,\mathbf{C})/\mathbf{P}_1$. Thus, the angles θ_{r0} parametrize a generic torus orbit in $SL(n,\mathbf{C})/\mathbf{P}_1$, while the 1-chop integrals I_{r1} index the family of such torus orbits. In fact, these integrals may be interpreted in terms of the Grassmannian "cross-ratios" introduced in [GM, Kap] to parametrize configurations of points in projective spaces; see Section 7 for an example.

Notation 1.1 The guiding idea is that the the k-chop integrals for $SL(n,\mathbf{C})$ are basically the 1-chop integrals for $SL(n-2k+2,\mathbf{C})$. We therefore introduce the Borel subgroups $\mathbf{B}_+^{(n-2k)}$ of $SL(n-2k,\mathbf{C})$. Furthermore, we will regard $SL(n-2k,\mathbf{C})$ as subgroup of $SL(n,\mathbf{C})$ via the obvious inclusion (as the middle block of an $n \times n$ matrix), and then view $\widetilde{\mathbf{P}}_{k'} \stackrel{\text{def}}{=} \mathbf{P}_{k'} \cap SL(n-2k,\mathbf{C})$, for $k' \geq k$, as parabolic subgroup of $SL(n-2k,\mathbf{C})$. ∎

At this stage, the partial flag $V_1 \subset V_{n-1}$ has been fixed. The 2-chop integrals I_{r2}, as functions on $\mathbf{Flag}_n$, depend only on the partial flag $V_2 \subset V_{n-2}$. Each fiber of $SL(n,\mathbf{C})/\mathbf{B}_+ \to \mathcal{F}_1$ has been identified with the manifold of shorter flags, $V_2 \subset V_3 \subset \ldots \subset V_{n-2}$, and we view it as the bundle

$$\begin{array}{ccc} \widetilde{\mathbf{P}}_2/\mathbf{B}_+^{(n-2)} & \longrightarrow & SL(n-2,\mathbf{C})/\mathbf{B}_+^{(n-2)} \\ & & \downarrow \\ & & SL(n-2,\mathbf{C})/\widetilde{\mathbf{P}}_2. \end{array}$$

The I_{r2} cut out, in each fiber, a subvariety $\mathcal{F}_2$ of $SL(n-2,\mathbf{C})/\widetilde{\mathbf{P}}_2$; it is a generic orbit of the action of a Cartan subgroup $\widetilde{\mathbf{H}}_1$ of $SL(n-2,\mathbf{C})$. And so on: the constructions repeat.

At the end of this process, the embedding of the level set variety in $\mathbf{Flag}_n$ is conjugate to another open embedding, into a product **Prod** of (partial) flag manifolds of decreasing dimension,

$$SL(n,\mathbf{C})/\mathbf{P}_1 \times \ldots \times SL(n-2k,\mathbf{C})/\widetilde{\mathbf{P}}_{k+1} \times \ldots \times SL(n-2m,\mathbf{C})/\widetilde{\mathbf{P}}_{m+1},$$

(with $m = [(n-1]/2])$. In **Prod**, the closure of the image of the level set variety is the closure of a product of generic torus orbits, one from each factor. This is the geometric expression of the linearization of the flows.

The rest of this paper is organized as follows. Notation and background material (about factorization, Thimm's method, and maps to flag manifolds) are collected in Section 2. The k-chop integrals are discussed in Section 3. We explain how representation theory leads to neat expressions like (6). Most results to this point hold for general semisimple Lie algebras, and are presented in that context. From Section 4 on, we return to

$\mathcal{SL}(n, \mathbf{C})$, because certain crucial facts have (to date) only a computational, rather than conceptual, explanation. In that section we give a new proof of the generic independence of the I_{rk}. Section 5 studies the subvariety $\mathcal{F}_1$ of $\mathbf{G}/\mathbf{P}_1$ defined by the 1-chop integrals for the case of $SL(n, \mathbf{C})$; we prove, as advertised earlier, that it is a generic torus orbit. In Section 6, we show first that the 1-chop flows generate a torus action on the smaller flag manifolds that are the fibers of (12), and then extend this result to all k-chop flows by induction. A fairly detailed description of the $\mathcal{SL}(4, \mathbf{C})$ Kostant-Toda lattice is provided in Section 7; the explicit formulas may serve to illustrate the general discussion in the body of the paper. The Conclusion mentions a few open problems and, of more interest, explains why our methods will not work for the orthogonal algebra D_n and the exceptional Lie algebra G_2, and shows how to get competing involutive integrals for $\mathcal{SL}(4, \mathbf{C})$. The Appendix deals with certain aspects of the relation between the full DLNT-Toda lattice and the Kostant-Toda version.

Acknowledgments. We thank Luen-chau Li for sending us his unpublished notes on the full Toda equations associated to the other classical Lie algebras, and Sam Evens for crucial explanations. Ercolani and Flaschka were supported by NSF grant DMS-9001897. Singer was supported by the Sloan Foundation and the AT&T Bell Laboratories.

2. Background

2.1 The Full Kostant-Toda Lattice

Let $\mathcal{G}$ be a complex semi-simple Lie algebra and $\mathbf{G}$ its simply connected Lie group. Choose a Cartan subalgebra $\mathcal{H}$ of $\mathcal{G}$, and a set Δ of positive roots for $\mathcal{H}$. Let $\mathcal{B}_+$ (resp. $\mathcal{N}_+$) be the Borel subalgebra for Δ (resp. its nilradical), and let $\mathcal{B}_-, \mathcal{N}_-$ be the opposite Borel and nilpotent algebras. Call the associated groups $\mathbf{B}_\pm$ and $\mathbf{N}_\pm$. The Killing form is denoted by $\kappa(\cdot\,,\,\cdot)$, and the adjoint action of $\mathbf{G}$ on $\mathcal{G}$ by $g \cdot X$.

We are interested in systems Hamiltonian with respect to the Lie-Poisson structure defined on $\mathcal{B}_+^*$ by

$$\{f, g\}(\beta) = \beta([\nabla f(\beta), \nabla g(\beta)]). \tag{13}$$

The gradient $\nabla f(\beta)$ belongs to $\mathcal{B}_+$: for all $\gamma \in \mathcal{B}_+^*$,

$$\lim_{\delta \to 0} \frac{f(\beta + \delta\gamma) - f(\beta)}{\delta} = \gamma(\nabla f(\beta)).$$

Every coadjoint orbit $\mathcal{O} \subset \mathcal{B}_+^*$ is a (complex) symplectic manifold. For simple complex Lie algebras, the dimensions of the generic coadjoint orbits in $\mathcal{B}_+^*$ are computed in [Ar, Trol, Tro2]. The paper [Tro2] announces the following elegant result. *Let s be the number of simple roots which are not mapped to their negatives by the longest element of the Weyl group of $\mathcal{G}$. The codimension of the generic orbit is $s/2$.* Unfortunately, the proof is not conceptual: the corank of the Poisson tensor is computed for each simple algebra, and the numbers obtained, miraculously, turn out to be the interesting numbers $s/2$.

For the case $\mathcal{G} = \mathcal{SL}(n,\mathbf{C})$, the codimension (found independently in [DLNT]) is $[(n-1)/2]$.

In this paper, we will realize $\mathcal{B}_+^*$ as a subspace of $\mathcal{G}$ parallel to $\mathcal{B}_-$. Let

$$\epsilon \stackrel{\text{def}}{=} \sum_{\alpha\in\Delta} x_\alpha,$$

(the x_α are fixed simple root vectors) and identify $\beta \in \mathcal{B}_+^*$ with $X \in \epsilon + \mathcal{B}_-$ when for all $Y \in \mathcal{B}_+$, $\beta(Y) = \kappa(X,Y)$.

The abstract coadjoint action of $\mathbf{B}_+$ on $\mathcal{B}_+^*$ now becomes an action of $\mathbf{B}_+$ on $\epsilon + \mathcal{B}_-$:

$$\mathrm{Ad}_b^* X = \epsilon + \mathbf{\Pi}_{\mathcal{B}_-} b^{-1} \cdot (X - \epsilon).$$

The Poisson bracket on $\epsilon + \mathcal{B}_-$ is given by (13), with the pairing provided by the Killing form. The functions of interest are usually restrictions to $\epsilon + \mathcal{B}_-$ of functions defined on all of $\mathcal{G}$. Write $\tilde{f} = f|_{\epsilon+\mathcal{B}_-}$. The Poisson bracket between two such functions can be found from

$$\{\tilde{f},\tilde{g}\}(X) = \kappa(X, [\mathbf{\Pi}_{\mathcal{B}_+} \nabla f(X), \mathbf{\Pi}_{\mathcal{B}_+} \nabla g(X)]), \tag{14}$$

where $X \in \epsilon + \mathcal{B}_-$ and $\nabla f, \nabla g$ are both computed in $\mathcal{G}$.

These constructions are somewhat unmotivated; they are best understood in terms of r-matrices and reduction [R], see the appendix.

Example. When $\mathcal{G} = \mathcal{SL}(n,\mathbf{C})$, $\mathcal{N}_+$ ($\mathcal{N}_-$) is the algebra of strictly upper (lower) triangular matrices while $\mathcal{B}_+$ ($\mathcal{B}_-$) is the algebra of upper (lower) triangular matrices with trace 0. The set $\epsilon + \mathcal{B}_-$ consists of matrices of the form (2). The *full Kostant-Toda lattice on $\mathcal{G}$* is the system of differential equations generated by the Hamiltonian $\frac{1}{2}$ Trace X^2 on $\epsilon + \mathcal{B}_-$. Hamilton's equations have the Lax form

$$\dot{X} = [X, \mathbf{\Pi}_{\mathcal{N}_-} X], \tag{15}$$

where $\mathbf{\Pi}_{\mathcal{N}_-}$ is projection onto $\mathcal{N}_-$ along $\mathcal{B}_+$. ∎

2.2 Integrability

A natural framework for seeking involutive families on $\epsilon + \mathcal{B}_-$ is provided by a general method due to Thimm [Th] and applied to full symmetric Toda in [DLNT].

Consider a nested chain of parabolic subalgebras of $\mathcal{G}$:

$$\mathcal{B}_+ = \mathcal{P}_{m+1} \subset \ldots \subset \mathcal{P}_k \subset \ldots \subset \ldots \subset \mathcal{P}_0 = \mathcal{G}.$$

The inclusion maps induce a sequence of maps of the vector space duals:

$$\mathcal{B}_+^* = \mathcal{P}_{m+1}^* \leftarrow \ldots \leftarrow \mathcal{P}_k^* \leftarrow \ldots \leftarrow \mathcal{P}_0^* = \mathcal{G}^* \cong \mathcal{G};$$

the projections

$$\rho_k : \mathcal{P}_k^* \to \mathcal{P}_{k+1}^*$$

are Poisson with respect to the Lie-Poisson structure on $\mathcal{P}_k^*$.

Let $\mathcal{CAS}_k$ denote the subspace of rational Casimirs in the function field $\mathrm{C}(\mathcal{P}_k^*)$. If $\mathcal{I}$ is an involutive family of rational functions in $\mathrm{C}(\mathcal{P}_{k+1}^*)$, then $\rho_k^*(\mathcal{I}) \cup \mathcal{CAS}_k$ generates an involutive family in $\mathrm{C}(\mathcal{P}_k^*)$. Proceeding inductively up the chain, one produces a large involutive family in $\mathcal{G}$.

Now let $I_1, \ldots, I_\ell$ be homogeneous generators of the invariant polynomials on $\mathcal{G}$. Since $\mathcal{CAS}_0 = \mathrm{C}(I_1, \ldots, I_\ell)$, the involutive families constructed by Thimm's method contain the Toda hamiltonian. One must remember that the functions obtained by this construction are in involution on all of $\mathcal{G}$, and the Hamiltonian vector fields generated by the invariant polynomials are trivial. By an application of the Adler - Kostant - Symes theorem, DLNT showed that the restrictions to $\mathcal{B}_+^*$ remain involutive, while the flows of the I_j become nontrivial.

The Thimm-DLNT method reduces the search for integrals to the construction of coadjoint invariants on $\mathcal{P}_k^*$. If the procedure leads to a maximal involutive family, complete integrability is established. If not, the system may still be integrable with integrals produced by other means. On the other hand, different parabolic chains can produce different involutive families which may not be mutually involutive. An example is given in the Conclusion.

Later, we shall need the *Levi decomposition* of a parabolic subalgebra (or subgroup); we fix our notation.

Definition 2.1 *A standard parabolic subalgebra $\mathcal{P}$ of $\mathcal{G}$ (i.e., one containing our fixed Borel subalgebra) has a canonical decomposition into a (vector space) direct sum of subalgebras*

$$\mathcal{L} \oplus \mathcal{R},$$

where $\mathcal{L}$ is reductive and $\mathcal{R}$ is the nilradical of $\mathcal{P}$. We call $\mathcal{L}$ the Levi summand*; it can be further decomposed as $\mathcal{L}^{ss} \oplus C$, where $\mathcal{L}^{ss}$ is semisimple and C is the center of $\mathcal{L}$. There is a corresponding decomposition $\mathbf{P} = \mathbf{LR}$ as semidirect product of groups and manifold product. For $\xi \in \mathcal{P}$, we write $[\xi]$ (resp. $[\xi]^{ss}$) for the $\mathcal{L}$ (resp. $\mathcal{L}^{ss}$) component. The same notation is used for the groups. We refer to $[\xi]$ (or $[p]$) as the* core *of ξ (resp. p). When the parabolic is $\mathcal{P}_k$ (or the group $\mathbf{P}_k$), the cores are denoted by $[\xi]_k$, $[\xi]_k^{ss}$, and so forth.*

Next, we introduce an explicit realization of the duals of parabolic subalgebras. Let $\mathcal{P}$ be a standard parabolic subalgebra of $\mathcal{G}$, with Lie subgroup $\mathbf{P}$ of $\mathbf{G}$. As with $\mathcal{B}_+^*$, we use the Killing form to represent $\mathcal{P}^*$ by a subspace of $\mathcal{G}$ parallel to $\mathcal{P}_-$ = transpose of $\mathcal{P}$ in $\mathcal{G}$. The shift $\tau \in \mathcal{R}$ is taken to be $\sum x_{\alpha_i}$, where the sum extends over the simple roots whose negatives do *not* belong to $\mathcal{P}$. The coadjoint action is realized by

$$\mathrm{Ad}_p^* X = \tau + \mathbf{\Pi}_{\mathcal{P}\_} p^{-1} \cdot (X - \tau), \quad X \in \tau + \mathcal{P},\ p \in \mathbf{P}.$$

Definition 2.2 *A function $f : \mathcal{P}^* \to \mathbf{C}$ is called a* coadjoint invariant *for* $\mathbf{P}$ *if*

$$f(\mathrm{Ad}_p^* X) = f(X), \quad X \in \mathcal{P}^*.$$

For variety, we also refer to such functions as parabolic Casimirs *or* parabolic invariants.

Proposition 2.1 *Under the restriction homomorphism:*

$$\begin{array}{rcl} \mathbf{C}(\mathcal{G}) & \to & \mathbf{C}(\tau + \mathcal{P}_-) \\ I(Y) & \mapsto & I(\tau + \mathbf{\Pi}_{\mathcal{P}\_} Y), \quad Y \in \mathcal{G}, \end{array}$$

the coadjoint invariants in $\mathbf{C}(\tau + \mathcal{P}_-)$ correspond to functions I in $\mathbf{C}(\mathcal{G})$ such that (i) $\nabla I \in \mathcal{P}$, (ii) $I(p^{-1} \cdot Y) = I(Y)$ for all $p \in \mathbf{P}$, $Y \in \mathcal{G}$.

The proof is easy.

Example 2.1 Let E_{rk}, I_{rk} and $\mathcal{P}_k$ be as in Definitions 1.1 and 1.2. By construction, the E_{rk} are independent of the coordinates in the nilradical $\mathcal{R}_k$, so that $\nabla E_{rk}(Y)$ (and hence $\nabla I_{rk}(Y)$) belongs to $\mathcal{P}_k$ for all $Y \in \mathcal{G}$. Semi-invariance of E_{rk} and invariance of I_{rk} under $\mathbf{P}_k$ is shown just as in [DLNT]. One can see easily that the character χ_k in (7) is

$$\chi_k(p) = p_{11} \cdots p_{kk} p_{n-k+1,n-k+1}^{-1} \cdots p_{nn}^{-1}. \quad \blacksquare \tag{16}$$

2.3 Factorization

We state some lemmas that are fundamental to the construction of explicit solutions of the Hamiltonian flows generated by parabolic Casimirs. The proofs are reasonably simple and can be found in [DLNT, S1]. Below, I is a coadjoint invariant for the standard parabolic subgroup $\mathbf{P}$, *and ∇I denotes its gradient computed in the whole Lie algebra $\mathcal{G}$.*

Definition 2.3 *Fix $X_0 \in \epsilon + \mathcal{B}_-$. Whenever the factorization in (17) is possible for $t \in \mathbf{C}$, define $n(t) \in \mathbf{N}_-$ and $b(t) \in \mathbf{B}_+$ by*

$$\exp t \nabla I(X_0) \stackrel{\text{def}}{=} n(t) b(t). \tag{17}$$

Lemma 2.1 *Let $Y \in \tau + \mathcal{P}_-$. If $p \in \mathbf{P}$, then $p \cdot \nabla I(Y) = \nabla I(p \cdot Y)$.*

Lemma 2.2 *Let $X_0 \in \epsilon + \mathcal{B}_-$. If $\exp t\nabla I(X_0) = n(t)b(t)$ as in (17), then $X(t) = n^{-1}(t) \cdot X_0 = b(t) \cdot X_0$ solves*

$$\dot{X} = [X, \Pi_{\mathcal{N}_-} \nabla I(X)], \quad X(0) = X_0.$$

Note that this lemma covers the case $\mathcal{P} = \mathcal{G}$, $I = (1/2)\,\mathrm{Trace}\, X^2$, i.e. the full Kostant-Toda equations.

Lemma 2.3 *With $X(t)$ as in the preceding lemma, $\nabla I(X(t)) = n^{-1}(t)\dot{n}(t) + \dot{b}(t)b^{-1}(t)$.*

When $g \in \mathbf{G}$ can be factored as $g = nb$, $n \in \mathbf{N}_-$, $b \in \mathbf{B}_+$, we write the lower unipotent factor as $n(g)$.

Lemma 2.4

$$[n(\exp t\nabla I(X))] = n(\exp t[\nabla I(X)]).$$

Proof. The key observation is that $[AB] = [A][B]$ if $A, B \in \mathbf{P}$. Since $\nabla I \in \mathcal{P}$, $[\exp t\nabla I(X)] = \exp t[\nabla I(X)]$. The lemma follows because the lower unipotent factor n commutes with the Levi projection $[\cdot]$ on $\mathbf{P}$. ∎

2.4 Map to the Flag Manifold

Our approach to the full Toda lattice centers on an embedding of the isospectral set $(\epsilon + \mathcal{B}_-)_{\Lambda_0}$ (Definition 1.3) into the flag manifold $\mathbf{G}/\mathbf{B}_+$. We now define the embedding for a general semisimple Lie algebra.

Let $\mathcal{S}$ denote a subspace of $\mathcal{B}_-$ such that $\mathcal{B}_- = [\epsilon, \mathcal{N}_-] \oplus \mathcal{S}$; one can choose $\mathcal{S}$ to be graded by the heights of the roots. The element $\epsilon \in \mathcal{B}_+$ is the sum of the simple positive roots.

Lemma 2.5 ([Ko3]) *Every regular[2] $X \in \epsilon + \mathcal{B}_-$ can be conjugated to an element $C \in \epsilon + \mathcal{S}$ by a unique unipotent matrix $L \in \mathbf{N}_-$: $X = L \cdot C$.*

When $\mathcal{G} = \mathcal{SL}(n, \mathrm{C})$, we may take $\epsilon + \mathcal{S}$ to be the plane of companion matrices as in Proposition 1.2.

Let Λ_0 be a regular element of the Cartan subalgebra $\mathcal{H}$, and as before let $(\epsilon + \mathcal{B}_-)_{\Lambda_0}$ denote the set of all $X \in \epsilon + \mathcal{B}_-$ that are conjugate to Λ_0. Lemma 2.5 provides a map from $(\epsilon + \mathcal{B}_-)_{\Lambda_0}$ to $\mathbf{G}/\mathbf{B}_+$:

$$\begin{aligned} K_{\Lambda_0} \ : \ (\epsilon + \mathcal{B}_-)_{\Lambda_0} &\longrightarrow \mathbf{G}/\mathbf{B}_+ \\ X &\longmapsto L^{-1} \bmod \mathbf{B}_+. \end{aligned} \tag{18}$$

[2] "Regular" means that the centralizer of X has minimal dimension = rank. If X is semisimple (meaning: it can be diagonalized), "regular" says that the eigenvalues are distinct.

Proposition 2.2 *(i) K_{Λ_0} is an embedding of the affine variety $(\epsilon+\mathcal{B}_-)_{\Lambda_0}$.*
(ii) The solution of the flow generated by a parabolic invariant I is mapped to

$$L^{-1}n(\exp t\nabla I(X)) \bmod \mathbf{B}_+ = L^{-1}(\exp t\nabla I(X)) \bmod \mathbf{B}_+.$$

(iii) When I is a Chevalley invariant ($I(g\cdot Y) = I(Y)$ for all $g \in \mathbf{G}$ and $Y \in \mathcal{G}$), then

$$L^{-1}\exp(t\nabla I(X)) \bmod \mathbf{B}_+ = \exp(t\nabla I(C_{\Lambda_0}))L^{-1} \bmod \mathbf{B}_+.$$

Proof. (i) is proved in [Ko3]. (ii) is implicit in [DLNT], and explicit in [S1]. (iii) is observed in [FH]. ∎

Remark. Part (iii) of this Proposition explains why we prefer $\epsilon + \mathcal{B}_-$ to the perhaps more appealing phase space of real symmetric matrices. To map an isospectral set $\mathcal{X}_{\Lambda_0}$ of symmetric matrices to the flag manifold, we should conjugate $X \in \mathcal{X}_{\Lambda_0}$ to a reference element, say the diagonal matrix Λ_0: $X = Q\Lambda_0 Q^{-1}$, with an orthogonal matrix Q that is determined only up to conjugation by an element from the group S of diagonal matrices with entries ± 1. Then we should send X to $Q^{-1} \bmod S$, in order to have the Toda flow act by exponentials on the left. But Q is not unique, and replacement of Q by Qs, $s \in S$, changes the coset $Q^{-1} \bmod S$. This problem can be overcome, with some effort, in the case of the tridiagonal Toda lattice, but it is not possible to make a smooth choice of Q over the larger isospectral set $\mathcal{X}_{\Lambda_0}$ [BFR]. ∎

A different embedding of $(\epsilon + \mathcal{B}_-)_{\Lambda_0}$ in $\mathbf{Flag}_n$ will be useful.

Definition 2.4 *Fix a $V \in \mathbf{G}$ for which $V \cdot C_{\Lambda_0} = \Lambda_0 \in \mathcal{H}$ (Λ_0 regular), and map*

$$X \mapsto V^{-1}L^{-1} \bmod \mathbf{B}_+. \tag{19}$$

We call this the torus embedding *of $(\epsilon + \mathcal{B}_-)_{\Lambda_0}$.*

One checks that the flows generated by the Chevalley invariants I_{r0} of X transform to the action of the diagonal Cartan subgroup $\mathbf{H}$ of $\mathbf{G}$. In particular, when $\mathbf{G} = SL(n,\mathbb{C})$, V may be taken to be a Vandermonde matrix, and the flow with Hamiltonian= Trace $X^2/2$ (*the* Toda flow) becomes $e^{t\Lambda_0}V^{-1}L^{-1} \bmod \mathbf{B}_+$.

3. Parabolic Casimirs

This section has several goals. We explain how parabolic semi-invariants produce Toda integrals. Several properties of the k-chop integrals in (5) were found by explicit calculation for $\mathcal{SL}(n,\mathbb{C})$, but fit more general patterns. In particular, we argue that parabolic Casimirs must assume the simple form (6) in the flag manifold. We also extend the chopped integral construction to other classical Lie algebras.

3.1 The Chops

This subsection deals only with $\mathcal{SL}(n, \mathrm{C})$. Definitions 1.1 and 1.2 introduced the k-chop integrals I_{rk}, $r = 1, \ldots, n-2k$, $k = 0, \ldots, [\frac{n-1}{2}]$ and the parabolic subgroups $\mathbf{P}_k$ for which they are coadjoint invariants (for $r = 1$, the I_{rk} are Casimirs on $\epsilon + \mathcal{B}_-$, and for $k = 0$, they are Chevalley invariants). These integrals are related in a remarkable way: for each k, all I_{rk} are functions of the eigenvalues of the Levi summand $[\nabla I_{2k}(X)]_k$ (in the Levi component $\mathcal{L}_k$ of $\mathcal{P}_k$). It is possible to construct $\nabla I_{2k}(X)$ directly from X, without ever taking a derivative. For example, when $k = 0$, then $I_{20}(X) = (\mathrm{Trace}\, X)^2/2 - (\mathrm{Trace}\, X^2)/2$, and $\nabla I_{20}(X) = -(X - \frac{1}{n}(\mathrm{Trace}\, X\, \mathrm{Id}))$. Proposition 3.1 generalizes this to $k > 0$.

Definition 3.1 *Fix $X \in \epsilon + \mathcal{B}_-$ and let k be an integer, $0 \leq k \leq [\frac{n-1}{2}]$. Break X into blocks of the indicated sizes:*

$$\begin{array}{cc} & \begin{array}{ccc} k & n-2k & k \end{array} \\ X = \begin{array}{r} k \\ n-2k \\ k \end{array} & \begin{pmatrix} X_1 & X_2 & X_3 \\ X_4 & X_5 & X_6 \\ X_7 & X_8 & X_9 \end{pmatrix} \end{array}$$

and, when $\det X_7 \neq 0$, *let* $\phi_k(X) \stackrel{\text{def}}{=} X_5 - X_4 X_7^{-1} X_8 \in \mathcal{GL}(n-2k, C)$. *Set* $\phi_0(X) \stackrel{\text{def}}{=} X$.

Proposition 3.1 ([S1, S2]) Id_j *is the* $j \times j$ *identity matrix. Convert*

$$(-1)\big(\phi_k(X) - (n-2k)^{-1}\, \mathrm{Trace}\, \phi_k(X)\, \mathrm{Id}_{n-2k}\big)$$

to an $n \times n$ *matrix by adding a border (of width* k*) of zeros. The result is precisely* $[\nabla I_{2k}(X)]_k^{ss}$.

The proof is by calculation; so far we have no Lie-algebraic explanation of this crucial fact. See Section 7 for an example.

Proposition 3.2 *The* I_{rk}, $r = 1, \ldots, n-2k$, *are the coefficients of* $\det(\lambda - \phi_k(X))$.

Proof. Letting $|\cdot|$ denote the determinant of a block matrix, we have

$$\widetilde{Q}_1(X, \lambda) = \begin{vmatrix} X_4 & X_5 - \lambda \\ X_7 & X_8 \end{vmatrix} = \begin{vmatrix} \mathrm{Id}_{n-2k} & -X_4 X_7^{-1} \\ 0 & \mathrm{Id}_k \end{vmatrix} \begin{vmatrix} X_4 & X_5 - \lambda \\ X_7 & X_8 \end{vmatrix},$$

which works out to

$$\begin{vmatrix} 0 & \phi_k(X)-\lambda \\ X_7 & X_8 \end{vmatrix} = |X_7|\,|\lambda-\phi_k(X)|. \blacksquare$$

In our application of these results, we will require of X that all $\phi_k(X)$ be generic in a sense defined later.

Example 3.1 When $(\epsilon+\mathcal{B}_-)_{\Lambda_0}$ is mapped to the flag manifold by the torus embedding (19), the I_{rk} assume the symmetric form announced in (6). We now show how this comes about. The calculation for $\mathcal{SL}(n,\mathbf{C})$ relies on the Laplace expansion of the determinant of a product of matrices; an intrinsic explanation is given in the next subsection.

Only the expansion of the 1-chop integrals I_{r1} will be done in detail. For a matrix A, denote by $A_{\hat{a}\hat{b}}$ the minor of the (a,b) element. With this notation, the generating polynomial (4) of the E_{r1} becomes

$$\widetilde{Q}_1(X,\lambda) = (X-\lambda)_{\hat{1}\hat{n}}.$$

Now

$$X = LV\Lambda_0V^{-1}L^{-1} \stackrel{\text{def}}{=} Z\Lambda_0Z^{-1}, \tag{20}$$

so

$$\widetilde{Q}_1(X,\lambda) = (Z\Lambda_0Z^{-1}-\lambda)_{\hat{1}\hat{n}} = \sum_{j=1}^{n} Z_{\hat{1}\hat{j}}(\Lambda_0-\lambda)_{\hat{j}\hat{j}}(Z^{-1})_{\hat{j}\hat{n}}$$

$$= (-1)^{\frac{n(n+1)}{2}} \sum_{j=1}^{n}(-1)^{j+1}(Z^{-1})_{j1}(\det Z)\prod_{s(\neq j)}(\lambda_{s0}-\lambda)\cdot(Z^{-1})_{\hat{j}\hat{n}}. \tag{21}$$

Recall that the torus embedding (19) maps $X \in (\epsilon+\mathcal{B}_-)_{\Lambda_0}$ to Z^{-1} mod $\mathbf{B}_+$. Two natural projections from $\mathbf{Flag}_n$ to the projective space $\mathbf{P}^{n-1}$ and its dual $(\mathbf{P}^{n-1})^*$ send a flag $V_1 \subset \ldots \subset V_{n-1}$ to the line V_1, respectively the hyperplane V_{n-1}. The element Z^{-1} mod $\mathbf{B}_+$ projects to the first column of Z^{-1}, respectively the collection of $(n-1)\times(n-1)$ minors $(Z^{-1})_{\hat{j}\hat{n}}$. Let π_j, π_j^* denote homogeneous coordinates on the projective spaces. Then (21) becomes (if, as we may, we choose Z to have determinant 1)

$$\widetilde{Q}_1(X,\lambda) = (-1)^{n(n+1)/2}\sum_{j=1}^{n}(-1)^{j+1}\prod_{s(\neq j)}(\lambda_{s0}-\lambda)\frac{\pi_j}{\pi_1}\frac{\pi_j^*}{\pi_1^*}. \tag{22}$$

One obtains expressions for the individual E_{r1} by expanding the product, and their ratios I_{r1} indeed have the form (6).

Analogously, one derives the formula for the k-chop integrals [S1]:

$$\tilde{Q}_k(X,\lambda) = (-1)^{k(k+1)/2} \sum_{|J|=n-k} (-1)^{\sum_{j\in J}} \prod_{j\in J} (\lambda_{j0}-\lambda) \frac{\pi_J}{\pi_{J_0}} \frac{\pi_J^*}{\pi_{J_0}^*}, \tag{23}$$

where: $J \subset \{1,\dots,n\}$, $J_0 = \{1,\dots,n-k\}$, π_J is the $k\times k$ minor of Z^{-1} formed from the first k columns and the rows from $\{1,\dots,n\}\setminus J$, and π_J^* is the minor formed from the first $n-k$ columns and the rows from J. ∎

The preceding example immediately leads to the following corollary:

Proposition 3.3 *As in (19), map $X \mapsto V^{-1}L^{-1} \bmod \mathbf{B}$. The locus $I_{r1} = c_{r1}, r = 1,\dots,n-2$, in $P^{n-1} \times (P^{n-1})^*$ is invariant under the action of the diagonal subgroup* $\mathbf{H}$.

Proof. Under the action of $h = \mathrm{diag}\,(h_1,\dots,h_n)$, $\pi_j \mapsto h_j\pi_j$ and $\pi_j^* \mapsto h_j^{-1}\pi_j^*$. ∎

3.2 Representation Theory

We now outline a less matrix-oriented derivation of the properties of the k-chop integrals. It will be seen that most of their important features have general analogs.

Let $\mathcal{G}$ be a complex semisimple Lie algebra. Let S denote the ring of polynomials on $\mathcal{G}$, and S^χ the C-subspace of semi-invariants transforming according to

$$f(p^{-1}\cdot Y) = \chi(p)f(Y), \quad Y \in \mathcal{G},\ p \in \mathbf{P}, \tag{24}$$

for a character χ of a parabolic subgroup $\mathbf{P}$.

Take two rational functions α, β on $\mathcal{G}$. In general, there is no simple relation between the restriction of the Poisson bracket $\{\alpha,\beta\}_{\mathcal{G}}$ to $\epsilon + \mathcal{B}_-$, and the Poisson bracket $\{\tilde{\alpha},\tilde{\beta}\}_{\epsilon+\mathcal{B}_-}$ of the restrictions (denoted by $\tilde{}$) of those functions to $\epsilon + \mathcal{B}_-$. For ratios of semi-invariants, however, the following holds (as follows from Proposition A.1 in the appendix):

Proposition 3.4 *Let $f_i, g_i \in S^{\chi_i}, i = 1,2$. Then*

$$\{\frac{\tilde{f}_1}{\tilde{g}_1}, \frac{\tilde{f}_2}{\tilde{g}_2}\}_{\epsilon+\mathcal{B}_-} = -\{\frac{f_1}{g_1}, \frac{f_2}{g_2}\}_{\mathcal{G}}\Big|_{\epsilon+\mathcal{B}_-}.$$

Since the Chevalley invariants of $\mathcal{G}$ are semi-invariant for the identity character, we deduce

Corollary 3.1 *Whenever $f, g \in S^\chi$ and J is a Chevalley invariant of $\mathcal{G}$, then*

$$\{\tilde{J}, \frac{\tilde{f}}{\tilde{g}}\}_{\epsilon+\mathcal{B}_-} = 0.$$

This prescription produces many (not necessarily independent or involutive) full Kostant-Toda constants of motion. Kostant's invariant theory of the adjoint representation suggests a construction of semi-invariants. We review the main features [Ko1, Dix].

The group $\mathbf{G}$ acts on the ring S of polynomials: $g \cdot f(Y) = f(g^{-1} \cdot Y)$. Let $\mathcal{CAS}$ be the ring of invariant polynomials, let J be the ideal generated by $\mathcal{CAS}$, and let J^+ be the set of $f \in J$ with zero constant term. The ring S is a product $J^+S \otimes H$, where H is a $\mathbf{G}$-module complement of J^+S, graded by degree. Under the action of $\mathbf{G}$, H decomposes into a sum of finite-dimensional irreducible (highest weight) representations of $\mathbf{G}$. The representation (ρ^χ, V^χ) with highest weight χ occurs ℓ^χ many times, where ℓ^χ is the multiplicity of the weight 0 in V^χ. In other words, one has the decomposition

$$H = \bigoplus_\chi \oplus_{j=1}^{\ell^\chi} V_j^\chi;$$

each V_j^χ is a copy of the abstract representation V^χ. The highest weight vectors (= functions) transform according to (24) for p in the appropriate parabolic. Thus, if f, g are highest weight functions for two copies of a representation V^χ, their ratio f/g, according to Corollary 3.1, will be a full Kostant-Toda integral.

Kostant's construction of highest weight functions will be important for us. Let (ρ_χ, V_χ) be the representation contragredient to (ρ^χ, V^χ). Let v^χ be the highest weight vector of (the abstract representation) V^χ, and let $w_i', i = 1, \ldots, \ell^\chi$ be a basis of the subspace V^{Λ_0} of V_χ annihilated by $\mathcal{G}^{C_{\Lambda_0}}$, the centralizer of the "companion" element C_{Λ_0} in Lemma 2.5. Define functions f_i on the conjugacy class $\{g \cdot C_{\Lambda_0} \mid g \in \mathbf{G}\}$ by

$$f_i(g \cdot C_{\Lambda_0}) = \langle v^\chi, \rho_\chi(g) w_i' \rangle. \tag{25}$$

Note that the definition is consistent: if $g \cdot C_{\Lambda_0} = C_{\Lambda_0}$, then $\rho_\chi(g) w_i' = w_i'$.

Example 3.2 We will compute (25) explicitly in a simple case, and show that it reproduces the generating function $\widetilde{Q}_1(X, \lambda)$ from equation (4).

Let $\mathcal{G} = \mathcal{SL}(n, \mathbf{C})$, and take ρ^χ to be the adjoint representation (whose character is given by (16) with $k = 1$); we realize this representation in

$$V^\chi = \mathbf{C}^n \otimes \bigwedge^{n-1} \mathbf{C}^n$$

(the one-dimensional subrepresentation will not affect the result). Let $\{\mathbf{e}_i\}$ be the standard basis of $\mathbf{C}^n$, and set

$$\mathbf{e}_i^* = \mathbf{e}_1 \wedge \ldots \wedge \widehat{\mathbf{e}_i} \wedge \ldots \wedge \mathbf{e}_n.$$

We identify C^n and $\bigwedge^{n-1} C^n$ with their duals so that $\langle \mathbf{e}_i, \mathbf{e}_j\rangle = \langle \mathbf{e}_i^*, \mathbf{e}_j^*\rangle = \delta_{ij}$.

For ease of calculation, we replace C_{Λ_0} by the (regular) diagonal matrix Λ_0; in formula (25), the w_i' (there are rank $= n-1$ many) must then be annihilated by the centralizer of Λ_0, i.e. by the Cartan subalgebra. Hence, they are linear combinations of $\mathbf{e}_j \otimes \mathbf{e}_j^*$, $j = 1, \ldots, n$. We take the w_i' to be the coefficients of the powers of λ in

$$w'(\lambda) \stackrel{\text{def}}{=} (-1)^{n(n+1)/2} \sum_{j=1}^{n} (-1)^{j+1} \prod_{s(\neq j)} (\lambda_{s0} - \lambda)\, \mathbf{e}_j \otimes \mathbf{e}_j^*;$$

the choice is suggested by equation (21). Finally, for $X \in (\epsilon + \mathcal{B}_-)_{\Lambda_0}$, take $Z = LU$ as in the torus embedding (19), so that $X = Z \cdot \Lambda_0 = Z\Lambda_0 Z^{-1}$.

The functions (25) are then encoded in

$$f(X,\lambda) \stackrel{\text{def}}{=} \langle \mathbf{e}_1 \otimes \mathbf{e}_n^*, \rho_\chi(Z) w'(\lambda)\rangle = \sum_{i=1}^{n-2} f_i(X)\lambda^i. \tag{26}$$

The action $\rho_\chi(Z)\, \mathbf{e}_j \otimes \mathbf{e}_j^*$ is $((Z^t)^{-1}\mathbf{e}_j) \otimes ((Z^t)^{-1}\mathbf{e}_j^*)$, so that, in the notation of (21),

$$\begin{aligned} \langle \mathbf{e}_1 \otimes \mathbf{e}_n^*, ((Z^t)^{-1}\mathbf{e}_j) \otimes ((Z^t)^{-1}\mathbf{e}_j^*)\rangle &= \langle \mathbf{e}_1, (Z^t)^{-1}\mathbf{e}_j\rangle \langle \mathbf{e}_n^*, (Z^t)^{-1}\mathbf{e}_j^*\rangle \\ &= (Z^{-1})_{j1} (Z^{-1})_{j\hat{n}}. \end{aligned}$$

The coefficient of λ^{n-1} in (26) vanishes, because the corresponding vector $\sum (-1)^j \mathbf{e}_j \otimes \mathbf{e}_j^*$ spans the trivial representation and is fixed under all of $SL(n, C)$. Hence, the function $f(X,\lambda)$ coincides with $\widetilde{Q}_1(X,\lambda)$ in (21), and the coefficient of λ^{n-2-r} is exactly E_{rk}. ∎

Proposition 3.5 *Let χ be a weight of the adjoint group of $\mathcal{G}$, and let $\mathbf{P}^\chi$ be the largest parabolic subgroup of $\mathbf{G}$ satisfying*

$$\rho^\chi(p) v^\chi = \chi(p) v^\chi, \quad p \in \mathbf{P}^\chi.$$

(i) For $p \in \mathbf{P}^\chi$ and $Y \in \mathbf{G} \cdot C_{\Lambda_0}$, $f_i(p^{-1} \cdot Y) = \chi(p) f_i(Y)$.

(ii) When $X \in (\epsilon + \mathcal{B}_-)_{\Lambda_0}$ is written $X = L \cdot C_{\Lambda_0}$, then the induced function on the big cell of $\mathbf{G}/\mathbf{B}_+$,

$$\tilde{f}_i(L^{-1}) \stackrel{\text{def}}{=} \langle v^\chi, \rho_\chi(L) w_i'\rangle,$$

depends only on L^{-1} mod $\mathbf{P}^\chi$.

(iii) The functions $\tilde{f}_i$ defined in (ii) extend to sections of a line bundle over $\mathbf{G}/\mathbf{P}^\chi$. The loci $\tilde{f}_i = c\tilde{f}_j$ are invariant under the action of the stabilizer subgroup $\mathbf{G}^{C_{\Lambda_0}}$ of C_{Λ_0}. (Thus, the sections $\tilde{f}_i$ are weight vectors for the weight zero in the representation of $\mathbf{G}$ on the space of holomorphic sections of this bundle.)

Proof. We use the following fact repeatedly: $\rho_\chi(g) = \rho^\chi(g^{-1})^t$. (i) $f_i(p^{-1} \cdot g \cdot C_{\Lambda_0}) = \langle v^\chi, \rho_\chi(p^{-1})\rho_\chi(g)w_i' \rangle = \langle \rho^\chi(p)v^\chi, \rho_\chi(g)w_i' \rangle$, but $\rho^\chi(p)v^\chi = \chi(p)v^\chi$. (ii) Write $L = L_1 L_2$, where $L_1 \in \mathbf{N}_- \cap \mathbf{P}$ and L_2 belongs to the opposite unipotent radical $\overline{\mathbf{R}}$ of $\mathbf{P}$. Then $\langle v^\chi, \rho_\chi(L_1)\rho_\chi(L_2)w_i' \rangle = \langle \rho^\chi(L_1^{-1})v^\chi, \rho_\chi(L_2)w_i' \rangle$. But $\rho^\chi(L_1^{-1})v^\chi = \chi(L_1^{-1})v^\chi = v^\chi$. Since $L^{-1} \bmod \mathbf{P} = L_2^{-1} \bmod \mathbf{P}$, the result follows. (iii) The linebundle is the one determined by the character χ; that the transition functions are correct follows from the semi-invariance in (i). Next, let h stabilize w_i', and map $L^{-1} \bmod \mathbf{B}_+ \mapsto h^{-1}L^{-1} \bmod \mathbf{B}_+$. We suppose that $h^{-1}L^{-1}$ admits a factorization $n((Lh)^{-1})b((Lh)^{-1})$, so that $\tilde{f}_i(L^{-1})$ is changed to $\tilde{f}_i(n((Lh)^{-1}))$. Then

$$\begin{aligned} \tilde{f}_i(n((Lh)^{-1})) &= \langle v^\chi, \rho_\chi(n((Lh)^{-1})^{-1})w_i' \rangle = \langle \rho^\chi(n((Lh)^{-1}))v^\chi, w_i' \rangle \\ &= \langle \rho^\chi((Lh)^{-1})\rho^\chi(b((Lh)^{-1}))v^\chi, w_i' \rangle \\ &= \chi(b((Lh)^{-1}))\langle v^\chi, \rho_\chi(Lh)w_i' \rangle \\ &= \chi(b((Lh)^{-1}))\langle v^\chi, \rho_\chi(L)w_i' \rangle, \end{aligned}$$

where the last equality follows because h stabilizes w_i'. Thus, $\tilde{f}_i(n((Lh)^{-1})) = \chi(b((Lh)^{-1}))\tilde{f}_i(L^{-1})$, and the conclusion follows. ∎

One sees from [Ko1] that on every orbit $\{g \cdot C_{\Lambda_0}\}$, all parabolic semi-invariants will have the form (25). Thus, all rational integrals provided by Corollary 3.1 will have the properties implied by the last proposition. Unfortunately, we do not know, except by explicit calculation for various algebras, when the gradient of such an integral will lie in $\mathcal{P}$; according to Proposition 2.1, this is necessary if we are to get an *involutive* family via Thimm's method. The w_i' in (25) must vary across orbits $\{g \cdot C_{\Lambda_0}\}$ in a very special way, as is seen in Example 3.2.

Proposition 3.2 about the k-chop integrals in $\mathcal{SL}(n,\mathbf{C})$ also has a very useful generalization. We state the result, and then give illustrations.

Proposition 3.6 *Let I be a coadjoint invariant for $\mathbf{P}$, satisfying the conditions in Proposition 2.1. (i) Let f be an invariant polynomial on the Levi component $\mathcal{L}$ of $\mathcal{P}$. Then $f \circ [\nabla I]$ is a coadjoint invariant for $\mathbf{P}$. (ii) Let $\mathbf{P}_0 \subset \mathbf{P}$ be parabolic and let χ be a character of $\mathbf{P}_0$. Let f be semi-invariant on $\mathcal{L}$ for the Levi factor $[\mathbf{P}_0]$ in $\mathbf{P}$. Then $f \circ [\nabla I]$ is semi-invariant on $\mathcal{G}$ for $\mathbf{P}_0$, with character χ.*

Proof. (i) Let $F(Y) = f([\nabla I(Y)])$. By assumption, $I(Y)$ depends only on the coordinates of Y in $\mathcal{P}_-$. Therefore, $\nabla F \in \mathcal{P}$. Now take $p \in \mathbf{P}$. Condition (ii) in Proposition 2.1 is checked as follows:

$$\begin{aligned} F(p \cdot Y) &= f([\nabla I(p \cdot Y)]) \stackrel{*}{=} f(p \cdot [\nabla I(Y)]) \\ &\stackrel{**}{=} f([p] \cdot [\nabla I(Y)]) \stackrel{***}{=} f([\nabla I(Y)]) = F(Y). \end{aligned}$$

Here $\stackrel{*}{=}$ comes from Lemma 2.1, $\stackrel{**}{=}$ from the properties of the Levi decomposition, and $\stackrel{***}{=}$ follows because f is an invariant polynomial on $\mathcal{L}$. (ii) The proof is the same until $\stackrel{***}{=}$; now a factor $\chi([p])$ is pulled out. ∎

Example 3.3 To illustrate part (i) of this proposition, take $\mathcal{G} = \mathcal{SL}(n, \mathrm{C})$ and $\mathbf{P} = \mathbf{P}_1$. Let $I = I_{21}$, and let f be one of the functions $\mathrm{Trace}\, Y^r$, which are certainly invariant on $\mathcal{L}$. Then the $n-3$ functions $\mathrm{Trace}\,([\nabla I_{21}(X)]^{ss})^r, r = 2, \ldots, n-2$, are parabolic invariants; in fact, as follows from Propositions 3.1 and 3.2, they are equivalent to the 1-chop integrals $I_{r1}, r = 2, \ldots, n-2$. The projection of $[\nabla I_{21}(X)]$ onto the center $\mathcal{C}$ of the Levi component is a multiple of the Casimir I_{11} on $\epsilon + \mathcal{B}_-$. See Section 7 for an example. ∎

Part (ii) of Proposition 3.6 is not quite as strong as part (i). There is *a priori* no reason for the induced semi-invariant $f \circ [\nabla I]$ on $\mathcal{G}$ to depend only on the coordinates of $\mathcal{P}_0^* \equiv (\mathcal{P}_0)_-$ (cf. Proposition 2.1 again). The general prescription is therefore not guaranteed to produce $\mathcal{P}_0$-coadjoint invariants. Explicit calculation is necessary, and may in fact lead to stronger conclusions than we have a right to expect. In $\mathcal{SL}(n, \mathrm{C})$, for example, we have the following remarkable fact [S2], proved by a brute-force calculation:

Proposition 3.7 *In the notation of Definition 3.1, $\phi_{k+1}(X) = \phi_1\big(\phi_k(X)\big)$ (when defined). From Proposition 3.6(ii), it follows that*

$$Q_{k+1}(X, \lambda) = Q_1(\phi_k(X), \lambda).$$

Hence, the $k+1$-chop integrals of X are equivalent to the 1-chop integrals of $\phi_k(X)$.

Remark 3.1 We will appeal to this Proposition in our inductive study of the k-chop flows. By Proposition 3.1, $\phi_k(X)$ is essentially the Levi component $[\nabla I_{2k}(X)]_k$ of the k-chop integral I_{2k}. The function $\mathrm{Trace}([\nabla I_{2k}(X)]_k)^2$ is a sort of "Killing form" which generates the k-chop flows, just as $\mathrm{Trace}\, X^2$ generates the basic Toda flows. Application of the map ϕ_1 to this k^{th} Levi component produces $[\nabla I_{2,k+1}(X)]_{k+1}$, which plays a similar role at the next level. ∎

Remark 3.2 The map ϕ_k can be defined for $X \in \mathcal{SL}(n, \mathrm{C})$, as long as $X_7 \neq 0$. It is shown in [S2] that $\tilde{\phi}_k \stackrel{\text{def}}{=} \phi_k - \frac{1}{n-2k} \mathrm{Trace}\, \phi_k \, \mathrm{Id}_{n-2k}$ is a *Poisson map* from $\mathcal{SL}(n, \mathrm{C})$ to $\mathcal{SL}(n-2k, \mathrm{C})$ in the r-matrix bracket on these algebras ($r = \Pi_{\mathcal{B}_+} - \Pi_{\mathcal{N}_-}$). ∎

3.3 Classical Algebras

Luen-chau Li, in unpublished notes, showed that the chop construction of Subsection 3.1 also works for the orthogonal and symplectic algebras (in the symmetric case). Several important properties of the chopped integrals can be deduced from the general results in Subsection 3.2.

Let $\mathcal{G}$ be a (complex) Lie algebra of type B_n, C_n, D_n. For each type, there is a standard representation of minimal dimension. The orthogonal algebras can be realized as matrices that are skew-symmetric with respect to the *anti*-diagonal. The symplectic algebra C_n consists of block matrices

$$\begin{pmatrix} A & B \\ C & -A' \end{pmatrix},$$

with the $n \times n$ matrices B, C symmetric about their antidiagonals, and A' denoting the antidiagonal transpose.

For $X \in \mathcal{G}$, define the chopped matrix $X_{(k)}$ as before (remove the first k rows and last k columns). The coefficients of the polynomials

$$\frac{\det(X - \lambda)_{(k)}}{\text{leading coefficient}}$$

will be parabolic Casimirs, analogous to the I_{rk} used for the $\mathcal{SL}(n,\mathbf{C})$ full Toda lattice. Because of the symmetries of X, certain coefficients of these polynomials will vanish. The results are summarized in the following theorem, which largely paraphrases L.-C. Li's notes. The dimension of the generic symplectic leaves was also computed by Trofimov [Tro1].

Theorem 3.1 *The k-chopped polynomials $\widetilde{Q}_k(X, \lambda)$ have $n - k + 1, k = 1, \ldots, n$, nontrivial coefficients for B_n, C_n, and $n+1-2[\frac{k+1}{2}], k = 1, \ldots, n-2$, for D_n. The coefficients are semi-invariants for the parabolics $\mathbf{P}_k$ that stabilize the fundamental weights ω_k; the characters are $2\omega_k$. Their ratios are involutive integrals of the full Toda equations. There are no Casimirs, save for D_n with odd n; in that case, the ratio of the first two nontrivial coefficients of $\widetilde{Q}_{n-1}(X, \lambda)$ (which are semi-invariant with character $2(\omega_{n-1} + \omega_n)$) is a Casimir. Let I denote a k-chop integral for B_n or C_n. The Chevalley invariants of the Levi component $[\nabla I(X)]_k$ of $\mathcal{P}_k$ are equivalent to* all *the k-chop integrals (in the case of $\mathcal{SL}(n,\mathbf{C})$, this includes the Casimirs). This is not true for D_n.*

4. Genericity and Independence

From now on we study the generic level set of the chopped constants of motion I_{rk}. We rely on explicit calculations with matrices. The Lie-algebraic significance of our computations is not understood, and while similar brute-force techniques appear to work with at least the classical algebras B_n and C_n, we do not report our incomplete results.

In this subsection, we define "generic", show that generic level sets exist, and use the map to the flag manifold to prove that the I_{rk} are independent. The first definition is designed for an induction argument.

Definition 4.1 *For each $k = 1, \ldots, m = \left[\frac{n-1}{2}\right]$, we say that X* has property $(\mathrm{A})_\mathrm{k}$, *if: (a) $\phi_j(X)$ exists and is regular semisimple for $j = 1, \ldots, k$, and (b) for $j = 1, \ldots, k$, the sets $\{\lambda_{1,j-1}, \ldots, \lambda_{n-2j+2,j-1}\}$ and $\{\lambda_{1j}, \ldots, \lambda_{n-2j,j}\}$ have no common elements. Define*

$$(\epsilon + \mathcal{B}_-)^k_{\Lambda_0} = \{X \in (\epsilon + \mathcal{B}_-)_{\Lambda_0} \mid X \text{ has property } (\mathrm{A})_\mathrm{k}\}. \qquad (27)$$

The largest possible value of k is $m = \left[\frac{n-1}{2}\right]$; we use the special symbol $(\epsilon+\mathcal{B}_-)^\diamondsuit_{\Lambda_0}$ to denote $(\epsilon+\mathcal{B}_-)^m_{\Lambda_0}$. This is our set of "generic" $X \in (\epsilon+\mathcal{B}_-)_{\Lambda_0}$. *A* level set *is a fiber of the rational map*

$$\begin{array}{rcl} (\epsilon + \mathcal{B}_-)^\diamondsuit_{\Lambda_0} & \to & C^d \\ X & \mapsto & (I_{rk}(X), k = 1, \ldots, m; r = 1, \ldots, n-2k) \end{array}$$

where $d = (n-2) + (n-4) + \ldots + (n - 2[\frac{n-1}{2}])$. The level set variety, *abbreviated $\mathcal{LSV}$, is the closure in $\epsilon + \mathcal{B}_-$ of a level set.*

We will show that $(\epsilon + \mathcal{B}_-)^\diamondsuit_{\Lambda_0}$ is dense in $(\epsilon + \mathcal{B}_-)_{\Lambda_0}$, and that the I_{rk} are independent on $(\epsilon + \mathcal{B}_-)^\diamondsuit_{\Lambda_0}$.

Recall from Definition 3.1 that $\phi_k(X)$ exists exactly when X_7 is invertible, that is, when $E_{0k} = \det X_7 \neq 0, k = 0, \ldots, \left[\frac{n-1}{2}\right]$ (cf. (4)). The complement of the set on which some E_{0k} vanishes is Zariski open in $(\epsilon + \mathcal{B}_-)_{\Lambda_0}$ and, under the Kostant map, in the flag manifold. This set has a certain stability property:

Lemma 4.1 *Let $M \in \mathbf{N}_- \cap \mathbf{P}_k$. If $\phi_k(X)$ exists, so does $\phi_k(M \cdot X)$.*

Proof. Write M in the form

$$M = \begin{array}{c} \\ k \\ n-2k \\ k \end{array} \begin{array}{c} \begin{array}{ccc} k & n-2k & k \end{array} \\ \begin{pmatrix} \mathrm{Id}_k & 0 & 0 \\ 0 & \widetilde{M} & 0 \\ 0 & 0 & \mathrm{Id}_k \end{pmatrix} \end{array}, \qquad (28)$$

where $\widetilde{M}$ belongs to $SL(n-2k, \mathrm{C})$. A short calculation using the definition of ϕ_k shows that

$$\phi_k(M \cdot X) = \widetilde{M} \cdot \phi_k(X). \qquad (29)$$

This fact not only proves the assertion, but will play an important role below. It is really a consequence of Lemma 2.1 combined with Proposition 3.1. ∎

The main step is to prove that for each $k = 1, \ldots, m = [\frac{n-1}{2}]$, the I_{rk} are independent on $(\epsilon + \mathcal{B}_-)^k_{\Lambda_0}$; the fact that this set is Zariski open is deduced simultaneously. Throughout, it is understood that the element X (in $(\epsilon + \mathcal{B}_-)_{\Lambda_0}$), or the corresponding $L^{-1} \bmod \mathbf{B}_+$ in $\mathbf{G}/\mathbf{B}_+$, are chosen from the open dense set where all ϕ_k are defined.

We exploit Proposition 3.5 (or more explicitly (23)) which says that the I_{rk}, for each k, descend to the partial flag manifold $SL(n,\mathbf{C})/\mathbf{P}_k$. To work with successive flag manifolds of this form, we need to factor the L in $L^{-1} \bmod \mathbf{B}_+$.

Notation 4.1 Let $n \in \mathbf{N}_-$. It admits a unique factorization $n = n_0 \ldots n_m$, where $n_j \in \mathbf{N}_- \cap \mathbf{P}_j$ has nonzero entries only in column j and row $n - j$. (For instance, n_0 has the form (11).) The factorization of L^{-1} will be written $L_0^{-1} \ldots L_m^{-1}$. ∎

The I_{rk} are functions of only some of the coordinates in $\mathbf{G}/\mathbf{B}_+$; precisely:

$$I_{rk}(L^{-1} \bmod \mathbf{B}_+) = I_{rk}(L_0^{-1} \ldots L_{k-1}^{-1} \bmod \mathbf{P}_k). \tag{30}$$

Of course, we think of the I_{rk} as functions on the flag manifold via the Kostant map (18). We will use the symbol I_{rk} to denote both the function $I_{rk}(X)$ and the function $I_{rk}(L^{-1} \bmod \mathbf{B}_+)$.

Lemma 4.2 *Fix k. Suppose that the functions*

$$M_{k-1}^{-1} \mapsto I_{rk}(L_0^{-1} \ldots L_{k-2}^{-1} M_{k-1}^{-1} \bmod \mathbf{P}_k), \; r = 1, \ldots, n-2k,$$

are independent on an open dense neighborhood of L_{k-1} in the $(2n-4k+1)$-dimensional affine space of M_{k-1}'s. Then the I_{rk} are independent on an open dense neighborhood of $L_0^{-1} \ldots L_{k-1}^{-1} \bmod \mathbf{P}_1$ in $\mathbf{G}/\mathbf{P}_1$.

Proof. The general situation is this: suppose we have a holomorphic map $f : \mathbf{C}^a \times \mathbf{C}^b \to \mathbf{C}^c$ with all functions $y \mapsto f_j(x_0, y)$ independent for a certain $x_0 \in \mathbf{C}^a$. (In our case, a is the dimension of the affine space of $L_0, \ldots, L_{k-2}$; b is the number of nonzero entries in M_{k-1}^{-1}, namely $2n - 4k + 1$; and c is the number of I_{rk}'s, namely $n - 2k$.)

The Jacobian $D_{(x,y)}f$ is a $c \times (a + b)$ matrix. If $c \leq b$, its rank is $\leq c$. According to our assumption, the Jacobian $D_y f$ with respect to y has maximal rank$= c$ at $x = x_0$. By analyticity, $D_{(x,y)}f$ has maximal rank in an open dense subset of $\mathbf{C}^a \times \mathbf{C}^b$. The condition $n - 2k = c \leq b = 2n - 4k + 1$ is satisfied because $k \leq [\frac{n-1}{2}]$. ∎

Lemma 4.3 *Under the hypotheses in the preceding lemma, the I_{rk} are independent on an open dense set in* $\mathbf{G}/\mathbf{B}_+$.

Proof. The map $L^{-1} \bmod \mathbf{B}_+ \mapsto L^{-1} \bmod \mathbf{P}_1$ is submersive, and by Lemma 4.2, so is $L^{-1} \bmod \mathbf{P}_1 \mapsto \mathrm{C}^{n-2}$, on an open dense set. Hence the composition is submersive on an open dense subset of $\mathbf{G}/\mathbf{B}_+$. ∎

These lemmas allow us to study the independence question in the smaller flag manifold that contains the "middle block" (see (28)) of L_{k-1}^{-1}.

Lemma 4.4 *The set* $(\epsilon + \mathcal{B}_-)^1_{\Lambda_0}$ *is Zariski open in* $(\epsilon + \mathcal{B}_-)_{\Lambda_0}$, *and the functions* I_{r1} *are independent on* $(\epsilon + \mathcal{B}_-)^1_{\Lambda_0}$.

Remark 4.1 The Plücker coordinates on $\mathbf{G}/\mathbf{P}_1$ will play a pivotal role here and later, so a brief review is in order. The map

$$g \bmod \mathbf{P}_1 \mapsto [g_{11} : \ldots : g_{n1}] \stackrel{\text{def}}{=} [\pi_1 : \ldots : \pi_n]$$

sends $g \bmod \mathbf{P}_1$ to $[\pi_1 : \cdots : \pi_n] \in \mathrm{P}^{n-1}$; this n-tuple—determined only up to an overall factor—describes a line V_1 in C^n, namely, the line through the vector represented by the first column of g. The map $g \bmod \mathbf{P}_1 \mapsto (i,n)$-*minors of* $g \stackrel{\text{def}}{=} \pi_i^*$ sends $g \bmod \mathbf{P}_1$ to a point $[\pi_1^* : \cdots : \pi_n^*] \in (\mathrm{P}^{n-1})^*$; this n-tuple—again, determined only up to an overall factor—describes a hyperplane V_{n-1} in C^n. One may think of V_{n-1} as being spanned by the first $n-1$ columns of g. The partial flag manifold $\mathbf{G}/\mathbf{P}_1$ consists of those pairs (V_1, V_{n-1}) for which $V_1 \subset V_{n-1}$; this condition is expressed by $\sum_{j-1}^{n}(-1)^j \pi_j \pi_j^* = 0$, the so-called Plücker relation. ∎

Proof. We map X to $V^{-1}L^{-1} \bmod \mathbf{B}_+$ via the torus embedding (19), restrict to the partial flag manifold $\mathbf{G}/\mathbf{P}_1$ (i.e., $X \mapsto V^{-1}L_0^{-1} \bmod \mathbf{P}_1$), and use explicit formulas for the functions $I_{r1}(V^{-1}L_0^{-1} \bmod \mathbf{P}_1)$ on $\mathbf{G}/\mathbf{P}_1$.

The equations for the I_{r1} on the partial flag manifold $\mathbf{G}/\mathbf{P}_1$ were found in Example 3.1; the final formula (22) can be rewritten as follows:

$$c\sum_{j=1}^{n}(-1)^{j+1}\frac{\pi_j\pi_j^*/\pi_1\pi_1^*}{\lambda - \lambda_{j0}} = \frac{\lambda^{n-2} + I_{11}\lambda^{n-3} + \ldots + I_{n-2,1}}{(\lambda - \lambda_{10})\cdots(\lambda - \lambda_{n0})}, \tag{31}$$

with a certain constant c. Elementary partial fractions calculus says that the I_{r1} are determined uniquely by the $\pi_i\pi_i^*/\pi_1\pi_1^*$, and vice versa. Hence the I_{rk} are independent exactly where the $\pi_i\pi_i^*$ are independent and $\pi_1\pi_1^* \neq 0$. The latter condition holds when $g \bmod \mathbf{P}_1$ can be written as $n \bmod \mathbf{P}_1, n \in \mathbf{N}_-$; this is the *big cell* of $\mathbf{G}/\mathbf{P}_1$. The function $\pi_i\pi_i^*/\pi_1\pi_1^*$ is submersive except where $\pi_i = \pi_i^* = 0$. Hence all values but zero are regular values.

Equation (31) shows that one or more of the products $\pi_i\pi_i^*$ vanish precisely when the numerator and denominator on the right have common roots. Thus, on the dense open set where all $\pi_i\pi_i^*$ are nonzero, the I_{rk} are independent, and moreover the roots λ_{r1} of the numerator are different from the roots λ_{r0} of the denominator. By Proposition 3.2, the I_{r1} are the coefficients of the characteristic polynomial of ϕ_1. Hence, independence certainly holds when property $(\mathrm{A})_1$ is satisfied, that is, when $X \in (\epsilon + \mathcal{B}_-)^1_{\Lambda_0}$. ∎

The proof of the next theorem introduces ideas and notation that will be crucial in the sequel; therefore, it will be interrupted with a Remark.

Theorem 4.1 *For every k, the set $(\epsilon+\mathcal{B}_-)^k_{\Lambda_0}$ is Zariski open in $(\epsilon+\mathcal{B}_-)_{\Lambda_0}$, and the I_{rk} are independent on $(\epsilon+\mathcal{B}_-)^k_{\Lambda_0}$.*

Proof. The proof is by induction. The case $k = 1$ was done in the preceding lemma. Suppose now that the proposition is true for $j = 1, \ldots, k-1$. Fix $L^{-1} \bmod \mathbf{B}_+ \in K_{\Lambda_0}((\epsilon+\mathcal{B}_-)^{k-1}_{\Lambda_0})$. We first show that the functions

$$M_{k-1}^{-1} \mapsto I_{rk}(L_0^{-1} \ldots L_{k-2}^{-1} M_{k-1}^{-1} \bmod \mathbf{P}_k)$$

are independent on a dense open set of M_{k-1}'s.

By assumption, $\phi_{k-1}(X)$ exists when $X = L \cdot C_{\Lambda_0}$. According to (29),

$$\phi_{k-1}(X) = (L_m \ldots L_{k-1}) \cdot \phi_k\big((L_{k-2} \ldots L_0) \cdot C_{\Lambda_0}\big),$$

and by Lemma 4.1,

$$C_{\Lambda_{k-1}}(L_0, \ldots, L_{k-2}) \stackrel{\text{def}}{=} \phi_{k-1}\big((L_{k-2} \ldots L_0) \cdot C_{\Lambda_0}\big) \tag{32}$$

exists as well (the notation is explained in a moment). In fact, the lemma says that if $M \in \mathbf{N}_- \cap \mathbf{P}_{k-1}$ is arbitrary and Y is defined by

$$Y \stackrel{\text{def}}{=} M \cdot (L_{k-2} \ldots L_0) \cdot C_{\Lambda_0}, \tag{33}$$

then $\phi_{k-1}(Y)$ exists and

$$\phi_{k-1}(Y) = \widetilde{M} \cdot C_{\Lambda_{k-1}}(L_0, \ldots, L_{k-2}). \tag{34}$$

Since $C_{\Lambda_{k-1}}$ is regular semisimple, there exists a $\widetilde{V} = \widetilde{V}(L_0, \ldots, L_{k-2}) \in SL(n-2k+2, \mathbb{C})$ that conjugates it to diagonal form, $\widetilde{V} \cdot C_{\Lambda_{k-1}} = \Lambda_{k-1} \in \mathcal{GL}(n-2k+2, \mathbb{C})$, and thus for the $\phi_{k-1}(Y)$ in (34),

$$\phi_{k-1}(Y) = \widetilde{M} \cdot \widetilde{V}^{-1} \cdot \Lambda_{k-1}. \tag{35}$$

Remark 4.2 We pause to explain the significance of these facts. In order to apply Lemma 4.2, we fix $L_0, \ldots, L_{k-2}$ but allow $M = L_{k-1} \cdots L_m$ to vary. In effect, we view $\mathbf{G}/\mathbf{B}_+$ as a fibration

$$\begin{array}{ccc} \mathbf{P}_{k-1}/\mathbf{B}_+ & \longrightarrow & \mathbf{G}/\mathbf{B}_+ \\ & & \Big\downarrow {\scriptstyle \pi_{k-1}} \\ & & \mathbf{G}/\mathbf{P}_{k-1} \end{array} \tag{36}$$

in which the fiber $\mathbf{P}_{k-1}/\mathbf{B}_+$ is isomorphic to the flag manifold

$$\mathbf{Flag}_{n-2k+2} = SL(n-2k+2, \mathrm{C})/\mathbf{B}_+^{(n-2k+2)}.$$

According to equations (34) and (35), every point $\widetilde{M}^{-1} \bmod \mathbf{B}_+^{(n-2k+2)}$ in the fiber over $L_0^{-1} \cdots L_m^{-1} \bmod \mathbf{P}_{k-1}$ corresponds to a $\phi_{k-1}(Y)$ via conjugation of $C_{\Lambda_{k-1}}(L_0, \ldots, L_{k-2})$ by $\widetilde{M}$, and moreover all those $\phi_{k-1}(Y)$ can be diagonalized through further conjugation by a $\widetilde{V}(L_0, \ldots, L_{k-2})$ which is *independent of* $\widetilde{M}$. This is exactly the situation we encountered at the outset: $X = L \cdot C_{\Lambda_0}$ and $X = L \cdot V \cdot \Lambda_0$. The notation $C_{\Lambda_{k-1}}$ is meant to suggest that this element does for the $(k-1)$-chop stage what Kostant's companion matrix did at the 0-chop stage. Note, however, that it no longer has "companion" form and that it varies with the basepoint in (36). ∎

We resume the proof. The object is to prove independence of the functions $I_{rk}(L_0^{-1} \cdots L_{k-2}^{-1} M_{k-1} \bmod \mathbf{P}_k)$ of M_{k-1}^{-1}, or equivalently, of the functions $I_{rk}(L_0^{-1} \cdots L_{k-2}^{-1} M^{-1} \bmod \mathbf{B}_+)$ of $M \in N_- \cap \mathbf{P}_{k-1}$. By Proposition 3.2, these are the coefficients of the characteristic polynomial of $\phi_k(Y)$, with Y as in (33); by Proposition 3.7, they are also the coefficients of the characteristic polynomial of $\phi_1(\phi_{k-1}(Y))$. This is the polynomial $Q_1(\phi_{k-1}(Y), \lambda)$, with Q_1 as in (5).

As just explained in Remark 4.2, there is a bijection between the big cell $\mathbf{N}_-^{(n-2k+2)}/\mathbf{B}_+^{(n-2k+2)}$ of $\mathbf{Flag}_{n-2k+2}$ and the set of all $\phi_{k-1}(Y)$. We can therefore compute $\widetilde{Q}_1(\phi_{k-1}(Y), \lambda)$ by the method of Example 3.1. We have

$$\widetilde{Q}_1(\phi_{k-1}(Y), \lambda) = \left(\phi_{k-1}(Y) - \lambda\, \mathrm{Id}_{n-2k+2}\right)_{\hat{1}, \widehat{(n-2k+2)}}.$$

Writing $\phi_{k-1}(Y) = \widetilde{M}\widetilde{V}\Lambda_{k-1}\widetilde{V}^{-1}\widetilde{M}^{-1}$ as in (35), we arrive at equation (20) in Example 3.1. The calculation in that example yields an analog of formula (22),

$$\widetilde{Q}_1(\phi_{k-1}(Y), \lambda) = c \sum_{j=1}^{n-2k+2} (-1)^{j+1} \prod_{s(\neq j)} (\lambda_{s,k-1} - \lambda) \frac{\pi_j \pi_j^*}{\pi_1 \pi_1^*}.$$

In this formula, of course, π_j and π_j^* denote the Plücker coordinates on the partial flag manifold $SL(n-2k+2,\mathbf{C})/\widetilde{\mathbf{P}}_k$ (see Notation 1.1). By our induction hypothesis, the eigenvalues $\lambda_{s,k-1}$ are distinct (i.e., $\phi_{k-1}(Y)$ is regular semisimple). The proof of Lemma 4.4 now applies verbatim and yields independence of the I_{rk} (thought of as functions of $\widetilde{M}_{k-1}$) on the dense open set where all $\pi_i\pi_i^*$ are nonzero. On this set, all roots λ_{rk} of $\widetilde{Q}_1(\phi_{k-1}(Y),\lambda)$ are different from $\lambda_{1,k-1},\ldots,\lambda_{n-2k+2,k-1}$. All the corresponding elements $Y = M\cdot L_{k-2}\cdots L_0\cdot C_{\Lambda_0}$ belong to $(\epsilon+\mathcal{B}_-)^k_{\Lambda_0}$.

The rest of the proof is routine. By induction hypothesis, the set $\{L^{-1} \bmod \mathbf{P}_{k-1} \mid L^{-1} \bmod \mathbf{B}_+ \in K_{\Lambda_0}((\epsilon+\mathcal{B}_-)^{k-1}_{\Lambda_0})\}$ is Zariski open in $\mathbf{G}/\mathbf{P}_{k-1}$, and in each fiber $SL(n-2k+2)/\mathbf{B}_+^{(n-2k+2)}$, the image of $K_{\Lambda_0}((\epsilon+\mathcal{B}_-)^k_{\Lambda_0})$ is also Zariski open. Hence, $K_{\Lambda_0}((\epsilon+\mathcal{B}_-)^k_{\Lambda_0})$ is Zariski open in all of $\mathbf{G}/\mathbf{B}_+$, and independence of the I_{rk} follows from Lemmas 4.2 and 4.3. ∎

Corollary 4.1 *All the I_{rk}, $r=1,\ldots,n-2k$; $k=1,\ldots,m$, are independent on $(\epsilon+\mathcal{B}_-)^{\diamond}_{\Lambda_0}$.*

Proof. The Jacobian of the I_{rk} with respect to $L_0,\ldots,L_m$ is block upper-triangular, and the diagonal blocks have full rank by Theorem 4.1. ∎

Remark 4.3 In [DLNT] and [S1], it is assumed that all eigenvalues λ_{rk}, $k=1,\ldots,m$; $r=1,\ldots,n-2k$, are distinct. This condition (called "simplicity" in those papers), is a bit stronger than the property $(\mathrm{A})_\mathrm{k}$ which we require to hold for $k=1,\ldots,m$. Our geometric picture helps one understand why some such restriction is necessary. Property $(\mathrm{A})_\mathrm{k}$ is relevant to the approach in [DLNT] in another way. The columns of the matrix Z introduced in the proof of 4.4 are the eigenvectors of X. The Plücker coordinates π_i^* of Z^{-1} are essentially the last entries of those eigenvectors, and they are used by DLNT to define the angle variables θ_{r0}. Our requirement that $\pi_i\pi_i^*$ be nonzero therefore guarantees that those entries do not vanish. A similar interpretation holds for other k. ∎

Remark 4.4 A less circuitous proof of independence could perhaps be based on the Plücker representations (23). We have not tried to do this, because the ideas developed in the present section will be important later. ∎

5. A Torus Orbit

Proposition 3.5 shows that an ideal generated by $\mathbf{P}^\chi$-parabolic invariants defines a torus invariant subvariety of $\mathbf{G}/\mathbf{P}^\chi$. In this section, we prove that (generically) the equations $I_{r1} = c_{r1}, r = 1, \ldots, n-2$, in fact cut out a single generic torus orbit $\mathcal{F}$ (defined below) in $SL(n, \mathrm{C})/\mathbf{P}_1$. (Corresponding results for the algebras B_n, C_n will be mentioned.) This torus orbit $\mathcal{F}$ will be the base for a bundle whose fibers are the flag manifolds $SL(n-2, \mathrm{C})/\mathbf{B}_+^{(n-2)}$ (see (12), and also Section 4); in the next section, we will see that the 1-chop flows are implemented by a torus action along those fibers.

On $(\epsilon + \mathcal{B}_-)^\diamond_{\Lambda_0}$, the independent equations $I_{r1} = c_{r1}, r = 1, \ldots, n-2$, cut out a submanifold of codimension $n-2$ which maps, under the embedding K_{Λ_0} in (18), to a smooth quasiprojective subvariety of $SL(n, \mathrm{C})/\mathbf{B}_+$, of the same codimension, contained in the torus invariant locus (cf. Proposition 3.3)

$$E_{r1} = c_{r1} E_{01}, \; r = 1, \ldots, n-2. \tag{37}$$

By Proposition 3.5(ii), these equations descend to $SL(n, \mathrm{C})/\mathbf{P}_1$. Since this homogeneous space has dimension $2n-3$ (cf. (11)), equations (37) cut out a variety $\mathcal{F}$ of pure dimension = rank = $n-1$.

Theorem 5.1 *The subvariety $\mathcal{F}$ of $SL(n, C)/\mathbf{P}_1$ cut out by the equations (37) is a generic closed torus orbit.*

We must define "generic closed torus orbit". A homogeneous space $\mathbf{G}/\mathbf{P}^\chi$ can be realized as the orbit $\mathcal{O}^\chi$ of $\mathbf{G}$ through the projectivized highest weight vector v^χ in the projectivized representation space $\mathrm{P}(V^\chi)$. A vector $v \in V^\chi$ can be written as $\sum_\mu \pi_\mu v^\mu$; the sum runs over a basis of weight vectors. The coefficients π_μ are called *Plücker coordinates*. Only the weight vectors $v^{w \cdot \chi}$, for w in the Weyl group $\mathbf{W}$, lie in the orbit $\mathcal{O}^\chi$.

Definition 5.1 *A* torus orbit *is the orbit of the action of a Cartan subgroup* $\mathbf{H}$ *of* $\mathbf{G}$ *on the homogeneous space* $\mathbf{G}/\mathbf{P}^\chi$. *Its (compact) closure is a* closed torus orbit. *A torus orbit is called* generic *if it contains a point p for which all Plücker coordinates $\pi_{w \cdot \chi}(p), w \in \mathbf{W}$, where $\mathbf{W}$ is the Weyl group of* $\mathbf{H}$, *are nonzero. The closure of a generic torus orbit is also called generic; it contains all fixed points of the torus action on* $\mathbf{G}/\mathbf{P}^\chi$. (All this is explained and proved in [GS].)

Proof of Theorem 5.1. According to Proposition 2.2(iii), the flows generated by the Chevalley invariants I_{r0} act on $SL(n, \mathrm{C})/\mathbf{P}_1$ via

$$\exp(t_r \nabla I_{r0}(C_{\Lambda_0}))\, L^{-1} \bmod \mathbf{P}_1.$$

Since C_{Λ_0} is regular, these flows are independent for $r = 1, \ldots, n-1$ ([Ko1]), and because C_{Λ_0} is assumed semisimple, they act as an $(n-1)$-dimensional algebraic torus. The torus is the centralizer $SL(n,\mathbf{C})^{C_{\Lambda_0}}$ of the companion matrix (9).

It is convenient, for the rest of the argument, to use the embedding (19). This is just a change of coordinates in the flag manifold which simplifies the analysis.

The variety $\mathcal{F}$ is a finite union of closed torus orbits of the maximal dimension, $n-1$. We now show that every irreducible component $\mathcal{F}_s$ of $\mathcal{F}$ is generic. Write the equations $E_{r1} = c_{r1}E_{01}$ in Plücker coordinates on $SL(n,\mathbf{C})/\mathbf{P}_1$ as in (22). If some π_i or π_i^* vanishes on an interior point of $\mathcal{F}_s$, it vanishes everywhere on $\mathcal{F}_s$ since the torus action scales the Plücker coordinates. The ideal of $\mathcal{F}_s$ must contain a torus invariant monomial forcing π_i, say, to vanish: that would be the equation $\pi_i\pi_i^* = 0$. According to the proof of Lemma 4.4, this can only happen when $\lambda_{r1} = \lambda_{s0}$ for some r and s, which is impossible by the definition of $(\epsilon + \mathcal{B}_-)^1_{\Lambda_0}$.

To complete the proof of the theorem, we will show that the degree of $\mathcal{F}$ and the degree of a generic closed torus orbit in $SL(n,\mathbf{C})/\mathbf{P}_1$ are equal.

There are natural mappings

$$SL(n,\mathbf{C})/\mathbf{P}_1 \xrightarrow{i} \mathbf{P}^{n-1} \times (\mathbf{P}^{n-1})^* \xrightarrow{S} \mathbf{P}(\mathbf{C}^n \otimes \bigwedge^{n-1} \mathbf{C}^n);$$

here i denotes the inclusion onto the hyperplane section defined by the Plücker relation

$$\sum_{i=1}^{n} (-1)^i \pi_i \pi_i^* = 0$$

(see Remark 4.1), and S is the Segre embedding

$$[\pi_1 : \ldots : \pi_n] \times [\pi_1^* : \ldots : \pi_n^*] \mapsto [\pi_i \pi_j^*].$$

In the Segre embedding,

$$\deg \mathcal{F} = \binom{2n-2}{n-1} = \deg SL(n,\mathbf{C})/\mathbf{P}_1.$$

This degree is computed as follows.

Since $\mathcal{F}$ is cut out in $SL(n,\mathbf{C})/\mathbf{P}_1$ by hyperplanes of the Segre embedding, its degree is the same as that of $SL(n,\mathbf{C})/\mathbf{P}_1$. But the latter is itself cut out by the (Plücker) hyperplane section in $\mathbf{P}^{n-1} \times (\mathbf{P}^{n-1})^*$. The degree of $\mathbf{P}^r \times \mathbf{P}^s$ under its Segre embedding, which can be computed from the leading coefficient of the corresponding Hilbert polynomial, is $\binom{r+s}{r}$.

For comparison, we now calculate the degree of a generic torus orbit in $SL(n,\mathbf{C})/\mathbf{P}_1$. By Proposition 2.10 in [O], this degree is $(n-1)!$ times the volume of a certain convex polytope $\square$ in $\mathbf{R}^{n-1}$, which, for *generic* torus orbits, is the Minkowski sum of the $(n-1)$-simplex $V_{n-1} \stackrel{\text{def}}{=} \{x_1 + \ldots + x_n =$

$1, x_1, \ldots, x_n \geq 0\}$ in R^n and its dual. The normalization is chosen so that V_{n-1} has volume $1/(n-1)!$.

The simplex V_{n-1} has $\binom{n}{j+1}$ j-dimensional faces, each being a simplex V_j. The polytope $\square$ is constructed by replacing each V_j by $V_j \times V_{n-j-2}$. This product contains $\binom{n-2}{j}$ $(n-2)$-simplices V_{n-2}, each of which serves as base for a V_{n-1}. Hence, $\square$ is made up of

$$\sum_{j=0}^{n-2} \binom{n-2}{j} \binom{n}{j+1} = \binom{2n-2}{n-1}$$

simplices V_{n-1}, each having volume $1/(n-1)!$. The degree of a generic closed torus orbit is therefore $\binom{2n-2}{n-1}$, which was also the degree of $\mathcal{F}$. It follows that $\mathcal{F}$ is a single closed, generic torus orbit. ∎

Remark 5.1 The closed torus orbit $\mathcal{F}$ is a singular variety. This follows from a count of the edges emanating from each vertex, combined with [O, Theorem 1.10]. ∎

Remark 5.2 A converse to Theorem 5.1 is also true: every generic closed torus orbit in $SL(n, C)/\mathbf{P}_1$ containing the image of an $X \in (\epsilon + \mathcal{B}_-)^{\diamond}_{\Lambda_0}$ is a level set of the I_{r1}. ∎

Remark 5.3 For the other classical Lie algebras, the number of 1-chop integrals is precisely what is needed to cut out a variety of dimension = rank in the appropriate partial flag manifold $\mathbf{G}/\mathbf{P}_1$ (cf. Theorem 3.1 and Proposition 3.5). A degree calculation as in the proof just concluded shows that for B_n, C_n this variety is again a single generic closed torus orbit (of degree 2^{2n+1}). We have not checked the situation for D_n, because some crucial constructions in the next few sections do *not* seem to work for it. ∎

6. Torus Action in the Fibers

Let LSV_1 denote the subvariety of $SL(n, C)/\mathbf{B}_+$ which is cut out by the 1-chop integrals (Roman LSV will refer to subsets of a flag manifold, while slanted $\mathcal{LSV}$ is used for subsets of $\epsilon + \mathcal{B}_-$). The results of the preceding section can be summarized in the diagram

$$\begin{array}{ccc} \mathbf{P}_1/\mathbf{B}_+ & \longrightarrow & LSV_1 \\ & & \downarrow \pi \\ & & \mathcal{F} \end{array} \tag{38}$$

where $\mathcal{F}$ is the generic closed torus orbit in $SL(n,\mathbf{C})/\mathbf{P}_1$ defined by the 1-chop integrals.

According to Proposition 2.2, the flow generated by a $\mathbf{P}_1$ invariant I acts on $\mathbf{Flag}_n$ as

$$L^{-1}n(\exp t\nabla I(X)) \bmod \mathbf{B}_+. \tag{39}$$

Because $\nabla I(X)$ belongs to the algebra $\mathcal{P}_1$, the exponential lies in the group $\mathbf{P}_1$, and therefore also $n(\exp t\nabla I(X)) \in \mathbf{P}_1$. It follows that such a flow fixes $L^{-1} \bmod \mathbf{P}_1 \in \mathcal{F}$, so that it must evolve along the fibers of the projection π in (38). These fibers are isomorphic to

$$\mathbf{P}_1/\mathbf{B}_+ = \mathbf{Flag}_{n-2} = SL(n-2,\mathbf{C})/\mathbf{B}_+^{(n-2)}.$$

We now show that the flows along the fibers are implemented by a torus action on $\mathbf{Flag}_{n-2}$.

6.1 1-Chop Flows in the Fibers

Remark 6.1 To understand the scope of subsequent results, one must keep two facts in mind. First, the 1-chop integrals are not defined for all $X \in (\epsilon + \mathcal{B}_-)_{\Lambda_0}$, since they are rational functions whose denominator may vanish. If $\phi_1(X)$ is not defined, then $X \in (\epsilon + \mathcal{B}_-)_{\Lambda_0} \setminus (\epsilon + \mathcal{B}_-)^{\diamond}_{\Lambda_0}$, and — as shown in Lemma 6.1 below — $K_{\Lambda_0}(X)$ lies in a fiber over the *boundary* of the open torus orbit in $\mathcal{F}$. Second, the factorization solution outlined in Subsection 2.3 does not always work. When lower-upper factorization fails, the solution of the Kostant-Toda equation has a pole. This occurs in the fibers over the so-called *Painlevé divisor*, which is the intersection of $\mathcal{F}$ with the complement of the big cell in $\mathbf{G}/\mathbf{P}_1$; such fibers are not in the image of $(\epsilon + \mathcal{B}_-)_{\Lambda_0}$ under the Kostant map K_{Λ_0}. We restrict our attention to the remaining fibers, which make up a bundle over a dense (quasi-affine) open subset $\mathcal{F}^0$ of the open torus orbit in $\mathcal{F}$. ∎

Lemma 6.1 *Let $\mathcal{F}$ be the closed torus orbit in $SL(n,\mathbf{C})/\mathbf{P}_1$ determined by the equations (37), where the c_{r1} are chosen so that equation (5) has distinct roots. Suppose that $K_{\Lambda_0}(X)$ lies in a fiber over $L_0^{-1} \bmod \mathbf{P}_1$, and $E_{01}(X) \neq 0$. Then the rational matrix $\phi_1(Y)$ in Definition 3.1 exists and is regular semisimple whenever $K_{\Lambda_0}(Y)$ lies in a fiber over $\mathcal{F}^0$.*

Proof. The Toda flows through X are described in Lemma 2.2, and clearly $X(t) = b(t) \cdot X$ implies that $X_{n1}(t) = b_{nn}(t)b_{11}(t)^{-1}X_{n1} \neq 0$. Thus, E_{01} is nonzero at one point in every fiber over $\mathcal{F}^0$. By Lemma 4.1, $\phi_1(Y)$ then exists at all points of the big cell over $\mathcal{F}^0$. ∎

Notation 6.1 In this subsection only, we modify the conventions of Notation 4.1. If $n \in \mathbf{N}_-$, it has a unique factorization $n = n_0 n_1$, where n_0 has the Heisenberg form (11) and (this is the change) $n_1 \in \mathbf{N}_- \cap \mathbf{P}_1$. When we think of n_1 as element of $SL(n-2, \mathbf{C})$, we write $\tilde{n}_1$.

Definition 6.1 *Define a map*

$$\jmath_1 : \mathbf{N}_- \bmod \mathbf{B}_+ \to SL(n, C)/\mathbf{P}_1 \times SL(n-2, C)/\mathbf{B}_+^{(n-2)}$$

on the big cell of $\mathbf{Flag}_n$ *by*

$$\jmath_1 : n \bmod \mathbf{B}_+ \mapsto n_0 \bmod \mathbf{P}_1 \times \tilde{n}_1 \bmod \mathbf{B}_+^{(n-2)}. \tag{40}$$

Theorem 6.1 *The map $\jmath_1$ of Definition 6.1 transforms the 1-chop flows in the generic fiber $\pi^{-1}(L_0^{-1} \bmod \mathbf{P}_1)$ into the action of a maximal torus of $SL(n-2, C)$ on the second component of the product. This torus is the centralizer of $C_{\Lambda_1}(L_0)$ (defined in (32)). The generic closed orbits of this torus in $SL(n-2, C)/\widetilde{\mathbf{P}}_2$ are precisely the (generic) level sets of the 2-chop integrals I_{r2}.*

Proof. When $K_{\Lambda_0}(X) = L^{-1} \bmod \mathbf{B}_+$, then the flow of X generated by a $\mathbf{P}_1$-invariant I is given by (39),

$$L_0^{-1} L_1^{-1} n(\exp(t\nabla I(X))) \bmod \mathbf{B}_+.$$

The first component in (40) is just $L_0^{-1} \bmod \mathbf{P}_-$, as was to be expected since the flows move in the fiber. The second component is $(L_1^{-1} n(\ldots))\tilde{} \bmod \mathbf{B}_+^{(n-2)}$. To simplify this expression, note first that $(\bmod\, \mathbf{B}_+)$

$$L_1^{-1} n(\ldots) = L_1^{-1} \exp(t\nabla I(L_1 \cdot L_0 \cdot C_{\Lambda_0})) = \exp(t\nabla I(L_0 \cdot C_{\Lambda_0}))L_1^{-1}. \tag{41}$$

This is an element of $\mathbf{P}_1$. By Lemma 2.4, the Levi factor of the exponential is the exponential of the Levi component of $t\nabla I$. Because the radical in the Levi decomposition is normal in $\mathbf{P}_1$, that factor of the exponential may be absorbed $\bmod\, \mathbf{B}_+$. Equation (41) then becomes $\exp(t[\nabla[I(L_0 \cdot C_{\Lambda_0})]_1^{ss})L_1^{-1} \bmod \mathbf{B}_+$, which can be thought of in a natural way as element of $SL(n-2, \mathbf{C})/\mathbf{B}_+^{(n-2)}$. When $I = I_{21}$, the exponential becomes $\exp(tC_{\Lambda_1}(L_0))$ by definition, and in general, by Propositions 3.1 and 3.2, it can be written $\exp(t\rho(C_{\Lambda_1}(L_0)))$ for some polynomial ρ. Thus, the second component of the map $\jmath_1$ is

$$\exp(t\rho(C_{\Lambda_1}(L_0))\widetilde{L}_1^{-1} \bmod \mathbf{B}_+^{(n-2)}. \tag{42}$$

The flows in (42) are precisely the action of the centralizer of $C_{\Lambda_1}(L_0)$.

As in (35), we may conjugate $C_{\Lambda_1}(L_0)$ to a diagonal matrix in $\mathcal{GL}(n-2,\mathbf{C})$ by means of a $\widetilde{V}(L_0) \in SL(n-2,\mathbf{C})$: $\widetilde{V}(L_0) \cdot C_{\Lambda_1}(L_0) = \Lambda_1$. This amounts to a change of coordinates in $\mathbf{Flag}_{n-2}$ that makes the 1-chop flows act by a diagonal torus. Now, we saw in the proof of Theorem 4.1 that the I_{r2} could be thought of as functions on the fiber $\pi^{-1}(L_0^{-1} \bmod \mathbf{P}_1) \equiv \mathbf{Flag}_{n-2}$. It was shown, moreover, that they depend only on the coordinates in the partial flag manifold $SL(n-2,\mathbf{C})/\widetilde{\mathbf{P}}_2$, where they have a Plücker coordinate expression analogous to (22). The proof of Theorem 5.1 is therefore applicable, and shows that the level set of the I_{r2} is a generic closed orbit of the 1-chop torus. ∎

6.2 The General Case

The general picture follows the pattern laid out in the preceding subsection. We return to the conventions of Notation 4.1.

The k-chop analog of Definition 6.1 will be needed. Introduce a map $\jmath_k$ from $\mathbf{N}_-$ mod $\mathbf{B}_+$ to a product of partial flag manifolds:

$$SL(n,\mathbf{C})/\mathbf{P}_1 \times SL(n-2,\mathbf{C})/\widetilde{\mathbf{P}}_2 \times \ldots \times SL(n-2k,\mathbf{C})/\mathbf{B}_+^{(n-2k)}$$

$$n \bmod \mathbf{B}_+ \mapsto n_0 \bmod \mathbf{P}_1 \times \tilde{n}_1 \bmod \widetilde{\mathbf{P}}_2 \times \ldots \times (n_k \ldots n_m)\tilde{} \bmod \mathbf{B}_+^{(n-2k)}.$$

The tilde means that we consider $\tilde{n}_j$ to be an element of $SL(n-2j,\mathbf{C})$.

Lemma 6.2 *Let I be a k-chop integral. Its flow acts on L^{-1} mod $\mathbf{B}_+$ by*

$$(L_0^{-1} \ldots L_{k-1}^{-1}) \exp\bigl(t\nabla I(L_{k-1} \cdot \ldots \cdot L_0 \cdot C_{\Lambda_0})\bigr)(L_k^{-1} \ldots L_m^{-1}) \bmod \mathbf{B}_+. \quad (43)$$

Proof. By part (ii) of Proposition 2.2, the flow moves L^{-1} mod $\mathbf{B}_+$ to $L^{-1}\exp\bigl(t\nabla I(X)\bigr)$ mod $\mathbf{B}_+$. By Lemma 2.1, since I is coadjoint invariant for $\mathbf{P}_k$ and the product $L_k^{-1} \ldots L_m^{-1} \stackrel{\text{def}}{=} n_k$ belongs to $\mathbf{P}_k$, we have

$$n_k \exp\bigl(t\nabla I(X)\bigr) = \exp\bigl(t\nabla I(n_k \cdot X)\bigr) n_k.$$

But

$$n_k \cdot X = (L_k^{-1} \ldots L_m^{-1}) \cdot (L_m \ldots L_0) \cdot C_{\Lambda_0},$$

and the result follows. ∎

Theorems 5.1 and 6.1 have the following k-chop analog, which is tedious to state but does summarize most of our work so far.

Theorem 6.2 *The map $\jmath_k$ takes the k-chop flows through L^{-1} mod $\mathbf{B}_+ \in K_{\Lambda_0}((\epsilon + \mathcal{B}_-)^{\diamondsuit}_{\Lambda_0})$ to the action of a maximal torus on the last factor in the* *range of $\jmath_k$, namely, the flag manifold $\mathbf{Flag}_{n-2k} = SL(n-2k)/\mathbf{B}_+^{(n-2k)}$. This torus is the centralizer of $C_{\Lambda_k}(L_0, \ldots, L_{k-1})$. Let LSV_k denote a generic level set of the I_{rk} in $\mathbf{Flag}_{n-2k+2}$. Then there is a fibration*

$$\begin{array}{ccc} \widetilde{\mathbf{P}}_k/\mathbf{B}_+^{(n-2k+2)} & \longrightarrow & LSV_k \\ & & \downarrow{\scriptstyle \pi_k} \\ & & \mathcal{F}_k \end{array} \tag{44}$$

where $\mathcal{F}_k$ is a generic closed torus orbit in $SL(n-2k+2)/\widetilde{\mathbf{P}}_k$. Let $\mathcal{F}_k^0$ denote the intersection of the open torus orbit in $\mathcal{F}$ with the big cell in $SL(n-2k, C)/\widetilde{\mathbf{P}}_{k+1}$. Then whenever $\jmath_k \circ K_{\Lambda_0}(X)$ lies in a fiber over $\mathcal{F}_k^0$, $\phi_{k+1}(X)$ is defined.

Proof. Apply $\jmath_k$ to (43), use (34), and follow the proof of Theorem 6.1. This will give the assertion about the torus action. The claim about the level set of the I_{rk} follows from the ideas in the proofs of Theorems 4.1 and 6.1. ∎

Finally, we want to look at all Kostant-Toda flows simultaneously.

Proposition 6.1 *Let $I_j, j = 0, \ldots, m$ be j-chop invariants. Their simultaneous action on $L^{-1} \bmod \mathbf{B}_+ \in K_{\Lambda_0}((\epsilon + \mathcal{B}_-)^{\diamond}_{\Lambda_0})$ is given by*

$$\prod_{j=0}^m \left(\exp(t_j \nabla I_j((L_0 \ldots L_{j-1}) \cdot C_{\Lambda_0})) L_j^{-1} \right) \bmod \mathbf{B}_+. \tag{45}$$

Here $L^{-1} = L_0^{-1} \ldots L_m^{-1}$ as in Notation 4.1, the t_j are independent complex parameters, and the product is ordered, with factors arranged (left to right) from $j = 0$ to $j = m$.

Proof. Lemma 6.2 is applied repeatedly. Because all the flows commute, the order in which they are applied is immaterial. First act by the I_0 flow; by the lemma, the exponential can be moved to the left of the entire product $L_0^{-1} \ldots L_m^{-1}$. Next, act by the I_1 flow; the corresponding exponential can be moved to the left of $L_1^{-1} \ldots L_m^{-1}$, and so forth. ∎

One consequence of formula (45) is perhaps disappointing. The fact that $\jmath_k$ realizes the k-chop flows as torus action might lead one to hope that $\jmath_m$ provides an equivariant map, of *all* flows simultaneously, to the product of partial flag manifolds

$$SL(n, C)/\mathbf{P}_1 \times SL(n-2, C)/\widetilde{\mathbf{P}}_2 \times \ldots \times SL(n-2m, C)/\mathbf{B}_+. \tag{46}$$

This is not the case. While

$$\jmath_m : L_0^{-1} \dots L_m^{-1} \bmod \mathbf{B}_+ \mapsto L_0^{-1} \bmod \mathbf{P}_1 \times \dots \widetilde{L}_m^{-1} \bmod \mathbf{B}_+,$$

one must re-factor the product in (45) into $M_0(t_1)^{-1} \dots M_m(t_m)^{-1}$ mod $\mathbf{B}_+$ before applying $\jmath_m$, and this destroys the toric nature of the action.

This difficulty seems to preclude simultaneous diagonalization of the flows on all the level sets *at once*. For a single level set LSV defined by fixed λ_{rk}, a geometric picture of linearization of the flows can be still be obtained. Because the I_{rk} are independent on $(\epsilon+\mathcal{B}_-)^{\diamond}_{\Lambda_0}$, one deduces that by starting at a generic L^{-1} mod $\mathbf{B}_+$ and acting by all possible flows, one can sweep out the open dense subset LSV^0 of LSV on which all flows are defined. Thus, one can coordinatize LSV^0 by an $(m+1)$-tuple $(h_0, \dots, h_m)$, where h_k is an element of the torus generated by $\nabla I_{rk}((L_0 \dots L_{k-1}) \cdot C_{\Lambda_0})$; the correspondence is

$$(h_0, \dots, h_m) \mapsto (h_0 L_0^{-1})(h_1 L_1^{-1}) \dots (h_m L_m^{-1}) \bmod \mathbf{B}_+.$$

This map can be composed with one to the product (46), with image

$$h_0 L_0^{-1} \bmod \mathbf{P}_1 \times \tilde{h}_1 \widetilde{L}_1^{-1} \bmod \widetilde{\mathbf{P}}_2 \times \dots,$$

where $\tilde{h}_k$ denotes the element of $SL(n-2k, \mathbb{C})$ corresponding to h_k of the torus centralizing $C_{\Lambda_k}(L_0, \dots, L_{k-1})$. It follows that LSV^0 is a (subset of) a product of open torus orbits in (46).

This "coordinatization" depends, of course, on the reference point L^{-1} mod $\mathbf{B}_+$. It is analogous to the standard parametrization of a level set by angle variables. There is always an arbitrary choice of origin, and the angles are then determined by flowing out from that origin.

7. An Example

We conclude by illustrating some of the earlier general results for the $\mathcal{SL}(4, \mathbb{C})$ Kostant-Toda lattice.

The elements $X \in (\epsilon + \mathcal{B}_-)_{\Lambda_0}$ have the form

$$X = \begin{pmatrix} f_1 & 1 & 0 & 0 \\ g_1 & f_2 & 1 & 0 \\ h_1 & g_2 & f_3 & 1 \\ k & h_2 & g_3 & f_4 \end{pmatrix}$$

with $\sum_i f_i = 0$ and fixed (distinct) eigenvalues $\lambda_{10}, \lambda_{20}, \lambda_{30}, \lambda_{40}$.

The 1-chop integrals I_{11} and I_{21} are the ratios of the coefficients in the polynomial

$$\det \begin{pmatrix} g_1 & f_2 - \lambda & 1 \\ h_1 & g_2 & f_3 - \lambda \\ k & h_2 & g_3 \end{pmatrix}.$$

The gradients, such as ∇I_{21}, are to be computed in the whole algebra, and then evaluated on $\epsilon + \mathcal{B}_-$. So, one must first replace the entry 1 by a variable u, take the gradient with respect to all the variables, and in the end set $u = 1$. The result has the form

$$\begin{pmatrix} 0 & * & * & * \\ 0 & f_3 - \frac{g_1 h_2}{k} & -u + \frac{g_1 g_3}{k} & * \\ 0 & -g_2 + \frac{h_1 h_2}{k} & f_2 - \frac{h_1 g_3}{k} & * \\ 0 & 0 & 0 & 0 \end{pmatrix} - \text{const Id}_4,$$

where the constant is chosen to make the trace zero, and u is set $= 1$.

A similar calculation shows $\nabla I_{11}(X)$ to be upper triangular. Therefore, $n(t) = n\big(\exp(t\nabla I_{11}(X))\big)$ is the identity, and the flow $X \mapsto n(t)^{-1} \cdot X$ is trivial. Hence, I_{11} is a Casimir.

According to Proposition 3.2, $\nabla I_{21}(X)$, or equivalently, $\phi_1(X)$, determines all the 1-chop integrals. In the present case, the central component of the Levi part is found to be $\frac{1}{4} I_{11} \text{diag}\,(1, -1, -1, 1)$; note that it determines the 1-chop Casimir I_{11}. The semisimple component $[\nabla I_{21}(X)]^{ss}$ of the Levi part is nonzero only in the middle 2×2 block, which is

$$\begin{pmatrix} \frac{1}{2}(f_3 - f_2 + \frac{h_1 g_3 - h_2 g_1}{k}) & -1 + \frac{g_1 g_3}{k} \\ -g_2 + \frac{h_1 h_2}{k} & -(\ldots) \end{pmatrix},$$

with trace zero. This 2×2 matrix is precisely $-\big(\phi_1(X) - \frac{1}{2}(\text{Trace}\,\phi_1(X))\,\text{Id}\big)$.

We turn now to the relation $X = L \cdot C_{\Lambda_0}$. The factorization in Notation 4.1 is

$$L = L_1 L_0 = \begin{pmatrix} 1 & 0 & 0 & 0 \\ 0 & 1 & 0 & 0 \\ 0 & \nu & 1 & 0 \\ 0 & 0 & 0 & 1 \end{pmatrix} \begin{pmatrix} 1 & 0 & 0 & 0 \\ a_1 & 1 & 0 & 0 \\ a_2 & 0 & 1 & 0 \\ c & b_2 & b_1 & 1 \end{pmatrix}. \tag{47}$$

Set $Z^{-1} = V^{-1}L^{-1} = (z_{ij})$, where (Definition 2.4) $V_{ij} = (\lambda_{j0})^{i-1}$ is a Vandermonde matrix. Introduce the Plücker coordinates of Z^{-1}, $\pi_i = z_{i1}$ and $\pi_i^* = (-)$ the cofactor of the $(i, 4)$ entry. (This differs from our earlier usage.) The 1-chop integrals have the Plücker expressions

$$I_{11} = \frac{\sum_i \sigma_2(\hat{\imath})\pi_i\pi_i^*}{\sum_i \sigma_1(\hat{\imath})\pi_i\pi_i^*} \tag{48}$$

$$I_{21} = \frac{\sum_i \sigma_3(\hat{\imath})\pi_i\pi_i^*}{\sum_i \sigma_1(\hat{\imath})\pi_i\pi_i^*} \tag{49}$$

where $\sigma_j(\hat{\imath})$ denotes the jth elementary symmetric function on the three entries of Λ_0 different from λ_{i0}. These formulas, involving only the fixed eigenvalues λ_{j0} and the Plücker coordinates of Z^{-1}, make clear once again that I_{11} and I_{21} depend on L_0 but not on L_1.

According to Proposition 2.2 and equation (19), the 0- and 1-chop flows act on $V^{-1}L^{-1}$ mod $\mathbf{B}_+$ by

$$HV^{-1}L_0^{-1}L_1^{-1}n \bmod \mathbf{B}_+.$$

Here $H = \operatorname{diag}(h_1, h_2, h_3, h_4)$ with $h_1h_2h_3h_4 = 1$. It is the torus generated by the Chevalley flows,

$$\exp\Big(\sum_{j=1}^{3} t_j(\Lambda_0{}^j - \frac{1}{4}\,\mathrm{Trace}\,\Lambda_0{}^j \cdot \mathrm{Id})\Big).$$

The matrix n is $n\big(\exp(t_{21}\nabla I_{21}(X))\big)$. It is clear from the expression for I_{21} that n has the same form as L_1 in (47), but of course the initial ν is replaced by a function $\mu(t_{21})$. This illustrates in a different way that the 1-chop flow affects L_1, but not L_0.

The separation of 0-chop and 1-chop flows according to the fibration (38) takes the following form. Modulo $\mathbf{P}_1$, $HZ^{-1}n$ becomes

$$\begin{pmatrix} 1 & 0 & 0 & 0 \\ h_2\pi_2/h_1\pi_1 & 1 & 0 & 0 \\ h_3\pi_3/h_1\pi_1 & 0 & 1 & 0 \\ h_4\pi_4/h_1\pi_1 & h_4\pi_2^*/h_2\pi_1^* & h_4\pi_3^*/h_3\pi_1^* & 1 \end{pmatrix}.$$

The fiber $\mathbf{P}_1/\mathbf{B}_+$, which is isomorphic to $\mathbf{P}^1$, was identified with $SL(2,\mathbf{C})/\mathbf{B}_+^{(2)}$ in Theorem 6.1. The 1-chop flow becomes (modulo $\mathbf{B}_+^{(2)}$)

$$\begin{pmatrix} 1 & 0 \\ \nu & 1 \end{pmatrix}\begin{pmatrix} 1 & 0 \\ \mu(t_{21}) & 1 \end{pmatrix}, \qquad \text{or} \qquad \exp\big(t_{21}C_{\Lambda_0}(L_0)\big)\begin{pmatrix} 1 & 0 \\ \nu & 1 \end{pmatrix}.$$

In each level set variety LSV (determined by Λ_1) in the flag manifold, the generic orbits of the diagonal torus of $SL(4,\mathbf{C})$ are parametrized by the 1-chop time variable t_{21}. A geometrically more natural local coordinate is the *cross-ratio* introduced in [GM]. Let $\pi_{ij} = z_{i1}z_{j2} - z_{i2}z_{j1} =$ the (i,j)-minor of the first two columns of Z_0^{-1}. The cross-ratio of Z_0^{-1} is defined to be $c = \pi_{12}\pi_{34}/\pi_{13}\pi_{24}$. The following proposition is easily verified [S1].

Proposition 7.1 *(i) The cross-ratio is constant on every torus orbit in* $\mathbf{G}/\mathbf{B}_+$. *(ii) The cross-ratio of a generic torus orbit is in* $\mathbf{P}^1 \setminus \{0,1,\infty\}$. *(iii) Each* $c \in \mathbf{P}^1 \setminus \{0,1,\infty\}$ *is the cross-ratio of precisely two generic torus orbits in* LSV. *(iv) The two torus orbits with cross-ratio*

$$(\lambda_{10} - \lambda_{20})(\lambda_{30} - \lambda_{40})/(\lambda_{10} - \lambda_{30})(\lambda_{20} - \lambda_{40})$$

are the sets of fixed points of the I_{21}*-flow.*

Part (iv) is the most interesting. A point X is fixed for the I_{21}–flow if $\nabla I_{21}(X)$ is upper triangular. The I_{21}-fixed-point sets are invariant under the 0-chop flows (because all flows commute). *Their cross-ratio is determined by* Λ_0, *and is independent of* Λ_1. This points to deeper geometric features of the level set varieties, which, however, we have not yet explored.

8. Conclusion

We want to provide some perspective on what we have done in this paper, and comment on what remains to be understood.

The level set of the constants of motion of a real integrable Hamiltonian system is (if all the commuting flows are complete) a product of lines and circles. When the system is complexified and the level set compactified, one often finds a mixture of torus orbits and abelian varieties. If the phase space lies in a Lie algebra, there are group-theoretic explanations of the properties of these geometric objects. Our goal is to understand the full Toda lattice for semisimple Lie algebras from this perspective.

The present paper is only a first step. We find that the integrals provided by Thimm's method applied to a parabolic chain have simple expressions when transferred to the flag manifold. The embedding of the level set in the flag manifold—a technique that is implicit or explicit in many papers treating many different systems— begins to become more accessible to analysis when the hierarchical structure of the constants of motion is unraveled and exploited. One can then see a complex analog of the Arnol'd-Liouville theorem: on generic points of the level set, the family of commuting flows is the action of an abelian group $(\mathrm{C}^*)^d$.

Several interesting questions were not (or barely) addressed in this paper. One would like to understand the geometry of the *compactified* level set in the flag manifold $\mathbf{G}/\mathbf{B}_+$. We have seen that it is a tower of fibrations. We know that they are nontrivial, i.e., not products, but can say little else.

The Hamiltonian structure of the full Toda lattice is also quite subtle. As explained in Section 3, there are many non-commuting integrals; their structure should be understood and used. We give an example.

In $\mathcal{SL}(4,\mathrm{C})$, the phase space $\epsilon+\mathcal{B}_-$ has dimension 9. There is one Casimir, making the generic symplectic leaf 8-dimensional, and 3 Chevalley invariants. One therefore needs one more integral. The integral I_{21} (see Section 7) is a parabolic Casimir for the subgroup $\mathbf{P}_{\omega_1+\omega_3}$ that stabilizes the weight $\omega_1+\omega_3$ (and I_{11}, arising from the same parabolic, is a Casimir for the Kostant-Toda lattice). If, however, one realizes $\mathcal{SL}(4,\mathrm{C})$ as the orthogonal algebra $\mathcal{SO}(6,\mathrm{C})$, it is more natural to use an integral J associated with the group $\mathbf{P}_{\omega_2}$, and get the Casimir as before. The Poisson bracket of I_{21} and J cannot be zero, and one has two different sets of involutive integrals, and two quite different level set varieties.

Finally, we want to comment on generalizations to other Lie algebras. Early in this paper, we mentioned several general results, because they provide strong evidence for the existence of a Lie-theoretic framework. Although our approach will work for the algebras B_n and C_n, we did not spend much space on the modifications, since they shed no light on the cases that require serious rethinking. Here are two examples.

Example. In D_4, the generic symplectic leaf in $\epsilon + \mathcal{B}_-$ has dimension 16. One needs 4 integrals, in addition to the 4 Chevalley invariants. Let us call them $I_{11}, I_{21}, I_{12}, I_{22}$; they are obtained by chopping, see Subsection 3.3. Let $\mathbf{P}_1$ be the parabolic that stabilizes the fundamental weight ω_1.

The equations $I_{11} = c_{11}, I_{21} = c_{21}$ cut out a 4-dimensional variety in the 6-dimensional homogeneous space $D_4/\mathbf{P}_1$, and it is the orbit of a 4-dimensional maximal torus of D_4. The fiber is the flag manifold of A_3. Now something new happens: the two 1-chop flows do *not* sweep out a generic orbit of A_3 in this fiber, and we do not "start over again". Calculations suggest that for the D_n series, the level set variety is a tower of fibrations whose bases and fibers are *nongeneric* torus orbits. ∎

Example. The exceptional algebra G_2 poses a different problem. A parabolic chain, as required by Thimm's argument, can contain just one parabolic (aside from $\mathbf{G}$ and $\mathbf{B}_+$). Both available parabolics have only one Casimir, and Thimm's method adds only one integral to the two Chevalley invariants. Unfortunately, *four* integrals are required in all. Thus, Thimm's method will not prove integrability. ∎

A. Appendix: Factorization and Reduction

As was mentioned at the outset, the Toda equations studied in [DLNT] are a flow on *symmetric* matrices. Our phase space, $\epsilon + \mathcal{B}_-$, looks very different. Nevertheless, the definition of the Casimirs and the chopped integrals is the same for both systems. We want to explain briefly why certain "kinematic" similarities are to be expected, while at the same time there remain significant "dynamical" differences. The reason, in brief, is this: the two systems are reductions, under different (but related) symmetry groups, of a single linear Hamiltonian system on $T^*\mathbf{G}$.

We are indebted to Percy Deift for nagging us until we made the effort to compare the symmetric and Kostant-Toda lattices. The techniques we need, it turned out, were already developed in a paper by Reiman [R].

A.1 Reduction by Left-Right Group Actions

We summarize the setup in [R], generally using Reiman's notation. Proofs of all assertions may be found in his paper.

Let $\mathcal{G}$ be a real split semisimple Lie algebra with Lie group $\mathbf{G}$, and suppose it can be written as vector space direct sum of two subalgebras, $\mathcal{G} = \mathcal{A} + \mathcal{B}$ (we now work over the reals to facilitate comparison with [DLNT]). Introduce the corresponding connected subgroups $\mathbf{A}$ and $\mathbf{B}$ of $\mathbf{G}$. It may or may not be true that $\mathbf{G} = \mathbf{AB}$. Let $\mathcal{G}_0 = \mathcal{A} \oplus \mathcal{B}$ be the Lie-algebra direct sum, and set $\mathbf{G}_0 = \mathbf{A} \times \mathbf{B}$.

Given a function $f : \mathcal{G}^* \to \mathbf{R}$, extend it to $\bar{f} : T^*\mathbf{G} \to \mathbf{R}$ by left translation. We are interested in the Hamiltonian system generated by the function $\overline{H}(\xi)$ obtained from the Killing form on $\mathcal{G}^*$ (identified with $\mathcal{G}$), $H(\xi) = \frac{1}{2}\kappa(\xi, \xi)$. In the right-trivialization of $T^*\mathbf{G}$, this system is linear:

$$\dot{g} = g\xi, \quad \dot{\xi} = 0, \tag{50}$$

with solution

$$g(t) = g(0)\exp(\xi(0)t), \quad \xi(t) = \xi(0). \tag{51}$$

Our aim is to reduce this system by a left action of $\mathbf{G}_0$ on $T^*\mathbf{G}$ in order to arrive at the Toda lattice.

Two cases will be of interest:

Case i): $\mathcal{A} = \mathcal{N}_-, \mathcal{B} = \mathcal{B}_+$;

Case ii): $\mathcal{A} = \mathcal{K}, \mathcal{B} = \mathcal{B}_+$, where $\mathcal{K}$ is the compact subalgebra in the Iwasawa decomposition.

In Case i), we will obtain (the real form of) our full Toda lattice, and in Case ii), the symmetric DLNT version.

The group $\mathbf{G}_0$ acts on $\mathbf{G}$ from the left according to $(a, b) \cdot g = agb^{-1}$. The action lifts to $T^*\mathbf{G}$, and it has a moment map

$$J : T^*\mathbf{G} \to \mathcal{G}_0^* \cong \mathcal{A}^\perp \oplus \mathcal{B}^\perp.$$

For $\xi \in \mathcal{G}_0^*$, we denote by $\overline{M}_\xi$ the reduced symplectic manifold $J^{-1}(\xi)/(\mathbf{G}_0)_\xi$.

Because $\overline{H}$ is left-invariant under $\mathbf{G}_0$, the system (50) descends to $\overline{M}_\xi$, and because the flow (51) is complete on $T^*\mathbf{G}$, the reduced flow is complete on $\overline{M}_\xi$.

For suitable choices of $\xi \in \mathcal{G}_0^*$, the reduced systems may be identified with full Toda lattices (see below). In Case ii), one obtains the DLNT system. In Case i), one gets a *completion* of our full Toda version, a manifold on which the Toda flows are complete.

The phase space $\epsilon + \mathcal{B}_-$ arises as follows. Define $\sigma : \mathbf{G}_0 \to \mathbf{G}$ by $\sigma(a, b) = ab^{-1}$. Then $T^*\sigma : T^*\mathbf{G} \to T^*\mathbf{G}_0$ is an isomorphism on fibers, and the fiberwise inverse $(T^*\sigma)^{-1} \overset{\text{def}}{=} \sigma^* : T^*\mathbf{G}_0 \to T^*\mathbf{G}$ makes sense.

For a function f on $T^*\mathbf{G}$, define $f^\sigma : T^* \to \mathbf{R}$ by $f^\sigma = f \circ \sigma^*$. One should think of f^σ as being obtained by translation from the identity fiber $\mathcal{G}_0^*$: left-translation by $\mathbf{A}$, right translation by $\mathbf{B}$. The map σ^* commutes with the left $\mathbf{G}_0$ actions on $T^*\mathbf{G}_0$ and $T^*\mathbf{G}$, and $f \mapsto f^\sigma$ is Poisson: $\{f^\sigma, g^\sigma\} = \{f,g\}^\sigma$.

The group $\mathbf{G}_0$ acts on $T^*\mathbf{G}_0$ by left translations, with moment map $J_0 : T^*\mathbf{G}_0 \to \mathcal{G}_0^*$. Since the Hamiltonian $\overline{H}^\sigma$ is left-invariant, it can be reduced to a coadjoint orbit $\mathcal{O}_\xi \subset \mathcal{G}_0^*$. (One such orbit will be $\epsilon + \mathcal{B}_-$.) Reiman connects these two reductions of (50) as follows.

Theorem. *The orbit $\mathcal{O}_\xi$ embeds into $\overline{M}_\xi$ as dense open subset. The reduced flow of $\overline{H}^\sigma$ maps to a (possibly not complete) subflow of the reduction of the flow (51) of $\overline{H}$.*

A.2 The Two Toda Lattices

The general theory is now illustrated with the two examples (Cases i) and ii) above). This is a small addition to Reiman's work; he concentrates on variants of tridiagonal Toda lattices. Except in the general Proposition A.1 below, we restrict the discussion to $\mathcal{G} = \mathcal{SL}(n, \mathbf{R})$.

In Case i), we have $\mathcal{G} = \mathcal{N}_- + \mathcal{B}_+$ and $\mathcal{G}^* = \mathcal{B}_- + \mathcal{N}_+$. Reduction takes place at a special element $\epsilon \oplus c \in \mathcal{G}_0^*$; here $\epsilon \in \mathcal{N}_+$ is the nilpotent element in (2), and $c \in \mathcal{B}_-$ is a convenient point in a generic coadjoint orbit of $\mathbf{B}_+$ on $\mathcal{B}_+^* \cong \mathcal{B}_-$. As shown in [Ar], one may take c to depend on m parameters $\vec{\alpha}$ as follows (when $n = 2m$):

$$c(\vec{\alpha})_{ij} \stackrel{\text{def}}{=} \begin{cases} \alpha_j, & \text{if } i = j = m+1, \ldots, 2m, \\ 1, & \text{if } i + j = 2m+1, j = 1, \ldots, m, \\ 0, & \text{otherwise.} \end{cases}$$

A similar formula holds for odd n. The $\mathbf{G}_0$-orbit through $\epsilon \oplus c$ is a symplectic leaf in the phase space $\epsilon + \mathcal{B}_-$. The reduction of the $\overline{H}^\sigma$-flow from $T^*\mathbf{G}_0$ to $\mathcal{O}_{\epsilon\oplus c}$ is precisely our Toda system. Its flow, as was pointed out repeatedly in the paper, is not complete. It would be complete on the larger reduced symplectic manifold $\overline{M}_{\epsilon\oplus c}$; our approach in effect builds this manifold through the map to the flag manifold.

In Case ii), $\mathcal{G} = \mathcal{K} + \mathcal{B}_+$ and $\mathcal{G}^* \cong \mathcal{S} + \mathcal{N}_+$, where $\mathcal{S}$ is the set of symmetric matrices. Reduction is performed at the element $0 \oplus c'$, where $c' = c(\vec{\alpha}/2) + c(\vec{\alpha}/2)^t$ (superscript ^{t} denotes transpose). Because the factorization $\mathbf{G} = \mathbf{KB}$ is global, one has $\mathcal{O}_{0\oplus c'} = \overline{M}_{0\oplus c'}$. The reduced space is a symplectic leaf in $\mathcal{S}$, and the reduced flow —the symmetric Toda flow— is complete on this leaf.

There is a natural symplectic diffeomorphism between the coadjoint orbits in $\epsilon + \mathcal{B}_-$ and $\mathcal{S}$. The orbits have the form

$$\{\epsilon + \textstyle\prod_{\mathcal{B}_-}(bcb^{-1}) \mid b \in \mathbf{B}_+\}$$

and

$$\{\prod_{\mathcal{S}}(bc'b^{-1}) \mid b \in \mathbf{B}_+\}.$$

It is easy to check that $\epsilon + X$ belongs to the first orbit if and only if, *for the same* $b \in \mathbf{B}_+$, $X_- + X_0 + X_-^t$ belongs to the second orbit (X_- and X_0 are the strictly lower and diagonal parts of $X \in \mathcal{B}_-$). Furthermore, the Poisson brackets in $\epsilon + \mathcal{B}_-$ and $\mathcal{S}$ are both given by

$$\{f, g\}(Y) = \text{Trace}\left(Y[\prod_{\mathcal{B}_+} \nabla f(Y), \prod_{\mathcal{B}_+} \nabla g(Y)]\right),$$

where the projections are along $\mathcal{N}_-$, resp. $\mathcal{K}$. The brackets of the coordinate functions $Y_{ij}, i \geq j$, on the two spaces are identical.

The Toda flows on orbits in $\epsilon + \mathcal{B}_-$ and $\mathcal{S}$, however, are not isomorphic. For one thing, one flow is incomplete, the other is complete. More concretely, the two reductions of the Hamiltonian $\overline{H}$ on $T^*\mathbf{G}$ yield distinctly different functions. Because of the isomorphism between $\epsilon + \mathcal{B}_-$ and $\mathcal{S}$, we get two quite different completely integrable systems of full Toda type on either of the two phase spaces. In the case of tridiagonal Toda, the Kostant and symmetric realizations are symplectically diffeomorphic [Ko2, R]. In the full case, such an isomorphism cannot exist globally; we do not know whether there exists a symplectic map defined at least on dense open subsets of the Kostant and symmetric orbits.

Finally, the reduction picture explains why the chopped constants of motion are obtained by similar prescriptions in the two cases. This follows from general considerations.

Proposition A.1 *With all notation as above, let* $f : \mathcal{G}^* \to \mathbf{R}$ *be invariant under the action of* $\mathbf{B}$, $f(\mathrm{Ad}_b^*\xi) = f(\xi), b \in \mathbf{B}$. *Let* $\tilde{f}$ *denote the projection of* f^σ *to* $\mathcal{O}_\xi \subset \mathcal{G}_0^*$. *Then* $\tilde{f}$ *is a constant of motion of the reduction of the system generated by* $\overline{H}^\sigma$.

Remark. In particular, since $\mathbf{B} = \mathbf{B}_+$ in both Case i) and Case ii), the class of $\mathbf{B}_+$-invariant functions (namely, ratios of semi-invariants with the same character) will induce constants of motion for both versions of the Toda lattice. ∎

Proof. Because H is a Casimir on $\mathcal{G}^*$, it Poisson-commutes with f. The pullbacks of H and f to $T^*\mathbf{G}$ under the momentum map for the right action of $\mathbf{G}$ on $T^*\mathbf{G}$ remain in involution; those pullbacks are just the left translates of H and f from the fiber over the identity in $\mathbf{G}$, which we have denoted by $\bar{f}$ and $\overline{H}$. In the left trivialization of $T^*\mathbf{G}$, we have $\bar{f}(g, \mu) = f(\mu)$, and similarly for $\overline{H}$. Now, the left action of $\mathbf{G}_0$ leaves these functions invariant. The group $\mathbf{G}_0$ acts on $T^*\mathbf{G}$ according to (a, b) : $(g, \mu) \mapsto (agb^{-1}, \mathrm{Ad}_{b^{-1}}^*\mu)$, but

$$\bar{f}\left((agb^{-1}, \mathrm{Ad}_{b^{-1}}^*\mu)\right) = f(\mathrm{Ad}_{b^{-1}}^*\mu) = f(\mu) = \bar{f}\left((g, \mu)\right).$$

Therefore, $\bar{f}$ and $\overline{H}$ can both be projected to the reduced phase space $\overline{M}_\xi$, and the resulting functions will again be in involution.

Similarly, $\bar{f}^\sigma$ and $\overline{H}^\sigma$ are invariant under the left action of $\mathbf{G}_0$ on $T^*\mathbf{G}_0$, and so their reductions to $\mathcal{O}_\xi$ are also in involution. ∎

The proposition shows that a thorough understanding of the *linear* problem (50) as a completely integrable Hamiltonian system is essential for a comprehensive theory of full Toda lattices.

REFERENCES

[Ar] A. A. Arhangelskiĭ, "Completely integrable Hamiltonian systems on the group of triangular matrices", *Mat. Sb.* **108**, No. 4 (1979) 134-142.

[BFR] A. M. Bloch, H. Flaschka, T. Ratiu, "A convexity theorem for isospectral manifolds of Jacobi matrices in a compact Lie algebra", *Duke Math. J.* **61** (1990) 41-65.

[DLNT] P. A. Deift, L.-C. Li, T. Nanda. C. Tomei, "The Toda lattice on a generic orbit is integrable", *Comm. Pure Appl. Math.* **39** (1986) 183-232.

[Dix] J. Dixmier, *Enveloping Algebras*, North-Holland 1977.

[FH] H. Flaschka, L. Haine, "Variétés de drapeaux et réseaux de Toda", *Math. Z.* **208** (1991) 545-556.

[GM] I. M. Gel'fand, R. W. MacPherson, "Geometry in Grassmannians and a generalization of the dilogarithm", *Adv. Math.* **44** (1982) 279-312.

[GS] I. M. Gel'fand, V. V. Serganova, "Combinatorial geometries and torus strata on homogeneous compact manifolds", *Uspehi Mat. Nauk* **42**:2 (1987) 107-134.

[Kap] M. M. Kapranov, "Chow quotients of Grassmannians, I", preprint 1992.

[Ko1] B. Kostant, "Lie group representations on polynomial rings", *Am. J. Math.* **85** (1963) 327-404.

[Ko2] B. Kostant, "The solution to a generalized Toda lattice and representation theory", *Adv. Math.* **34** (1979) 195-338.

[Ko3] B. Kostant, "On Whittaker vectors and representation theory", *Invent. Math.* **48** (1978) 101-184.

[Mo] J. Moser, "Finitely many point masses on the line under the influence of an exponential potential", in *Springer Lecture Notes in Physics* **38** (1975) 467-497.

[O] T. Oda, *Convex Bodies and Algebraic Geometry*, Springer-Verlag 1988.

[R] A. G. Reiman, "Integrable Hamiltonian systems connected with graded Lie algebras", *Zap. Nauch. Sem. LOMI* **95** (1980) 3-54.

[RSTS] A. G. Reiman, M. S. Semënov-Tian-Shantsky, "Reduction of Hamiltonian systems, affine Lie algebras and Lax equations", *Invent. Math.* **54** (1979) 81-100.

[S1] S. Singer, Ph. D. Dissertation, Courant Institute, 1991.

[S2] S. Singer, "Some maps from the full Toda lattice are Poisson," *Phys. Lett.* **A 174** (1993), 66-70.

[Th] A. Thimm, "Integrable geodesic flows on homogeneous spaces", *Ergod. Th. & Dynam. Sys.* **1** (1982) 495-517.

[Tro1] V. V. Trofimov, "Euler equations on Borel subalgebras of semisimple Lie algebras", *Izvestija* **43** (1979) 714-732.

[Tro2] V. V. Trofimov, "Finite-dimensional representations of Lie algebras and completely integrable systems", *Mat. Sb.* **11** (1980) 610-621.

N. M. Ercolani
Department of Mathematics
The University of Arizona
Tucson, AZ 85721, USA

H. Flaschka
Department of Mathematics
University of Arizona
Tucson, AZ 85721, USA

S. Singer
Department of Mathematics
Haverford College
Haverford, PA 19041, USA

Deformations of a Hamiltonian Action of a Compact Lie Group

Victor Guillemin*

1. Introduction

Let (M, ω) be a compact $2n$-dimensional symplectic manifold, G a commutative connected compact Lie group, and $\tau : G \times M \longrightarrow M$ a Hamiltonian action of G on M with moment map, $J : M \longrightarrow \mathfrak{g}^*$. Since M is compact, J can be normalized by requiring that

$$\int J\omega^n = 0. \tag{1.1}$$

I will assume (1.1) to be in effect from now on. Also to simplify the exposition below I will make the following three assumptions:

(1.2)
1. G acts freely on an open dense set
2. M^G is finite
3. $J : M^G \longrightarrow \mathfrak{g}^*$ is injective.

I will call the image of the map in part 3 of (1.2) the *vertex set* of (ω, τ) and denote it by $\Delta = \Delta(\omega, \tau)$. The main question I want to address in this paper is to what extent is (ω, τ) *determined* by its vertex set? I will only attempt to answer this question at the deformation level. That is, I will describe how small deformations of (ω, τ) affect the vertex set and prove a local rigidity theorem which says, roughly speaking, that one can't deform the pair (ω, τ) non-trivially without deforming its vertex set. To be more specific the main result of this paper will be the following:

Theorem. *The moduli space of Hamiltonian actions of G on M is a manifold of dimensions, $m = \beta_2 =$ the second Betti number of M. Moreover, one can find $2m$ points*

$$q_1, q_1', q_2, q_2', \ldots, q_m, q_m' \tag{1.3}$$

in $\Delta(\omega, \tau)$ having the property that, as one varies (ω, τ) the distances between q_i and q_i' vary smoothly with respect to (ω, τ) and are a system of coordinates on this moduli space.

---

*Supported by NSF grant DMS 890771.

Let me now describe how to prove this result (and, in particular, how to choose the q_i's and q_i''s). Since $J: M^G \longrightarrow \mathfrak{g}^*$ is injective each $q \in \Delta$ has a unique pre-image, p, in M^G. Let

$$\Sigma_q = \{\alpha_{i,q};\ i = 1,\dots,n\} \tag{1.4}$$

be the weights of the isotropy representation of G on T_pM. It is clear that, under deformation of (ω,τ), the q's can deform; however, the $\alpha_{i,q}$'s, being points of the weight lattice of G, can't. Thus, the following lemma shows that when one deforms (ω,τ) the relative positions of the points in the vertex set can't vary arbitrarily.

Lemma 1. *For every $\alpha \in \Sigma_q$ the half-line*

$$q + t\alpha, \qquad t > 0, \tag{1.5}$$

contains at least one other vertex.

Thus, for instance, if the vertex, q, is *simple* (i.e. if none of the $\alpha_{i,q}$'s are collinear) there will be n vertices on the rays emanating from q in the directions, $\alpha_{i,q}, i = 1,\dots,n$; and, as (ω,τ) varies the positions of these vertices relative to q can only vary by moving along these rays.

Next I will describe how to choose the points (1.3). let W be a fixed connected component of the set:

$$\{\xi \in \mathfrak{g}, \alpha_{i,q}(\xi) \neq 0 \text{ for all } i \text{ and } q\} \tag{1.6}$$

(i.e. let W be the choice of a "positive Weyl chamber" relative to the set of weights, $\alpha_{i,q}$). Given any ξ in W and any point, $q \in \Delta$, I will define the *index*, σ_q, to be the number of weights, $\alpha_{i,q}$, in Σ_q such that $\alpha_{i,q}(\xi) < 0$. (Note that this number depends on the choice of W but not on the choice of ξ in W.) These indices turn out, modulo some Morse theoretic considerations which I'll discuss in the next section, to determine the Betti numbers of M :

Lemma 2. *The i-th Betti number of M is zero if i is odd, and if i is even, i.e. $i = 2k$, the i-th Betti number of M is equal to the number of vertices q, with $\sigma_q = k$.*

I will prove, in fact, a fairly constructive version of this result. I will associate to each vertex, q, of index k an element c_q of $H_{2k}(M,\mathbb{R})$ and show that the set

$$\{c_q, \quad \sigma_q = k\} \tag{1.7}$$

forms a *basis* of $H_{2k}(M,\mathbb{R})$.

Now suppose, in particular, that $\sigma_q = 1$. Then, by definition, there exists a unique weight, $\alpha_{i,q}$, in the set, Σ_q, with the property that $\alpha_{i,q}(\xi) < 0$. Moreover, Lemma 1 says that there exist one or more vertices on the ray

$$q + t\alpha_{i,q}, \quad 0 < t < \infty. \tag{1.8}$$

The following will be a more precise version of this result:

Lemma 3. *Let $t_0 = [\omega](c_q)$. Then the point, $q' = q + t_0\alpha_{i,q}$ on the ray (1.8) is a vertex.*

Thus, in particular, if $q_1, \dots, q_m$ are the vertices of index one, the distances between q_1 and q_1' q_2 and q_2' etc. can be computed by evaluating $[\omega]$ on the basis vectors, $c_{q_1} \dots, c_{q_m}$ of $H_2(M, \mathbb{R})$; so one gets the following corollary of Lemma 3.

Lemma 4. *The distances between q_i and q_i' $i = 1, \dots, m$, determine the cohomology class of ω in $H^2(M, \mathbb{R})$.*

To explore this fact we need next a result which says, roughly speaking, that the action of a compact group on a compact manifold is indeformable. For the moment let G be an arbitrary compact Lie group (not necessarily commutative or connected). Let S be an open convex neighborhood of 0 in $\mathbb{R}^k$, M a compact manifold and τ_s, $s \in S$, an action of G on M. We will say that τ_s depends smoothly on s if the action of G on $M \times S$ defined by

$$g(m, s) = (\tau_s(g, m), s)$$

is a smooth action. For the proof of the following see the appendix:

Lemma 5. *If τ_s depends smoothly on S there exists a diffeomorphism, $f_s : M \longrightarrow M$, depending smoothly on s, such that $f_s \tau_s f_s^{-1} = \tau_0$.*

In other words all the τ_s's are isomorphic. Thus, in proving the theorem one can assume that the action, τ, is fixed and that the only deformation that takes place is that of the symplectic form, ω. Moreover, by Moser's theorem the only way to deform the Hamiltonian pair, (ω, τ), leaving τ fixed, is to deform $[\omega]$.

Now equip M with a G-invariant metric and consider the affine map,

$$H^2(M, \mathbb{R}) \longrightarrow \Omega^2_{\text{DeRham}}(M)$$

which maps the cohomology class, c, onto the two-form

$$\omega_c = \omega + \mu_{c-c_0}$$

c_o being the cohomology class of ω and μ_{c-c_0} the unique harmonic representative of $c - c_0$. It is clear that ω_c is G-invariant, closed, represents c and is equal to ω when $c = c_0$. In particular, if c is close to $[\omega]$, ω_c will be symplectic. Therefore, since $H^1(M, \mathbb{R}) = 0$ (see, for instance, Lemma 2), the action, τ, will be Hamiltonian on the symplectic manifold (M, ω_c); and hence the set

$$\{(\omega_c, \tau), c \;\text{ close to } [\omega]\}$$

will be an open subset of the moduli space of Hamiltonian actions of G on M. Thus, by Lemma 4, the distance between the q_i's and q_i''s will be a system of coordinates on this space, concluding the proof of the theorem.

Comments

1. If $\dim M = 2\dim G$, Delzant's theorem [D] says that the vertex set of the triple (M,ω,τ) determines it up to an equivariant diffeomorphism. Thus the theorem above can be viewed as a local version of Delzant's theorem for the case, $\dim M > 2\dim G$.
2. There is an analogue of the theorem above for G non-abelian; however, in the statement of this theorem, the set of vertices, Δ, has to be replaced by the set of vertices of the Kirwan polytope, and the hypotheses (1.2) have to be changed somewhat. (Details will appear in another article.)

2. The proofs of lemmas 1 and 2.

Proof of Lemma 2. Fix, once and for all, a G-invariant Riemannian metric on M having the property that for every $p \in M^G$ the inner product on T_p defined by this metric is Kaehlerian, and the decomposition of T_p into two dimensional weight space is orthogonal. Now let ξ be a point in W, let J^ξ be the ξ-component of the moment map and let $\mathfrak{v}$ be the gradient of J^ξ. Note that $\mathfrak{v}$ is G-invariant since the metric is and also J^ξ itself.

The proof of Lemma 2 is an elementary application of Morse theory: It is easy to see that J^ξ is a Morse function and that its critical points are the fixed points of G. Moreover, if p is a critical point its Morse index is $2\sigma_q$ where q is the vertex corresponding to it. In particular, the critical points are all of even index, and the Morse inequalities (which, in this case, are equalities since there are no critical points of odd index) imply Lemma 2. (For more details see [A] or [GS]).

Finally, by inspecting the Morse-Whitney decomposition of M associated with the gradient flow, $\exp t\mathfrak{v}$, one sees that the unstable manifold, $M^-(p)$, of $\exp t\mathfrak{v}$ through p is a "pseudo-cycle", i.e. the closure of $M^-(p)$, minus $M^-(p)$ itself, is of dimension two less than $M^-(p)$. Thus it supports a homology class, c_q, in $H_{2k}(M,\mathbb{R})$ (where $k = \sigma_1$) and these homology classes form a basis of $H_{2k}(M,\mathbb{R})$.

Proof of Lemma 1. Let α be in Σ_q, and let $\mathfrak{h} = \{\eta \in \mathfrak{g}, \alpha(\eta) = 0\}$. Since α is in the weight lattice of G, $\mathfrak{h}$ is the Lie algebra of a *closed* connected subgroup, H, of G; and the quotient group, G/H, is a circle group with Lie algebra, $\mathfrak{g}/\mathfrak{h}$. Let X be the connected component of M^H containing the pre-image, p, of q in M^G. X is a G-invariant symplectic submanifold of M (on which H acts trivially) and J maps X into the line

$$q + t\alpha, \quad -\infty < t < \infty. \tag{2.1}$$

Therefore the images of the G-fixed points in X are vertices lying on this line. Let us now prove that there are vertices on the interval, $t > 0$: Note that since H acts trivially on X, the action of G on X can be viewed as an action of the circle group, G/H; and this action is Hamiltonian since the action of G is. Moreover, if one identifies $(\mathfrak{g}/\mathfrak{h})^*$ with $\mathfrak{h}^\circ$ and identifies $\mathfrak{h}^\circ$ with $\mathbb{R}$ by means of α, the moment map associated with this action is the composition of J with the projection map, $q + t\alpha \longrightarrow t$. Let us denote this map by ϕ. A simple computation shows that p is a critical point of ϕ and that the Hessian of ϕ at p is positive definite on the two-dimensional subspace of T_pX associated with the weight, α. Thus, in particular, p is not a local maximum of ϕ; and, hence, there exists another critical point, p', with $\phi(p') > \phi(p)$. It follows that the vertex, $q' = J(p')$, lies to the right of q on the line (2.1). Q.E.D.

Finally to prove Lemma 3, it suffices to prove Lemma 3 with M replaced by X, G replaced by G/H and J replaced by ϕ. We will deal with this special case of Lemma 3 in the next section.

3. The proof of Lemma 3

Let (X, ω) be a compact symplectic manifold and τ a Hamiltonian action of S^1 on X with moment map, $\phi : X \longrightarrow \mathbb{R}$. I will assume below that the fixed point set of τ is finite or, alternatively, that ϕ is a Morse function. Equip M with a G-invariant Riemannian metric and let $\mathfrak{v}$ be the gradient of ϕ. Given a critical point of ϕ of index 2, say p, let $X^-(p)$ be the unstable manifold of p (i.e. the set of all points, $x \in X$, with the property that the trajectory of $\mathfrak{v}$ through x has p as its omega-limit point). This manifold is two-dimensional and S^1-invariant. Moreover, by a theorem of Kirwan, [K], it is a symplectic submanifold of X. The main result of this section* is the following:

Theorem. *The closure of $X^-(p)$ in X consists of $X^-(p)$ plus one additional point, p'. This point is a critical point of ϕ and*

$$\phi(p) - \phi(p') = \int_{X^-(p)} \omega \tag{3.1}$$

Proof. Since $\exp t\mathfrak{v}$ is a gradient flow, every trajectory of $\exp t\mathfrak{v}$ on $X^-(p)$ has a unique alpha-limit point. Each of these α-limit points is a critical point of ϕ, and hence a fixed point of S^1. Therefore, since the action of S^1 on $X^-(p)$ permutes these trajectories, these α-limit points have to be the same. This proves the first assertion in the theorem above. To prove the second assertion, i.e. to verify (3.1), one first notes that, since $X^-(p)$ is topologically just a cell, the restriction of $-\omega$ to $X^-(p)$ is the exterior derivative of a G-invariant one-form, ν. Now let $\mathfrak{w}$ be the infinitesimal generator of the circle group action. Then:

$$d\phi = \iota(\mathfrak{w})\omega = -\iota(\mathfrak{w})d\nu = d\iota(\mathfrak{w})\nu.$$

---

*From which one easily deduces Lemma 3. See the comments at the end of the last section.

Thus

$$\phi = \iota(\mathfrak{w})\nu + c, \qquad c \in \mathbb{R};$$

and noting that $\mathfrak{w} = 0$ at p, one gets $c = \phi(p)$; or, in other words, one gets the formula:

$$\iota(\mathfrak{w})\nu = \phi - \phi(p) \tag{3.2}$$

Now, for every a on the interval,

$$\phi(p') < a < \phi(p), \tag{3.3}$$

let X_a^- be the set of points, $x \in X^-(p)$, where $\phi(x) \geq a$. Let p_0 be any point on ∂X_a^-. Then the map,

$$\gamma : \mathbb{R}/\mathbb{Z} \longrightarrow \partial X_a^- \tag{3.4}$$

sending s onto $(\exp ts\mathfrak{w})(p_0)$ is a diffeomorphism and, by Stokes theorem,

$$\int_{X_a^-} \omega = -\int_{\mathbb{R}/\mathbb{Z}} \gamma^* \nu.$$

But $\gamma^*\nu = \gamma^*(\iota(\mathfrak{w})\nu)ds = (a - \phi(p))ds$ by (3.2) so

$$\int_{X_a^-} \omega = \phi(p) - a$$

and, letting a tend to $\phi(p')$, one obtains (3.1).

Appendix: *The action of a compact Lie group on a compact manifold is indeformable.*

Let G be a compact Lie group, and M a compact manifold, Let S be an open convex neighborhood of the origin in $\mathbb{R}^k$ and, for each $s \in S$, let

$$\tau_s : G \times M \longrightarrow M \tag{A.1}$$

be an action of G on M. τ_s is said to *depend smoothly on s* if the action

$$\tau : G \times M \times S \longrightarrow M \times S \tag{A.2}$$

of G on $M \times S$ defined by

$$\tau(g, m, s) = (\tau_s(g, m), s) \tag{A.3}$$

is smooth. Notice that the action defined by (A.3) leaves invariant the projection

$$\psi : M \times S \longrightarrow S \tag{A.4}$$

sending (m, s) onto s. We will prove below the following indeformability theorem:

Theorem. *If τ_s depends smoothly on s there exists a family of diffeomorphisms of M : $f_s, s \in S$, depending smoothly on s, with the property that for all $(g, m, s) \in G \times M \times S$*

$$f_s \circ \tau_s(g, m) = \tau_0(g, f_s(m)). \tag{A.5}$$

Proof. Without loss of generality one can assume that S is one-dimensional and is a subinterval of the real line containing the origin. Since G is compact one can equip $M \times S$ with a Riemannian metric which is invariant under the action (A.3). Since the function (A.4) is proper and G-invariant, its gradient flow can be integrated; and, for all $s \in S$, provides one with a diffeomorphism of $M \times \{0\}$ onto $M \times \{s\}$ which intertwines the actions, τ_0 and τ_s.

REFERENCES

[A] M. Atiyah, *Convexity and commuting Hamiltonians*, Bull. Lond. Math. Soc. **14** (1982), 1–15.

[D] T. Delzant, *Hamiltoniens périodiques et images convexes de l'application moment*, Bull. Soc. Math. France **116** (1988), 315–339.

[GS] V. Guillemin and S. Sternberg, *Convexity properties of the moment mapping*, Invent. Math. **67** (1982), 491–513.

[K] F. Kirwan, *Convexity properties of the moment mapping III*, Invent. Math. **77** (1984), 547–552.

Department of Mathematics
Massachusetts Institute of Technology
Cambridge, MA 02139
USA

Linear-Quadratic Metrics "Approximate" any Nondegenerate, Integrable Riemannian Metric on the 2-Sphere and the 2-Torus

A.T. Fomenko

A nondegenerate, Riemannian metric is called *integrable* if its geodesic flow is integrable. An integrable nondegenerate Riemannian metric is called *linear-quadratic* if its geodesic flow admits an additional integral which is linear or quadratic in the momenta.

Conjecture. Let g_{ij} be an arbitrary integrable nondegenerate orientable Riemannian metric on 2-sphere or 2-torus. Then the complexity (see [1]) of its geodesic flow coincides with the complexity of some linear-quadratic Riemannian metric.

The conjecture states that, from the point of view of their complexity, integrable geodesic flows with an additional integral which is linear or quadratic in momenta exhaust all integrable nondegenerate geodesic flows on 2-sphere and 2-torus. This means that the integrability of geodesic flows on these 2-surfaces is of a "linear or quadratic" nature, in the sense of complexity.

Note that for geodesic flows on 2-sphere and 2-torus the isoenergy surfaces are $\mathbb{R}P^3$ and T^3, respectively. All possible Hamiltonian systems on $\mathbb{R}P^3$ and T^3 from the viewpoint of their complexity were completely investigated by Nguyen Tien Zung [3]. If the conjecture above is true then all points of the complexity table [1] corresponding to integrable metrics must belong to a particular region discovered by E. N. Selivanova, T. Z. Nguyen and L.S. Polyakova [2], [4] and represented in Figs. 1,2 by small disks inside of the $\mathbb{R}P^3$- and T^3-regions.

An integrable Riemannian metric g_{ij} on a 2-dimensional manifold is said to be *orientable* if all saddle critical circles of the corresponding geodesic flow possess orientable separatrix diagrams.

It has been shown using V.V. Kalashnikov's results [5] that, for the class of orientable, integrable metrics on the 2-sphere, the above conjecture is "almost correct":

Proposition (T.Z. Nguyen, L. S. Polyakova): The number of critical circles for an arbitrary integrable nondegenerate orientable Riemannian metric g_{ij} on the 2-sphere equals $4k + 2$ for some integer k.

Corollary (T.Z. Nguyen, L.S. Polyakova): The integrable nondegenerate orientable geodesic flows on the 2-sphere form the proper subset in the $\mathbb{R}P^3$-region represented in Figure 1 by white disks, black disks and white disks with dots.

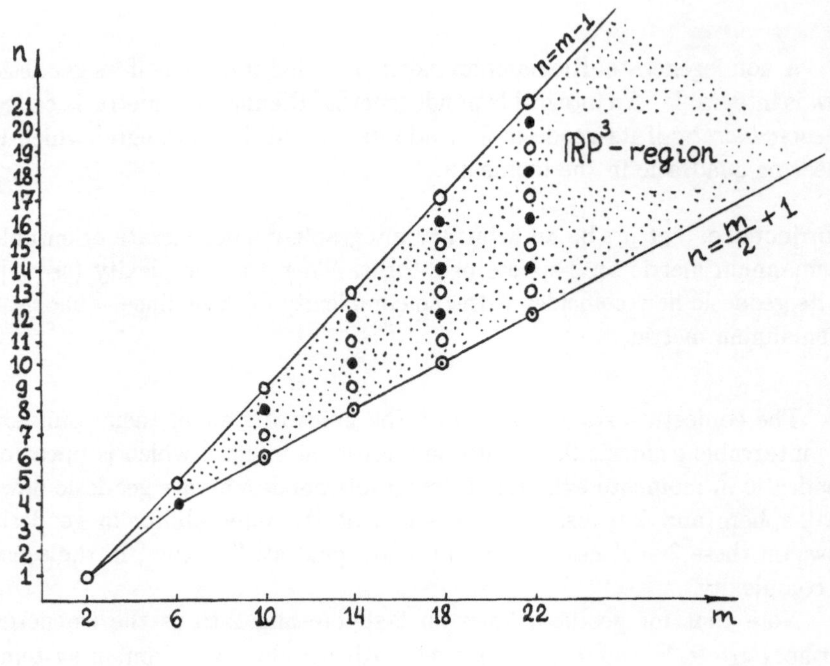

Figure 1.

Here the white (resp. black) disks represent geodesic flows with a linear (resp. quadratic) additional integral. The white disks with dots represent integrable Riemannian metrics which have not yet been identified and may even not exist.

Thus, the "zone of all integrable metrics" practically coincides with the "zone of linear-quadratic integrable metrics," except for a set of points on the lower edge of the region in Fig. 1. It is an extremely interesting problem to complete the analysis and investigate this special line. Fig. 2 represents integrable geodesic flows in the case of the 2-torus.

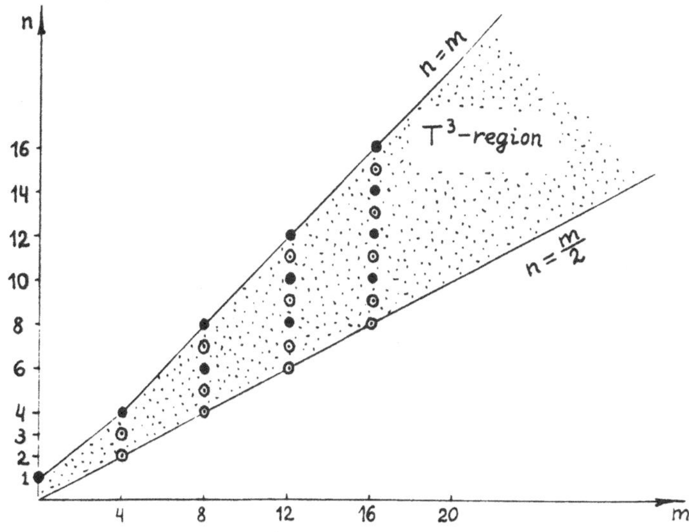

Figure 2.

REFERENCES

[1] A.T. Fomenko, "Topological classification of all Hamiltonian differential equations of general type with two degrees of freedom", in *The Geometry of Hamiltonian systems.* Proceedings of a workshop held in Berkeley, June 5–16, 1989. Springer-Verlag, New York, 1991, pp. 131–339.

[2] E.N. Selivanova, "Topological classification of integrable Bott geodesic flows on the two-dimensional torus", in *Advances in Soviet mathematics.* Amer. Math. Soc., **6** (1991), pp. 209–228.

[3] T.Z. Nguyen, "On the complexity of integrable Hamiltonian systems on three-dimensional isoenergy submanifolds", in *Advances in Soviet mathematics.* Amer. Math. Soc., **6** (1991), pp. 229–255.

[4] T.Z. Nguyen, T.S. Polyakova, *A topological classification of integrable geodesic flows on two-dimensional sphere with quadratic in momenta additional integral,* Jour. of Non-linear Sciences, **3**:1 (1993), in press.

[5] V.V. Kalashnikov, *Description of the Structure of Fomenko invariants on the boundary and inside Q-domains, estimates of their number on the lower boundary for the manifolds S^3, $\mathbb{R}P^3$, $S^1 \times S^2$, and T^3,* Advances in Soviet Mathematics, Amer. Math. Soc., **6** (1991), pp. 297–304.

Department of Mathematics
MGU
Moscow, Russia

Canonical Forms for Bihamiltonian Systems

Peter J. Olver

Dedicated to the Memory of Jean-Louis Verdier

BiHamiltonian systems were first defined in the fundamental paper of Magri, [**5**], which deduced the integrability of many soliton equations from the fact that they could be written in Hamiltonian form in two distinct ways. More recently, the classical completely integrable Hamiltonian systems of ordinary differential equations, such as the Toda lattice and rigid body, have been shown to be biHamiltonian systems. However, recent results of Brouzet, [**1**], extended by Fernandes, [**3**], indicate that there are global, topological obstructions to the existence of a biHamiltonian structure for a general completely integrable Hamiltonian system. The connection between biHamiltonian structures and R-matrices, [**10**], which provide solutions to the classical Yang-Baxter equation, has given additional impetus to their study.

Magri's Theorem demonstrates the existence of an infinite hierarchy of commuting Hamiltonians and flows, provided that the two Hamiltonian structures are compatible, in the sense to be defined below. Therefore, any biHamiltonian system of ordinary differential equations will be completely integrable, as long as a sufficient number of the integrals are functionally independent. Explicitly, *a priori* conditions guaranteeing the independence of the integrals are not so evident, and one method of elucidating such conditions is to determine the possible canonical forms for biHamiltonian systems. In [**11**], Turiel gave a complete classification of "generic" compatible non-degenerate biHamiltonian structures — a "double Darboux Theorem". In [**8**], this classification was used to find the associated canonical forms for such biHamiltonian systems, and their complete integrability, or lack thereof. The integrability depends on the algebraic structure of the biHamiltonian structure. The main result is that any biHamiltonian system associated with a nondegenerate biHamiltonian structure, each of whose eigenvalues appear in just one irreducible substructure, is necessarily completely integrable; in all other cases, there do exist "non-integrable" biHamiltonian systems. (See below for the precise terminology.)

In this brief survey, I will first review these results on the canonical forms for compatible, nondegenerate complex-analytic biHamiltonian systems. Details, as well as some preliminary extensions to the classification of incompatible biHamiltonian structures, can be found in the author's paper [**8**]. Secondly, I will briefly describe the results of Brouzet and Fernandes on integrable Hamiltonian systems without biHamiltonian structures. An outstanding problem in this area is to determine similar canonical forms for degenerate (and compatible) biHamiltonian systems. Unfortunately, Turiel's approach, which is fundamentally tied to the covariant differential form framework for symplectic structures, does not appear to readily generalize, since degenerate Poisson structures can only be readily expressed in the contravariant language of bi-vector fields, [**7**].

A system of differential equations is called *biHamiltonian* if it can be written in Hamiltonian form in two distinct ways:

$$\frac{dx}{dt} = J_1 \nabla H_1 = J_2 \nabla H_0. \tag{1}$$

Here $J_1(x), J_2(x)$ are Hamiltonian operators, not constant multiples of each other, determining Poisson brackets: $\{F, G\}_\nu = \nabla F^T J_\nu(x) \nabla G$. The biHamiltonian structure detemined by J_1, J_2 is *compatible* if the sum $J_1 + J_2$ is also Hamiltonian. The biHamiltonian structure is *nondegenerate* if the first Hamiltonian operator J_1 is nonsingular.

Theorem. *Suppose J_1, J_2 determine a nondegenerate, compatible biHamiltonian structure. For any associated biHamiltonian system (1), there exists a hierarchy of Hamiltonian functions H_0, H_1, H_2, ..., all in involution with respect to either Poisson bracket, $\{H_j, H_k\}_\nu = 0$, and generating mutually commuting biHamiltonian flows*

$$\frac{dx}{dt} = J_1 \nabla H_k = J_2 \nabla H_{k-1}. \tag{2}$$

We classify biHamiltonian structures pointwise according to the algebraic invariants of the skew-symmetric matrix pencil $\lambda J_1(x) + \mu J_2(x)$ at each x. According to the Weierstrass theory, *cf.* [**2**], the complete system of algebraic invariants of a non- degenerate matrix pencil consists of the eigenvalues, the elementary divisors, and the Segre characteristic. (Degenerate pairs of skew-symmetric matrices are handled by the more detailed Kronecker theory.) A pencil is called *elementary* if it has just one complex eigenvalue, and *irreducible* if it has Segre characteristic $[(nn)]$, analogous to a single Jordan block. Every non-degenerate complex matrix pencil can, algebraically, be decomposed into a direct sum of irreducible matrix pencils. (For simplicity, we restrict our attention to complex-analytic systems, although the real case offers little additional difficulty.) The algebraic invariants of a biHamiltonian structure are invariant under the flow of any associated biHamiltonian system. A biHamiltonian structure is *generic* on a domain M if it has constant Segre characteristic, and the number of functionally independent eigenvalues does not change on M.

Theorem. *Every generic non-degenerate, compatible biHamiltonian structure can be locally expressed as a Cartesian product of elementary biHamiltonian structures. Every associated biHamiltonian system decomposes into independent subsystems corresponding to the elementary substructures, each of which consists of an autonomous Hamiltonian system whose dimension is twice the number of irreducible substructures for the given eigenvalue, coupled with a sequence of linear, non-autonomous Hamiltonian systems. In particular, the biHamiltonian system is completely integrable if and only if there is just one irreducible substructure for each eigenvalue.*

When an eigenvalue is constant, the elementary substructure decomposes into a Cartesian product of irreducible substructures; however, this decomposition does *not* hold in the case of non-constant eigenvalues. We will now present the details of the Turiel classification and the structure of associated biHamiltonian systems.

Without loss of generality, we may assume that neither 0 nor ∞ is an eigenvalue, so that the biHamiltonian structure is determined by two compatible symplectic Hamiltonian operators. (Otherwise, replace J_1, J_2 by two other linearly independent members of the corresponding pencil.) Darboux' theorem, [**7**; Theorem 6.22], implies that we can write the first Hamiltonian operator in canonical form

$$J_1 = \begin{pmatrix} 0 & I \\ -I & 0 \end{pmatrix}, \tag{3}$$

relative to canonically conjugate coordinates $x = (p, q)$. Therefore, only the canonical form of the second Hamiltonian operator needs to be explicitly indicated.

Given a Hamiltonian pair J_1, J_2, any associated biHamiltonian system must be a solution to the linear system of partial differential equations

$$\nabla H_1 = M \nabla H_0, \qquad M = J_1^{-1} \cdot J_2, \tag{4}$$

where M is the transpose of the recursion operator, [**7**]. We remark here that the simple system of differential equations (4), which arises in a surprising number of different contexts, is not well understood, except when the matrix M is constant, in which case the general solution can be found in [**4**]. In the present case, the solutions all have a similar pattern. On any convex open subdomain, the two Hamiltonians H_0, H_1 are expressed as a sum of "basic" Hamiltonians $H_0^{(k)}, H_1^{(k)}$, which are individually solutions to (4):

$$H_i(x) = H_i^{(0)}(x) + H_i^{(1)}(x) + \ldots + H_i^{(n)}(x), \qquad i = 0, 1.$$

Moreover, each basic pair $H_0^{(k)}, H_1^{(k)}$, can be most simply expressed in terms of the derivatives with respect to a parameter s evaluated at $s=0$ of a single arbitrary analytic function $F(\xi_1(x,s),\ldots,\xi_m(x,s))$ depending on certain parameterized variables $\xi_j(x,s)$. We can therefore summarize the general classification results in this convenient form.

I. Irreducible, Constant Eigenvalue Pairs.

Canonical coordinates:

$$(p,q)=(p_0,p_1,\ldots,p_n,q_0,q_1,\ldots,q_n),\qquad n\geq 0.$$

Second Hamiltonian operator:

$$J_2=\begin{pmatrix}0 & \lambda\mathrm{I}+U\\ -\lambda\mathrm{I}-U^T & 0\end{pmatrix}.$$

Here $\lambda\mathrm{I}+U$ denotes an irreducible $(n+1)\times(n+1)$ Jordan block matrix with eigenvalue λ.

Parametrized variables:

$$\pi(s)=p_0+sp_1+s^2p_2+\cdots+s^np_n,\qquad \varpi(s)=q_n+sq_{n-1}+s^2q_{n-2}+\cdots+s^nq_0.$$

Basic Hamiltonians:

$$H_0^{(k)}(x)=\frac{1}{\lambda}\,\frac{\partial^k}{\partial s^k}\,F_k(\pi(s),\varpi(s))\bigg|_{s=0}+k\,\frac{\partial^{k-1}}{\partial s^{k-1}}\,F_k(\pi(s),\varpi(s)))\bigg|_{s=0},$$
$$H_1^{(k)}(x)=\frac{1}{\lambda}\,\frac{\partial^k}{\partial s^k}\,F_k(\pi(s),\varpi(s))\bigg|_{s=0},\qquad 0\leq k\leq n.$$

The Hamiltonians are polynomials in the "minor variables" $p_1,\ldots,p_n$, $q_0,\ldots,q_{n-1}$, whose coefficients are certain derivatives of the arbitrary smooth functions $F_k(p_0,q_n)$ of the remaining two "major variables" p_0, q_n. This implies, *cf.* [**8**], that any biHamiltonian system corresponding to an irreducible, constant eigenvalue biHamiltonian structure is completely integrable, since it can be reduced to a single two-dimensional (planar) autonomous Hamiltonian system for the major variables, with Hamiltonian $n!F_n(p_0,q_n)$. (Curiously, the major variables are *not* canonically conjugate for any of the Hamiltonian structures in the pencil determined by J_1 and J_2.) The time evolution of the minor variables is then determined by successively solving a sequence of forced planar, linear Hamiltonian systems in the variables p_k, q_{n-k}.

II. Elementary, Constant Eigenvalue Pairs.

Canonical coordinates:

$$(p,q) = (p^1,\dots,p^m,q^1,\dots,q^m),$$

$$p^i = (p_0^i,\dots,p_{n_i}^i), \quad q^i = (q_0^i,\dots,q_{n_i}^i),$$

where $n_1 \geq n_2 \geq \cdots \geq n_m \geq 0$.

Second Hamiltonian operator:

$$J_2 = \begin{pmatrix} & & & \lambda \mathrm{I} + U_1 & & \\ & & & & \ddots & \\ & & & & & \lambda \mathrm{I} + U_m \\ -\lambda \mathrm{I} - U_1^T & & & & & \\ & \ddots & & & & \\ & & -\lambda \mathrm{I} - U_m^T & & & \end{pmatrix},$$

where $\lambda \mathrm{I} + U_i$ denotes an irreducible $(n_i+1)\times(n_i+1)$ Jordan block as above.

Parametrized variables:

$$\begin{aligned} \pi^i(s) &= p_0^i + sp_1^i + s^2p_2^i + \cdots + s^{n_i}p_{n_i}^i, \\ \varpi^i(s) &= q_{n_i}^i + sq_{n_i-1}^i + s^2q_{n_i-2}^i + \cdots + s^{n_i}q_0^i. \end{aligned}$$

We define

$$\pi^{(k)}(s) = (\pi^1(s),\dots,\pi^{m_k}(s)), \qquad \varpi^{(k)}(s) = (\varpi^1(s),\dots,\varpi^{m_k}(s)),$$

where m_k denotes the number of indices n_i with $n_i \geq k$, *i.e.*, the number of irreducible substructures of dimension $\geq 2k+2$; in particular $m_0 = m$.

Basic Hamiltonians:

$$\begin{aligned} H_0^{(k)}(x) = \frac{1}{\lambda}\,\frac{\partial^k}{\partial s^k}\,F_k(\pi^{(k)}(s),\varpi^{(k)}(s))\bigg|_{s=0} &+ \\ + k\,\frac{\partial^{k-1}}{\partial s^{k-1}}\,F_k(\pi^{(k)}(s),\varpi^{(k)}(s))\bigg|_{s=0}, &\qquad 0 \leq k \leq n_1. \end{aligned}$$

$$H_1^{(k)}(x) = \frac{1}{\lambda}\,\frac{\partial^k}{\partial s^k}\,F_k(\pi^{(k)}(s),\varpi^{(k)}(s))\bigg|_{s=0},$$

As in the irreducible case, the Hamiltonians are polynomials in the minor variables p^i_j, $q^i_{n_i-j}$, $j \geq 1$, whose coefficients are certain derivatives of arbitrary functions of the major variables $p^i_0, q^i_{n_i}$. Thus, such a biHamiltonian system reduces to an autonomous $(2m)$-dimensional Hamiltonian system in the major variables, coupled with a sequence of linear non-autonomous Hamiltonian systems in the appropriate minor variables p^i_k, $q^i_{n_i-k}$, $n_i \geq k \geq 1$.

III. Irreducible, Non-constant Eigenvalue Pairs.

Canonical coordinates:

$$(p,q) = (p_0, p_1, \ldots, p_n, q_0, q_1, \ldots, q_n), \qquad n \geq 0.$$

Second Hamiltonian operator:

$$J_2 = \begin{pmatrix} 0 & P(p)^{-1} \\ -P(p)^{-T} & 0 \end{pmatrix},$$

where $P(p)$ denotes the $(n+1) \times (n+1)$ banded upper triangular matrix

$$P_n(p) = P(p) = \begin{pmatrix} p_0 & p_1 & p_2 & & \cdots & p_n \\ & p_0 & p_1 & p_2 & & \\ & & p_0 & p_1 & & \vdots \\ & & & \ddots & \ddots & \\ & & & & p_0 & p_1 \\ & & & & & p_0 \end{pmatrix}. \tag{5}$$

Here p_0 is the eigenvalue. The explicit formula for the Hamiltonian operator J_2 in terms of $p_0, \ldots, p_n$ is quite complicated. However, remarkably, the inverse matrix J_2^{-1} is also Hamiltonian, and, in fact, isomorphic to the Hamiltonian structure determined by J_2; see [8] for an explicit change of variables mapping the one Hamiltonian structure to the other.

Parametrized variables:

$$\pi(s) = p_0 + sp_1 + s^2 p_2 + \cdots + s^n p_n, \qquad \varpi(s) = q_n + sq_{n-1} + s^2 q_{n-2} + \cdots + s^n q_0.$$

Basic Hamiltonians:

$$H_0^{(-1)} = \widetilde{h}(p_0), \qquad H_1^{(-1)} = h(p_0), \qquad \text{where} \qquad \widetilde{h}'(s) = sh(s),$$

$$H_0^{(k)}(x) = \frac{\partial^k}{\partial s^k} \{\pi(s)\pi'(s)F_k(\pi(s), \varpi(s))\}\Big|_{s=0}, \qquad 0 \le k \le n-1.$$

$$H_1^{(k)}(x) = \frac{\partial^k}{\partial s^k} \{\pi'(s)F_k(\pi(s), \varpi(s))\}\Big|_{s=0},$$

Here $\pi'(s)$ is the derivative of π with respect to s.

In this case, the eigenvalue is a constant, hence p_0 is a first integral. Once its value is fixed, the other minor variable q_n is determined by solving a single autonomous ordinary differential equation. The remaining minor variables $p_1, \ldots, p_n, q_0, \ldots, q_{n-1}$ satisfy a sequence of forced, linear planar Hamiltonian systems.

IV. Elementary, Non-constant Eigenvalue Pairs.

Canonical coordinates:

$$(p, q) = (p_0, p^1, \ldots, p^m, q_0, q^1, \ldots, q^m), \qquad m \ge 2,$$
$$p^i = (p_1^i, \ldots, p_{n_i}^i), \qquad q^i = (q_1^i, \ldots, q_{n_i}^i), \qquad 1 \le i \le m,$$

where $n_1 \ge n_2 \ge \ldots \ge n_k \ge 1$.

Second Hamiltonian operator:

$$J_2 = \begin{pmatrix} 0 & \widehat{P}(p)^{-1} \\ -\widehat{P}(p)^{-T} & 0 \end{pmatrix},$$

where

$$\widehat{P}(p) = \begin{pmatrix} p_0 & p^1 & p^2 & \cdots & & p^m \\ & P_{n_1-1}(\hat{p}^1) & 0 & \cdots & & 0 \\ & & \ddots & & & \vdots \\ & & & P_{n_{m-1}-1}(\hat{p}^{m-1}) & & 0 \\ & & & & & P_{n_m-1}(\hat{p}^m) \end{pmatrix}.$$

Here $\hat{p}^i = (p_0, p_1^i, \ldots, p_{n_i-1}^i)$, and the P_{n_i-1}'s are as given in (5). Again p_0 is the eigenvalue. Note that this particular biHamiltonian structure is pointwise algebraically reducible, but cannot be decoupled using canonical transformations.

Parametrized variables:

$$\pi^i(s) = p_0^i + sp_1^i + \cdots + s^{n_i} p_{n_i}^i, \qquad \varpi^i(s) = q_{n_i}^i + sq_{n_i-1}^i + \cdots + s^{n_i-1} q_1^i, \quad i \geq 1.$$

We further define

$$\mu^i(s) = \frac{\tau^i(s)}{s}, \qquad \sigma^i(s) = \varpi(\tau^i(s)),$$

where $\tau^i(s)$ solves the implicit series equation

$$\pi^i(\tau^i(s)) = \pi^1(s), \qquad i \geq 2.$$

Let

$$\mu^{(k)}(s) = (\mu^1(s), \ldots, \mu^{m_k}(s), \qquad \sigma^{(k)}(s) = (\sigma^1(s), \ldots, \sigma^{m_k}(s)),$$

where m_k denotes the number of indices n_i with $n_i \geq k$.

The Lagrange inversion formula, [**6**], implies that the latter two parametrized variables have the alternative expansions

$$\begin{aligned} \mu^i(s) &= \sum_{n=0}^{n_i-1} \frac{s^n(\zeta^1(s))^{n+1}}{(n+1)!} \frac{d^n}{dt^n} \left[\frac{1}{(\zeta^i(t))^{n+1}} \right] \Bigg|_{t=0}, \\ \sigma^i(s) &= q_{n_i}^i + \sum_{n=0}^{n_i-1} \frac{s^n(\zeta^1(s))^n}{n!} \frac{d^n}{dt^n} \left[\frac{1}{(\zeta^i(t))^{n+1}} \frac{d\omega^i(t)}{dt} \right] \Bigg|_{t=0}, \end{aligned} \tag{6}$$

where $\zeta^i(s) = (\pi^i(s) - p_0)/s$. The expansions (6) can be expressed in terms of the remarkable nonlinear series differential operator

$$\mathcal{D} = D^{-1} : e^{sDu} : D = 1 + \sum_{n=1}^{\infty} \frac{s^n}{n!} D^{n-1} u^n D, \qquad D = \frac{d}{dt}, \quad u = u(t), \tag{7}$$

with s replaced by $s\zeta^1(s)$. In (7), the colons denote *normal ordering* of the non-commuting operators D and u, which is analogous to the so-called "Wick ordering" in quantum mechanics. The operator $\mathcal{D}$ has the surprising property that it commutes with *any* analytic function $\Phi(u)$, *i.e.*, $\mathcal{D}\Phi(u) = \Phi(\mathcal{D}u)$. See [**9**] for details and applications of this operator in combinatorics, orthogonal polynomials and new higher order derivative identities.

Basic Hamiltonians:

$$H_0^{(-1)} = \widetilde{h}(p_0), \qquad H_1^{(-1)} = h(p_0), \qquad \text{where} \qquad \widetilde{h}'(s) = sh(s),$$
$$H_0^{(k)}(x) = \frac{\partial^k}{\partial s^k} \left\{ s\zeta^1(s) \, \frac{d\pi^1}{ds} \, F_k(\pi^1(s), \mu^{(k)}(s), \omega^1(s), \sigma^{(k)}(s)) \right\} \bigg|_{s=0},$$
$$H_1^{(k)}(x) = \frac{\partial^k}{\partial s^k} \left\{ \frac{d\pi^1}{ds} \, F_k(\pi^1(s), \mu^{(k)}(s), \omega^1(s), \sigma^{(k)}(s)) \right\} \bigg|_{s=0},$$

where $0 \leq k \leq n_1 - 1$. In general, such biHamiltonian systems reduce to the integration of a $(2m-2)$-dimensional autonomous Hamiltonian system for the coordinates p_1^i, $q_{n_i}^i$, $i = 1, \ldots, m$, followed by a sequence of forced linear Hamiltonian systems. The eigenvalue p_0 is constant, and the final coordinate q_0 is determined by quadrature. Actually, the initial Hamiltonian system can be reduced in order to $2m - 3$ since it only involves the homogeneous ratios of momenta $r^i = p_1^i / p_1^1$, $i \geq 2$, as can be seen from the second formula (6) for μ^i.

The converse problem is whether every completely integrable Hamiltonian system admits a biHamiltonian structure. One must be careful how to formulate this question, since, away from an equilibrium point, every vector field can be locally straightened out, with flow merely given by a translation, and so, locally, every system of ordinary differential equations is trivially "integrable" and trivially biHamiltonian. Consequently, one should only expect obstructions to the existence of a biHamiltonian structure globally, or, at the very least, in a neighborhood of an invariant torus. If we require that the integrals of the system be expressed as functions of the eigenvalues of the biHamiltonian structure, then, in a neighborhood of an invariant torus, the answer to the above question is "no", as first shown by Brouzet, [1], for $n = 2$, and generalized to arbitrary n by Fernandes, [3].

Theorem. *Let $\dot{u} = J\nabla H$ be a completely integrable Hamiltonian system possessing n functionally independent integrals in involution, whose level sets are compact (and hence n-dimensional tori), and whose Hamiltonian has nondegenerate Hessian. Then the system is biHamiltonian with functionally independent real eigenvalues if and only if the graph of the Hamiltonian H is a hypersurface of translation relative to the affine structure determined by the action variables.*

If $(s^1, \ldots, s^n)$ are the action variables, then the graph of the Hamiltonian satisfies the hypothesis of the Theorem if and only if it admits a parametrization of the form

$$s^i = a_1^i(y^1) + \cdots + a_n^i(y^n), \quad i = 1, \ldots, n,$$
$$H(s^1, \ldots, s^n) = \phi_1(y^1) + \cdots + \phi_n(y^n).$$

Thus, to produce examples of completely integrable systems without bi-Hamiltonian structure, it suffices to devise Hamiltonians whose graph is never a hypersurface of translation. A simple example is

$$H(p,q) = p^1(1 + (p^2)^2) + (p^3)^2 + \cdots + (p^n)^2$$

on the space $(p,q) \in \mathbf{R}^n \times \mathbf{T}^n$, where $\mathbf{T}^n$ denotes the n-dimensional torus. Brouzet's original example corresponds to the case $n = 2$. A more interesting example is provided by the perturbed Kepler problem

$$H = \frac{1}{2}\left(p_r^2 + \frac{p_\theta^2}{r^2} + \frac{p_\phi^2}{r^2 \sin^2\theta}\right) - \frac{1}{r} + \frac{\varepsilon}{2r^2},$$

expressed in spherical coordinates, which, for $\varepsilon \neq 0$, remains integrable, but loses its biHamiltonian structure, [**3**].

REFERENCES

1. Brouzet, R., Systèmes bihamiltoniens et complète intégrabilité en dimension 4, *Comptes Rendus Acad. Sci. Paris* **311** (1990) 895–898.
2. Gantmacher, F.R., *The Theory of Matrices*, vol. 2, Chelsea Publ. Co., New York, 1959.
3. Fernandes, R., Completely integrable biHamiltonian systems, University of Minnesota, 1991.
4. Jodeit, M. and Olver, P.J., On the equation $\nabla f = M \nabla g$, *Proc. Roy. Soc. Edinburgh* **116** (1990), 341–358.
5. Magri, F., A simple model of the integrable Hamiltonian equation, *J. Math. Phys.* **19** (1978), 1156–1162.
6. Melzak, Z.A., *Companion to Concrete Mathematics*, Wiley-Interscience, New York, 1973.
7. Olver, P.J., *Applications of Lie Groups to Differential Equations*, Graduate Texts in Mathematics, vol. 107, Springer-Verlag, New York, 1986.
8. Olver, P.J., Canonical forms and integrability of biHamiltonian systems, *Phys. Lett.* **148A** (1990) 177–187.
9. Olver, P.J., A nonlinear differential operator series which commutes with any function, *SIAM J. Math. Anal.* **23** (1992), 209–221.
10. Semenov-Tian-Shanskii, M.A., What is a classical R-matrix? *Func. Anal. Appl.* **17** (1983), 259–272.

11. Turiel, F.-J., Classification locale d'un couple de formes symplectiques Poisson-compatibles, *Comptes Rendus Acad. Sci. Paris* **308** (1989), 575–578.

Peter J. Olver,
School of Mathematics
University of Minnesota
Minneapolis, Minnesota, 55455
USA

Research supported in Part by NSF Grant DMS 89-01600.

Bihamiltonian Manifolds And Sato's Equations

Paolo Casati, Franco Magri, and Marco Pedroni

ABSTRACT. This paper is a concise introduction to Sato's equations from the point of view of Hamiltonian mechanics. It aims to show that the theory of soliton equations may be completely built on the study of the Casimir functions of a pencil of Poisson brackets on a Poisson manifold.

1. Introduction

This paper is the second report on work that is still in progress (see [1]), and which seeks to give a coherent picture of the theory of soliton equations [2] from the point of view of Hamiltonian mechanics. We recall that, according to Sato's standard approach [3], soliton equations are defined as follows: let L be the monic n–th order differential operator,

$$L = \partial^n - \sum_{k=0}^{n-2} u_k \partial^k, \tag{1.1}$$

and let Q be its n–th root in the ring of pseudodifferential operators,

$$Q^n = L. \tag{1.2}$$

This operator obviously commutes with L, and it may be considered to be the fundamental solution of the equation,

$$[Q, L] = 0, \tag{1.3}$$

since any other solution can be written as a linear combination of powers of Q with constant coefficients. By means of Q, one constructs either Sato's equations,

$$\frac{\partial Q}{\partial t_k} = [Q, (Q^k)_+], \tag{1.4}$$

or the Lax equations,

$$\frac{\partial L}{\partial t_k} = [L, (L^{k/n})_+]. \tag{1.5}$$

This work has been supported by the Italian M.U.R.S.T. and by the G.N.F.M. of the Italian C.N.R.

They define a hierarchy of compatible systems of partial differential equations in two independent variables, x and t, and in $(n-1)$ field functions, u_k (the coefficients of L). These equations are called soliton equations.

This formulation is rather abstract and, apparently, without relation with classical mechanics. Despite this impression, our aim is to show that the equations of Sato's theory have a clear, terse, and illuminate meaning from the point of view of Hamiltonian mechanics. To show that, we have to construct the theory of soliton equations on a different basis. An earlier attempt in this direction was made in [4]. Ours is different: we believe that this theory can be (almost) completely derived from a single basic principle: to construct integrable Hamiltonian systems one has to consider the Casimir functions of a pencil of Poisson brackets on a bihamiltonian manifold. This principle, and its related concepts, are presented in Sec.2. To keep the exposition as simple as possible, in this section we have adopted the language and the techniques of classical Hamiltonian mechanics. That allows us to give the theory a concise and neat form. The applications, on the contrary, may become quite involved. For this reason, in Sec.3, we have restricted ourselves to the simplest example of soliton equation, the well-known Korteweg-de Vries equation,

$$u_t = \tfrac{1}{4}u_{xxx} - \tfrac{3}{2}uu_x. \tag{1.6}$$

It is sufficiently general to display (almost) all the features of the theory, and yet sufficiently simple to be worked out explicitly. The last section, finally, provides the connection between Sato's approach and the Hamiltonian approach.

2. The bihamiltonian approach

This section seeks to introduce the basic ideas of the bihamiltonian approach to the integrable systems. This is a variant of the classical Liouville theory, where a family of first integrals which are in involution is replaced by a second Poisson bracket on the manifold. In our opinion, this variant is particularly useful for the study of soliton equations. The language is purely geometrical and stresses the general nature of the concepts that surpass the limits of soliton theory. The manifolds are supposed to be smooth and finite-dimensional, and the maps are of class C^∞. However, many results can be extended to infinite-dimensional manifolds, as will be shown by an example in the next section.

2.1. Poisson pencil. Let M be a manifold endowed with two Poisson brackets, $\{\cdot,\cdot\}_0$ and $\{\cdot,\cdot\}_1$. We say that M is a *bihamiltonian manifold* if the linear combination,

$$\{f,g\}_\lambda := \{f,g\}_0 - \lambda\{f,g\}_1, \tag{2.1}$$

of these brackets satisfies the Jacobi identity for any value of the real or complex parameter, λ. This means that the cyclic compatibility condition,

$$\begin{aligned} \{f,\{g,h\}_0\}_1+\{h,\{f,g\}_0\}_1+\{g,\{h,f\}_0\}_1+ \\ +\{f,\{g,h\}_1\}_0+\{h,\{f,g\}_1\}_0+\{g,\{h,f\}_1\}_0=0, \end{aligned} \tag{2.2}$$

is valid for any triple of functions, (f,g,h), on M. In this case, the bracket, $\{\cdot,\cdot\}_\lambda$, is called the *Poisson pencil*, defined by $\{\cdot,\cdot\}_0$ and $\{\cdot,\cdot\}_1$ on M.

Let us consider $\{\cdot,\cdot\}_\lambda$ as a deformation of $\{\cdot,\cdot\}_0$, and let us assume that there exist Casimir functions of the Poisson pencil which are deformations of the Casimir functions of $\{\cdot,\cdot\}_0$. This means that we assume the existence of one–parameter families of functions, $h(\lambda)$, such that

$$\{h(\lambda),f\}_\lambda=0 \tag{2.3}$$

for any function, f, and for any value of λ, which, for $\lambda\to\infty$, tend to a Casimir function, h_0, of $\{\cdot,\cdot\}_0$. Accordingly, we assume that $h(\lambda)$ admits a Laurent expansion in λ,

$$h(\lambda)=\sum_{k\geq 0}h_k\lambda^{-k}, \tag{2.4}$$

starting from h_0. Eq.(2.3) immediately implies the recurrence relation,

$$\{h_k,\cdot\}_1=\{h_{k+1},\cdot\}_0, \tag{2.5}$$

on the coefficients, h_k. In this paper we shall avoid discussing the existence of such Casimir functions from a theoretical point of view, being satisfied to demonstrate their existence in applications.

Let $h(\lambda)$ and $l(\lambda)$ be two Casimir functions. Then Eq.(2.5) implies

$$\{h_k,l_j\}_0=\{h_{k-1},l_j\}_1=\{h_{k-1},l_{j+1}\}_0=\ldots=\{h_0,l_{j+k}\}_0=0, \tag{2.6}$$

and, similarly,

$$\{h_k,l_j\}_1=\{h_{k+1},l_j\}_0=0. \tag{2.7}$$

Therefore, we conclude that the coefficients of the Laurent expansion of the Casimir functions of the pencil are in involution with respect to both brackets generating the pencil. This remark is the basis of the bihamiltonian approach to integrable systems. It suggests considering the equations,

$$\dot f=\{h_k,f\}_1=\{h_{k+1},f\}_0, \tag{2.8}$$

as good candidates for descriptions of integrable systems. Their integrability depends on the number and on the independence of the first integrals, h_k, which will not be considered here. We shall conclude these remarks by showing that Eq.(2.8) can also be written in the form,

$$\dot f=\{h^k(\lambda),f\}_\lambda, \tag{2.9}$$

where

$$h^k(\lambda):=\sum_{j=0}^{k}h_j\lambda^{k-j}. \tag{2.10}$$

They are, consequently, simultaneously Hamiltonian with respect to all the Poisson brackets of the pencil.

2.2. Reduction. We now recall an interesting Poisson reduction canonically defined on every bihamiltonian manifold (see [1], Sec.1, for the proof). It follows from the observation that the Casimir functions of $\{\cdot,\cdot\}_0$ generate a Poisson subalgebra with respect to $\{\cdot,\cdot\}_1$, because of compatibility condition (2.2). Thus we can use the functions of the center, K_0, of $\{\cdot,\cdot\}_0$ in two distinct ways:

(1) to define the symplectic leaves of $\{\cdot,\cdot\}_0$ as level sets of these functions;
(2) to define an integrable distribution, D, on M, generated by the vector fields $\{k,\cdot\}_1$ where $k \in K_0$.

Let us choose a specific, symplectic leaf, S, of $\{\cdot,\cdot\}_0$, and let us denote the foliation induced on S by D by E. The leaves of E are the intersections of S with the leaves of D. We shall assume that E is sufficiently regular and that there exists a quotient space, $N = S/E$, and we denote the canonical immersion of S in M and the canonical projection of S onto N by $i : S \to M$ and $\pi : S \to N$ respectively. Then

Proposition 2.1. *The quotient space, $N = S/E$, is a bihamiltonian manifold. On N there exists a unique Poisson pencil, $\{\cdot,\cdot\}_N^\lambda$, such that*

$$\{f,g\}_N^\lambda \circ \pi = \{F,G\}_M^\lambda \circ i \tag{2.11}$$

for any pair of functions, F and G, which extend the functions f and g of N into M, and are constant on D. Technically, this means that F satisfies the conditions,

$$F \circ i = f \circ \pi \tag{2.12}$$

$$\{F,k\}_1 = 0, \tag{2.13}$$

for any function $k \in K_0$.

As a simple corollary of this theorem one easily proves

Proposition 2.2. *Let h_k the coefficients of the series* (2.4). *The bihamiltonian vector fields,*

$$\dot{f} = \{h_k, f\}_1 = \{h_{k+1}, f\}_0, \tag{2.14}$$

are tangent to the symplectic leaf S and are projectable on the quotient space $N = S/E$. The projected vector fields are bihamiltonian with respect to the reduced Poisson pencil on N.

In the next section we shall use this proposition to derive the KdV hierarchy.

2.3. The Lax representation. In soliton theory it is easier to work with forms than with functions. Consequently, it is worthwhile to formulate the equations of motion and commutativity conditions (2.7) using 1–forms instead of functions. This is easily done by introducing the Poisson brackets on forms as in [5]. Let P be the Poisson tensor associated with any given Poisson bracket, $\{\cdot,\cdot\}$, through the relation,

$$\{f,g\} = \langle df, Pdg\rangle . \tag{2.15}$$

A bracket on 1–forms, v_1 and v_2, is then defined by setting

$$\langle\{v_1,v_2\},X\rangle = \frac{\partial}{\partial t_2}\langle v_1,X\rangle - \frac{\partial}{\partial t_1}\langle v_2,X\rangle + \langle v_1, L_X(P)v_2\rangle , \tag{2.16}$$

where X is any vector field, $L_X(P)$ is the Lie derivative of P along X, and $\frac{\partial}{\partial t_i}$ is the derivative along Pv_i. On exact 1–forms,

$$\{df_1, df_2\} = d\{f_1,f_2\}. \tag{2.17}$$

If we denote the Lie derivative of an arbitrary 1–form along a Hamiltonian vector field Pdh by $\dot{v}$, we can write

$$\dot{v} = \{v, dh\}. \tag{2.18}$$

This is the form assumed by the equations of motion when they are formulated on forms rather than on functions.

Let now $v_k = dh_k$ be the differentials of the coefficients, h_k, of a Casimir function of the Poisson pencil. Let

$$v^k(\lambda) = \sum_{j=0}^{k} v_j\lambda^{k-j} = dh^k(\lambda) \tag{2.19}$$

$$v(\lambda) = \sum_{j\geq 0} v_j\lambda^{-j} = dh(\lambda). \tag{2.20}$$

Since the coefficients, h_k, commute in pairs with respect to both Poisson brackets, $\{\cdot,\cdot\}_0$ and $\{\cdot,\cdot\}_1$,

$$\{v_k, v_j\}_0 = 0 \tag{2.21}$$

$$\{v_k, v_j\}_1 = 0 \tag{2.22}$$

$$\{v(\lambda), v^k(\lambda)\}_\lambda = 0. \tag{2.23}$$

These three simple relations play so prominent a role in the theory of soliton equation that they ought to be written explicitly.

Proposition 2.3 (Lax representation). *The exact 1-forms, $v_k = dh_k$, associated with the coefficients, h_k, of a Casimir function of the Poisson pencil, satisfy the relations,*

(2.24)
$$\frac{\partial}{\partial t_k}\langle v_{j+1}, X\rangle - \frac{\partial}{\partial t_j}\langle v_{k+1}, X\rangle + \langle v_{j+1}, L_X(P_0)v_{k+1}\rangle = 0,$$

(2.25)
$$\frac{\partial}{\partial t_k}\langle v_j, X\rangle - \frac{\partial}{\partial t_j}\langle v_k, X\rangle + \langle v_j, L_X(P_1)v_k\rangle = 0,$$

where X is any vector field on M, and the symbol $\frac{\partial}{\partial t_k}$ denotes the Lie derivative along the k–th flow of the hierarchy (2.8). *The associated polynomials, $v^k(\lambda)$, satisfy the conditions*

$$\frac{\partial}{\partial t_k}\langle v^j(\lambda), X\rangle - \frac{\partial}{\partial t_j}\langle v^k(\lambda), X\rangle + \langle v^j(\lambda), L_X(P_\lambda)v^k(\lambda)\rangle = 0, \tag{2.26}$$

called the zero–curvature representation *of the hierarchy* (2.8), *while the sum, $v(\lambda)$, of the series* (2.20) *satisfies the conditions*

$$\frac{\partial}{\partial t_k}\langle v(\lambda), X\rangle + \langle v(\lambda), L_X(P_\lambda)v^k(\lambda)\rangle = 0, \tag{2.27}$$

called the Lax representation *of the hierarchy.*

3. KdV hierarchy

This section is devoted to the study of a special class of bihamiltonian manifolds. The purpose is to derive the well–known KdV hierarchy according to the geometric scheme of the previous section.

3.1. Lie–Poisson manifolds. Let $\mathfrak{g}$ be a simple Lie algebra, and let M be the space of C^∞–maps from S^1 into $\mathfrak{g}$. We denote by x the coordinate on S^1, and either by $S(x)$ or simply by S a map from S^1 into $\mathfrak{g}$. The main example dealt with in this paper is $\mathfrak{g} = \mathfrak{sl}(2, \mathbb{C})$. Accordingly, we write

$$S = \begin{pmatrix} p & r \\ q & -p \end{pmatrix}, \tag{3.1}$$

where p, q, r are three arbitrary periodic functions playing the role of global coordinates on our infinite dimensional manifold. A curve, $S(t)$, in M is a one–parameter family of maps from S^1 into $\mathfrak{g}$. The tangent vector is denoted by

$$\dot S = \begin{pmatrix} \dot p & \dot r \\ \dot q & -\dot p \end{pmatrix}, \tag{3.2}$$

where the dot means differentiation with respect to the parameter t. A covector is a map,

$$V = \begin{pmatrix} v_1 & v_2 \\ v_3 & -v_1 \end{pmatrix}, \tag{3.3}$$

from S^1 into $\mathfrak{g}$, whose value on the tangent vector, $\dot{S}$, is given by

$$\langle V, \dot{S}\rangle = \int_{S^1} (V(x), \dot{S}(x))_{\mathfrak{g}} dx = \int_{S^1} (2\dot{p}v_1 + \dot{q}v_2 + \dot{r}v_3) dx. \tag{3.4}$$

The space M inherits a Lie algebra structure from $\mathfrak{g}$, and a cocycle, ω, from S^1:

$$[S_1, S_2]_M (x) = [S_1(x), S_2(x)]_{\mathfrak{g}} \tag{3.5}$$

and

$$\omega(\dot{S}_1, \dot{S}_2) = \int_{S^1} (\dot{S}_1, \frac{d}{dx}\dot{S}_2) dx. \tag{3.6}$$

They allow us to define two Poisson brackets on M as follows: the first bracket is defined by

$$\{f, g\}_0(S) = -(A, [df(S), dg(S)]) \tag{3.7}$$

where A is a fixed element in $\mathfrak{g}$, a constant loop, to be chosen below, $f : M \to \mathbb{R}$ and $g : M \to \mathbb{R}$ are arbitrary functionals on M, and $df(S)$ and $dg(S)$ are their differentials, regarded as elements of the loop–algebra itself. The second bracket is defined by

$$\{f, g\}_1(S) = \omega(df(S), dg(S)) + (S, [df(S), dg(S)]). \tag{3.8}$$

It is a standard result that these brackets are compatible, and that M is, consequently, a bihamiltonian manifold for any choice of A. The corresponding Poisson tensors are given by

$$(P_0)_S V = [A, V] \tag{3.9}$$

and

$$(P_1)_S V = V_x + [V, S] \tag{3.10}$$

where V_x denotes the partial derivative of the loop V along the circle S^1. Our problem is to find the Casimir functions of the corresponding pencil.

3.2. Casimir functions. We first consider the kernel of the pencil, i.e. the matrices, V, which solve the equation,

$$V_x + [V, S + \lambda A] = 0. \tag{3.11}$$

We observe that the spectrum of V is independent of x since

$$\frac{d}{dx}\mathrm{Tr}V^k = 0. \tag{3.12}$$

Hence, we can characterize the solutions of Eq.(3.11) by fixing their spectrum. We set

$$\mathrm{Tr}\frac{V^2}{2} = \lambda, \tag{3.13}$$

selecting in this way a basic solution of Eq.(3.11). Any other solution can be obtained from it by multiplication by a constant Laurent series, at least for a suitable choice of A. We remark that $V(\lambda)$ has the same spectrum as

$$\Lambda = \begin{pmatrix} 0 & 1 \\ \lambda & 0 \end{pmatrix}. \tag{3.14}$$

Therefore there exists a nonsingular matrix, $K(\lambda, S)$, depending on λ and on the point S of the manifold, such that

$$V(\lambda) = K(\lambda)\Lambda K(\lambda)^{-1}. \tag{3.15}$$

This relation, which may be called a *dressing transformation*, has been studied in [1]. It is the basic element in the demonstration that $V(\lambda)$ is an exact 1–form. Indeed, let us introduce the matrix,

$$M(S,\lambda) := K^{-1}(S + \lambda A)K - K^{-1}K_x, \tag{3.16}$$

and let us define the function

$$H(\lambda) = \langle M, \Lambda\rangle = \int_{S^1} (M, \Lambda)\, dx. \tag{3.17}$$

Proposition 3.1. *For every constant matrix Λ, $V(\lambda)$ is an exact 1–form, with potential $H(\lambda)$,*

$$V(\lambda) = dH(\lambda). \tag{3.18}$$

The function $H(\lambda)$ is therefore a Casimir function of the Poisson pencil (3.7–8).

Proof. Since V and $S + \lambda A$ satisfy Eq.(3.11), matrices Λ and M satisfy the equation, $\Lambda_x + [\Lambda, M] = 0$. Since $\Lambda_x = 0$, then $[\Lambda, M] = 0$. Let $\dot{S}$ be any tangent vector to M,

$$\begin{aligned} \left\langle dH(S), \dot{S}\right\rangle &= \frac{d}{dt}H(S + t\dot{S})|_{t=0} = \left\langle \frac{d}{dt}M(S + t\dot{S})|_{t=0}, \Lambda\right\rangle \\ &= \int_{S^1} \left(K^{-1}\dot{S}K + [M, K^{-1}\dot{K}], \Lambda\right) dx, \end{aligned} \tag{3.19}$$

where we have integrated by parts, and where $\dot{K} = \frac{d}{dt}K(S+t\dot{S})|_{t=0}$. Since $[M,\Lambda]=0$,

$$\begin{aligned} \left\langle dH(S),\dot{S}\right\rangle &= \int_{S^1}\left(K^{-1}\dot{S}K,\Lambda\right)dx = \int_{S^1}\left(\dot{S},K\Lambda K^{-1}\right)dx \\ &= \left\langle V,\dot{S}\right\rangle. \end{aligned} \tag{3.20}$$

□

This result solves the first part of our problem. We still have to characterize the Casimir functions which are deformations of the Casimir functions of $\{\cdot,\cdot\}_0$. The solution of this problem depends on the choice of A. Following a suggestion of Drinfeld and Sokolov ([4], p.2010), we choose

$$A = \begin{pmatrix} 0 & 0 \\ 1 & 0 \end{pmatrix}. \tag{3.21}$$

The meaning of this choice in the framework of the theory of Poisson reduction has been discussed in [7]. We can now explicitly solve matrix equation (3.11): the solution is

$$V(\lambda) = \begin{bmatrix} r^{-1}(-\frac{1}{2}v_x + pv) & v \\ \begin{array}{l} r^{-2}(-\frac{1}{2}v_{xx} + (pv)_x)+ \\ \quad + r^{-1}(q+\lambda)v + r^{-3}r_x(\frac{1}{2}v_x - pv) \end{array} & r^{-1}(\frac{1}{2}v_x - pv) \end{bmatrix} \tag{3.22}$$

if v is a solution of

(3.23)
$$-\tfrac{1}{2}r^{-2}v_{xxx} + \tfrac{3}{2}r^{-3}r_xv_{xx} + \left[2r^{-1}(u+\lambda) - \tfrac{3}{2}r^{-4}{r_x}^2 + \tfrac{1}{2}r^{-3}r_{xx}\right]v_x + \left[r^{-1}u_x - r^{-2}r_x(u+\lambda)\right]v = 0,$$

where

$$u = q + p^2r^{-1} + p_xr^{-1} - pr^{-2}r_x. \tag{3.24}$$

The meaning of Eq.(3.23) is easily discovered. Since it can also be written as

(3.25)
$$\frac{d}{dx}\left[\tfrac{1}{4}r^{-2}{v_x}^2 - \tfrac{1}{2}r^{-2}vv_{xx} + \tfrac{1}{2}r^{-3}r_xvv_x + r^{-1}(u+\lambda)v^2\right] = \frac{d}{dx}\mathrm{Tr}\frac{V(\lambda)^2}{2} = 0,$$

we see that it states the isospectrality of $V(\lambda)$. To satisfy condition (3.13), we have to choose $v(\lambda)$ in such a way that

$$\tfrac{1}{4}r^{-2}{v_x}^2 - \tfrac{1}{2}r^{-2}vv_{xx} + \tfrac{1}{2}r^{-3}r_xvv_x + r^{-1}(u+\lambda)v^2 = \lambda. \tag{3.26}$$

Let us try to solve this equation by a series expansion,

$$v(\lambda) = \sum_{k\geq 0} v_k \lambda^{-k}. \tag{3.27}$$

By equating the coefficients of the powers of λ we get the system,

$$\begin{cases} r^{-1}{v_0}^2 = 1 \\ \frac{1}{4}r^{-2}{v_{0x}}^2 - \frac{1}{2}r^{-2}v_0v_{0xx} + \frac{1}{2}r^{-3}r_xv_0v_{0x} + r^{-1}u{v_0}^2 + 2r^{-1}v_0v_1 = 0 \\ \frac{1}{2}r^{-2}v_{0x}v_{1x} - \frac{1}{2}r^{-2}(v_0v_{1xx} + v_1v_{0xx}) + \frac{1}{2}r^{-3}r_x(v_0v_{1x} + v_1v_{0x}) + \\ +r^{-1}uv_0v_1 + r^{-1}({v_1}^2 + 2v_0v_2) = 0 \\ \cdots\cdots \end{cases} \tag{3.28}$$

which recursively yields v_0, v_1, ... and so on. This proves that function (3.17) is a deformation of a Casimir function of $\{\cdot,\cdot\}_0$.

Finally, to explicitly compute this Casimir function, we observe that a possible choice for K is

$$K = \begin{bmatrix} v^{\frac{1}{2}} & 0 \\ r^{-1}v^{-\frac{1}{2}}\left(\frac{1}{2}v_x - pv\right) & v^{-\frac{1}{2}} \end{bmatrix}. \tag{3.29}$$

Then

$$M = K^{-1}(S + \lambda A)K - K^{-1}K_x = \frac{r}{v}\Lambda, \tag{3.30}$$

and

$$H(\lambda) = \langle M, \Lambda\rangle = \int_{S^1} 2\lambda\frac{r}{v}dx. \tag{3.31}$$

This solves our problem. We now introduce the bihamiltonian hierarchy associated with $H(\lambda)$.

Definition 3.2 (matrix KdV hierarchy). *The* matrix KdV hierarchy *is the bihamiltonian hierarchy defined by the Casimir function* (3.31) *on the loop–algebra M endowed with the Poisson pencil* (3.7–8). *The equations of the hierarchy are explicitly given by*

$$\dot{S}_k = V_{kx} + [V_k, S] = [A, V_{k+1}], \tag{3.32}$$

where the matrices V_k are the coefficients of the Laurent expansion of solution (3.22) *of Eq.*(3.11), *which satisfies the isospectrality condition* (3.13). *They can also be written in the form,*

$$\dot{S}_k = \left(V^{2k+1}\right)_{+x} + [\left(V^{2k+1}\right)_+, S + \lambda A], \tag{3.33}$$

where

$$(V^{2k+1})_+ = \sum_{j=0}^{k} V_j \lambda^{k-j} \tag{3.34}$$

is the positive part of the Laurent expansion of the power V^{2k+1} of V. The last representation may be called the rule of fractional powers.

We need only justify the last representation. We observe that, by Eq.(3.13) or Eq.(3.14), V verifies the condition

$$V^2 = \lambda I, \tag{3.35}$$

and, thus, it is a square root of matrix, λI. Consequently,

$$V^k(\lambda) := \sum_{j=-1}^{k} V_j \lambda^{k-j} = (\lambda^k V(\lambda))_+ = (V^{2k+1})_+, \tag{3.36}$$

where V^{2k+1} is the $(2k+1)$–th power of V. Then Eq.(3.33) coincides with Eq.(2.5) of the previous section.

3.3. The Poisson reduction. We now perform the Poisson reduction described in subsection 2.2, in order to obtain the usual KdV hierarchy. This reduction has already been studied in [1] so it will suffice to give the results without proofs.

The first step is the choice of a symplectic leaf of P_0. Let us choose

$$S = \begin{pmatrix} p & 1 \\ q & -p \end{pmatrix} \tag{3.37}$$

corresponding to the constraint, $r = 1$. This is strongly suggested by the form of Eq.(3.28), where r appears as an annoying coefficient. The second step is to find the distribution D associated with the Casimir functions of $\{\cdot,\cdot\}_0$. It is given by

$$\dot{S} = \begin{pmatrix} -w & 0 \\ w_x + 2pw & w \end{pmatrix}. \tag{3.38}$$

By eliminating the arbitrary function, w, between the relations,

$$\dot{p} = -w \qquad \text{and} \qquad \dot{q} = w_x + 2pw, \tag{3.39}$$

and by integrating the resulting constraint, we obtain the projection, $\pi : S \to N$,

$$u = p_x + p^2 + q. \tag{3.40}$$

The function, u, plays the role of a global coordinate on the quotient space, N. The Poisson pencil on N is obtained by standard computations. The result is

$$\{f_1, f_2\}_\lambda = \omega(v_1, v_2) + (u, [v_1, v_2]_N) + \lambda(1, [v_1, v_2]_N), \tag{3.41}$$

where $v_j = df_j$ and

$$[v_1, v_2]_N = v_1 v_{2x} - v_2 v_{1x} \tag{3.42}$$

$$\omega(v_1, v_2) = \tfrac{1}{2} \int_{S^1} v_{1x} v_{2xx} dx. \tag{3.43}$$

The whole result of the process of reduction is summarized in the following

Proposition and Definition 3.3 (scalar KdV hierarchy). *The quotient space, N, canonically associated with the symplectic leaf* (3.37), *is the Virasoro algebra of vector fields on the circle, endowed with the Poisson pencil* (3.41). *The Casimir function on N is the projection of the Casimir function* (3.31) *on M, along the canonical projection* (3.40). *Its differential, $v(\lambda) = dh(u, \lambda)$, is the unique solution of the equation,*

$$\tfrac{1}{4}{v_x}^2 - \tfrac{1}{2} v v_{xx} + (u + \lambda) v^2 = \lambda. \tag{3.44}$$

The coefficients v_k of its Laurent expansion are given by the recursive system (3.28), *with $r = 1$, and are called the* Gelfand–Dickey polynomials. *The equations of the hierarchy associated with this Casimir function have the explicit form,*

$$\dot{u}_k = -\tfrac{1}{2} v_{kxxx} + 2u v_{kx} + u_x v_k = -2 {v_{k+1}}_x. \tag{3.45}$$

They can also be written in the form

$$\dot{u}_k = -\tfrac{1}{2} v^k(\lambda)_{xxx} + 2u v^k(\lambda)_x + u_x v^k(\lambda), \tag{3.46}$$

where $v^k(\lambda) = \sum_{j=0}^{k} v_k \lambda^{k-j}$. They are the equations of the usual KdV hierarchy.

3.4. The Lax representation. To obtain the 0–curvature representation and the Lax representation of the KdV hierarchy, we have to compute the Poisson brackets on 1–forms. We begin with the matrix case. By using the definition given in subsection 2.3, we find that

$$\{V_1, V_2\}_0 = \frac{\partial V_1}{\partial t_2} - \frac{\partial V_2}{\partial t_1} \tag{3.47}$$

$$\{V_1, V_2\}_1 = \frac{\partial V_1}{\partial t_2} - \frac{\partial V_2}{\partial t_1} + [V_1, V_2], \tag{3.48}$$

for any pair of matrices, V_1, V_2, where $\frac{\partial}{\partial t_1}$ and $\frac{\partial}{\partial t_2}$ are the time derivatives along the vector fields associated with V_1 and V_2 respectively. If we denote the time derivative along the k–th flow of the hierarchy by $\frac{\partial}{\partial t_k}$, and if we recall that $P_1 V_k = P_0 V_{k+1} = P_\lambda V^k(\lambda)$, from Prop.2.3, we immediately get

Proposition 3.4 (matrix Lax representations). *The 1–forms, V_k, defining the matrix KdV hierarchy, satisfy the conditions*

$$\frac{\partial V_{j+1}}{\partial t_k} - \frac{\partial V_{k+1}}{\partial t_j} = 0 \tag{3.49}$$

$$\frac{\partial V_j}{\partial t_k} - \frac{\partial V_k}{\partial t_j} + [V_j, V_k] = 0, \tag{3.50}$$

and the polynomials $V^k(\lambda)$ verify the 0–curvature representations,

$$\frac{\partial V^j(\lambda)}{\partial t_k} - \frac{\partial V^k(\lambda)}{\partial t_j} + [V^j(\lambda), V^k(\lambda)] = 0, \tag{3.51}$$

and the series, $V(\lambda)$, verifies the Lax representation,

$$\frac{\partial V(\lambda)}{\partial t_k} + [V(\lambda), V^k(\lambda)] = 0. \tag{3.52}$$

Eq.(3.49) and Eq.(3.50) express the fact that the 1–forms, V_k, commute with respect to the pair of Poisson brackets defined on M.

The same Proposition is valid in the scalar case, if we replace the commutator of matrices by the commutator in the Virasoro algebra (3.42). The reason is that the Poisson brackets on N have the form (3.9–10), up to replacement of the commutators.

Proposition 3.5 (scalar Lax representation) The 1–forms, v_k, $v^k(\lambda)$ and $v(\lambda)$, satisfy the equations on N,

(3.53)
$$\frac{\partial v_{j+1}}{\partial t_k} - \frac{\partial v_{k+1}}{\partial t_j} = 0$$

(3.54)
$$\frac{\partial v_j}{\partial t_k} - \frac{\partial v_k}{\partial t_j} + [v_j, v_k]_N = 0$$

(3.55)
$$\frac{\partial v^j(\lambda)}{\partial t_k} - \frac{\partial v^k(\lambda)}{\partial t_j} + [v^j(\lambda), v^k(\lambda)]_N = 0 \quad \text{(0–curvature)}$$

(3.56)
$$\frac{\partial v(\lambda)}{\partial t_k} + [v(\lambda), v^k(\lambda)]_N = 0 \qquad \text{(Lax representation)},$$

where $\frac{\partial}{\partial t_k}$ is the time derivative along the k–th flow of the hierarchy (3.45).

In the next section, we shall show that Eqs. (3.52) and Eq.(3.56) coincide, up to an isomorphism, with the famous Sato equations (1.4). As far as Eq.(3.53) is concerned, we observe that it implies the existence of an exact 1–form

$$d\theta = \sum_{j\geq 0} v_{j+1} dt_j. \tag{3.57}$$

The potential θ is related to the well–known τ–function [3] of the KdV hierarchy by

$$\theta = \frac{d}{dx} \log \tau, \tag{3.58}$$

as shown in [1].

4. The spectral problem and Sato's approach

In this section we construct the link between the geometric approach and Sato's approach to the soliton equations. The starting point is the study of the spectral problem for the matrix, $V(\lambda)$, defining the matrix KdV hierarchy (evaluated for $r = 1$).

4.1. The spectral problem. We have already pointed out that $V(\lambda)$ is an isospectral matrix whose characteristic polynomial is

$$z^2 = \lambda, \tag{4.1}$$

by Eq.(3.35). Therefore, all the relevant information on this matrix is encoded in its eigenvectors which are defined by the linear system,

$$\begin{cases} -\frac{1}{2}v_x\psi_1 + v(\psi_2 + p\psi_1) = z\psi_1 \\ [-\frac{1}{2}v_{xx} + (u+\lambda)v]\psi_1 + \frac{1}{2}v_x(\psi_2 + p\psi_1) = z(\psi_2 + p\psi_1), \end{cases} \tag{4.2}$$

up to a multiplicative factor. The crucial point is the choice of a suitable normalization condition. We choose

$$\begin{cases} \psi_1 = \psi \\ \psi_2 + p\psi_1 = \psi_x \end{cases} \tag{4.3}$$

on the experimental evidence that it provides the quickest way of obtaining Sato's formalism. A deeper understanding of this condition is still lacking. With this choice, the eigenvalue problem (4.2) reduces to the single equation,

$$-\tfrac{1}{2}v(\lambda)_x\psi + v(\lambda)\psi_x = z\psi, \tag{4.4}$$

provided that z obeys Eq.(4.1). The following observation is important.

Lemma 4.1 (the Schrödinger equation). *Any solution of Eq.(4.4) is a solution of the Schrödinger equation,*

$$\psi_{xx} - u\psi = \lambda\psi. \tag{4.5}$$

Proof. Eq.(4.4) yields:

$$\begin{aligned} z^2\psi &= -\tfrac{1}{2}v(\lambda)_x(z\psi) + v(\lambda)(z\psi)_x \\ &= -\tfrac{1}{2}v(\lambda)_x(-\tfrac{1}{2}v_x\psi + v\psi_x) + v(\lambda)(-\tfrac{1}{2}v_{xx}\psi + \tfrac{1}{2}v_x\psi_x + v\psi_{xx}) \\ &= (\tfrac{1}{4}v_x^2 - \tfrac{1}{2}vv_{xx})\psi + v^2\psi_{xx} \\ &= \lambda\psi + v^2[\psi_{xx} - (u+\lambda)\psi], \end{aligned} \tag{4.6}$$

since, by Eq.(3.44),

$$\tfrac{1}{4}v_x^2 - \tfrac{1}{2}vv_{xx} + (u+\lambda)v^2 = \lambda. \tag{4.7}$$

Hence, Eq.(4.5) follows from Eq.(4.1). □

This result can be employed in the study of the eigenvalue problem (4.4), in order to eliminate the spectral parameter, z, on the left–hand side of the equation, and to obtain a true eigenvalue problem.

Indeed we can write,

$$z\psi = \sum_{k\geq 0}(-\tfrac{1}{2}v_{kx}\lambda^{-k})\psi + (v_k\lambda^{-k})\psi_x = \sum_{k\geq 0}(-\tfrac{1}{2}v_{kx}I + v_k\partial)(\lambda^{-k}\psi). \tag{4.8}$$

This suggests that we may introduce the operators

$$P_k := -\tfrac{1}{2}v_{kx}I + v_k\partial, \tag{4.9}$$

$$L := \partial^2 - u, \tag{4.10}$$

and

$$Q := \sum_{k\geq 0} P_k L^{-k}, \tag{4.11}$$

and that we may state the following result,

Lemma 4.2 (Sato's eigenvalue problem). *The eigenvectors of matrix, $V(\lambda)$, which satisfy the normalization condition (4.3) are in one-to-one correspondence with the eigenfunctions ψ of the Sato operator, Q,*

$$Q\psi = z\psi. \tag{4.12}$$

4.2. The Poisson pencil. The previous discussion strongly suggests the existence of a morphism allowing us to represent the geometrical objects of the theory as pseudodifferential operators. The operator Q, for instance, is the natural candidate to represent the matrix, $V(\lambda)$, that is the Casimir function of the Poisson pencil. The basic object of this construction is to understand how to represent the Poisson pencil on M,

$$\dot{S} = V_x + [V, S + \lambda A], \tag{4.13}$$

in terms of pseudodifferential operators. We shall now offer some arguments to show that this pencil is described by the commutator $[\cdot, L]$. Preliminarily, let us recall the three different types of representation of the matrix KdV equations:

$$\dot{S}_k = V_{kx} + [V_k, S] \tag{4.14}$$

$$\dot{S}_k = [A, V_{k+1}] \tag{4.15}$$

$$\dot{S}_k = V^k(\lambda)_x + [V^k(\lambda), S + \lambda A]. \tag{4.16}$$

Moreover, we observe that we can obtain the fourth representation,

$$\dot{S}_{k+1} - \dot{S}_k\lambda = V_{k+1\ x} + [V_{k+1}, S + \lambda A], \tag{4.17}$$

by comparing two successive equations of the hierarchy. All these representations are equivalent, as it is easily shown. We now obtain the corresponding operator–representations.

Lemma 4.3. *The operators, P_k and L, defined by Eq.(4.9–10), satisfy the equations,*

$$\frac{\partial L}{\partial t_{k+1}} - \frac{\partial L}{\partial t_k} \cdot L = [P_{k+1}, L] \tag{4.18}$$

$$\frac{\partial L}{\partial t_k} = [P^k, L] \tag{4.19}$$

$$\frac{\partial L}{\partial t_k} = [P_{k+1}L^{-1}, L]_R \tag{4.20}$$

$$\frac{\partial L}{\partial t_{k+1}} = [P_{k+1}L^{-1}, L]_S, \tag{4.21}$$

where $[\ ,\ ]_R$ and $[\ ,\ ]_S$ are the Poisson tensors associated with the so-called R–bracket and Sklyanin bracket, respectively [8]. *They are defined by*

$$[P, L]_R = [R(P), L] - R([P, L]) \tag{4.22}$$

$$[P, L]_S = R(L \cdot P)L - LR(P \cdot L), \tag{4.23}$$

where $R(P) = \frac{1}{2}(P_+ - P_-)$ *is the R–matrix. Moreover*

$$P^k := \sum_{j=0}^{k} P_j L^{k-j}. \tag{4.24}$$

Proof. First, we prove the basic Eq.(4.18):

(4.25)
$$\begin{aligned}[L, P_k] =& L \cdot P_k - P_k \cdot L = \\ =& -\tfrac{1}{2}v_{kxxx} + u_x v_k + 2v_{kx}\partial^2 = \\ =& -\tfrac{1}{2}v_{kxxx} + 2uv_{kx} + u_x v_k + 2v_{kx}(\partial^2 - u) = \\ =& \quad \dot{u}_k - \dot{u}_{k-1}L = \\ =& -\frac{\partial L}{\partial t_k} + \frac{\partial L}{\partial t_{k-1}}L.\end{aligned}$$

By setting

$$P_{k+1} = P^{k+1} - P^k L \tag{4.26}$$

from Eq.(4.18) we get

$$\frac{\partial L}{\partial t_{k+1}} - [P^{k+1}, L] = (\frac{\partial L}{\partial t_k} - [P^k, L])L = (\frac{\partial L}{\partial t_0} - [P^0, L])L^{k+1} = 0, \tag{4.27}$$

since

$$\frac{\partial L}{\partial t_0} = [P^0, L] \tag{4.28}$$

as it is easily shown. Finally, Eq.(4.20) and Eq.(4.21) are easily obtained by solving Eq.(4.18) with respect to $\frac{\partial L}{\partial t_{k+1}}$ and $\frac{\partial L}{\partial t_k}$. Indeed it implies

$$\frac{\partial L}{\partial t_{k+1}}L^{-1} - \frac{\partial L}{\partial t_k} = [P_{k+1}, L]L^{-1} \tag{4.29}$$

and therefore

$$\begin{aligned}([P_{k+1}, L]L^{-1})_+ &= -\frac{\partial L}{\partial t_k} \\ ([P_{k+1}, L]L^{-1})_- &= \frac{\partial L}{\partial t_{k+1}}L^{-1}.\end{aligned} \tag{4.30}$$

Finally

$$\begin{aligned}\frac{\partial L}{\partial t_k} &= ([L, P_{k+1}L^{-1}])_+ \\ \frac{\partial L}{\partial t_{k+1}} &= -([L, P_{k+1}L^{-1}])_- L.\end{aligned} \tag{4.31}$$

As $P_+ = R + \frac{1}{2}I$ and $P_- = -R + \frac{1}{2}I$, we can write Eq.(4.31) in terms of R:

(4.32)
$$\begin{aligned}\frac{\partial L}{\partial t_k} &= P_+([L, P_{k+1}L^{-1}]) = R([L, P_{k+1}L^{-1}]) + \tfrac{1}{2}[L, P_{k+1}L^{-1}] = \\ &= R([L, P_{k+1}L^{-1}]) - [L, R(P_{k+1}L^{-1})],\end{aligned}$$

since $(P_{k+1}L^{-1})_+ = 0$ and $R(P_{k+1}L^{-1}) = -\frac{1}{2}(P_{k+1}L^{-1})$. Hence Eq.(4.20) is proved. On the other hand

(4.33)
$$\begin{aligned}\frac{\partial L}{\partial t_{k+1}} &= -P_-([L, P_{k+1}L^{-1}])L = R([L, P_{k+1}L^{-1}])L - \tfrac{1}{2}[L, P_{k+1}L^{-1}]L = \\ &= R(L \cdot P_{k+1}L^{-1})L - R(P_{k+1})L - \tfrac{1}{2}LP_{k+1} + \tfrac{1}{2}P_{k+1}L = \\ &= R(L \cdot P_{k+1}L^{-1})L - \tfrac{1}{2}LP_{k+1} = R(L \cdot P_{k+1}L^{-1})L - LR(P_{k+1}) = \\ &= R(L \cdot P_{k+1}L^{-1})L - LR(P_{k+1}L^{-1} \cdot L) = [P_{k+1}L^{-1}, L]_S,\end{aligned}$$

and also Eq.(4.21) is proved. □

The comparison of Eq.(4.14–17) with Eq.(4.18–21) clearly points out two simple rules of the morphism we are looking for:

(1) the Poisson pencil, $V_{kx} + [V_k, S + \lambda A]$, is systematically replaced by the commutator, $[P_k, L]$,
(2) the expansions in powers of λ are replaced by expansions in powers of L.

Moreover, the simple splitting of the Poisson pencil on M,

$$V_x + [V, S + \lambda A] = V_x + [V, S] - [A, V]\lambda, \tag{4.34}$$

characteristic of the matrix formulation, is replaced by the more intriguing splitting,

$$[P, L] = [P \cdot L^{-1}, L]_S - [P \cdot L^{-1}, L]_R \cdot L \tag{4.35}$$

of the operator approach. This provides a hint for the geometrical interpretation of the classical R–matrix.

4.3. Sato's approach. We can finally come back to Sato's approach considered in Sec. 1. Our aim is to show that the operator, Q, associated with the Casimir function of the Poisson pencil is indeed Sato's operator.

Proposition 4.4. *The operator, Q, defined by Eq.(4.11), satisfies the equations,*

$$[Q, L] = 0 \tag{4.36}$$

and

$$Q^2 = L. \tag{4.37}$$

They correspond exactly to the equations

$$V_x + [V, S + \lambda A] = 0 \tag{4.38}$$

and

$$V^2 = \lambda I \tag{4.39}$$

characterizing the Casimir functions of the Poisson pencil. Thus, Sato's operator is the operator form of the differential of the Casimir function of the Poisson pencil.

Proof Eq.(4.36) follows from

$$[Q, L] = \sum_{j\geq 0}[P_j, L]L^{-j} = \sum_{j\geq 0}(\frac{\partial L}{\partial t_j} - \frac{\partial L}{\partial t_{j-1}}L)L^{-j} = -\frac{\partial L}{\partial t_{-1}}L = 0, \tag{4.40}$$

since $\frac{\partial L}{\partial t_{-1}} = 0$. Eq.(4.37) is proved by observing that Q is a monic pseudodifferential operator of the form,

$$Q = \partial + \sum_{j\geq 1} q_j \partial^{-j}, \tag{4.41}$$

commuting with L. It coincides with the square root of L by Schur's lemma (see [9], p. 3.86). □

It is now really easy to obtain Sato's equations (1.4). As an intermediate step, we remark that the operators, P_k, satisfy the equations,

$$\frac{\partial P_{k+1}}{\partial t_j} - \frac{\partial P_{j+1}}{\partial t_k} = 0 \tag{4.42}$$

$$\frac{\partial P_k}{\partial t_j} - \frac{\partial P_j}{\partial t_k} + [P_k, P_j] = 0, \tag{4.43}$$

as a consequence of the equations,

$$\frac{\partial V_{k+1}}{\partial t_j} - \frac{\partial V_{j+1}}{\partial t_k} = 0 \tag{4.44}$$

$$\frac{\partial V_k}{\partial t_j} - \frac{\partial V_j}{\partial t_k} + [V_k, V_j] = 0, \tag{4.45}$$

of the matrix formulation. This can be easily seen by a direct inspection. This observation is another important piece of information concerning the Hamiltonian interpretation of Sato's approach. It gives a further support to the idea that the operators, P_k, are the operator representation of the differentials of the Hamiltonians of the KdV hierarchy. Then, we prove the following:

Proposition 4.5. *Sato's operator, Q, satisfies the equations,*

$$\frac{\partial Q}{\partial t_{k+1}} - \frac{\partial Q}{\partial t_k} L = [P_{k+1}, Q] \tag{4.46}$$

$$\frac{\partial Q}{\partial t_k} = [P^k, Q]. \tag{4.47}$$

Proof. We first prove Eq.(4.46):

(4.48)

$$\begin{aligned}
&\frac{\partial Q}{\partial t_{k+1}} - \frac{\partial Q}{\partial t_k} L - [P_{k+1}, Q] = \\
&= \sum_{j\geq 0} \left(\frac{\partial P_j}{\partial t_{k+1}} - \frac{\partial P_j}{\partial t_k} L - [P_{k+1}, P_j] \right) L^{-j} + \\
&+ \sum_{j\geq 0} P_j \left(\frac{\partial L^{-j}}{\partial t_{k+1}} - \frac{\partial L^{-j}}{\partial t_k} L - [P_{k+1}, L^{-j}] \right) = \\
&= -\frac{\partial P_0}{\partial t_k} + \sum_{j\geq 0} \left(\frac{\partial P_j}{\partial t_{k+1}} - \frac{\partial P_{j+1}}{\partial t_k} - [P_{k+1}, P_j] \right) L^{-j} = \\
&= \sum_{j\geq 0} \left(\frac{\partial P_j}{\partial t_{k+1}} - \frac{\partial P_{k+1}}{\partial t_j} + [P_j, P_{k+1}] \right) L^{-j} + \sum_{j\geq 0} \left(\frac{\partial P_{k+1}}{\partial t_j} - \frac{\partial P_{j+1}}{\partial t_k} \right) L^{-j} = \\
&= 0.
\end{aligned}$$

Then, Eq.(4.47) is proved exactly as in Lemma 4.3. □

We have thus obtained the result we were looking for: starting from the Hamiltonian approach to the KdV hierarchy we have recovered Sato's operator, Q, and we have shown that it obeys Sato's equations (1.4) as a consequence of the equations,

$$\frac{\partial V}{\partial t_k} = [V^k, V], \tag{4.49}$$

of the matrix formulation. Sato's equations are thus a way of writing the vanishing of certain Poisson brackets on forms by using pseudodifferential operators. This result justifies the name of "Lax representation" given to Eq.(2.27) of Sec. 2.

5. Final comments

This paper is centered around one simple idea: the deformation of a Poisson bracket and of its Casimir functions provides a way of constructing an algebra of functions which are in involution. This idea has been proved to be effective in the paradigmatic example of the KdV hierarchy. To be worked out, this technique has required a completely new approach to the theory of soliton equations. In particular, we have been obliged to regard the KdV hierarchy as a special reduction of a simpler theory in a bigger space, which we called the matrix–KdV hierarchy. At the matrix level, the theory is elementary, and it perfectly reproduces the geometrical scheme of Hamiltonian mechanics. After the reduction, the theory becomes more and more algebraic, up to Sato's approach, where every trace of the geometrical origin seems to be lost. This paper is an attempt to display as clearly as possible the hidden connections between Sato's theory and the Hamiltonian properties of soliton equations. It seems to us interesting that it has been possible to deduce Sato's operator from the simple concept of the Casimir functions of a Poisson pencil. In the last section we initiated the construction of a code of transcription, where every formula of Sato's approach may be compared with a corresponding formula of the geometrical approach. Admittedly, a lot of work is still necessary to make this code complete and convincing.

REFERENCES

1. P. Casati, F. Magri, M. Pedroni, *Bihamiltonian Manifolds and τ-function*, Mathematical Aspects of Classical Field Theory 1991 (M.J. Gotay, J.E. Mardsen, V.E. Moncrief, eds.), Contemporary Mathematics **132**, American Mathematical Society, Providence.

2. A. C. Newell, *Solitons in Mathematics and Physics*, S.I.A.M., Philadelphia, 1985.

3. E. Date, M. Jimbo, M. Kashiwara, T. Miwa, *Transformation Groups for Soliton Equations*, Proceedings of R.I.M.S. Symposium on "Nonlinear Integrable Systems–Classical Theory and Quantum Theory" (1983), World Scientific, Singapore.

4. H. Flaschka, A. C. Newell, T. Ratiu, *Kac–Moody Lie Algebras and Soliton Equations*, Physica 9D (1983), 300–323.

5. Y. Kosmann–Schwarzbach, F. Magri, *Poisson–Nijenhuis Structures*, Ann. Inst. Poincaré **53** (1990), 35–81.

6. V. G. Drinfeld, V. V. Sokolov, *Lie Algebras and Equations of Korteweg-de Vries Type*, J. Sov. Math. **30** (1985), 1975–2036.

7. P. Casati, M. Pedroni, *Drinfeld Sokolov Reduction on a Simple Lie Algebra from the Bihamiltonian Point of View*, Lett. Math. Phys. **25** (1992), 89–101.

8. M. A. Semenov–Tian–Shansky, *What is the classical r–matrix?*, Funct. Anal. and Appl. **17** (1983), 259–272.

9. D. Mumford, *Tata Lectures on Theta II*, Birkhäuser, Boston, 1984.

Paolo Casati
Dottorato in Matematica
Università di Torino
Via Principe Amedeo 8
I-10123 Torino, Italy

Franco Magri
Dipartimento di Matematica dell'Università di Milano
Via C. Saldini 50
I-20133 Milano, Italy

Marco Pedroni
Dottorato in Matematica
Università di Milano,
Via C. Saldini 50
I-20133 Milano, Italy

PART III
Solvable Lattice Models

Generalized chiral Potts models and minimal cyclic representations of $U_q(\hat{\mathfrak{gl}}(n, C))$.

E. Date

Finite-dimensional cyclic representations of the quantized universal enveloping algebra $U_q(\mathfrak{g})$ exist only when the deformation parameter q is a root of unity, $q^N = 1$. In such representations, the N-th power of each Chevalley generator acts as a non-zero scalar. These values of e_i^N, f_i^N and t_i^N are part of a continuous family of parameters upon which the cyclic representations depend.

The chiral Potts model, which is the first example of a 2-dimensional solvable lattice model whose Boltzmann weights are parametrized by points on algebraic curves of genus greater than 1, is now treated in the framework of the cyclic representations of $U_q(\hat{\mathfrak{gl}}(2, C))$. The Boltzmann weights constitute an R-matrix for a cyclic representation. In order for R-matrices of cyclic representations to exist, the continuous parameters of the representations have to satisfy some algebraic relations, and this fact explains the origin of the non-trivial algebraic curves in this theory.

In the talk at the symposium in memory of J.-L. Verdier, a generalization of the chiral Potts model based on minimal cyclic representations of $U_q(\hat{\mathfrak{gl}}(n, C))$ was presented. The content was based on our joint paper of the same title with M. Jimbo, K. Miki and T. Miwa which appeared in *Comm. Math. Phys* **137**, 133–147 (1991). For earlier works, see the references cited in that paper.

In relation with this work, we presented a kind of branching rule for tensor products of cyclic and (finite-dimensional) highest-weight representations of $U_q(\hat{\mathfrak{gl}}(n, C))$, which will appear in the proceedings of the NATO ARW on "Quantum Field Theory, Statistical Mechanics, Quantum Groups and Topology", Miami 1991.

Solvable models based on more general cyclic representations are presented in the recent preprint of Kashaev, Mangazeev and Stroganov "N^3-state R-matrix related with the $U_q(\hat{\mathfrak{sl}}(3))$ algebra at $q^{2N} = 1$".

E. Date
Osaka University
Faculty of Engineering Science
Toyonaka, Osaka 560
Japan

Infinite Discrete Symmetry Group for the Yang–Baxter Equations and their Higher Dimensional Generalizations [1]

M. Bellon, J.-M. Maillard, and C. Viallet

Abstract. We show that the Yang-Baxter equations for two dimensional vertex models admit as a group of symmetry the infinite discrete group $A_2^{(1)}$. The existence of this symmetry explains the presence of a spectral parameter in the solutions of the equations. We show that similarly, for three-dimensional vertex models and the associated tetrahedron equations, there also exists an infinite discrete group of symmetry. Although generalizing naturally the previous one, it is a much bigger hyperbolic Coxeter group. We indicate how this symmetry can help to resolve the Yang-Baxter equations and their higher-dimensional generalizations and initiate the study of three-dimensional vertex models. These symmetries are naturally represented as birational projective transformations. They may preserve non trivial algebraic varieties.

Key-words: Yang-Baxter equations, Star-triangle relations, Tetrahedron equations, Inversion relations, Integrable models, Coxeter groups, Weyl group, Automorphisms of algebraic varieties, Birational transformations, Cremona transformations, Iteration of mappings.

1. Introduction

The Yang-Baxter equations, which appeared twenty years ago[2], have acquired a predominant role in the theory of integrable two-dimensional models in statistical mechanics [6, 7] and field theory (quantum or classical). They have actually outpassed the borders of physics and have become fashionable in some parts of the mathematics literature. They in particular support the construction of quantum groups [8, 9].

The Yang-Baxter equations [7] and their higher dimensional generalizations are now considered as the defining relations of integrability. They are the "Deus ex machina" in a number of domains of Mathematics and Physics

---

[1] work supported by CNRS

[2] In fact, fifty years ago, Lars Onsager was totally aware of the key role played by the star-triangle relation in solving the two-dimensional Ising model, but he preferred to give an algebraic solution emphasizing Clifford algebras [1, 2, 3, 4, 5].

(Knot Theory [10], Quantum Inverse Scattering [11], S-Matrix Factorization, Exactly Solvable Models in Statistical Mechanics, Bethe Ansatz [12], Quantum Groups [13, 9], Chromatic Polynomials [14] and more awaited deformation theories). The appeal of these equations comes from their ability to *give global results from local ones.* For instance, they are a sufficient and, to some extent, necessary [15] condition for the commutation of families of transfer matrices of arbitrary size and even of corner transfer matrices. From the point of view of topology, one may understand these relations by considering them as the generators of a large set of *discrete deformations of the lattice.* This point of view underlies most studies in knot theory [10] and statistical mechanics ($\mathbb{Z}$-invariance [16, 17]).

We want to analyze the Yang-Baxter equations and their higher dimensional generalizations [18, 19, 20, 21] without prejudice about what should be a solution, that is to say proceed by *necessary* conditions.

We will exhibit an infinite discrete group of transformations acting on the Yang-Baxter equations or their higher dimensional generalizations (tetrahedron, hyper-simplicial equations).

These transformations act as an automorphy group of various quantities of interest in Statistical Mechanics (partition function,...), and are of great help for calculations, even outside the domain of integrability (critical manifolds, phase diagram,...) [22].

We show here is that *they form a group of symmetries of the equations defining integrability.* They consequently appear as a group of automorphisms of the algebraic varieties parametrizing the solutions of the Yang-Baxter or tetrahedron equations. We will denote this group *Aut*.

The existence of *Aut* drastically constrains the varieties where solutions may be found. In the general case, it has *infinite* orbits and gives *severe constraints* on the algebraic varieties which parametrize the possible solutions (genus zero or one curves, algebraic varieties which are not of the general type [23]). In the non-generic case, when *Aut* has finite order orbits, the algebraic varieties can be of general type, but *the very finiteness condition allows for their determination* [24].

In the framework of infinite group representations, it is crucial to recognize the *essential difference* between what these symmetry groups are for the Yang-Baxter equations and what they are for the higher dimensional tetrahedron and hyper-simplicial relations: the number of involutions generating our groups increases from 2 to 2^{d-1} when passing from two-dimensional to d-dimensional models and the group jumps from the semi-direct product $\mathbb{Z} \ltimes \mathbb{Z}_2$ to a much larger group, i.e., a *group with an exponential growth* with the length of the word[3].

[3]It is worth recalling that for the Zamolodchikov solution [19, 21] of the tetrahedron relation, the partition function is similar to the one of the two-dimensional checkerboard Ising model. This example seems to indicate that three-dimensional integrability can only occur when the 2^{d-1} generators of the group satisfy additional relations allowing

The existence of *Aut* as a symmetry of the Yang-Baxter equations has the following consequence: we may say that solving the Yang-Baxter equation is equivalent to solving all its images by *Aut*. These images *generically tend to proliferate*, simply because *Aut* is infinite. Considering that the equations form an *overdetermined set*, it is easy to believe that the total set of equations is "less overdetermined" when the orbits of *Aut* are of finite order. One can therefore imagine that the *best candidates* for the integrability varieties are *precisely* the ones where the symmetry group possesses *finite orbits*: the solutions of Au-Yang et al. [25, 26, 27] seem to confirm this point of view [28, 29].

A contrario, if one gets hold of an apparently isolated solution, the action of *Aut* will multiply it until building up, in experimentally not so rare cases, a continuous family of solutions from the original one. This is the solution to the so-called baxterization problem [30].

We first show that the simplest example of Yang-Baxter relation which is the star-triangle relation [7] has an *infinite discrete group* of symmetries generated by three involutions. These involutions are deeply linked with the so-called *inversion relations* [31, 32, 33, 34].

This analysis can be extended to the "generalized star-triangle relation" for Interaction aRound the Face models without any major difficulties [6, 35].

2. The star-triangle relations

2.1 The setting

We consider a spin model with nearest neighbour interactions on square lattice. The spins σ_i can take q values. The Boltzmann weight for an oriented bond $\langle ij\rangle$ will be denoted hereafter by $w(\sigma_i, \sigma_j)$. The weights $w(\sigma_i, \sigma_j)$ can be seen as the entries of a $q \times q$ matrix. In the following we will introduce a pictorial representation of the star-triangle relation. An arrow is associated to the oriented bond $\langle ij\rangle$. The arrow from i to j indicates that the argument of the Boltzmann weight w is (σ_i, σ_j) rather than (σ_j, σ_i). This arrow is relevant only for the so-called chiral models [25], that is to say that the $q \times q$ matrix describing w is not symmetric. An interesting class of $q \times q$ matrices has been extensively investigated in the last few years [25, 27, 26]: the general cyclic matrices. It is important to note that we *do not* restrict ourselves to this particular class of matrices. Let us give the following non cyclic nor symmetric 6×6 matrix as another illustrative

for a mere polynomial growth of the size, and possibly reducing to a semi-direct product of finite groups and $\mathbb{Z}$ factors.

example:

$$\begin{pmatrix} x & y & z & y & z & z \\ z & x & y & z & y & z \\ y & z & x & z & z & y \\ y & z & z & x & z & y \\ z & y & z & y & x & z \\ z & z & y & z & y & x \end{pmatrix} \tag{1}$$

2.2 The relations

We introduce the star-triangle equations both analytically[4] and pictorially:

$$\sum_{\sigma} w_1(\sigma_1, \sigma) \cdot w_2(\sigma, \sigma_2) \cdot w_3(\sigma, \sigma_3) = \lambda\, \overline{w}_1(\sigma_2, \sigma_3) \cdot \overline{w}_2(\sigma_1, \sigma_3) \cdot \overline{w}_3(\sigma_1, \sigma_2). \tag{2}$$

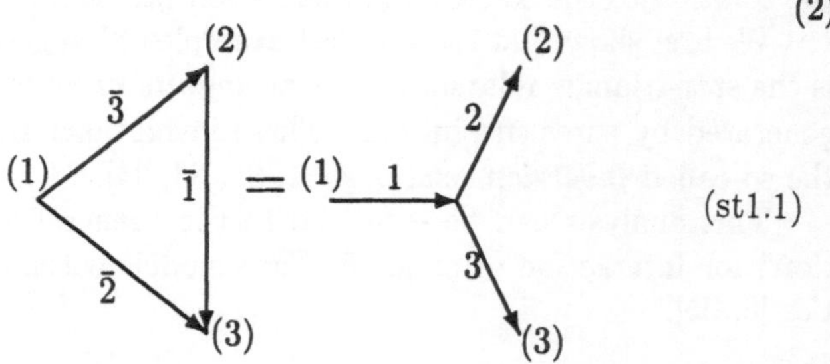

One should note that satisfying equation (2) *together with* the relation (st1.2) obtained by reversing all arrows, is a *sufficient* condition for the commutation of the diagonal transfer matrices of *arbitrary size M* with periodic boundary conditions $\mathbb{T}_M(w_2, \overline{w}_2)$ and $\mathbb{T}_M(\overline{w}_3, w_3)$:

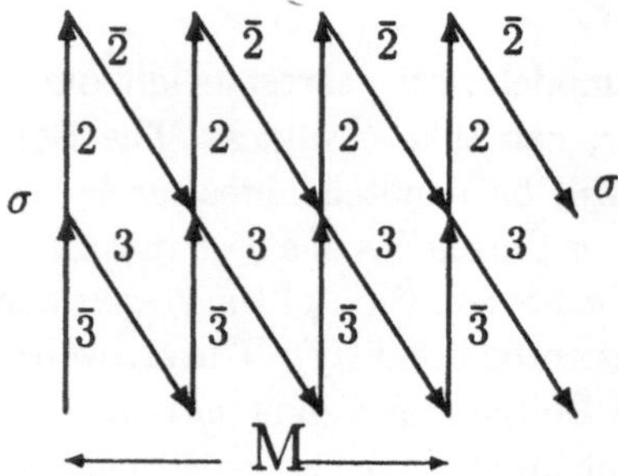

Note that for cyclic matrices ([25, 27, 26]) the star-triangle relations (st1.1) and (st1.2) give the same equations since one exchanges (st1.2) and (st1.1) by spin reversal.

One could obviously imagine many other choices for the arrows on the six bonds, however only three of them lead to the commutation of

[4]Since the w_i and $\overline{w}_i$ are homogeneous variables, there will always be a global multiplicative factor λ floating around in the star-triangle equations.

diagonal transfer matrices. We therefore have three systems of equations to study. For example, if the Boltzmann weights are given by the 6×6 matrix (1), these three systems of equations are respectively made of 20 different equations or 35 or 36.

3. The Yang-Baxter relation for vertex models

We shall not get here into the arcanes of this relation, which appears in the theory of integrable models [9], the theory of factorizable S-matrix in two-dimensional field theory, the quantum inverse scattering method [11], knot theory and has been given a canonical meaning in terms of Hopf algebras [36] (quantum groups [8, 9, 37, 38, 39]) and the list is far from exhaustive. We just want to fix some notations for later use.

We consider a vertex model on a two-dimensional square lattice. To each bond is associated a variable with q possible states and a Boltzmann weight $w(i,j,k,l)$ is assigned to each vertex

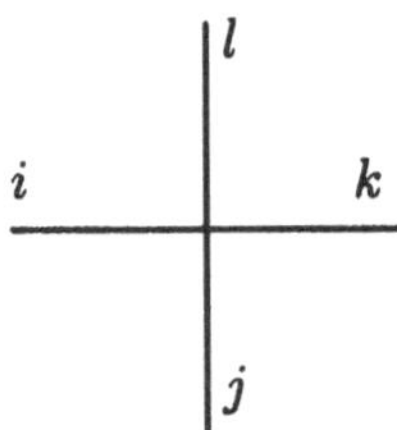

In order to write the Yang-Baxter relation, the q^4 homogeneous weights $w(i,j,k,l)$ are first arranged in a $q^2 \times q^2$ matrix R:

$$R^{ij}_{kl} = w(i,j,k,l). \tag{3}$$

The Yang-Baxter relation is a trilinear relation between three matrices $R(1,2)$, $R(2,3)$ and $R(1,3)$:

$$\sum_{\alpha_1,\alpha_2,\alpha_3} R^{i_1 i_2}_{\alpha_1 \alpha_2}(1,2) R^{\alpha_1 i_3}_{j_1 \alpha_3}(1,3) R^{\alpha_2 \alpha_3}_{j_2 j_3}(2,3) = \sum_{\beta_1,\beta_2,\beta_3} R^{i_2 i_3}_{\beta_2 \beta_3}(2,3) R^{i_1 \beta_3}_{\beta_1 j_3}(1,3) R^{\beta_1 \beta_2}_{j_1 j_2}(1,2). \tag{4}$$

The assignation (3) is arbitrary and we may specify it by complementing the vertex with an arrow and attributing numbers to the lines

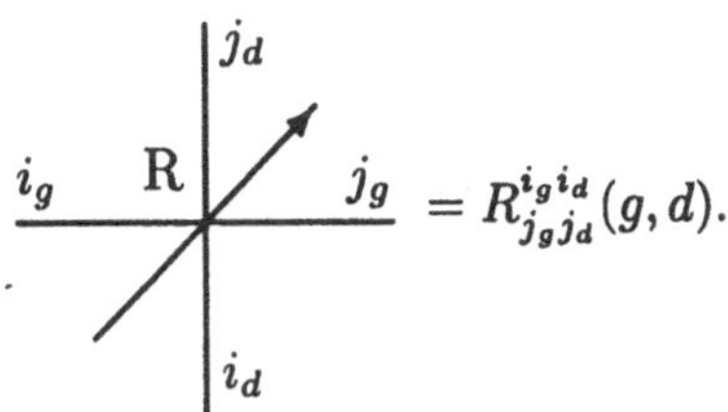

With these rules relation (4) has the following graphical representation

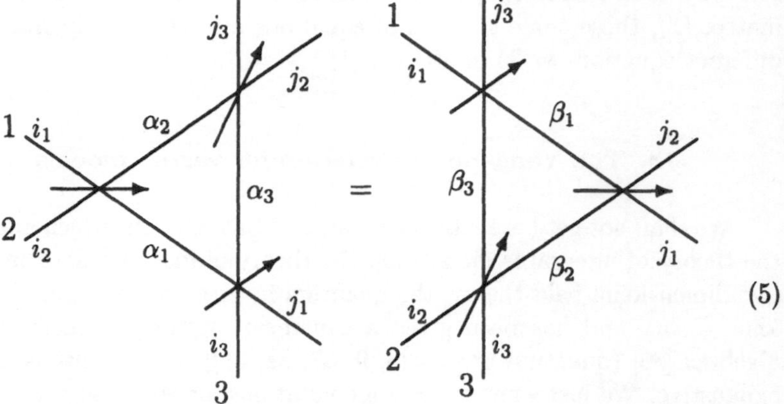

The lines carry indices 1,2,3.

Some especially interesting solutions depend on a continuous parameter called the "spectral parameter". The presence of this parameter is fundamental for many applications in physics, as for example the Bethe Ansatz method [40, 5, 11, 12]. One of the main issues in the full resolution of (4) is precisely to describe what is this parameter and the algebraic variety on which it lives, although its presence may obscure the algebraic structures underlying the Yang-Baxter equation (*the discovery of quantum groups was allowed by forgetting this parameter* [39, 8, 41, 9]). The problem of building up continuous families of solutions from an isolated one, known as the *baxterization* [10], is made straightforward by our study. Indeed our results *explain the presence of the spectral parameter* in the solution of the equation (see also [24]).

4. Infinite discrete symmetry group for the star-triangle relation

4.1 The inversion relation

Two distinct inverses act on the matrix of nearest neighbour spin interactions: the matrix inverse I and the dyadic (element by element) inverse J. We write down the inversion relations both analytically and pictorially:

$$\sum_{\sigma} w(\sigma_i, \sigma) \cdot I(w)(\sigma, \sigma_j) \quad = \quad \mu\, \delta_{\sigma_i \sigma_j}, \tag{6}$$

$$w(\sigma_i, \sigma_j) \cdot J(w)(\sigma_i, \sigma_j) \quad = \quad 1. \tag{7}$$

where $\delta_{\sigma_i \sigma_j}$ denotes the usual Kronecker delta.

w $I(w)$

σ_i σ σ_j = $\sigma_i = \sigma_j$

w

and σ_i σ_j $=$ σ_i σ_j

$J(w)$

The two involutions I and J generate an *infinite discrete group* Γ (Coxeter group) isomorphic to the infinite dihedral group $\mathbb{Z}_2 \ltimes \mathbb{Z}$. The $\mathbb{Z}$ part of Γ is generated by IJ. In the parameter space of the model, that is to say some projective space $\mathbb{C}\mathbf{P}_{n-1}$ (n homogeneous parameters), I and J are birational involutions. They give a *non-linear representation of this Coxeter group by an infinite set of birational transformations* [24]. It may happen that the action of Γ on specific subvarieties *yields a finite orbit.* This means that the representation of Γ identifies with the p-dihedral group $\mathbb{Z}_2 \ltimes \mathbb{Z}_p$.

4.2 The symmetries of the star-triangle relations

The two inversions I and J act on the star-triangle relation. Let us give a pictorial representation of this action, starting from (st1.1) as an example:

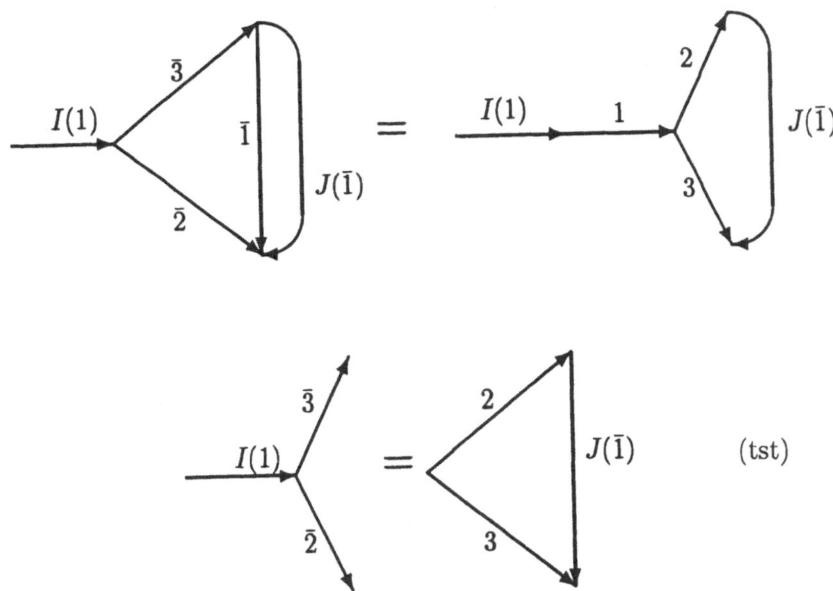

The transformed equation reads:

$$\lambda \sum_{\sigma_1} I(w_1)(\tau,\sigma_1)\cdot\overline{w}_2(\sigma_1,\sigma_3)\cdot\overline{w}_3(\sigma_1,\sigma_2) = w_2(\tau,\sigma_2)\cdot w_3(\tau,\sigma_3)\cdot J(\overline{w}_1)(\sigma_2,\sigma_3). \tag{8}$$

We get an action on the space of solutions of the star-triangle relation. If $(w_1, w_2, w_3, \overline{w}_1, \overline{w}_2, \overline{w}_3)$ is a solution of eq(2) (see picture (st1.1) for the specific arrangement of arrows), then $(I(w_1), \overline{w}_3, \overline{w}_2, J(\overline{w}_1), w_3, w_2)$ is also a solution of eq(2), at the price of a permitted redefinition of λ. In this transformation, the weights w_1 and $\overline{w}_1$ play a special role.

At this point, it is better to formalize this action by introducing some notations. We may choose as a reference star-triangle relation $\mathcal{ST}$, the symmetric configuration:

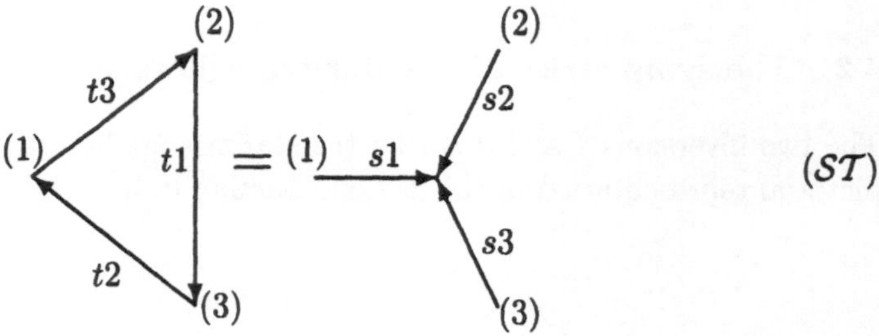

Any configuration may be obtained by reversing some arrows and permuting some bonds. With evident notations, we will denote by $R_{s1}, R_{s2}, R_{s3}, R_{t1}, R_{t2}, R_{t3}$ the reversals of arrows, and by $P_{si,sj}, P_{si,tj}, P_{ti,tj}$ the permutations of bonds. Moreover I and J act on the bonds as $I_{s1}, I_{s2}, \ldots$ The action of I and J described above (where 1 was playing a special role) identifies with the action of

$$\mathcal{K}_1 = R_{s2} R_{t3} I_{s1} J_{t1} P_{s2,t3} P_{s3,t2}. \tag{9}$$

It is easy to check that $\mathcal{K}_1$ is an involution.

We may construct two similar involutions $\mathcal{K}_2$ and $\mathcal{K}_3$, obtained by cyclic permutation of the indices 1, 2, 3. The involutions $\mathcal{K}_i (i = 1, 2, 3)$ verify the defining relations of the *Weyl group of an affine algebra of type* $A_2^{(1)}$ [42]:

$$(\mathcal{K}_1\mathcal{K}_2)^3 = (\mathcal{K}_2\mathcal{K}_3)^3 = (\mathcal{K}_3\mathcal{K}_1)^3 = 1. \tag{10}$$

We denote $\mathcal{A}ut$ the group generated by the three involutions $\mathcal{K}_i$ $(i = 1, 2, 3)$.

5. Infinite discrete symmetry group for the Yang-Baxter equation

5.1 The inversion relations.

The R-matrix appears naturally as a representation of an element of the tensor product $\mathcal{A} \otimes \mathcal{A}$ of some algebra $\mathcal{A}$ with itself. This algebra is a nice Hopf algebra in the context of quantum groups. We shall not dwell on this here but recall some simple operations on R.

In $\mathcal{A} \otimes \mathcal{A}$ we have a product inherited from the product in $\mathcal{A}$:

$$(a \otimes b)(c \otimes d) = ac \otimes bd. \tag{11}$$

R is an invertible element of $\mathcal{A} \otimes \mathcal{A}$ for this product and we shall denote by $I(R)$ the inverse for this product:

$$R \cdot I(R) = I(R) \cdot R = 1 \otimes 1. \tag{12}$$

In terms of the representative matrix this reads:

$$\sum_{\alpha,\beta} R^{ij}_{\alpha\beta}\, I(R)^{\alpha\beta}_{uv} = \delta^i_u\, \delta^j_v = \sum_{\alpha,\beta} I(R)^{ij}_{\alpha\beta}\, R^{\alpha\beta}_{uv}. \tag{13}$$

This is nothing else but the so-called *inversion relation* for *vertex* models [31, 32, 35, 43, 23]. On $\mathcal{A} \otimes \mathcal{A}$ we have a permutation operator σ:

$$\sigma(a \otimes b) = b \otimes a, \tag{14}$$

$$(\sigma R)^{ij}_{uv} = R^{ji}_{vu}, \quad \text{for the matrix } R. \tag{15}$$

Note that the representation of σ is just the conjugation by the permutation matrix P:

$$P^{ij}_{kl} = \delta_{il}\delta_{jk}, \tag{16}$$

$$\sigma R = PRP. \tag{17}$$

In the language of matrices we have a notion of transposition. Let us define partial transpositions t_g and t_d by:

$$(t_g R)^{ij}_{uv} = R^{uj}_{iv}, \tag{18}$$

$$(t_d R)^{ij}_{uv} = R^{iv}_{uj}, \tag{19}$$

and the full transposition

$$t = t_g t_d = t_d t_g. \tag{20}$$

We shall in the sequel use another inversion J defined by:

$$J = t_g I t_d = t_d I t_g, \tag{21}$$

or equivalently:

$$\sum_{\alpha,\beta} R^{\alpha u}_{v\beta}\, J(R)^{\alpha i}_{j\beta} = \delta^i_u\, \delta^j_v = \sum_{\alpha,\beta} J(R)^{i\beta}_{\alpha j}\, R^{u\beta}_{\alpha v} \tag{22}$$

These operators verify straightforwardly:

$$\begin{aligned} I^2 &= J^2 = 1, \quad It = tI, \quad Jt = tJ, \\ \sigma^2 &= t^2 = 1, \quad \sigma I = I\sigma, \quad \sigma J = J\sigma, \\ (\sigma t_g)^2 &= (\sigma t_d)^2 = t, \quad \sigma t_g \sigma t_d = 1. \end{aligned} \tag{23}$$

Each of these operations has a graphical representation. For the inversion I or more precisely for σI it is:

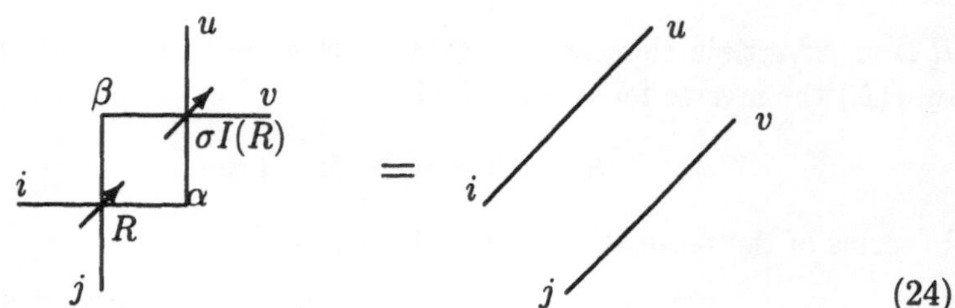

(24)

the inversion J reads:

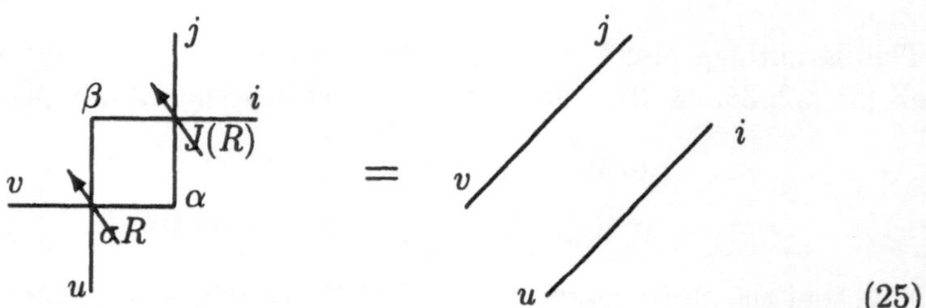

(25)

and the transposition reads:

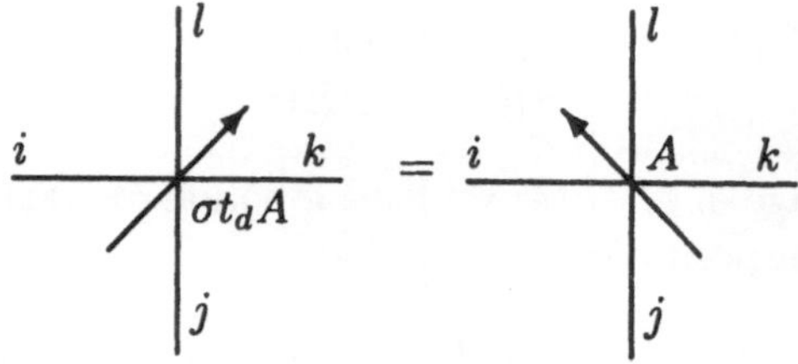

Note that the two inversions I and J do not commute. They generate an infinite discrete group Γ, the infinite dihedral group, isomorphic to the semi-direct product $\mathbb{Z} \ltimes \mathbb{Z}_2$. This group is represented on the matrix elements by *birational transformations* [24, 44, 45] acting on the projective space of the entries of the matrix R. Remark that for the *vertex models*, the birational transformations associated to the two involutions I and J are naturally related by collineations (see (21): this should be compared with the situation for nearest neighbour interaction spin models [24, 46].

5.2 The symmetries of the Yang-Baxter equations.

At the price of the redefinitions:

$$A = tR(2,3), \tag{26}$$
$$B = \sigma t_d R(1,3), \tag{27}$$
$$C = R(1,2), \tag{28}$$

we may picture the Yang-Baxter relation in a more symmetric way:

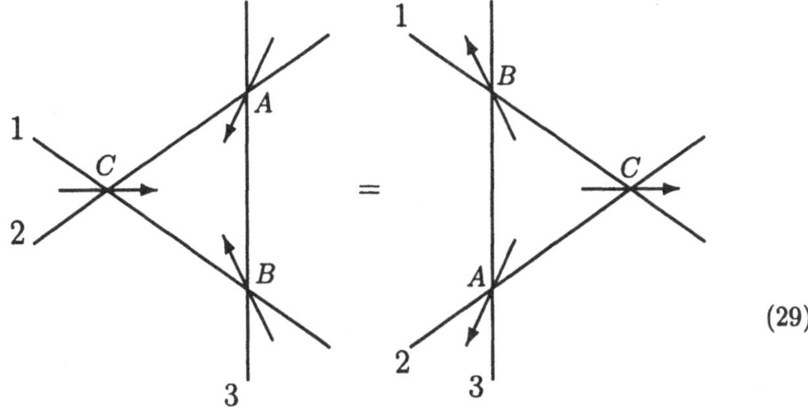

(29)

We may bracket (29) with 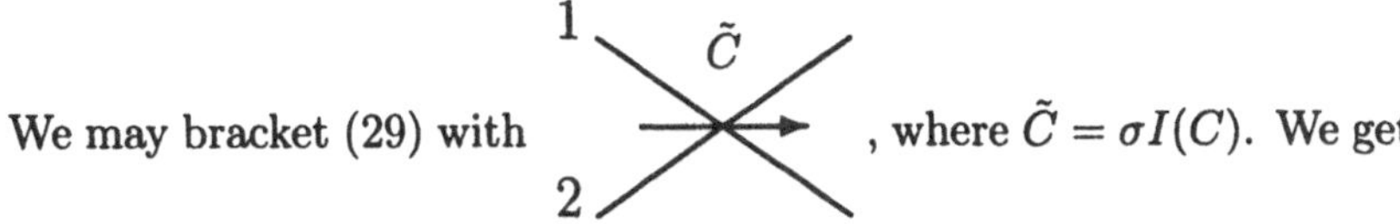

, where $\tilde{C} = \sigma I(C)$. We get

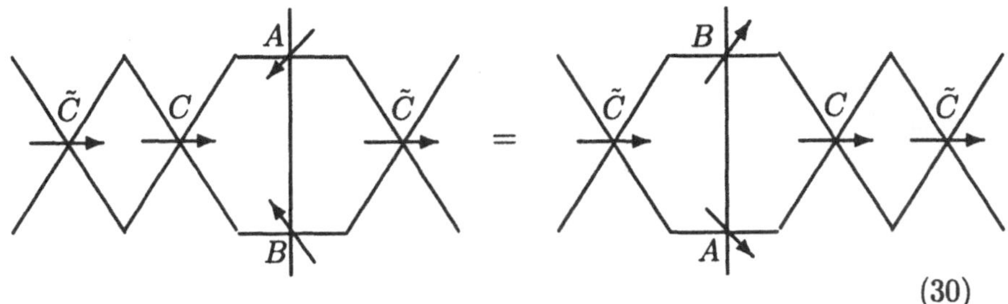

(30)

that is to say

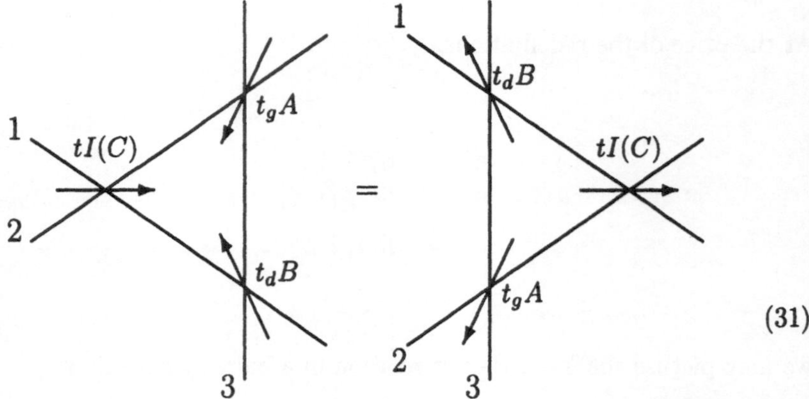

(31)

This relation is nothing but (29) after the redefinitions

$$\begin{aligned} A &\rightarrow t_g A, \\ B &\rightarrow t_d B, \\ C &\rightarrow tI\, C. \end{aligned} \tag{32}$$

We may denote by K_3 the operation (32). We have two other similar operations K_1 and K_2

$$K_1: \begin{array}{l} A \rightarrow tI\,A \\ B \rightarrow t_g B \\ C \rightarrow t_d C \end{array} \quad , \qquad K_2: \begin{array}{l} A \rightarrow t_d A \\ B \rightarrow tI\,B \\ C \rightarrow t_g C \end{array} \quad .$$

The discrete group $\mathcal{A}ut$ generated by the K_i's ($i = 1, 2, 3$) is a symmetry group of the Yang-Baxter equations. These generators K_i ($i = 1, 2, 3$) are involutions. The K_i's satisfy the relation $(K_1K_2K_3)^2 = 1$. Actually, the operation $K_1K_2K_3$ is just the inversion I on R. Among the elements of the discrete group generated by the K_i's we have in particular:

$$(K_1K_2)^2: \quad A \rightarrow It_gIt_gA = tIJA, \tag{33}$$

$$B \rightarrow t_dIt_dIB = tJIB, \tag{34}$$

$$C \rightarrow C. \tag{35}$$

Since IJ is of infinite order, we have generated an *infinite discrete group* of symmetries. This is exactly the phenomenon that we described in section 4.2 for the star-triangle equations.

Under this form it is not so evident to find the actual structure of the group. Let us introduce K_A, K_B and K_C, which are simply related to the K_i's by the transposition of two vertices:

$$K_A: \begin{array}{l} A \rightarrow \sigma tIA \\ B \rightarrow t_g\sigma C \\ C \rightarrow \sigma t_g B \end{array} \quad , \qquad K_B: \begin{array}{l} A \rightarrow \sigma t_g C \\ B \rightarrow \sigma tIB \\ C \rightarrow t_g\sigma A \end{array} \quad , \qquad K_C: \begin{array}{l} A \rightarrow t_g\sigma B \\ B \rightarrow \sigma t_g A \\ C \rightarrow \sigma tIC \end{array} \quad .$$

It is easily verified that:

$$K_A^2 = K_B^2 = K_C^2 = 1, \tag{36}$$

and

$$(K_A K_B)^3 = (K_B K_C)^3 = (K_C K_A)^3 = 1, \tag{37}$$

with no other relations. We recover the affine Coxeter group $A_2^{(1)}$ we already encountered in section 4.2.

A fundamental remark: Beware that, due to the different arrangement of indices, the relations we consider are not the Yang-Baxter equations that one considers in the study of quantum groups (shortly $RRR = RRR$) but rather its avatar $ABC = CBA$. The relevance of these relations is detailed in the standard literature on integrable models [9] and quantum groups [37, 38, 39].

We have here a very powerful instrument for two purposes: it defines *adequate patterns* for the matrix R [47]. It permits the so-called *baxterization of an isolated solution* just acting with tIJ. Indeed if a set of relations among the entries of R are preserved by IJ (or at least by tIJ), they will stay for every transforms of the initial Yang-Baxter relation. We shall illustrate in section 7.1.1 the baxterization on the Baxter eight-vertex model [48, 16] and show in section 7.1.2 how to introduce a spectral parameter for the solutions of the Yang-Baxter equations associated to $sl(n)$ algebras.

6. The tetrahedron equations and their symmetries

This equation is a generalization of the Yang-Baxter equation to three dimensional vertex models [19, 18, 21]. We give a pictorial representation of the three-dimensional vertex by

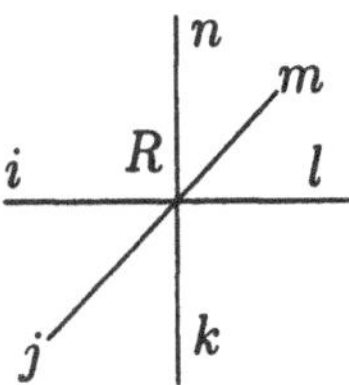

The Boltzmann weights of the vertex are denoted $w(i, j, k, l, m, n)$ and may be arranged in a matrix of entries

$$R^{ijk}_{lmn} = w(i, j, k, l, m, n). \tag{38}$$

The tetrahedron equation has a pictorial representation:

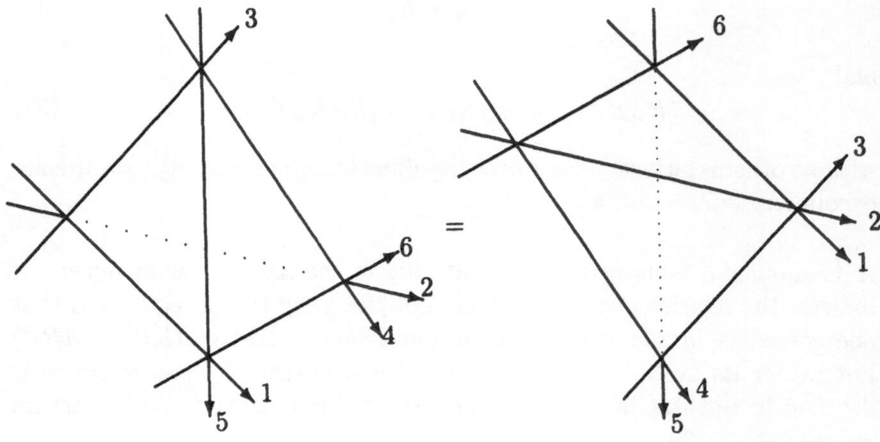

The algebraic form is

$$R_{123}R_{543}R_{516}R_{426} = R_{426}R_{516}R_{543}R_{123}. \tag{39}$$

We may here again introduce an inverse I

$$\sum_{\alpha_g,\alpha_m,\alpha_d} (IR)^{i_g i_m i_d}_{\alpha_g \alpha_m \alpha_d} \cdot R^{\alpha_g \alpha_m \alpha_d}_{j_g j_m j_d} = \delta^{i_g}_{j_g} \delta^{i_m}_{j_m} \delta^{i_d}_{j_d}. \tag{40}$$

We also introduce the partial transpositions t_g, t_m and t_d with

$$(t_g R)^{i_g i_m i_d}_{j_g j_m j_d} = R^{j_g i_m i_d}_{i_g j_m j_d}, \tag{41}$$

and similar definitions for t_m and t_d.

We redefine

$$A = R_{123}, \quad B = t_d R_{543}, \quad C = t_g t_m R_{516}, \quad D = t R_{426}, \tag{42}$$

where t is the full transposition $t_g t_m t_d$. Equation (39) then takes the more symmetric form

$$\sum_{s_1,\ldots,s_6} A^{i_1 i_2 i_3}_{s_1 s_2 s_3} B^{i_5 i_4 j_3}_{s_5 s_4 s_3} C^{j_5 j_1 i_6}_{s_5 s_1 s_6} D^{j_4 j_2 j_6}_{s_4 s_2 s_6} = \sum_{r_1,\ldots,r_6} D^{r_4 r_2 r_6}_{i_4 i_2 i_6} C^{r_5 r_1 r_6}_{i_5 i_1 j_6} B^{r_5 r_4 r_3}_{j_5 j_4 i_3} A^{r_1 r_2 r_3}_{j_1 j_2 j_3}. \tag{43}$$

We may multiply the previous equation by $(IA)^{u_1 u_2 u_3}_{i_1 i_2 i_3}$ and $(tIA)^{v_1 v_2 v_3}_{j_1 j_2 j_3}$ and sum over (i_1, i_2, i_3) and (j_1, j_2, j_3). This amounts to a bracketing of the tetrahedron equations by two times the same vertex, in a procedure trivially generalizing the one for the Yang-Baxter equation (30). We recover (43) with A, B, C and D transformed by

$$\begin{aligned} K_1 : A &\to tIA \\ B &\to t_dB \\ C &\to t_mC \\ D &\to t_mD. \end{aligned} \tag{44}$$

We have in a similar way the operations

$$K_2 : \begin{array}{l} A \to t_dA \\ B \to tIB \\ C \to t_gC \\ D \to t_gD \end{array}, \qquad K_3 : \begin{array}{l} A \to t_gA \\ B \to t_gB \\ C \to tIC \\ D \to t_dD \end{array}, \qquad K_4 : \begin{array}{l} A \to t_mA \\ B \to t_mB \\ C \to t_dC \\ D \to tID \end{array}.$$

Each of these four operations is an involution. They satisfy various relations, for instance $(K_1K_2K_3K_4)^2 = 1$. *The K_i's generate a group Aut_3 which is a symmetry group of the tetrahedron equations.* This group is "monstrous" since the number of elements of length smaller than l is of exponential growth with respect to l, unlike the case of the affine Coxeter groups (as $A_2^{(1)}$ for the Yang-Baxter equation) where this number is of polynomial growth.

The operations playing a role similar to the one of I and J in the two-dimensional Yang-Baxter equations are the *four* involutions

$$I, \quad J = t_gIt_mt_d, \quad K = t_mIt_dt_g, \quad L = t_dIt_gt_m. \tag{45}$$

We call Γ_3 the group generated by these four involutions. Γ_3 is also a symmetry group for the three dimensional vertex model *even if* [33] the model *does not* satisfy the tetrahedron equation.

In order to precise the algebraic structure of the group Γ_3 generated by I, J, K and L, it is simpler to consider as generators two of the partial transpositions t_g and t_d, I and the full transposition t. The third partial transposition can be recovered as the product tt_gt_d and t commutes with all other generators and so contributes a mere $\mathbb{Z}_2$ factor in the group. We are thus considering the Coxeter group generated by three involutions t_g, t_d and I, with two of them commuting: this is represented by the following Dynkin diagram

$$\underset{t_g}{\bullet} \overset{\infty}{\text{———}} \underset{I}{\bullet} \overset{\infty}{\text{———}} \underset{t_d}{\bullet}$$

For this group again, the number of elements of length smaller than l is greater than $2^{l/2}$. This is in fact a *hyperbolic* Coxeter group [49].

7. Consequences of this symmetry group

7.1 The baxterization

The problem of the baxterization is to introduce a spectral parameter into an isolated solution of the Yang-Baxter equations [10]. We have solutions of this problem by acting with the symmetry group Γ.

7.1.1 Baxterization of the Baxter model

Consider the matrix of the symmetric eight vertex model

$$R = \begin{pmatrix} a & 0 & 0 & d \\ 0 & b & c & 0 \\ 0 & c & b & 0 \\ d & 0 & 0 & a \end{pmatrix}. \tag{46}$$

Notice that this form is preserved by I and J and that $tR = R$. The action of I is

$$a \longrightarrow \frac{a}{a^2 - d^2} \qquad b \longrightarrow \frac{b}{b^2 - c^2} \tag{47}$$

$$c \longrightarrow \frac{-c}{b^2 - c^2} \qquad d \longrightarrow \frac{-d}{a^2 - d^2} \tag{48}$$

and the action of J is

$$a \longrightarrow \frac{a}{a^2 - c^2} \qquad b \longrightarrow \frac{b}{b^2 - d^2} \tag{49}$$

$$c \longrightarrow \frac{-c}{a^2 - c^2} \qquad d \longrightarrow \frac{-d}{b^2 - d^2} \tag{50}$$

We shall look at the solutions of the Yang-Baxter equations for matrices R of the form (46). The leading idea is that the parametrization of the solutions is just the parametrization of the algebraic varieties preserved by tIJ in the projective space $\mathbb{C}P_3$ of the homogenous parameters (a, b, c, d). The remarkable fact is that not only these varieties exist but can be completely described. We use the visualization method we have already used [24, 50] for spin models, that is to say just draw the orbits obtained by numerical iteration and look.

This is best illustrated by Figure 1. This figure shows the orbit of point (*), which is a matrix of the form (46). It is drawn by the iteration of IJ acting on the initial point (*). The resulting points densify on the elliptic curve given by the intersection of the two quadrics Δ_1 = constant and Δ_2 = constant (Clebsch's biquadratic), with Δ_1 and Δ_2 the Γ invariants

$$\Delta_1 = \frac{a^2 + b^2 - c^2 - d^2}{ab + cd},$$

$$\Delta_2 = \frac{ab - cd}{ab + cd}. \tag{51}$$

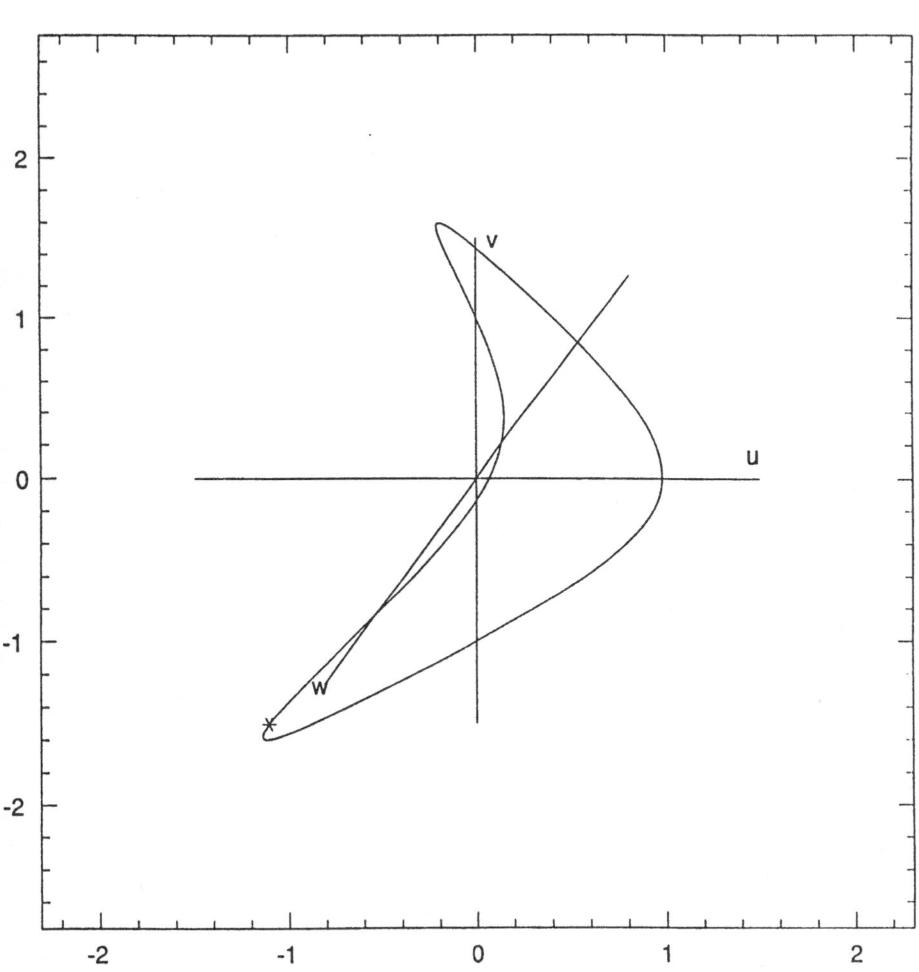

Figure 1. Baxterization of the point $*$

Similar calculations can be performed for a very general 16-vertex model for which the 4×4 R-matrix is symmetric:

$$R = \begin{pmatrix} a & e & f & d \\ e & b & c & g \\ f & c & b' & h \\ d & g & h & a' \end{pmatrix}. \tag{52}$$

Amazingly the baxterization of this 16-vertex model leads to *curves*. These curves are also *intersection of quadrics* (even in the general case for which their is no solution for the Yang-Baxter equations) [51].

7.1.2 Baxterization of the R matrix of $sl_q(n)$

Another example corresponds to the baxterization of solutions associated to $sl(n)$ algebras [13]. There are special solutions generally denoted R_+ and R_-. For the simplest four-dimensional representation of the $sl(2)$ case, we have

$$R_+ = \begin{pmatrix} q & 0 & 0 & 0 \\ 0 & 1 & q-q^{-1} & 0 \\ 0 & 0 & 1 & 0 \\ 0 & 0 & 0 & q \end{pmatrix}. \tag{53}$$

and a similar expression for R_- [13]. Looking for a family containing both R_+ and R_- our baxterization procedure leads to the well-known [5] six-vertex model R-matrix $R = \lambda R_+ + 1/\lambda R_-$.

We let as an exercise for the reader to treat the $sl(3)$ case. In a forthcoming publication we will show that these ideas can be generalized to all the universal R-matrices [9] for every representation [52]. This group appears in field theory, in the analysis of classical R-matrices [53].

7.2 A strategy for the resolution of a star-triangle equation

We may use the symmetry group *Aut* (or more simply Γ on each Boltzmann weight) to find the integrability varieties. In general, Γ points towards specific algebraic varieties in the parameter space (the varieties where the orbits lie), and *Aut* sometimes allows to reduce the number of unknowns in the original equations (sections 4.2 and 5.2). We may in certain instance bring these equations to a handable "isotropic" form ($ABC = CBA \rightarrow RRR = RRR$ form), and find particular isotropic solution of the equations.

This is best exemplified [54] with the chiral model, with five homogeneous parameters $w(k)$, $k = 0, 1, 2, 3, 4$ and weight matrix

$$\begin{pmatrix} w(0) & w(1) & w(2) & w(3) & w(4) \\ w(4) & w(0) & w(1) & w(2) & w(3) \\ w(3) & w(4) & w(0) & w(1) & w(2) \\ w(2) & w(3) & w(4) & w(0) & w(1) \\ w(1) & w(2) & w(3) & w(4) & w(0) \end{pmatrix} \tag{54}$$

For the non-chiral model obtained by setting $w(1) = w(4)$ and $w(2) = w(3)$, the exact result of Fateev-Zamolodchikov [55] is recovered in [54] *without prejudice* on the properties of the solution (e.g. self-duality). One can go further and look at the integrability varieties of the general chiral Potts model [27].

The above (non-chiral) "isotropic" point is a particular solution of the star-triangle relation for this model (isotropic star-triangle). We use the non-homogeneous variables $x(k) = w(k)/w(0)$, $k = 1, \ldots, 4$. We introduce the infinitesimal perturbation $X_i(k)$ of $x_i(k)$, and $\overline{X}_i(k)$ of $\overline{x}_i(k)$, with obvious notations.

The linearized star-triangle relations yield a homogeneous linear system for $X_i(k), \overline{X}_i(k)$. This system is not only compatible, but it has *a four dimensional space of solutions.*

The solutions verify:

$$X_1(k)+X_2(k)+X_3(k) = \overline{X}_1(k)+\overline{X}_2(k)+\overline{X}_3(k) = 0, \quad k = 1, \ldots, 4. \tag{55}$$

If we introduce the symmetric and antisymmetric vectors X^s and X^a (resp. $\overline{X}^s$ and $\overline{X}^a$)

$$X^s = \begin{pmatrix} 1 \\ s \\ s \\ 1 \end{pmatrix} \qquad X^a = \begin{pmatrix} 1 \\ a \\ -a \\ -1 \end{pmatrix} \qquad \overline{X}^s = \begin{pmatrix} 1 \\ \overline{s} \\ \overline{s} \\ 1 \end{pmatrix} \qquad \overline{X}^a = \begin{pmatrix} 1 \\ \overline{a} \\ -\overline{a} \\ -1 \end{pmatrix} \tag{56}$$

with

$$s = \frac{2 - 4c(2) - c(4) + c(6) + 7c(8) - 7c(10) - c(12) - 3c(14)}{-2 - 4c(2) + 8c(4) - c(6) + 2c(8) - 4c(10) + c(12) - 2c(14)}$$

$$a = \frac{-2 + 4c(2) + c(4) - c(6) - 9c(8) + 9c(10) + c(12) - c(14)}{-6 + 2c(2) + 8c(4) + 17c(6) - 42c(8) + 2c(10) + 23c(12) + 22c(14)}$$

$$\overline{s} = \frac{-2 + 10c(2) - 12c(4) + 7c(6) - 7c(8) + 3c(10) + 4c(12) - 2c(14)}{10 - 15c(2) + 11c(4) - 9c(6) + 2c(10) + c(12) + 4c(14)}$$

$$\overline{a} = \frac{10 - 4c(2) - 12c(4) + 18c(6) - 26c(8) + 14c(10) + 8c(12) - 8c(14)}{11 + 20c(2) - 60c(4) + 68c(6) - 90c(8) + 50c(10) + 32c(12) - 50c(14)}$$

with $c(p) = cos(p\theta)$, i.e., numerically:

$$s \simeq -58.28463... \quad a \simeq -1.0308189..., \quad \overline{s} \simeq -1.834537..., \quad \overline{a} \simeq 4.543390...$$

and set

$$X_i = s_i X^s + a_i X^a \tag{57}$$
$$\overline{X}_i = \overline{s}_i \overline{X}^s + \overline{a}_i \overline{X}^a \qquad i = 1, 2, 3, \tag{58}$$

we get

$$\begin{aligned} s_1 + s_2 + s_3 &= a_1 + a_2 + a_3 = 0, \\ \overline{s}_i &= -\alpha\, s_i, \\ \overline{a}_1 &= \beta\,(a_2 - a_3), \\ \overline{a}_2 &= \beta\,(a_3 - a_1), \\ \overline{a}_3 &= \beta\,(a_1 - a_2), \end{aligned} \tag{59}$$

with

$$\alpha = -\frac{1}{4\cos^2(\theta)} \simeq -.2527617250... \tag{60}$$
$$\beta = \frac{28}{31} - \frac{38}{31}c(1) + \frac{20}{93}c(3) + \frac{118}{93}c(7) \simeq .0108158287... \tag{61}$$

This proves[5] the existence of *a four parameter family of solutions* of the star-triangle equations containing the isotropic point. This family contains in particular the previously mentioned Fateev-Zamolodchikov solution [55]. It is remarkable that IJ is of *finite order* on this curve (the order is five). We have the prejudice that the integrability surface is of the same nature, i.e. is a locus of points where $(IJ)^5 = 1$. *Notice that such a locus is automatically invariant by both I and J.* Such a surface is given by the two equations:

$$\begin{aligned} A\ &\Big(w(1)w(3)w(4)^2 - 3w(2)^2w(4)^2 + 2w(1)w(2)w(4)w(0) \\ &+ w(3)w(2)^2w(0) - 3w(1)w(3)^2w(0) + 2w(4)w(2)w(3)^2 \Big) \\ &- 4w(2)^2w(4)^2 - 4w(1)w(3)^2w(0) + 3w(3)w(2)^2w(0) \\ &+ 3w(1)w(3)w(4)^2 + w(4)w(2)w(3)^2 + w(1)w(2)w(4)w(0) = 0 \end{aligned} \tag{62}$$

and

$$A\ \Big(-2w(3)^2w(0)^2 - 2w(1)w(2)w(4)^2 - w(4)w(3)^2w(1) \tag{63}$$

[5] As a consequence of the implicit function theorem and the algebraicity of the solutions of the star-triangle equations

$$
\begin{aligned}
&+\ 3w(0)w(3)w(4)^2+3w(2)w(1)w(3)w(0)-w(2)w(0)^2w(4)\Big)\\
&-\ w(1)w(2)w(4)^2+2w(2)w(0)^2w(4)-w(0)w(3)w(4)^2\\
&+\ 2w(4)w(3)^2w(1)-w(3)^2w(0)^2-w(2)w(1)w(3)w(0)=0
\end{aligned}
$$

with $A=\frac{1}{2}(-1\pm\sqrt{5})$. The vectors $X^s, \overline{X}^s, X^a, \overline{X}^a$ are tangent to this surface for $A=\frac{1}{2}(-1+\sqrt{5})$.

The consequences of equation (59) on the commutation of transfer matrices $T_i=\mathbb{T}_M(\overline{w}_i, w_i)$ are the following: locally near the isotropic point T_i depends on $(s_i, a_i, \overline{s}_i, \overline{a}_i)$. The commutation of T_1 and T_2 is obtained by imposing relations (59). We first need $\overline{s}_1=-\alpha\, s_1$ which allows three parameters for T_1. At this point (59) fixes a_2 and $\overline{a}_2$ and the only free parameter for T_2 is s_2, giving a *one parameter family of commuting transfer matrices.*

The integrability surface is actually the locus of points where $(IJ)^5 = e$ [28, 56, 57, 58, 59, 60]. In general, we believe that the varieties corresponding to *finite dimensional orbits* of the group Γ are *good candidates for integrability* [54].

In support of our inclination for finite orbits, we recall the asymmetric eight-vertex model ((52) with $e=f=g=h=0$) for which the *free fermion condition* $aa'+bb'-c^2-d^2=0$ implies *both the integrability and the finiteness of the orbits* of Γ (hint: I and J reduce to *linear permutations* up to signs). It is actually easy to find explicitly the subvarieties where Γ is of finite order [24]

When searching for isolated solutions of the Yang-Baxter equations, the group Γ can be used to get necessary conditions on the R-matrices. From the equation $RRR=RRR$, we deduce an infinite set of other relations of the kind $Rg(R)h(R)=h(R)g(R)R$, where g and h belong to Γ, leading to the commutation of the transfer matrices with periodic boundary conditions $T_N(R)$ and $T_N(g(R))$. Even for $N=1$ and $g=I$, this leads to non-trivial *necessary conditions.* For the 16 vertex model (52), this gives one *quartic* condition with 72 terms.

7.3 Three-dimensional models

Our strategy for finding solutions of the tetrahedron equations is to seek for patterns of the Boltzmann weights of the three dimensional vertex *compatible with the symmetry group* Γ_3. By this we mean that its form should be preserved by Γ_3.

7.3.1 A first model

We will therefore consider a simple model where i, j, k, l, m and n take only two values $+1$ and -1. The matrix (38) is an 8×8 matrix. We will require that its pattern is invariant under the inverse I [47] and the various partial transpositions t_g, t_m and t_d. We aim at having a generalization of the Baxter eight-vertex model and we impose the following restrictions:

$$\begin{aligned} w(i,j,k,l,m,n) &= w(-i,-j,-k,-l,-m,-n), & (64)\\ w(i,j,k,l,m,n) &= 0 \quad \text{ if } ijklmn = -1. & (65) \end{aligned}$$

These constraints amount to saying that the 8×8 matrix splits into two times the same 4×4 matrix. It is further possible to impose that this matrix is symmetric since, in this case, $t_g R$ (and any other partial transpose) is also symmetric. Let us introduce the following notations for the entries of the 4×4 block of the R matrix

$$\begin{pmatrix} a & d_1 & d_2 & d_3 \\ d_1 & b_1 & c_3 & c_2 \\ d_2 & c_3 & b_2 & c_1 \\ d_3 & c_2 & c_1 & b_3 \end{pmatrix}. \tag{66}$$

The four rows and columns of this matrix correspond to the states $(+,+,+)$, $(+,-,-)$, $(-,+,-)$ and $(-,-,+)$ for the triplets (i,j,k) or (l,m,n). The R-matrix can be completed by spin reversal, according to the rule (64). t_g simply exchanges c_2 with d_2 and c_3 with d_3, t_m and t_d can be similarly defined and I acts as the inversion of this 4×4 matrix.

For this three dimensional model, the coefficients of the characteristic polynomial of the 4×4 matrix (66) give a good hint for invariants under Γ_3. They are

$$\begin{aligned} \sigma_1^{(3d)} &= a + b_1 + b_2 + b_3, & (67)\\ \sigma_2^{(3d)} &= a(b_1+b_2+b_3) + b_1 b_2 + b_2 b_3 + b_3 b_1 & (68)\\ &- (c_1^2 + c_2^2 + c_3^2 + d_1^2 + d_2^2 + d_3^2), \ldots . \end{aligned}$$

Since $\sigma_2^{(3d)}$ is invariant by t_g, t_m amd t_d and takes a simple factor (the inverse of the determinant) under the action of I, the variety $\sigma_2^{(3d)} = 0$ *is invariant under* Γ_3. Given the hugeness of the group Γ_3, it is already an astonishing fact to have such a covariant expression. In fact we can exhibit *five linearly independent polynomials* with the same covariance, which give *four invariants*, as follows:

$$a b_1 + b_2 b_3 - c_1^2 - d_1^2, \qquad c_2 d_2 - c_3 d_3, \tag{69}$$

and the ones deduced by permutations of 1, 2 and 3. They form a five dimensional space of polynomials. Any ratio of the five independant polynomials is invariant under all the four generating involutions. In other

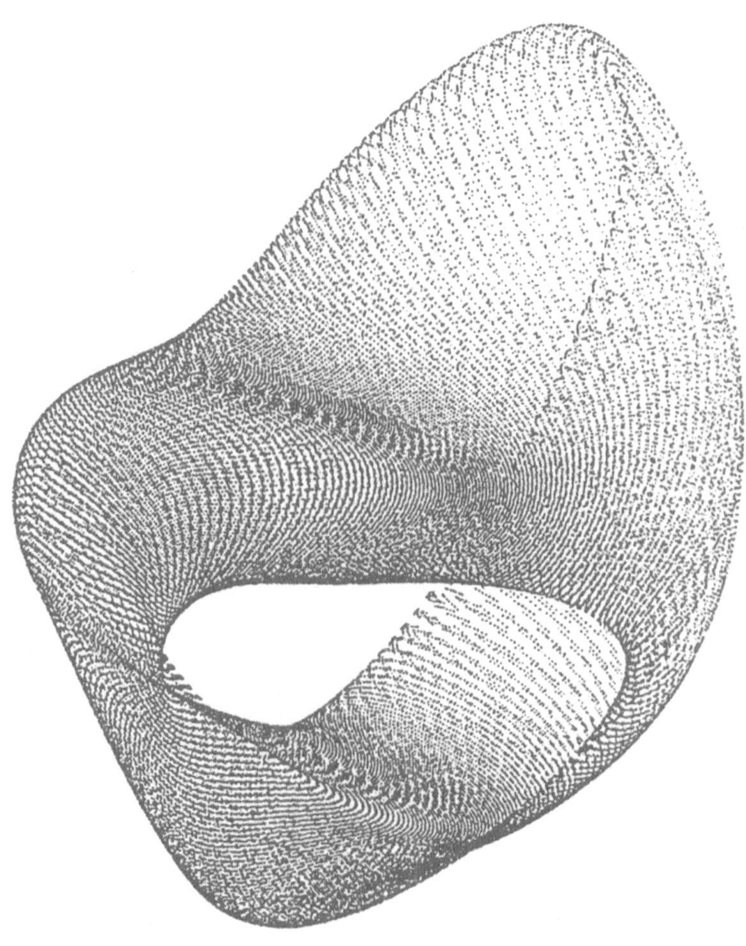

Figure 2.

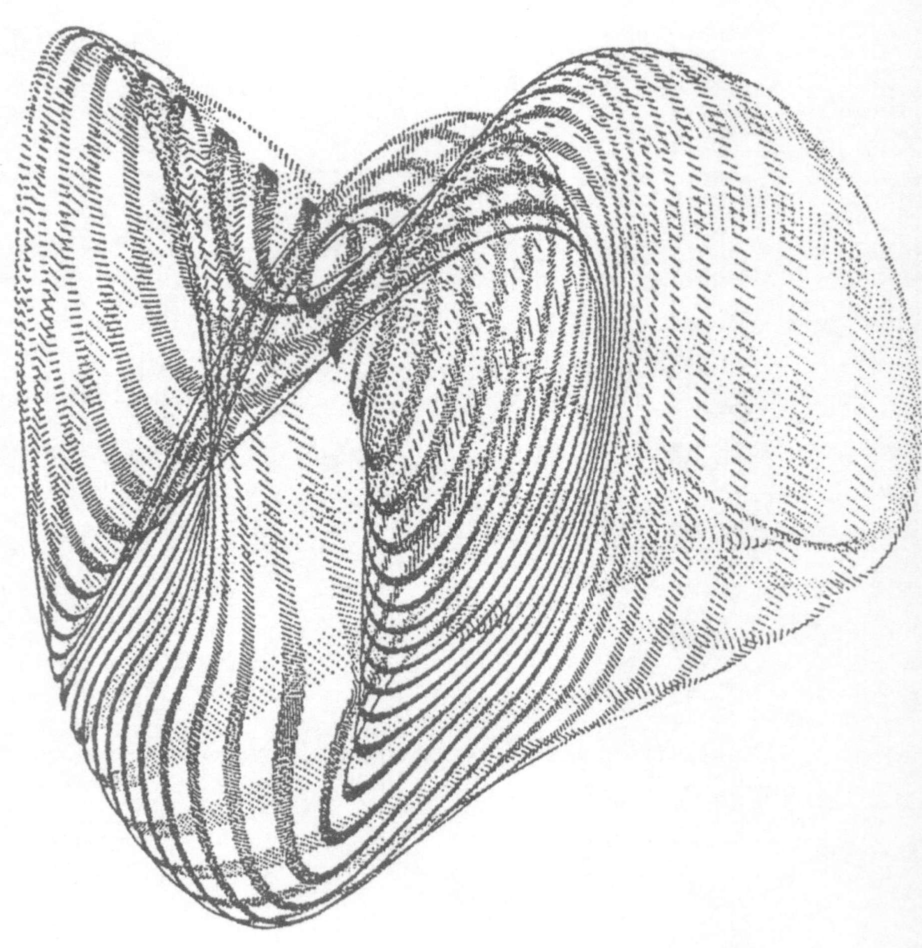

Figure 3.

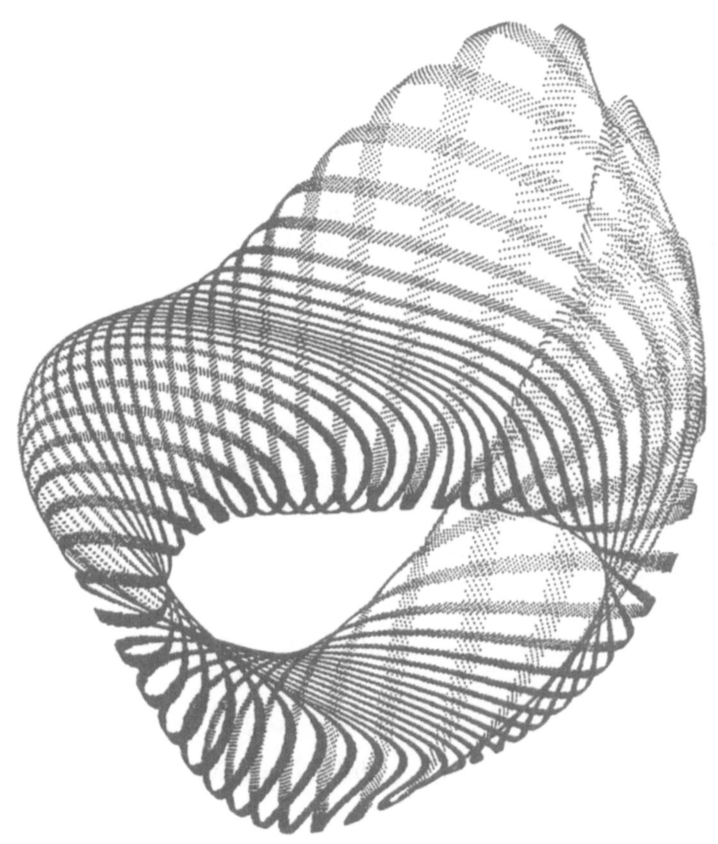

Figure 4.

words $\mathbb{CP}_9$ is foliated by five dimensional algebraic varieties invariant under Γ_3.

To have some flavour of the possible (integrable ?) algebraic varieties invariant under Γ_3, we study its orbits [24, 50]. We start with the study of the subgroup generated by some infinite order element namely IJ. This element gives a special role to axis 1. The transformation IJ *does preserve the symmetry under the exchange of 2 and 3.* If the initial point is symmetric under the exchange of 2 and 3, the orbit under IJ is thus a *curve.* Other starting points lead to orbits lying on a *two dimensional* variety given by the intersection of seven quadrics (see figure 2,3,4). However, what we are interested in are the orbits of the *whole* Γ_3 group. The size of this group prevent us from studying exhaustively the full set of group elements of a given length even for quite small values of this length. We have nevertheless explored the group by a random construction of typical elements of increasingly large length [30]. This confirms that we generically only have the four invariants described previously.

7.3.2 A second model

We also consider a simple model where i, j, k, l, m and n take only two values $+1$ and -1 and which is also a generalization of the Baxter eight vertex model. The Boltzmann weights $w(i,j,k,l,m,n)$ are given by:

$$w(i,j,k,l,m,n) = f(i,j,k)\,\delta_l^i\,\delta_m^j\,\delta_n^k + g(i,j,k)\,\delta_{-l}^i\,\delta_{-m}^j\,\delta_{-n}^k \tag{70}$$

$$f(i,j,k) = f(-i,-j,-k) \text{ and } g(i,j,k) = g(-i,-j,-k) \tag{71}$$

Equations (71) are symmetry conditions reducing the numbers of homogeneous parameters from 16 to 8.

As for the previous model, there exists an invariant of the action of the whole group Γ_3:

$$\frac{f(+,+,+)f(+,-,-)f(-,+,-)f(-,-,+)}{g(+,+,+)g(+,-,-)g(-,+,-)g(-,-,+)} \tag{72}$$

Considering the subgroup of Γ_3 generated by the infinite order element IJ, one can easily find other invariants, namely

$$\frac{f(+,+,+)f(+,-,-)}{g(+,+,+)g(+,-,-)} \tag{73}$$

and

$$\frac{f(+,+,+)^2 + f(+,-,-)^2 - g(+,+,+)^2 - g(+,-,-)^2}{g(+,+,+)g(+,-,-)} \tag{74}$$

For this model [61], the trajectories under IJ are *curves* in $\mathbb{CP}_7$.

8. Conclusion

An important problem in statistical mechanics and field theory, is the understanding of the role of the dimension of the lattice on both the algebraic aspects and the topological aspects. All this touches various fields of mathematics and physics: algebraic geometry, algebraic topology, quantum algebra. Indeed the Coxeter groups we use are at the same time groups of automorphisms of algebraic varieties, symmetries of quantum Yang-Baxter equations (and their higher dimensional avatars). They also provide an *extension to several complex variables functions of the notion of the fundamental group* Π_1 of a Riemann surface, with of course a much more involved covering structure [33, 50].

We believe moreover that the *space of parameters* seen as a projective space is the appropriate place to look at, if one wants to substantiate the deep topological notion embodied in the notion of **Z**-invariance [16] and free the models from the details of the lattice shape.

Actually, we have exhibited an infinite discrete symmetry group for the Yang-Baxter equations and their higher dimensional generalization acting on this parameter space. This group is the Coxeter group $A_2^{(1)}$ (semi-direct product of $\mathbb{Z} \times \mathbb{Z}$ by some finite group). We have shown that this symmetry is responsible for the presence of the spectral parameter. In other words, the *discrete* symmetry gives rise to a *continuous* one (see [54]). A similar study for the generalized star-triangle relation of the Interaction aRound a Face model, sketched in [35], can be performed rigorously along the same lines, leading to the same result. Also note that the same groups generated by involutions appear in the study of semi-classical r-matrices [53]. An interesting point will be to exhibit the action of our symmetry group on the underlying quantum group for the Yang-Baxter equations [52].

Our symmetry group is a group of automorphisms of the integrability varieties. This should give precious informations on these varieties. In particular one should decide if, up to Lie groups factors (which cannot be excluded because of the existence of "gauge" symmetries, weak graph duality [62], ...), these varieties can be anything else than abelian varieties, or even product of curves: can they be for example K_3 surfaces, are they homological obstructions to the occurence of anything but curves ?

For three-dimensional vertex models, the symmetry group, though generalizing very naturally the previous group (generated by four involutions with similar relations) is drastically different: it is so "large"[6] that the chances are quite small that it leaves enough room for any invariant integrability varieties. It is not useless to recall the unique non-trivial known solution of the tetrahedron equations (Zamolodchikov's solution) [19, 18, 21].

[6] One should keep in mind that very "large" sets of rational transformations may preserve algebraic curve of genus zero or one. Just think of the transformations on the circle generated by $\{\theta \rightarrow \theta + \lambda,\ \theta \rightarrow 2\theta,\ \theta \rightarrow 3\theta\}$ [63]

For this model the three axes are not on the same footing, so that we do not have a "true" three dimensional symmetry for the model (two-dimensional checkerboard models coupled together). Is there still any hope for a three-dimensional exactly solvable model with *genuine three-dimensional symmetry*? We think that the group of symmetries we have described gives the best line of attack to this problem. We will show that Γ_3 and even more $\mathcal{A}ut_3$ are generically too "large" to allow any non-trivial solution of the tetrahedron equations with genuine three dimensional symmetry [64].

Acknowledgments: We would like to thank J. Avan, O. Babelon, and M. Talon, for very stimulating discussions and comments.

L'un de nous (JMM), désire rendre hommage à la mémoire de Jean-Louis Verdier. Il y a plus de dix ans de cela, travailler sur les modèles intégrables était plutôt mal vu dans la communauté de la physique théorique française. Alors que je travaillais avec M.-T. Jaekel, Jean-Louis Verdier nous consacra une après-midi de discussion chaque semaine et ce, durant des années. Nous discutions d'équation de Yang-Baxter, de ses avatars (groupes de tresses, algèbre de Hecke, relations d'entrelacement, ...), de relations tetraèdre, de relations d'inverse, de modèle de Potts, toujours dans un esprit de géométrie algébrique.

J'ai un merveilleux souvenir de ces discussions: il y avait de part et d'autre un réel effort pour communiquer et pour montrer à l'autre, au delà des considérations adventices qui trop souvent encombrent, l'idée la plus intrinsèque, le concept, la structure qui font réellement marcher les choses. Il y avait le dialogue profond et honnête de gens qui ont compris que derrière ces idées se trouve quelque chose de passionnant. Je peux témoigner de la patience, de la générosité, du désintéressement de Jean-Louis Verdier: tout ce temps consacré à deux jeunes chercheurs plutôt marginaux ne servait en rien à sa carrière.

A l'heure où l'intégrabilité reçoit une reconnaissance institutionnelle à travers trois médailles Fields, mais où les phénomènes de mode tendent à remplacer les idées par des campagnes d'influence et le savoir-faire par le faire-savoir, à cette heure je tiens à dire que Jean-Louis Verdier me manque profondément.

Appendix: Symmetry group of the star-triangle relation

The three involutions generating the symmetry group of the star-triangle relations read:

$$\mathcal{K}_1 = R_{s2}R_{t3}I_{s1}J_{t1}P_{s2,t3}P_{s3,t2}\,, \quad \mathcal{K}_1^2 = 1 \tag{1}$$
$$\mathcal{K}_2 = R_{s3}R_{t1}I_{s2}J_{t2}P_{s3,t1}P_{s1,t3}\,, \quad \mathcal{K}_2^2 = 1 \tag{2}$$
$$\mathcal{K}_3 = R_{s1}R_{t2}I_{s3}J_{t3}P_{s1,t2}P_{s1,t1}\,, \quad \mathcal{K}_3^2 = 1 \tag{3}$$

If σ is the cyclic permutation, $\sigma = \sigma_s\sigma_t$ with $\sigma_s = P_{s_2,s_3}P_{s_1,s_2}$ and $\sigma_t = P_{t_2,t_3}P_{t_1,t_2}$, the involutions $\mathcal{K}_i$ are related by

$$\mathcal{K}_2 = \sigma^2\mathcal{K}_1\sigma, \quad \mathcal{K}_3 = \sigma^2\mathcal{K}_2\sigma \tag{4}$$

This symmetry group contains an action of IJ . It may be obtained by successively operating with the previous involutions: first act with $\mathcal{K}_1$, then with $\mathcal{K}_3$, then operate with $\mathcal{K}_2$ and finally with $\mathcal{K}_1$. This sequence of operations, when used on relations (st1.1), yields:

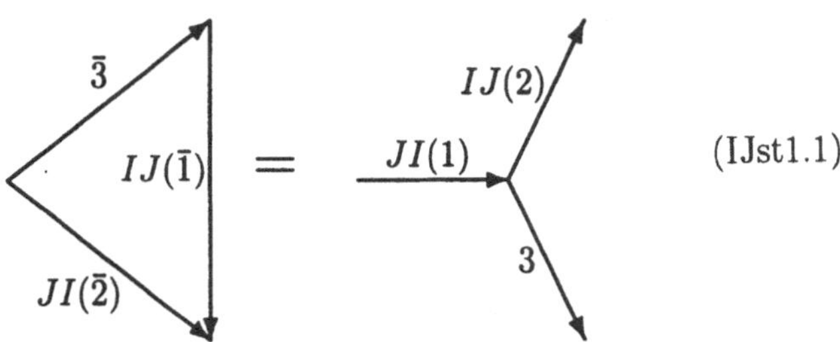

(IJst1.1)

This sequence of transformations amounts to acting with the product

$$G_3 = \sigma\mathcal{K}_1\mathcal{K}_2\mathcal{K}_3\mathcal{K}_1 = R_{s1}R_{s2}R_{t1}R_{t2}(JI)_{s1}(IJ)_{s2}(IJ)_{t1}(JI)_{t2}. \tag{5}$$

We may define similarly

$$G_2 = \sigma G_3\sigma^2 = R_{s3}R_{s1}R_{t3}R_{t1}(JI)_{s3}(IJ)_{s1}(IJ)_{t3}(JI)_{t1} \tag{6}$$
$$G_1 = \sigma G_2\sigma^2 = R_{s3}R_{s2}R_{t3}R_{t2}(IJ)_{s3}(JI)_{s2}(JI)_{t3}(IJ)_{t2} \tag{7}$$

We have the relations:

$$G_1G_2G_3 = 1 \tag{8}$$
$$G_iG_j = G_jG_i \qquad \forall i,j = 1,2,3. \tag{9}$$

The symmetry group $\mathcal{A}ut$ is the semi-direct product of the Weyl group of an A_2 (finite dimensional simple of rank 2) Lie algebra by a bidimensional lattice translation group $\mathbb{Z}\times\mathbb{Z}$.

REFERENCES

[1] L. Onsager. The Ising model in two dimensions. In R.E. Mills, E. Ascher, and R.I. Jaffee, editors, *Critical Phenomena in Alloys, Magnets and Superconductors*, New York, (1971). MacGraw-Hill.

[2] L. Onsager. Phys. Rev. **65** (1944), pp. 117–149.

[3] M.J. Stephen and L. Mittag. J. Math. Phys. **13** (1972), pp. 1944–1951.

[4] R.J. Baxter and Enting, 399th *Solution of the Ising Model.* J. Phys. **A11** (1978), pp. 2463–2473.

[5] P.W. Kasteleyn. Exactly solvable lattice models. In *Proc. of the 1974 Wageningen Summer School: Fundamental problems in statistical mechanics III*, Amsterdam, (1974). North–Holland.

[6] R. J. Baxter. In *Proc. of the 1980 Enschede Summer School: Fundamental problems in statistical mechanics V*, Amsterdam, (1981). North–Holland.

[7] R.J. Baxter. *Exactly solved models in statistical mechanics.* London Acad. Press, (1981).

[8] V.G. Drinfel'd. Quantum groups. In *Proceedings of the International Congress of Mathematicians.* Berkeley, (1986).

[9] O. Babelon and C-M. Viallet. *Integrable Models, Yang-Baxter Equation, and Quantum Groups, Part I.* preprint SISSA 54 EP, (1989).

[10] V.F.R Jones. *Baxterization.* preprint CMA-R23-89, (1989).

[11] L.D. Faddeev. Integrable models in $1+1$ dimensional quantum field theory. In *Les Houches Lectures (1982)*, Amsterdam, (1984). Elsevier.

[12] M. Gaudin. *La fonction d'onde de Bethe.* Collection du C.E.A. Série Scientifique. Masson, Paris, (1983).

[13] L.D. Faddeev. Quantum groups. In *Tel-Aviv Landau conference*, (1990).

[14] J. M. Maillard and R. Rammal, *Some analytical consequences of the inverse relation for the Potts model.* J. Phys. **A16** (1983), p. 353.

[15] P. Lochak and J. M. Maillard, *Necessary versus sufficient conditions for exact solubility.* Journ. Math. Phys. **27** (1986), p. 593.

[16] R.J. Baxter, *Solvable eight-vertex model on an arbitrary planar lattice.* Phil. Trans. R. Soc. London **289** (1978), p. 315.

[17] H. Au-Yang and J.H.H Perk, *Critical correlations in a $\mathbb{Z}$ invariant inhomogeneous Ising model.* Physica **A144** (1987), pp. 44–104.

[18] R.J. Baxter, *On Zamolodchikov's Solution of the Tetrahedron Equations.* Comm. Math. Phys. **88** (1983), pp. 185–205.

[19] A.B. Zamolodchikov, *Tetrahedron equations and the relativistic S-matrix of straight-strings in 2+1 dimensions.* Comm. Math. Phys. **79** (1981), p. 489.

[20] M. T. Jaekel and J. M. Maillard, *Symmetry relations in exactly soluble models.* J. Phys. **A15** (1982), pp. 1309–1325.

[21] R.J. Baxter, *The Yang-Baxter equations and the Zamolodchikov model.* Physica **D18** (1986), pp. 321–347.

[22] D. Hansel, J. M. Maillard, J. Oitmaa, and M. Velgakis, *Analytical properties of the anisotropic cubic Ising model.* J. Stat. Phys. **48** (1987), p. 69.

[23] J. M. Maillard, *Automorphism of algebraic varieties and Yang–Baxter equations.* Journ. Math. Phys. **27** (1986), p. 2776.

[24] M.P. Bellon, J-M. Maillard, and C-M. Viallet, *Integrable Coxeter Groups.* Physics Letters **A 159** (1991), pp. 221–232.

[25] H. Au-Yang, B.M. Mc Coy, J.H.H. Perk, S. Tang, and M.L. Yan, *Commuting transfer matrices in the chiral Potts models: solutions of the star–triangle equations with genus* ≥ 1. Phys. Lett. **A123** (1987), p. 219.

[26] B.M. Mc Coy, J.H.H. Perk, S. Tang, and C-H. Sah, *Commuting transfer matrices for the four–state self–dual chiral Potts model with genus three uniformizing Fermat curve.* Phys. Lett. **A125** (1987), p. 9.

[27] R.J. Baxter, J.H.H. Perk, and H. Au-Yang, *New solutions of the star–triangle relations for the chiral Potts model.* Phys. Lett. **A128** (1988), p. 138.

[28] D. Hansel and J. M. Maillard, *Symmetries of models with genus* > 1. Phys. Lett. A **133** (1988), p. 11.

[29] J. Avan, J-M. Maillard, M. Talon, and C-M. Viallet, *Algebraic varieties for the chiral Potts model.* Int. Journ. Mod Phys. **B4** (1990), pp. 1743–1762.

[30] M.P. Bellon, J-M. Maillard, and C-M. Viallet, *Infinite Discrete Symmetry Group for the Yang-Baxter Equations: Vertex Models.* Phys. Lett. B **260** (1991), pp. 87–100.

[31] Y.G. Stroganov. Phys. Lett. **A74** (1979), p. 116.

[32] R.J. Baxter, *The Inversion Relation Method for Some Two-dimensional Exactly Solved Models in Lattice Statistics.* J. Stat. Phys. **28** (1982), pp. 1–41.

[33] M. T. Jaekel and J. M. Maillard, *Inverse functional relations on the Potts model.* J. Phys. **A15** (1982), p. 2241.

[34] J. M. Maillard, *The inversion relation : some simple examples (I) and (II).* Journal de Physique **46** (1984), p. 329.

[35] J. M. Maillard and T. Garel, *Towards an exhaustive classification of the star triangle relation (I) and (II).* J. Phys. **A17** (1984), p. 1251.

[36] E. Abe. *Hopf Algebras.* Cambridge University Press, Cambridge, (1977).

[37] O. Babelon. Integrable systems associated to the lattice version of the Virasoro algebra. I: the classical open chain. In *Integrable systems and quantum groups*, March 1990. Pavia.

[38] J.L. Verdier, *Groupes quantiques [d'après V.G. Drinfel'd].* Séminaire Bourbaki, 39ème année **685** (1987), pp. 1–15.

[39] M. Jimbo, *A q–difference analogue of* $U(\mathcal{G})$ *and the Yang–Baxter equation.* Lett. Math. Phys. **10** (1985), p. 63.

[40] E.H. Lieb and F.Y. Wu. Two dimensional ferroelectric models. In C. Domb and M.S. Green, editors, *Phase Transitions and Critical Phenomena*, volume 1, pages 331–490, New York, (1972). Academic Press.

[41] S.L. Woronowicz, *Twisted* $SU(2)$ *Group. An Example of Noncommutative Differential Calculus.* Publ. Res. Inst. Math. Sci. **23** (1987), pp. 117–181.

[42] V. G. Kac. *Infinite dimensional Lie algebras.* Cambridge University Press, second edition, (1985).

[43] J.-M. Maillard. The star–triangle relation and the inversion relation in statistical mechanics. In *Brasov International Summer School on Critical Phenomena, Theoretical Aspects*, (1983).

[44] M.P. Bellon, J.-M. Maillard, and C-M. Viallet. *Higher Dimensional Mappings with Exact Properties.* Phys. Lett. A159 (1991) p. 233.

[45] M.P. Bellon, J.-M. Maillard, and C-M. Viallet. *Mappings for the* $\mathbb{Z}_N$ *chiral Potts Model.* in preparation.

[46] I.R. Shafarevich. *Basic algebraic geometry.* Springer study. Springer, Berlin, (1977). page 216.

[47] M.P. Bellon, J.-M. Maillard, and C-M. Viallet. *Matrix Patterns for Integrability.* in preparation.

[48] R.J. Baxter, *Partition Function of the Eight-Vertex Lattice Model.* Ann. Phys. **70** (1972), p. 193.

[49] J.E. Humphreys. *Reflection Groups and Coxeter Groups.* Cambridge University Press, Cambridge, (1990).

[50] M.P. Bellon, J.-M. Maillard, and C-M. Viallet, *Higher dimensional mappings.* Physics Letters **A 159** (1991), pp. 233–244.

[51] M.P. Bellon, J.-M. Maillard, and C-M. Viallet. *Infinite Discrete Group of Symmetries for the Sixteen Vertex Model.* Phys. Lett. B281 (1992), p. 315.

[52] M.P. Bellon, J.-M. Maillard, and C-M. Viallet. *Beyond q-analogs.* in preparation.

[53] J. Avan. *Classical R-matrices with general automorphism groups.* Preprint Brown HET 781, (1991).

[54] M.P. Bellon, J-M. Maillard, and C-M. Viallet, *Infinite Discrete Symmetry Group for the Yang-Baxter Equations: Spin models.* Physics Letters **A 157** (1991), pp. 343–353.

[55] V.A. Fateev and A.B. Zamolodchikov, *Self-dual solutions of the star-triangle relations in $\mathbb{Z}_N$ models.* Phys. Lett. **A92** (1982), p. 37.

[56] V.V. Bazhanov and Yu.G. Stroganoff. J. Stat. Phys. **59** (1990), p. 799.

[57] V.V. Bazhanov and R.M. Kashaev. *Cyclic L-opreators related with a 3-state R matrix.* preprint RIMS-702, Kyoto (to appear in Comm. Math. Phys.), (1990).

[58] V.V. Bazhanov, R.M. Kashaev, V.V. Mangazeev, and Yu.G. Stroganoff. $Z_N^{\times\, n-1}$ *Generalization of the chiral Potts model.* preprint IHEP-90-137, Protvino (to appear in Comm. Math. Phys.), (1990).

[59] E. Date, M. Jimbo, K. Miki, and T. Miwa. *New R matrices associated with cyclic representations of $U_q(A_2^{(2)})$.* preprint RIMS-706, July 1990.

[60] E. Date, M. Jimbo, K. Miki, and T. Miwa. *Generalized chiral Potts model and minimal cyclic representations of $U_q(\widehat{gl}(n,C)$.* preprint RIMS-715, (1990).

[61] M.P. Bellon, J-M. Maillard, and C-M. Viallet, *Rational Mappings, Arborescent Iterations, and the Symmetries of Integrability.* Physical Review Letters **67** (1991), pp. 1373–1376.

[62] A. Gaaf and J. Hijmans, *Symmetry relations in the sixteen-vertex model.* Physica **A80** (1975), pp. 149–171.

[63] M.P. Bellon, J.-M. Maillard, and C-M. Viallet. *Mappings In Higher Dimensions.* Preprint PAR-LPTHE-91-31.

[64] M.P. Bellon, J.-M. Maillard, and C-M. Viallet. *A No-Go Theorem for the Tetrahedron Equations.* in preparation.

Laboratoire de Physique Théorique et des Hautes Energies
Université Paris VI, Tour 16, 1er étage
4 Place Jussieu
F–75252 Paris Cedex 05
France

PART IV
Topological Field Theory

Integrable Systems and Classification of 2-Dimensional Topological Field Theories

B. Dubrovin

Abstract. In this paper we consider the so-called WDVV equations from the point of view of differential geometry and of the theory of integrable systems as defining relations of 2-dimensional topological field theory. A complete classification of massive topological conformal field theories (TCFT) is obtained in terms of the monodromy data of an auxillary linear operator with rational coefficients. The procedure of coupling a TCFT to topological gravity is described (at tree level) via certain integrable bihamiltonian hierarchies of hydrodynamic type and their τ-functions. A possible role for the bihamiltonian formalism in the calculation of higher genus corrections is discussed. As a biproduct of this discussion, new examples of infinite dimensional Virasoro-type Lie algebras and their nonlinear analogues are constructed. As an algebro-geometrical application it is shown that WDVV is just the universal system of integrable differential equations (higher order analogue of the Painlevé-VI equation) specifying the periods of the Abelian differentials on Riemann surfaces as functions on moduli of these surfaces.

This paper is an extended version of the talk given at the Conference on Integrable Systems (Luminy, July 1991) and further results [39] have been included. I dedicate it to the memory of J.-L. Verdier.

Introduction

A quantum field theory (QFT) on a D-dimensional manifold M consists of:

1) a family of local fields $\phi_\alpha(x)$, $x \in M$ (functions or sections of a fiber bundle over M). A metric $g_{ij}(x)$ on M usually is one of the fields (the gravity field).

2) A Lagrangian $L = L(\phi, \phi_x, \dots)$. Classical field theory is determined by the Euler–Lagrange equations

$$\frac{\delta S}{\delta \phi_\alpha(x)} = 0, \quad S[\phi] = \int L(\phi, \phi_x, \dots).$$

3) A procedure of quantization is usually based on the construction of an appropriate path integration measure $[d\phi]$. The partition function is a result of the path integration over the space of all fields $\phi(x)$

$$Z_M = \int [d\phi] e^{-S[\phi]}.$$

Correlation functions (non normalized) are defined by a similar path integral

$$< \phi_\alpha(x)\phi_\beta(y)\ldots >_M = \int [d\phi]\phi_\alpha(x)\phi_\beta(y)\ldots e^{-S[\phi]}.$$

Since the path integration measure is almost never well-defined (and also taking into account the fact that different Lagrangians could give equivalent QFT), an old idea was to construct a self-consistent QFT by solving a system of differential equations for correlation functions. These equations were scrutinized in 2D conformal field theories where D=2 and Lagrangians are invariant with respect to conformal transformations

$$\delta g_{ij}(x) = \epsilon g_{ij}(x), \quad \delta S = 0.$$

This theory is still far from complete because of the complexity (and, probably, nonintegrability) of the differential equations determining correlators.

Here I will consider another class of solvable QFT: topological field theories. These theories admit *topological invariance.* They are invariant with respect to an arbitrary change of the metric $g_{ij}(x)$ on M

$$\delta g_{ij}(x) = \text{arbitrary}, \quad \delta S = 0.$$

On the quantum level that means that the partition function Z_M depends only on the topology of M. All the correlation functions also are topological creatures; they depend only on the labels of the operators and on the topology of M but not on the positions of the operators

$$< \phi_\alpha(x)\phi_\beta(y)\ldots >_M \equiv < \phi_\alpha\phi_\beta\cdots >_M .$$

The simplest example is 2D gravity with the Hilbert – Einstein action

$$S = \int R\sqrt{g}d^2x = \text{Euler characteristic of } M.$$

There are two ways of quantizing this functional. The first one is based on an appropriate discrete version of the model ($M \to$ polyhedron). This leads to a consideration of matrix integrals of the form [55]

$$Z_N(t) = \int_{X^*=X} \exp\{-\mathrm{tr}(X^2 + t_1X^4 + t_2X^6 + \ldots)\}dX,$$

where the integral is taken over the space of all $N \times N$ Hermitian matrices X. Here t_1, $t_2, \ldots$, are called *coupling constants.* A solution of 2D gravity [1] is based on the observation that after an appropriate limiting procedure $N \to \infty$, and a renormalization of t, the limiting partition function coincides with the τ-function of the KdV hierarchy.

The second approach to 2D gravity is based on an appropriate supersymmetric extension of the Hilbert – Einstein Lagrangian [2]. This reduces the path integral over the space of all metrics $g_{ij}(x)$ on a surface M of the given genus g to an integral over the finite-dimensional space of conformal classes of these metrics, i.e., over the moduli space $\mathcal{M}_g$ of Riemann surfaces of genus g. The correlation functions of the model are expressed via intersection numbers of certain cycles on the moduli space [2-4, 46, 49]

$$\phi_\alpha \leftrightarrow c_\alpha \in H_*(\mathcal{M}_g), \quad \alpha \in \mathbf{N}$$

$$< \phi_\alpha \phi_\beta \ldots >_g = \#(c_\alpha \cap c_\beta \cap \ldots)$$

here the subscript g designates the correlators on a surface of genus g. This approach is often called *cohomological field theory.*

It was conjectured by Witten that both approaches to 2D quantum gravity should give the same results. This conjecture was proved by Kontsevich [42-43], who showed that the generating function

$$F(t) = \sum_{g,n} \sum_{\alpha_1,\ldots,\alpha_n} \frac{t_{\alpha_1} \ldots t_{\alpha_n}}{n!} < \phi_{\alpha_1} \ldots \phi_{\alpha_n} >_g,$$

the free energy of 2D gravity, is a logarithm of the τ-function of a solution of the KdV hierarchy. This was the original form of Witten's conjecture. The τ-function is specified by the string equation (see eq. (3.16b) below).

Other examples of 2D TFT constructed in [2-6, 8-9, 46-49, 57] proved that they could have important mathematical applications, and may be the best tool for treating sophisticated topological objects. In these examples correlators can be expressed via intersection numbers on moduli spaces or their coverings [48] of holomorphic maps from Riemann surfaces to a complex or even almost complex variety (*topological sigma-models* [2]) or via the intersection-form of a singularity in catastrophe theory (*topological Landau – Ginsburg models* [9, 7]; see also [10]). (We shall not discuss here the interesting relations between these models.) This gives rise to the following

Problems. What could be the intrinsic origin of integrability in 2D TFT? How can one classify 2D TFT? Is it possible to find an analogue of the KdV hierarchy in order to calculate the partition function of a given TFT model?

In this paper an approach to these problems is proposed based on differential geometry and on the theory of classical integrable systems of the KdV type. The main ingredient of my approach is the Hamiltonian formalism of integrable hierarchies of KdV type (see, e.g., [21, 25, 29]) and, especially, the Hamiltonian analysis of the semi-classical limits of these systems [18-21].

Let me start by considering the *matter sector* of a 2D topological field theory. That means that the set of local fields $\phi_1(x), \ldots, \phi_n(x)$ (the so-called *primary fields* of the model) does not contain the metric. (Afterwards one should integrate over the space of metrics. This should give rise to a procedure of *coupling to topological gravity* that will be described below.) Then the correlators of the fields $\phi_1(x), \ldots, \phi_n(x)$ obey very simple algebraic axioms (a consequence [51] of Atiyah's general axioms [50] of topological field theory).

Let

$$\eta_{\alpha\beta} = < \phi_\alpha \phi_\beta >_0$$

(0 means genus zero correlator),

$$c_{\alpha\beta\gamma} = < \phi_\alpha \phi_\beta \phi_\gamma >_0 .$$

Then

1) These tensors are symmetric and $\det(\eta_{\alpha\beta}) \neq 0$. I will use the tensor $\eta_{\alpha\beta}$ and the inverse $(\eta^{\alpha\beta}) = (\eta_{\alpha\beta})^{-1}$ for lowering and raising indices.

2) $c_{\alpha\beta}^\gamma = \eta^{\gamma\epsilon} c_{\alpha\beta\epsilon}$ is the tensor of structure constants of a commutative associative algebra A with a unity. That means that, for a basis $e_1, \ldots, e_n$ in A, the multiplication law has the form

$$e_\alpha e_\beta = c_{\alpha\beta}^\gamma e_\gamma.$$

(We will normalise a basis in such a way that e_1 = the unity of A. So $c_{1\alpha}^\beta = \delta_\alpha^\beta$.)

3) Let $H = \eta^{\alpha\beta} e_\alpha e_\beta \in A$. Then, for correlators of genus g, the following formula holds

$$< \phi_\alpha \ldots \phi_\gamma >_g = < e_\alpha \ldots e_\gamma, H^g > .$$

In this way

Topologically invariant Lagrangian → correlators of local physical fields

we lose too much relevant information. To capture more information about a topological Lagrangian, we will consider a topological field theory, together with its deformations preserving topological invariance

$$L \to L + \sum t^\alpha L_\alpha^{(pert)}$$

(t^α are coupling constants). Here we use the ideas and results of [8, 51]. In these papers a general construction of a class of 2D TFT by the twisting of $N = 2$ superconformal field theories was proposed. The so-called *topological conformal field theories* (TCFT) are obtained by this procedure. For any TCFT with n local observables (primary operators) a *canonical* n-parameter deformation preserving topological invariance was constructed. All the correlators of the primary fields $\phi_1, \dots, \phi_n$ in the perturbed TCFT now depend on coupling parameters $t_1, \dots, t_n$. This dependence is not arbitrary but obeys the following equations:

1) $\eta_{\alpha\beta} \equiv \text{const in } t \ (t^1, \dots, t^n)$

2) $c^{\alpha}_{1\beta} \equiv \delta^{\alpha}_{\beta}$

3) $c_{\alpha\beta\gamma} = \frac{\partial^3 F(t)}{\partial t^\alpha \partial t^\beta \partial t^\gamma}$ for some function $F(t)$ (primary free energy).

Equations of associativity give a system of nonlinear PDE for $F(t)$

$$\frac{\partial^3 F(t)}{\partial t^\alpha \partial t^\beta \partial t^\lambda} \eta^{\lambda\mu} \frac{\partial^3 F(t)}{\partial t^\mu \partial t^\gamma \partial t^\sigma} = \frac{\partial^3 F(t)}{\partial t^\alpha \partial t^\gamma \partial t^\lambda} \eta^{\lambda\mu} \frac{\partial^3 F(t)}{\partial t^\mu \partial t^\beta \partial t^\sigma} \tag{0.1}$$

with the constraint

$$\frac{\partial^3 F(t)}{\partial t^1 \partial t^\alpha \partial t^\beta} = \eta_{\alpha\beta}. \tag{0.2}$$

These equations were called in [39] *the Witten – Dijkgraaf – E. Verlinde – H. Verlinde* (WDVV) equations. In fact in TCFT one should assume invariance of a solution with respect to scaling transformations of the form

$$t^\alpha \mapsto c^{1-q_\alpha} t^\alpha$$

$$\eta_{\alpha\beta} \mapsto c^{q_\alpha + q_\beta - d} \eta_{\alpha\beta}$$

$$c^{\gamma}_{\alpha\beta} \mapsto c^{q_\alpha + q_\beta - q_\gamma} c^{\gamma}_{\alpha\beta}$$

for some numbers q_α (*charges* of the fields ϕ_α) and d (*dimension* of the model).

My program now is:

1. To classify 2D TFT as solutions of WDVV equations, and

2. For any solution of WDVV (I recall that this describes the matter sector of a TFT model) to construct (i.e., to calculate the partition function and correlators) the complete TFT model (coupling of the given matter sector to topological gravity).

Problem 1 was investigated in [39]. A Lax pair for the WDVV equations was constructed. For the so-called *massive* TCFT models, where the algebra $e_\alpha e_\beta = c^{\gamma}_{\alpha\beta}(t) e_\gamma$, for almost all t, has no nilpotents, it was shown that solutions of the WDVV equations form a $(\frac{n(n-1)}{2} + 1)$-dimensional family (where n is the number of primaries). It turns out that the WDVV equations for the massive case are equivalent to the equations of isomonodromy deformations of an ordinary linear differential operator with rational coefficients. These isomonodromy deformation equations coincide with the Painlevé-VI equation (for $n = 3$) and with higher order analogues of the Painlevé-VI for $n > 3$. Monodromy data (i.e., Stokes matrices) of the operator with rational coefficients serve as parameters of massive TCFT-models.

Concerning the second problem my conjecture is that the set of solutions of WDVV parametrizes a large class of hierarchies of (1+1)-integrable systems. All the well-known hierarchies are in this class but they are only isolated points in it.

The basic idea behind the construction of these integrable hierarchies comes from the Feynmann diagram expansion machinery which is standard in the quantum field theory. In 2D QFT it becomes a representation of the partition function and correlators as a sum of contributions from surfaces of different genera g (*genus expansion*). The idea is that the genus expansion of the partition function should coincide with the small dispersion expansion (see below) of the τ-function of some integrable hierarchy.

A first step in this direction has been taken in [39]: for any solution of WDVV, a hierarchy of integrable Hamiltonian equations of hydrodynamic type was constructed such that the τ-function of a particular solution of this hierarchy coincides with the genus zero approximation of the corresponding TFT model coupled to gravity (see Section 3 below).

From the point of view of the WDVV equations, the hierarchy determines a family of symmetries of the equation (0.1) (see below, Proposition 3.1). To go further one should solve a non-standard "inverse problem": to reconstruct an integrable hierarchy from its zero-dispersion limit. Some examples of such a reconstruction are discussed in Sections 3, 4. Probably, the bihamiltonian formalism could be useful in order to complete the solution of this problem.

Examples of solutions of WDVV and of the corresponding integrable hierarchies are described in Section 4. Almost all the known examples are obtained as a result of analysis of the semiclassical (particularly, dispersionless) limiting procedure in integrable hierarchies of KdV type. More precisely, let

$$\partial_{t_k} y^a = f_k^a(y, \partial_x y, \partial_x^2 y, \ldots),\ a = 1, \ldots, l,\ k = 0, 1, \ldots$$

be a commutative hierarchy of Hamiltonian integrable systems of the KdV type. "Hierarchy" means that the systems are ordered, say, by action of a recursion operator. The number of recursions determines the level of a system in the hierarchy. Systems of the level zero form a primary part of the hierarchy (these correspond to the primary operators in TFT); others can be obtained from the primaries by recursions. The hierarchy posesses a rich family of finite-dimensional invariant manifolds. Some of them can be found in a straightforward way; one needs to apply sophisticated algebraic geometry methods [28] to construct a wider class of invariant manifolds. Any of these manifolds, after an extension to the complex domain, turns out to be fibered over some base M (a complex manifold of some dimension n) with m-dimensional tori as the fibers (common invariant tori of the hierarchy). For $m = 0$, M is nothing but the family of common stationary points of the hierarchy. For $m > 0$, M is a moduli space of Riemann surfaces of some genus g with certain additional structures: marked points, marked meromorphic functions, etc. These are the families of finite-gap ("g-gap") solutions of the hierarchy. The main observation is that any such M determines a solution of the WDVV equation (M is a Frobenius manifold in the terminology of Section 1 below, or the "small phase space" of a TFT theory, in the terminology of [3-4, 46]). For $m = 0$ and the set M of stationary points of the Gelfand – Dickey hierarchy this essentially follows from [8, 11]; for the general case (including arbitrary genera g) a construction of a solution of WDVV was given in [13-14] (see also the recent preprint [44]).

To give an idea of how an integrable Hamiltonian hierarchy of the above form induces tensors $c_{\alpha\beta}^{\gamma}$, $\eta_{\alpha\beta}$ on a finite dimensional invariant manifold M, I need to introduce the notion of the semiclassical limit of a hierarchy near a family M of invariant tori (sometimes it is called also a *dispersionless limit* or *Whitham averaging* of the hierarchy; see details in [15-21]). In the simplest case of the family of stationary solutions the semiclassical limit is defined as follows: one should substitute in the equations of the hierarchy

$$x \mapsto \epsilon x = X, \ t_k \mapsto \epsilon t_k = T_k$$

and let ϵ tend to zero. For more general M (family of invariant tori) one should add averaging over the tori. As a result one obtains a new integrable Hamiltonian hierarchy where the dependent variables are coordinates v^1, ..., v^n on M and the independent variables are the slow variables X and T_0, T_1, ... This new hierarchy always has the form of a quasilinear system of PDE of the first order

$$\partial_{T_k} v^p = c_{k}{}_{q}^{p}(v)\partial_X v^q, \ k = 0, 1, \dots$$

for some matrices of coefficients $c_{k}{}_{q}^{p}(v)$. One can keep in mind the simplest example of the semiclassical limit (just the dispersionless limit) of the KdV hierarchy. Here M is the one-dimensional family of constant solutions of the KdV hierarchy. For example, rescaling the KdV equation, one obtains

$$u_T = uu_X + \epsilon^2 u_{XXX}$$

(KdV with small dispersion). After $\epsilon \to 0$ one obtains

$$u_T = uu_X.$$

The semiclassical limit of each equation in the KdV hierarchy has the form

$$\partial_{T_k} u = \frac{u^k}{k!}\partial_X u, \; k = 0, 1, \ldots.$$

A semiclassical limit of spatially discrete hierarchies (like the Toda system) is obtained in a similar way. It still is a system of quasilinear PDEs of the first order.

Let us come back to the determination of tensors $\eta_{\alpha\beta}$, $c^\gamma_{\alpha\beta}$ on M. To introduce $\eta_{\alpha\beta}$ we need to use the Hamiltonian structure of the original hierarchy. A semiclassical limit (or "averaging") of this Hamiltonian structure in the sense of the general construction of S.P. Novikov and the author induces a Hamiltonian structure of the semiclassical hierarchy: a Poisson bracket of the form

$$\{v^p(X), v^q(Y)\}_{\text{semiclassical}} = g^{ps}(v(X))[\delta^q_s \partial_X \delta(X-Y) - \Gamma^q_{sr}(v) v^r_X \delta(X-Y)]$$

where $g^{pq}(v)$ are the contravariant components of a metric on M and $\Gamma^q_{pr}(v)$ are the Christoffel symbols of the Levi-Cività connection for $g^{pq}(v)$ (the so-called *Poisson brackets of hydrodynamic type*). (Strictly speaking the metric and the connection are defined on a real part of M that parametrizes smooth solutions of the original hierarchy with some reality constraints. But the formulae for the metric and the connection admit an extension onto all M.) From the general theory of Poisson brackets of hydrodynamic type [18-21] one concludes that the metric $g^{pq}(v)$ on M should have zero curvature. So local flat coordinates $t^1, \ldots, t^n$ on M exist such that the metric in these coordinates is constant

$$\frac{\partial t^\alpha}{\partial v^p}\frac{\partial t^\beta}{\partial v^q} g^{pq}(v) = \eta^{\alpha\beta} = \text{const}.$$

The Poisson bracket $\{\;,\;\}_{\text{semiclassical}}$ in these coordinates has the form

$$\{t^\alpha(X), t^\beta(Y)\}_{\text{semiclassical}} = \eta^{\alpha\beta}\delta'(X-Y).$$

The tensor $(\eta_{\alpha\beta}) = (\eta^{\alpha\beta})^{-1}$ together with the flat coordinates t^α is the first part of a structure we want to construct. (The flat coordinates t^1, ..., t^n can be expressed via Casimirs of the original Poisson bracket and action variables and wave numbers along the invariant tori; see details in [18-21].)

To define a tensor $c^\gamma_{\alpha\beta}(t)$ on M (or, equivalently, the "primary free energy" $F(t)$) we need to use a semiclassical limit of the τ-function of the original hierarchy [11, 53-54, 61]

$$\log \tau_{\text{semiclassical}}(T_0, T_1, \ldots) = \lim_{\epsilon \to 0} \epsilon^{-2} \log \tau(\epsilon t_0, \epsilon t_1, \ldots).$$

Then

$$F = \log \tau_{\text{semiclassical}}$$

where $\tau_{\text{semiclassical}}$ should be considered as a function only of n of the slow variables of the same level. (To satisfy the normalization (0.2), one should properly choose these n slow variables and normalize the values of the others. This specifies uniquely the semiclassical τ-function.) The semiclassical τ-function as the function of all the slow variables coincides with the tree-level partition function of the matter sector $\eta_{\alpha\beta}$, $c_{\alpha\beta}^{\gamma}$ coupled to topological gravity.

Summarizing, we can say that a structure of a Frobenius manifold (i.e., a solution of WDVV) on an invariant manifold M of an integrable Hamiltonian hierarchy is induced by a semiclassical limit of the Poisson bracket of the hierarchy and of the τ-function of the hierarchy. So the above conjecture can be reformulated as follows: the WDVV equations just specify the semiclassical limits of τ-functions of Hamiltonian integrable hierarchies.

I shall not consider in this paper another type of integrable system involved in 2D TFT: the so-called equations of *topological-antitopological fusion* proposed in [40]. These equations describe the ground state metric on a given 2D TFT model. See [45] concerning the theory of integrability of these equations. An interesting relation of these equations with the theory of harmonic maps was also found in [45].

1. Geometry of Frobenius manifolds

I recall that A is called a Frobenius algebra (over **R** or **C**) if it is a commutative associative algebra with a unity and with a nondegenerate invariant inner product

$$< ab, c >=< a, bc > . \tag{1.1}$$

If e is the unity of A then the invariant inner product on A can be written in the form

$$< a, b >= \omega_e(ab) \tag{1.2a}$$

where

$$\omega_e(a) =< e, a > . \tag{1.2b}$$

Moreover, for any linear functional $\omega \in A^*$ the inner product

$$< a, b >_\omega = \omega(ab) \tag{1.3}$$

is invariant. It is nondegenerate for generic ω. Any invariant inner product on a finite-dimensional Frobenius algebra A (only finite-dimensional algebras will be considered) can be represented in the form (1.3).

If e_i, $i = 1, \ldots, n$, is a basis in A, then the structure of a Frobenius algebra on $A$is specified by the coefficients η_{ij}, c_{ij}^k, where

$$< e_i, e_j >= \eta_{ij} \tag{1.4a}$$

$$e_i e_j = c_{ij}^k e_k \tag{1.4b}$$

(summation over repeated indices will be assumed). The matrix η_{ij} and the structure constants c_{ij}^k satisfy the following conditions:

$$\eta_{ji} = \eta_{ij}, \quad \det(\eta_{ij}) \neq 0 \tag{1.5a}$$

$$c_{ij}^s c_{sk}^l = c_{is}^l c_{jk}^s \tag{1.5b}$$

(associativity),

$$c_{ijk} = \eta_{is} c_{jk}^s = c_{jik} = c_{ikj} \tag{1.5c}$$

(commutativity and invariance of the inner product). If $e = (e^i)$ is the unity of A then

$$e^s c_{sj}^i = \delta_j^i \tag{1.5d}$$

(the Kronecker delta).

One-dimensional Frobenius algebras are parametrized by one number (length of the unity). Any semisimple n-dimensional Frobenius algebra is isomorphic to the direct sum of n one-dimensional Frobenius algebras

$$f_i f_j = \delta_{ij} f_i, \quad < f_i, f_j >= \eta_{ii} \delta_{ij}. \tag{1.6}$$

Moreover, any Frobenius algebra without nilpotents is semisimple.

Let us consider a particular class of deformations of Frobenius algebras.

Definition 1.1. A manifold M is called *quasi-Frobenius* if it is equipped with three tensors, $c = (c_{ij}^k(x))$, $\eta = (\eta_{ij}(x))$, $e = (e^i(x))$, satisfying (1.5) for any $x \in M$.

In other words, these three tensors define the structure of a Frobenius algebra in the space of smooth vector fields $Vect(M)$ over the ring $\mathcal{F}(M)$ of smooth functions on M:

$$[v \cdot w]^k(x) = c_{ij}^k(x) v^i(x) w^j(x), \tag{1.7a}$$

$$< v, w > (x) = \eta_{ij}(x) v^i(x) w^j(x) \tag{1.7b}$$

for any v, $w \in Vect(M)$.

Complex quasi-Frobenius manifolds are defined in a similar way but the tensors c, η, e should be holomorphic. They define the structure of a Frobenius algebra in the space of holomorphic vector fields over the ring of holomorphic functions.

Informally speaking, n-dimensional quasi-Frobenius manifolds are n-parameter deformations of n-dimensional Frobenius algebras. For any $x \in M$, the tangent space T_xM is a Frobenius algebra with the structure constants $c_{ij}^k(x)$, invariant inner product $\eta_{ij}(x)$, and unity $e^i(x)$.

As was explained above, in physical applications there are additional restrictions on quasi-Frobenius manifolds.

Definition 1.2. A quasi-Frobenius M is called a *Frobenius manifold* if the invariant metric

$$ds^2 = \eta_{ij}(x)dx^i dx^j \tag{1.8a}$$

is flat, the unity vector field e is covariantly constant

$$\nabla e = 0 \tag{1.8b}$$

(here ∇ is the Levi-Cività connection associated with the metric ds^2), and the tensor

$$(\nabla_z c) < u, v, w > \tag{1.8c}$$

is a symmetric tensor in the vectors $u, \ldots, z$.

Locally. Frobenius manifolds are in 1-1 correspondence with solutions of WDVV equations (i.e., with 2D TFTs). Indeed, for the flat metric (1.8a), locally flat coordinates t^α exist such that the metric is constant in these coordinates, $ds^2 = \eta_{\alpha\beta}dt^\alpha dt^\beta$, $\eta_{\alpha\beta} = \text{const}$. The covariantly constant vector field e in the flat coordinates has constant components; using a linear change of coordinates, one can obtain $e^\alpha = \delta_1^\alpha$. The tensor $c_{\alpha\beta\gamma}(t)$ in these coordinates satisfies the condition

$$\partial_\delta c_{\alpha\beta\gamma} = \partial_\gamma c_{\alpha\beta\delta}. \tag{1.9a}$$

This means that $c_{\alpha\beta\gamma}(t)$ can be represented in the form

$$c_{\alpha\beta\gamma}(t) = \partial_\alpha \partial_\beta \partial_\gamma F(t) \tag{1.9b}$$

for some function $F(t)$ satisfying the WDVV equations.

The first step in solving WDVV is to obtain a "Lax pair" for these equations. The most convenient way is to represent them as the compatibility conditions of an overdetermined linear system depending on a spectral parameter λ.

Proposition 1.1. *A quasi-Frobenius manifold is Frobenius iff the unity e is covariantly constant and the pencil of connections*

$$\tilde{\nabla}_u(\lambda)v = \nabla_u v + \lambda u \cdot v, \quad u,\ v \in Vect(M) \tag{1.10}$$

is flat identically in λ.

Flatness of the pencil of connections (1.10) is equivalent to the flatness of the metric η and to the equation

$$\nabla_u(v \cdot w) - \nabla_v(u \cdot w) + u \cdot \nabla_v w - v \cdot \nabla_u w = [u, v] \cdot w \tag{1.11}$$

for any three vector fields u, v, w. Here $[u, v]$ means the commutator of the vector fields. This equation is equivalent to the symmetry of the tensor (1.8c).

Corollary. *WDVV is an integrable system.*

Indeed, WDVV is equivalent to the compatibility of the following linear system

$$\tilde{\nabla}_\alpha(\lambda)\xi = 0, \quad \alpha = 1, ..., n, \tag{1.12a}$$

(here ξ is a covector field), or, equivalently, in the flat coordinates t^α,

$$\partial_\alpha \xi_\beta = \lambda c_{\alpha\beta}^\gamma(t)\xi_\gamma. \tag{1.12b}$$

The compatibility of the system (1.12) (identically in the spectral parameter λ) together with the symmetry of the tensor $c_{\alpha\beta\gamma} = \eta_{\alpha\epsilon}c_{\beta\gamma}^\epsilon$ is equivalent to WDVV.

It turns out that symmetries of Frobenius manifolds play an important role in the geometrical foundations of TFT. We start with the notion of *algebraic symmetry* of a Frobenius manifold.

Definition 1.3. A diffeomorphism $f : M \to M$ of a Frobenius manifold is called an algebraic symmetry if it preserves the multiplication law of vector fields:

$$f_*(u \cdot v) = f_*(u) \cdot f_*(v) \tag{1.13}$$

(here f_* is the induced linear map $f_* : T_xM \to T_{f(x)}M$).

Proposition 1.2. *The Algebraic symmetries of a Frobenius manifold form a finite-dimensional Lie group $G(M)$.*

The generators of the action of $G(M)$ on M (i.e., the representation of the Lie algebra of $G(M)$ in the Lie algebra of vector fields on M) are the vector fields w such that

$$[w, u \cdot v] = [w, u] \cdot v + [w, v] \cdot u \tag{1.14}$$

for any vector fields u, v.

Note that the group $G(M)$ is always nontrivial: it contains the one-parameter subgroup of shifts along the coordinate t^1. The generator of this subgroup coincides with the unity vector field e.

The group $G(M)$ can be calculated for the important class of *massive* Frobenius manifolds.

Definition 1.4. A Frobenius manifold is called massive if the algebra T_xM is semisimple for any $x \in M$.

In physical language, massive Frobenius manifolds are coupling spaces of massive TFT models.

Main lemma. *The connected component of the identity in the group $G(M)$ of algebraic symmetries of an n-dimensional massive Frobenius manifold is an n-dimensional commutative Lie group that acts locally transitively on M.*

This is a reformulation of the main lemma of [39].

Action of the group of algebraic symmetries provides a new affine structure on a massive Frobenius manifold. The structure tensor c_{ij}^k is constant in this affine structure.

From the main lemma the following statement follows.

Theorem 1.1. [39] *On a massive Frobenius manifold there exist local coordinates $u^1, \ldots, u^n$ such that the multiplication law of vector fields in these coordinates has the form*

$$\partial_i \cdot \partial_j = \delta_{ij}\partial_i, \tag{1.15}$$

where $\partial_i = \partial/\partial u^i$. The invariant metric η in these coordinates has a diagonal form

$$\eta_{\alpha\beta}dt^\alpha dt^\beta = \sum_{i=1}^n \eta_{ii}(u)(du^i)^2 \tag{1.16}$$

satisfying the equations

$$d\left(\sum_{i=1}^n \eta_{ii}(u)du^i\right) = 0, \tag{1.17a}$$

$$\sum_{k=1}^n \partial_k \eta_{ii} = 0. \tag{1.17b}$$

Conversely, for a flat diagonal metric with the properties (1.17) and $t^\alpha = t^\alpha(u)$, $\alpha = 1, \ldots, n$, being the flat coordinates for the metric, the formulae

$$\eta_{\alpha\beta} = \sum_{i=1}^n \eta_{ii}(u)\frac{\partial u^i}{\partial t^\alpha}\frac{\partial u^i}{\partial t^\beta}, \tag{1.18a}$$

$$c_{\alpha\beta}^\gamma(t) = \sum_{i=1}^n \frac{\partial u^i}{\partial t^\alpha}\frac{\partial u^i}{\partial t^\beta}\frac{\partial t^\gamma}{\partial u^i}, \tag{1.18b}$$

$$e^\alpha = \sum_{i=1}^n \frac{\partial t^\alpha}{\partial u^i} \tag{1.18c}$$

determine (locally) a massive Frobenius manifold.

The above coordinates, $u^1, \ldots, u^n$, on a massive Frobenius manifold are determined uniquely up to permutations and shifts. They are called *canonical coordinates* on the massive Frobenius manifold M. The tensor c of structure constants in these coordinates has the following canonical constant form

$$c_{ij}^k = \delta_{ij}\delta_j^k. \tag{1.19}$$

The canonical coordinates u^i can be found by solving an overdetermined system of differential equations

$$\frac{\partial t^\alpha}{\partial u^i}\frac{\partial t^\beta}{\partial u^j}c_{\alpha\beta}^\gamma = \delta_{ij}\frac{\partial t^\gamma}{\partial u^i}.$$

For massive conformally invariant Frobenius manifolds (see the next section) they can be found in a purely algebraic way (see Proposition 2.4 below).

To complete the local classification of massive TFT one has to classify flat diagonal metrics with the properties (1.17). This class of metrics was studied by Darboux [33] and Egoroff. Following Darboux, I will call them *Egoroff metrics.* The vanishing of the curvature of these metrics can be written in the form of the following system of PDE's (*Darboux–Egoroff system*) for the *rotation coefficients*

$$\gamma_{ij}(u) = \frac{\partial_j\sqrt{\eta_{ii}(u)}}{\sqrt{\eta_{jj}(u)}},\ i \neq j \tag{1.20}$$

$$\partial_k\gamma_{ij} = \gamma_{ik}\gamma_{kj},\quad i,\ j,\ k \text{ are distinct,} \tag{1.21a}$$

$$\sum_{k=1}^n \partial_k\gamma_{ij} = 0,\ i \neq j \tag{1.21b}$$

$$\gamma_{ji} = \gamma_{ij}. \tag{1.21c}$$

It is interesting to observe that the same equations (for even n) arise in the calculation [34] of multipoint correlators in an impenetrable Bose-gas, see Appendix to [45].

The integrability of the Darboux–Egoroff system was observed in [23]. It essentially coincides with the "pure imaginary reduction" of the n-wave system (see [26, 25]). This can be represented as the compatibility conditions of the following linear system (depending on a spectral parameter λ)

$$\partial_j\psi_i = \gamma_{ij}\psi_j,\quad i \neq j \tag{1.22a}$$

$$\sum_{k=1}^{n} \partial_k \psi_i = \lambda \psi_i. \tag{1.22b}$$

To complete the local classification of massive Frobenius manifolds one first should apply an appropriate version of the inverse spectral transform (IST) to solve the Darboux–Egoroff system (1.21). Below I will give an example of IST for the important case of self-similar solutions of (1.21) (so called topological conformal field theories [8, 51]). To find the metric (1.16) and the flat coordinates $t^\alpha = t^\alpha(u)$ for a given solution $\gamma_{ij}(u)$ one has to fix a basis $\psi_{i\alpha}(u)$, $\alpha = 1, \ldots, n$, in the space of solutions of the system (1.22) for $\lambda = 0$

$$\partial_j \psi_{i\alpha} = \gamma_{ij} \psi_{j\alpha}, \quad i \neq j, \tag{1.23a}$$

$$\sum_{k} \partial_k \psi_{i\alpha} = 0, \tag{1.23b}$$

$\alpha = 1, \ldots, n$. Then we set

$$\eta_{ii}(u) = \psi_{i1}^2(u), \tag{1.24a}$$

$$\eta_{\alpha\beta} = \sum_{i=1}^{n} \psi_{i\alpha}(u) \psi_{i\beta}(u), \tag{1.24b}$$

$$\frac{\partial t_\alpha}{\partial u^i} = \psi_{i1}(u) \psi_{i\alpha}(u), \tag{1.24c}$$

$$c_{\alpha\beta\gamma}(t(u)) = \sum_{i=1}^{n} \frac{\psi_{i\alpha} \psi_{i\beta} \psi_{i\gamma}}{\psi_{i1}}. \tag{1.24d}$$

These formulae complete the local classification of complex massive Frobenius manifolds. They are parametrized (locally) by n arbitrary functions of one variable (the parametrization of solutions of the Darboux–Egoroff system) and also by n complex parameters because of the ambiguity in the choice of solutions ψ_{i1} in formulae (1.24).

To classify real Frobenius manifolds one should apply IST to various real forms of the Darboux–Egoroff system. We will not do it here (see [27] for discussion of real forms of the system (1.21) in algebraic-geometry IST).

Global topology of massive Frobenius manifolds is rather poor. We say that an n-dimensional manifold M admits an S_n-structure if the structure group of the tangent bundle TM can be reduced to the symmetric group S_n. An atlas of coordinates charts on M is *compatible* with the given S_n-structure if differentials of the transition functions are the corresponding elements of S_n (in the standard n-dimensional representation). Globally, an S_n-manifold M with a compatible atlas is determined by an affine representation of $\pi_1(M) \to S_n \to Aff_n$, i.e., the transition functions have the form

$$u^i \mapsto u^{\sigma(i)} + a^i_\sigma, \tag{1.25a}$$

$$a^i_{\sigma'\sigma} = a^{\sigma'(i)}_{\sigma} + a^i_{\sigma'}, \tag{1.25b}$$

for σ, $\sigma' \in S_n$. As an example of an S_n-manifold one can bear in mind the space of all polynomials $M = \{P(u) = u^n + a_1 u^{n-1} + \cdots + a_n |\, a_1, \ldots, a_n \in \mathbf{C}\}$ without multiple roots. Compatible coordinates are the roots of $P(u)$. The transition functions (1.25) are given by the standard n-dimensional representation of the braid group $\pi_1(M) = B_n$.

The Darboux–Egoroff system is well-defined on any S_n-manifold M with a marked compatible atlas. To obtain a massive Frobenius structure on M one should find a solution $\gamma_{ij}(u)$ that is covariant with respect to transformations of the form (1.25). This "boundary value problem" seems to be more complicated.

In all the examples (below) of massive TFT, the coupling space M (massive Frobenius manifold) can be extended by adding a certain locus M_{sing} of at least real codimension 2. The structure of the Frobenius manifold can be extended on $\widehat{M} = M \cup M_{sing}$ but the algebra structure on the tangent spaces $T_x\widehat{M}$ for $x \in M_{sing}$ has nilpotents. The flat metric $\eta_{\alpha\beta}$ is extended on $\widehat{M}$ without degeneration. So $\hat{M}$ is still a locally Euclidean manifold.

Remark. The notion of the Frobenius manifold admits algebraic formalization in terms of the ring of functions on a manifold. More precisely, let R be a commutative associative algebra with unity over a field k of characteristics $\neq 2$. We are interested in structures of the Frobenius algebra over R in the R-module of k-derivations $Der(R)$ (i.e., $u(\kappa) = 0$ for $\kappa \in k, u \in Der(R)$) satisfying

$$\tilde{\nabla}_u(\lambda)\tilde{\nabla}_v(\lambda) - \tilde{\nabla}_v(\lambda)\tilde{\nabla}_u(\lambda) = \tilde{\nabla}_{[u,v]}(\lambda) \text{ identically in } \lambda \tag{1.26a}$$

$$\text{for } \tilde{\nabla}_u(\lambda)v = \nabla_u v + \lambda u \cdot v, \tag{1.26b}$$

$$\nabla_u e = 0 \text{ for all } u \in Der(R) \tag{1.26c}$$

where e is the unity of the Frobenius algebra $Der(R)$. The non-degeneracy of the symmetric inner product

$$<\ ,\ >: Der(R) \times Der(R) \to R$$

means that it defines an isomorphism $\mathrm{Hom}_R(Der(R), R) \to Der(R)$. I recall that the covariant derivative is a derivation $\nabla_u v \in Der(R)$ defined for any u, $v \in Der(R)$, which is determined from the equation

$$\begin{aligned} &< \nabla_u v, w >= \\ &\frac{1}{2}[u < v, w > + v < w, u > - w < u, v > + < [u, v], w > \\ &\qquad + < [w, u], v > + < [w, v], u >] \end{aligned} \tag{1.27}$$

for any $w \in Der(R)$ (here [,] denotes the commutator of derivations). Observe that the notion of infinitesimal algebraic symmetry also can be algebraically formalized in a similar way. It would be interesting to find a purely algebraic version of Theorem 1.1. This could give an algebraic approach to the problem of the classification of Frobenius manifolds.

In the conclusion of this section we consider a closure of the class of massive Frobenius manifolds to be the set of all Frobenius manifolds with an n-dimensional commutative group of algebraic symmetries. Let A be a fixed n-dimensional Frobenius algebra with structure constants c_{ij}^k and an invariant nondegenerate inner product $\epsilon = (\epsilon_{ij})$. Let us introduce the matrices

$$C_i = (c_{ij}^k). \tag{1.28}$$

An analogue of the Darboux–Egoroff system (1.21) for an operator-valued function

$$\gamma(u) : A \to A, \ \gamma = (\gamma_i^j(u)), \ u = (u^1, \dots, u^n) \tag{1.29a}$$

(an analogue of the rotation coefficients) where the operator γ is symmetric with respect to ϵ,

$$\epsilon\gamma = \gamma^{\mathrm{T}}\epsilon \tag{1.29b}$$

has the form

$$[C_i, \partial_j\gamma] - [C_j, \partial_i\gamma] + [[C_i, \gamma], [C_j, \gamma]] = 0, \ i, j = 1, \dots, n, \tag{1.30}$$

$\partial_i = \partial/\partial u^i$. This is an integrable system with the Lax representation

$$\partial_i\Psi = \Psi(\lambda C_i + [C_i, \gamma]), \ i = 1, \dots, n. \tag{1.31}$$

It is convenient to consider $\Psi = (\psi_1(u), \dots, \psi_n(u))$ as a function with values in the dual space A^*. Note that A^* also is a Frobenius algebra with the structure constants $c_k^{ij} = c_{ks}^i\epsilon^{sj}$ and the invariant inner product $< , >_*$ determined by $(\epsilon^{ij}) = (\epsilon_{ij})^{-1}$.

Let $\Psi_\alpha(u)$, $\alpha = 1, \dots, n$, be a basis of solutions of (1.31) for $\lambda = 0$,

$$\partial\Psi_\alpha = \Psi_\alpha[C_i, \gamma], \ \alpha = 1, \dots, n, \tag{1.32a}$$

such that the vector $\Psi_1(u)$ is invertible in A^*. We put

$$\eta_{\alpha\beta} = < \Psi_\alpha(u), \Psi_\beta(u) >_* \tag{1.32b}$$

$$\mathrm{grad}_u t_\alpha = \Psi_\alpha(u) \cdot \Psi_1(u) \tag{1.32c}$$

$$c_{\alpha\beta\gamma}(t(u)) = \frac{\Psi_\alpha(u) \cdot \Psi_\beta(u) \cdot \Psi_\gamma(u)}{\Psi_1(u)}. \tag{1.32d}$$

Theorem 1.2. *Formulae (1.32) for an arbitrary Frobenius algebra A locally parametrize all Frobenius manifolds with an n-dimensional commutative group of algebraic symmetries.*

Considering u as a vector in A and $\Psi_1^2 = \Psi_1 \cdot \Psi_1$ as a linear function on A one obtains the following analogue of Egoroff metrics (on A)

$$ds^2 = \Psi_1^2(du \cdot du). \tag{1.33}$$

2. Conformal invariant Frobenius manifolds and isomonodromy deformations

Definition 2.1. A diffeomorphism $f : M \to M$ is called a *conformal symmetry* if

$$f_*(u \cdot v) = \mu_f^c f_*(u) \cdot f_*(v) \tag{2.1a}$$

$$< f_*(u), f_*(v) >= \mu_f^\eta < u, v > \tag{2.1b}$$

$$f_*(e) = \mu_f^e e \tag{2.1c}$$

for some functions μ_f^c, μ_f^η, μ_f^e. A Frobenius manifold M is called *conformal invariant* if it admits a one-parameter group of conformal symmetries $f^{(\tau)}$ such that the tensors $f_*^{(\tau)}(c)$, $f_*^{(\tau)}(\eta)$, $f_*^{(\tau)}(e)$ determine a Frobenius structure on M for any τ.

Let v be the generator of the one-parameter group of conformal symmetries on a conformal invariant Frobenius manifold.

Proposition 2.1. *On a massive conformal invariant Frobenius manifold an action of the one-parameter group of conformal symmetries is generated by the field*

$$v = \sum_{i=1}^n u^i \partial_i \tag{2.2}$$

(modulo obvious transformations $v \mapsto av + be$ for constant a and b). It acts on the tensors c, η, e by the following formulae

$$\mathcal{L}_v c = c \tag{2.3a}$$

$$\mathcal{L}_v e = -e \tag{2.3b}$$

$$\mathcal{L}_v \eta = (2 - d)\eta \tag{2.3c}$$

where d is a constant.

Here $\mathcal{L}_v$ means the Lie derivative along the vector field v.

Corollary. *For a massive conformal invariant Frobenius manifold, the rotation coefficients* $\gamma_{ij}(u)$ *satisfy the similarity condition*

$$\gamma_{ij}(cu) = c^{-1}\gamma_{ij}(u) \tag{2.4a}$$

or, equivalently

$$\sum_{k=1}^{n} u^k \partial_k \gamma_{ij}(u) = -\gamma_{ij}(u). \tag{2.4b}$$

For $n = 2$ the similarity reduction (2.4) of the Darboux–Egoroff system can be solved immediately:

$$\gamma_{12} = \gamma_{21} = \frac{id}{2} \frac{1}{u^1 - u^2}. \tag{2.5}$$

For the first nontrivial case $n = 3$ the system (1.21), (2.4) reads

$$\Gamma_1' = \Gamma_2 \Gamma_3 \tag{2.6a}$$

$$(z\Gamma_2)' = -\Gamma_1 \Gamma_3 \tag{2.6b}$$

$$((z-1)\Gamma_3)' = \Gamma_1 \Gamma_2 \tag{2.6c}$$

where

$$\gamma_{ij}(u) = \frac{1}{u^2 - u^3} \Gamma_k(z), \ i, \ j, \ k \text{ are distinct} \tag{2.7a}$$

$$z = \frac{u^1 - u^3}{u^2 - u^3}. \tag{2.7b}$$

It has an obvious first integral

$$\Gamma_1^2 + (z\Gamma_2)^2 + ((z-1)\Gamma_3)^2 = \text{const.} \tag{2.8}$$

Using this integral one can reduce [30] system (2.6) to a particular case of the Painlevé-VI equation.

For $n > 3$ system (1.21), (2.4) can be considered as a higher order analogue of the Painlevé-VI. To find solutions of this system one can use an appropriate version of IST: the so-called method of isomonodromy deformations [31]. This gives a parametrization of the solutions of the system (1.21), (2.4) by monodromy data of the following system of linear ODEs with rational coefficients:

$$\lambda \frac{d\psi}{d\lambda} = (\lambda U - [U, \gamma])\psi. \tag{2.9a}$$

Here

$$U = \operatorname{diag}(u^1, \ldots, u^n), \tag{2.9b}$$

$$\gamma = (\gamma_{ij}(u)). \tag{2.9c}$$

Solutions of this linear ODE have some monodromy properties, i.e., they are multivalued functions in the complex λ-plane.

Proposition 2.2. *The monodromy transformations of solutions of system (2.9) do not depend on the parameters u iff the matrix $\gamma_{ij}(u)$ is a solution of system (1.21), (2.4).*

The linear system (2.9) has two singular points: a regular singularity at $\lambda = 0$ and an irregular one at $\lambda = \infty$. Monodromy transformations of solutions of the system near $\lambda = 0$ have the form

$$\psi \mapsto \exp(-2\pi i[U, \gamma])\psi. \tag{2.10}$$

So the eigenvalues of the matrix $[U, \gamma]$ are first integrals of the system (1.21), (2.4). They generalise the first integral (2.8). Monodromy at infinity is determined by a $n \times n$ *Stokes matrix* S (see [31] for details). The diagonal terms of S equal 1; $n(n-1)/2$ of the other entries of matrix S vanish. Other matrix elements of S can be used as local parameters of massive conformal-invariant Frobenius manifolds (just $n(n-1)/2$ arbitrary complex parameters; one should add one more parameter: a norming constant of a solution $\psi_{i1}(u)$ in (1.24) which is an eigenvector of matrix $[U, \gamma]$). The monodromy at $\lambda = 0$ can be expressed via S using cyclic relations (see [39]). If the Stokes matrix S is sufficiently close to the unity matrix, then the inverse problem of the monodromy theory (i.e., to determine the coefficients of the linear operator (2.9) from the given monodromy data) is always solvable. The solution can be obtained by solving linear integral equations [39].

Let us assume that the monodromy of the operator (2.9) at the origin is semisimple. That means that matrix $[U, \gamma]$ has pairwise different eigenvalues $\mu_1, \ldots, \mu_n$. Let us order them in such a way that

$$\mu_\alpha + \mu_{n-\alpha+1} = 0. \tag{2.11}$$

Proposition 2.3. *Flat coordinates on a massive conformal invariant Frobenius manifold with semisimple monodromy of (2.10) at $\lambda = 0$ can be chosen in such a way that the generator v of conformal symmetries takes the form*

$$v = \sum (1 - q_\alpha) t^\alpha \partial_\alpha + \sum r_\alpha \partial_\alpha \tag{2.12}$$

for

$$q_\alpha = \mu_1 - \mu_\alpha \tag{2.13a}$$

where μ_α are the eigenvalues of matrix $[U, \gamma]$ ordered as in (2.11) and for some numbers r_α.

In other words, the tensors c, η, e should be conformally covariant with respect to the following transformations

$$t^\alpha \mapsto c^{1-q_\alpha} t^\alpha + r_\alpha \tag{2.14a}$$

$$c_{\alpha\beta}^\gamma \mapsto c^{q_\alpha + q_\beta - q_\gamma} c_{\alpha\beta}^\gamma \tag{2.14b}$$

$$\eta_{\alpha\beta} \mapsto c^{q_\alpha + q_\beta - d} \eta_{\alpha\beta} \tag{2.14c}$$

where

$$d = q_n = 2\mu_1 \tag{2.13b}$$

is the same as in (2.3c),

$$e \mapsto c^{-1} e. \tag{2.14d}$$

Equation (2.14c) means that $\eta_{\alpha\beta} \neq 0$ only for $q_\alpha + q_\beta = d$. The second sum in (2.12) can be killed by a shift if all $q_\alpha \neq 1$.

The numbers q_α are called the *charges* of the TCFT model, d is called *dimension* of the model. For topological sigma-models it coincides with the complex dimension of the target-space. Scaling laws (2.14) with $r_\alpha = 0$ were obtained in [8] using the assumption that the TCFT model is obtained by twisting of a N=2 supersymmetric model of QFT. They imply superselection rules for tree-level correlators in the conformal point $t = 0$ (the stationary point of the vector field v (2.3)). In our approach the scaling laws follow from the simple symmetry assumption on the Frobenius manifold. The scaling laws (2.3) for $r_\alpha \neq 0$ (being relevant only for $q_\alpha = 1$) look like an artifact of the geometrical approach (though they select the only physically interesting solution (4.5) for the case $n = 2$, $d = 1$). Classification of TCFT models with non semi-simple monodromy still is an open problem.

Summarizing we obtain

Theorem 2.1. *All massive conformal invariant Frobenius manifolds are parametrized by monodromy data of the linear operator*

$$\Lambda = \lambda\partial_\lambda - \lambda U + M(u) \tag{2.15a}$$

$$U = \operatorname{diag}(u^1, \ldots, u^n). \tag{2.15b}$$

$$M^{\mathrm{T}} = -M. \tag{2.15c}$$

Manifolds with semisimple monodromy at the origin $\lambda = 0$ form a $[\frac{n(n-1)}{2} + 1]$-parameter family. The free energy $F(t)$ of such a Frobenius manifold can be expressed via quadratures of a higher order analogue of the Painlevé-VI transcendents, i.e., solutions of the equations of isomonodromy deformations of (2.15).

For nonresonant conformal invariant Frobenius manifolds (see (3.19) below) with a semisimple monodromy at the origin, the structure functions $c_{\alpha\beta}^{\gamma}(t)$ can be expressed algebraically (i.e., without quadratures) via the above higher order analogue of the Painlevé-VI transcendents. Also one has

Proposition 2.4. *The canonical coordinates $u^1, \dots, u^n$ on a massive conformal invariant Frobenius manifold coincide with the eigenvalues of the matrix*

$$\tilde{U} = (\tilde{U}_\beta^\gamma(t)) = ((1 + q_\beta - q_\gamma) F_\beta^\gamma(t)) \tag{2.16a}$$

$$F_\beta^\gamma(t) = \eta^{\gamma\epsilon} \partial_\beta \partial_\epsilon F(t). \tag{2.16b}$$

It would be interesting to have an understanding of the physical sense of the operator $\tilde{U}$ for TCFT models.

Remark. We saw that monodromy is an important invariant of a massive conformal invariant Frobenius manifold. It can also be defined for an arbitrary conformal invariant Frobenius manifold by considering the linear operator

$$\tilde{\Lambda} = \lambda \partial_\lambda - \lambda \tilde{U} + \tilde{M}, \tag{2.17a}$$

where

$$\tilde{M} = (\tilde{M}_\beta^\gamma) = (q_\beta \delta_\beta^\gamma), \tag{2.17b}$$

the matrix $\tilde{U}$ has the form (2.16). WDVV equations determine isomonodromy deformations of $\tilde{\Lambda}$.

Monodromy properties of eigenfunctions of $\tilde{\Lambda}$ near the irregular singularity $\lambda = \infty$ (i.e., Stokes matrices) strongly depend on the algebraic structure of the multiplication on TM. These Stokes matrices are constrained by cyclic relations since monodromy near the origin $\lambda = 0$ is fixed by the given charges q_α. An advantage of the isomonodromy problem (2.15) for massive Frobenius manifolds is its universality (independence of the charges; the charges can be expressed via an arbitrary Stokes matrix of (2.15)). Note that a basis of common eigenfunctions of $\tilde{\Lambda}$,

$$\tilde{\Lambda}\xi = \kappa \xi, \tag{2.18a}$$

and (1.12) take the form

$$\xi_\beta(t, \lambda) = \partial_\beta h_\alpha(t, \lambda), \ \kappa = d - q_\alpha, \ \text{for any } \alpha = 1, \dots, n \tag{2.18b}$$

where the solutions $h_\alpha(t, \lambda)$ of (3.5) are normalized by (3.6).

3. Coupling to gravity. Systems of hydrodynamic type: their Hamiltonian formalism, solutions, and τ-functions

Let us fix a Frobenius manifold (i.e., a solution of the WDVV equations). Considering this as the primary free energy of the matter sector of a 2D TFT model, let us try to calculate the tree-level (i.e., the zero-genus) approximation of the complete model obtained by coupling of the matter sector to topological gravity. The idea of using hierarchies of Hamiltonian systems of hydrodynamic type for such a calculation was proposed by E. Witten [46] for the case of topological sigma-models. An advantage of my approach is that it provides an effective construction of these hierarchies for any solution of WDVV. The tree-level free energy of the model will be identified with the τ-function of a particular solution of the hierarchy. For a TCFT-model (i.e., for a conformal invariant Frobenius manifold) the hierarchy carries a bihamiltonian structure under a non-resonance assumption for the charges and dimension of the model (this bihamiltonian structure was constructed in [39] for the case of massive perturbations of a TCFT model; here I generalize it for an arbitrary TCFT model). This gives an answer to a question of [46] (see p.283). As it was mentioned in the Introduction, the bihamiltonian structure could be useful for the calculation of higher genus corrections.

So let $c_{\alpha\beta}^{\gamma}(t)$, $\eta_{\alpha\beta}$ be a solution of WDVV, $t = (t^1, \ldots, t^n)$. I will construct a hierarchy of systems of first-order PDEs linear in the derivatives (*systems of hydrodynamic type*) for functions $t^\alpha(T)$, T is an infinite vector

$$T = (T^{\alpha,p}), \quad \alpha = 1, \ \ldots, \ n, \quad p = 0, \ 1, \ \ldots; \ T^{1,0} = X,$$

$$\partial_{T^{\alpha,p}} t^\beta = c_{(\alpha,p)}{}_{\gamma}^{\beta}(t) \partial_X t^\gamma \tag{3.1a}$$

for some matrices of coefficients $c_{(\alpha,p)}{}_{\gamma}^{\beta}(t)$. The marked variable $X = T^{1,0}$ usually is called the *cosmological constant.*

I will consider the equations (3.1) as dynamical systems (for any (α, p)) on the space of functions $t = t(X)$ with values in the Frobenius manifold M.

A. Construction of the systems. I define a Poisson bracket on the space of functions $t = t(X)$ (i.e., on the loop space $\mathcal{L}(M)$) by the formula

$$\{t^\alpha(X), t^\beta(Y)\} = \eta^{\alpha\beta}\delta'(X - Y). \tag{3.2}$$

All the systems (3.1a) have the Hamiltonian form

$$\partial_{T^{\alpha,p}} t^\beta = \{t^\beta(X), H_{\alpha,p}\} \tag{3.1b}$$

with Hamiltonians of the form

$$H_{\alpha,p} = \int h_{\alpha,p+1}(t(X))dX. \tag{3.3}$$

The generating functions of densities of the Hamiltonians

$$h_\alpha(t,\lambda)=\sum_{p=0}^{\infty}h_{\alpha,p}(t)\lambda^p,\ \alpha=1,\dots,n, \tag{3.4}$$

coincide with the flat coordinates of the perturbed connection $\tilde\nabla(\lambda)$ (see (1.10)). That means that they are determined by the system (cf. (1.12))

$$\partial_\beta\partial_\gamma h_\alpha(t,\lambda)=\lambda c_{\beta\gamma}^\epsilon(t)\partial_\epsilon h_\alpha(t,\lambda). \tag{3.5}$$

This gives simple recurrence relations for the densities $h_{\alpha,p}$. Solutions of (3.5) can be normalized in such a way that

$$h_\alpha(t,0)=t_\alpha=\eta_{\alpha\beta}t^\beta, \tag{3.6a}$$

$$<\nabla h_\alpha(t,\lambda),\nabla h_\beta(t,-\lambda)>=\eta_{\alpha\beta}. \tag{3.6b}$$

Here ∇ is the gradient (in t). It can be shown that the Hamiltonians (3.3) are in involution. So all the systems of the hierarchy (3.1) commute pairwise.

B. Specification of a solution $t=t(T)$. Hierarchy (3.1) admits an obvious scaling group

$$T^{\alpha,p}\mapsto cT^{\alpha,p},\ \ t\mapsto t. \tag{3.7}$$

Let us take the nonconstant invariant solution for the symmetry

$$(\partial_{T^{1,1}}-\sum T^{\alpha,p}\partial_{T^{\alpha,p}})t(T)=0. \tag{3.8}$$

(I identify $T^{1,0}$ and X. So the variable X is suppressed in the formulae.) This solution can be found without quadratures from a fixed point equation for the gradient map

$$t=\nabla\Phi_T(t), \tag{3.9}$$

$$\Phi_T(t)=\sum_{\alpha,p}T^{\alpha,p}h_{\alpha,p}(t). \tag{3.10}$$

The existence and uniqueness of such a fixed point can be proved for sufficiently small $T^{\alpha,p}$ for $p>0$ (more precisely, in the domain where $T^{\alpha,0}$ is arbitrary, $T^{1,1}=o(1)$, $T^{\alpha,p}=o(T^{1,1})$ for $p>0$).

C. τ-function. Let us define coefficients $V_{(\alpha,p),(\beta,q)}(t)$ from the expansion

$$(\lambda+\mu)^{-1}(<\nabla h_\alpha(t,\lambda),\nabla h_\beta(t,\mu)>-\eta_{\alpha\beta})=\sum_{p,q=0}^{\infty}V_{(\alpha,p),(\beta,q)}(t)\lambda^p\mu^q$$
$$\equiv V_{\alpha\beta}(t,\lambda,\mu). \tag{3.11}$$

The infinite matrix of coefficients $V_{(\alpha,p),(\beta,q)}(t)$ has a simple meaning: it is the energy-momentum tensor of the commutative Hamiltonian hierarchy (3.1). That means that the matrix entry $V_{(\alpha,p),(\beta,q)}(t)$ is the density of flux of the Hamiltonian $H_{\alpha,p-1}$ along the flow $T^{\beta,q}$:

$$\partial_{T^{\beta,q}} h_{\alpha,p}(t) = \partial_X V_{(\alpha,p),(\beta,q)}(t). \tag{3.12}$$

Then

$$\log\tau(T) = \frac{1}{2}\sum V_{(\alpha,p),(\beta,q)}(t(T))T^{\alpha,p}T^{\beta,q} + \sum V_{(\alpha,p),(1,1)}(t(T))T^{\alpha,p} + \frac{1}{2}V_{(1,1),(1,1)}(t(T)). \tag{3.13}$$

Remark. More generally, a family of solutions of (3.1) has the form

$$\nabla[\Phi_T(t) - \Phi_{T_0}(t)] = 0 \tag{3.14}$$

for an arbitrary constant vector $T_0 = T_0^{\alpha,p}$. For massive Frobenius manifolds they form a dense subset in the space of all solutions of (3.1) (see [22, 23, 39]). Formally they can be obtained from the solution (3.9) by a shift of the arguments $T^{\alpha,p}$. The τ-function of the solution (3.14) can be formally obtained from (3.13) by the same shift. For the example of topological gravity [3, 46] such a shift is just the operation that relates the tree-level free energies of the topological phase of 2D gravity and of the matrix model. It should be taken into account that the operation of such a time shift in systems of hydrodynamic type is a subtle one: it can pass through a point of gradient catastrophe where derivatives become infinite. The corresponding solution of the KdV hierarchy has no gradient catastrophes but oscillating zones arise (see [32] for details).

Theorem 3.1. *Let*

$$\mathcal{F}(T) = \log\tau(T), \tag{3.15a}$$

$$< \phi_{\alpha,p}\phi_{\beta,q}\ldots >_0 = \partial_{T^{\alpha,p}}\partial_{T^{\beta,q}}\ldots\mathcal{F}(T). \tag{3.15b}$$

Then the following relations hold:

$$\mathcal{F}(T)\big|_{T^{\alpha,p}=0 \text{ for } p>0,\ T^{\alpha,0}=t^\alpha} = F(t) \tag{3.16a}$$

$$\partial_X\mathcal{F}(T) = \sum T^{\alpha,p}\partial_{T^{\alpha,p-1}}\mathcal{F}(T) + \frac{1}{2}\eta_{\alpha\beta}T^{\alpha,0}T^{\beta,0} \tag{3.16b}$$

$$< \phi_{\alpha,p}\phi_{\beta,q}\phi_{\gamma,r} >_0 = < \phi_{\alpha,p-1}\phi_{\lambda,0} >_0 \eta^{\lambda\mu} < \phi_{\mu,0}\phi_{\beta,q}\phi_{\gamma,r} >_0 . \tag{3.16c}$$

Let me establish now a 1-1 correspondence between the statements of the theorem and the standard terminology of QFT. In a complete model of 2D TFT (i.e., a matter sector coupled to topological gravity) there are an infinite number of operators that are usually denoted by $\phi_{\alpha,p}$ or $\sigma_p(\phi_\alpha)$. The operators $\phi_{\alpha,0}$ can be identified with the primary operators ϕ_α; the operators $\phi_{\alpha,p}$ for $p > 0$ are called *gravitational descendants* of ϕ_α. Respectively, one has an infinite number of coupling constants $T^{\alpha,p}$. Formula (3.15a) expresses the tree-level (i.e., genus zero) partition function of the model of 2D TFT via the logarithm of the τ-function (3.13). Equation (3.15b) is the standard relation between the correlators (of genus zero) in the model and the free energy. Equation (3.16a) shows that before coupling to gravity the partition function (3.15a) coincides with the primary partition function of the given matter sector. Equation (3.16b) is the string equation for the free energy [3, 4, 8, 46]. And equations (3.16c) coincide with the genus zero recursion relations for correlators of a TFT [4, 46].

In particular, from (3.15) one obtains

$$< \phi_{\alpha,p}\phi_{\beta,q} >_0 = V_{(\alpha,p),(\beta,q)}(t(T)), \tag{3.17a}$$

$$< \phi_{\alpha,p}\phi_{1,0} >_0 = h_{\alpha,p}(t(T)), \tag{3.17b}$$

$$< \phi_{\alpha,p}\phi_{\beta,q}\phi_{\gamma,r} >_0 = < \nabla h_{\alpha,p} \cdot \nabla h_{\beta,q} \cdot \nabla h_{\gamma,r}, [e - \sum T^{\alpha,p}\nabla h_{\alpha,p-1}]^{-1} > . \tag{3.17c}$$

The second factor of the inner product in the r.h.s. of (3.17c) is an invertible element (in the Frobenius algebra of vector fields on M) for sufficiently small $T^{\alpha,p}$, $p > 0$. From the last formula one obtains

Proposition 3.1. *The coefficients*

$$c_{p,\alpha\beta}^{\ \ \gamma}(T) = \eta^{\gamma\mu}\partial_{T^{\alpha,p}}\partial_{T^{\beta,p}}\partial_{T^{\mu,p}} \log \tau(T) \tag{3.18}$$

for any p and any T are structure constants of a commutative associative algebra with the invariant inner product $\eta_{\alpha\beta}$.

As a rule, such an algebra has no unity.

In fact the proposition also holds for a τ-function of an arbitrary solution of the form (3.14).

We see that hierarchy (3.1) determines a family of Bäcklund transforms of the WDVV equation (0.1)

$$F(t) \mapsto \tilde{F}(\tilde{t}),$$

$$\tilde{F} = \log \tau, \ \tilde{t}^\alpha = T^{\alpha,p}$$

for a fixed p and for an arbitrary τ-function of (3.1). So it is natural to consider equations of the hierarchy as Lie – Bäcklund symmetries of WDVV.

Up to now I did not even use the scaling invariance (2.14). It turns out that this gives rise to a bihamiltonian structure of the hierarchy (3.1).

Let us consider a conformal invariant Frobenius manifold, i.e., a TCFT model with charges q_α and dimension d. We say that a pair α, p is *resonant* if

$$\frac{d+1}{2} - q_\alpha + p = 0. \tag{3.19}$$

Here p is a nonnegative integer. The TCFT model is *nonresonant* if all pairs α, p are nonresonant. For example, models satisfying the inequalities

$$0 = q_1 \leq q_2 \leq \ldots \leq q_n = d < 1 \tag{3.20}$$

are nonresonant.

Theorem 3.2. *1) For a conformal invariant Frobenius manifold with charges q_α and dimension d, the formula*

$$\begin{aligned}\{t^\alpha(X), t^\beta(Y)\}_1 = [(&\frac{d+1}{2} - q_\alpha)F^{\alpha\beta}(t(X)) \\ &+ (\frac{d+1}{2} - q_\beta)F^{\alpha\beta}(t(Y))]\delta'(X-Y)\end{aligned} \tag{3.21}$$

$$F^{\alpha\beta}(t) = \eta^{\alpha\alpha'}\eta^{\beta\beta'}\partial_{\alpha'}\partial_{\beta'}F(t)$$

determines a Poisson bracket compatible with the Poisson bracket (3.2). 2) For a nonresonant TCFT model all the equations of the hierarchy (3.1) are Hamiltonian equations also with respect to the Poisson bracket (3.21).

The nonresonancy condition is essential: equations (3.1) with resonant numbers (α, p) do not admit another Poisson structure.

Remark. According to the theory [18-21] of Poisson brackets of hydrodynamic type, any such bracket is determined by a flat Riemannian (or pseudo-Riemannian) metric $g_{\alpha\beta}(t)$ on the target space M (more precisely, one needs a metric $g^{\alpha\beta}(t)$ on the cotangent bundle to M). In our case the target space is the Frobenius manifold M. The first Poisson structure (3.2) is determined by the metric being specified by the double-point correlators $\eta_{\alpha\beta}$. The second flat metric for the Poisson bracket (3.21) on a conformal invariant Frobenius manifold M has the following geometrical interpretation. Let ω_1 and ω_2 be two 2-forms on M. We can multiply them $\omega_1,\ \omega_2 \mapsto \omega_1 \cdot \omega_2$ using the multiplication of tangent vectors and the isomorphism η between tangent and cotangent spaces. Then the new inner product $< \, , \, >_1$ is defined by the formula

$$< \omega_1, \omega_2 >_1 = \mathrm{i}_v(\omega_1 \cdot \omega_2). \tag{3.22}$$

Here i_v is the operator of contraction with the vector field v (the generator of conformal symmetries (2.3)). The metric (3.22) can be degenerate.

The theorem states that, nevertheless, the Jacobi identity for the Poisson bracket (3.21) holds.

The main examples of solutions of WDVV and of corresponding hierarchies will be given in the next section. Here I will consider the simplest class of examples where $c_{\alpha\beta}^{\gamma}$ does not depend on t. They form the structure constants of a Frobenius algebra A with an invariant inner product $< \, , \, >$ ($\eta_{\alpha\beta}$ in a basis $e_1 = 1, \ldots, e_n$). Let

$$\mathbf{t} = t^{\alpha} e_{\alpha} \in A. \tag{3.23}$$

The linear system (3.5) can be solved easily:

$$h_{\alpha}(t, \lambda) = \lambda^{-1} < e_{\alpha}, e^{\lambda \mathbf{t}} - 1 > .$$

This gives the following form of hierarchy (3.1)

$$\partial_{T^{\alpha,p}} \mathbf{t} = \frac{1}{p!} e_{\alpha} \mathbf{t}^{p} \partial_X \mathbf{t}. \tag{3.24}$$

The solution (3.9) is specified as the fixed point

$$G(\mathbf{t}) = \mathbf{t}, \tag{3.25a}$$

$$G(\mathbf{t}) = \sum_{p=0}^{\infty} \frac{\mathbf{T}_p}{p!} \mathbf{t}^{p}. \tag{3.25b}$$

Here I introduce A-valued coupling constants

$$\mathbf{T}_p = T^{\alpha,p} e_{\alpha} \in A, \; p = 0, \; 1, \ldots. \tag{3.26}$$

The solution of (3.25) has the well-known form

$$\mathbf{t} = G(G(G(\ldots))) \tag{3.27}$$

(infinite number of iterations). The τ-function of solution (3.27) has the form

$$\log \tau = \frac{1}{6} < 1, \mathbf{t}^3 > - \sum_{p} \frac{< \mathbf{T}_p, \mathbf{t}^{p+2} >}{(p+2)p!} + \frac{1}{2} \sum_{p,q} \frac{< \mathbf{T}_p \mathbf{T}_q, \mathbf{t}^{p+q+1} >}{(p+q+1)p!q!}. \tag{3.28}$$

For the tree-level correlation functions of a TFT-model with constant primary correlators one immediately obtains

$$< \phi_{\alpha,p} \phi_{\beta,q} >_0 = \frac{< e_{\alpha} e_{\beta}, \mathbf{t}^{p+q+1} >}{(p+q+1)p!q!}, \tag{3.29a}$$

$$< \phi_{\alpha,p}\phi_{\beta,q}\phi_{\gamma,r} >_0 = \frac{1}{p!q!r!} < e_\alpha e_\beta e_\gamma, \frac{t^{p+q+r}}{1 - \sum_{s\geq 1} \frac{T_s t^{s-1}}{(s-1)!}} > . \tag{3.29b}$$

For $n = 1$ the formulae (3.29) give the tree-level correlators of the topological gravity (see [3, 46]). For $n = 24$ one obtains the tree-level correlators of the topological sigma-model with a K3-surface as the target space. Here the algebra $A = H^*(K3)$ is a graded one: it has a basis P, $Q_1, \ldots, Q_{22}$, R of degrees 0, 1 (all the Q's) and 2 respectively. The multiplication has the form

$$P \text{ is the unity, } Q_iQ_j = \eta_{ij}R, \; Q_iR = R^2 = 0 \tag{3.30}$$

for a nondegenerate symmetric matrix η_{ij}. The scalar product (the intersection number) has the form

$$\eta_{PR} = 1, \; \eta_{Q_iQ_j} = \eta_{ij}.$$

Let us consider now the second hamiltonian structure (3.21). I start with the most elementary case $n = 1$ (the pure gravity). Let me redenote the coupling constant by

$$u = t^1.$$

The Poisson bracket (3.21) in this case reads

$$\{u(X), u(Y)\}_1 = \frac{1}{2}(u(X) + u(Y))\delta'(X - Y). \tag{3.31}$$

This is nothing but the Lie – Poisson bracket on the dual space to the Lie algebra of one-dimensional vector fields.

For an arbitrary graded Frobenius algebra A, the Poisson bracket (3.21) is also linear in the coordinates t^α

$$\{t^\alpha(X), t^\beta(Y)\}_1 = [(\frac{d+1}{2} - q_\alpha)c_\gamma^{\alpha\beta}t^\gamma(X) + (\frac{d+1}{2} - q_\beta)c_\gamma^{\alpha\beta}t^\gamma(Y)]\delta'(X - Y). \tag{3.32}$$

It therefore determines the structure of an infinite dimensional Lie algebra on the loop space $\mathcal{L}(A^*)$ where A^* is the dual space to the graded Frobenius algebra A. Theory of linear Poisson brackets of hydrodynamic type and of corresponding infinite dimensional Lie algebras was constructed in [34] (see also [18]). But the class of examples (3.32) is a new one. Observe that the case $A = H^*(K3)$ is a nonresonant one.

Let us come back to the general (i.e., nonlinear) case of a TCFT model. I will assume that the charges and the dimension are ordered in such a way that

$$0 = q_1 < q_2 \leq \ldots \leq q_{n-1} < q_n = d. \tag{3.33}$$

Then from (3.21) one obtains

$$\{t^n(X), t^n(Y)\}_1 = \frac{1-d}{2}(t^n(X) + t^n(Y))\delta'(X - Y). \tag{3.34}$$

Since

$$\{t^\alpha(X), t^n(Y)\}_1 = [(\frac{d+1}{2} - q_\alpha)t^\alpha(X) + \frac{1-d}{2}t^\alpha(Y)]\delta'(X-Y), \quad (3.35)$$

the functional

$$P = \frac{2}{1-d}\int t^n(X)dX \quad (3.36)$$

generates spatial translations. We see that for $d \neq 1$ the Poisson bracket (3.21) can be considered as a nonlinear extension of the Lie algebra of one-dimensional vector fields. An interesting question is to find an analogue of the Gelfand – Fuchs cocycle for this bracket. I found such a cocycle for a more special class of TCFT models. We say that a TCFT-model is *graded* if, for any t, the Frobenius algebra $c_{\alpha\beta}^\gamma(t)$, $\eta_{\alpha\beta}$ is graded.

Theorem 3.3. *For a graded TCFT-model the formula*

$$\{t^\alpha(X), t^\beta(Y)\}\hat{}_1 = \{t^\alpha(X), t^\beta(Y)\}_1 + \epsilon^2\eta^{1\alpha}\eta^{1\beta}\delta'''(X-Y) \quad (3.37)$$

determines a Poisson bracket compatible with (3.2) and (3.21) for arbitrary ϵ^2 (the central charge). For a generic graded TCFT model this is the only one deformation of the Poisson bracket (3.21) proportional to $\delta'''(X-Y)$.

For $n = 1$, equation (3.37) determines nothing but the Lie – Poisson bracket on the dual space to the Virasoro algebra

$$\{u(X), u(Y)\}\hat{}_1 = \frac{1}{2}[u(X) + u(Y)]\delta'(X-Y) + \epsilon^2\delta'''(X-Y) \quad (3.38)$$

(the second Poisson structure of the KdV hierarchy). For $n > 1$ and constant primary correlators (i.e., for a constant graded Frobenius algebra A) the Poisson bracket (3.37) can be considered as a vector-valued extension (for $d \neq 1$) of that of the Virasoro.

Graded TCFT models occur as the topological sigma-models with a Calabi – Yau manifold of (complex) dimension d as the target space [2, 46, 57]. They are nonresonant where d is even. In particular, for $d = 2$ one obtains the K3-models where the primary correlators are constant. For $d > 2$ they are not constant because of instanton corrections [46, 47, 57]. As it was explained in [57], finding of these primary correlators for the Calabi – Yau models (and, therefore, graded solutions of WDVV) could be a crucial point in solving the problem of mirror symmetry.

The compatible pair of the Poisson brackets (3.2) and (3.37) generates an integrable hierarchy of PDEs for a nonresonant graded TCFT using the standard machinery of the bihamiltonian formalism [52]

$$\partial_{T^{\alpha,p}}t^\beta = \{t^\beta(X), \hat{H}_{\alpha,p}\} = [\frac{d+1}{2} - q_\alpha + p]^{-1}\{t^\beta(X), \hat{H}_{\alpha,p-1}\}\hat{}_1. \quad (3.39)$$

Here the Hamiltonians have the form

$$\hat{H}_{\alpha,p} = \int \hat{h}_{\alpha,p+1} dX, \tag{3.40a}$$

$$\hat{h}_{\alpha,p+1} = h_{\alpha,p+1}(t) + \epsilon^2 \Delta \hat{h}_{\alpha,p+1}(t, \partial_X t, \ldots, \partial_X^p t; \epsilon^2) \tag{3.40b}$$

where $\Delta \hat{h}_{\alpha,p+1}$ are some polynomials determined by (3.39). They are graded-homogeneous of degree 2 where $\deg \partial_X^k t = k$, $\deg \epsilon = -1$. It is clear that hierarchy (3.1) is the zero-dispersion limit of this hierarchy. For $n = 1$ using the pair (3.2) and (3.38) one immediately obtains the KdV hierarchy. Note that this describes the topological gravity. It would be interesting to investigate the relation of the hierarchies determined by the pair (3.2) and (3.37) to a nonperturbative (i.e., for all genera) description of the the K3 models coupled to gravity. For a model with constant correlators (for a graded Frobenius algebra A) the first nontrivial equations of the hierarchy are

$$\partial_{T^{\alpha,1}} \mathbf{t} = e_\alpha \mathbf{t}\mathbf{t}_X + \frac{2\epsilon^2}{3-d} e_\alpha e_n \mathbf{t}_{XXX}. \tag{3.41}$$

For non-graded TCFT models it could be of interest to find nonlinear analogues of the cocycle (3.37). These should be differential geometric Poisson brackets of the third order [58, 18] of the form

$$\begin{gathered} \{t^\alpha(X), t^\beta(Y)\}\hat{}_1 = \{t^\alpha(X), t^\beta(Y)\}_1 + \\ \epsilon^2 \{ g^{\alpha\beta}(t(X))\delta'''(X-Y) + b_\gamma^{\alpha\beta}(t(X)) t_X^\gamma \delta''(X-Y) + \\ [f_\gamma^{\alpha\beta}(t(X)) t_{XX}^\gamma + h_{\gamma\delta}^{\alpha\beta}(t(X)) t_X^\gamma t_X^\delta]\delta'(X-Y) + \\ [p_\gamma^{\alpha\beta}(t) t_{XXX}^\gamma + q_{\gamma\delta}^{\alpha\beta}(t) t_{XX}^\gamma t_X^\delta + r_{\gamma\delta\lambda}^{\alpha\beta}(t) t_X^\gamma t_X^\delta t_X^\lambda]\delta(X-Y)\}. \end{gathered} \tag{3.42}$$

I recall (see [58, 18]) that the form (3.42) of the Poisson bracket should be invariant with respect to nonlinear changes of coordinates in the manifold M. This implies that the leading term $g^{\alpha\beta}(t)$ transforms as a metric (that may be degenerate) on the cotangent bundle T_*M, and that $b_\gamma^{\alpha\beta}(t)$ are contravariant components of a connection on M etc. The Poisson bracket (3.42) is assumed to be compatible with (3.2). Then the compatible pair (3.2), (3.42) of the Poisson brackets generates an integrable hierarchy of the same structure (3.39), (3.40).

4. Examples

Example 1. I start with the most elementary examples of solutions of WDVV for $n = 2$. Only massive solutions are of interest here (a 2-dimensional nilpotent Frobenius algebra has no nontrivial deformations).

The Darboux–Egoroff equations in this case are linear. I consider only TCFT case (the similarity reduction of WDVV). Let us redenote the coupling constants

$$t^1 = u,\ t^2 = \rho. \tag{4.1}$$

For $d \neq 1$ the primary free energy F has the form

$$F = \frac{1}{2}\rho u^2 + \frac{g}{a(a+2)}\rho^{a+2}, \tag{4.2}$$

$$a = \frac{1+d}{1-d}, \tag{4.3}$$

where g is an arbitrary constant. The second term in the formula for the free energy should be understood as

$$\frac{g}{a(a+2)}\rho^{a+2} = \int\int\int g(a+1)\rho^{a-1}.$$

The linear system (3.5) can be solved via Bessel functions [39]. Let me take the $T = T^{1,1}$ flow as an example of equations of hierarchy (3.1),

$$u_T + uu_X + g\rho^a\rho_X = 0 \tag{4.4a}$$

$$\rho_T + (\rho u)_X = 0. \tag{4.4b}$$

These are the equations of isentropic motion of a one-dimensional fluid with the dependence of the pressure on the density of the form $p = \frac{g}{a+2}\rho^{a+2}$. The Poisson structure (3.2) for these equations was proposed in [37]. For $a = 0$ (equivalently $d = -1$) the system coincides with the equations of waves on shallow water (the dispersionless limit [59] of the nonlinear Schrödinger equation (NLS)).

For $d = 1$ the primary free energy has the form

$$F = \frac{1}{2}\rho u^2 + ge^\rho. \tag{4.5}$$

This coincides with the free energy of the topological sigma-model with CP^1 as the target space. Note that this can be obtained from the same solution of the Darboux–Egoroff system as the semiclassical limit of the NLS (the case $d = -1$ above) for different choices of the eigenfunction ψ_{i1} (in the notations of (1.24)). The corresponding $T = T^{2,0}$-system of the hierarchy (3.1) reads

$$u_T = g(e^\rho)_X$$

$$\rho_T = u_X.$$

Eliminating u one obtains the long wave limit

$$\rho_{TT} = g(e^{\rho})_{XX} \tag{4.6}$$

of the Toda system

$$\rho_{n\,tt} = e^{\rho_{n+1}} - 2e^{\rho_n} + e^{\rho_{n-1}}. \tag{4.7}$$

(The 2-dimensional version of (4.6) was obtained in the formalism of Whitham-type equations in [44].) It would be interesting to prove that the nonperturbative free energy of the CP^1-model coincides with the τ-function of the Toda hierarchy.

Example 2. Topological minimal models. I consider here the A_n-series models only. The Frobenius manifold M here is the set of all polynomials (*Landau - Ginsburg superpotentials*) of the form

$$M = \{w(p) = p^{n+1} + a_1 p^{n-1} + \ldots + a_n |\, a_1, \ldots, a_n \in \mathbf{C}\}. \tag{4.8}$$

For any $w \in M$ the Frobenius algebra $A = A_w$ is the algebra of truncated polynomials

$$A_w = \mathbf{C}[p]/(w'(p) = 0) \tag{4.9}$$

(the prime means the derivative with respect to p) with the invariant inner product

$$< f, g >= \mathrm{res}_{p=\infty} \frac{f(p)g(p)}{w'(p)}. \tag{4.10}$$

Algebra A_w is semisimple if the polynomial $w'(p)$ has simple roots. The canonical coordinates (1.15) $u^1, \ldots, u^n$ are the critical values of the polynomial $w(p)$

$$u^i = w(p_i), \text{ where } w'(p_i) = 0, \; i = 1, \ldots, n. \tag{4.11}$$

Let us take the following diagonal metric on M

$$\sum_{i=1}^{n} \eta_{ii}(u)(du^i)^2, \quad \eta_{ii}(u) = [w''(p_i)]^{-1}. \tag{4.12}$$

It can be proved that this is a flat Egoroff metric on M. The corresponding flat coordinates on M have the form

$$t^\alpha = -\frac{n+1}{n-\alpha+1} \mathrm{res}_{p=\infty} w^{\frac{n-\alpha+1}{n+1}}(p)dp, \; \alpha = 1, \ldots, n. \tag{4.13}$$

The metric (4.12) in these coordinates has the constant form

$$\sum_{i=1}^{n} \eta_{ii}(u)(du^i)^2 = \eta_{\alpha\beta} dt^\alpha dt^\beta, \quad \eta_{\alpha\beta} = \delta_{n+1, \alpha+\beta}. \tag{4.14}$$

The orthonormal basis in A_w with respect to this metric consists of the polynomials $\phi_1(p),\dots,\phi_n(p)$ of degrees 0, 1, ... , $n-1$, resp., where

$$\phi_\alpha(p) = \frac{d}{dp}[w^{\frac{\alpha}{n+1}}]_+, \quad \alpha = 1,\dots,n. \tag{4.15}$$

Here $[\ ]_+$ means the polynomial part of the power series in p. This is a TCFT model with the charges and dimension

$$q_\alpha = \frac{\alpha-1}{n+1},\ d = q_n = \frac{n-1}{n+1}. \tag{4.16}$$

In fact one obtains a n-parameter family of TFT models with the same canonical coordinates u^i of the form (4.11) where

$$\eta_{ii}(u) \mapsto \eta_{ii}(u,c) = [w''(p_i)]^{-1}[\sum c_\alpha\phi_\alpha(p_i)]^2, \tag{4.17a}$$

$$t^\alpha \mapsto t^\alpha(c) = -\frac{n+1}{n-\alpha+1}\operatorname{res}_{p=\infty} w^{\frac{n-\alpha+1}{n+1}}(p)[\sum c_\gamma\phi_\gamma(p)]dp \tag{4.17b}$$

depending on arbitrary parameters c_1, ..., c_n. This reflects an ambiguity in the choice of the solution ψ_{i1} in the formulae (1.24). These models are conformally invariant if only one of the coefficients c_γ is nonzero.

The corresponding hierarchy of the systems of the hydrodynamic type (3.1) coincides with the dispersionless limit of the Gelfand–Dickey hierarchy for the scalar Lax operator of order $n+1$. This essentially follows from [8, 11]. I recall that the Gelfand–Dickey hierarchy for an operator

$$L = \partial^{n+1} + a_1(x)\partial^{n-1} + \dots + a_n(x)$$

$$\partial = d/dx$$

has the form

$$\partial_{t^{\alpha,p}} L = c_{\alpha,p}[L, [L^{\frac{\alpha}{n+1}+p}]_+],\ \alpha = 1,\dots,n,\ p = 0,1,\dots \tag{4.18}$$

for some constants $c_{\alpha,p}$. Here $[\ ]_+$ denotes differential part of the pseudo-differential operator. The dispersionless limit of the hierarchy is defined as follows: one should substitute

$$x \mapsto \epsilon x = X,\ t^{\alpha,p} \mapsto \epsilon t^{\alpha,p} = T^{\alpha,p} \tag{4.19}$$

and let ϵ tend to zero. The dispersionless limit of the τ-function of the hierarchy is defined [11, 53-54, 61] as

$$\log\tau_{\text{dispersionless}}(T) = \lim_{\epsilon\to 0}\epsilon^{-2}\log\tau(\epsilon t). \tag{4.20}$$

The modified minimal model (4.17) is related to the same Gelfand – Dickey hierarchy with the following modification of the L-operator

$$L \mapsto \tilde{L} = \sum c_\gamma [L^{\frac{\gamma}{n+1}}]_+. \tag{4.21}$$

The linear equation (3.5) for the minimal model can be solved in the form [39]

$$h_\alpha(t;\lambda) = -\frac{n+1}{\alpha}\mathrm{res}_{p=\infty} w^{\frac{\alpha}{n+1}}\, {}_1F_1(1; 1+\frac{\alpha}{n+1}; \lambda w(p))dp. \tag{4.22}$$

Here ${}_1F_1(a;c;z)$ is the Kummer (or confluent hypergeometric) function [35]

$${}_1F_1(a;c;z) = \sum_{m=o}^{\infty} \frac{(a)_m}{(c)_m}\frac{z^m}{m!}, \tag{4.23a}$$

$$(a)_m = a(a+1)\dots(a+m-1). \tag{4.23b}$$

The generating function (3.11) has the form

$$V_{\alpha\beta}(t;\lambda,\mu) = (\lambda+\mu)^{-1}[\eta^{\mu\nu}(\mathrm{res}_{p=\infty} w^{\frac{\alpha}{n+1}-1}\, {}_1F_1(1;\frac{\alpha}{n+1};\lambda w(p))\phi_\mu(p)dp)\times$$

$$(\mathrm{res}_{p=\infty} w^{\frac{\beta}{n+1}-1}\, {}_1F_1(1;\frac{\beta}{n+1};\mu w(p))\phi_\nu(p)dp) - \eta_{\alpha\beta}]. \tag{4.24}$$

From this one obtains formulae for the τ-function.

Example 3. $M_{g;n_0,\dots,n_m}$-models [13, 14]. Let $M = M_{g;n_0,\dots,n_m}$ be a moduli space of dimension

$$n = 2g + n_0 + \dots + n_m + 2m \tag{4.25}$$

of sets

$$(C;\infty_0,\dots,\infty_m;w;k_0,\dots,k_m;a_1,\dots,a_g,b_1,\dots,b_g) \in M_{g;n_0,\dots,n_m} \tag{4.26}$$

where C is a Riemann surface with marked points ∞_0, ..., ∞_m, and a marked meromorphic function

$$w : C \to CP^1, \;\; w^{-1}(\infty) = \infty_0 \cup \dots \cup \infty_m \tag{4.27}$$

having a degree $n_i + 1$ near the point ∞_i, and a marked symplectic basis $a_1,\dots,a_g$, $b_1,\dots,b_g \in H_1(C,\mathbf{Z})$, and marked branches of roots of w near ∞_0, ..., ∞_m of the orders $n_0+1,\dots,n_m+1$, resp.,

$$k_i^{n_i+1}(P) = w(P), \;\; P \text{ near } \infty_i. \tag{4.28}$$

(This is a connected manifold as it follows from [56].) We need the critical values of w

$$u^j = w(P_j),\ dw|_{P_j} = 0,\ j = 1, \dots, n \tag{4.29}$$

(i.e., the ramification points of the Riemann surface (4.27)) to be local coordinates in open domains in M where

$$u^i \neq u^j \text{ for } i \neq j \tag{4.30}$$

(by the Riemann existence theorem). Another assumption is that the one-dimensional affine group acts on M as

$$(C; \infty_0, \dots, \infty_m; w; \dots) \mapsto (C; \infty_0, \dots, \infty_m; aw + b; \dots) \tag{4.31a}$$

$$u^i \mapsto au^i + b,\ i = 1, \dots, n. \tag{4.31b}$$

Let dp be the normalized Abelian differential of the second kind on C with a double pole at ∞_0

$$dp = dk_0 + \text{ regular terms} \tag{4.32a}$$

$$\oint_{a_i} dp = 0,\ i = 1, \dots, g. \tag{4.32b}$$

Using u^i as the canonical coordinates (1.15) I define a flat Egoroff metric on M by the formula

$$ds^2 = \sum_{i=1}^{n} \eta_{ii}(u)(du^i)^2, \tag{4.33a}$$

$$\eta_{ii}(u) = \text{res}_{P_i} \frac{(dp)^2}{dw}. \tag{4.33b}$$

It can be extended globally on M. The corresponding flat coordinates are

$$t^{i;\alpha} = -\frac{n_i + 1}{n_i - \alpha + 1} \text{res}_{\infty_i} k_i^{n_i - \alpha + 1} dp,\ i = 0, \dots, m,\ \alpha = 1, \dots, n_i; \tag{4.34a}$$

$$p^i = \text{v.p.} \int_{\infty_0}^{\infty_i} dp = \lim_{Q \to \infty_0} \Big(\int_Q^{\infty_i} dp + k_0(Q)\Big),\ i = 1, \dots, m; \tag{4.34b}$$

$$q^i = -\text{res}_{\infty_i} w dp,\ i = 1, \dots, m; \tag{4.34c}$$

$$r^i = \oint_{b_i} dp,\ s^i = -\frac{1}{2\pi i} \oint_{a_i} w dp,\ i = 1, \dots, g. \tag{4.34d}$$

The metric (4.33) in these coordinates has the following form

$$\eta_{t^{i;\alpha} t^{i;\beta}} = \frac{1}{n_i + 1} \delta_{ij} \delta_{\alpha + \beta, n_i + 1} \tag{4.35a}$$

$$\eta_{p^i q^j} = \delta_{ij} \tag{4.35b}$$

$$\eta_{r^i s^j} = \delta_{ij}, \tag{4.35c}$$

and the other components of η vanish. The unity vector field is a unit vector along the coordinate $t^{0;1}$.

Proposition 4.1. *The flat metric (4.35) is well-defined globally on M and the flat coordinates (4.34) are globally independent analytic functions on M.*

As a consequence we see that the moduli space M is an unramified covering over a domain in $\mathbf{C}^n$ (see [13, 14]).

Let us introduce primary differentials on C (or on a universal covering $\tilde{C}$ of $C \setminus \infty_0 \cup \ldots \cup \infty_m$) of the form

$$\phi_{t^A} = \partial_{t^A}(pdw)_{w=\text{const}} \tag{4.36}$$

where

$$p(P) = \int_{Q_0}^{P} dp, \tag{4.37a}$$

$$Q_0 \in C,\ w(Q_0) = 0, \tag{4.37b}$$

t^A is one of the flat coordinates (4.34). Note that the definition (4.36) of the primary differentials can be rewritten as

$$\phi_{t^A} = -\partial_{t^A}(wdp)_{p=\text{const}} \tag{4.38}$$

where the multivalued coordinate p on C is defined in (4.37). So $w(p)$ plays the role of the Landau – Ginsburg superpotential for the $M_{g;n_0,\ldots,n_m}$-models. More explicitly, $\phi_{t^{i;\alpha}}$ is a normalized Abelian differential of the second kind with a pole in ∞_i,

$$\phi_{t^{i;\alpha}} = -\frac{1}{\alpha} dk_i^\alpha + \text{ regular terms} \quad \text{near } \infty_i,$$

$$\oint_{a_j} \phi_{t^{i;\alpha}} = 0; \tag{4.39a}$$

ϕ_{p^i} is a normalized Abelian differential of the second kind on C with a pole only at ∞_i, with the principal part of the form

$$\phi_{p^i} = dw + \text{ regular terms} \quad \text{near } \infty_i,$$

$$\oint_{a_j} \phi_{p^i} = 0; \tag{4.39b}$$

ϕ_{q^i} is a normalized Abelian differential of the third kind with simple poles at ∞_0 and ∞_i with residues -11 and $+1$ resp.; ϕ_{r^i} is a normalized multivalued differential on C with increments along the cycles b_i of the form

$$\phi_{r^i}(P+b_j)-\phi_{r^i}(P)=-\delta_{ij}dw,$$

$$\oint_{a_j}\phi_{r^i}=0; \tag{4.39c}$$

ϕ_{s^i} are the basic holomorphic differentials* on C normalized by the condition

$$\oint_{a_j}\phi_{s^i}=2\pi i\delta_{ij}. \tag{4.39d}$$

The inner product (4.33) in terms of the primary differentials ϕ_{t^A} reads

$$\eta_{AB}=\sum_{i=1}^{n}\mathrm{res}_{P_i}\frac{\phi_{t^A}\phi_{t^B}}{dw}. \tag{4.40}$$

The structure functions $c_{ABC}(t)$ can be calculated by

$$c_{ABC}(t)=\sum_{i=1}^{n}\mathrm{res}_{P_i}\frac{\phi_{t^A}\phi_{t^B}\phi_{t^C}}{dw\,dp}. \tag{4.41}$$

The extension of the Frobenius structure on the entire moduli space M is given by the condition that the differential

$$\frac{\phi_{t^A}\phi_{t^B}-c_{AB}^{C}\phi_{t^C}dp}{dw} \tag{4.42}$$

be holomorphic for $|w|<\infty$. The Frobenius algebra on T_tM will be nilpotent for Riemann surfaces $w: C\to CP^1$ with branch points of order greater than 2. This is a conformally invariant Frobenius manifold with the dimension

$$d=\frac{n_0-1}{n_0+1} \tag{4.43a}$$

and charges

$$q_{t^{i;\alpha}}=\frac{\alpha}{n_i+1}-\frac{1}{n_0+1} \tag{4.43b}$$

$$q_{r^i}=q_{p^i}=\frac{n_0}{n_0+1} \tag{4.43c}$$

---

* The 1-form pdw was used by Novikov and Veselov in their theory of algebro-geometric Poisson brackets [60]. The coordinates s^i are the algebro-geometric action variables of [60]. In [60] it was also an important point that derivatives $\partial_{s^i}(pdw)$ are the normalized holomorphic differentials.

$$q_{s^i} = q_{q^i} = -\frac{1}{n_0+1}. \tag{4.43d}$$

For the particular case $g = m = 0$ we obtain the Frobenius manifolds of the minimal models (the previous example). For $g = 0$, $m > 0$ we obtain models with rational functions as superpotentials.

The generating functions $h_{t^A}(t;\lambda)$ (3.4) have the form

$$h_{t^{i;\alpha}}(t;\lambda) = -\frac{n_i+1}{\alpha}\mathrm{res}_{p=\infty_i} k_i^\alpha {}_1F_1(1;1+\frac{\alpha}{n_i+1};\lambda w(p))dp. \tag{4.44a}$$

$$h_{p^i} = \mathrm{v.p.}\int_{\infty_0}^{\infty_i} e^{\lambda w}dp \tag{4.44b}$$

$$h_{q^i} = \mathrm{res}_{\infty_i}\frac{e^{\lambda w}-1}{\lambda}dp \tag{4.44c}$$

$$h_{r^i} = \oint_{b_i} e^{\lambda w}dp \tag{4.44d}$$

$$h_{s^i} = \frac{1}{2\pi i}\oint_{a_i} pe^{\lambda w}dw. \tag{4.44e}$$

Remark. Integrals of the form (4.44) seem to be interesting functions on the moduli space of the form $M_{g;n_0,\ldots,n_m}$. The simplest example of such an integral for a family of elliptic curves reads

$$\int_0^\omega e^{\lambda\wp(z)}dz \tag{4.45}$$

where $\wp(z)$ is the Weierstrass function with periods 2ω, $2\omega'$. For real negative λ a degeneration of the elliptic curve ($\omega\to\infty$) reduces (4.45) to the standard probability integral $\int_0^\infty e^{\lambda x^2}dx$. So the integral (4.45) is an analogue of the probability integral as a function on λ and on moduli of the elliptic curve. I recall that dependence on these parameters is specified by the equations (3.5), (2.17).

Gradients of this functions on the moduli space M have the form

$$\partial_{t^A}h_{t^{i;\alpha}} = \mathrm{res}_{\infty_i}k_i^{\alpha-n_i-1}{}_1F_1(1;\frac{\alpha}{n_i+1};\lambda w(p))\phi_{t^A}, \tag{4.46a}$$

$$\partial_{t^A}h_{p^i} = \eta_{t^A p^i} - \lambda\mathrm{v.p.}\int_{\infty_0}^{\infty_i} e^{\lambda w}\phi_{t^A} \tag{4.46b}$$

$$\partial_{t^A}h_{q^i} = \mathrm{res}_{\infty_i}e^{\lambda w}\phi_{t^A} \tag{4.46c}$$

$$\partial_{t^A}h_{r^i} = \eta_{t^A r^i} - \lambda\oint_{b_i} e^{\lambda w}\phi_{t^A} \tag{4.46d}$$

$$\partial_{t^A} h_{s^i} = \frac{1}{2\pi i} \oint_{a_i} e^{\lambda w} \phi_{t^A}. \tag{4.46e}$$

The generating function $V_{\alpha\beta}(t; \lambda, \mu)$ of coefficients of the τ-function (3.13) can be calculated via inner products (with respect to the matrix (4.35)) of (4.46). In particular, for a part of the Hessian of the primary free energy $F(t)$ (a function on M) one obtains [13, 14]

$$\frac{\partial^2 F}{\partial s^i \partial s^j} = -\tau_{ij} = -\oint_{b_j} \phi_{s^i}. \tag{4.47}$$

This is nothing but the matrix of periods of holomorphic differentials on M. Other second derivatives of F also turn out to be certain periods of some Abelian differentials on C.

Conclusion

WDVV is a universal system of integrable differential equations for periods of Abelian differentials on Riemann surfaces.

I recall that this system is a higher order analogue of the Painlevé-VI equation (i.e., equations of isomonodromy deformations of (2.15)). To specify the solution of WDVV one needs to find the monodromy matrix of the linear operator (2.15) for the eigenfunctions of the form (4.44). I will do it in a forthcoming publication.

We obtain the following picture of "Painlevé uniformisation" of the moduli spaces $M_{g;n_0,\ldots,n_m}$: (1) a global system of analytic coordinates on $M_{g;n_0,\ldots,n_m}$; (2) periods of Abelian differentials on curves $C \in M_{g;n_0,\ldots,n_m}$ are certain higher order Painlevé transcendents as functions of these coordinates.

Remark. For any Hamiltonian $H_{A,p}$ of the form (3.3), (4.44) one can construct a differential $\Omega_{A,p}$ on C or on the covering $\tilde{C}$ with singularities only at the marked infinite points such that

$$\frac{\partial}{\partial u^i} h_{t^A,p} = \text{res}_{P_i} \frac{\Omega_{A,p}dp}{dw}, \; i = 1, \ldots, n. \tag{4.48}$$

See[13] for an explicit form of these differentials (for $m = 0$ also see [14]).

Using these differentials the hierarchy (3.1) can be written in the Flaschka – Forest –McLaughlin form [16]

$$\partial_{T^{A,p}} dp = \partial_X \Omega_{A,p} \tag{4.49}$$

(derivatives of the differentials are to be calculated with w =const.).

The matrix $V_{(A,p),(B,q)}(t)$ determines a pairing of these differentials with values in functions on the moduli space

$$(\Omega_{A,p}, \Omega_{B,q}) = V_{(A,p),(B,q)}(t) \tag{4.50}$$

Particularly, the primary free energy F as a function on M can be written in the form [13, 14]

$$F = -\frac{1}{2}(pdw, pdw). \tag{4.51}$$

Note that the differential pdw can be written in the form

$$pdw = \sum \frac{n_i+1}{n_i+2}\Omega_{\infty_i}^{(n_i+2)} + \sum t^A \phi_{t^A} \tag{4.52}$$

where $\Omega_{\infty_i}^{(n_i+2)}$ is the Abelian differential of the second kind with a pole at ∞_i of the form

$$\Omega_{\infty_i}^{(n_i+2)} = dk_i^{n_i+2} + \text{ regular terms} \quad \text{near } \infty_i. \tag{4.53}$$

For the pairing (4.50) one can obtain from [44] the following formula

$$(f_1 dw, f_2 dw) = \frac{1}{2}\int\int_C (\bar\partial f_1 \partial f_2 + \partial f_1 \bar\partial f_2) \tag{4.54}$$

where the differentials ∂ and $\bar\partial$ along the Riemann surface should be understood in the distribution sense. The meromorphic differentials $f_1 dw$ and $f_2 dw$ on the covering $\tilde C$ should be considered to be piecewise meromorphic differentials on C with jumps on some cuts.

The corresponding hierarchy (3.1) is obtained by averaging along invariant tori of a family of g-gap solutions of a KdV-type hierarchy related to a matrix operator L of the matrix order $m+1$ and of orders $n_0+1, \dots, n_m+1$, in $\partial/\partial x$. The example $m=0$ (the averaged Gelfand–Dickey hierarchy) was considered in more details in [14]. The Poisson bracket (3.2) is a result of semiclassical limit (or averaging) [18-21] of the first Hamiltonian structure of the Gelfand–Dickey hierarchy; averaging of the second Hamiltonian structure (the classical W-algebra) gives the Poisson structure (3.21). Therefore, (3.21) can be considered to be the semiclassical limit of the classical W-algebras. The corresponding flat metric (3.22) on the moduli space $M_{g;n_0,\dots,n_m}$ is well-defined on a subset of Riemann surfaces having $w=0$, a non-ramifying point.

Also, for $g+m>0$ one needs to extend the KdV-type hierarchy to obtain (3.1) (see [13-14]). To explain the nature of such an extension let us consider the simplest example of $m=0$, $n_0=1$. The moduli space M consists of hyperelliptic curves of genus g with marked homology basis

$$y^2 = \prod_{i=1}^{2g+1}(w-w_i). \tag{4.55}$$

This parametrizes the family of g-gap solutions of the KdV. The L operator has the well-known form

$$L = -\partial_x^2 + u. \tag{4.56}$$

In real smooth periodic case $u(x+T)=u(x)$ the quasimomentum $p(w)$ is defined by the formula

$$\psi(x+T,w)=e^{ip(w)T}\psi(x,w) \tag{4.57}$$

for a solution $\psi(x,w)$ of the equation

$$L\psi=w\psi \tag{4.58}$$

(the Bloch – Floquet eigenfunction). The differential dp can be extended onto the family of all (i.e., quasiperiodic complex meromorphic) g-gap operators (4.55) as a normalized Abelian differential of the second kind with a double pole at the infinity $w=\infty$. (So the above superpotential (4.38) has the sense of the Bloch dispersion law, i.e., the dependence of the energy w on the quasimomentum p.) The Hamiltonians of the KdV hierarchy can be obtained as coefficients of the expansion of dp near the infinity. To obtain a complete family of conservation laws of the averaged hierarchy (3.1) one needs to extend the family of the KdV integrals by adding nonlocal functionals of u of the form

$$\oint_{a_i} w^k dp,\ \oint_{b_i} w^{k-1}dp,\ k=1,2,\dots. \tag{4.59}$$

As in (4.17) one can deform the above Frobenius structure on the moduli space $M=M_{g;n_0,\dots,n_m}$ by changing the differential dp,

$$dp\mapsto \tilde{dp}=\sum c_A\phi_{t^A} \tag{4.60}$$

for arbitrary constant coefficients. (The deformed Frobenius structure genericaly is well-defined only on a subset of M.) Particularly, if $\tilde{dp}$ is a differential of the third kind on C then the "dimension" d of this model always equals 1. The corresponding hierarchy (3.1) is obtained by averaging a Toda-type system.

Here I consider the simplest example of such a deformation. Let M be the 3-dimensional family of elliptic curves

$$y^2=4(w-c)^2-g_2(w-c)-g_3=4(w-c-e_1)(w-c-e_2)(w-c-e_3) \tag{4.61}$$

with ordered roots e_1, e_2, e_3. It is convenient to use the Weierstrass uniformization of (4.63)

$$w=\wp(z)+c \tag{4.62a}$$

$$y=\wp'(z) \tag{4.62b}$$

(I will use the standard notations [35] of the theory of elliptic functions). Let us use the holomorphic differential

$$dp = \frac{\pi i dz}{\omega} \tag{4.63}$$

to construct a Frobenius structure on M (here $\wp(\omega) = e_1$). The corresponding Landau – Ginsburg superpotential is the Weierstrass function (4.62a) where one should substitute $z = \omega p/\pi i$. The flat coordinates t^1, t^2, t^3 for the superpotential read

$$t^1 = -c + \frac{\eta}{\omega} \tag{4.64a}$$

$$t^2 = -1/\omega \tag{4.64b}$$

$$t^3 = 2\pi i \tau \quad \text{where } \tau = \omega'/\omega, \tag{4.64c}$$

$\wp(\omega') = e_3$, $\eta = -\int_0^\omega \wp(z)dz$. The charges of this manifold are $q_1 = 0$, $q_2 = \frac{1}{2}$, $q_3 = d = 1$.

Remark. The above models with $m = 0$, $g > 0$ can be obtained [39] in a semiclassical description of the correlators of multimatrix models (at the tree-level approximation for small couplings they correspond to various self-similar solutions of the hierarchy (3.1)) as functions of the couplings after passing through a point of gradient catastrophe. The idea of such a description originated in the theory of a dispersive analogue of shock waves [32]; see also [18].

More general algebraic-geometrical examples of solutions of WDVV were constructed in [44]. In these examples M is a moduli space of Riemann surfaces of genus g with a marked normalized Abelian differential of the second kind dw with poles at marked points and with fixed b-periods

$$\oint_{b_i} = B_i, \; i = 1, \dots, g.$$

For $B_i = 0$ one obtains the above Frobenius structures on $M_{g;n_0,\dots,n_m}$. Unfortunately, for $B \neq 0$ the Frobenius structures of [44] do not admit a conformal invariance.

Acknowledgments. I wish to thank E. Witten and C. Vafa for instructive and stimulating discussions. I acknowledge the support of INFN, Sez. di Napoli, where this paper was completed.

REFERENCES

[1] F. Brézin and V. Kazakov, *Phys. Lett.* **B 236** (1990) 144.
M. Douglas and S. Shenker, *Nucl. Phys.* **B 335** (1990) 635.
D.J. Gross and A. Migdal, *Phys. Rev. Lett.* **64** (1990) 127.
D.J. Gross and A. Migdal, *Nucl. Phys.* **B 340** (1990) 333.
T. Banks, M. Douglas, N. Seiberg, and S. Shenker, *Phys. Lett.* **B 238** (1990) 279.
M. Douglas, *Phys. Lett.* **B 238** (1990) 176.

[2] E. Witten, *Commun. Math. Phys.* **117** (1988) 353; **118** (1988) 411.

[3] E. Witten, *Nucl. Phys.* **B 340** (1990) 281.

[4] R. Dijkgraaf and E. Witten, *Nucl. Phys.* **B 342** (1990) 486.

[5] J. Distler, *Nucl. Phys.* **B 342** (1990) 523.

[6] K. Li, *Topological gravity and minimal matter*, CALT-68-1662, August 1990; *Recursion relations in topological gravity with minimal matter*, CALT-68-1670, September 1990.

[7] E. Martinec, *Phys. Lett.* **B 217** (1989) 431; *Criticality, Catastrophe and Compactifications*, V.G. Knizhnik memorial volume, 1989.
C. Vafa and N.P. Warner, *Nucl. Phys.* **B 324** (1989) 427.
S. Cecotti, L. Girardello and A. Pasquinucci, *Nucl. Phys.* **B 328** (1989) 701;
S. Cecotti, L. Girardello and A. Pasquinucci, *Int. J. Mod. Phys.* **A 6** (1991) 2427.

[8] R. Dijkgraaf. E. Verlinde and H. Verlinde, *Nucl. Phys.* **B 352** (1991) 59; *Notes on topological string theory and 2D quantum gravity*, PUPT-1217, IASSNS-HEP-90/80, November 1990.

[9] C. Vafa, *Mod. Phys. Lett.* **A 6** (1991) 337.

10. B. Blok and A. Varchenko, *Int. J. Mod. Phys.* **A7** (1992) 1467.

[11] I. Krichever, *Comm. Math. Phys.* **143** (1992) 415.

[12] I. Krichever, *Whitham theory for integrable systems and topological field theories.* To appear in Proceedings of Summer Corgese School, July 1991.

[13] B. Dubrovin, *Differential geometry of moduli spaces and its application to soliton equations and to topological conformal field theory*, Preprint No. 117 of Scuola Normale Superiore, Pisa, November 1991.

[14] B. Dubrovin, *Comm. Math. Phys.* **145** (1992) 195.

[15] G.B. Whitham, *Linear and Nonlinear Waves*, Wiley Intersci., New York - London - Sydney, 1974.
S. Yu. Dobrokhotov and V.P. Maslov, *Multiphase asymptotics of nonlinear PDE with a small parameter, Sov. Sci. Rev.: Math. Phys. Rev.* **3** (1982) 221.

[16]. H. Flaschka, M.G. Forest and D.W. McLaughlin, *Comm. Pure Appl. Math.* **33** (1980) 739.

[17] I. Krichever, *Funct. Anal. Appl.* **22** (1988) 200; *Russ. Math. Surveys*

44 (1989) 145.
[18] B. Dubrovin and S. Novikov, *Russ. Math. Surveys* **44**:6 (1989) 35.
[19] B. Dubrovin and S. Novikov, *Sov. Math. Doklady* **27** (1983) 665.
[20] S. Novikov, *Russ. Math. Surveys* **40**:4 (1985) 85.
[21] B. Dubrovin, *Geometry of Hamiltonian Evolutionary Systems*, Bibliopolis, Naples 1991.
[22] S. Tsarev, *Math. USSR Izvestija* **36** (1991); *Sov. Math. Dokl.* **34** (1985) 534.
[23] B. Dubrovin, *Funct. Anal. Appl.* **24** (1990).
[24] B. Dubrovin, *Funct. Anal. Appl.* **11** (1977) 265.
[25] S.P. Novikov (Ed.), *The Theory of Solitons: the Inverse Problem Method*, Nauka, Moscow, 1980. Translation: Plenum Press, N.Y., 1984.
[26] V. Zakharov and S. Manakov, *Sov. Phys. JETP* **42** (1976) 842.
[27] B. Dubrovin, *J. Sov. Math.* **28** (1985) 20; *Theory of operators and real algebraic geometry*, In: *Lecture Notes in Mathematics* **1334** (1988) 42.
[28] B. Dubrovin, I. Krichever and S. Novikov, *Integrable Systems*. I. Encyclopaedia of Mathematical Sciences, vol.4 (1985) 173, Springer-Verlag.
[29] L.D. Faddeev and L.A. Takhtajan, *Hamiltonian Methods in the Theory of Solitons*. Springer-Verlag, 1987.
[30] A.S. Fokas, R.A. Leo, L. Martina, and G. Soliani, *Phys. Lett.* **A 115** (1986) 329.
[31] B.M. McCoy, C.A. Tracy, T.T. Wu, *J. Math. Phys.* **18** (1977) 1058; M. Sato, T. Miwa, and M. Jimbo, *Publ. RIMS* **14** (1978) 223; **15** (1979) 201, 577, 871; **16** (1980) 531.
A. Jimbo, T. Miwa, Y. Mori, M. Sato, Physica **1D** (1980) 80.
H. Flaschka and A.C. Newell, *Comm. Math. Phys.* **76** (1980) 65.
A.R. Its and V. Yu. Novokshenov, *The Isomonodromic Deformation Method in the Theory of Painlevé Equations*, Lecture Notes in Mathematics 1191, Springer-Verlag, Berlin 1986.
[32] A.V. Gurevich and L.P. Pitaevskii, *Sov. Phys. JETP* **38** (1974) 291; *ibid.*, **93** (1987) 871; *JETP Letters* **17** (1973) 193.
V. Avilov and S. Novikov, *Sov. Phys. Dokl.* **32** (1987) 366.
V. Avilov, I. Krichever and S. Novikov, *Sov. Phys. Dokl.* **32** (1987) 564.
[33] G. Darboux, *Leçons sur les systèmes ortHogonaux et les cordonnées curvilignes*, Paris, 1897.
D. Th. Egoroff, *Collected papers on differential geometry*, Nauka, Moscow (1970) (in Russian).
[34] A. Balinskii and S. Novikov, *Sov. Math. Dokl.* **32** (1985) 228.
[35] W. Magnus, F. Oberhettinger and R.P. Soni, *Formulas and Theorems for the Special Functions of Mathematical Physics*, Springer-Verlag, Berlin-Heidelberg, New York, 1966.
[36] I.M. Gelfand and L.A. Dickey, *A Family of Hamilton Structures Related*

to Integrable Systems, preprint IPM/136 (1978) (in Russian).
M. Adler, *Invent. Math.* **50** (1979) 219.
I. Gelfand and I. Dorfman, *Funct. Anal. Appl.* **14** (1980) 223.

[37] P.J. Olver, *Math. Proc. Cambridge Philos. Soc.* **88** (1980) 71.

[38] R. Dijkgraaf, *Intersection Theory, Integrable Hierarchies and Topological Field Theory*, Preprint IASSNS-HEP-91/91, December 1991.

[39] B. Dubrovin, *Nucl. Phys.* **B 379** (1992) 627.

[40] S. Cecotti, C.Vafa, *Nucl. Phys.* **B367** (1991) 359.

[41] A.R. Its, A.G.Izergin, and V.E.Korepin, *Comm. Math. Phys.* **129** (1990) 205.

[42] M. Kontsevich, *Funct. Anal. Appl.* **25** (1991) 50.

[43] M. Kontsevich, *Comm. Math. Phys.* **147** (1992) 1.

[44] I. Krichever, *The τ-Function of the Universal Whitham Hierarchy, Matrix Models and Topological Field Theories*, Preprint LPTENS-92/18.

[45] B. Dubrovin, *Comm. Math. Phys.* **152** (1993), 539.

[46] E. Witten, *Surv. Diff. Geom.***1** (1991) 243.

[47] C. Vafa, *Topological Mirrors and Quantum Rings*, Preprint HUTP-91/A059.

[48] E. Witten, *Algebraic Geometry Associated with Matrix Models of Two-Dimensional Gravity*, Preprint IASSNS-HEP-91/74.

[49] E. Witten, *Int. J. Mod. Phys.* **A6** (1991) 2775.

[50] M.F. Atiyah, *Topological Quantum Field Theories*, Publ. Math. I.H.E.S. **68** (1988) 175.

[51] R. Dijkgraaf, *A Geometrical Approach to Two-Dimensional Conformal Field Theory*, Ph.D. Thesis (Utrecht, 1989).

[52] F. Magri, *J. Math. Phys.* **19** (1978) 1156.

[53] K. Takasaki and T. Takebe, *Quasi-classical Limit of KP Hierarchy, W-Symmetries and Free Fermions*, Preprint KUCP-0050/92; *W-Algebra, Twistor, and Nonlinear Integrable Systems*, Preprint KUCP-0049/92.

[54] Y. Kodama, *Phys. Lett.* **129A** (1988) 223; *Phys. Lett.* **147A** (1990) 477;
Y. Kodama and J. Gibbons, *Phys. Lett.***135A** (1989) 167.

[55] E. Brezin, C. Itzykson, G. Parisi, and J.-B.Zuber, *Comm. Math. Phys.* **59** (1978) 35.
D. Bessis, C. Itzykson, and J.-B. Zuber, *Adv. Appl. Math.* **1** (1980) 109.
M.L. Mehta, *Comm. Math. Phys.* **79** (1981) 327;
S. Chadha, G. Mahoux, and M.L.Mehta, *J.Phys.* **A14** (1981) 579.

[56] S. Natanzon, *Sov. Math. Dokl.* **30** (1984) 724.

[57] E. Witten, *Lectures on Mirror Symmetry*, In: *Proceedings MSRI Conference on Mirror Symmetry*, March 1991, Berkeley.

[58] B. Dubrovin and S. Novikov, *Sov. Math. Doklady* **279** (1984) 294.

[59] V. Zakharov, *Funct. Anal. Appl.* **14** (1980).

[60] S. Novikov and A.Veselov, *Sov. Math. Doklady* (1981); *Proceedings of Steklov Institute* (1984).

[61] P.D. Lax, C.D. Levermore, and S. Venakides, *The Generation and Propagation of Oscillations in Dispersive IVPs and their Limiting Behaviour*, In: *Important Developments in Soliton Theory 1980 - 1990*, Eds. A. Fokas and V. Zakharov.

On leave from Department of Mechanics and Mathematics, Moscow State University, 119899 Moscow, Russia

Present address:
SISSA
via Beirut, 2
I-34013 Trieste, Italy

Participants

M.A. Annamalai
Pondicherry University
Department of Mathematics
Kalapet, Pondicherry 605014
India

O. Babelon
CNRS -Université de Paris VI
L.P.T.H.E.
4, place Jussieu
75252 - Paris Cedex 05
France

D. Bennequin
Université de Strasbourg I
Laboratoire de Mathématiques
7, rue René Descartes
67084 - Strasbourg Cedex,
France

A. Bruguières
Université de Paris VII
UFR de Mathématiques
2, place Jussieu
75251 Paris Cedex 05
France

P. Cartier
I.H.E.S.
35, route de Chartres
91440 Bures-sur-Yvette,
France

A. Chenciner
Université de Paris VII
UFR de Mathématiques
2, place Jussieu
75251 Paris Cedex 05
France

A. Cohen
Université de Paris XIII
Mathématiques
Avenue J.-B. Clément
93430 Villetaneuse
France

P. Damianou
P.O. Box 6601
Limassol
Cyprus

E. Date
Osaka University
Faculty of Engineering Science
Toyonaka, Osaka 560
Japan

L. Dickey
University of Oklahoma
Department of Mathematics
601 Elm Avenue
Norman, Oklahoma 73019
USA

B. Dubrovin
SISSA
Via Beirut,2
34013 Trieste
Italy

N. M. Ercolani
The University of Arizona
Department of Mathematics
Tucson, Arizona 85721
USA

H. Flaschka
The University of Arizona
Department of Mathematics
Tucson, Arizona 85721
USA

J.-P Françoise
Université de Paris VI
4, place Jussieu
75252 Paris Cedex 05
France

L. Freidel
E.N.S. Lyon
Département de Physique Théorique
46, allée d'Italie
69364 Lyon Cedex 07
France

P. Gauduchon
Ecole Polytechnique
URA 169 CNRS
91128 Palaiseau
France

R. Gergondey
Université de Lille I
UFR de Mathématiques
59655 Villeneuve d'Ascq Cedex
France

D. Gurevitch
Max Planck Inst. für Mathematik
Gottfried Claren Str. 26
5300 Bonn 3
Germany

A.B. Guerrero
National University of Colombia
Department of Mathematics
Bogota
Colombia

V. Guillemin
MIT
Department of Mathematics
Cambridge, Massachusetts 02139
USA

L. Haine
Université Catholique de Louvain
Institut de Mathématique
Pure et Appliquée
Chemin de Cyclotron, 2
1348, Louvain-la-Neuve
Belgium

N. Kamran
Mc Gill University
Department of Mathematics
Montreal, Quebec H3A 2K6
Canada

T. Kohno
Kyushu University 33
Department of Mathematics
Hakozaki, Fukyoka, 812
Japan

H. Knörrer
ETH Zentrum
8092 Zürich
Switzerland

Y. Kosmann-Schwarzbach
Université de Lille I
UFR de Mathématiques
59655 Villeneuve d'Ascq Cedex
France

Lê Dung Trang
Université de Paris VII
UFR de Mathématiques
2, place Jussieu
75251 Paris Cedex 05
France

Lê Ngoc Chuyen
Institute of Mathematics
PO Box 631, Bo Ho
10 000 Hanoi
Vietnam

F. Magri
Università di Milano
Dipart. di Mat. "F. Enriques"
Via Saldini 50
20133 Milano
Italy

J.-M. Maillard
Université de Paris VI
L.P.T.H. E., CNRS
4, place Jussieu
75252 Paris Cedex 05
France

G. Maltsiniotis
Université de Paris VII
UFR de Mathématiques
2, place Jussieu
75251 Paris Cedex, France

Nguyen Viet Dung
Institute of Mathematics
P.O. Box 631, Bo Ho
10 000 Hanoi
Vietnam

S.P. Novikov
Academy of Sciences
L.D. Landau Inst. for Theor. Physics
Kosygina, 2
117940, GSP-1, Moscow, V-334
Russia

P. Olver
School of Mathematics
University of Minnesota
Minneapolis, Minnesota 55455
USA

M. Pedroni
Università di Milano
Dipart. di Mat. "F. Enriques"
Via Saldini, 50
20133 Milano
Italy

L.A. Piovan
Universidad Nacional del Sur
Departemento de Matematica
Av. Alem 1253
8000 Bahia Blanca
Argentina

E. Previato
Boston University
Department of Mathematics
Boston, Massachusetts 02215
USA

T. Ratiu
University of California
Department of Mathematics
Santa Cruz, California 95064
USA

M. Rosso
Université de Strasbourg I
Laboratoire de Mathématiques
7, rue René Descartes
67084 Strasbourg Cedex
France

M.A. Semenov-Tian-Shansky
LOMI
Fontanka 27
Saint-Petersburg, 191011
Russia

L.A. Takhtajan
SUNY at Stony Brook
Department of Mathematics
Stony Brook, NY 11794-3651
USA

K.M. Tamizhmani
Pondicherry University
Department of Mathematics
Kalapet, Pondicherry 650001
India

H. Torriani
Univ. Estadual de Campinas
Inst. de Mat. Est.
e Ciencias de Computacao
13081 Campinas SP
Brazil

A. Treibich
Université de Lille I
UFR de Mathématiques
59655 Villeneuve d'Ascq Cedex
France

H. Umemura
Kumamoto University
Department of Mathematics
Kumamoto 860
Japan

J.F. van Diejen
University of Amsterdam
Fac. Math/Computer Science
Plantage Muidergracht 24
1018 TV Amsterdam
The Netherlands

P. Vanhaecke
Katholiecke Universiteit Leuven
Department of Mathematics
Celestijnenlaan 200B
3001 Leuven
Belgium

P. van Moerbeke
Université Catholique de Louvain
Institut de Mathématique
Pure et Appliquée
Chemin du Cyclotron, 2
1348, Louvain-la-Neuve
Belgium

G. Wilson
London University
Imperial Coll. Sci. Techn.
Department of Mathematics
London, SW7 2BZ
Great Britain

J.B. Zuber
C.E.A. Saclay
Service Physique Théorique
91191 Gif-sur-Yvette Cedex
France

Index

AKNS equations, 131, 132, 133, 140, 147, 153, 157, 159, 161
Algebraically integrable equations, 121

Baker (wave) functions, 43, 44, 116, 120, 126, 147, 149, 150, 152, 154, 155, 157, 158, 161, 163
Bäcklund transformations, 163, 164, 165, 167, 168, 171
Baxterization, 292, 294
Betti numbers, 227
Bihamiltonian structures, 239, 252, 335, 338, 342
Birational transformations, 48, 283, 286
Boltzmann weights, 279

Casimir functions, 182, 194, 195, 202, 217, 219, 223, 253
Central extensions, 136
Calabi-Yau models, 342, 343
Character formula, 62
Chazy equation, 122, 123, 125, 126
Chern-Simons action, 116
Chiral Potts model, 275
Complexity, 235
Conformal field theory, 62, 313, 314, 316
Correlation functions, 314, 315
Coxeter groups, 283
Cyclic representations, 275

Deformation of Hamiltonian actions, 227
Dressing transformations, 133, 134, 258

Eigenvalue pairs, 242, 243, 244, 245
Elliptic solitons, 45, 46, 47, 50

Factorization method, 182, 192, 220
Fay identity, 168
Flag manifolds, 183, 196, 203, 204, 205, 211, 214, 215, 219
Flat connections, 133
Fractional powers, 261
Fredholm
determinants, 117
operators, 148
Frobenius
algebras, 329
manifolds, 321, 323, 324, 325, 330
Fuchsian groups, 124, 125
Full Kostant-Toda lattice, 181
Fusion algebras, 62

Gelfand-Dickey polynomials, 262
Grassmannians, 116, 126, 147, 148, 154, 155, 157, 158, 159, 165, 166, 167

Halphen system, 118, 120, 121, 122, 126
Heat equation, 73
Heisenberg group, 64
Hirota's direct method, 131
Hyperelliptic curves, 75, 78, 81, 82, 84

Intersection numbers, 315, 341
Isomonodromy deformations, 317, 330, 331, 333, 334, 352
Isospectral equations, 164
Iteration of mappings, 292

Jacobians, 39, 40, 41, 47, 52, 58, 76, 173

Kac-Moody algebras, 62, 126
Korteweg-de Vries equation, 163, 165, 252
KP hierarchy, 39, 44, 147, 155, 156, 158, 159
Krichever correspondence, 39, 42, 43, 44, 126

Lax equations, 116, 119, 255, 256, 263
Levi decomposition, 191, 213
Lie-Poisson structures, 189, 191
Loop
  algebras, 133
  groups, 132

Mean curvature, 81, 82, 83, 87, 90
Modular forms, 115, 120, 121, 123, 124, 126, 127
Monodromy transformations, 317

Painlevé equations, 117, 317, 331, 333, 352
Parabolic subalgebras, 190, 191
Path integrals, 313, 314
Poisson pencils, 252, 254, 258, 266
Pseudodifferential operators, 165, 251, 266, 269, 270

Quadrics, 69, 73
Quantized universal algebras, 275
Quartics, 69, 73
Quasiperiodic solutions, 82

Reduction, 116, 183, 220, 254, 261
R-matrices, 275, 286, 289, 294
Resolvents, 153

Sato's equations, 265, 268, 270
Schur polynomials, 168, 169
Schottky problem, 61
Self-dual Yang-Mills equations, 115, 116, 117, 126
Sinh-Gordon equation, 81
Sklyanin brackets, 266
Solitons, 147, 264
Spectral curves, 77, 172
Spectral parameters, 133, 134, 163, 277, 282, 289, 291, 303
Star-triangle relations, 279, 282, 283
Symplectic leaves, 187, 254
Systems of hydrodynamic type, 318, 334

τ-functions, 39, 44, 115, 116, 117, 122, 126, 127, 131, 132, 139, 140, 147, 150, 156, 157, 158, 161, 163, 170, 264
Tetrahedron equations, 289, 291, 297
Theta
  characteristics, 64, 65, 67, 68, 75
  divisors, 39, 41, 45, 63, 65, 67
  functions, 39, 41, 44, 56, 61, 64, 69, 70, 73, 74, 83, 117
Toda lattice, 116, 181
Topological
  classification, 237
  field theory, 314, 315, 316, 335, 337, 342
Torelli theorem, 42
Tori, 81, 90
Torus orbits, 209, 210, 211, 212, 215, 219, 220

Verlinde numbers, 62, 63, 67, 69, 70
Vertex
  models, 281
  sets, 227

Weyl groups, 284

Yang-Baxter equations, 281, 285, 287
Young diagrams, 167, 168, 169, 170

Progress in Mathematics

Edited by:

J. Oesterlé
Département de Mathématiques
Université de Paris VI
4, Place Jussieu
75230 Paris Cedex 05, France

A. Weinstein
Department of Mathematics
University of California
Berkeley, CA 94720
U.S.A.

Progress in Mathematics is a series of books intended for professional mathematicians and scientists, encompassing all areas of pure mathematics. This distinguished series, which began in 1979, includes authored monographs and edited collections of papers on important research developments as well as expositions of particular subject areas.

We encourage preparation of manuscripts in some form of TeX for delivery in camera-ready copy which leads to rapid publication, or in electronic form for interfacing with laser printers or typesetters.

Proposals should be sent directly to the editors or to: Birkhäuser Boston, 675 Massachusetts Avenue, Cambridge, MA 02139, U. S. A.

1 GROSS. Quadratic Forms in Infinite-Dimensional Vector Spaces
2 PHAM. Singularités des Systèmes Différentiels de Gauss-Manin
3 OKONEK/SCHNEIDER/SPINDLER. Vector Bundles on Complex Projective Spaces
4 AUPETIT. Complex Approximation, Proceedings, Quebec, Canada, July 3-8, 1978
5 HELGASON. The Radon Transform
6 LION/VERGNE. The Weil Representation, Maslov Index and Theta Series
7 HIRSCHOWITZ. Vector Bundles and Differential Equations Proceedings. Nice, France, June 12-17, 1979
8 GUCKENHEIMER/MOSER/NEWHOUSE. Dynamical Systems, C.I.M.E. Lectures. Bressanone, Italy, June, 1978
9 SPRINGER. Linear Algebraic Groups
10 KATOK. Ergodic Theory and Dynamical Systems I
11 BALSEV. 18th Scandinavian Conferess of Mathematicians, Aarhus, Denmark, 1980
12 BERTIN. Séminaire de Théorie des Nombres, Paris 1979-80
13 HELGASON. Topics in Harmonic Analysis on Homogeneous Spaces
14 HANO/MARIMOTO/MURAKAMI/ OKAMOTO/OZEKI. Manifolds and Lie Groups: Papers in Honor of Yozo Matsushima
15 VOGAN. Representations of Real Reductive Lie Groups
16 GRIFFITHS/MORGAN. Rational Homotopy Theory and Differential Forms
17 VOVSI. Triangular Products of Group Representations and Their Applications
18 FRESNEL/VAN DER PUT. Géométrie Analytique Rigide et Applications
19 ODA. Periods of Hilbert Modular Surfaces
20 STEVENS. Arithmetic on Modular Curves

21 KATOK. Ergodic Theory and Dynamical Systems II
22 BERTIN. Séminaire de Théorie des Nombres, Paris 1980-81
23 WEIL. Adeles and Algebraic Groups
24 LE BARZ/HERVIER. Enumerative Geometry and Classical Algebraic Geometry
25 GRIFFITHS. Exterior Differential Systems and the Calculus of Variations
26 KOBLITZ. Number Theory Related to Fermat's Last Theorem
27 BROCKETT/MILLMAN/SUSSMAN. Differential Geometric Control Theory
28 MUMFORD. Tata Lectures on Theta I
29 FRIEDMAN/MORRISON. Birational Geometry of Degenrations
30 YANO/KON. CR Submanifolds of Kaehlerian and Sasakian Manifolds
31 BERTRAND/WALDSCHMIDT. Approximations Diophantiennes et Nombres Transcendants
32 BOOKS/GRAY/REINHART. Differential Geometry
33 ZUILY. Uniqueness and Non-Uniqueness in the Cauchy Problem
34 KASHIWARA. Systems of Microdifferential Equations
35 ARTIN/TATE. Arithmetic and Geometry: Papers Dedicated to I. R. Shafarevich on the Occasion of His Sixtieth Birthday. Vol. 1
36 ARTIN/TATE. Arithmetic and Geometry: Papers Dedicated to I. R. Shafarevich on the Occasion of His Sixtieth Birthday. Vol. 1I
37 DE MONVEL. Mathématique et Physique
38 BERTIN. Séminaire de Théorie des Nombres, Paris 1981-82
39 UENO. Classification of Algebraic and Analytic Manifolds
40 TROMBI. Representation Theory of Reductive Groups
41 STANLEY. Combinatorics and Commutative Algebra
42 JOUANOLOU. Théorèmes de Bertini et Applications
43 MUMFORD. Tata Lectures on Theta II
44 KAC. Infitine Dimensional Lie Algebras
45 BISMUT. Large deviations and the Malliavin Calculus
46 SATAKE/MORITA. Automorphic Forms of Several Variables, Taniguchi Symposium, Katata, 1983
47 TATE. Les Conjectures de Stark sur les Fonctions L d'Artin en $s = 0$
48 FRÖLICH. Classgroups and Hermitian Modules
49 SCHLICHTKRULL. Hyperfunctions and Harmonic Analysis on Symmetric Spaces
50 BOREL ET AL. Intersection Cohomology
51 BERTIN/GOLDSTEIN. Séminaire de Théorie des Nombres, Paris 1982-83
52 GASQUI/GOLDSCHMIDT. Déformations Infinitesimales des Structures Conformes Plates
53 LAURENT. Théorie de la Deuxième Microlocalisation dans le Domaine Complexe
54 VERDIER/LE POTIER. Module des Fibres Stables sur les Courbes Algébriques: Notes de l'Ecole Normale Supérieure, Printemps, 1983
55 EICHLER/ZAGIER. The Theory of Jacobi Forms
56 SHIFFMAN /SOMMESE. Vanishing Theorems on Complex Manifolds
57 RIESEL. Prime Numbers and Computer Methods for Factorization
58 HELFFER/NOURRIGAT. Hypoellipticité Maximale pour des Opérateurs Polynomes de Champs de Vecteurs
59 GOLDSTEIN. Séminaire de Théorie des Nombres, Paris 1983–84
60 PROCESI. Geometry Today: Giornate Di Geometria, Roma. 1984

61 BALLMANN/GROMOV/SCHROEDER. Manifolds of Nonpositive Curvature
62 GUILLOU/MARIN. A la Recherche de la Topologie Perdue
63 GOLDSTEIN. Séminaire de Théorie des Nombres, Paris 1984–85
64 MYUNG. Malcev-Admissible Algebras
65 GRUBB. Functional Calculus of Pseudo-Differential Boundary Problems
66 CASSOU-NOGUES/TAYLOR. Elliptic Functions and Rings and Integers
67 HOWE. Discrete Groups in Geometry and Analysis: Papers in Honor of G.D. Mostow on His Sixtieth Birthday
68 ROBERT. Autour de L'Approximation Semi-Classique
69 FARAUT/HARZALLAH. Deux Cours d'Analyse Harmonique
70 ADOLPHSON/CONREY/GHOSH/YAGER. Analytic Number Theory and Diophantine Problems: Proceedings of a Conference at Oklahoma State University
71 GOLDSTEIN. Séminaire de Théorie des Nombres, Paris 1985–86
72 VAISMAN. Symplectic Geometry and Secondary Characteristic Classes
73 MOLINO. Riemannian Foliations
74 HENKIN/LEITERER. Andreotti-Grauert Theory by Integral Formulas
75 GOLDSTEIN. Séminaire de Théorie des Nombres, Paris 1986–87
76 COSSEC/DOLGACHEV. Enriques Surfaces I
77 REYSSAT. Quelques Aspects des Surfaces de Riemann
78 BORHO/BRYLINSKI/MACPHERSON. Nilpotent Orbits, Primitive Ideals, and Characteristic Classes
79 MCKENZIE/VALERIOTE. The Structure of Decidable Locally Finite Varieties
80 KRAFT/PETRIE/SCHWARZ. Topological Methods in Algebraic Transformation Groups
81 GOLDSTEIN. Séminaire de Théorie des Nombres, Paris 1987–88
82 DUFLO/PEDERSEN/VERGNE. The Orbit Method in Representation Theory: Proceedings of a Conference held in Copenhagen, August to September 1988
83 GHYS/DE LA HARPE. Sur les Groupes Hyperboliques d'après Mikhael Gromov
84 ARAKI/KADISON. Mappings of Operator Algebras: Proceedings of the Japan-U.S. Joint Seminar, University of Pennsylvania, Philadelphia, Pennsylvania, 1988
85 BERNDT/DIAMOND/HALBERSTAM/HILDEBRAND. Analytic Number Theory: Proceedings of a Conference in Honor of Paul T. Bateman
86 CARTIER/ILLUSIE/KATZ/LAUMON/MANIN/RIBET. The Grothendieck Festschrift: A Collection of Articles Written in Honor of the 60th Birthday of Alexander Grothendieck. Vol. I
87 CARTIER/ILLUSIE/KATZ/LAUMON/MANIN/RIBET. The Grothendieck Festschrift: A Collection of Articles Written in Honor of the 60th Birthday of Alexander Grothendieck. Volume II
88 CARTIER/ILLUSIE/KATZ/LAUMON/MANIN/RIBET. The Grothendieck Festschrift: A Collection of Articles Written in Honor of the 60th Birthday of Alexander Grothendieck. Volume III
89 VAN DER GEER/OORT / STEENBRINK. Arithmetic Algebraic Geometry
90 SRINIVAS. Algebraic K-Theory
91 GOLDSTEIN. Séminaire de Théorie des Nombres, Paris 1988–89
92 CONNES/DUFLO/JOSEPH/RENTSCHLER. Operator Algebras, Unitary Representations, Enveloping Algebras, and Invariant Theory. A Collection of Articles in Honor of the 65th Birthday of Jacques Dixmier

93 AUDIN. The Topology of Torus Actions on Symplectic Manifolds
94 MORA/TRAVERSO (eds.) Effective Methods in Algebraic Geometry
95 MICHLER/RINGEL (eds.) Representation Theory of Finite Groups and Finite Dimensional Algebras
96 MALGRANGE. Equations Différentielles à Coefficients Polynomiaux
97 MUMFORD/NORI/NORMAN. Tata Lectures on Theta III
98 GODBILLON. Feuilletages, Etudes géométriques
99 DONATO /DUVAL/ELHADAD/TUYNMAN. Symplectic Geometry and Mathematical Physics. A Collection of Articles in Honor of J.-M. Souriau
100 TAYLOR. Pseudodifferential Operators and Nonlinear PDE
101 BARKER/SALLY. Harmonic Analysis on Reductive Groups
102 DAVID. Séminaire de Théorie des Nombres, Paris 1989-90
103 ANGER /PORTENIER. Radon Integrals
104 ADAMS /BARBASCH/VOGAN. The Langlands Classification and Irreducible Characters for Real Reductive Groups
105 TIRAO/WALLACH. New Developments in Lie Theory and Their Applications
106 BUSER. Geometry and Spectra of Compact Riemann Surfaces
108 BRYLINSKI. Loop Spaces, Characteristic Classes and Geometric Quantization
108 DAVID. Séminaire de Théorie des Nombres, Paris 1990-91
109 EYSSETTE/GALLIGO. Computational Algebraic Geometry
110 LUSZTIG. Introduction to Quantum Groups
111 SCHWARZ. Morse Homology
112 DONG/LEPOWSKY. Generalized Vertex Algebras and Relative Vertex Operators
113 MOEGLIN/WALDSPURGER. Décomposition spectrale et séries d'Eisenstein
114 BERENSTEIN/GAY/VIDRAS/YGER. Residue Currents and Bezout Identities
115 BABELON/CARTIER/KOSMANN-SCHWARZBACH. Integrable Systems, The Verdier Memorial Conference: Actes du Colloque International de Luminy
116 DAVID. Séminaire de Théorie des Nombres, Paris 1991-92

THERAPEUTIC CHANGE WITH DIFFICULT CLIENTS

Second Edition

Precursors and Techniques in the CHANGES Model

BRETT D. WILKINSON | FRED J. HANNA

Published by
American Psychological Association
750 First Street, NE
Washington, DC 20002
https://www.apa.org

Order Department
https://www.apa.org/pubs/books
order@apa.org

Typeset in Meridien and Ortodoxa by Lumina Datamatics, India

Printer: Sheridan Books, Chelsea, MI
Cover Designer: Anthony Paular Design, Newbury Park, CA

Library of Congress Cataloging-in-Publication Data

Names: Wilkinson, Brett D. author | Hanna, Fred J. author
Title: Therapeutic change with difficult clients : precursors and techniques in the CHANGES model / by Brett D. Wilkinson and Fred J. Hanna.
Other titles: Therapy with difficult clients
Description: Second edition. | Washington, DC : American Psychological Association, [2025] | Revised edition of: Therapy with difficult clients : using the precursors model to awaken change / Fred J. Hanna. Washington, DC : American Psychological Association, c2002. | Includes bibliographical references and index.
Identifiers: LCCN 2024061452 (print) | LCCN 2024061453 (ebook) | ISBN 9781433843167 paperback | ISBN 9781433843174 ebook
Subjects: LCSH: Impasse (Psychotherapy) | Attitude change
Classification: LCC RC489.I45 H36 2025 (print) | LCC RC489.I45 (ebook) | DDC 616.8914—dc23/eng/20250307
LC record available at https://lccn.loc.gov/2024061452
LC ebook record available at https://lccn.loc.gov/2024061453

https://doi.org/10.1037/0000451-000

Printed in the United States of America

10 9 8 7 6 5 4 3 2 1

For Katherine: my confidante, my muse, my love.

—BW

To my wife, Constance Hanna, for her profound insight, knowledge, compassion, and understanding.

—FH

CONTENTS

Preface ix
Acknowledgments xi
Introduction 3

I. ACTIVATING CHANGE WITH DIFFICULT CLIENTS 13

1. Toward a Model of Change for Difficult Clients 15
2. What Makes a Client Difficult? 35

II. THE CHANGES MODEL, STRATEGIES, AND TECHNIQUES 49

3. Orienting the Relationship Around Change 51
4. C: Confronting the Problem 69
5. H: Hope for Change 89
6. A: Awareness of the Problem 105
7. N: Necessity for Change 127
8. G: Grit, or Willingness to Experience Anxiety or Difficulty 145
9. E: Effort, or Will to Change 167
10. S: Social Support for Change 183

III. THE CHANGES ASSESSMENT 195

11. Rating Client Potential for Therapeutic Change 197
12. Rating Therapist Interference in the Change Process 209

IV. SPECIALIZED APPLICATIONS **223**

13. Guidance for Advanced Training and Supervision **225**

14. Oppression, Perspicacity, and Liberation **239**

15. Addiction and Substance Use **255**

V. CONCLUSION **271**

16. At the Horizon of Change **273**

References 289
Index 317
About the Authors 327

PREFACE

We have spent our careers seeing clients in therapy, training and supervising graduate students, providing consultations in the community, and conducting research on therapy processes. Across these avenues, our study of therapeutic change has been a persistent theme. We consider the subject of therapeutic change to be one of the most intriguing and extraordinary topics among the vast range of phenomena that constitute human life. Change remains ever at the heart of our work as therapists and, indeed, at the root of our personal experiences in navigating a fulfilling life.

Well-established techniques quite often hit the mark with relative ease among those clients who are willing and able to intentionally engage in the therapeutic change process. However, one of the most challenging aspects of therapy is working with clients who have little interest in change, or who think that change is a waste of time, or who somehow have come to believe that change is a threat to their personal freedom or sense of being. The limits of psychotherapy effectiveness become starkly evident when we look steadfastly at the difficulty of creating change for such clients, making it one of the most important areas of consideration for ongoing research efforts.

The many theories of psychotherapy, from behavioral to psychodynamic and from existential to family systems, are remarkably instructive and descriptive of the subtleties and complexities of human thought, feeling, behavior, and relationships. There is no substitute for the study of the great insights these theories catalog and represent. Add to those the integrative and eclectic approaches to psychotherapy, and one has a rich and wide spectrum of perspectives on human problems. Nevertheless, classic theories and innovative integrative approaches do not go far enough in capturing the essential structures and dynamics of change. We remain in need of the knowledge of what makes people change. We also believe there is a need for more techniques and approaches to help such clients and the therapists who work with them.

Our investigations of therapeutic change with difficult clients have revealed a host of subtle processes that affect change. Such subtle processes often go unacknowledged in routine psychotherapy, not because they are hidden but because they are pervasive. Such pervasiveness often makes them difficult to identify and articulate. For example, what are the subtle processes that move a callous, cruel man from nearly exclusive self-centeredness toward opening up to the feelings and views of others? What are the dynamics that make an involuntary, defiant, court-ordered client wake up to become aware of a problem and involved in therapy? What brings a helplessly depressed person to the point of realizing that depression is something that can be influenced and relieved? Similarly, when a client is sitting on the fence between taking responsibility and blaming others, what nudges the person toward responsibility? These are difficult questions and certainly require further research, but we now know that such changes result not from a single process but a combination of many.

A major purpose of this book is to identify seven of these processes and move them from the vague to the defined. We have called the processes precursors because they herald the arrival of change. When the precursors dawn, change is on the horizon. We hope to outline the intricate and variable manner in which the precursors act interdependently to produce change. To help therapists use the CHANGES model, we propose a clinical assessment tool that reveals which of the precursors is missing in the client who is not achieving change. Finally, we provide techniques for implementing each precursor in accordance with the assessment. This is the essence of the precursors approach to therapeutic change. The precursors are offered as an aid to clients for whom therapy is a painful and fruitless struggle.

From the outset, we want to make it clear that the precursors are neither our invention nor our discovery. They were recognized long before the dawn of psychotherapy. We have only shown how they interact and described an array of techniques that stimulate the precursors in clients who may not otherwise achieve positive change. We share these techniques and approaches in the hope they will help therapists work with difficult clients to achieve change sooner rather than later.

ACKNOWLEDGMENTS

I (Brett) have had the privilege of learning from and connecting with so many inspiring figures, wise friends, and good colleagues over the years. I would like to first thank Fred Hanna, who has been an inspirational mentor and good friend. The opportunity to collaborate on this important work has been a pleasure, and I am grateful for the chance to contribute to the field in this way.

To my wife, Katherine, whose insightfulness as a therapist, patience as my life partner, and boundless love as a mother, has been my lodestar. To our children, Elianora and Rowan, whose earnestness and curiosity are a welcomed reminder of what matters most in life. To my father, stepmother, and sister, for their unwavering love and support, I am forever grateful. And to my late mother, of course, whose heart endures.

I would also like to share my gratitude for the mentorships of the late counselor Tim Robertson, the evolutionary psychologist Harmon Holcomb, and the Husserlian scholar Ronald Bruzina, all of whom fostered my curiosity and lent compassionate guidance at critical stages in my journey. Appreciation also extends to my friend of 30 years, Kusha Sefat, whose intellectual virtuosity is equaled only by his humor, adaptability, and brilliance at the piano. And to my many colleagues, graduate students, and former clients: I have gained so much from my relations with each of you, and I am thankful for the chance to have played a small part in your big life story.

I (Fred) have been inspired by many people in this field. Some I have met in person, and others I have not. I am grateful to them all for providing insights that have helped me find my path in this fascinating and gratifying work. First, I want to thank my wife, Constance Hanna, a wonderfully effective and empathic therapist. Her wisdom and advice have kept me practical and concrete.

I also thank two professional philosophers from whom I benefited enormously. Ramakrishna Puligandla taught me the subtleties and liberating potential of dialectical philosophy from the viewpoints of both Kant and Nagarjuna. James Daley, friend and mentor, validated my ideas and taught me the insider's perspective on phenomenology as a philosophical discipline and its implications for psychology. Studying with him brought Husserl and Heidegger to life, especially in terms of consciousness, dialectics, and ontology. My study with both of these men afforded great insight into the nature and goals of psychotherapy from a philosophical perspective.

My collaboration with Kaisa Puhakka has been nothing less than extraordinary. I am deeply thankful to her for teaching me the subtleties of object relations therapy. Most of all, she has always been willing to take part in a vital interchange of ideas on Asian philosophy and psychology that was part of our perennial quest to understand the human predicament.

The pioneering insights of Jerome Frank are now legendary, and I am deeply grateful to him for the extraordinarily helpful conversations we had during my time at Johns Hopkins. I thank George Howard for his wisdom, innovative research, and progressive ideas concerning psychotherapy. When I was a graduate student, he took me under his wing and patiently answered dozens of questions. He helped shed light on a path that would have taken me many years to find on my own.

I also thank Hal Arkowitz, who encouraged me to write the original article on which this book is based. Michael Mahoney's comments on my work have been of great value. His deep and global view of psychotherapy and therapeutic change has been a rich source of insight for me for many years. I extend my heartfelt thanks to him for his invaluable suggestions concerning the original draft of this book. I have also found the work of Arnold Lazarus to be a source of instruction and learning. I have greatly admired his teachings, bold thinking, and commentary on psychotherapy as well as his therapeutic wisdom. I owe a debt of gratitude to Al Ottens, who has always been there for perceptive comments, support, and reassurance. I also thank Martin Ritchie, Chris Aanstoos, and Lorean Roberts for their professional and personal support. Finally, I want to acknowledge and extend deep thanks to my friend and coauthor, Brett Wilkinson. He is a person of great insight, creativity, and vision. Our nearly 15-year association has been most rewarding, supportive, and fortuitous.

In terms of the manuscript itself, we would like to thank Susan Reynolds, Senior Acquisitions Editor at American Psychological Association Books, for her patience and encouragement in this long process. We thank Molly Gage for her guidance in development.

THERAPEUTIC CHANGE WITH DIFFICULT CLIENTS

Introduction

In this book, we endeavor to clarify that which brings about beneficial change in human beings. Its promise lies in the fact that it concentrates on within-individual catalysts that bring therapeutic change into being. Psychotherapy has grown and evolved immensely in the intervening 140 years since Sigmund Freud studied hypnosis under Jean-Martin Charcot at the Pitié-Salpêtrière Hospital in Paris. Nowhere is this more evident than in the area of theory-into-practice, as more than 500 theories of, and approaches to, psychotherapy have been proposed as an efficacious means to facilitate change with clients (Goldfried, 2019). Yet we still know relatively little about the intricacies of therapeutic change itself; much of this essential knowledge remains to be discovered.

Psychotherapy researchers have expended considerable effort demonstrating that psychotherapy facilitates change, whereas mechanisms or processes by which such change is produced remain a bit of a mystery, which is not to say that valuable inroads have not been made. Outcome research does not always present a sufficiently detailed or accurate picture of therapeutic change in practice, which led Gendlin (1986) to promote the value of microprocess moments in therapy. Such change process research has long been a valuable route to "identifying, describing, explaining, and predicting the effects of the processes that bring about therapeutic change" (Greenberg, 1986, p. 4) that complements process-outcome research, randomized control trials, and experimental designs.

Yet, even within the tighter sphere of change process research, the primary methods of analysis tend to emphasize change only within the context of therapeutic practice. As outlined by Elliott (2010), qualitative helpful factors design asks clients to identify what in the therapeutic process helped facilitate

https://doi.org/10.1037/0000451-001
Therapeutic Change With Difficult Clients: Precursors and Techniques in the CHANGES Model, Second Edition, by B. D. Wilkinson and F. J. Hanna

change, microanalytic sequential process design tends to closely scrutinize coded therapist–client interactions to identify change catalysts, and the significant events approach examines major transition points in the therapeutic process. Taking a closer look at research that employs such designs, the focal points of analysis tend to be therapist behaviors, attitudes, and dispositions; therapist-led interventions, techniques, and methods; and therapist–client relationship factors.

The CHANGES model is an outgrowth of qualitative research in the spirit of a significant events approach but without the within-session task analysis features often associated with such designs (Pascual-Leone et al., 2009). Instead, Hanna and Ritchie (1995) compiled 32 variables cited in the psychotherapy literature as having the capacity for producing change. Participants were screened for having undergone a major, significant moment of therapeutic change and were then asked to rate those 32 variables on a "perceived potency scale" to gauge perceived causal relationships between variables and events. Ratings on the 5-point scale included: 0 = *not present*; 1 = *present but not a factor in change*; 2 = *somewhat of a factor*; 3 = *a definite factor*; 4 = *a necessary condition for change*; and 5 = *a sufficient condition for change*. The study concluded that several of the 32 variables may regulate both the rate and magnitude of therapeutic change, seven of which have been identified as fundamental precursors of therapeutic change in the CHANGES model.

FUNDAMENTAL QUESTIONS

A more thorough understanding of the intricacies of therapeutic change may be the most important step toward improving the effectiveness of psychotherapy. Procedures can then be developed that derive from that new and vital knowledge. When it comes to using techniques, therapists still rely, for the most part, on those that have been cataloged in alignment with the various theories, such as behavioral, Adlerian, gestalt, or cognitive. Are there undiscovered change principles that can lead to the development of more effective techniques and more efficient use of the techniques we have?

There are some fundamental questions about change that need to be addressed in this regard. Why is it that some clients change relatively quickly in therapy while others make little, if any, progress? Why is change painfully difficult for some clients and comparatively easy for others? Why do some clients welcome change, but others resist and struggle against it every inch of the way? Why do some clients achieve core personality changes, whereas others make relatively minor, linear adjustments? Why do some clients recognize that therapeutic change is important, but others see it as threatening? And how can the most beneficial change be produced in the shortest amount of time, especially in this age of managed care?

Each of these questions emphasizes client-specific, within-individual factors for which we have an incomplete understanding. We hope to persuade therapists that a better understanding of within-individual change factors is crucial to fostering change, particularly for difficult clients. The latter point is of universal concern in psychotherapy as we continue to wrangle with how to help clients who are disinterested in change. Tried and true techniques, rigorous treatment methods, and every bit of efficacy evidence born of research are of little use to the therapist who sits across from a client who has no interest in changing. As the American journalist Sydney J. Harris (1986) wrote, "Our dilemma is that we hate change and love it at the same time. What we really want is for things to remain the same but get better" (p. 36). As a principle, this is perhaps what makes therapy one of the most difficult professions there is. Like chess, it is relatively easy to learn principles and maneuvers, but it is incredibly difficult to master and more difficult still when working with a disengaged client.

PRECURSORS OF CHANGE

This book offers one approach to these fundamental questions by suggesting a set of seven critical variables called "precursors" that are conducive to psychotherapeutic change. A *precursor* is generally defined as that which precedes and, so to speak, "announces" the arrival of something else. We refer to these seven variables as precursors because their presence indicates the imminent manifestation of change. The seven precursors are not focused on the therapist, theories, or techniques. Each has to do with client-specific factors, that is, what the client brings to the session.

The seven precursors are concerned with pivotal within-individual processes upon which change depends. Each has empirical validation. The precursors are not arranged in terms of potency or order of implementation, but rather to produce the acronym *CHANGES* as a mnemonic device:

- **C**onfronting the problem
- **H**ope for change
- **A**wareness of the problem
- **N**ecessity for change
- **G**rit, or a willingness or readiness to experience anxiety or difficulty
- **E**ffort or will toward change
- **S**ocial support for change

Each precursor is a transient and conditional state that may be present, or not, within a person at any given time for a particular problem or issue. They are not dispositions, personality traits, or characteristics, such as ego strength or hardiness. Confronting the problem, one of the most powerful of the precursors, is something that a person does, or not, in real time. It is not a part of a person's personality makeup. Even if a person is particularly skilled or attentive to the process of confronting many of their life problems, there may still be problems about which they are unaware, or are aware but would rather not be, or are aware and would simply prefer to put off considering until some later date or set of conditions. Anyone with any kind of personality, disordered or not, can activate precursors, assuming there are no neurophysiological impediments.

The seven precursors are a set of interdependent, necessary conditions for therapeutic change. When precursors are not activated in a client, change is unlikely to occur regardless of how great the therapist is, how potent the theory is, how close the relationship is, or how capable the person is. These client-specific ingredients, working in various combinations, seem to regulate the speed, intensity, and magnitude of change and can be considered regulators of the change process (see Hanna & Ritchie, 1995). In other words, the more they are present in a person, the more quickly change will occur and, in some cases, the deeper that change will be in the person's psyche.

Each precursor can be formulated in different ways and with different terms and jargon, but we have tried to keep them as theory-free as possible. In fact, they are assumed by all the theories and techniques of counseling and therapy. The precursors are pervasive; one can pick up virtually any book on therapy that contains successful case examples, and the precursors will be very much in evidence in each case. In unsuccessful cases, the precursors will be absent. There may be more precursors than those discussed in this book, but these seven seem to be key ingredients of change.

UNIQUE ASPECTS OF THE CHANGES MODEL

Several aspects of the CHANGES model are rather unique. One unique aspect of the CHANGES model is that it concentrates solely on a set of within-individual change factors without focusing on stages, theories, or personality traits. Additionally, the CHANGES model is a taxonomy of client-specific processes with corollary interventions and techniques developed across virtually all the schools of psychotherapy. The activation of precursors can be brought about by myriad, well-established therapy practices regardless of whether or not the therapist or client is aware that the activation of an underlying precursor was responsible for the change. Finally, focusing on client-specific change factors means that therapeutic change can occur outside of therapy, as maintained by Gendlin (1986). As such, therapeutic change does not require therapy; techniques to activate precursors abound, and the CHANGES model addresses the common factors of change itself.

Therapeutic Change as a Metacognitive Skill

The CHANGES model is also unique in its consideration of metacognitive aspects of change. Rather than addressing first-order thought processes, metacognition has to do with thinking about thinking, intentional psychological acts, or what were once called "acts of consciousness" (Husserl, 1913/1982), those inner decisions about where to direct awareness, the act of regulating one's thoughts and behaviors, or degree of awareness of awareness (Hanna et al., 1995). Barring a few exceptions, metacognition has seldom been considered in relation to change processes.

Most research tends to focus on the role that metacognition plays in maladaptive behaviors and associated beliefs; for instance, in how worrying arises from a metacognitive belief that worry is an effective means of coping, while the idea that "people are dangerous" is an instance of cognitive belief that feeds into that metacognitive style (Wells, 2007). However, when therapeutic change itself is examined from a metacognitive perspective, it begins to look more and more like a skill.

Behavior therapists have long emphasized the importance of skill building. In the metacognitive context, change itself is a skill, and mastery of the precursors is tantamount to mastery of therapeutic change itself (Hanna, 1996). As noted earlier, precursors are not traits or dispositions. Each precursor is a transient, conditional state that may or may not be present in a person at a given time for a given problem. However, having an awareness of the role that the precursors play in the change process can profoundly impact one's relationship to, and understanding of, change itself. If one understands that grit, confronting, and a sense of necessity are required catalysts for change to take place, then it becomes possible to orient oneself toward the activation of these core states. In this respect, the intentional, metacognitive activation of precursors is a skill that can be learned.

The Role of Interpersonal Savvy in Facilitating Change

Although therapeutic change is a metacognitive skill that can be learned and engaged without the need for therapy, therapists obviously play a central role in the activation of precursors and, thus, the facilitation of therapeutic change for clients. While the value of using various techniques and strategies to facilitate precursor activation and client change is a founding premise of this book, the pivotal role of interpersonal dynamics must also be underscored, particularly in therapy with difficult clients. Insofar as the CHANGES model broadly aligns with the common factors model, a therapist's ability to successfully activate precursors is oft contingent upon a strong therapeutic relationship. As such, Chapter 3 is dedicated to techniques for both building the relationship and building the therapeutic encounter around the principle of therapeutic change itself.

However, a considerable portion of the content within this book implicitly—and in some cases, explicitly—suggests that therapists with a degree of interpersonal savvy may be particularly well suited to the task of activating precursors in work with difficult clients. It is basically understood that it takes a certain degree of interpersonal skill, ability, or savvy to convince someone that the challenge of change is a worthwhile endeavor. If change were mundane or simple, then everyone would do it whenever necessary, and there would be absolutely no need for psychotherapists.

The truth is, change is hard. When a client shows up for therapy and makes immediate progress toward their goals, that client likely already had many, if not all, established precursors in place. Difficult clients are, as discussed in Chapter 2, only "difficult" insofar as they do not have activated precursors in place. The task of establishing those precursors is often not an easy one, and the barriers presented by a client to their activation are often complex. In the end, it often takes a certain degree of interpersonal savvy, and perhaps perspicacity (see Chapter 14), to break through with such difficult clients, making relational contact and getting the client onboard with the idea that therapy is about change and that change itself is worthwhile.

Techniques Oriented Around Change Principles

Perhaps the most unique aspect of this book is the vast array of techniques that have been compiled and described for activating each of the precursors. It is unusual for techniques to be cataloged in terms of change principles. In Part II, each chapter includes a subsection that lists and describes various techniques to activate each particular precursor. Many of the techniques have not been previously described, while others are more familiar and well-established techniques that have been adapted specifically for bringing about the presence of precursors among difficult clients.

Precursors in Group and Family Therapies

The CHANGES model is not limited to work with individuals. It also applies to group and family therapy. The precursors can be assessed and identified not only with each individual member of the group or family but also with the group or family as a whole. A group will display a configuration of precursors of its own. Whitaker held that a family has its own character as an entity in its own right (Simon, 1985). Therapy groups can be seen in the same way. Thus, the CHANGES model and assessment can help a therapist working with difficult groups and families.

Human Being as Active Agent

The CHANGES model makes great use of the concept of *active agency*, the view that human beings function as agents, actively influencing both their own minds and their environments (see Avdi et al., 2015; Gorlin & Békés, 2021; Harré, 1984). The behaviorist view of human beings leaned more toward a mechanistic or deterministic conception of human beings as merely effects of and responses to environmental forces. The latter perspective did not allow for the idea that human beings actually determined or shaped events in their lives, only that their environments and events determined and shaped them. This perspective, albeit useful, is obviously limited and has steadily diminished in popularity across decades of advancement in the cognitive sciences.

In parallel with the decline of this behaviorist view, the active agency conception that was central to the approaches of Rogers and Adler has gained considerable attention and widespread acceptance (Hoener et al., 2012). Led by deepening philosophical and neuroscientific discussions on the mind–body problem, the rise of influential research based on principles of embodied cognition and enactivism reflects a significant advancement in phenomenological discourse on the relation between self and world (Chemero, 2013; Gallagher, 2017; Wilkinson & Wilkinson, 2024). People are indeed influenced by the world just as much as they are agents influencing that same world in return (Zahavi, 2008).

In truth, the idea of active agency in psychology has been around for over a century. It was the basis of the psychology of William James, for example, and is assumed in almost all the major schools of psychotherapy today. In fact, many historians of psychology believe that contemporary psychology in general is now returning to the approaches and conceptions found in James's (1890/1981) classic *Principles of Psychology* (Hergenhahn, 1996). Metacognition also rests comfortably within the active agency conception, especially with regard to the realms of awareness, making decisions, consciously changing thoughts, exerting effort, forming and reforming mental images, and deliberately acting on the body and its impulses to enact new behaviors. Nowhere are such processes better described than in James's classic book. The CHANGES model shows how research confirms active agency and how therapeutic change depends to a large degree on it.

DEGREE OF CHANGE: FIRST ORDER AND SECOND ORDER

Both within and outside of the therapeutic encounter, the depth of change can vary. Although first-order and second-order change are defined a bit differently by different authors, the essential meaning is consistent (Sperry, 2022; Watzlawick et al., 1974). *First-order change* is linear, surface-level, uncomplicated, straightforward change that takes the form of an adaptation or adjustment. For example, one client may learn new communication skills to better get along with her daughter, while another client may learn to be assertive with his supervisor.

Second-order change, on the other hand, is more profound. It is a sweeping, deep structure or core change within an individual (Lyddon, 1990; Sperry, 2022). This kind of change fundamentally alters a person's core sense of self, mode of being, or essential worldview.

Second-order change radiates into and transfers across the wide array of a person's personality traits, activities, and interests. According to the CHANGES model, second-order change often involves intense initial turmoil or stress that is directly related to internal conflicts of considerable magnitude. These initial conditions are enough to threaten the person's equilibrium and psychological stability prior to the change itself. In other words, second-order change arises, often but not always, out of a crisis. It is also associated with advancement across stages of development (Gilligan, 1982; Hanna et al., 1995). Research has indicated that the precursors are intimately involved with the occurrence of both first-order and second-order change (Hanna & Ritchie, 1995).

An important goal for the fields of counseling and psychotherapy should be to develop new therapies that lead to greater magnitude and intensity of therapeutic change. When psychotherapy evolves to the point where almost any client who is free of organic brain damage can derive benefit relatively quickly, deeply, and with relative ease, the discipline will have taken a quantum leap. It is, arguably, our ethical duty to explore and develop new approaches that can lead to such results. The CHANGES model outlined in this book, applied to difficult clients, is not a quantum leap in itself but is based on the research of second-order, client-specific change experiences. It is offered to point the way forward to one possible, perhaps promising avenue of approach toward that goal.

MAJOR POINTS OF THE BOOK

Readers should consider the following central points as preparation for the chapters that follow:

1. The presence of precursors makes change possible.
2. The absence of even one precursor can inhibit therapeutic progress. When the missing precursor is activated, therapeutic progress can be made.
3. Client resistance, no matter how one defines it, indicates the absence of precursors.
4. Precursors regulate the rate, intensity, and magnitude of therapeutic change. The more they are present, the more likely change is to occur.
5. Therapy with difficult clients often involves a different set of skills than therapy with clients who are motivated or involved in the therapy process.
6. With many difficult clients, it is helpful to first establish precursors that are missing or deficient before proceeding to routine therapy approaches.
7. Difficult clients need a therapist with particularly effective relationship skills as well as a depth of empathy that surpasses what is necessary for willing and involved clients.
8. A therapist with inactive precursors can negatively affect therapeutic progress, inadvertently blocking client progress in a number of ways.
9. A remarkable number of techniques can be used to activate and enhance precursors.
10. Therapeutic change is a life skill that can be learned, practiced, and taught.

ORGANIZATION OF THE BOOK

This book is divided into five sections. It should be noted that all client names and other identifying information have been altered throughout the book to preserve confidentiality. Part I introduces the model. Chapter 1 introduces the technical, philosophical, and practical foundations of the CHANGES model.

Chapter 2 examines and reframes the basic idea of client resistance to treatment approaches, interventions, and strategies.

Part II is arguably the heart of this book, with eight chapters that describe clinical applications of the CHANGES model. Chapter 3 lists important guidelines and tips for relationship building with difficult clients. In Chapters 4 through 10, each precursor is defined and described according to how it brings about change. The clinical markers of each precursor are provided to help therapists detect their presence and absence in clients. Finally, each chapter includes techniques and practices for precursor activation, with case vignettes and transcriptions used to show techniques in action.

Part III describes the use of the CHANGES assessment form. Chapter 11 is devoted to the use of the form with difficult clients and provides several examples. Chapter 12 describes the use of the form to determine the level of precursors of a therapist working with a particular difficult client. In this chapter, examples are given of how a lack of activated precursors for the therapist in relation to a particular client can have a direct bearing on the climate, tone, and success of therapy.

Part IV is the specialized clinical applications section. Chapter 13 looks at therapist precursors in light of supervision practices, reframing the idea of countertransference as therapist interference. We endeavor to demonstrate how therapist reactions to clients can be leveraged for therapist insight and growth and how therapists can more effectively manage personal issues that inevitably arise in work with difficult clients. Chapter 14 discusses multicultural issues from the viewpoint of oppression. An oft-overlooked characterological outgrowth of oppression is discussed with the goal of supporting liberation among clients who identify with oppressed groups rather than facilitating mere adaptation or adjustment to oppressive conditions and environments. Chapter 15 reviews applications of the CHANGES model related to addiction and substance use, introducing the grand reframe of addictive behaviors and providing a litany of techniques and strategies for use with difficult clients in both individual and group substance use treatment.

Finally, Part V reviews the major advantages of the CHANGES model, proposes a metatheoretical framework based on the principles set forth in this book, identifies potential business applications using the proposed CHANGES Assessment for Businesses and Corporations, and examines some avenues for future research.

We are honored to present this material to you, dear reader, in hopes that the CHANGES model lends further insight into therapeutic change processes along with tools to facilitate said change.

ACTIVATING CHANGE WITH DIFFICULT CLIENTS

1

Toward a Model of Change for Difficult Clients

Therapeutic change is the raison d'être of psychotherapy, serving as its most defining characteristic and its primary criterion for gauging success. Bereft of the concept of change, therapy practice loses its meaningfulness. Psychotherapists are first and foremost agents of therapeutic change, integrating relational skills with strategies and interventions to help clients attain beneficial ends. However, too much attention has been historically devoted to theories rather than what facilitates change itself (Gaines & Goldfried, 2021). If therapeutic change is such a crucial aspect of psychotherapy in practice, then it is extremely important to understand its basic nature.

This book examines what brings about beneficial change in human beings, regardless of whether one participates in therapy or not. Its promise lies in the fact that it concentrates on therapeutic change itself by centralizing client-focused factors that can be understood, identified, monitored, discussed, evaluated, and activated in the therapeutic encounter. As such, this book examines the capacity of individuals to activate therapeutic change, the many barriers to doing so, and the power of psychotherapists to catalyze this process using a wide variety of well-established techniques.

Lazarus (1990) opined years ago that psychotherapy was "still in the dark ages" (p. 356). In terms of what we know about the nature of change and how to help people achieve it as quickly as possible, this assessment remains more or less accurate today (Kramer et al., 2024; Silberschatz, 2017). Our collective difficulty in identifying mechanisms of therapeutic change has led some to quite appropriately label it the "black box of psychotherapy" (Zilcha-Mano, 2021, p. 516). Consequently, for psychotherapy to rise to a new level of effectiveness, it is

https://doi.org/10.1037/0000451-002
Therapeutic Change With Difficult Clients: Precursors and Techniques in the CHANGES Model, Second Edition, by B. D. Wilkinson and F. J. Hanna

necessary to consider a range of potential paths that can lead to a more complete and dynamic view of how people change.

None of this is to suggest the field has not made significant advancements. In recent years, data-driven researchers seeking to advance psychotherapy have emphasized the value of randomized control trials to identify empirically supported treatments for specified symptoms and disorders (Philips & Falkenström, 2021). Neuroscientific developments have resulted in new psychotherapy intervention models (Cozolino, 2017; Smith et al., 2020), while integration-focused scholars have proposed a variety of unified psychotherapy frameworks based on the premise that metatheoretical modeling is needed to advance the field beyond its current, preparadigmatic state (Constantino et al., 2021; Marquis et al., 2021; Schiepek & Pincus, 2023).

Yet there is no solid evidence that psychotherapy outcomes are improving for clients in a consistent manner despite such advancements (Insel, 2022). The relative value of empirically supported treatments over myriad other approaches to psychotherapy remains questionable, as meta-analyses and metascientific reviews highlight concerns about replicability, effect size, and general efficacy (Machado & Beutler, 2016; Sakaluk et al., 2019). Significant questions remain as to the limits of neuroscience in the social sciences (De Vos & Pluth, 2016) and its applicability to psychotherapy practice (Insel, 2022; Paris, 2017; Wilkinson, 2019). In a tone quite reminiscent of Lazarus (1990), the psychotherapy integration movement has been described as "very much in its infancy" (O'Leary, 2021, p. 3).

There are, of course, many ways to approach and scrutinize these complex concerns, and there are no easy answers to the question of what makes psychotherapy work, or not work as the case may be. We do not claim to provide any definitive answers herein. However, we do wish to suggest that amid endless empirical, neurobiological, and integrative arguments about what precisely makes psychotherapy work, there seems to be considerably more emphasis on what therapists do than on what clients do. Scholarly interest in measuring therapist factors tends to outweigh emphasis on client factors (Fuertes & Nutt Williams, 2017). Wachtel (2018) argued that it is time to overcome our theoretical tribalism and "focus on principles and processes of change rather than branded packages" (p. 202). We agree; thus, we focus on change as a client-specific factor.

THERAPEUTIC CHANGE AND CLIENT-SPECIFIC FACTORS

Although therapeutic change is relatively easy to define at a superficial level, it is a remarkably complex, dynamic, and intricate phenomenon. As a working definition, *therapeutic change* is a beneficial, positive alteration in thoughts, behaviors, feelings, or interpersonal interaction that leads to improved or more effective coping or functioning and greater satisfaction with one's outer and inner life. In general, it is anything that constitutes an improvement to a person's life in terms of feeling, thinking, behaving, or relating. Therapeutic change is associated with simply learning to get along with one's supervisor, better managing one's time, or feeling better after talking about a problem. It can also be as profound as conquering major depression, recovering from addiction, overcoming the devastating effects of abuse, or finding meaning in a seemingly empty existence.

The motivation for therapeutic change can drive us to pursue love, travel the world, start a family, attend college, launch a career, pick up a hobby, or do myriad other things that human beings tend to believe will lead to greater satisfaction or fulfillment in life. It can also drive us to garner power, seek fame, acquire unnecessary possessions, and even use alcohol or drugs under the misguided premise that substance-fueled changes in feeling, perceiving, thinking, and behaving are a useful, if transient, form of therapeutic change. While the desire for therapeutic change is a fundamental aspect of human experience, not all paths to it are equal in effectiveness.

It is a mistake to discuss therapeutic change only in the context of psychotherapy and counseling. The fact is, people change all the time without therapy. We have known for nearly 75 years that a third of troubled individuals will improve on their own with no therapeutic intervention at all (Eysenck, 1952). We have known for nearly 40 years that about 15% of clients improve before engaging in their first session (Howard et al., 1986). If change occurs without an empathic therapist and without an identifiable theoretical approach, then change is largely in the domain of the person (Cuijpers et al., 2019; Hanna & Ritchie, 1995; Norcross & Prochaska, 1986a, 1986b). Therefore, a model of therapeutic change ought to account for and describe change outside of psychotherapy.

Common factors research has long demonstrated that extratherapeutic variables, or client-specific factors, account for 20%–40% of positive change outcomes in psychotherapy (Lambert, 1992; Peterson, 2019; Wampold, 2001). Client-specific factors include personal qualities, environmental influences, and what the client actually does in therapy and in life. There is considerable evidence that these are also the most important factors on which to focus a model of change (Bohart, 2000; Swift et al., 2023). After all, it is the client who does the work and who makes the changes. So, what kind of client is most likely to change? If we knew the answer to this pivotal question, we could likely predict who will respond to therapeutic interventions and who will not.

THE SEVEN PRECURSORS OF CHANGE

According to the CHANGES model, the client who is most likely to change is the client who has established the precursors of change in relation to a specified problem. A *precursor* is defined as a prerequisite or precondition that indicates an impending phenomenon. Therefore, a precursor of change is a specified precondition by which change may occur. The seven precursors of the CHANGES model seem to be present in clients who are actively engaged in the therapeutic change process. The seven precursors of change are as follows:

- *Confronting the problem* involves the steady and deliberate attending to and observing of anything intimidating, painful, or confusing, in spite of the inclination to avoid, shun, or act out. It is operative when a client looks at a problem squarely and directly and continues to observe, explore, or investigate it until the client grasps its essence.

- *Hope for change* is the client's realistic expectation that change can, and will probably, occur. It is not wishing, longing, desiring, or yearning. The hopeful client sees the possibility of change and the path to accomplish it. This recognition has the power to motivate even an apathetic client, especially if that client also has a sense of necessity.

- *Awareness of the problem* involves knowing that a problem exists and having a good sense of what that problem or issue is, as well as of the thoughts, feelings, and behaviors connected with it. Awareness is the opposite of denial or obliviousness. Without it, a person has no idea where to direct resources for change. With awareness, a client can pinpoint areas of dysfunction or need and identify relevant thoughts, feelings, and behaviors.

- *Necessity for change* is a felt sense of urgency or need that change takes place on the part of the client. In the person's assessment, current conditions are not at all satisfactory and must give way to a different set of circumstances.

- *Grit, or a willingness to experience anxiety or difficulty* indicates a readiness to feel the discomfort that accompanies change. Defensiveness, the diametric opposite to this precursor, is usually defined as an attempt to avoid anxiety. When grit is present, anxiety or difficulty is not resisted but directly experienced, with the knowledge that doing so is necessary for change to occur.

- *Effort or will toward change* indicates action actually taken toward solving the problem. It is the expending of energy and the movement made. It also involves will in the sense of making a commitment, coming to a decision, and initiating action. Effort manifests in two domains: the mind, in changing one's thoughts and attitudes, or the world, in coping with real-life situations.

- *Social support for change* consists of being engaged in confiding, supportive relationships dedicated to the client's well-being and improvement. Social support paves the path toward therapeutic change when those relationships function to enhance and inspire the presence of each of the other precursors.

The "CHANGES" acronym is designed to enhance therapist recall of the seven precursors. The model is rather like a display case, as the primary objects of attention are the precursors themselves alongside their interactions. As such, the CHANGES model does not sequentially arrange the seven precursors in their order of potency or in the order in which they occur. The precursors are present in clients in different configurations for different moments of change, and for the same person, different precursors may have varying degrees of influence from one instance to the next.

Since the precursors are a client variable, they are active in the change process both in and out of therapy. For instance, people who improve without therapy have often accurately identified a problem (awareness), become dedicated to overcoming that problem (sense of necessity), read a book (confronting), talked to a friend (social support) who made things seem not so bad (hope), and engaged in actions to facilitate change (effort) despite their discomfort (grit). As Prochaska et al. (1994) once put it, "It can be argued that all change is self-change, and that therapy is simply professionally coached self-change" (p. 17).

The CHANGES model maintains that every person has a complex configuration of waxing and waning precursors related to any given problem. It does not matter if the person is in therapy or not. If a client is in therapy, the counselor's theoretical orientation is inconsequential so long as the precursors are activated. Contrary to popular belief, talking about a problem does not always constitute "the talking cure"; talking is not always enough to produce change. Many clients talk about problems indefinitely, session after session, but nothing seems to change. This is not so much the fault of the client as a faulty understanding of therapeutic change itself.

If people possess, develop, or implement the precursors in the CHANGES model, then change will occur with or without a therapist. Logically, if change can occur without an empathic therapist who employs a psychological theory in practice, then (a) change is largely in the domain of the person and (b) it is the primary responsibility of therapists to identify and facilitate the activation of precursors. The CHANGES model thus represents a common factor approach to identifying the principles of therapeutic change itself rather than the more widely recognizable common factors of therapy. Armed with a well-founded grasp of therapeutic change itself, the role of the therapist is to help clients activate the seven precursors of the CHANGES model related to a given problem.

Notably, the precursors also precede, regulate, and influence the better-known change processes cataloged in psychotherapy. Most of these change processes are associated with classical theories; examples include insight, catharsis, changes in beliefs or thoughts, reinforcement, exposure, desensitization, problem solving, the act of acceptance, working through transference, and the corrective emotional experience. Without the precursors, these intervention-based change processes are unlikely to result in lasting therapeutic change. Client variables such as motivation and involvement are also widely regarded as important to the therapeutic change process, yet these too are dependent upon the precursors in many circumstances. The precursors can give definition, detail, and applicability to research findings on motivation, involvement, and other client variables.

VALUE OF THE MODEL FOR THERAPY WITH DIFFICULT CLIENTS

The test of any model of therapeutic change is whether or not it will apply to difficult clients. In the context of this book, a *difficult client* is a person for whom change is not forthcoming. The CHANGES model is presented precisely for the purpose of working with people who find change to be difficult, intolerable, or even a source of pain, inconvenience, or failure. When change is not forthcoming, the CHANGES model suggests this occurs because there are too few precursors present and operative in a client. As such, a difficult client displays what might be called *change deficits*, which indicates some lack of skill, ability, desire, and/or willingness to enact therapeutic change processes. A therapist using the CHANGES model can help difficult clients by providing education about the precursors of change and using techniques designed to activate the precursors.

Of course, there are many other ways to define "difficult," some of which do not fit with our approach in this book. Many difficult clients have been referred to as "resistant," a carry-over from psychoanalytic explanations of unconscious transference issues. When Freud (1910) said, "The patient who comes seeking desperately for help soon bends every effort to defeat help being given" (p. 54), he meant that resistance is an inner dynamic that leads people to repress and avoid painful or uncomfortable material in the unconscious. In such a context, virtually everyone displays some kind of resistance, and we maintain that this is a valuable way to understand some difficult clients.

However, a less specialized and more widespread use of the term insinuates that a lack of therapeutic change is due to client recalcitrance, active/passive aggressiveness, or ambivalence. A good working definition of *resistance* in this vein is "The patient's efforts to obstruct the aims and process of treatment" (Walrond-Skinner, 1986, p. 298). Although this is certainly accurate for some clients in some cases, therapists run the risk of impairing their empathy by framing difficult clients as having a characterological deficit in general. There is a long history of attributing poor client outcomes to self-sabotage rather than the complexities of sociocultural and relational factors. It is also far easier to blame clients than to recognize and acknowledge our culpability as therapists.

Change is difficult. Attempting to keep things the same is generally easier than making things different. While some people may view change as a worthy challenge to overcome, many others find the prospect of change threatening, upsetting, disturbing, overwhelming, painful, ill-advised, impossible, or even just plain silly. Some clients are uninformed about the benefits of change, feel undeserving of positive life changes, or vigorously engage in self-protective maneuvers to prevent change. Like communication or social skills, some people never learned how to change or to accept help to change. We have encountered some clients who never learned the fundamental lesson that therapeutic change is possible, much less valuable and helpful. Whatever the reason, if a person does not find the prospect of change appealing, then change is unlikely to be forthcoming.

Interaction and Interdependence of the Seven Precursors

The major psychotherapy theories identify the following as primary human motivations: reducing anxiety (Freud), attaining a life goal (Adler), self-actualization (Rogers), making sense of the world (Kelly), meeting needs (Murray), and reinforcement (Skinner). Even a casual study of these motivations reveals the importance of each and the liability of excluding any. Attempts to take something enormously complex and reduce it to one basic cause or principle, once referred to as "single principle imperialism" (Koch, 1981), is not a viable option in our search for understanding the dynamic nature of therapeutic change.

Human beings are too complicated to assign single principles to explain behaviors and motivations. In kind, therapeutic change should not be unduly oversimplified as if it operates in a linear, stepwise fashion. The precursors are not symmetrical, proportional, or equal in potency. They are also not sequential. Any given precursor can take the lead in one change experience, only to be far less prominent and consequential in another. The CHANGES model operates with a rotating form of circular causality. Instead of a static, two-dimensional patchwork quilt, the model is closer to a dynamic three-dimensional quilt where each patch changes in size, volume, area, and color in accord with the idiosyncrasies of a given client in a given situation (see Figure 1.1).

As common factors of change, the precursors are interdependent variables that overlap in both meaning and function. They can combine and recombine in

FIGURE 1.1. The Precursors and Their Interconnection, Interaction, and Interdependence on Each Other and Therapeutic Change

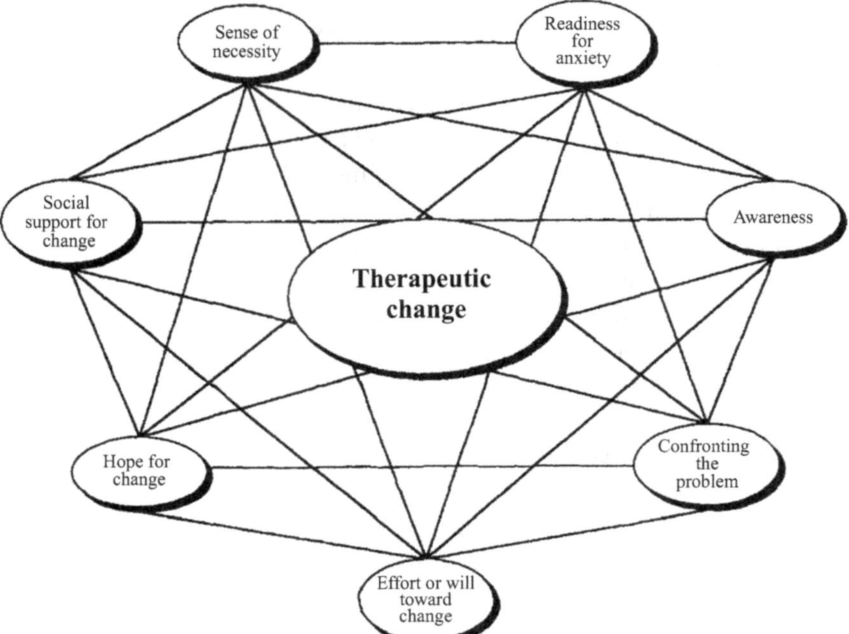

an extraordinary number of ways, and they vary in degree of presence from case to case or problem to problem. For example, a client who has reached a therapeutic impasse has an adequate presence of two precursors, awareness and a sense of necessity, but is largely lacking in all the others. Clients with this combination are quite common, and their lack of progress can be confusing without reference to the precursors. Yet when other change ingredients, such as confronting and being willing to experience anxiety, are addressed, change may occur. In a similar vein, deficits in awareness and necessity can be bolstered by social support and hope, while adding a sense of necessity can drive confronting the problem to greater depths and inspire grit, or a willingness to experience anxiety and difficulty.

Precursors are not static but can wax and wane throughout various stages of therapy. They can be powerfully present one day and nowhere to be found the next. They can vary in strength from minute to minute in the same session, which supports the need to study microprocesses in research. If a client does not trust their therapist, the precursors can almost literally be observed to diminish when that client enters the presence of the distrusted therapist. Many teenagers find it difficult to engage in therapy and act out or protest when they are told they must work with a counselor with whom they feel no rapport. Precursors vanish on the spot. Even with therapists whom they do trust, clients can bring a different level of precursors each day to therapy.

In addition, different precursors can vary in intensity and magnitude in different areas of a person's life. For instance, a person may be high in awareness but low in the sense of necessity when it comes to the need to quit drinking. On the other hand, when it comes to saving their marriage, that same person might have a high sense of necessity but very little awareness of the problems of the marriage. Each person can have a different configuration of precursors for each particular problem, which is one reason therapeutic change in general is so difficult to reliably map.

Finally, the precursors include, undercut, and incorporate the major theories of therapy. They are both transtheoretical and integrative without being dependent on the major theories, although our understanding of the precursors is certainly enhanced by theories and vice versa. These precursors are functions or conditions that can be found in a person's mindset (in the case of hope, grit, and sense of necessity), in one's environment (in the case of social support), or as metacognitive acts (in the case of awareness, effort or will toward change, and confronting the problem).

AT THE BASIS OF THEORY

Goldfried (1980) once wrote, "There exist certain timeless truths, consisting of common observations of how people change. These observations date back to early philosophers and are reflected in great works of literature" (p. 996). He further noted that these are "robust phenomena, as they have managed to survive the distortions imposed by the therapists' varying theoretical biases" (p. 996). The CHANGES model represents a compiled set of such change characteristics. The precursors are prior to and underlie all of the major theories of psychotherapy, making them applicable across theories in a manner that aligns with the psychotherapy integration movement. Researchers and clinicians long sought to improve psychotherapy along the lines of theory. In search of a new, global theoretical perspective, the field has accumulated over 500 schools of thought (Goldfried, 2019), of which 50 or so might be considered major approaches.

Eventually, the lofty ideal of the global theory of psychotherapy lost its luster, and a shift toward theoretical integration was underway (Stricker & Gold, 2013). The shift probably began with the classic study by Smith and Glass (1977), which found no evidence that any one of the major theories is more effective than any of the others. Soon after, the field entered a posttheoretical era more akin to "a maturing, scientifically based art rather than of an ideologically based sectarian mission" (Omer & London, 1988, p. 178). There has been little evidence to dissuade integrative researchers from the idea that all major schools of thought, from psychoanalysis to existential to cognitive behavioral, are equivalent in effectiveness under appropriate conditions (Wampold & Imel, 2015).

As a result, we now have a host of approaches that are characterized by categories and terms such as eclectic, integrative, transtheoretical, and containing the common factors of therapy. The early integrative movement revolutionized the landscape of psychotherapy. Wachtel's (1977) theoretical integration of behavioral and psychoanalytic therapies showed how two seemingly incompatible theories can work together in relative harmony. The development of technical eclecticism (Lazarus, 1976, 1996) streamlined psychotherapy by formalizing an emphasis on effective techniques with a minimum of theory. The common factors approach identified what each major theory has in common, seeking the middle path between advocating techniques and combining theories (see Frank & Frank, 1991; Grencavage & Norcross, 1990). Additionally, the transtheoretical model set forth by Prochaska et al. (1994) indicated how therapeutic change takes place in and across various stages of therapy.

Yet this is just the tip of the iceberg. The process of integrating or combining theories and treatments has added greater complexity to the theoretical picture and, in some ways, rendered the original problem worse than ever. To be clear, we maintain that original theories and integrative approaches continue to provide a treasure trove of insights concerning psychotherapy and human behavior. The CHANGES model is neither antitheoretical nor anti-intellectual. However, an emphasis on theories, whether in isolation or in combination, tends to inadvertently sustain the clinical fixation on the common factors of therapy rather than the common factors of change itself.

The CHANGES model encourages a therapist to incorporate any or all theories and techniques, as appropriate, into practice. Each theory presupposes the activation of the precursors in clients, although they seldom recognize or identify the role of precursors in the process of change. Across all psychotherapy theories, the precursors of change are always presupposed. Read the hundreds of successful case studies cited by therapists in books and articles, and the precursors are observably operative within and across all of those clients, regardless of the espoused theory.

Psychological theories play a critical role in the common factor process of providing an adaptive, psychologically derived, and culturally embedded explanation for distress, as well as a guide for implementing rituals and methods to enact something helpful (Frank & Frank, 1991). However, such explanations and methods are only of value when the client is invested in the change process. Confronted with a difficult client who has no precursors established, the finest therapists in the world will find their most advanced methods and explanations falling on deaf ears.

Realigning Psychotherapy Theories Around Change

It is well recognized that dogmatically insisting upon the "Truth" of any psychotherapy theory is a mistake born of philosophical naivete (Wampold & Imel, 2015; Yadlin-Gadot, 2016). Many theories take positions that were abandoned by the discipline of philosophy long ago. Traditional cognitive therapy, for example, subscribes to the view that reasoning is more important than behavior or affect in approaching various psychological problems. Such a view has its roots in the centuries-old philosophy of rationalism, the idea that the world is based on rational principles and amenable to rational speculations and explanations, as held by the likes of Descartes and Spinoza.

In terms of neurobiology, the assumption that affect and cognition engage independent brain regions or unique neural circuitry has long been framed by affective neuroscientists as one of "seven sins in the study of emotion" (Davidson, 2003, p. 129). The modern neuroscientific understanding that the brain is a large-scale, coordinated network means that we can no longer naively suggest that cognition, emotion, and perception are independent faculties (Pessoa, 2023). In cognitive therapy, dialogues on the role of emotion in the therapeutic process began to shift at the turn of the century (Samoilov & Goldfried, 2000), and the cognitive primacy hypothesis (i.e., the view that access to affective content is contingent upon cognitive categorization) is now widely acknowledged as an insufficient thesis at best (Reisenzein, 2019; Stevens, 2024).

The ostensible opposite of rationalism is empiricism, or the idea that we must rely on the senses to understand the world. Ultimately, empiricism led to a philosophy known as *logical positivism*, in which its original intent was to abandon any metaphysical and theoretical speculations that could not be verified by scientific inquiry. Popper (1963) once told Nobel Laureate Peter Medawar that logical positivism takes the naive position that "The world is all surface" (Medawar, 1984, p. 101), indicating that mental states need not be considered. For years, the use of the term *mind* was looked down upon in behaviorally oriented psychology circles as shortsighted and unenlightened

(Hackert & Weger, 2018). Contemporary research, however, has made it ever clearer that our perceptions and sense experiences affect, and are interdependently affected by, our thoughts and beliefs (Pessoa, 2022). Philosophy abandoned logical positivism decades ago (BonJour, 2009), but experimental psychology remains bound to positivist assumptions (Mayrhofer et al., 2021).

The problem herein is not with therapeutic practices per se but dogmatic adherence to particular theoretical vantage points as "right," "true," or "correct" assessments of the human condition. Current theories of psychotherapy neither capture nor contain truth any better than older ones. They contain too many elements of metaphysics. Both religion and psychotherapy theories are inextricable from metaphysics, one of the most perpetually perplexing domains of human knowledge (O'Donohue, 1989; van Inwagen, 2024).

Metaphysics is, simply stated, the seeking or study of knowledge that is beyond the capacity of science to verify. Most theories of psychotherapy contain metaphysical elements that cannot be verified by science, such as the id, the self-actualizing tendency, or the collective unconscious. The self is also a metaphysical construct that Adler (1956), as well as the philosopher Hume (1739/1978), poignantly questioned. Kant (1787/1929) showed in his classic *Critique of Pure Reason* that reason alone cannot solve metaphysical problems. James (1907) repeatedly showed in his treatises that reason could not solve metaphysical problems. James, whom Whitehead (1925, 1938) hailed as one of the four most important philosophers in history, developed pragmatism as a dialectical means of coping with and transcending metaphysically driven theoretical and conceptual conundrums.

Pragmatism, in a psychotherapy context, is a dialectical mode of thinking that fluidly moves between and among theories without any attachment beyond the immediate therapeutic situation. In the Jamesian tradition, what is true is so in terms of its pragmatic value. For instance, the highly praised idea of human free will is a metaphysical position that has pragmatic value in therapy. The problem is that our fascination with theoretical explanations—which are inextricably bound up in metaphysical presuppositions—delimits our focus on what produces change. It is in James's pragmatism and radical empiricism that the CHANGES model approach finds its truest inspiration.

James (1904/1977) wrote a brilliant series of articles in the first decade of the 20th century called *Essays in Radical Empiricism*. Both Russell (1972) and Whitehead (1925) believed that in those essays, James had discovered the long-awaited solution to the centuries-old Cartesian dualism of mind versus matter. Essentially, James showed how all of life, space, time, relationships, objects, and even awareness boil down to pure experience. James (1965) was clear that radical empiricism involves taking a phenomenological stance, such that "The only things that shall be debatable among philosophers shall be things definable in terms drawn from experience" (p. 105).

Rather than continuously engage in and elevate speculative metaphysics (see Henriques, 2019; Hibberd, 2014, for examples), an approach based on James' philosophy of radical empiricism might liberate psychotherapy from its longstanding conceptual entrenchment (Hanna, 1994). From a Jamesian perspective, if therapeutic change is at or near the essence of psychotherapy, perhaps there is good reason to focus the development of therapy models and speculation around therapeutic change itself, what produces it, and what enhances it (Goldfried, 2019). This is the heart of the precursors approach, and it is no surprise that some researchers have been advocating this for decades. However, the integrative question remains: What do we do with all of these theories?

Theories Are for Clients

The renowned family therapist Whitaker (1976) believed that theories were ultimately destructive to therapy, claiming that "My theory is that all theories are bad except for the beginner's game playing, until he gets the courage to give up theories and just live" (p. 154). Whitaker believed that adherence to a theory is a way of avoiding the basic anxiety of not knowing what the truth about human beings really is. Strean (1993) was even more direct:

> It is helpful for practitioners to study carefully their affinity to a particular theoretical perspective or therapeutic model as well as their abhorrence of other perspectives and models. When our clients idealize and/or denounce certain individuals, or certain "isms" with a great deal of affect, we try to help them resolve their infantile attachments and overdetermined hatred. (p. 14)

The CHANGES model aligns with common factors and regards classic theories of psychotherapy as better reserved for use by clients than by therapists. Clients, for example, have much to gain by understanding that thinking influences feeling and behavior. Of course, cognitive theory is not absolute truth, but many clients have used its germane ideas to effectively overcome difficulties.

As Frank and Frank (1991) observed, the healing process seems to require some kind of rationale or myth to serve as an explanation of a psychological problem, and the theory of each therapeutic approach provides a different rationale or myth. Thus, depressed clients can benefit, for example, from the 19th-century James–Lange theory of emotion, which states that if one acts confidently or happily, one eventually begins to think and feel that way. Similarly, clients can respond positively to existential and gestalt explanations, just as they do with those of Adlerian and family systems.

Theoretical explanations have healing value in themselves—as reframes—and should be readily available for clients who might be able to benefit from them when their situation warrants. Of course, students and practicing therapists should also study them, but not to the degree that they would actually believe in the ultimate truth of any one of them. Dialectical philosophers such as James, Heidegger, and Husserl have shown that theories possess no truth value in and of themselves. Theories are for clients, and the realization that the so-called truths of psychotherapy can be framed and jargonized from a variety of perspectives can liberate therapists from entrenchment in a given theory. That liberation allows for a fluid, dialectical movement between and among therapy approaches, with the needs of the client dictating the therapeutic approach chosen—a primary assumption in many psychotherapy integration circles.

The precursors underlie and run through the background of therapy theories, but they are seldom focused upon as goals in and of themselves. We are not suggesting that the CHANGES model should be the central focus of all approaches; that would be quite inaccurate. However, emphasis on the CHANGES model is particularly relevant when working with difficult clients who lack the precursors necessary to make standard therapy approaches successful, as is discussed in the next chapter. In ongoing efforts to improve therapeutic outcomes, therapists and researchers alike would do well to recognize the critical role played by the precursors as common factors of change.

AT THE BASIS OF TECHNIQUES

Techniques abound in psychotherapy. Role-playing, guided imagery, the empty chair, identifying and disputing dysfunctional cognitions, behavioral contracting, systematic desensitization, flooding, and self-monitoring are just a few. Just as the various theories assume the existence of the precursors, psychotherapy techniques are implemented with the assumption that the precursors are present in the clients who engage in techniques. However, this assumption is almost never stated as such. In the broad psychotherapy literature, the precursors are only seldom mentioned in the context of being necessary for techniques to work. Implementing techniques with a client who has no precursors in place is not much different from doing therapy with a sleeping client.

Just as theories might be oriented around change, it may also be advantageous to orient techniques around change itself in the form of functions such as the precursors. As in the case of theories, the precursors are also at the basis of techniques: constantly assumed but seldom acknowledged.

Techniques are thus dependent on precursors of change for their effectiveness. If a client has no sense of necessity, is not willing to experience any difficulty in therapy, thinks that there is no problem at all, and will not look at it or do anything about the problem or issue, it is naive to assume that a technique will lead to therapeutic change. The therapist must shift to helping the client activate precursors so that the technique in question can take hold. If precursors are present and the relationship is sound, then various techniques are likely to be effective.

One must also keep in mind that techniques used to activate precursors will work with some clients some of the time, but they will not work with all clients all of the time. They also will not work with the same client at all times. It is important that therapists not be bound by a manual or formula when working with difficult clients. Therapist creativity and spontaneity are crucial. One's therapeutic toolbox should be filled with options suitable for a wide variety of clients and situations. Knowing as many techniques as possible is optimum, as is being able to deliver them with timing, finesse, wisdom, and the proper amount of indifference or enthusiasm according to the needs and dictates of the moment. Of course, any reasonable techniques from the major theories may be called upon if and when they can be of therapeutic service to a client.

FRAMING TECHNIQUES AS EXPERIMENTS

Some difficult clients are hesitant to follow a therapist in a certain direction, particularly when the therapist frames it as a technique. The idea of being "therapized" is anathema to many. In gestalt therapy, techniques are typically framed as experiments (Polster & Polster, 1973) that allow therapists to be creative in the therapeutic process, launching techniques based upon the spontaneous identification of client needs, relational dynamics, and the climate of that particular moment. We have found that clients tend to be more willing to try something new when it has been framed as an experiment rather than a technique, although a client's level of commitment or engagement may remain low. Most techniques in this book can be framed as an experiment to encourage client involvement. To set up experiments, the therapist should always ask the client's permission to try something new and inform the client that they may end it at any time.

CHANGE AS A SKILL

The active agent conception of the human being views therapeutic change as something intentional and purposeful on the part of the client. Change is not necessarily something that occurs in a person according to a mathematical equation. It does not proceed in the same fashion as a physics experiment or the motion of billiard balls. Of course, mathematical equations and physics formulas do indeed describe much of the activity of the world, but they do not fully or adequately describe human beings. Specifically, they do not account for the fact that human beings are active agents who write the equations and the formulas, do the experiments, and set the billiard balls in motion. We understand only half of the process of change if we do not attend to the powers that set change into motion intentionally, purposefully, and with choices.

In psychotherapy, some clients are more than willing to talk about problems and describe their feelings but never seem to make a move toward change. In the midst of their dissatisfaction and unhappiness, one might ask why they are not willing to change. Yet sometimes, it is not a lack of willingness at all but rather a lack of knowledge or practice. We have found that people can learn how to change like they learn how to communicate or to have gratifying relationships. In other words, therapeutic change can be taught, developed, and honed as a skill.

Like nearly any other skill, some people have learned it and others have not, but everyone can learn how to cultivate it. Alternatively, some people have it but do not seem to know that they do, whereas others know they have it but are unsure how to intentionally engage it. Some people have learned the skill of change through hard-earned life lessons, accomplished in spite of low self-esteem and high adversity. Other people, unaware and unknowing, await change as something to magically arrive and are disappointed to discover that change is their own responsibility. These are all symptoms of a lack of precursors.

A primary message of the CHANGES model is that education is often necessary with difficult clients so they can know how therapeutic change works. Education about the change process itself provides a map and a rationale and is often effective (Bohart & Tallman, 1999; Eubanks & Goldfried, 2019). One of the fundamental mistakes made in psychotherapy and counseling is to assume that clients understand change processes. If they did, change might be accomplished much more quickly, easily, and on a routine basis. At first glance, it may seem silly to educate a difficult client about change when some may not care in the slightest. No doubt this is true in some instances. However, therapists using an active agent approach treat human beings as if they have choices and options. Change is one such choice, and if change looks to a client like it is in their best interest, that change will have a better chance of manifesting via psychotherapy.

METACOGNITIVE ASPECTS OF CHANGE

Many of the precursors of change are neither cognitive, affective, or behavioral. They are metacognitive. *Metacognition* can be defined as the psychological acts of recognizing, deciding, monitoring, and attending to self-regulation, self-instruction, and the processing of experience, memory, and attention (Flavell, 1979; Rhodes, 2019). It is knowing about knowing, thinking about thinking, or, in the case of memory, remembering that one once remembered. Most important for the investigation of therapeutic change, it can also be defined as the skill or ability "to be aware of being aware" (Singer, 2017, p. 608).

As a domain-general psychological skill tied to accelerated learning outcomes, metacognition includes processes like constructive and de-automated self-talk, intentional and self-directed goal formation, along with the vital areas of self-awareness and the de-automatization of habits and reactions (Eccles & Feltovich, 2008; Meichenbaum & Asarnow, 1979; Singer, 2017). Many metacognitive functions involve subtle acts of *willing*—intending the exertion of mental effort. The precursors most associated with metacognitive functions are awareness, confronting the problem, effort or will toward change, and grit, or the willingness to experience anxiety or difficulty. However, all precursors are arguably influenced by metacognitive beliefs.

Metacognition has been identified as a predictor of change across a wide variety of issues, from depression and anxiety to domestic violence to addiction to personality disorders (Carcione et al., 2019; Hamonniere & Varescon, 2018;

Seow et al., 2021). Insofar as metacognitive skills are not the same as cognitive skills, one person can be extremely intelligent but less capable of change than someone far less intellectually gifted but more metacognitively skilled. Perhaps this is why clients who intellectualize often find it difficult to change (Di Giuseppe et al., 2021). Metacognition is involved with other classic defense mechanisms, as well. Most therapists have seen instances in which deflection or repression is momentarily suspended, resulting in honest client engagement.

A metacognitive process is arguably occurring in such instances, as the cognitive beliefs connected with protecting the person's psychological system are metacognitively suspended or held in abeyance through an implicit decision process. Structural and functional explanations for such a process are plentiful in cognitive psychology (Jankowski & Holas, 2014; Kuhn, 2022; Leschziner & Brett, 2019), and alternative framings have been advanced in cognitive phenomenology (Arango-Muñoz, 2019; Norman & Furnes, 2016; Reggia et al., 2016; Wilkinson, 2023). Regardless of any such explanations, the CHANGES model supports the suspension or abeyance process so that therapeutic change can be launched through the activation of specific precursors.

The awareness precursor is among the most obvious of the metacognitive aspects of change. Sometimes a client decides to allow their perceptions to penetrate and enter the field of consciousness; this is essentially the decision to become aware of a problem, and it is highly metacognitive. For instance, a client may be surrounded by indications that their negative statements and behaviors are harming loved ones. That data does not register, however, until the person allows that data into awareness by selectively attending to that part of their life. That profound and fascinating process of allowing such data to register can sometimes be crucial to manifesting cognitive dissonance and the decision to stop destructive acts.

In terms of grit, the willingness to experience anxiety or difficulty is also a metacognitive process. Clients who are motivated to change must will themselves to experience the anxiety that accompanies the change process, as well as monitor their anxiety levels to avoid becoming overwhelmed. Willing and monitoring are inextricably linked in the process of enduring anxiety or difficulty, with research pointing to the interconnected nature of grit and metacognition in terms of achievement-oriented processes (Wang et al., 2023; Weintraub et al., 2023).

There is another highly relevant and important lesson that metacognition can provide to therapists. In many cases, clients are quite capable of actively influencing and changing their feelings and moods. Change of feelings can come about not only through the usual, well-documented cognitive and behavioral processes but also through direct and deliberate attention to feelings or emotions. Research on active–passive emotion regulation strategies consistently indicate that an active strategy leads to superior mental health outcomes (Birditt et al., 2020; Hipson et al., 2019; Silk et al., 2006) and that an active strategy enhances one's ability to directly modify emotions (Rholes et al., 1989).

The CHANGES model accounts for the possibility of such active and direct change. However, the term metacognition has its problems in this instance, and the phenomenon might be better described by terms such as dialectical thinking (see Slife, 1987; Wilken & Miyamoto, 2018) or wisdom (see Ardelt & Ferrari, 2014; Hanna & Ottens, 1995). No matter what we call it, this area has great potential for further research.

THE TRANSTHEORETICAL MODEL AND MOTIVATIONAL INTERVIEWING

The transtheoretical model of change (TTM; Prochaska et al., 1994) delineates five major stages of change: precontemplation, contemplation, preparation, action, and maintenance. Each stage requires the presence of the precursors to make progression to the next stage possible. This is part of how the CHANGES model actively complements the TTM's stages. The precursors are implicit and necessary not only in the beginning but during each stage of change. Prochaska et al. (1992) once noted that the stages are nonlinear and spirallike, involving regression back to earlier stages and launches forward to latter stages. When the precursors have waned, a client is likely to regress toward earlier stages. Overall, the precursors are critical in the process of moving clients from the stages of precontemplation and contemplation into the stages of preparation and action.

The TTM also identifies 10 change processes—five cognitive/affective and five behavioral—each of which is a categorical framing for relevant stage-based "strategies that may be composed of a seemingly endless number of techniques" (Gutierrez & Czerny, 2018, p. 203). In other words, each TTM process lets a therapist know what type of activity might best be used to facilitate movement from one stage to the next. For instance, consciousness-raising strategies like bibliotherapy might help move a client from precontemplation to contemplation or from contemplation to preparation.

While this is obviously a useful approach, TTM processes are fundamentally different from the precursors because processes only progress when certain preconditions are met. The precursors are preconditions of change itself. Processes are complex, iterative, and rarely linear. The process of reevaluating one's sense of identity in relation to a problematic behavior (e.g., learning to regard oneself as a "healthy nonsmoker") tends to occur in fits and starts, two steps forward and one step back, as they say. A single moment of insight rarely leads to lasting and transformational change.

When progress is being made, the precursors are in place. When progress is not being made, the precursors have waned. As clients move in and out of various processes, those shifts can be attributed to the flux and flow of the seven precursors from moment to moment or problem to problem. As such, the CHANGES model asks an important additional question: What client-specific factors must be activated for that client to consistently make therapeutic progress by steadfastly engaging in change-fostering therapeutic activities?

For example, imagine a therapist asking a client to read a self-help book. Upon finishing the book, the client moves from contemplation to preparation, stating to the therapist, "I'm fully aware of the problem now, and I'm ready to talk about what to do next." According to the TTM, we might say the consciousness-raising strategy of bibliotherapy successfully moved the client into the preparation stage of change. It is also fair to assume that the reading might have facilitated other processes, such as self-reevaluation, environmental reevaluation, and emotional arousal.

However, per the TTM, we can only say that reading the book either did or did not serve its purpose as a tool for facilitating those four processes of change. The CHANGES model pushes us to further distinguish between the client for whom that bibliotherapy strategy was effective and the client for whom it was not. To address that difference, we must examine the flux in specific client precursors during the period in which the therapeutic change process actually took place.

For the client who read the entire book voraciously over one weekend, perhaps there was already some basic sense of necessity that grew when she read its first few inspiring pages. The book then deepened her sense of necessity while also activating awareness and fostering hope. Fortunately, that client had deep reserves of grit that helped her keep reading through a tough chapter that "hit a bit too close to home," and so she continued to confront the problem by continuing to read despite her ongoing discomfort. Perhaps her effort was also being actively supported by a caring partner.

But what of the client who could not read a full chapter? Perhaps he did not really believe that the problem was so urgent that it required reading an entire book. Even though there was a sufficient sense of necessity for change that he sought therapy, there was not enough to make reading uncomfortable materials that "hit a bit too close to home" a worthwhile endeavor. As a result, there was no catalyst for hope to grow, as the client did not have the willingness to experience anxiety, confront the problem, and exert real effort. Without any social support, he continues to attend therapy but does not finish the book, telling his therapist he will get around to it next week.

Thus, the CHANGES model captures those client-specific factors that underlie strategy-specific processes within the TTM. Just as a change process has been defined in TTM as "any activity that you initiate to help modify your thinking, feeling, or behavior" (Prochaska et al. 1994, p. 25), the precursors in the CHANGES model can be defined as the client-specific preconditions that must be in place in order for any such activities to successfully modify thinking, feeling, or behavior such that therapeutic change occurs. Both the CHANGES model and the TTM identify the value of using techniques to facilitate change. However, the CHANGES model is unique in terms of identifying barriers to therapeutic change plus specific techniques to overcome those barriers.

While the TTM and the CHANGES model are models of change, motivational interviewing (MI) is not a model at all. It is a method used to engage clients in "a collaborative conversation style for strengthening . . . motivation and commitment to change" (W. R. Miller & Rollnick, 2012, p. 29). Leaning into the client-centered principle of unconditional positive regard and the use of reflecting skills, MI is an evidence-based approach to working with ambivalent clients (Rosengren, 2017). MI and the TTM both foster therapeutic change for difficult clients, with MI serving as an intervention style that can be applied within the TTM framework.

ASSESSMENT OF THE SEVEN PRECURSORS

As presented in Section III, the CHANGES assessment is an application of the CHANGES model and a therapy tool that helps clinicians determine which precursors need to be activated to bring about change. However, the assessment is secondary in importance to a deep understanding of the CHANGES model as a conceptual framework and the various techniques to activate missing precursors or to grow underactivated ones. The CHANGES assessment is a practical tool for gauging and monitoring degrees of precursor activation. For experienced therapists who grasp the fundamental nature and function of the CHANGES model, the assessment is easily internalized.

However, therapists who are becoming newly acquainted with the CHANGES model may find it helpful to use the assessment form in sessions. The assessment is rated by the therapist but may be completed with the help of the client when appropriate. It enables the therapist to see how the precursors overlap and interact and provides a relatively clear picture of what might be needed for a particular client to achieve change. Therapist and client can then work toward implementing and increasing missing precursors using the treatment strategy or technique suggestions provided throughout Section II. Again, the conceptual basis of the CHANGES model and its techniques take precedence, so we reserve a detailed discussion of the assessment for Section III.

CONCLUSION

As examined in this chapter, the CHANGES model is an example of how specific change processes can be identified in a context that owes allegiance to no particular theory. It is an example of how a therapy approach can focus on change itself and orient techniques, theories, procedures, and stages around specific change functions. The seven precursors are so fundamental that they lie at the core of theories, manuals, approaches, and procedures. This is not a global model of therapy, nor is it meant to be. But it may serve as an example of how techniques and manuals can be oriented around the therapeutic change process rather than treated as ends in and of themselves.

Consequently, one reason major schools of therapy are more or less equivalent is rather simple: They all have a fairly equivalent grasp on change itself. The same is generally true for techniques. Both theories and techniques require, draw from, and make use of the precursors. The absence of precursors means the absence of change. Any therapy, any procedure, or any approach that evokes, inspires, or brings to emergence the primary prerequisites of change will tend to be a valid therapy. And the therapist who encourages the presence of precursors will be an effective therapist regardless of training and theoretical background.

Insofar as up to 85% of what transpires in psychotherapy is common to all approaches (Cuijpers et al., 2019; Peterson, 2019; Strupp, 1996), researchers who compare therapies are not examining apples and oranges so much as tangerines and clementines. If psychotherapy is to advance as a discipline, it must be able to achieve therapeutic change more quickly, deeply, and efficiently. Advancements in theory and technique, while important, must be accompanied by a clearer and more complete understanding of what brings about change. To that end, therapists have a duty to understand why a client is not changing. Therapists must also be able to identify missing change factors in a client and then prescribe targeted strategies or techniques that can activate those change factors. Although it can be used with any client, the CHANGES model seems particularly valuable with difficult clients because a difficult client is one with few, if any, activated precursors.

2

What Makes a Client Difficult?

The primary duty of a therapeutic change model is to elucidate the process of how people change as well as how people remain the same. According to the CHANGES model, people change when the precursors of change are activated, and people stay the same when the precursors of change are inactive or underdeveloped. This means that when a client is responding well to therapy and undergoing change, there is little or no need to use the CHANGES model. Whatever approach the therapist is using is probably fine. We know from common factors research that all of the major approaches seem to work about equally well (Wampold & Ulvenes, 2019), with the exception of particularly effective applications for specified disorders, such as exposure therapy for phobias.

The secondary duty of a therapeutic change model is to show how change can be achieved where there are factors working against it. In other words, any model of change worth its therapeutic salt must highlight new ways to work with clients who have difficulty creating change. If precursors of change are absent or underdeveloped, the client may be regarded as "difficult." Our use of that term is not meant to pejoratively suggest a characterological deficit. Rather, it means there is a deficit in terms of the client's skill, ability, desire, or willingness to enact therapeutic change. A client is considered difficult when therapeutic change is not forthcoming. One difficult client may present as unmotivated and indifferent, another as aggressive and indignant, and still another as deferential and agreeable. However, all such clients may be considered difficult insofar as change is not occurring, and such change deficits are occurring because the precursors are not established.

Case examples have been disguised to protect client confidentiality.

https://doi.org/10.1037/0000451-003
Therapeutic Change With Difficult Clients: Precursors and Techniques in the CHANGES Model, Second Edition, by B. D. Wilkinson and F. J. Hanna

The irony is that if not for difficult clients, therapy could become quite automatized, and virtually any client would benefit from just about any competently applied theory. Indeed, chatbots trained using large language models to disclose humanlike emotions have proven successful with clients (Park et al., 2023). Such outcomes are not solely due to advancements in artificial intelligence, as people reported experiencing "therapist understanding" from mental health software prototypes long before the advent of large language models (Selmi et al., 1990). Similarly, bibliotherapy can be effective and helpful in a way that does not require a therapist (Christensen & Jacobson, 1994; Liu et al., 2022).

We demonstrated in the previous chapter that the reason for success when therapeutic change occurs is that the precursors are present and operational. Such change-activated clients can make substantial progress with people, chatbots, or even self-directed bibliotherapy. However, with difficult, self-undermining, manipulative, unwilling, or involuntary clients, the precursors are hardly present, if at all. Computer programs and bibliotherapy simply will not do the job. As such, the therapeutic mandate proposed in this book is to implement the CHANGES model with difficult clients so that more conventional therapeutic approaches can eventually be used once again.

DIFFICULTY AND RESISTANCE AS SELF-PROTECTION

The CHANGES model views resistance to change primarily as a mode of self-protection. Some people are so afraid of the world that they are hesitant to make any contact with it, let alone effect change within it. Perhaps this signals a kind of existential agoraphobia (Vahdani & Phillips, 2021). At some deep existential level, they believe making direct contact with the world is tantamount to experiencing pain and hold corresponding maladaptive beliefs that lead to further alienation from the world. Other people seem to believe that they do not have the capacity to alter either the world or their mind. Still others believe that they do not deserve the benefit of change, and through a lack of effort, they deliberately deprive themselves of change. Many passive clients are in this group.

Yet resistance can also be quite creative, such as when a client is perceived as difficult and disinterested in change in therapy but actively seeks therapeutic change through some misguided means. In other words, a client may honestly be trying to improve their conditions or fix a problem but is using flawed methods or inaccurate knowledge, as in the case of abusing drugs or joining a cult.

Thus, change is not always therapeutic. For some difficult clients, almost every experience of change they have ever undergone is associated with pain, suffering, loss, or oppression. Abuse causes change, as does violence and neglect. Even though psychotherapy holds no intention to produce these negative experiences, in a difficult client's mind, psychotherapy's tacit promise of change might be tantamount to the threat of pain by association. Therefore, in this private logic, all change (even positive change) can carry the implicit guarantee of being a painful, unbearable experience that should be avoided or deflected to protect oneself. Such a mindset is devoid of the precursors of change. Part of the process of establishing the precursors involves overriding such negative associations. Many techniques identified in this book are designed to do that precisely.

In the classic *Human Change Processes*, Mahoney (1991) noted that people who actively fight beneficial change do so because their beliefs and general outlook are so delicately in balance that change appears as a threat so great it could cause collapse and with it the loss of functioning and mental equilibrium. Interpreted in a different way, to the client for whom change is not appealing, the prospect of change, therapeutic or otherwise, represents the onset of a crisis. *Crisis* has been defined as an event or events that cause cognitive, emotional, and environmental turmoil so great that the person goes into a type of shock called "disequilibrium" (Janosik, 1986; Magnavita, 2006). Examples of such events are the death of a child, abuse, suicidal ideation, or tragedies caused by natural disasters. Crisis work consists not of enhancing the person's coping skills and range of functioning but of restoring that person to their precrisis level of coping and functioning.

Some clients experience a different kind of disequilibrium as a result of the perceived threat of change. They have feelings of instability and uncertainty, intense emotional swings, and a loss of perceived control (Pagnini et al., 2016). Clients of this bent understandably get angry, oppositional, and spiteful toward a therapist or anyone else who pushes them to change. Assuming that people are, to some degree, free agents, imposing change on a person, regardless of professional credentials, looks like punishment to a client who is not convinced that it will help. Such clients often fight hard to preserve whatever precious freedom and self-determination they have left. They dig in their heels and engage in a power struggle with the therapist, seeking to avoid any sort of surrender to a person they do not fundamentally trust. We have found that some of these clients are suffering from a previously unidentified trauma and are still trying to hold things together to get along in the world. From this perspective, it is wise to frame difficulty with change as a form of self-protection rather than implying that such clients are stubborn, defiant, or misguided.

Clients "Trained" to Be Difficult by Their Therapists

When a client is not changing, it is often for an excellent reason, and it could be that a therapist is indirectly bringing about the problem. The following case descriptions present a few examples of clients whose therapists played a key role in their difficulty. Two of these case examples will be reexamined in Chapter 12 in this volume, where the therapists are rated on the CHANGES assessment form.

Beth was a White, cisgender woman in her late 30s who was losing her home because of financial issues and losing her children to an ex-husband, who was falsely accusing her of abusing them. She was frantic. Her previous therapist had insisted that she explore issues of childhood, which were causing her current problems. The implicit promise was that if she resolved her childhood issues, her current problems would also resolve. Beth was an angry, suspicious, and bitter client who was still desperate for help.

Her new therapist found it perfectly reasonable that Beth was unwilling to explore her childhood and her emotions. What she needed from therapy was a problem-solving approach. Beth had been "resistant" to anything else and was very difficult to work with. But with the new therapist, she worked hard to resolve the real-time problems in which she was immersed. Beth's resistance was caused by her previous therapist's misguided insistence on the "proper" treatment.

A different example of therapist-initiated "resistance" is the case of Kurt, a 31-year-old male therapist, and his client Janey, a female university senior in her mid 20s. Janey was highly cooperative in answering questions in her early sessions with Kurt and initially seemed to value her therapy sessions. She freely and openly discussed her issues and problems. Her major complaint was that the men in her life verbally mistreated and did not respect her. She claimed to be worried about this issue and wondered if she would ever find someone who really loved her. She was articulate, easily identified her beliefs and feelings, and reported several seemingly important insights. On the other hand, there was a certain naivete about her that was reflected in her agreeableness as well as her tendency to smile while in pain and to laugh in a shrill, tinny tone.

After 20 therapy sessions, change was not apparent. Janey still allowed her current male partner to verbally mistreat her, and she complained of the same problems from session to session. Finally, she stopped attending sessions and did not return phone calls. In his final case note, Kurt identified that she had most likely terminated therapy due to her own "resistance to change."

In attempting to understand all this in supervision, Kurt discovered, reluctantly and with no small amount of embarrassment, that he harbored a hidden belief, formed in high school, that "attractive girls with fake laughter" who "smiled for no reason" were "stupid" and unlikely to achieve real significance in their lives. He realized that this implicit bias slipped out in various ways in his interactions with Janey, and although he never did or said anything overtly disrespectful, he never really empathized with her. Kurt had quietly minimized Janey's problems, delimited her capacity for change in his own mind, and failed to challenge her surface-level agreeableness. Janey had been an eager client in need of real support, but Kurt had created barriers to potential change.

A quite different example involves Jerry, a pleasant 30-year-old man who reported feeling "lost" and "wayward" about the lack of direction in his life. Although quite intelligent, he had no career plans, and nothing seemed to inspire him to want to do anything with his life. Nancy, his therapist, had 9 years of experience working in a university counseling center. In supervision, she reported that after 16 sessions, he was so "totally passive" and so "resistant" to help that she could not get him "to do anything" at all about his problem.

When asked to report her real feelings about the client in supervision, she was embarrassed to report that she found him "irritating" and "maddening" to work with, and she disliked the way he talked so slowly and deliberately, as though he were "picking each word, one by one." To her credit, she felt quite badly about her feelings toward him, but she also said that she was not able to help and suggested referring Jerry to another therapist.

The supervisor recommended that she write down all the people in her life of whom Jerry reminded her. The next week, she reported with a sense of both relief and astonishment that Jerry reminded her of her sexually abusive brother, from whom she had cut herself off 20 years earlier. After this supervision, she no longer considered Jerry to be resistant or difficult, and he began to make progress. In fact, she began to like him. At one point, Jerry told her that she seemed "nicer." This therapist learned a valuable lesson.

Such examples show how clients can be difficult because of the therapist's faulty perceptions or an inappropriate or ill-advised approach. The therapists were not immediately aware of their own contributions to their clients' perceived resistance. It is extremely important to emphasize this aspect of difficulty and resistance in therapy.

Clients Difficult All on Their Own

In many cases, the client is difficult despite the best efforts of the most competent therapists, as described in the following three case examples. Diagnoses from the *Diagnostic and Statistical Manual of Mental Disorders* are deliberately omitted in what follows so that a distinct picture of the client can emerge without clinical labels. We reexamine these cases in Chapter 12 in this volume.

The Story of Tommy

Tommy, age 29, was tall, handsome, charming, humorous, and thoroughly self-centered. He was married to Julie, whom he described as a loving and devoted wife and caring mother of their three children. He and his wife had been "fighting," and she insisted that he go to therapy.

Tommy hated himself for cheating on her at every opportunity, but he seemingly would not, or could not, stop. He would sometimes criticize her for being "oblivious" and "foolish," but when challenged, he would say, "My wife is a saint." He also said it was "only natural" for men to cheat on their wives. He reported that one of his friends had advised him to seek therapy because he was so unhappy and because he was "throwing away a good woman." A salesperson who regularly made house calls as part of his work, Tommy was so good at concealing his activities that Julie had no inkling of his promiscuity. Tommy, meanwhile, showed neither preference nor prejudice regarding the women he seduced. He was addicted to charming and exploiting women for sex.

Tommy boasted of his sexual prowess and said that such talents were wasted on a wife. Then, in almost the same breath and in an explosion of guilt, he ruthlessly criticized and condemned himself for being a liar and a phony.

He believed, as a Christian, that he was headed "straight to hell" for his many affairs. His fear of hell seemed to trouble him more than his infidelity, much more so, it appeared, than any potential harm done to his family or marriage. Tommy had been to three therapists trying to rid himself of his anxiety. He had apparently competed with and fought his therapists at every turn. In a tone of voice that combined both bravado and shame, he claimed that none of his therapists were of any help at all. His current therapist deduced that his apparent interest in therapy amounted to the improbable goal of reducing his anxiety without changing his behaviors. Beyond what he stated, his chief goal for therapy seemed to be reassurance that he would indeed make it to heaven, that his behavior was really okay, and that he was a good person. He had also confided in his minister but found no support for his promiscuity.

The Story of Joy

Joy identified herself as an "asexual former showgirl" from the Lake Tahoe region in Nevada. She was in her late 30s with dark hair and bright blue eyes. She was heavily into the occult and new age pursuits, believing that she was clairvoyant and psychically gifted. She had been involved with several new age cults. She was actively and avowedly celibate and freely admitted to a contempt for men. She also claimed that she was destined to find her "soul mate" sometime in the next year. She had been to many psychics and astrologers but had decided to seek therapy as a way of opening the psychological blocks that stood in the way of her realizing her "destiny" as an occult "adept."

In therapy, her insistence on staying "positive" was so strong that she avoided any disclosure that was associated with negative emotion or inferred weakness, as such things ran counter to the list of new age affirmations that she repeated to herself each day. She seemed interested only in discovering how she could undo the lingering negative "vibrations" of the destructive people of her past. These included her family, ex-husband, and many of what she called "false friends." She assumed that psychotherapy could teach her how to completely cut herself off from all negative emotions and influences, especially her parents. Joy was also highly demanding and wanted instant results. Quite rigid in her thinking, she often lectured her therapist on new age principles, which she said were superior to those of psychotherapy. Not surprisingly, Joy reached a rather dramatic impasse in her therapy after eight sessions, showing considerable impatience and correctly claiming to have derived no benefit.

The Story of Ricky

Ricky was 18 years old and a member of a street gang. He was tall, handsome, and muscular and had been placed in an outpatient adolescent drug and alcohol treatment program when he was 17. He was addicted to crack cocaine and was on probation for its possession and sale. He was a father to at least three children born of adolescent girls. He reported matter-of-factly that all women loved him, and he blatantly and proudly admitted, after some coaxing, that he liked to impregnate young girls. It was also clear that he had supplied many young people with drugs and had committed other crimes, although, according to his probation officer, he had not been brought up on charges.

He had an outwardly pleasant demeanor and smiled in a cocky, self-assured way. He had a coldness about him—a deeply disturbing lack of warmth or engagement that oftentimes evoked visceral reactions of intimidation among the therapists in the program. Ricky never exhibited any overt violence. However, with any kind of confrontation of his behaviors or attitudes came a subtle nonverbal threat of intimidation. He had a way of cocking his head and curling his lip in a silent, smiling snarl whenever questioned about his lifestyle or attitudes. These defiant behaviors seemed to open a window into a deep, smoldering hostility. Ricky was highly intelligent and manipulative. He displayed an extraordinary sense of entitlement that was, of course, part of his criminal mindset. In therapy, it soon became clear that he was immersed in a deep hatred of the world and people in general. Therapy to him was a form of control and punishment to be outwitted and thwarted at all costs. He often told his therapist, "You best not mess with my head."

Difficult Indeed

The examples in the previous section describe people who were extraordinarily difficult to engage in therapy. Standard therapeutic approaches were of no help, and each client was adept at strangling the therapeutic change process. Even those who recognized that change was necessary did not want the type of change that psychotherapy offered.

ROOTS OF DIFFICULTY: IMPLICIT LEARNING AND MALADAPTIVE BELIEFS

Difficulty or resistance to change is a consequence of beliefs that arise through implicit learning and are typically formed so early in life that the seeds of its formulation are sown before one learns to speak. *Implicit learning* involves the preconscious, nonintentional formation of knowledge, concepts, and beliefs about the environment as well as causal-relational structures between objects and events (Reber et al., 2019). Through implicit learning, people formulate assumptions upon which an entire meaning system can rest (Cleeremans, 2019). These assumptions are often highly emotionally charged and, if challenged, evoke vehement reactions ranging from protest and defiance to betrayal and hurt. If those primary, fundamental assumptions do not change, then neither will the person.

It is generally understood that the experiential, preverbal nature of implicit learning makes it "verbally inaccessible" (Hartmann, 2019, p. 264) and thus difficult to articulate via standard talk therapy. Trauma-focused somatic therapies have arisen to address how embodied, implicit learnings interfere with healthy functioning (van der Kolk, 2014). The fact that so much psychotherapy is narratively driven may explain why therapeutic change is so often difficult for clients with early life traumas. Fortunately, implicit learnings also tend to manifest cognitively as maladaptive, irrational, or pathogenic beliefs about the nature of reality and fundamental aspects of living (David et al., 2009; Gazzillo, 2023). They take the form of preverbally formed yet verbally identifiable judgments on subjects like awareness, love, people, self, problems, the world, or life.

Examples of some of these fundamental preverbal beliefs that directly affect the psychotherapy process are "interactions with people cause pain," "awareness of a problem only makes it worse," and "people only care about themselves." There is an old notion that one should never discuss religion or politics in casual conversation; such topics, like many discussed in therapy, evoke deep emotional responses or what is generally recognized in cognitive behavioral therapy as "hot" thoughts (McKay et al., 2021). In our experience, implicit knowledge can be changed but only with care and extreme courtesy, and usually, permission from the client must be deliberately and openly sought beforehand.

The Story of Kathy

Kathy had a doctoral degree in psychology and ran her own private psychotherapy practice. She was extremely self-aware and articulate. Somehow, she came to be convinced that no one would ever be there for her when she really needed them, not even her best friends. When asked, she said that even when her friends had been supportive and there for her, she always considered it surprising and unlikely. She had felt deep emotional anxiety over this issue and had consistently framed and described it as a problem in trusting others. Being familiar with her highly admirable qualities and knowing her to be a trusting person, the therapist eventually became highly suspicious of this view. When the therapist asked if trust might not be the real problem, she became upset and insisted that trust had been a problem all her life.

The therapist first asked Kathy if she noticed how upset she became when asked the question. Kathy nodded. The therapist then asked her if she would be willing to explore that but noted that if at any point she thought it was a waste of time to pursue, to make it known and they would move on. She agreed. The therapist asked her to hold on tightly to that feeling of upset and protest toward the question and to follow it into her mind. Being highly aware, Kathy was easily capable of such a directive, so the therapist encouraged her to allow any memory images to emerge that might be related to this feeling. She began to cry. She was soon relating memory images in which she was in her crib, crying and waiting for her mother to come to her. But her mother never came, and she remembered being horribly alone and deeply despondent, abandoned, and unloved. It was during this time, she said, that she formed a belief that no one is reliable and that people cannot be counted on to be there in times of need. She estimated that she was about a year to 18 months old.

Of course, it is not important whether the memory is accurate, if it actually occurred, or if that was her actual age at the time. Adler (1956) noted that early memories need not be verifiably true for a person to obtain therapeutic benefit from examining them. It is the maladaptive belief about people that was at the heart of the problem, regardless of the story content.

After this session, Kathy called her mother and told her about this early memory. Her mother, as Kathy later related the story, was quite surprised and expressed how bad she had felt listening to her baby cry for long periods. She said that she did not attend to her because the conventional wisdom at the time was to avoid spoiling the child by deliberately letting her cry, which allegedly would make the child more self-reliant. Seeing and dismissing the maladaptive belief was of value to Kathy but having her mother more or less verify the memory was of greater significance.

THERAPIST VARIABLES INFLUENCING THERAPEUTIC SUCCESS

Research into differential therapist effectiveness has consistently demonstrated that some therapists are more effective than others when it comes to creating positive therapeutic outcomes. Evidence first surfaced when Strupp and Hadley (1979) found that college professors untrained in psychotherapy were just as effective as experienced professional therapists. Many of the clients in this study had challenging disorders, yet there were no appreciable differences in effectiveness. Five years later, a meta-analysis comparing paraprofessionals to professionals (i.e., licensed psychologists, psychiatrists, social workers) found that "paraprofessionals must be considered as effective additions to the helping services, and in many cases they are more effective than professional counselors" (Hattie et al., 1984, p. 540). Level of education and years of professional experience were not significant in determining effectiveness.

Years later, Stein and Lambert (1995) found that professionals fared only slightly better than paraprofessionals and stated that "given the enormous, national investment of physical and human resources in graduate programs, it is quite remarkable that more compelling evidence is not available that demonstrates that graduate training directly relates to enhanced therapy outcomes" (p. 194). Such classic studies have withstood criticism over many years, and the results remain relevant: People who have only 20 or so hours of therapy training can be nearly as effective as professionals with advanced degrees, hundreds of hours of training, and years of experience.

Studies demonstrating a measurable impact of therapist experience on client outcomes have tended to have negligible effect sizes (Kraus et al., 2016; Walsh et al., 2019). Instead, research shows that the most effective psychotherapists facilitate change regardless of the presenting concern (Nissen-Lie et al., 2016), consistently over time (Kraus et al., 2016), regardless of years of experience in the field (Delgadillo et al., 2020), and that they might be identifiable from the earliest point of training (Edmondstone et al., 2023).

What do such studies really tell us? According to the CHANGES model, the dilemma boils down to two essential points. The first is that some people are somehow more fundamentally capable of helping clients change, regardless of education or training. In other words, professional training seems limited in its capacity to train some people to be more effective change agents. The second point is that if we more fully understand therapeutic change as a process in and of itself, perhaps we can better grasp what therapists must do to help make it happen.

Therapist effectiveness has little to do with GRE scores, high grade point averages, number of publications, research skills, or the ability to pass exams or write dissertations (Castonguay & Hill, 2017). Therapist effectiveness is not dependent on the university attended or the type of training program but rather on the personal characteristics of the therapist (Edmondstone et al., 2023; Firth et al., 2015; Goldfried et al., 1990; Peterson, 1995). Of course, this raises the question of what it takes to not only facilitate change but also to train and prepare professionals to be effective change facilitators.

Wisdom and the Wise Therapist

There is indeed a concept that accounts for the mostly unspoken and largely unspecified skills that differentiate the average therapist from the highly effective therapist. The concept is called wisdom (Grossman et al., 2013; Hanna & Ottens, 1995; Sternberg & Glück, 2021). This ancient concept is imported from research in developmental psychology and has great explanatory power. Researchers generally describe *wisdom* as a high degree of knowledge and expertise regarding the living of life itself. It is not the same as intelligence, although the two complement each other quite well. Characteristics of wisdom include empathy, compassion, metacognitive skills, perspicacity, self-awareness, and self-transcendence, among many others (see Table 2.1).

It may well be that lack of wisdom is at the core of the puzzle concerning why years of experience as a therapist does not seem to reliably predict effectiveness (Skovholt, 2017). One of the characteristics of wisdom is to resist habit in or automatization of thought and behavior routines (Sternberg & Glück, 2021). Take two therapists with 20 years of experience. One continuously self-reflected, learned, and improved therapy skills for a full, rich 20 years of experience, whereas the other had 1 year of therapy experience that was then automatized and repeated 20 times. In the latter case, the therapist more or less went through the motions of therapy on autopilot, with the net result of no longer improving their skills (Dawson, 2018; Dumont, 1991).

The effective therapist tends to be a wise therapist. Of course, intelligence is certainly handy and helpful, but when it comes to the skills necessary to be effective, the characteristics and thinking styles that make up wisdom seem more relevant than those of intelligence (Levitt & Piazza-Bonin, 2016). Yet we accept people into graduate training programs with criteria based on intelligence rather than wisdom. Could it be that graduate-level trainees end up with the same general level of wisdom as paraprofessionals because wisdom is not taught in graduate schools? After all, if wisdom is randomly distributed across the

TABLE 2.1. Characteristics of Wisdom

Wisdom characteristic	Definitional element
Empathy	• strong sense of the feelings and outlooks of others • understands others from their subjective point of view and from a perspective that is not self-centered
Concern	• compassion for others • cares for the welfare of living beings and the environment
Recognition of affect	• recognizes the interdependence of cognition and affect • aware of own emotions and feeling states
De-automatization	• resists habitual, automatic behavior and thinking patterns • emphasis on awareness of actions and responsible choice
Sagacity	• strong listening skills • deep insight and awareness into human beings and relationships • self-knowledge and capacity for self-transcendence • ability to learn from mistakes
Dialectical reasoning	• recognizes the power of context and the interplay of opposing views • fluid and intuitive reasoning • ability to consider all sides of an issue • orientation toward beneficial change
Efficient coping skills	• copes smoothly and efficiently across a range of people and situations • ability to find fulfillment and meaning in life
Tolerance of ambiguity	• recognizes ambiguity as intrinsic to the nature of being in the world • ability to perceive, appreciate, integrate, and use shades of gray
Perspicacity	• ability to "see through" situations • capacity to avoid being fooled or deceived • intuitive understanding and accurate interpretations of their environment • ability to look beyond appearances
Problem-solving skills	• identifies and frames a problem such that the solution is efficient • capacity to reframe problems and situations • expertise in the use of transferable metaphors
Metacognitive capacity	• ability to recognize presuppositions and assumptions • awareness of awareness; knowing of knowing; thinking about thinking • capacity to direct awareness, behavior, and emotion toward change

Note. From "The Role of Wisdom in Psychotherapy," by F. J. Hanna and A. J. Ottens, 1995, *Journal of Psychotherapy Integration*, *5*(3), p.199 (https://doi.org/10.1037/h0101273). Copyright 1995 by the American Psychological Association.

population, then it should be rather evenly distributed among professional and paraprofessional therapist samples as well (see Hanna & Ottens, 1995).

Regardless, a wise therapist may well be the most capable of inspiring and implementing the CHANGES model. Pure intelligence is not oriented around skills such as listening, empathy, and interpersonal expertise (Skovholt, 2017). Thus, the crucial difference on which to place a research focus may not be the professional and paraprofessional, but the wise and not so wise. The difference may well lie in the ability to facilitate the precursors. With regard to research, it may be time to explore some research avenues implied or indicated by the CHANGES model.

Training the Wisdom Out of the Therapist

Lazarus (1989b) observed that formal education and training in psychotherapy and counseling can take a person's natural therapeutic warmth, wisdom, genuineness, and empathy and degrade them into a stylized professionalism that is detrimental to a would-be therapist's natural skills. He used the metaphor of "taking a can of spray paint to an artistic masterpiece" (Lazarus, 1990, p. 352) to describe how academic training can spoil a student's natural helping skills and suggested that much of what is taught to graduate students about doing therapy amounts to "deadwood, superstition, and plain rubbish" (p. 352) rather than useful knowledge about what is actually effective.

From the perspective of the CHANGES model, graduates of many training programs primarily learn a professional language; an array of theories, research findings, research methods, and ethical principles and behaviors; and a carefully cultivated professional demeanor, but they are not necessarily taught how to facilitate therapeutic change itself. There are therapists and trainees at all levels who, for example, do not use session time in ways conducive to arriving at solutions or strengthening therapeutic relationships with the goal of change. There are people working in community agencies without even a bachelor's degree who are natural therapists with a knack for helping clients change. Such natural therapists tend to have wonderful relationship skills, make perceptive decisions with a sense of precise timing, and tailor their interventions to client needs.

None of this is to suggest that graduate training is unimportant, as effective training programs can and do enhance psychological mindedness, refine presence, and provide a launch point for skilled interpersonal work with clients. However, it seems that what is most important is quite simple: The therapist must have the skills to help bring about change. According to the CHANGES model, the skillfulness of a therapist may be little more than the ability to inspire, engage, and enhance the presence of precursors in clients. Such therapists can orchestrate and implement change ingredients with a client through empathy, warmth, persuasiveness, techniques, reframes, metaphors, and general wisdom about human beings and life. The therapist could be the family doctor or the professional therapist in private practice. It could be the school counselor at the local high school or the recovering alcoholic who works for near minimum wage at the local community substance use treatment center. Whether a professional or paraprofessional, the presence of these characteristics in a person indicates a capacity to become an agent of change and a fine therapist.

Fundamentally, there are a certain number of well-off people who can be helped by almost any well-intended therapist using almost any sound approach (Wampold & Imel, 2015). Bartenders and hairdressers routinely perform this function. Some astrologers, psychics, and channelers act as therapists as well (Lester, 1982). If all that is needed, according to Frank and Frank (1991), is a myth to explain healing, astrology and channeling can certainly serve that function. People claim to gain benefit from these offbeat, at best, and at worst harmful therapies (Singer & Lalich, 1996). Why? That is the point of this book. If the precursors of change are present, that person is likely to change. If the precursors are not present, change is unlikely. As such, the most important skill of a therapist is the ability to activate the precursors within the CHANGES model.

CONCLUSION

Study the vast number of theories, procedures, and techniques used in psychotherapy, and one finds that most assume some degree of motivation and involvement on the part of the client. In the previous cases of Tommy, Joy, and Ricky, involvement was lacking in all three, and motivation was either absent, in the case of Ricky, or alloyed and misdirected despite a stated interest in change by Tommy and Joy. Techniques typically used with clients, such as role-playing, the empty chair, identifying and disputing irrational beliefs or dysfunctional cognitions, behavioral contracting, desensitization, and so on usually will not work with a client who is unmotivated and uninvolved. Without involvement and motivation, few techniques are likely to be effective, in spite of the therapist's level of skill. That is why positive outcomes seem to be related to these variables.

Clients who steadfastly refuse to engage in role-plays often judge such an exercise as useless. The empty chair is viewed with particular disdain by such clients, and the same can be said for many other techniques. The potency of powerful change-producing techniques such as these is irrelevant when a client is uninvolved or unmotivated. Many of these clients are tremendously difficult to engage. A few are almost inaccessible. The skills needed to work with such clients and engage them in therapy are seldom taught in graduate programs and are almost always learned on the job. Thus, the ability of psychotherapy to reach such clients more quickly and effectively represents one of the growth paths of the entire discipline.

There are a host of styles of being difficult. Many clients desperately want to change but find the process too painful and fight it at each juncture. Others are not forthrightly resistant or difficult at all; they would change, but they just do not see the point. For yet others, change seems a pleasant prospect but involves too many sacrifices to appear desirable. Another style is to see change as a hindrance to an established way of being. These people often see the therapist as an enemy intruder, there to confuse and confound. For many people with criminal histories, change is the same as an admission of weakness and guilt and is to be avoided at all costs. It can also be a threat to their "freedom" to do whatever they want, whenever they want, to any victim they choose. Another style of difficult client simply does not believe that change is possible—for anyone—and therefore devotes nothing to it, other than lip service, to a judge, employer, or spouse, as the case may be.

In view of all of these issues, the field is clearly in need of a broader and deeper understanding of why some people change and others do not. This a primary goal of the CHANGES model. The following section is dedicated to examining how therapists can accomplish this task, by examining each of the seven precursors in the CHANGES model along with myriad techniques to systematically grow client motivation and involvement in the therapeutic change process.

II

THE CHANGES MODEL, STRATEGIES, AND TECHNIQUES

3

Orienting the Relationship Around Change

It would be a mistake to place one's trust solely in techniques when working with difficult clients because without a strong therapeutic relationship, techniques are likely to fall flat or not take hold (Cochran & Cochran, 2015). When client motivation is absent, the therapist's primary task is to get the client to participate or cooperate in therapy. Getting an unmotivated, disinterested client engaged in therapy requires a different set of skills than applying the steps of a particular theoretical approach, technique, or manual. In such cases, the ability to be persuasive and influential is often necessary to get a client on track. Empathy must be sufficiently advanced that the client feels understood by the therapist, even when the therapist does not at all agree with the client. The therapist may even experience nausea or visceral discomfort while empathizing with clients who display unsavory or destructive attitudes. Nevertheless, without the ability to reach and make that vital connection with a difficult client, little or no change is likely.

If a client does not respect a therapist, then premature termination is likely. That respect depends on various factors, but a major factor seems to be whether the therapist truly understands the client. Therapists have known for a long time that "feeling understood" is at or near the heart of how a person experiences empathy from another (Van Kaam, 1966). Adler (1927) gave perhaps the most descriptive definition of empathy when he portrayed it as seeing with the eyes of another, hearing with the ears of another, and feeling with the heart of another. Research indicates that empathy, warmth, and genuineness are related to a large percentage of the gains attained in therapy regardless of the therapist's

Case examples have been disguised to protect client confidentiality.

https://doi.org/10.1037/0000451-004
Therapeutic Change With Difficult Clients: Precursors and Techniques in the CHANGES Model, Second Edition, by B. D. Wilkinson and F. J. Hanna

theoretical orientation (Elliott et al., 2018; W. R. Miller & Moyers, 2021). When it happens, in therapy or out, the empathic person is valued by the other.

Thus, it is of central importance to respect and work from within the client's frame of reference, as Rogers (1951, 1957) repeatedly emphasized. Difficult clients who feel understood by a therapist no longer find it easy to summarily dismiss the approaches or observations of the therapist. Conversely, it is likely that many people can accomplish change faster on their own than with a therapist who is not empathic and does not engage a client in a viable working relationship. A therapist can know all the most innovative and brilliant techniques, but if they are not adept at empathy, they are likely to lose clients (W. R. Miller & Moyers, 2021) or have clients who do not change. We have talked with many therapy dropouts, and many of the complaints went like this:

- "My therapist never saw where I was coming from."
- "My therapist kept telling me what was wrong with me and never understood me."
- "My therapist's 'know-it-all' attitude bugged me."
- "My therapist made me feel stupid."
- "We just never connected."

The therapeutic relationship is filled with complexities and subtleties that research still seeks to understand. Many mysterious aspects remain. For starters, we do not know what empathy really is beyond a few surface definitions (see Maibom, 2017, 2020). Is it the process of experiencing another's inner world through a tangible and embodied connection (Manganaro, 2017)? Or is it a simulation-based, cognitive representation that we use to imagine what it would be like to have this person's experience (Decety & Jackson, 2004; Marsh, 2018)? One mode involves phenomenological contact, whereas the other is inferential. Is it both or neither? Each of these perspectives on empathy is based on a different set of metaphysical assumptions (Hanna & Shank, 1995; Maibom, 2017), and at this point, we really do not know. But empathy is so crucial to successful therapy that a more complete understanding could make therapy more effective.

INCREASING THE PRECURSORS THROUGH THE RELATIONSHIP

It would be simplistic to view the precursors as separate from the therapeutic relationship because they are interdependent. According to the CHANGES model, the therapeutic relationship is a necessary precondition for the activation of precursors, which in turn support therapeutic change. Therapeutic core conditions alone are insufficient to facilitate change until precursors are activated in the client. In this respect, the purpose of the therapeutic relationship is to provide a setting in which the precursors can be established and stabilized in a client, difficult or otherwise.

Of course, this is not as simple as it might initially seem. Weaving change through the relationship requires tremendous skill, particularly when a client is ambivalent about change. We have repeatedly found that techniques only occasionally work with a guarded and defensive client who still does not trust the therapist or the therapy process. The most enthusiastic and otherwise competent therapist can try a potent therapy technique on a difficult client, only to have it land with a "thud," to no effect at all. When this happens, it is often the relationship that is lacking.

Successful therapeutic relationships involve more than interpersonal trust. Again, inspiration and persuasion (see Afonseca et al., 2023; Frank & Frank, 1991) are also involved. Clients who do not feel supported and encouraged by a therapist are more likely to drop out of therapy (Friedlander, 2015), seek and accomplish change on their own, or find another therapist. Furthermore, a client can have a wildly different configuration of precursors with different therapists. It is the therapist's duty to tailor the relationship to provide an environment conducive to developing such functions as hope, effort, and confronting the problem.

As far as the CHANGES model is concerned, if the relationship is not established as a working alliance, the precursors may never develop or stabilize. The therapist must develop and refine the relationship so the precursors can incubate and grow to the point where the client can exit therapy in an improved state of coping, health, or functioning. What makes the therapeutic relationship therapeutic at all may well be that it activates the seven precursors, and a relationship that does not activate the precursors is not therapeutic.

Lazarus (1993) called on therapists to be an "authentic chameleon"—in other words, to provide a relationship that meets the client's needs without losing one's own genuineness or authenticity in the process. Some clients require a businesslike collaboration, whereas others need a lot of encouragement and nurturing. Either of these groups would probably find the other style unsatisfying. However, being an authentic chameleon goes beyond merely supplying the appropriate style. In the context of contemporary research on the therapeutic relationship, it also involves being able to match a client's worldview in an empathic and understanding way so there is as little "personality clash" as possible, in spite of interpersonal differences (Clark, 2023).

PURPOSE OF THE RELATIONSHIP: ACHIEVING CHANGE

Linehan (1993) often observed that successful relationships and interactions with difficult clients are brought about by teaching them what kinds of behaviors lead to therapeutic change. She referred to such behaviors as "therapy enhancing" and noted that they should not be expected of clients but should instead be taught. In this same vein, when a client learns that change comes about as a result of cultivating the precursors, the relationship becomes based on change principles. Therefore, a client's agreement can be procured to allow the therapist to gently or firmly point out when one or more of the precursors are being ignored or avoided. Timing is crucial, of course, but it does help to keep a client on task.

In addition, if the client knows and is reminded that the entire interaction is based on change, much less ambiguity is allowed into the relationship. There will be fewer opportunities to lead therapy astray by initiating casual conversation designed to avoid therapy involvement. The purpose of therapy is not often readily apparent to difficult clients. This is largely because the difficult client is not thinking about therapeutic change and therefore sees a relationship with a therapist in ways that often have little or nothing to do with psychotherapy being a means of facilitating change.

The easiest way around this is to clearly establish that the purpose of the relationship is to achieve change, even if change is framed as the goal of happiness for the client. Anything that strays from this purpose is seen as interfering with therapy and can be named as such. So, when clients suggest business dealings, casual meetings, romantic interludes, or conversation about sports, the purpose of therapy can be gently reiterated. In extreme cases, the therapist may even suggest that therapy be terminated on the grounds that the purpose of change is not being served and make a referral if appropriate.

ESTABLISHING THE THERAPEUTIC RELATIONSHIP WITH DIFFICULT CLIENTS

The following sections outline strategies for increasing empathy with a wide variety of difficult clients. Although some of these may be obvious to certain readers, the list forms a comprehensive overview of how to work with clients who care little for therapy, are not interested in seeking help, or are so fragile they are intimidated or threatened by the very prospect of change, no matter how much they might benefit from it.

There are many styles and patterns of difficult clients, so these strategies are not appropriate for all clients at all times. Therapists can refer to this list if needed when the appropriate moment arises and as appropriate for the client's needs at that time. We make no claim that any of these suggestions are original to this book; many therapists have adhered to and followed these suggestions for decades.

Being Courteous and Requesting Permission

Even though therapists and counselors are professionals, many difficult clients see routine actions by therapists as intrusive or even rude. From their perspectives, therapists regularly engage in meddling and prying behaviors. Courtesy can clear the way for probes that might otherwise be rebuffed. In many instances, it can be considered a basic courtesy to ask permission before posing a question or making a statement. Clients will usually agree, thus entitling and empowering the therapist to do so. One might also ask a client if there is a particular way to be challenging or confrontational that they would not perceive as rude or intrusive. In response, clients often give clear instructions that make the job much easier.

Asking permission is best done before asking deeply sensitive questions, and the therapist should add that the client does not have to answer. Clients nearly always agree. Asking permission has two advantages. One is that the person becomes prepared or ready for the question, which increases their willingness to experience the anxiety involved. The other advantage is that the client notes that the therapist respects the client's inner world. This is a form of communicating empathy in and of itself.

Part of asking permission is to be courteous about how one goes about it. This does not sound like much, but some difficult clients are going out of their way to answer such questions at all. For many, answering deeply personal questions involves a lot of effort in an area in which they have little practice or experience. Statements such as, "Do you mind if I ask a question you might find sensitive?" prepare the client for the question and show respect for their dignity and right to privacy.

Being Persuasive

Beyond core therapeutic conditions, persuasion in the service of therapeutic change is a uniquely crucial therapeutic skill in work with difficult clients (Anderson et al., 2020). Frank and Frank (1991) observed that persuasion is a common factor of all therapy approaches, operating in diverse ways but always present. Rhetoric plays a necessary role in facilitating change with difficult clients, as the stylistic use of language and its delivery influence how clients view problems, situations, and events (Frank, 1987). Compelling psychotherapy research in discourse and conversation analysis serves to highlight the legitimate value of persuasive rhetoric (Avdi & Georgaca, 2007; Peräkylä, 2019; Smoliak & Strong, 2018). As Jerome Frank told me (FH) in 1996, persuasion is inseparable from effective therapy, and methods of rhetorical influence should be taught in graduate school.

However, misperceptions of what constitutes persuasion may serve as a significant barrier to using the CHANGES model, which maintains that it is the role and responsibility of therapists to help clients activate precursors and facilitate change. If a therapist is dissuaded of the value of persuasion or otherwise regards persuasion as mere advice-giving or, worse yet, manipulation, then it will be difficult for them to make use of the model and techniques in this book with intentionality. Such negative reactions to a "call for persuasiveness" might be born out of recognizing that power in the therapeutic encounter is inherently imbalanced and that it is the therapist's responsibility to empower clients (Totton, 2018). Therapists must indeed avoid the needless power struggle that occurs when a client perceives the therapist as taking a "one-up" position that conveys an unspoken "I know better than you" message.

Yet persuasion is different from manipulation, just as rhetoric is different from grandstanding. The difference stems from intent, as guided by our ethical codes and professional norms. If therapists are trying to facilitate change, it is incumbent upon them to grow their skill as artful rhetoricians who can persuasively articulate the value of the change process itself (Anderson et al., 2020). Through the art of persuasion, therapists can encourage clients to do more of what is working, deemphasize what is not working, and promote the value of personal growth, mental health, and well-being (Anderson et al., 2016). Persuasion also plays a pivotal role in expectancy effects (Goodwin et al., 2018), hope (Griffith & Dsouza, 2012), and psychoeducation (Stice et al., 2008).

Most transcribed client–therapist dialogues in this book demonstrate persuasiveness, and many of the techniques herein require artful use of persuasion to be effective. Good timing, tact, and word choice all require a capacity to "read the room" in a manner that captures the context or prevailing mood of a particular moment. Facilitating client movement toward change without impinging on freedom of thought, identity, or autonomy is thus a fine art indeed. Being persuasive is not about getting your way or acting as though you know better than others. As such, therapists would do well to follow the proofs set forth by Aristotle on three modes of persuasion: personal character (ethos), stirring of emotions (pathos), and coherent or logical argumentation (logos; Bartlett, 2019). In alignment with our professional responsibility as ethical stewards of therapeutic change, the artful use of rhetoric in psychotherapy should proceed in a wise, caring, and truthful manner.

Validating Positive Qualities

A client does not have to have a high IQ to be smart. Clients who are perceptive, shrewd, or have street savvy should be validated for it. For example, when a therapist encourages a client to explain how a problem came into their life, they can say, "Come on, you are very smart. You know what's going on here." Any positive quality that the client displays can be used to facilitate therapeutic progress. Much of this approach to therapy involves discovering and revealing those positive qualities and genuinely admiring them in a client. Such positive qualities can form the platform upon which the precursors rest and find stability.

Additionally, positive qualities should be expounded upon at reasonable length whenever possible so the therapist does not come off as merely ingratiating. For example, if a client says, "I'm a bad person. I always hurt the people around me," then simply saying, "I think you're good" is not enough. The quality of goodness is more fully demonstrated when a therapist responds,

> Sounds to me like you're actually quite a good person. Bad people don't care about whether they hurt others. They just hurt them and don't bother to think about it. I would say that you're actually a good person trying to figure out how to be better. What do you think?

The therapist then waits for an answer so the client has a chance to vocalize agreement or otherwise engage in the process.

Such validation can have extraordinary effects in terms of stabilization, removing self-doubt, building the relationship, and helping a client feel more deserving of positive changes. This practice should be done constantly in therapy to create an ever-increasing list of positive qualities. Some clients initially object, but it can be helpful and quite humorous to all when they acquiesce.

Giving the Client the Option of Telling the Therapist to Back Off

After getting permission to intrude or confront a client, a therapist can say, "If I am bothering you, tell me to back off, okay?" This will often ease the tension of the question or the therapist's persistence by returning a sense of control to a threatened client. This can also be effective with clients who are intimidating, some of whom, when they hear the option of telling the therapist to back off, will consider it beneath them to admit that mere questions could ever bother them.

A client who is manipulative may indeed tell the therapist to back off. When this happens, the therapist readily agrees to do so but then asks if the client is getting uncomfortable. The therapist can also ask if this discomfort occurs often, and when it does, in what kinds of situations and with what kinds of people it occurs. This must be done casually, as these moments can be delicate, requiring good timing and proper phrasing based on client needs and the situation.

Attending to Metalogue, Then Bringing It Into Dialogue

Metalogue is a term we have adapted from its original use by Gregory Bateson (Bateson & Bateson, 1987) to refer to the unspoken conversation that the client may be having with a therapist. Because human beings have only one tongue but process their thoughts in parallel—that is, in many separate streams at a time (Metzinger, 2004; Ornstein, 2003)—many of these thought streams are never given voice. Clients are unlikely to express silent, self-statement thoughts, such as, "I wonder if I can trust this person," or "Why is she looking at me like that? She doesn't like me," or "I wonder if I can fool this guy." These thoughts may be occurring simultaneously, even as the client answers the therapist's questions.

With particularly difficult clients, there may be a different metalogue containing self-statements such as, "Therapy's a bunch of crap," "I wish I didn't have to be here," or "I hope she doesn't say anything about my drinking." These latter statements can be clear thought content to the client, even if they are simultaneously pledging to be cooperative with a reassuring smile.

Attending to metalogue statements has two advantages. The first is that it saves valuable session time by directly addressing not only the presenting complaints but deeper, unspoken issues. The second is that by doing so, a client will come to respect the therapist in a way that facilitates disclosure and cooperation.

The therapist attends to the metalogue through an inferential process. For example, a client may reassure the therapist that she wants to get better yet is not showing effort, confronting the problem presented, or being willing to experience anxiety. The therapist might ask the client, "Do you mind if I ask you a question?" The client will usually say something like, "Sure, go ahead." The therapist then says, "Please tell me if I am wrong, but I get the sense that you really don't want anything to do with this therapy thing. Is that right?" The therapist does not have to be right but may just be close enough that, for the first time, a real and genuine dialogue takes place. By bringing the metalogue into dialogue, therapy becomes more vital and interesting as well.

Reflecting Meanings Before Feelings

It is easy to plunge right into exploring a client's feelings, but this may not be appropriate with clients who are unwilling to experience anxiety or unpleasant affect. For such clients, feelings are not particularly accessible. It is often helpful to initially reflect meanings instead. Frank (1987) observed that "All psychotherapeutic endeavors, whatever their form, transpire entirely in the realm of meanings" (p. 293). Getting a sense of the meaning system of that client can help in accessing feelings later. Casual expressions of opinions can be extrapolated and subjected to an inductive reasoning process that leads to a person's core cognitions. This is also a way of being able to match a client's worldview in a way that expedites the process of self-disclosure and the formation of a viable relationship.

For example, a male client once made a sexist statement about women, smiling a knowing, "brotherhood" type of smile to subtly suggest some form of comradery with the therapist in the moment. Rather than confront the client on his sexism, the therapist found it more effective to explore and discover his attitude toward people in general. The therapist's response was, "Do you mind if I try and sort out where you are coming from?" Following the client's hesitant nod, the therapist continued: "If you think that about women, then is it fair to say that you generally believe some people are meant to be taken advantage of by others?" This client said, "Well, yeah, I do, sort of." The therapist paused and said, "Well, I guess my question for you then is to which group you figure you belong: those who get used or those who do the using?"

This opened up a fruitful exploration of how this client had been used by others and how all his life he strove to become one of the users. This then led to a discussion of whether being a user was admirable and who the users were in his life that he strove to be like. Of course, it turned out that the people he was striving to imitate were people that he despised. This explained a lot of the self-hate that he disclosed later in the therapeutic process.

It is often true that even casual statements on mundane subjects can activate an inductive reasoning process, which in turn can support the isolation and identification of maladaptive beliefs about self, people, problems, or life in general. With a client who withholds information, it sometimes helps to evoke brief opinion statements on politics, sports, or even religion and from there, move deeper toward the underlying presuppositions of the statement itself. Whether one is a conservative, a Democrat, or a Dallas Cowboys fan, casual opinions can open windows into a client's deeper issues and problems. If a person says they are a conservative because their father was, the therapist can issue a challenge aimed at discovering whether the client ever made any independent decisions or whether it is a good idea to be one's own person.

This approach is sometimes highly valuable in easing a client toward detaching from sources of harmful or dysfunctional beliefs.

Increasing Therapist Capacity for Empathy

Increasing one's capacity for empathy seems to be more of a therapeutic duty than a mere suggestion. However, one particular angle on the topic of empathy is seldom discussed in the literature: How much is enough? Does a great therapist have a great amount of empathy? How much empathy does a client need for change to be accomplished?

Certain exercises can help a therapist increase empathy. One is to practice assuming the roles of other people, specifically those one knows or has known. It is also important to be able to assume the role of people with personality disorders such as narcissistic, borderline, or antisocial disorders. Personality disorders that have delusional aspects, such as schizotypal and paranoid, also make for excellent practice. Therapists can demonstrate their ability to think in a disordered way without agreeing with that way of thinking. Clients usually know this and can use the therapist as a model.

Carl Whitaker, the renowned family therapist, often advised being able to think like schizophrenic clients to be better able to work with them. North (1987), a psychiatrist who was once diagnosed with catatonic schizophrenia, stated that this experience made her better able to help such clients because of her ability to empathize with them. Of course, being able to assume the viewpoints of people with these disorders is not the same as having had them, but it certainly can help a client to feel understood. It is also important to clients to know that the therapist is at least attempting to see things from their perspective.

It is often helpful to a therapist to do an empathy check by simply asking the client if they feel understood. This can be done in general or specific ways on specific points. Statements like, "I am trying to see where you are coming from here," and adding, "Am I getting it?" serve this purpose. Simply asking, "Do you think I understand you?" is another empathy check. These serve a dual purpose: to help the therapist ensure that they understand the client and to communicate to the client that the therapist is attempting to truly understand.

Telling the client things like, "You are ultimately the expert on you" also serves this purpose by acknowledging that clients have access to information about themselves that is not available to anyone else. If the client informs the therapist that they do not feel understood, this can open the door to honest conversation and a potential disclosure of information that enhances the relationship.

Another exercise is to role-play a difficult client with a colleague or supervisor by fully assuming the client's meaning and value system and answering questions while playing that client's identity. A therapist should be able to role-play that client so thoroughly that they eventually begin to have a feel for what it is to be that person, in addition to the diagnosis or problem.

A final exercise involves the therapist role-playing the client during the session and checking with the client to determine whether the therapist is really grasping their perspective on life and living. As part of this role-play, it may help to have the client assume the role of therapist. This can help raise self-awareness in the client. Although some difficult clients will eschew this kind of approach, some will take to it, and it can be quite humorous if done lightheartedly. To the degree a therapist can empathically act as a client in role-play, it is important not to slip into caricature. Therapists should be thoughtful about how they convey the expressions, mannerisms, and vocalizations of a client. The same goes for trainees, who often need guidance on the subtleties of character development in role-play exercises.

Establishing the Client's Meaning System or Worldview

This strategy is an extension of the previous one and is related to a little-known school of cognitive therapy called *philosophical psychotherapy* (Mills, 2001; Sahakian, 1976). Based on what clients tell a therapist, even casually, an entire meaning structure can be deduced and proposed to the client (see McMullin, 1986). For example, a client casually declares, "You know you have to look out for number one in this world." A deductive response might be, "So if you are number one, does that mean that no one else in the entire world is as important as you?" A few clients will agree, and others may respond with a "not quite" answer. These are easy enough to handle with standard cognitive approaches. However, many difficult clients will boldly and shamelessly answer, "Of course I'm number one; who else could be?" In this case, an entire meaning system can be structured in a hierarchy.

Consider the example of William. He was 35, depressed, and said in the first session that looking out for himself was his primary task in life. His therapist drew a pyramid on paper, drew in various levels, and placed William's name at the apex or top level. Then the therapist asked who would be number two and so on down the levels. The therapist asked what sort of treatment people deserved at each level of the hierarchy. This helped to discern William's entire moral or ethical philosophy.

Like many clients of the narcissistic style, William calmly and directly said that people low in the hierarchy can be manipulated, cheated, and otherwise mistreated. The therapist asked if he himself was ever low on someone else's pyramid. He nodded hesitantly but seriously. This opened an inroad into his world. The therapist inquired whether he was ever manipulated, cheated, or otherwise mistreated by that person or persons. When he nodded affirmatively, the therapist asked if it hurt or caused him anxiety or difficulty. Once again, William nodded. From this point, the therapist was able to construct his meaning system, which included a view of the world based on contempt for and resentment toward others.

The therapist asked permission to reflect that viewpoint back to him for verification. With William's consent, the therapist stated,

> Please tell me if I'm wrong, but it seems that for you, life seems to be played in terms of who outsmarts and outmaneuvers the other and is based on each person causing

> pain and hurt to others in the process. And whoever does it best wins. So, if no one else is looking out for you, then you have to make yourself number one in order to survive. Is that close?

William replied, "I guess so." Because William had already admitted that he was depressed, the therapist asked if this lifestyle worked to make him happy. When William said, "Not really," the therapist asked if they could use therapy to find a lifestyle that could. Although not productive of change on the spot, this empathic approach placed William on the path to awareness and a sense of necessity for change.

Operating Within the Client's Meaning System or Framework

Once the client's worldview is somewhat defined and outlined, it can become a pivot for various interventions. Techniques can be introduced and framed from within that perspective. It is not helpful to expect a difficult client, who is minimally involved in therapy anyway, to learn a new lexicon. From the perspective of many difficult clients, a therapist's use of jargon-filled explanations can look and feel suspiciously like an attempt to convert them to a new religion.

For example, the same "number one on the pyramid" technique was once used on a different male client who was highly self-centered and self-serving. The therapist got a good idea of the client's meaning system and began to work within it. Later in therapy, the therapist seriously and solemnly reframed the client's existence as the "Unknown King," noting that because he was the most important person in the world, he should be ruling the human race and should be given anything and everything he desired. The client looked at the therapist long and hard. After a pause of several seconds, the therapist said,

> And if that's true, then you must be a tremendously unhappy person who is never respected, meets frustration and disappointment at every turn in life, and believes other people are ignorant, foolish, and petty. After all, they don't recognize that you are the king.

From that point forward, the client began to make progress slowly toward change.

Redefining the Problem

There is always a fine balance between addressing a client's presenting concerns and resolving the almost inevitable deeper issues underlying them. Contrary to some views on this, it can be done briefly. When a client is low in precursors, their explanation of the problem is usually naive and inaccurate. A young woman being physically abused by her husband says it was her fault. A man with antisocial personality disorder says people do not like him because he is "too intense."

On the other hand, there is often a grain of truth in the explanation. A battered woman says that she cannot leave her husband because he needs her. Surely, he does, but that is not the problem. A man who stalks a woman says that he knows that if only the woman would get to know him, she would come to love him. Maybe in theory, but that part of him is so hidden by pathology that the real "him" may never emerge for the woman to see. A kernel of truth is the seed of delusion.

Difficult clients should indeed be acknowledged and understood from within their meaning system, but therapists must be aware of the fact that the meaning system is probably deeply flawed. Clients in general, and especially those who are difficult, fail to account for unconscious motives and beliefs.

Bringing Down Defense Systems From Within

When encountering a well-defended client, it sometimes helps to envision gently bypassing their walls of defense and peacefully entering the city. The goal is to be given an entrance point into the city itself. It will usually be given only to someone who is perceived as a friend. A therapist is the enemy of what Winnicott (1960/2018) called the "false self." In this context, the false self is what erects those defenses and guards the walls. The therapist's purpose is to restore or establish the greatness of the client's "real self," that which is open to experience and not entrenched, bunkered down, or ready for war. Once inside, the therapist usually perceives that the outwardly forbidding ramparts and walls nearly always look delicate and fragile from the inside.

To get past the walls, it sometimes helps to present oneself as a consultant to help improve the design and engineering of the walls. Validate defenses, even hostile ones, by pointing them out and asking if they are effective and what could make them more so. For example, one could say, "I have noticed that you do not like anyone being too inquisitive toward you and that you highly value your privacy. I respect that. Do you find that you like keeping people at a distance?"

People have excellent reasons for erecting defenses. Paradoxically, if the reasons are validated and encouraged, it sometimes reduces the need to erect them. One can also point out the energy drain that comes with keeping the defenses active so much of the time. If a person is tired, that could be reframed as due to their devoting a considerable amount of mental energy to defending instead of living.

Respecting and Encouraging the Client's Autonomy and Freedom

Respecting and encouraging autonomy and freedom is helpful with nearly all clients, but especially with criminal or antisocial clients. Regardless of any debate on determinism and free will, research has shown that appealing to the prospect of change as a free choice has considerable therapeutic value for clients (Kanfer & Grimm, 1978; Lavik et al., 2018). In fact, this has long been shown to be one of the few effective strategies with antisocial clients (see Kierulff, 1988; Samenow, 1998; Simourd et al., 2016). There is a catch, however. Criminals have learned they can violate a wide range of societal norms and dictates. They know they have the freedom to do so, and they live for it. This seems to be not so much a cognitive calculation as a palpable thrill or sensation that reinforces antisocial behaviors. It is as important to address the thrill or sensation of freedom, and the beliefs or attitudes connected with it, as it is to examine their irrational beliefs about criminality in general (see Benn, 2021; Samenow, 1998).

Perhaps the most important step toward working with the criminal mindset is to understand two things. First, the criminal seeks the admirable goal of absolute freedom of behavior—the ability to do whatever, whenever, and however. A criminal client will often combine the first goal with the contemptible goal of being able to have complete disregard for responsibility, accountability, and the well-being of others. This is often a core aspect of their flawed meaning system (Benn, 2021).

Encouraging Uncooperativeness as a Protection of Freedom

This is paradoxical, but it is also true. If a client with limited precursor activation steadfastly refuses to disclose anything of real substance, it might be of help to momentarily suspend the effort to get the client to self-disclose and instead talk about why the client likes to be uncooperative. It often turns out that some clients are so protective of their freedom and autonomy that being uncooperative is merely an attempt to retain the dignity and integrity that comes with being free. If this is the case, the therapist can offer to make an agreement with the client:

> I'll make a deal with you. The minute you feel that I am taking away your freedom in any way, please tell me. I have no interest in doing anything of the sort. If anything, therapy is meant to strengthen and increase your freedom. What do you think?

Once this is done, it is surprising how much a client will allow a therapist to probe and challenge.

Answering the "What's-in-It-for-Me?" Question

Difficult clients are often uninterested in therapy until they can answer the question, What's in it for me? Whenever possible, engage the client in the first session. There are many ways to do this, depending on the style and disorders of clients themselves. Maybe the best option is to offer a sense of hope, even if the client does not admit to lacking it. It is not necessary to promise any resolution of problems, of course, but it often helps to present the possibility of change or better conditions.

Many difficult clients view therapy with more than a little trepidation or suspicion. They are not likely to become involved in the process until they answer the What's-in-it-for-me? question. When it becomes apparent that they can gain some benefit from participating, their interest level will pick up to that degree. In the beginning, the answer to the question might be horribly mundane, such as "to satisfy my probation officer," but even that is a start.

Displaying Compassion and Caring

Compassion may be the most underrated element in all of psychotherapy (Gilbert, 2020; Lewin, 1936). Many clients have told me that the only thing that kept them going was knowing that someone cared about what was going to happen to them. If a client knows someone genuinely cares, this functions as a ticket or sanction to improve and get better. When someone else cares, it allows a person to feel deserving of improvement. So many people believe that if they are not loved, they are unworthy of happiness or joy. Thus, a demonstration of care by others can serve as an indication of a person's significance. This could be one explanation for why people with borderline and narcissistic disorders are obsessed with obtaining the love and admiration of others.

Alternatively, some difficult clients are extremely adept at getting people to hate them. It serves the inverse purpose of finding significance, as the false self further solidifies the core belief of unworthiness or insignificance (Masterson, 1988; Oberst & Stewart, 2014). A savvy therapist will be on the lookout for this unconscious tactic since it shuts down the change process immediately.

Moderating Compassion With Antisocial Clients

With clients diagnosed as antisocial, compassion and caring require a slightly different approach. These people are often accustomed to manipulating people who care about them. Many believe that showing love or caring is a sign of weakness. Thus, they tend to peg a warm, caring therapist as someone who is easily manipulated and unworthy of respect. For a therapist to openly display warmth and caring can be a mistake with some of these clients. With others, it is a wise and important approach. Clinical judgment needs to be exercised, but a fairly reliable indicator is showing some genuine warmth and watching to see how the client responds.

Addressing Problems in Chunks

If too much contradictory, destabilizing, or painful material is opened too soon in sessions, clients will tend to withdraw and feel disempowered and threatened by the depth and extent of the problems. It gives the "rug-out-from-under" feeling, and one's inner world appears to be poised at the brink of a void. The problem no longer seems manageable and is a cause of reduced hope.

The solution is to address things in chunks. The aim is to handle only what the client can confront without being overwhelmed. Sometimes, a client is difficult precisely because their tolerance of pain is so low. Additionally, clients can be "trained" to be difficult when they are asked to confront or change too much at once. Asking a client to confront too much material actually reduces the confronting precursor. Finally, a client can feel degraded or demeaned by a therapist who outlines a vast map of psychopathology. The relationship can be severely damaged, and a premature termination may be imminent.

Expecting Client Hesitancy After Great Effort

Nearly every therapist will recognize this phenomenon: A great session finally occurs with a difficult client, and you are pleased that a breakthrough has occurred, yet in the next session, the client is withdrawn, uncooperative, or perhaps even belligerent. In some cases, especially with people diagnosed as borderline, there may be extreme hostility and defiance toward the therapist for "tricking" them into disclosing or working. This can be upsetting to a therapist who does not know that such behaviors should be expected.

Thus, after a session in which a client does much confronting and exerts great effort, do not plunge further ahead but assess where the client stands with what was done. Enjoy the plateau. The client may not be ready for more just yet. Pacing in such situations is of the utmost importance.

Setting Boundaries

Each therapist has their own limits on what they will accept from clients in terms of middle-of-the-night telephone calls and the like. It is important to define limits that neither impede nor threaten therapeutic change. Therapists should instruct clients that they will not always have time to talk, that taking "no" for an answer will sometimes be necessary, and that keeping agreements and agreeing to cancel rather than skipping sessions is desirable behavior (Linehan, 1993; Pope & Keith-Spiegel, 2008).

Although extremely rare, threats of physical harm from a client—whether thinly or thickly veiled—should not be tolerated under any circumstances. If the client is already on probation or parole, therapists should inform the police of any threats (the therapist should be in contact with the probation or parole officer anyway). It should be made firmly clear to the client that despite compassion, empathy, caring, and the like, no boundary violations of this variety will be tolerated, especially those of security.

It is equally important for female therapists to make it clear that sexual comments or innuendoes will not be allowed. Some difficult male clients are adept at using graphic sexual language in a session to give themselves a cheap thrill and make the therapist uncomfortable. Any actions like this require immediate attention, and termination may be necessary if it continues. Grounds for termination should be made clear ahead of time with some difficult clients. With borderline clients especially, the range of acceptable and unacceptable behaviors needs to be set clearly and concisely so the client can be reminded, with as little ambiguity as possible, when they have gone too far (Linehan, 1993; McCloskey et al., 2021). At the same time, it is important to soothe a client's feelings about observing limits while not compromising the limits themselves.

Matching the Client's Emotional State

Match the client's emotional state but with one vital additive—interest. For example, if a client is sad or dejected, the therapist does not have to be melancholy and helpless along with the client. However, if empathic contact is present, the therapist will resonate with the client's emotions enough to reflect back that sadness. This empathic resonance helps the client accept the therapist and their input.

The danger is that the therapist can sink into sympathy and lose the empathic contact. After all, some difficult clients have backgrounds so devastating and tragic that a therapist's empathy can easily degrade into sympathy. One route of escape from this is through the element of interest. We have found that a strong and powerful sense of genuine interest on the part of a therapist, focused on a client and their condition, can be contagious. A therapist's interest can stimulate a client to also be interested without any verbal persuasion. In many cases, especially when coupled with humor, a therapist's contagious interest can help a client to be lifted, or at least dislodged, from entrenched emotions or mindsets.

Opening a Therapeutic Window

Much of working with difficult clients is probing here and there, trying this approach and that, searching for an entry point, an inroad, or a window to open that makes the client accessible. This is a difficult phenomenon to describe, as it is highly subjective. In any case, in working with difficult clients, one becomes familiar with canned responses, disingenuous behaviors, and flat, feigned, or exaggerated affect. Many clients have become hardened to any external inspection and have a response for just about any potential intrusion. But every now and then, a probe or a confrontation will bring a hesitance, a response that reveals a soft spot in the armor. A client may look at you with a sensitivity or genuineness that was not present and indeed may not have ever been there before.

When a therapeutic window is opened, a therapist must enter through it soon, for the client may close it with a new defense formed in a matter of moments. For example, some angry, defiant teenagers are proud of their anger. Although anger management strategies can get them to talk about their anger, many such clients will change little, if anything, about it. The teenager might even brag about their anger as though it is a prized possession. An example of opening a therapeutic window with this type of client is to ask, "Have you ever been hurt?" The answer is typically yes. This makes the client pause. The window can then be further opened by asking, "Is that hurt related to your anger?" and then, "If you didn't have the hurt, what would happen to the anger?" This approach will often open a window that allows entrance beyond anger into the client's inner world. It changes the relationship and initiates the change process (see Hanna & Hunt, 1999), as opening the window and entering smoothly are part of bringing down defenses from within.

Looking for and Connecting With the "I Behind the Eye"

In instances where a therapist has experienced a sense of connection with a client, it is quite revealing to ask oneself what got connected. Was it two brains? Two worldviews? Two selves? Two multigenerational family systems? Phenomenologically, the therapist may perceive something in a client that forms the basis for a relational exchange. It may appear as though one has made subtle yet deep contact with a client through a maze of turbulent thought patterns, raging emotions, and destructive behaviors.

Thinking too much in respect of clinical labels sometimes risks obscuring the actual person behind a series of professional templates and categories. When it comes to establishing relationships, it is probably better to suspend the labels and templates so one can get a sense of connection with the person. Gendlin (1992) described the person as the "I who looks at you from behind the eyes" (p. 453). That "I"—the real person—can be accepted unconditionally, even though negative behaviors and thoughts are seen as needing change. Unfortunately, sometimes it takes a year or so for that "I behind the eye" to show up in a therapy session, and sometimes it may not happen at all, especially with antisocial clients. But for the therapist who knows how to locate that "I," it may happen sooner.

Developing Perspicacity as a Wisdom Characteristic

Clients with antisocial or conduct disorders often respect therapists who are suspicious. "Bullshit" is a common term in casual conversation in this culture (Frankfurt, 2005). Its use is ubiquitous among antisocial clients, and most learn at an early age how to dole it out. Their peers ridicule them when they are susceptible to it, saying, "You idiot, you believed that bullshit?!" They will test therapists by tossing out falsehoods to gauge their sharpness or gullibility, and a degree of tacit respect is given to a person who can recognize it. One of the characteristics of an effective therapist, especially with this population, is what Sternberg (1990) called "perspicacity," or the capacity to see beyond appearances, to not be easily deceived or fooled.

Antisocial clients will be forced to respect a therapist who has a well-functioning, reasonably accurate "bullshit detector." This may not be sufficient by itself to get such a client to be compliant or to work in therapy, but it is essential to establishing a relationship and cannot be underestimated. The same is true in working with clients who use alcohol and drugs, another population notorious for dishonesty and deception concerning drug use and behaviors. It is difficult to teach, but I have observed that even initially naive therapists can cultivate it through experience.

Matching the Therapist With the Client

An intangible therapeutic factor largely unaccounted for by research is that some clients do better with particular therapists. Much of it seems to be related to shared worldviews (Davis et al., 2018; B. S. Kim et al., 2005; Lyddon, 1989), as clients and therapists who share general outlooks or beliefs seem to do better together in therapy. Of course, the therapist's personal responses and issues also play an important role, as will be explored in the next chapter. Yet there are simply times when one counselor will develop natural rapport with a client whereas another therapist will not. In such cases, switching therapists for the client's sake will likely help bring about change sooner.

Using the Concentric Circle Technique

When working with difficult clients, it is common for therapists to wonder how close they really are to a particular client as well as how deeply such clients are allowing the therapist into their confidence. In other words, it is helpful to know the level or degree of a client's self-disclosure. Lazarus (1989b) described a remarkably helpful and effective technique for finding out how far a client has "let one in." This adaptation is done by drawing five concentric circles with the core or innermost circle labeled as Circle 1, and the outermost perimeter labeled as Circle 5. The therapist then asks the client to conceive of the core self as being in the center, at Circle 1, where their most private and personal information is held. Circle 5, at the outer perimeter, is where the most superficial and insignificant information is—that which everyone knows. The therapist can ask the client directly, "Into which circle have you allowed me?" After finding out, the therapist can ask, "What would it take for me to get into the next circle?" or "Whom have you allowed into circle 1?" or "For a person to get into circle 1 with you, what do they have to do or be like?" Clients can be surprisingly honest in using this technique. It takes little effort but can be highly productive.

Using Different Therapy Modalities

The three major therapeutic modalities are individual, group, and family therapies. Each has its own particular and peculiar set of benefits not readily available from the others. Many difficult clients need all three, if and when appropriate. Individual therapy may not be the mode a particular difficult client needs most. For example, an alcohol abuser may be more in need of a group with similar issues and problems than an individual setting. Chances are they may need both—and family therapy as well. Children and adolescents can greatly benefit from family therapy, not to mention parents. Each modality provides a unique opportunity to build relationships with a therapist, with peers, and with family members. These skills can then be used in everyday life to build a social support network, possibly leading to a logarithmic effect on change.

A CLOSING NOTE ON THE THERAPEUTIC RELATIONSHIP

A strong therapeutic relationship is the foundation for positive changes to take place in psychotherapy. The CHANGES model requires the establishment of relational factors such as empathy, warmth, and genuineness in order to work with difficult clients. However, there are a number of interpersonal skills and communication techniques that therapists can utilize to foster engagement and enhance rapport, as seen in this chapter. Therapists should keep the importance of the relationship foremost in their minds in sessions, as it can be all too easy to forego empathic contact with a difficult client once the therapist initiates a challenging technical intervention. In moving from contact to intervention, therapists must not lose sight of the subjective personhood of the client.

4

C: Confronting the Problem

The term "confronting," as a precursor of change, has nothing to do with hostility, opposition, antagonism, or provocation. It also has nothing to do with challenging a client on an inconsistency, hesitancy, or incongruency. Like most therapy techniques, challenging a client is an attempt to implement this precursor, but the challenge is not the precursor itself.

In the CHANGES model, *confronting* is defined as the active, intentional, sustained, and deliberate directing of attention or awareness toward anything that is painful, intimidating, or stultifying. Confronting thus involves a continuous examination or investigation of something in spite of fear, confusion, or any tendency toward avoidance or acting out. Almost anything at all can be confronted: mental images, memories, emotional pain, behaviors of all varieties, thoughts, thought patterns, beliefs, persons, places, objects, and relationships.

Confronting extends well beyond mere acknowledgment of a problem, as in the case of the awareness precursor (see Chapter 6, this volume). Confronting the problem is a dramatic and radical extension of awareness, involving sustained attention on a problem with the intent to penetrate and understand it. When one actively sustains and deepens awareness, one might be said to "enter into" the problem, looking and observing in a manner undeterred by confusion and undaunted by pain or intimidation. One essentially uses their attention and powers of perception to look into, through, and perhaps even beyond a problem or issue rather than skirting or dancing around it.

Case examples have been disguised to protect client confidentiality.

https://doi.org/10.1037/0000451-005
Therapeutic Change With Difficult Clients: Precursors and Techniques in the CHANGES Model, Second Edition, by B. D. Wilkinson and F. J. Hanna

METACOGNITIVELY "HOLDING THE PROBLEM STEADY"

Confronting is a metacognitive process that involves directing and regulating awareness in the context of choice. We have known for a long time that attention is selective (James, 1890/1981; van Ede & Nobre, 2021), and that it does not randomly come to rest on various phenomena. It attaches to and apprehends certain objects and things for the purposes of survival, pleasure, interest, and other influences. This metacognitive aspect of confronting is not found among the other precursors.

Confronting is thus uniquely metacognitive in that it requires sustaining one's attention in spite of the impulse to avoid, give in to confusion, or act out. It involves holding the problem steady in the mind such that it does not waver, wander, or disperse. When a person confronts an intimidating, painful, or confusing phenomenon, there is a natural tendency to turn away or avoid (Scalabrini et al., 2020). Whether the phenomenon resides within one's mind or in the world, the person must hold it steady in conscious awareness and closely contemplate it. If the person is unable to sustain active attention, the degree and depth of confronting will be severely diminished or impaired.

Research on sustained attention in cognitive science and neuroscience distinguishes attentional arousal from attentional allocation, with the latter informing opportunity cost models and information processing stances (Esterman & Rothlein, 2019). Opportunity cost models suggest it is only difficult to sustain attention insofar as there is a subjective cost: It feels better to focus on good things rather than bad things. As such, confronting a negatively valenced stimulus (e.g., personal problems) is generally less rewarding and more subjectively difficult than engaging a positively valenced stimulus (e.g., fantasizing). Significant motivation is required to offset this cost. In neuroscientific terms, enhanced communication between the default mode and attentional networks is observed during rewarding activities, indicating "that when motivated, participants' internal thoughts may in fact be more stimulus-related" (Esterman & Rothlein, 2019, p. 177).

HOW CONFRONTING THE PROBLEM LEADS TO THERAPEUTIC CHANGE

As far as therapeutic change is concerned, holding the problem or issue steady in one's mind begins a fundamental shift from being at the mercy of the problem to being in control of it. While the mechanism at work herein is subtle, it is pivotal to various effective interventions and practices. If a person can exert control over the problem enough to hold it steady in the mind, they are more likely to be able to influence or affect it in the world. Consequently, some remarkable change processes occur as a result of confronting a problem. These processes can be considered from a number of perspectives, as outlined in this section.

Confronting and Systematic Desensitization

One of the most powerful procedures in psychotherapy is systematic desensitization, a cornerstone of behavior therapies. It is based on the idea that if a client is exposed to repulsive or frightening stimuli gradually and with increasing intensity, anxiety responses to the stimuli will eventually reduce in intensity or extinguish altogether. This procedure is especially effective in treating phobias (Fear, 2018). Wolpe (1958) attributed the effectiveness of systematic desensitization to what he called *reciprocal inhibition*, or the idea that one cannot be relaxed and anxious at the same time. Thus, if a person can relax in the presence of noxious or repulsive objects or scenes, such as a snake or a great height, the phobic reaction will diminish and extinguish. Whether done via imagination (in vitro) or real life (in vivo), research has demonstrated that systematic desensitization produces therapeutic change for symptoms of anxiety, depression, posttraumatic stress, dissociation, and psychosis, to name a few (L. A. Brown et al., 2019).

While reciprocal inhibition and exposure involve confronting to good effect, these processes are not synonymous with confronting as an act. What helps a client conquer paralyzing fear is the continued, steady, and deliberate perception of the noxious object or situation. One must be willing to maintain engagement with the phobic stimulus, which has been referred to in other quarters as "resolute perception" (Hanna & Puhakka, 1991; Vickery et al., 2023). According to the CHANGES model, therapy procedures involving exposure and reciprocal inhibition require confronting as a necessary condition for change. The confronting precursor is an essential change mechanism underlying exposure, reciprocal inhibition, systematic desensitization, and flooding.

Cutting to the Essence of a Problem

Husserl's (1913/1982, 1936/1970) philosophy of phenomenology is particularly insightful when applied to the precursor of confronting and neatly explains several aspects of the process of confronting in therapy. In the early 1900s, Husserl developed a methodology of seeing and observing that was meant to free a person of their preconceptions. His method was probably the most fundamental, basic, rigorous discipline for observation and awareness ever developed in the Western world (Herrnstein & Boring, 1965; Zahavi, 2017).

The method was, at its core, quite simple. One puts aside as many of one's preconceptions and assumptions as possible and views the issue or object with sustained awareness. The item or issue can be any object (mental or physical) or relationship or any other worldly or mental experience or phenomenon. This is done until a person reaches or penetrates the core or essence of that phenomenon, with the consequence of thoroughly understanding it. Husserl said this is no mere intellectual process, nor is it a theory. It is an experiential process that Husserl called the *eidetic reduction*, meaning that one arrives at the essence of the phenomenon under investigation.

In the philosophy of Husserl and others, such as Sartre, Heidegger, and Merleau-Ponty, phenomenology is an extraordinarily complex subject with an intricate language for describing various aspects and modes of consciousness. It involves a method of seeing and observing that is as primary and unbiased as is possible for human beings to achieve (Herrnstein & Boring, 1965; Zahavi, 2017). In the phenomenological method, one observes any mental or physical phenomenon or experience openly and freely, with minimal expectations and preconceptions, with the intent of understanding it at its roots. In a therapy context, a remarkably similar method is often unwittingly applied to such problems as a cruel boss, childhood abuse, depression, anxiety, and rage. As one continues to look and purely observe, eventually the phenomenon becomes clearer until one begins to have insights into its nature and character. This is a fundamental aspect of the method of scientific observation, and as Nobel Prize laureate Sir Peter Medawar (1984) noted, it is a process that seems to be as intuitive as it is intellectual; certainly, Husserl would have agreed.

Husserl's method is near to the essence of what occurs in many forms of psychotherapy. Therapists and counselors routinely and regularly ask clients to delve into an area of pain or uncertainty with the intent of coming to a more complete understanding of it. This is the case even if the understanding arrived at is that the phenomenon itself is too vague to be understood, as in attempts to figure out another person's intentions. Continued, intense, sustained inquiry and observation has profound effects in terms of change. Carl Rogers and other experiential therapists have referred to the continuous process of examining feelings and views of a problem as "peeling the onion." This is a fitting metaphor that describes the method of cutting to the essence of an issue, and it describes the process made possible by confronting a problem.

Cutting to the essence of a problem or issue does not have to occur verbally. It can be accomplished without putting words to experience. Many clients do this routinely, even if they may be thought of as having "cognitive deficits." In such cases, behavior change comes with the confronting but may not be verbalized in the form of an insight. As Arkowitz (1989) once noted, insight often comes after behavior change. Once again, it is the confronting that seems to be the catalyst.

Ancient Approaches to Confronting

Confronting is at the basis of Indian psychology, being well expressed in the ancient text of yoga psychology, the *Yoga Sutras of Patanjali* (Aranya, 1983; Feuerstein, 1989; Nguyen, 2016). Patanjali's yoga is widely recognized as the chief application of the principles of various psychologies of India. *Samyama* is the name of a technique used in yoga that makes direct use of intense and highly concentrated confronting of various phenomena to arrive at deep insight and transcendence of self. Similarly, confronting has been a cornerstone of Buddhism for 2,500 years (Rahula, 1978). Buddhist monks have been practicing a wide variety of cognitive–behavioral techniques for the same length of time (de Silva, 1985; Tirch et al., 2015). Confronting is a crucial aspect of these practices (Pandita, 1991), although in Buddhism it is more commonly referred to as *mindfulness*. Mindfulness is fundamental to the Buddhist concept of therapeutic change and is a fascinating subject in and of itself. In many ways, this precursor can be understood as the Western equivalent.

Simplicity and Complexity

A fascinating phenomenological aspect of confronting is how the more one confronts a problem, the clearer and less mysterious it becomes. At first glance, the issue becomes simpler, but this is misleading. What is more likely is that the problem, as it becomes more thoroughly explored, examined, and investigated, becomes less and less obscure and vague, so that its intricacies stand out in greater detail.

For example, in the case of a relationship that has gone sour, a client explores and examines all aspects of it in therapy. The relationship is quite involved and has many dynamics, outside influences and pressures, and personal and interpersonal perspectives. In therapy, the person examines the relationship in terms of empathy, love, roles, social interactions, mutual friends, parental influences, sexual relations, met and unmet needs, spiritual beliefs, and children, if any. In the process, the relationship becomes more and more crystallized until the person achieves an overarching view of it accompanied by a more thorough understanding. The relationship is still quite complex, but now the complexities are mapped and stand out with more clarity.

Reality Testing

Another aspect of confronting as part of the change process involves differentiating reality from illusion or, in some cases, delusion. Confronting is intrinsic to *reality testing*, a continuous process of verifying the accuracy of our perceptions (Goldstein, 2013). Reality testing is related to but moves beyond simple clarification of the problem. It is the work done to test whether perceptions and beliefs hold up to the proving ground of the experiential world.

Setting aside metaphysical speculations as to the nature of reality, it is often important for clients to compare what is actually present in the world to what is in their minds. For instance, if a client is suspicious that their spouse is cheating on them, it is useful to determine whether this is true, that is, whether the spouse is cheating in fact. Otherwise, the client may construct a false reality forever, leading to further dysfunction, despite the fact their belief is not accurate.

From a pragmatic perspective, reality could be said to be what remains after all our theories, beliefs, constructions, and perceptual filters about it are stripped away. This was precisely Husserl's goal in phenomenology and is often what happens in psychotherapy. It is far more functional to cope with a situation that is accurately perceived.

A Smaller Problem Space

Another curious, highly subjective phenomenon associated with confronting the problem is consistent with Lewin's (1935, 1936) conception of what he called the life space. The *life space* is that part of the person's perceived world that is taken up by events, circumstances, or issues that are absorbing attention. According to Lewin, some of these issues can appear large in a spatial sense, whereas others can appear small. In everyday language, we speak of "big" problems and "little" problems.

The act of thoroughly confronting usually makes a problem or issue seem less significant. Where a problem once loomed large and foreboding, it can seem smaller and more manageable after thoroughly confronting it. The problem becomes less oppressive as it is progressively confronted. Some clients convey that they feel "bigger" after directly confronting a problem. A problem can seem smaller, or the "I" of a person can seem bigger. Either way, the implications for therapeutic change are obvious. For example, resolving a conflict with one's teenage son or daughter might initially seem quite daunting, but after confronting the problem, it feels more like a difficult but manageable task. When a person becomes competent at confronting, depression or anxiety can be viewed in this same fashion. This is part of developing the ability to confront as a skill of change.

Removing the Power of the Problem

When a problem or issue is ignored, resisted, or denied, it takes on what can appear to be a life of its own, influencing the person in a variety of ways. For instance, a 45-year-old client named Dominica believed that she had a lump in her breast, which she feared could be malignant. She did all she could to divert her attention from it, avoid talking about it, and pretend it was not an issue. Meanwhile, when she did think about it, she would cry, thinking that her life might be over. When she entered therapy for another reason, she had been ignoring this problem for well over a year and a half, fully aware that things could get worse if not treated early. Little by little over the course of therapy, she began to realize that ignoring the problem was actually lending it power over her. She saw how it intimidated her and exerted influence over her in various ways. It affected how she treated her children, how she viewed her job, and how she saw herself. When she finally looked at the problem directly and thoroughly, she was, by degrees, no longer intimidated and immersed in it and acquired some overview of her life again. The mere act of deciding to see a physician gave her a sense of control, although it did not remove her fear. The lump turned out to be benign.

This aspect of confronting is related to the popular notion of empowerment. When the problem occupies a smaller region in the life space and begins to lose its ominous or intimidating quality, one becomes empowered to move out from under the burden of worldly or psychological oppression. It also enables one to extend one's range of influence, scope, or outlook beyond a painfully limiting set of circumstances to more comfortably deal with the challenges at hand. There is little question that such conditions contribute to the well-being of a person.

The more intensively something is confronted, the more it diminishes in size and power. When a problem or issue is continually contemplated, a person seems better able to endure it without becoming overwhelmed. This is at the heart of the change principle of exposure. The key is to be able to look and observe even while caught up in painful emotions or profoundly confusing circumstances. Confronting is a skill that can be taught, trained, developed, and practiced.

The "Vaccine Effect" of Confronting

Confronting the problem, especially when combined with awareness and grit, appears to have a mental effect similar to the effect that a vaccine has on the body. With a vaccine, patients are exposed to a harmless form of a virus under conditions the body can control. In psychotherapy, clients are placed in conscious contact with an area of emotional pain or other difficulty under controlled conditions so that harm will not result. Confronting the phenomenon can have a healing or strengthening effect. It seems to reduce anxiety, allow for insight and personal growth, give more clarity to thought, release painful emotions, and improve interpersonal functioning. Confronting adverse conditions in a controlled setting also allows a person to better adapt to and, by degrees, master their environment.

SIGNIFICANT MOMENTS OF CHANGE INVOLVING CONFRONTING

John had been in therapy for 10 months, attempting to resolve the loss of a relationship with a partner he had loved very deeply. During therapy, feelings of depression would only slightly lift before returning. Cognitive approaches were used in an attempt to change John's thinking, but change was not forthcoming. Meanwhile, he was obsessed with his former partner, thinking about her continuously. He reported that he had been depressed for as long as he could remember, saying that his life had been "dominated by sadness and anxiety" and that he had been to many therapists. Therapy had "only helped slightly," he said. Then came a session in which the therapist was questioning him on what aspects of the loss of his relationship were the most difficult.

Suddenly, for no apparent reason, John was aware of a vivid memory from his childhood, around age 6. He "was lying in bed" and "heard an airplane flying overhead." He was "terrified" that the plane was going to drop a nuclear bomb. As he relayed this to his therapist, he realized that his lifelong sadness and depression somehow stemmed from this moment. He remembered how frightened he was of a nuclear holocaust as a child during the Cold War and how this had negatively affected his entire outlook on the world. He almost instantly realized that the strong anxiety he felt about the woman was of the exact quality as the anxiety he felt about nuclear war as a child. He saw that his fear of nuclear war was remarkably the same as his fear of losing his romantic partner.

On seeing how silly this was, he said that his "sadness and depression disappeared and never came back." Following this, he no longer obsessed about his lost love and stopped compulsively talking about it. He said, "She wasn't the problem; this incident was." He claimed it was this session that, as he put it, "freed up my emotions" so that he could "live a more real life." John's story is a classic example of peeling the onion until the actual problem emerges.

In another example, 13-year-old Bobby developed a phobia of dogs. His therapist asked him about a particular dog he found to be scary. He shared that the dog, named Sam, belonged to a neighbor and that this apparently ill-tempered animal was constantly trying to bite him. The therapist asked Bobby to hold a picture of Sam in his mind and "not let him get away or bite you; just look at him." Bobby was initially reluctant to try, but with some therapeutic encouragement, he eventually experimented with the visualization and did quite well. As Bobby did so, his fear began to diminish. He was amazed that the longer he imagined Sam, the less fearful he felt. Bobby's struggle was a genuinely heroic demonstration of courageous confronting in therapy.

CLINICAL MARKERS: HOW TO RECOGNIZE CONFRONTING THE PROBLEM

Clinical indicators of the presence of confronting are fairly straightforward and, in many cases, immediately evident. When a client is actively confronting, there is a sense of perseverance in facing up to and continuing to observe. The person displays an animated excitement or composure in meeting the challenges of life or therapy. There is a steadfastness, alternatively described as a "doggedness" or "digging in one's heels" in seeing and studying an issue to its resolution. Such a person will show a spark of determination in holding attention on a problem. The person is honest and direct, and defensiveness is overridden by the desire to overcome obstacles and barriers. They accept feedback and often ask questions of the therapist that take the therapy to a deeper level.

Metacognitively, what to look for in clients confronting the problem is the ability to hold the problem steady in their minds to be inspected and explored. Such clients seem to have attention sufficiently free so they can direct it to virtually any area of their lives or life histories. They engage in little or no deflection and do not turn away from or avoid an issue, emotion, or problem. The person is able to selectively view virtually any phenomenon.

Cognitively, there is a confidence in a confronting client that says, "I can look at anything at all in my mind" without fear of being harmed, consumed, destroyed, or disintegrated. The person believes that confronting is an important act and part of the healing process and that studying a problem or issue will enable them to understand and resolve it. This person is convinced that the way out of a problem or situation is by working one's way to the core or essence of problems, issues, emotions, beliefs, and so on. But the client does not necessarily have to articulate insights so much as be inclined toward behavioral change.

The confronting person or client is a confident person who is stable in the ability to maintain perspective. Confronting ability is a sign of high ego strength, whereby a person does not easily wallow or stew in their emotions but is consistently able to keep a sense of awareness and presence even in the face or threat of pain, discomfort, or deeply disturbing circumstances. This person is not intimidated by strong feelings or emotions and often displays a sense of calm, composure, or even poise in the face of adversity or hardship. From the therapist's perspective, it is common to find oneself deeply admiring such a client, appreciating and perhaps even being moved or touched by the client's courage, fortitude, and perseverance.

HOW TO DETECT THE ABSENCE OF CONFRONTING THE PROBLEM

Clients who do not confront will protect themselves from looking at or contemplating an issue or problem in the belief that harm will result. They will be adept at deflecting challenges from the therapist by changing the subject or implying or directly telling the therapist that looking at or knowing about a problem is neither necessary nor helpful. Some of these clients may be adept at "talking around the problem" or "skirting the issue."

Such clients may wallow in their problems, becoming overwhelmed and ensnared. They hesitate or waver in the act of viewing and find it difficult to maintain focus. They tend to offer or even prefer superficial or shallow solutions and minimize the importance of examining an issue or aspect of their lives. Such clients often need a great deal of coaxing or encouragement and may be easily intimidated by the magnitude of even small problems. Clients who seldom achieve closure on issues may be lacking in this precursor. Cognitively, a client who does not confront has not learned that confronting a problem leads to its resolution.

One of the most difficult of all difficult client styles is the intellectualizing client. This client is often highly intelligent and possibly highly educated and can simulate confronting and make it look like they are deeply examining issues. These clients often mistakenly believe that confronting is one of their greatest strengths and will offer a host of reasons for why the problem developed. Such clients will use sophisticated vocabulary and even therapy jargon. They may be successful at "snowing" the therapist into believing they are in control when in fact they are using their intellects to protect themselves from exposure to—or in gestalt language, contact with—the painful emotions, difficulties, or problems present.

In such cases, the client is not confronting at all. Instead, they are compulsively analyzing or thinking about an issue or problem. Indirectly, intellectualizing can anesthetize emotional pain to a small extent, providing the secondary gain or reinforcer that perpetuates the dysfunctional behavior.

One of Perls's (1973; Perls et al., 1951) best known and important contributions to experiential therapy is his therapeutic notion that we ought to "lose our minds and come to our senses." The intellectualizing client has a thinking mind so hyperactive that it has stepped beyond its boundaries and taken over the functions of being and observing. Such clients' high intelligence and compulsive thinking sabotage their ability to penetrate the essence of a problem and achieve a viable resolution. For such clients to achieve change, they may need to learn that there is a difference between thinking and looking and that analyzing and contemplating are not the same endeavor (see Heidegger, 1927/1962). Fortunately, many intellectualizers will readily respond to a range of techniques that enhance this precursor.

TECHNIQUES: ENHANCING CONFRONTING

Virtually all of the techniques used in psychotherapy, with the exception of medications and pure operant conditioning, use the confronting precursor. A vast array of procedures, such as identifying and disputing dysfunctional beliefs, role-playing, the empty chair, social skills and assertiveness training, working through transference, interpretation, paradox, reframing, the use of metaphor, and the talking cure itself, all require the problem to be confronted directly or indirectly. In essence, any approach that calls for sustained, continuous, and deliberate attention to a situation, whether to discuss or cope with it, uses confronting. The techniques provided in this chapter specifically support confrontation with difficult clients.

There are some important points to bear in mind regarding confronting. Forcing a person to confront too much, too soon, is traumatic by definition. Across myriad forms of exposure therapy, it is well understood that bringing a client into contact with painful phenomena must be done with care and attention to the person's level of tolerance (see Richard & Lauterbach, 2011). Overwhelming a client with mental, emotional, or environmental material that is too much to confront will not only bring about early termination; it will cause harm. Thus, exposure to sensitive memories, feelings, or beliefs should be done gradually so the client will be successful and not view therapy as a source of failure and pain. The techniques and strategies in this chapter are designed to encourage, implement, and strengthen the confronting precursor.

The Strengths Metaphor

Confronting is not only a psychological act; it is also an ability. Like muscles with weight lifting, it becomes stronger with practice and discipline. Yet some clients seem to find confronting their problems not only difficult but abhorrent and repulsive. For many, the act itself is painful. This is especially the case with clients with personality disorders. Although such clients are often almost vilified as being uncooperative, manipulative, malicious, and intrinsically flawed, the truth may be that, at the core, they lack the skill and ability to confront such things as unpleasant thoughts, feelings, behaviors, and events. These clients are often in need of education concerning this and other precursors. Many have never learned that confronting, rather than being harmful, is the beginning of healing and a source of exhilaration.

Thus, the metaphor of strength can be used to show clients that the more one confronts issues, the better at it one gets. Although some people can confront a vast range of psychopathology in one grand sweep of consciousness, others can take in only small bits and portions at a time (and even then, with great difficulty). The difference, as in lifting weights or exercising, lies in understanding that the more confronting one does, the more capable one is of doing more of it and the more stamina and strength one develops.

Confronting the Hesitancy to Confront

Often, clients are aware of a problem area, such as alcoholism, depression, anxiety, or some compulsive behavior, but are unwilling to confront anything about it. In such cases, the therapist can offer an approach framed as an experiment. The therapist can openly inform the client that confronting the problem is not necessary. Next, they can ask the client to think about the problem and report all the thoughts and feelings that arise. The therapist again asks the client not to talk about the problem itself but rather whatever springs to mind when thinking about it. The client will often report disliking the subject and simply not wanting to think about it or any other problems. These statements can then be fully acknowledged and reflected.

This technique aims to identify the automatic thoughts, beliefs, and feelings that stand in the way of confronting. The therapist can divert the client and ask them to talk about not wanting to talk about or explore anything, fully understanding and reflecting back everything the client says. Eventually, the client may be more likely to confront the problem or issues and may do so spontaneously.

Confronting the Idea of Change Itself

The technique described previously can also be applied to difficult clients who find the idea of change itself uncomfortable or objectionable. The client can be asked, "Consider the idea of making changes [or one particular change] in your life." Next, the therapist can ask the client about what kinds of thoughts and feelings rise to awareness, including thought streams and unspoken verbal responses such as, "Go to hell," "That's a load of bull," or similar statements. It will often turn out that the idea of therapeutic change insults the person's character or dignity and that the mere suggestion of its need is a kind of criticism. Furthermore, the contemplation of change by a difficult client can bring up a variety of beliefs about change, such as "To change is to fail," "To change is an admission of being wrong," or "To change is to give up." When disputed, some fascinating clinical material that accompanies these beliefs is often encountered.

In Vivo Confronting

Talking in therapy does not always lead to the talking cure. Talk can be used to circumvent and even avoid problems and issues, resulting in the illusion of progress but never any real resolution. In many cases, the emotions, passions, and desires involved with dysfunctional behaviors must be aroused so they can be addressed in the present moment. Talk therapy does not always accomplish this. In vivo confronting is a variation of the well-tested and venerable techniques derived from the behavioral principles of exposure and reciprocal inhibition, which also recognize the limits of the talking cure. This technique is equally well explained by the concept of the resolute perception of particular issues and problems. The problem can be recreated in the therapy room in real time with therapeutic benefit, invoking the vaccine effect.

In vivo confronting is an adaptation of the classical behavioral techniques of flooding and systematic desensitization, as well as resolute perception. Of course, the classical application of these techniques is with phobias, and it is so well documented there is no need for elaboration here. However, these principles, all based on the confronting precursor, can be applied to compulsive behaviors, such as stealing, substance use, and compulsive sexual behaviors. In applying this approach, it is wise to inform the client of the procedure and frame it as an experiment that can be backed out of at any time.

The Story of Marvin

Marvin was a difficult, conduct-disordered teenager convicted of theft and placed in an outpatient program for adolescent criminal offenders. Because stealing was the problem that landed Marvin in the outpatient treatment program, the consulting therapist decided to use a technique designed to get Marvin to confront his compulsive stealing behaviors, which he had willingly discussed in group.

The therapist asked Marvin if he was willing to try an experiment regarding his urge to take things that belonged to others. Marvin agreed. At that point, the therapist pulled out a $20 bill and placed it in front of the client, turned his back on him (he was sitting between the client and the door, of course), and asked him to report his thoughts and feelings as he looked at the money. Marvin immediately said, "I want to take it, man." The therapist asked him, "What are you feeling in your body right now, Marvin?" He replied that he was feeling excited and pointed to his stomach. When asked, he added that this was associated with the challenge of getting away with something and doing something at which he felt competent (stealing).

The therapist asked him, "Marvin, is this the same feeling you get when you are in the street or in a store and you think about stealing something?" He replied somewhat contemplatively, "Yeah, it's the same feeling." The therapist asked, "When you have this feeling, are you feeling pretty good about yourself?" Marvin looked at the therapist and said, "Yeah, it's like I feel really cool, and I can take anything." A discussion then ensued concerning the relationship between stealing and his sense of self-esteem and self-worth and how his parents and uncle had tied these together. Marvin recognized this and was by now capable of elaborating on it. The therapist asked if this was how he gained respect from his peers. Marvin replied affirmatively to this, saying that among his friends, he was acknowledged as "the best."

The term "stealing" was now being used directly. Importantly, it soon became clear that, to Marvin, the most appealing aspect of stealing was the feeling, the excitement, and the challenge of taking something not his own. He reported that the biggest obstacle to quitting stealing was not the value of the item stolen but the appeal of the thrill and excitement that came with the mere prospect of stealing it. The therapist asked him, "Do you feel more alive when you are thinking about stealing something?" He quietly said, "Yeah." As a result of this experiment, the therapist asked if it would be okay to get a little more serious about producing that feeling, to which Marvin agreed.

The therapist placed a credit card in front of him (turned upside down and face down) and once again turned to the door. With his back to the client, the therapist asked if the feeling of excitement was there again. Marvin replied that it was indeed, and even stronger this time. He was asked to describe the feeling in detail, outlining where in his stomach he felt the feeling. Marvin did so. The therapist asked Marvin, "When you have the feeling, do you feel like you *have* to take something?" He thought a minute and said, "Yeah, it's like it's there, and I gotta do it." The therapist asked if he could feel good by not giving in and being ruled by the feeling. Marvin had never considered this.

The therapist took Marvin to a local store, walked around with him, and asked if he felt the excitement in his stomach. Marvin reported that, like always, he did. There was a pocketknife that appealed to him. Marvin was asked to describe the feeling and what it was like to watch the feeling and not be taken in by it or drawn into it. "Weird" was the word he first used to describe this new feeling. "Good weird or bad weird?" asked the therapist. "Kinda good," he said, "Like I don't have to give in to it just because I feel it." The therapist asked if part of this new feeling was a sense of strength. Marvin responded positively, and the therapist pointed out that perhaps the real strength was not in the guts to steal but in the strength to resist. He had already proved that he had the former but not the latter.

This was the beginning of Marvin's rehabilitation. In his case, it was not only his thinking but his need for sensation that needed attention. The success of this technique in Marvin's case was through establishing the confronting precursor by getting him accustomed to looking at and examining issues. After this, he responded more positively to both group and individual therapy. When dealing with people who engage in criminal behaviors, therapy is probably incomplete without addressing the sensations and feelings that accompany the acts (Samenow, 1998).

The Story of Rusty

Rusty, age 17, was addicted to crystal meth. He had been arrested several times for possession and had just spent 6 months in a youth detention center for dealing meth and other drugs. He had been expelled from three schools for behavior problems. His father was an angry man who weighed nearly 500 pounds. He was habitually abusive, both physically and verbally, toward Rusty; was on disability; and rarely left the house in their poor neighborhood.

In treatment, Rusty often said that he loved meth, could do nothing about quitting, and did not want to quit. The director of the program described Rusty to the consulting therapist as a noncompliant borderline client for whom the treatment program had been so far ineffective. He had already been in treatment in this outpatient substance use program for about 4 months, and his urine screens had consistently come back dirty. Rusty sometimes acquired money by breaking into cars or mugging people for their purses and wallets.

In the first session, the consulting therapist asked Rusty if he was willing to try an experiment. Rusty agreed, probably because of the novelty of having a new therapist and because of the chance to get out of group. After establishing an initial rapport, the therapist offered a small piece of chalk and asked Rusty to pretend that it was "crystal." Rusty was surprised at this but also excited to get the chance to talk about meth. "When you look at this crystal, Rusty, what are you thinking?" Rusty unhesitatingly answered, "I want some right now, man. You have no idea how bad." The therapist acknowledged this and said, "So you're craving the high at this very moment?" Once again, the immediate answer was, "Yeah, bad."

"Is there a place in your body where you feel the craving, Rusty?" Rusty considered this for a moment, looked up, and said, "Yeah, here," and pointed to the area in the upper part of his stomach. "When you look at that part of your body, what feeling is in there?" Rusty again considered the question for a moment and said, "I don't know ... emptiness." The therapist was encouraged by this answer, which indicated an unexpected degree of awareness. He asked, "Is that emptiness a good feeling?" to which Rusty replied, "No. I don't like it." The therapist went on, "And when you're high, what does the crystal do to that emptiness feeling?" Rusty said, "It's like it fills it up." The therapist asked, "So is it fair to say that you use crystal to fill up that feeling of emptiness you don't like?" Rusty thought about it momentarily, looked up thoughtfully, and said, "Yeah, I guess I do."

Continuing, the therapist asked, "How much does smoking meth actually fill up the emptiness feeling? All of it, or only part?" "Only part of it," was Rusty's honest reply. "So, you always have some emptiness inside you, and it bothers you. Is that fair to say?" Rusty agreed. "You know, Rusty," the therapist said,

> Counseling can show you how to fill up the emptiness in ways that won't land you in jail or in programs like this one, and there are no cravings like the ones you have right now. All you have to do is keep looking at and talking about your problems just like you did here. What do you think?

Rusty, deep in thought, said, "Yeah, why not."

This was the beginning of Rusty's road to recovery from his use of crystal meth. His urine screens were clean from that session onward. His continued abstinence was documented for 2 years after his graduation from the program, which occurred 90 days after that session. Of course, there is far more to this story.

This conversation was just the beginning, and many other precursors needed to be established in Rusty before his recovery strengthened and stabilized. However, the establishment of the confronting precursor in this session was what placed him on the road to recovery and was the key to all of his successful treatment. This case example shows how it is possible for confronting to be used to an advantage even with clients known for being difficult.

The Onion Peeler

The onion peeler, a powerful technique that uses direct confronting, is derived from Husserl's (1913/1982) phenomenological method and from various applications of existential and gestalt therapies (Perls et al., 1951). It can be quite intense and should be used only with a client who has a rating of at least 2 for awareness on the assessment form, and the grit precursor also contributes to its success. Additionally, it is powerful in enhancing the change process among clients who are not difficult at all. It is called the onion peeler because it strips away automatic thoughts, self-talk, beliefs, defenses, and feelings and helps a client arrive at the heart of the problem or issue. It evokes deep emotions and can produce insights that are central to the nature of the problem.

The onion peeler is a cyclical question that is repeated over and over again after getting an answer to each cycle. The central question is, "How does [fill in the blank with whatever the problem is] appear to you now?" Variations such as "How does it look to you at this point?" or "What seems most obvious about it now?" can also be used occasionally. Another variation is, "How are you experiencing [the problem] now?" These questions require the client to observe in order to answer the question. Remarkably, the problem will appear to change after several cycles, and possibly it will change again and again until the core of the onion is reached.

Before beginning the technique, it is important to identify or name the problem or issue with the client's agreement that this indeed characterizes the problem itself. This is where awareness is established. Once the problem has been identified, the characterization can be inserted into the body of the cyclical question. The client should be fully informed of the nature of the procedure, its purpose, and what to expect in terms of how it is done.

The problem itself can be almost anything. For example, the problem could be a person, such as an ex-spouse, current spouse, son, daughter, employer, probation officer, teacher, betraying friend, or abusive parent or relative. It can be a condition or feeling, such as depression, anxiety, loneliness, or helplessness, or it can be a situation, such as a lack of romance, homelessness, or an unpleasant marital or work situation. One can fill in the blank with alcohol, cocaine, or other drug of choice. "How does your body appear to you now?" can be used with some people with eating disorders. For clients who are on probation or parole, "How does [the particular crime] appear to you now?" can be used. Of course, therapeutic judgment should be exercised about the proper timing of these techniques.

As the onion peels, the question can be further refined as the problem appears to change or become reshaped, allowing for more focus. For instance, after asking, "How does your depression appear to you now?" 15 times or so, the problem reshapes itself, and the question becomes, "How does being alone appear to you now?" or "How does being a failure appear to you now?" With the client's approval, the refined version is then used. The technique itself can be used at many levels. At much higher developmental levels, well beyond the focus of this book, the sentence can be filled in with such items as the ego, life, meaning, god, or the world in general.

Every now and then, in between the cycling questions, the therapist is advised to add a processing question, such as "What's happening?" or "What are you experiencing?" or "How are you feeling about the problem?" This helps maintain the communication link between therapist and client and is often where the client will report insights, progress, or how the problem seems different.

Initial responses to the onion peeler question are often defensive, with answers ranging from "Who cares?" to "It sucks." The therapist acknowledges these responses, saying "okay" or "alright" and perhaps reflecting on the meaning or feeling involved. However, to be effective, the question should be returned to as soon as possible to continue peeling the onion. Later answers tend to be characterized by more intense feelings, such as deep concern or helplessness.

In the case of a highly defensive male alcohol abuser, for example, his initial answers to "How does drinking appear to you now?" were along the lines of not having a problem or not needing to drink. After about 20 repetitions, answers revolved around his loving to drink, and after about 60 or 70 questions, he was worried that he could not quit. Obviously, the technique can last several sessions, and if done smoothly, it should not be a grinding or abrasive process. It is always important to check with the client to make sure their interest in continuing the procedure is still present or if the client feels like progress is being made. If the client is not interested, the therapist should not continue.

How Could the Problem Be Worse?

We have focused on clients who deny their problem behaviors or attitudes and who avoid awareness or confronting by minimizing their dysfunctionality and presenting it as normal or even advantageous. However, there are clients at the other end of the scale who will not confront an issue or problem simply because it is deemed so terrible that it is unconfrontable and beyond their capacity to view, explore, or otherwise tolerate. With such clients, a technique can be used that develops the confronting precursor indirectly by imagining something even worse. By talking about how things could have been worse, the client is more or less forced to contextualize the event or situation. In seeing how things could have turned out worse than they did, the client may begin to build tolerance for the current condition as it is.

This is delicate work and appropriate only in particular situations where the client will not think their problem is being invalidated or rendered insignificant.

Thus, while fully and rightfully respecting the severity of the problem, the therapist can ask the client, "As terrible as this is, it seems you were fortunate that [a worse circumstance] did not occur." This can sometimes be appropriate for clients who have suffered a disability or family tragedy.

However, this technique is contraindicated for victims of physical abuse and sexual assault. For instance, rape has been minimized throughout history and across cultures (Daly, 1978; Feinstein, 2018). To lessen or dismiss the profound tragedy of rape by finding something worse would be destructive and harmful, only serving to perpetuate the ill-founded, ignorant, patriarchal tradition that contributes to ongoing violence against women. Thus, it is better to avoid this technique than to inadvertently do harm with it.

Mirroring

The mirroring technique involves presenting a physical mirror and asking the client to dialogue with it or to comment on what they perceive in it (Mahoney, 1991). Although its potential might not be immediately accessible with difficult clients, it can be used as a means to get a client to begin to confront the self and how they regard it. It has the advantage of showing a client how little actual confronting of their life has taken place. If done correctly, it can get a client to begin to confront dysfunctional traits or aspects of the self that are a source of shame, self-blame, guilt, or regret. It helps a person take stock of their life and goals and can be used to assess degrees of self-loathing.

Representational or Concretized Confronting

Some difficult clients find the prospect of confronting thoughts or problems to be too abstract. There are times when it is easier to treat problems and emotions as concretely as possible using objects or props to represent thoughts and images to enhance a client's understanding of confronting. This is often helpful with children but can be highly effective with adults as well. It is especially helpful for clients who are not adept at holding a problem steady in their minds so that it can be more readily and fully studied or examined. For example, if a particular client tends to deflect or avoid issues and problems, the therapist can point this out to them and represent the thought, idea, or image as a physical object to be used as a prop.

For instance, Elizabeth, age 45, was wealthy and unhappy and blamed a wide range of people for her "unhappiness," including her children and her parents. In her descriptions of the people in her life, she perceived them as wanting her to be unhappy and treating her in a way that was disrespectful and insensitive to her needs. Her depression was "their fault," and attempts to isolate beliefs related to her unhappiness had failed. Unfortunately, she repeatedly avoided her therapist's attempts to get her to confront her own actions as part of her situation. Other approaches to blaming had also been ineffective. Rather than pursuing this line further, the therapist pointed out her avoidance and attempted to concretize the problem to help her hold the issue steady.

THERAPIST: I've noticed that as soon as I bring up the possibility of how your attitude might play some small part in your unhappiness, you immediately change the subject or blame someone, like your husband or your children. Please tell me if I am wrong.

CLIENT: You don't understand—they are all mean and cruel to me. They hurt me over and over again.

THERAPIST: Yes, you feel that all they cared about was making you hurt and that you did nothing to bring this on. It just wasn't your fault at all.

CLIENT: Yes, that's it. I don't deserve this!

THERAPIST: I don't blame you for feeling that way. Isn't it interesting, though, that every time I mention the possibility that your own attitude might have some part in your unhappiness, you immediately change the subject?

CLIENT: But they were the ones ...

THERAPIST: Please pardon me for interrupting you, but do you see, you are doing it again. You don't have to agree with me. Just tell me if I am wrong.

CLIENT: About what?

THERAPIST: Do you change the subject every time I ask if your own attitude might play a part in your unhappiness?

CLIENT: Because it's not true.

THERAPIST: Fair enough. Can I have your permission to go on with this just a bit more?

CLIENT: I suppose.

THERAPIST: Please tell me if I am wrong. Is it fair to say you don't like the idea that at least part of your unhappiness has to do with how you might be thinking and acting at times?

CLIENT: I don't like that idea at all.

THERAPIST: I understand. Would you mind trying an experiment?

CLIENT: I guess so.

THERAPIST: (offering the client a pen) Let's pretend for just a moment, Elizabeth, that this pen is the thought that your attitude might have some role in your own unhappiness, okay?

CLIENT: Okay ...

THERAPIST: I am going to hand the pen to you as though I were saying that thought to you again. As I do, please show me, with the pen, what you do with that thought.

CLIENT: (throws that pen to the side of the room and looks at the therapist)

THERAPIST: Is it like you don't want anything to do with that thought at all?

CLIENT: Nothing.

THERAPIST: (retrieving the pen and handing it back to her) Would you mind just holding it in your hand and telling me what feelings and thoughts you have?

CLIENT: (looking at the pen and beginning to cry) Sure, I have made mistakes, but they are mean to me, and then I do things that I know aren't right, but I just can't help it.

In this fashion, the technique broke through some of Elizabeth's strong deflections to the point of opening a productive dialogue about mistakes she had made and her regrets and guilt about them. She slowly progressed toward examining and changing her beliefs and behaviors.

Concentration-Based Techniques

As with representational confronting, concentration-based techniques help clients hold the problem steady in their mind. One technique adapted from gestalt therapy (Perls et al., 1951) is a form of in vitro exposure that involves asking the client to hold a thought, image, feeling, emotion, problem, or situation in the mind. As with concretized confronting, the therapist asks the client to continue to look at the object in the mind, holding it steady and reporting on what occurs as they do so. The image of a boss, parent, or coworker are fairly standard examples. Third-wave mindfulness-based cognitive therapies regularly use similar techniques to reduce experiential avoidance and encourage varying iterations of acceptance (see Hayes & Linehan, 2018).

Such moments call for effective psychoeducation as well as conveying empathy and understanding to clients because holding a problem steady in the mind can seem contrary to common sense. As a defense, some clients believe that unconsciously jettisoning a thought or image from the mind is exactly what will relieve them of anxiety. They believe that if it does not stay in the mind, it won't hurt. In a similar vein, Gendlin's (1981) focusing technique is a form of confronting that helps clients direct sustained attention toward implicit and often murky, indistinct bodily sensations. In all such cases, confronting as sustained attention on the problem is the central precursor at work.

The Miracle Question

The miracle question technique originated with solution-focused brief therapy (de Shazer, 1985) but was actually used by Adler (1956) many decades earlier. Assuming that the person has a small degree of awareness of a problem, this technique can be a bridge to greater confronting. The miracle question goes like this: "Suppose one night, while you were asleep, there was a miracle and this problem was solved. How would you know? What would be different?" To be answered, this well-crafted question requires the act of confronting and clarifies the problem even further by requiring the client to look at the problem itself, then look through it and beyond it. It can also build the hope precursor.

A CLOSING NOTE ON CONFRONTING

At the very heart and soul of what brings about positive change as well as mental and emotional healing, confronting may be the most powerful therapeutic change mechanism. Virtually every therapy technique makes use of confronting the problem as a precursor of change. What we have presented here are strategies and techniques to stimulate the confronting of problems with difficult clients. However, the effective implementation of such techniques to activate confronting also enables standard treatment approaches to work more effectively. When clients confront a problem, dedication to the therapeutic process tends to increase and success is far more likely in the end.

5

H: Hope for Change

It is no surprise that Dante described hell as a realm bereft of hope. Without hope, life holds little in the way of inspiration or motivation. *Hope for change* is the realistic expectation that the future will be positive and experienceable (Snyder, 1994). Hope may be said to be a precursor not only of change but possibility. It inspires action and courage and paves the way for the realization of dreams, whether simple or sublime. Unfortunately, hope is widely mischaracterized as something akin to longing, desiring, and especially wishing. In our professional context, wishing carries the overtones of a collapsed sense of hope, in which one has, to some degree, bid farewell to the real world and has become resigned to an inert fantasy that derives its power from yearning.

Hope is about what can occur in reality, what can truly be accomplished, even if the odds are long or daunting. Competitive athletes, for example, live on a diet of hope that involves envisioning a realistic form of success, not some idle fantasy. The only thing that hope shares with qualities such as longing, wishing, or yearning is a concern for a better future. At its peak, hope involves not only seeing the future as experienceable but inviting or welcoming the future as preferable to the present (Aubuchon-Endsley et al., 2015).

Thus, hope is different from blind optimism, which is often unrealistic and can take the form of fantasy (Schmid, 2019). Hope is based on realistic vision and probability, replete with options and a plan to meet the future. Hope is intimately related to the ability to solve and frame problems in ways that lend themselves to solutions (Oettingen & Chromik, 2017). What Seligman (2006) called *learned optimism* is closely related to hope, which can be practiced as a skill. What may look like a fantasy future to one person may be a calculated, realistic possibility to another who has learned that no matter how bleak the outlook, there may be a

Case examples have been disguised to protect client confidentiality.

https://doi.org/10.1037/0000451-006
Therapeutic Change With Difficult Clients: Precursors and Techniques in the CHANGES Model, Second Edition, by B. D. Wilkinson and F. J. Hanna

hidden solution. Without a realistic element of calculation and discernment, however, dysfunctional fantasy is always a pitfall.

Hope seems to have profound effects on the human psyche. It can inspire a person to not only survive but to live more fully. It has been described as the activator of the motivational system in human beings (Seligman, 2006) that grows investment in the outcome of one's actions. Hope has been associated with enhancing a person's coping ability from a wide variety of perspectives (Schmid, 2019). Just as this book conceives of it as a precursor of change, hope has been described as a prerequisite for effective coping (Al-Yagon & Margalit, 2017; Ong et al., 2017). Jerome Frank (Frank & Frank, 1991) identified hope as the operative factor in the placebo effect: Symptom alleviation by placebo is common in studies of depression, and hope may be the active ingredient.

Conversely, a lack of hope can delay recovery and even hasten death (Frank, 1961). Cannon's (1942) classic anthropological study of "voodoo death" in such places as Haiti and Australia found that voodoo deaths occurred due to the expectancy effect, when fearful beliefs about "black magic" resulted in a complete loss of hope (Hahn & Kleinman, 1983). In the Nazi concentration camps of World War II, Frankl (1992) and Bettelheim (1960) identified the disastrous effects of giving up, or the absence of hope, on survival rates among the prisoners under extreme conditions. Nardini's (1952) research on American soldiers in Japanese prison camps during World War II showed that those who managed to survive demonstrated certain dispositional tendencies, including the hopeful expectation that conditions could be tolerated and that, eventually, they would be set free.

HOW HOPE LEADS TO THERAPEUTIC CHANGE

Hope's role in change is subtle, without the drama of some of the other precursors, such as confronting or effort. While hope and self-efficacy are often grouped together under the heading of client expectancy factors (Constantino et al., 2023), it has alternatively been framed in terms of perceived possibility rather than expected probability (Nelissen, 2017). In this respect, hope has an indirect or subtle influence on therapeutic change outcomes: It is responsible for alighting the path to change rather than for precipitating the walking of it.

In the CHANGES model, hope's real power is in its widespread effects on other precursors. Hope can enhance the intensity of other precursors and, in some cases, can be a catalyst that activates other precursors. The opposite is also true, of course, since the precursors are profoundly interdependent. Hope makes experiencing anxiety or difficulty more tolerable by indicating the positive payoff for the discomfort. It makes confronting easier through the knowledge that one will not become lost in a tangle of confusion and darkness. Hope jump-starts effort by making a positive outcome appear viable so that the end seems imminent and the effort expended seems well spent. Hope can be the wellspring that arises from, and even further inspires, genuine and avid social supports.

Hope can enhance a sense of necessity by bringing a person out of a state of apathy to the point of recognizing the urgency of the current predicament. A situation that once looked overwhelming and discouraging, with no solution or egress in sight, is radically altered with the introduction of hope. With hope, it still appears as a problem, but not a problem that is "tightly packed" and impenetrable. It is now "porous" with the possibility of options and alternatives and, most important, the potential for resolution. Yet a sense of necessity can also inspire hope when the necessity is so urgent that it forces a person to confront the problem and envision hopeful solutions.

Any problem or issue can be perceived as a threat or as a challenge (Oettingen & Chromik, 2017). When a problem is seen as a threat, one is more likely to withdraw, avoid, or otherwise defend against it. When a problem is seen as a challenge, one is more likely to approach it with interest or excitement. Challenges can be framed as opportunities. As a precursor, hope can transform a threat into a challenge. For example, the loss of a job might look like a threat through the lens of hopelessness and an opportunity through the lens of hope. A hopeful future is perceived as fluid rather than static, and problems are considered a source of potential gain rather than probable loss. Many have pointed out the remarkable way in which the word "crisis" is written in Chinese, where it is represented by two characters: one denoting danger and the other, opportunity.

SIGNIFICANT MOMENTS OF CHANGE INVOLVING HOPE

Anthony, age 17, was well over 6 feet tall, handsome, and charming. Unfortunately, he was also a member of a gang heavily involved with drugs and violence. His fellow gang members respected him for his fearlessness, ability to fight, and loyalty. Like so many adolescents in his situation, he said that he wanted to improve his life and help his little brother, who idolized him, as well as the aunt who had loved him and raised him after his mother was killed. He said he was tired of the fighting and "all the hard-ass bullshit." But he reported that it really didn't matter: "I'll be dead before I turn 21," he told his therapist softly and with great conviction. Then he added, "Besides, if I ever try to leave the gang, they will f—up my world."

The therapist told him three stories of other gang members who were able to leave their gangs and make something of their lives, and he became visibly more animated. The therapist also promised to "hook him up" with one of those men who had made great strides in improving his life. Anthony jumped at the chance to speak with this man. The possibility that he could help himself and his family escape a desperate situation inspired him. What was once a threat had become a challenge.

Tina, age 36, was held in psychological bondage by her husband. She was mentally immobilized by his dominance and verbal abuse. Her extreme dependence appeared to be induced by a brainwashing campaign carried out from the time they were "high school sweethearts." Having given up on her own chances for happiness and fulfillment, her chief concern was for her three children, who ranged from ages 8 to 14. During their 13 years of marriage, Tina's husband had convinced her that she was incompetent, powerless, insignificant, and incapable of surviving without him. She had been in therapy for six sessions, originally for her oldest child, a highly intelligent boy who was flunking out of his freshman year in high school. She had been extremely careful to avoid any mention of her relationship with her husband.

In the seventh session, which she attended alone, she tearfully spoke of her sense of being trapped by her husband and how she needed to "get away" from him but just did not know how. She was convinced, however, that this was not easily done. "I don't think I can survive on my own," she said fearfully but also with a strange sense of calm. She also stated with grim satisfaction, "He needs me."

In time, Tina eventually achieved therapeutic change when she saw that she did not have to be dependent on her husband. Hope dawned when she learned of other women who had moved out of similar oppressive circumstances and spoke with a couple of divorced friends who had been through similar situations. As she began to see solutions to her problem, she eventually gained the confidence to move out on her own. Her husband did not allow this without a lot of anger and fuss, but she eventually reported, "It's okay. I can handle him now."

Both Anthony and Tina developed a sense of hope when they recognized that other people in similar situations to their own had successfully created change. They subsequently explored their problems with a greater focus on solutions, and the possibility of change became increasingly realistic. They each began to imagine a future in which they were no longer helplessly trapped, and that future appeared experienceable, valuable, and worthy of pursuit.

CLINICAL MARKERS: HOW TO RECOGNIZE HOPE FOR CHANGE

The most important clinical indicator of the presence of hope is a sense of confidence that conditions will improve (Larsen & Stege, 2010). Change, in the form of therapeutic goals, is seen as genuinely attainable. The outlook will be realistically positive and, even in the face of adversity, the client will not become discouraged or apathetic. A hopeful client perceives a problem in terms of solutions rather than becoming entrenched in "awfulizing" the problem. That perception is accompanied by enthusiasm or even excitement about life (Schmid, 2019; Snyder, 1994). Perhaps the ultimate sense of hope occurs when a client grows thankful for lessons learned from a problem, and demoralization or distress is seen as merely temporary.

The following statements are clues to the presence of hope:

- "I know things will eventually be okay, but things are really difficult right now."
- "I know I will get through this, but it is so hard and I feel so terrible."
- "I have been through tough times before, and I can get through this, too."

A therapist can also test for hope by gently probing with alternative views. For example, if a client remarks, "I have absolutely no idea how to deal with this situation," a therapist can reply, "Do you have a sense that you will eventually figure it out?" A client inclined toward hope will indeed have that sense, knowing intuitively that a way exists but has not yet been discovered.

Hope and Humor

Hope also gives rise to *therapeutic humor* (Gladding & Drake Wallace, 2016; Sultanoff, 2013; Snyder, 1994), or humor that is insightful and uplifting and that transports a person out of a burdened mindset and into alternative perspectives that cause some degree of amusement. For instance, a male client with a verbally abusive wife once said with a sense of trepidation, "My wife said she's going to leave. I don't know what I'll do without her." The therapist smiled and replied with a somewhat satirical tone, "That's quite a problem. What will you do with all of that freedom and well-being?" The client suddenly laughed, and the seriousness of the moment eased a bit. He was surprised that he did so, and this fostered further discussion. The client's response also hinted that he was hopeful enough to eventually be able to see a future without her.

Humor, in this sense, can function as a therapeutic reframe of an otherwise oppressive problem. Ideally, therapeutic humor is directed at the ambiguity or quirks of life or the foibles of human nature. The kind of humor to avoid is that which is at the expense of another person, such as gloating over the misfortunes of others. Other types of destructive humor involve ridiculing another about looks, mannerisms, or lack of intelligence or skills. Humor based on stereotyping is, of course, another nontherapeutic variety. In any case, hope can give rise to humor and using humor properly can give rise to hope, reframing the problem just enough to lift the seriousness and rigidity of a problem to allow a healthy escape from an otherwise discouraging situation.

A therapist who is genuinely lighthearted and can tactfully integrate humor has a real talent. Clients generally look forward to sessions where the work is hard but fun and done in a setting that combines lightheartedness with learning. Timing and appropriateness are key, as it is important to fit humor into an acceptable framework for the client. While humor cannot be taught to therapists in any consistent way, modeling it in supervision may be a good approach (Gladding & Drake Wallace, 2016). In some cases, therapeutic humor can be conducive to an increase in awareness.

HOW TO DETECT THE ABSENCE OF HOPE FOR CHANGE

A client lacking hope will display an almost palpable sense of despondency or despair and probably a sense of apathy as well. They may believe that life is pointless and there is no significance in anything. Other common emotions are resentment and deep self-pity. Sometimes, a hopeless person displays a pessimistic humor that conveys bitterness or biting sarcasm. Such a client will question the worthiness of life and living. A fatalism, as though life is out of human control, may manifest and pervade statements and beliefs. The person may say in one way or another, "Things never turn out well for me," or "It doesn't matter what I do; it's always a disaster."

Such a person displays little confidence in their ability to solve problems. Discouragement and disappointment are always just around the corner for these unfortunate people, and their inclination toward resilience in difficult experiences is low. A client low in hope may display some degree of suicidal ideation without necessarily ever intending suicide. On the other hand, suicide may be imminent, and resolving this crisis will become the first duty of therapy. Clients low in hope may also harass themselves with self-doubt and express a general sense of protest against the seeming unfairness of life and the impossibility of ever being happy. To the truly hopeless person, life is a painful process and staying alive means the pain will continue. When hopelessness reaches its greatest depths, each passing moment is a kind of torment.

Farber (1968) suggested that suicide is an inverse function of hope. He described it as a disease of hope that proceeds from a "no exit" belief, wherein a person believes that death is the only escape from an intolerable situation or set of conditions. Ensuing research determined that suicide is more closely associated with hopelessness than depression (A. T. Beck et al., 1974, 1979), and the Beck Hopelessness Scale (A. T. Beck et al., 1985) developed to measure the likelihood of suicide remains in widespread use today (Kocalevent et al., 2017; Marchetti, 2019). Difficulties in social problem solving have also been linked to suicide risk, with perceived burdensomeness serving as a primary mediating factor (Chu et al., 2018).

TECHNIQUES: BUILDING HOPE

When the precursor of hope is activated, the vision of a bright, promising future brings with it the love of life and the joy of simply being. When coupled with other precursors, such as awareness, effort, and social support, the scenario begins to look like a poet's portrait of the undaunted human spirit. It does seem to be true that where there is hope, there is life. It is simply a matter of degree.

If a client leaves the first session with a sense of hope, the odds of their attending another session are much higher. Building hope involves directly empowering a person, helping them to become more capable and confident as part of perceiving a favorable, realistic future. The enterprise is largely dedicated to helping a client become more stable in the face of adversity and better able to focus on positive outcomes with the realistic expectation that problems and situations will turn out well. The client who is hopeful has sufficient coping skills to deal with problems and issues, as well as that fascinating quality that psychodynamic therapists refer to as ego strength. The hopeful person also has the capacity to see beyond problems toward a bright future.

The techniques and strategies in this section are designed to instill hope as a precursor to change by strengthening and empowering the person. When a client has hope, the other precursors can be positively affected by its pervasive influence throughout the range of that person's life. The gift of hope is the gift of a desirable future. Thus, hope building is done through any approach that can make the future more tolerable. The following techniques can help the therapist build the hope precursor in clients.

Overcoming the Influence of Negative Role Models

This technique addresses the impact of negative role models on client behavior, particularly among adolescents and young adults who are abusing drugs or alcohol. According to Bandura (1977), people tend to imitate or take on the roles and behaviors of the people whom they admire. This technique seeks to reduce the internal power that the influential person exerts on the client's mindset and provide a sense of hope for living more congruently. The strategy is simple since such role models tend to hold significant weight in the client's mind and lead the client to imitate particular roles, mannerisms, or behaviors.

Begin by asking the client about the most influential people in their life, with an emphasis on those people the client "most wants to be like" in terms of certain behaviors. For instance, if a teen client is getting into fistfights at school, ask about their role models for being tough and winning fights. If a client is using drugs, ask about their role models for drug use. Have the client list these people and, one at a time, provide a detailed description of each. Be sure to have the client include musicians, actors, and other media, popular, or subcultural figures of relevance.

Monitor what the client identifies as admirable about their role models, empathize with the client about those points of commonality, and ask the client if there have been negative outcomes or life struggles for any of those role models. In many cases, the client will hope for the accolades or perks that seem to accompany the behavior of their role model, such as power, status, money, or influence. Peer recognition is a powerful influence on behavior among adolescents and young adults. However, in other cases, the client will identify with the "crash-and-burn" aspect of a role model's life simply because this is how they seem to feel themselves. A 19-year-old budding alcoholic once told his therapist, "Kurt Cobain's my hero. He died, like, forever ago, but I understand him. No one got him, you know? But I do. I'll probably be dead before 30, too."

Whether related to drug and alcohol use or not, it is important to validate the perceived connection between the client and their role model before examining potential negative outcomes. Trying to undermine the influence of the role model can be difficult. It takes a certain finesse. Work with the client to examine the similarities and differences between themselves and the selected role model. Look for any differences that the client perceives as a personal deficit. It can be helpful to then note areas in which the client has a strength that the role model does not or to highlight behaviors of, and outcomes for, a role model that the client might find uncomfortable or distasteful.

At each step in the process, ask the client if they still choose to have the person be influential. Note any dysfunctional cognitions or beliefs that are connected to this person. Use a cognitive approach, including the cognitive therapy of oppression if appropriate, to refute and reformulate any harmful beliefs. For situations in which the role model has died or has been incarcerated, it can be beneficial to ask the client what they know that the role model did not know. Additionally, it can be helpful to put the admired person in the empty chair and conduct a therapeutic dialogue. Regardless, this technique has the potential to free the client from negative influences and move the client toward empowerment through an enhanced sense of agency and authenticity. Most importantly, the technique can have a profound impact on hope, as the client comes to recognize that their own fate does not have to align with the deleterious outcomes experienced by certain negative role models.

The Jamesian Device

James (1907) developed a theory of truth that is suited to personal knowing that does not rely only on objective criteria. This has an application in therapy that can greatly help some clients. Many difficult clients are remarkably hesitant to believe that anything positive about them may be true. James' definition of truth can provide clients with an internal measuring device that can indicate whether or not a particular belief or statement has any truth value. The therapist explains to the client that when a statement or belief is true for a person, it produces positive feelings and a sense of well-being and resonates and harmonizes within oneself. Even if the statement is negative, that resonance will take place, producing a feeling of peaceful acceptance "deep down inside." On hearing this idea, many clients immediately discount it as wishful thinking. However, when told it was originated by America's greatest psychologist (Korn et al., 1991), it is easier to accept, or at least less easy to dismiss.

For example, Michelle, a woman in her early 20s, was depressed and lacking in self-efficacy. She found it exceedingly difficult to accept any positive statements about herself from her therapist and would not engage in disputing negative beliefs. She told her therapist several times that it was nice to suggest such positive things, but she knew they simply were not true, and for that reason, she would not believe any of them. "I know," she said, "that there is nothing positive about me and that all the people who have hurt me were right."

At that point, the therapist introduced James's theory of truth and explained it in language that was meaningful to Michelle. The therapist then asked her if she would like to try an experiment. With a sigh, Michelle agreed. The therapist said,

> I am going to give you a statement to repeat to yourself out loud. When you repeat it, rather than think about whether it is true or false, I would like you just to watch and see how the statement makes you feel inside.

The therapist gave Michelle the self-statement, "I am a good person," a declaration she would not even consider previously. Michelle repeated it to herself a couple of times, looked up at the therapist hesitantly, and said, "It makes me feel kind of good to say that, but I am just kidding myself." The therapist asked, "Did it feel good deep down and give you a sense of inner peace?" Michelle, now on new territory, only nodded. "Well, Michelle," said the therapist matter-of-factly, "according to William James, you may very well be a good person."

Michelle was forced to consider this and, in the following sessions, tried out many other self-statements using her new internal measuring device. The therapist also told her she could use this tool in everyday life when someone criticized her or forced a belief on her. Such affirmations and positive self-statements had not worked previously. This experiential technique bypassed the difficulty she was having in terms of disputing her cognitive distortions. Of course, this internal measuring device can only gauge the validity of self-evaluative statements and not of environmental conditions or other people. Otherwise, it would convince paranoid or narcissistic clients, for example, that their ideas of reference are correct, making such clients even more difficult to treat. Examples of statements that should not be used in this technique are, "They are out to destroy my life," "I'm going to win the lottery," or "[fill in the blank] is in love with me."

Empowerment Strategies

In the enterprise of hope building, it is wise to take every opportunity to help a client feel capable and empowered to make changes. Many therapists have told me they sometimes do not know how to empower dysfunctional clients who seem to have little going for them. As part of empowerment, it helps to recall that even the most dysfunctional people occasionally do something functional and to concentrate on reinforcing that fact, rather than implying that they lack skills and have to learn new skills. In many cases, new skills do need to be learned, but it is easier to do so when these are built on a foundation of established skills. Two modes of empowerment are especially helpful in this regard: reframing negative behaviors as skills and converting a threat into a challenge.

Reflecting the Hopelessness

It is often a mistake for a therapist to refuse to accept the client's conviction of the hopelessness of a situation or condition. To work with a client who is bereft of hope, it is often vital for the therapist to reflect that hopelessness, conveying that it is fully understood and why. A client may be more inclined to listen to messages of hope when it appears that the therapist truly understands the grimness of the situation. Many hopeless clients seem to demand that their therapists see the hopelessness and feel it as well. To meet this demand, it is helpful to empathically paraphrase and summarize exactly why things cannot and will not improve. The therapist then invites the client to comment on the degree of the therapist's understanding.

For example, a therapist might say to such a client,

> Please tell me if I am tracking with you here. You figure that no matter what you do, things are going to be a disaster, and you will end up being a failure. There is no point in trying, and the only way out is by ending your life. Am I close to understanding how you see all this?

When clients are convinced that a therapist truly understands the depth and degree of their hopelessness and the internal logic of that hopelessness, then they may be more inclined to explore or consider alternative ideas or options.

Reframing Negative Behaviors as Skills

One of the most helpful ways of empowering clients is to demonstrate to them that their dysfunctional, negative behaviors can be reframed into positive skills that are available to them at their own choosing. All too often, the message received by people who behave in maladaptive ways is that the behaviors are "bad," or something that no good or worthy person would ever do. The key is to reframe negative behaviors as skills. For example, rather than inform a manipulative client that their behavior is destructive, the therapist can reframe it as a kind of skill that involves accurate perception and persuasiveness in getting people to do what they want.

Clients are far more willing to talk about their manipulative behavior when it is framed as a skill. The therapist can admire it as a hard earned and well-practiced skill that has served a useful purpose. The therapist can then suggest that the skill is actually underused and can be used in other, more positive ways. If one is a selfish manipulator, one can redirect or reverse that very same skill and become a helper. The same perception that is used to exploit a person can also be used to help. The same is true of the skill of counseling: If that skill were reversed or redirected, an effective therapist could become an extraordinarily effective verbal abuser.

Thus, the therapist can inform the client that up until now, that skill at manipulation has been used for selfish purposes, but if the client so wished, they could use the same skill in a positive way to become a corporate manager, salesperson, or counselor. When a client realizes that they already have the blueprint of positive skills, it builds hope by making the refinement of those skills seem like a realistic possibility. This technique can also be used with other "skills." Lying, for example, presupposes a creative ability to reshape and reformulate information. The negative behavior of verbal abuse, when reversed, can be validating, encouraging, and even inspiring. The blueprint is obviously therein. The challenge is in identifying how to constructively adjust its use.

Converting Threats Into Challenges

When a person is threatened by circumstances, problems, or conditions, hope can be adversely and sometimes seriously affected. The greater the perceived threat, the lower the degree of hope for a positive outcome. One strategy is to reframe a perceived threat into what looks like a challenge, thus rendering it more likely to be handled successfully (Oettingen & Chromik, 2017). If a threatening problem is perceived as a challenge, hope can be sustained or even enhanced.

The procedure involves two steps. The first is to find and identify the client's skills, of which they may not even be aware. If the client was a victim of oppression, for example, they may be highly perceptive of others, even if easily exploited. Similarly, being a survivor of abuse can be reframed as an ability to endure hardship as part of solving a problem. Ingratiating or people-pleasing behaviors require the skill of perception as part of knowing what people want. A host of skills can be recognized in clients, ranging from patience and strength to assertiveness and intelligence. A skill is a skill, no matter how developed it is. If a client recognizes it as a skill, hope may be enhanced.

After skills are identified, the problem itself can be addressed. The therapist suggests that the client's skills are especially suited to the problem at hand and are what is needed to solve it. When clients see their skills in other domains can be transferred to the current problem, hope can be increased, and the effort precursor may be enhanced as well.

The Story of Gwen

Gwen, age 38, was a salesperson for a major pharmaceutical company. She presented with what appeared to be a career issue of being bored with her work, and she was no longer willing to continue her career. She referred to her job as a "drug pusher" and said that it was not in harmony with her true nature. In spite of her depression, she had an inauthentic way of smiling through it and saying that everything was as it should be. In the second session, she hesitantly though enthusiastically revealed that she was studying with a group whose guru had magical powers over her well-being and spiritual development. She also revealed that the real reason she was in therapy was that her husband, Bill, had insisted on it. Bill believed that the guru was crazy and was convinced Gwen had fallen prey to his manipulations and that it was ruining their marriage. The guru himself, she said, was opposed to her receiving therapy and was not happy with it.

Gwen maintained a schizotypal belief system based on and reinforced by her guru, who had convinced her that even her dreams and moods were a result of his control over her mind. Part of this belief system was that the guru was putting her through uncertainty and depression for the purpose of spiritual "testing" and "cleansing" for her eventual enlightenment. Gwen had little self-efficacy and almost no faith in her own ability to achieve happiness and fulfillment. Her faith in the magical powers of the guru was a way of bypassing her helplessness so he could achieve happiness for her. Only he could do it. She vigorously defended against any perceived criticism of the guru.

In the third session, Gwen admitted that even though she had great faith in her guru, he was not making her as happy as she would have liked. However, she anxiously added that this was a "test" of her loyalty and resolve and that, in due time, she would be rewarded with a mystical experience of supreme peace and tranquility.

In the next session, the therapist recognized and reframed her resolve and her faith as skills. Taking great care not to threaten or attack her delusional system, the therapist reframed Gwen's belief in the guru and her presenting problem of wanting to quit her job as related to the deeper, more pervasive, and admirable goal of attaining self-knowledge and self-mastery. Gwen agreed that this was indeed her ultimate goal. The guru's test that had caused her so much distress was reframed as part of the challenge for her to come to know herself.

Gwen accepted the reframe, as it fit rather easily and comfortably into her delusional system. The therapist further informed Gwen that therapy was another path to self-knowledge and self-mastery that did not have to conflict with her spiritual beliefs and that therapy had the capacity to help her be happy and find fulfillment. The therapist did not challenge her spiritual beliefs; instead, they reflected to Gwen that her beliefs were from ancient sources and were not dependent on the guru for their validity. The therapist took great care not to invalidate the guru himself because it was clear that Gwen would likely terminate therapy if this were to happen.

This approach was consistent with her beliefs and delusions about her guru and allowed egress from her attachment to the guru and the cult in which she was entrenched. The basis of hope provided to Gwen early in the therapeutic relationship was the prospect that therapy could lead to self-knowledge and self-mastery. Crucial to her liberation from the guru was the recognition that the vast power she had attributed to him was evidence of her own power of belief and that she could begin to use some of it now instead of waiting. This was another great source of hope for her.

In time, she began to see that her intense dislike of her job was a source of self-knowledge in itself, and she no longer linked it with her guru's ridicule of such "lowly" occupations. She also took heart in her absolute faith in the guru, having reframed it as an insight into her admirable qualities of faithfulness, loyalty, and friendship. Most important, she began to see the guru's narcissistic need for admiration and control. Her faith in him progressively weakened as she regarded him in this realistic fashion. After about 15 months, she left her job and entered law school full-time, and she gradually ended her involvement with the guru and his circle of admirers. Gwen attributed her change to the promise of hope, the various reframes, and the relationship that kept her engaged in therapy in spite of her guru's admonitions to terminate.

Hope as Contagion

We know that a therapist's positive expectations of a client can impact the client's success in therapy (Sperry, 2022). The therapist's confidence can be contagious. When working with a client to build hope through empowerment, just believing a client can indeed get through an issue or successfully solve a problem will indirectly affect the client. This belief is communicated through the therapist's metalogue and dialogue; of course, it also helps if the therapist directly and often voices their belief in a client's skills or ability to deal with an issue.

Many clients have told their therapists that they successfully handled a problem because of the faith or confidence the therapist had in them. In this way, hope is contagious.

Hope Requires a Sense of Worthiness

It is important to ask clients whether or not they feel worthy of having good things in life, as unworthiness can act as a powerful obstacle to therapeutic change. Many difficult clients, when asked, report that they do not deserve to be happy and do not feel good about themselves or their lives because of mistakes they have made in the past. Ironically, such a stance serves as evidence of goodness or worthiness, while for many clients, it is perceived as the path to eventual redemption.

Regardless, if a client deems themselves unworthy of positive life experiences, change will not occur because the client has lost hope in the possibility of redemption. Clients in recovery and in treatment often say this is why they did not previously respond to therapy after years of treatment across multiple treatment programs. Again, the main reason clients feel unworthy is because they have harmed others in the past and punish themselves accordingly. Although this is sometimes an unconscious phenomenon, many clients are distinctly aware of the issue when pressed to discuss it:

THERAPIST: Do you feel like you deserve to be happy?

CLIENT: (shaking head) No, not really.

THERAPIST: Would you like to feel deserving and worthy of a happy life?

CLIENT: Yeah, sure. But I don't know what that even means.

THERAPIST: If you did feel worthy, how do you think your life might be different?

CLIENT: I don't know. I probably wouldn't be so depressed all the time.

THERAPIST: That sounds pretty great, not being depressed. So, when therapists suggest things to do to reduce your depression, there's a big part of you inside that says, "Nope, not going to do it because I deserve to be sad." Is that right?

CLIENT: Exactly. I don't deserve to be happy. I've done too many bad things.

THERAPIST: So, you're not allowing yourself to feel happy?

CLIENT: Man, you don't know me as well as you think. I've done *really* bad things.

THERAPIST: Do you feel bad about the bad things you've done?

CLIENT: Yeah, I do. I think about it a lot.

THERAPIST: You know, the fact that you feel bad is just proof that you're a good person.

CLIENT: (rolling eyes) Whatever.

THERAPIST: Do you think a truly bad person would feel bad about hurting others?

CLIENT: Probably not.

THERAPIST: Bad people don't feel remorse or guilt. Good people do. You feel guilty, which means you're a good person. You've made big mistakes, but you're obviously still a good person because you clearly feel remorseful.

CLIENT: I guess. It still doesn't change what I did, though.

THERAPIST: No, but maybe you still deserve some good things in life. Maybe your path to redemption can include positive changes. Besides, it's hard to overcome our mistakes and do better by others when we're depressed, don't you think?

CLIENT: Yeah, maybe.

THERAPIST: So, I'll ask again: if you did feel worthy, how would your life be different?

CLIENT: Maybe I'd do more good things for the world if I was just a bit happier.

THERAPIST: Okay. What do you think about exploring the good things you could do?

CLIENT: Alright, yeah. Let's give it a shot.

Hope is the wellspring from which all self-transformation occurs. If a client believes they do not deserve to transform, therapists need to resuscitate hope. Doing self-forgiveness work can help get a client to let go of the past and start to forgive themselves for the harmful acts committed toward others. Simply acknowledging that one is a good person can be transformative. In terms of addiction, therapists can ask a client whether their drug or alcohol use was a response to feeling worthless or undeserving of a good life. Self-forgiveness can be a powerful catalyst for sobriety. Other operative precursors in this technique are grit, confronting, awareness, and effort.

Creative Narratives: Relating and Rewriting

In some cases, therapists can begin work on the hope precursor by asking a client to tell the story of their life. There is no need to focus on tragedy or trauma, but only to get a view of life as the client perceives it. If the story is told tragically, the therapist can help the client reconstruct the story to retell it with a sense of hope, empathy, and understanding—as a story still in the making. If the story is told in a narcissistic fashion, complete with fantasies of brilliance or superiority, the story can be retold to show how difficult it must have been for others to appreciate such a great person. The retelling could include how important the development of empathy and compassion for others might have been so that the other people in the client's life would have been more communicative and appreciative. The purpose, again, is to open a therapeutic window.

The retelling of a life story can be powerful for some clients. It allows a client to become liberated from a stuck or fixed pattern of conceiving the self and surrounding contexts. When creatively and therapeutically retold, the story can allow therapeutic change to become part of the story itself, supporting the emergence of precursors such as willingness, awareness, and necessity.

Telling Stories of Recovery

Like Anthony, who struggled to imagine a future for himself, and Tina, who could not envision leaving her husband, many difficult clients have great difficulty conceiving the possibility of change. Telling stories of clients and people with similar problems who made positive changes in their lives can be helpful. Clients will often inquire about the person or client mentioned by the therapist, sometimes bringing them up months or weeks after the story was told. Stories can be introduced by saying, "You know, I once knew someone who was similar to you," or "You remind me of someone who was in a predicament similar to yours." Clients have sometimes told me later that a particular story about a person with a similar dilemma kept them going during difficult times.

Exposure to Others Who Solved the Same Problem

Group therapy can be remarkably powerful in building hope, with unique benefits not found in individual or family therapy. Yalom (1995) noted that hope is often instilled when clients observe their peers making therapeutic changes. Group therapy can communicate a client's potential for change in a way that goes far beyond encouraging messages from a therapist. In addition to groups, hearing speakers, watching films, or reading books that contain stories of change can also be a powerful source of hope. In treatment programs for adults or adolescents, bringing in people who have worked on similar issues to tell their stories of change is another effective strategy. I have observed several instances in which a sense of necessity arose in a client when a person inspired hope by relating a personal story. This is especially helpful, for example, with adolescent gang members who believe they have no future and no escape from their circumstances. It is also helpful for substance users.

Examining Maladaptive Beliefs About the Future

Because hope intertwines with positive expectations for the future (Wampold & Flückiger, 2023), hope building can take the form of examining maladaptive beliefs about the future itself and the context of those beliefs. If a client is habitually inclined to regard the future with suspicion, fear, apprehension, and anxiety, that client's capacity for hope will diminish. In many cases, if maladaptive beliefs are not addressed and disputed, the techniques in this chapter may not bear fruit. Distorted views and beliefs about the future can affect the smooth implementation of any program of therapeutic change, regardless of the theoretical approach. It is the therapist's job to help identify and properly phrase these beliefs with the client's assurance of accuracy before disputing them. The following are examples of maladaptive beliefs about the future:

- "The future holds nothing for me."
- "Only bad things are waiting for me."
- "The world is a cruel place to live."
- "No matter what you do, things will never really get better."
- "Life is a process of failing to realize one's dreams."
- "The future holds only what others want."
- "Other people dictate my future."
- "The passing of time is painful."
- "The future provides nothing but anxiety."
- "The future is filled with continuous disappointment."
- "The future is filled with unforeseen catastrophes."
- "I am unable to affect the future."
- "The future is a continuous threat to my well-being."
- "Hope is for fools."
- "Nobody has any idea of the future."
- "My life has been disappointing, and the future will be more of the same."

A CLOSING NOTE ON HOPE

As we have stated previously, all of the precursors are interdependent and interactive, and thus, the increase of one precursor can increase the presence of others depending on the circumstances. Of all the precursors, hope demonstrates this interactive quality as much or more than the others. For example, a client may lack a sense of necessity, but that lack may be due to apathy, specifically, in the belief that things cannot and will not ever get better anyway. Thus, for such a person, there is no point in entertaining any sense of necessity because hopelessness has rendered such action a waste of time. Hope can also stimulate the precursors of grit and confronting the problem where these qualities did not exist previously. It can be surprising and encouraging to client and therapist alike when motivation to change emerges alongside the rediscovery of hope. If there is anything at all magical about the therapeutic encounter, it is to be found in the moment of hope's resuscitation.

6

A: Awareness of the Problem

Many philosophers have framed awareness as the most fundamental aspect of life, its qualities extolled in wonder and awe since the time of the ancients. Generally speaking, *awareness* is the cognizance, recognition, or knowledge of any mental or physical object, relation, or event. As a precursor of change, awareness involves a client's recognition of, or clarity of perception about, a problem. Put another way, awareness is the identification or pinpointing of issues or relationships that need to be addressed as part of the therapy process and its tasks. It is the function that moves issues from the edge or ground of consciousness into figural focus. Plucking a problem out of the realm of the obscure and dropping it into the light, awareness illuminates and enables genuine recognition of the potential for change. It gives substance to shadow and form to oblivion.

People lacking in the awareness precursor have, for the most part, shut off the process of knowing and growing. When there is no awareness of the existence of a problem, therapeutic change occurs only by accident or through a process that does not require consciousness raising, such as pharmacotherapy or pure operant conditioning. It is therefore a vital element of the psychotherapy process. Without awareness, self-determined change would probably never directly take place. When awareness is present, only denial itself can be denied. Unsurprisingly, awareness has been identified as a common factor tied to the change process across the established psychotherapies (Drozd & Goldfried, 1996; Høglend & Hagtvet, 2019).

Case examples have been disguised to protect client confidentiality.

https://doi.org/10.1037/0000451-007
Therapeutic Change With Difficult Clients: Precursors and Techniques in the CHANGES Model, Second Edition, by B. D. Wilkinson and F. J. Hanna

DEGREES OF CONSCIOUS AWARENESS

Awareness is not an all-or-nothing phenomenon. It can be active to varying degrees, reflecting the extent to which an individual is consciously aware of a problem. The first degree involves uncertainty as to whether there is a problem at all. Whether framed in terms of defense mechanisms such as denial and repression or simply in terms of obliviousness, many clients are unaware and unsuspecting that a problem exists, even when the problem is severe. Few people who lack any awareness of a problem or issue seek therapy on their own, and most such clients have been referred by a spouse, judge, employer, or friend. For those clients, the first order of business is usually to identify a problem that needs attention.

The second degree of awareness denotes ambiguity as to the nature of the problem. The person may be aware that something is wrong but has no idea what that something is. Such awareness is often accompanied by a cognitive dissonance that initiates the change process. In such cases, a person may have a high sense of necessity and grit. However, without awareness to pinpoint the area on which to focus, the client will demonstrate a certain amount of vagueness about what to address. Such a person may come to therapy to try to determine if there is indeed anything to address at all. Often, this kind of client secretly wants the therapist to say, "You're fine. There is really nothing wrong and you can go home."

The third degree of awareness denotes a genuine clarity about the nature of a problem, including the ability to identify specific behaviors, emotions, beliefs, cognitions, and interpersonal issues in need of remedy. The more details a client recognizes in a problem, the easier it is to manage it and maneuver around potential pitfalls. Such awareness can include recognizing the origins of a problem and properly attributing responsibility for a problem without blaming.

HOW AWARENESS OF THE PROBLEM LEADS TO THERAPEUTIC CHANGE

Awareness is fundamental and integral to all the other precursors. To the degree that a client can learn to detect such things as emotional difficulties, unresolved issues, problematic behaviors, harmful relationships, dysfunctional thoughts, anxieties, and negative environmental stressors, the potential for change gains entrance into the realm of the imminently real via awareness. The capacity for awareness can be enhanced through therapy to great effect, even if it is seldom sufficient by itself for change to occur. To better understand how awareness leads to change, we look to various processes to better understand and support the advancement of awareness.

Metacognitive Aspects of Awareness and Change

Metacognition involves two primary functions: monitoring and control (Norman et al., 2019). Awareness is obviously fundamental to the monitoring function, which includes detecting, seeing, and observing, in one's "inner eye," cognitions, beliefs, and thoughts. Perhaps the most important metacognitive aspects of awareness are what needs to be observed and where to direct attention. Whether a client is avoiding or dealing with an issue hinges on awareness. In this regard, the awareness precursor is tremendously powerful.

Another metacognitive process related to awareness, known as *decentering*, involves shifting from being within subjective experience to reflecting upon that experience (Bernstein et al., 2015). Many clients are so enmeshed and immersed in a situation, behavior, thought, feeling, or sensation that they can contemplate or think of little else. One of the most valuable aspects of awareness involves the action of taking a step back and viewing a problem, emotion, or obsessive or compulsive pattern dispassionately (Bernstein et al., 2015; Goldfried, 1995). This essential action of stepping back is a prime ingredient of change, creativity, and mental health (Coffey et al., 2010).

A. T. Beck (1976) referred to such stepping back as "distancing," not in the negative sense of avoidance but in the sense of getting an overview. This has also been described as taking a broad or overarching viewpoint. By removing oneself from confusions and mental entanglements, one can get perspective. For cognitive therapists, it is the act of stepping back in one's mind and viewing thoughts or beliefs connected with an event or emotion from a self-distanced stance (Travers-Hill et al., 2017). Self-distancing is experientially synonymous with concepts such as cognitive defusion within third-wave cognitive behavioral therapy approaches (Hayes & Linehan, 2018).

In many instances, increased awareness is the metacognitive equivalent of learning how to negotiate one's way through a cognitive and interpersonal minefield. Each successive increase in awareness reveals the position and placement of other mines, which can then be avoided on the way toward freedom. Just as the shunning of awareness is the essence of defensiveness, the strengthening of awareness opens the door to exquisite feelings of vitality. The key to that vitality seems to be the belief that even the awareness of pain is better than no awareness at all.

Consciousness Raising

Increasing awareness is often referred to in therapy and counseling literature as *consciousness raising*, which has been cited as a common factor across therapy approaches (Prochaska & Norcross, 2018). Freud attributed awareness to the functioning of the ego. Of course, the concept of the ego and awareness was around long before Freud, and he himself studied it under Franz Brentano (Hergenhahn, 1996), who also taught the founder of phenomenology, Edmund Husserl. The philosophical discipline of phenomenology made consciousness its central focus of study.

Awareness was praised and celebrated in the *Upanishads* (Easwaran, 2007), one of the world's oldest known religio-philosophical works. It was also emphasized by other ancient Asian philosopher–psychologists such as Buddha, Shankara, and Patanjali (Gupta, 1998; Hanna, 1993, 1995; Theise & Kafatos,

2016), who regarded conscious awareness as the passive witnessing of all phenomena of mind and the world. Western philosophers have long written about awareness in the form of consciousness; it is a theme in the writings of Plotinus, Kant, Husserl, and Sartre.

Consciousness raising is present in virtually every form of psychotherapy and counseling other than pure operant conditioning. It involves attaining clarity of perception of thought processes, mental images, feelings, emotions, interpersonal interactions, and environmental conditions. It also involves recognizing the effects of one's actions upon any of these aspects of living. It plays a critical role in existential and humanistic therapies, both of which claim a long tradition of emphasizing this phenomenon. Awareness as consciousness raising is an indispensable part of behavior therapy as well, which utilizes such techniques as self-monitoring and self-observation for the purpose of bringing a person to discern more fully and precisely the nature of behaviors or reactions, ranging from problem drinking and smoking to irrational fears or phobias.

Behavioral techniques that make use of exposure would hardly be workable if not for awareness, and it is here that much of therapeutic change originates. However, while it is true that behaviorists can manipulate change in a person through conditioned responses to stimuli, that change is entirely dependent on a largely unchanging environment to maintain new behaviors. For example, many clients change in inpatient drug and alcohol treatment programs, only to relapse when exposed to old environmental cues due to impaired meta-awareness (Ruimi et al., 2018) or lack of cognitive integration (Vafaie & Kober, 2022). In awareness, change only becomes stable and lasting when new cognitive structures are successfully integrated with older knowledge and experience.

Awareness Remains Consistent

One of the most remarkable things phenomenologists have discovered about awareness is its amazing consistency. When present, awareness itself seems not to change in its essential character, regardless of what is being contemplated (Merleau-Ponty, 1962), even during drug-induced psychedelic experiences (Yaden et al., 2021). In other words, the content of awareness changes and can be altered by various mental states and drug influence. However, as a function, awareness itself remains consistent. In the Tibetan Buddhist tradition of Dzogchen, this stream of consciousness is referred to as *Rigpa*, "which continues with no beginning and no end, without any break ... and constitutes the mental continuum of each Being" (Klein & Wangyal, 2006, p. 73). In the field of experimental psychology, Powers (1973) made a similar discovery about the consistency of awareness and expressed it well:

> Awareness seems to have the same character whether one is being aware of his finger or of his faults, his present automobile or the one he wishes Detroit would build, the automobile's hubcap or its environmental impact. Perception changes like a kaleidoscope, while that sense of being aware remains quite unchanged. (p. 200)

As we know, awareness can be directed toward the contemplation of a tremendous range of physical and psychological phenomena. Psychological phenomena include mental images, thoughts, desires, intentions, and even hallucinations and delusions. They include emotions as well, from sadness and grief to enthusiasm, joy, and happiness. Awareness of physical phenomena requires the senses to contemplate the sights, sounds, smells, tastes, and touch of, for example, music, mountains, oceans, city traffic, living beings, and inanimate objects. Physical sensations, for instance, headaches, dizziness, nausea, motion, goosebumps, and sexuality are other examples. Thus, awareness is tremendously consistent yet versatile, capable of attending to and beholding a fantastic array of phenomena without itself being altered in its essential character.

Empathy as a Form of Awareness

From the perspective of psychotherapy, empathy is intimately related to awareness. *Empathy,* or the inferential knowledge or perception of how or what another person is perceiving, thinking, or feeling, is a form of awareness. In the CHANGES model, empathy falls under the awareness category. People who have a reasonable amount of it are more likely to engage others in harmonious social interactions than those who lack the awareness necessary to recognize the cues and indicators of the thoughts, feelings, and responses of others. This has important implications for people with antisocial, narcissistic, borderline, and histrionic personality disorders, where lack of empathy serves as a fundamental aspect of psychopathology (Kajonius & Dåderman, 2017).

In some people, transformational change arises from sudden empathic awareness. When a client becomes aware that their temper scares and alienates their young children, that may be enough to enable them to start managing their outbursts. In such cases, awareness catalyzes all of the other precursors and launches the change process. But for most clients who find change particularly difficult, awareness is a helpful but insufficient catalyst for activating the other precursors.

SIGNIFICANT MOMENTS OF CHANGE INVOLVING AWARENESS

The following example of Claire's significant moment of change illustrates the mechanism of coming to awareness and the accompanying action of stepping back from the problem. As she put it, her awareness came "in a flash." After 9 years of marriage, Claire began to notice that she was becoming increasingly depressed. As time went by, she told her therapist, "It just became so obvious that I was in a state of depression." This depression was accompanied by a series of illnesses in the form of infections that were perplexing to several doctors. Although these illnesses were "more annoying than life-threatening," one particular infection did become rather serious. The combination of medicine she was taking as well as her depression had made her lethargic. Her husband was noncommunicative, and she had been consistently "rebuffed" whenever she tried to talk to him. He had also consistently refused her requests to attend marital therapy.

Her change began right after her most serious bout with an infection. "I was standing in the kitchen next to my kitchen stove—I remember staring at one of the burners," Claire reported. The "realization hit me that because my life was in such a disarray that I was dying ... not so much physically dying; mentally and emotionally I was dying." She went on: "I realized I had to take any step necessary to change my situation. I felt certain that divorce was the answer." Soon after that realization, she initiated divorce proceedings, and "all my physical problems dissipated ... the medication started to work." She reported that she "healed and felt great within 1 to 2 months." She had not had a similar illness since that time.

In summarizing her change, Claire explained, "I was denying myself an opportunity for true happiness. I couldn't say that I didn't matter anymore—life and being happy was very important to me." She also reported that she realized, "I'm the only one who can create my happiness." She arranged her life to maintain some control over it from that moment on. She recognized that the change was sparked by the initial awareness of her circumstances, which in turn stimulated the presence of other precursors, for instance, a sense of necessity for change. All of this culminated in her determined, focused, and intensely inspiring movement toward change.

CLINICAL MARKERS: HOW TO RECOGNIZE AWARENESS

Awareness of the problem is relatively easy to detect among clients. When awareness is amply present, the client is able to identify thoughts and feelings with ease. Such a client is alert and catches on to where a therapist may be going with a particular technique or series of questions. There will be certainty about the person, a poise that manifests as an air of assurance and confidence. The person may show that assurance even when depressed, discouraged, or stressed. The person will generally respond quickly to queries regarding behaviors, feelings, or thoughts and will often offer helpful details above and beyond what was requested.

However, they do not always change, so change itself should not be looked for as a clinical marker.

In this regard, the aware client can sometimes be among the trickiest of all difficult clients. When working with a client high in awareness, therapists are likely to believe that change is imminent. However, many articulate, self-disclosing, and insightful clients do not take any steps toward change. Such clients may be remarkably skilled at pointing out their issues and explaining their maladies, which can seem like a gift to the inexperienced therapist until it becomes apparent that change is nowhere in sight and may never occur. Some of these clients have been validated by previous therapists for their awareness but not for making changes in their lives. As has been noted, awareness, like the other precursors, is a necessary yet insufficient condition for change by itself.

Alternatively, a person with high awareness may not be particularly verbal or articulate but may demonstrate considerable awareness nonetheless. In such cases, the therapist should help articulate certain details that the client may recognize but need assistance in describing. With an empathic understanding of what the client is experiencing, the therapist may articulate that understanding while simultaneously asking the client to modify or verify, as the case may require.

The certainty accompanying this precursor manifests in the cognitive context as well. When awareness is high, the person can be certain of being uncertain. The person can easily identify thoughts, beliefs, conclusions, schemas, memories, perceptions, and most other cognitive phenomena. They will find it easy to see how cognition affects behavior and feelings and vice versa. In cases of high awareness, thoughts are generally clear, attention is focused, and the person can step back from a problem or complex of emotions relatively easily. This client will also be likely to identify not only the problem but also the circumstances surrounding the problem and, perhaps, the source of the problem. They will likely be adept at assigning or assessing responsibility for a given situation or set of circumstances and be able to identify contributing factors in interpersonal dynamics as well. It is often helpful to observe the degree of detail, coherence, or clarity in the stories the person tells related to the presenting problem. The way a person tells such stories can give clues as to the degree of their awareness.

People with high awareness are not likely to get lost in feelings or emotions. With little prompting or encouragement, they can extract themselves from entanglement in strong emotions and return to viewing the problem or issue. Clients with a high degree of awareness may report a variety of mixed emotions, such as simultaneous love and hate or concurrent sadness, helplessness, and anger. A person with a high degree of awareness will also be likely to accurately and poignantly describe inner conflicts about decisions as well as opposing intentions and purposes.

While aware clients who are not verbally inclined are unlikely to spontaneously articulate experiences, they will readily relate to reframes or reflections made by the therapist. Overall, a client with awareness is every therapist's desire. Even when change is not occurring, session time passes in a deceptively smooth way—a note of caution to not sing the praises of awareness in therapy too loudly.

HOW TO DETECT THE ABSENCE OF AWARENESS

The absence of awareness is one of the most frustrating conditions a therapist encounters outside of deliberate recalcitrance or deception. It is a form of psychological obliviousness, the hallmark of a difficult client. Vagueness and obscurity define the client's relation to their inner and outer worlds. Such clients will be consistently unclear, unsure, and, at worst, oblivious to all that surrounds them. They typically answer the therapist's questions with brief, noncommittal statements like "I don't know" or "I guess." Blaming is common among people lacking in awareness, not so much because of an inclination toward avoidance as because of an inability to properly designate influences and causes.

A client can be burdened by an array of issues and yet be largely unaware of how profoundly those issues negatively affect their life. To such clients, the mind and the world are shrouded in haze and obscurity. An example is the obsessive–compulsive client who repeats and checks behaviors, knowing only that "it has to be done." Another example is the criminal who has spent a lifetime learning to victimize others. His behaviors have been validated and reinforced repeatedly by peers and the environment, to the point where he believes the only improvement needed in his life is in his skills at deception and intimidation. Clients lack awareness in different ways, but in each, the pattern is similar: confusion, ignorance, or dogmatic beliefs dominate. Another indicator of low awareness is when a clearly problematic belief or behavior is regarded by the client as an asset. This is evident when the client takes an "ignorance is bliss" kind of stance.

Additionally, it is common for a client to have acute awareness of problems in some aspects of their life while being oblivious to others. This is sometimes referred to as *splitting* or *compartmentalizing*. For instance, a client may be quite aware of her feelings about a romantic relationship but only dimly aware of the gravitas of her behavior outbursts when arguing with her partner in public. Another client may be sensitive to his partner's emotions and reactions to various situations yet have little awareness of how his own behaviors and beliefs affect his relationship. Another common example of lack of awareness is the employer who is convinced that she is loved by all but is, in reality, poorly regarded by nearly every staff member. Such people are excellent candidates for group therapy since receiving interpersonal feedback on how one is perceived by others is among its greatest therapeutic benefits (Yalom, 1995).

Another important manifestation of the lack of awareness is a general pattern of minimizing, which involves behaving as though a genuinely important event or experience is inconsequential. Minimizing typically occurs when the client claims that a problem behavior such as lying or drinking is of little or no consequence, resulting in accusations that their spouse or employer, for example, are "overreacting," "blowing things out of proportion," or "barking up the wrong tree." Both minimizing and exaggerating a problem can indicate that awareness is limited.

Without the precursor of awareness, a presenting problem will seem vague, overly generalized, or indistinct, and therapy progress will likely be slow, tedious, and difficult. More importantly, therapy techniques are likely to be ineffective to the degree that clarity of the problem is not attained. Without a focal point for techniques, the entire exercise may appear pointless to difficult clients. Thus, it is often wise, in the initial stages, to devote some time to defining and clarifying the presenting complaint, problem, or symptom.

TECHNIQUES: CULTIVATING AWARENESS

If awareness is not present and operative to some degree, the other precursors will lack focus and power in producing change. Therefore, when working with clients low in this precursor, the therapeutic task is to build or cultivate awareness so that confronting and effort resources can be brought to bear on a known problem or concern. Bringing about awareness in clients who receive low ratings in this precursor can be challenging and demands considerable creativity on the part of the therapist. This section contains several approaches to cultivating awareness in the form of metaphors, reframes, and techniques.

Metaphors That Illustrate Lack of Awareness

As with the willingness precursor, metaphors can help a client recognize that a problem or an area of life needs attention, even when the client does not initially perceive the problem. Some of the metaphors that follow are rather graphic, even off-color. However, metaphors that are used effectively will illustrate how a client has overlooked or ignored important issues. Metaphors are also meant to normalize a lack of awareness so the client does not feel criticized. A difficult client will often perceive a direct statement about lack of awareness as an insult delivered by an arrogant therapist. Metaphors help avoid this pitfall when used with tact and timing in the proper context.

Successully using metaphors in this context is limited only by a therapist's wisdom and imagination. We have seen many memorable cases in which therapeutic change emerged through the use of a simple but powerful metaphor. To produce metaphors with great change potential, understanding the client's frame of reference is essential. Clients often provide the best metaphors themselves, whether knowingly or unknowingly, and therapists should draw upon such insights in abundance.

Blind Spots in the Rearview Mirror

This metaphor is helpful for any client who has a driver's license. The therapist tells the client that people do not see everything on the road, and that is why we have rearview mirrors: so we can be more aware of what traffic surrounds us. Unfortunately, rearview mirrors have blind spots. Relying only on the mirrors will not reveal a nearby vehicle at certain angles. Sometimes, we have to turn around and look at the road directly.

Awareness also has blind spots, and to compensate for those blind spots, we sometimes have to turn around and look; that is, inspect our minds—for thoughts, intentions, and feelings—and our environment for what is happening around and within us. This metaphor is effective in getting a client to see that there may be things they are doing or thinking that have negative consequences and to indicate that it is the right time to check things out.

Bad Breath

For some clients, this graphic, albeit crude, metaphor can be highly effective in communicating how a person can have issues or problems they are unaware of. The therapist can ask if the client has ever known someone with bad breath who had no idea of it and needed to be told. The therapist can make the point that all people have blind spots. They can then seek permission to point out a possible blind spot for the client's consideration. If the relationship is reasonably secure, the client will often give permission, even though they may not agree that they have the issue. But by that point, the door has been opened for discussion through further use of the metaphor. Another point the therapist can make is that all of us rely on someone else for feedback at certain times in our lives. In the context of therapy, a therapist's function is to provide that feedback in a trusting, confidential climate.

Body Odor

This metaphor serves the same purpose as the bad breath metaphor and can be used if and when appropriate with clients for whom it is suitable. It can be effective in conveying the idea of how a person can have a problem but lack awareness of it, especially if it affects relationships with others. Body odor can be used as a metaphor for a "behavior odor," that is, a behavior such as being obnoxious or arrogant that can get a person into difficulty, but the person is so used to it that they can no longer "smell it" or be aware of it. It is a primitive but often useful metaphor.

The Mountain Overlook

Part of attaining awareness is to step back to gain a more global view of the situation or problem. A. T. Beck (1976) referred to this psychological act as self-distancing, while proponents of acceptance and commitment therapy discuss it in the context of cognitive defusion (Hayes et al., 2011). It is often needed when a client is so deeply committed to a particular viewpoint that they will not reconsider it. Stepping back can be suggested in therapy with the help of the mountain overlook metaphor. It is easier to understand a city by looking down at it from the top of a mountain than by walking a single street or alley. The therapist can ask the client to step back and provide what might be called "the big picture" of a situation. This process can be enhanced by outlining, mapping, or diagramming the problem to get a more complete perspective. The therapist can sometimes add new aspects to the map or diagram for client consideration as part of attaining an overarching view of the problem or issue.

Finding an Example of an Unaware Person

Nearly everyone knows someone with quirks, odd habits, or idiosyncrasies, of which, it seems, the person is completely unaware. It is sometimes helpful to use this observation to create an opening through which clients low in awareness can take a fresh look at themselves. The first step is to get the client to recall someone who had strange, harmful, or otherwise eccentric habits. Once recalled, it can be pointed out that virtually all of us have quirks we are not aware of. At that point, the therapist can respectfully ask if it is possible that the client might have some habits or actions they are not aware of. The therapist then asks permission to point one out for the client's consideration. This will often help initiate the process of enhancing the awareness precursor.

Reframes That Raise Awareness

Reframes are invaluable when trying to build client awareness of a problem. As a technique, reframes have been praised as a foundational element of all psychotherapies (Barker, 2013). The most effective reframes arise from empathic attunement and can be particularly powerful when formulated in accordance with the client's language, experience, and understanding.

Reframing Negative Behaviors in a Positive Light

Marvin, age 17, was court-ordered into an outpatient treatment program for adolescent criminal offenders after a theft conviction. After only a month in the program, he had acquired the reputation of being difficult and incorrigible. He was described to the consulting therapist as a "compulsive thief and liar." Marvin had little regard for boundaries and felt entitled to own any object he perceived. This client's father, apparently a career criminal, was currently serving time in prison, and his mother's whereabouts were unknown. He was being raised by his favorite uncle, who also appeared to be a habitual criminal offender. Marvin had little or no awareness of the consequences of his actions and would not tolerate any form of discussion about theft or stealing.

It eventually became clear that, as a child, Marvin's father, mother, and uncle consistently validated and encouraged him when he came home with something stolen. Stealing made his parents proud of him. It was a source of pride and esteem for Marvin. Once aware of the historical reinforcement of stealing behaviors, the therapist began to reframe his stealing as "being a good boy." Marvin found this amusing and often laughed out loud when the therapist said it. The therapist asked Marvin if he could use that phrase in place of the word "stealing." Marvin agreed. From his perspective, the reframe was much closer to reality and greatly contributed to the fast formation of a therapeutic relationship.

Eventually, awareness of his problem emerged as he explored the contradictory messages he had received from his parents, schools, courts, and society and the consequences of those contradictions. He also developed awareness of the problem as a multigenerational family issue. Marvin's story is further explored in relation to the confronting precursor.

Awareness as Savoir Faire

For many adolescents and even some adults, awareness can be a foreign term. A solid goal in educating clients on the benefits of awareness is to help them see it as a desirable and worthy pursuit. It can help to point out that many of the people the client admires already possess awareness in some form or another, even if it is in terms of knowing what people want or "what's cool." Reframing awareness as knowing "what's happening" or "what's going down" places it in a more desirable light for many clients. Eventually, the client learns that awareness is related to therapeutic change and leads to better coping strategies and success in life. When awareness is aligned with the client's personal goals, a greater level of hope develops as the future is perceived with increased clarity. The trick is to help the client see that greater awareness is what has really been desired all along anyway.

A Situation Rather Than a Problem

Some difficult clients respond negatively at even the hint that their issue is a problem. Some boldly proclaim, "This is not a problem!" It can be a mistake to assert that a client has a problem, as a common response is for clients to stiffen or become defiant or indignant. Although many clients with a lack of awareness will not admit to a problem, they may be amenable to reframing it as "a situation that needs attention." Thus, terms like "situation," "challenge," or "puzzle" can substitute for "problem" quite nicely. One can introduce the idea by questioning and probing the consequences of the situation if it were to continue unchecked. Of course, the initial response would be a superficial denial, but this will usually diminish to the degree the situation is defined.

Getting Lost in the Thought Stream

People who are subject to compulsive behaviors such as excessive shopping, sexual acting out, checking, and substance use often find themselves caught up in escalating desires and thoughts that lead to repetition of the unwanted behavior. Therapists can help clients enhance their awareness of the phenomenon by reframing these automatic thoughts as being "lost in the thought stream." This can be combined with a metaphor of being caught up in the river rapids of the mind, which carry one away helplessly. The metaphor can then be extended by introducing the idea that it is possible to swim to the shore as well as to remain on the shore of the river, observing the rapids from a safe, calm vantage point.

The client can be guided to identify the pattern of cascading thoughts, desires, and emotions that consume them, sweeping them away in the stream. It also helps for the client to observe and identify the environmental triggers that lead to an onslaught of desires, emotions, and thoughts. The therapist may even, in a controlled fashion, recreate the cascade in the session and help the client "swim to the shore," so to speak. The metaphor of "keeping one's head above water" can further symbolize how the client avoids becoming submerged in the rapids of desires and thoughts and reaches the stability of the shore, which is, of course, sustained awareness.

Confrontation

Confrontation is perhaps as old as therapy itself and requires no explanation other than to say that it is completely different from the confronting precursor. Virtually every school of psychotherapy has a version of it, except perhaps "pure" client-centered therapy in the vein of Brodley (2002), although even therein, it has been argued that confrontation takes place in its own form and fashion (see Kensit, 2000). Regardless, confrontations remain a classically effective way to bring a client to awareness. The key to confrontation is that it be done with compassion and empathy. Otherwise, unnecessary power struggles may occur.

One example of using confrontation to catalyze the awareness precursor involved Jeff, a 24-year-old man who had just completed a year-long prison stint for assault. He was emotionally volatile, with a history of escalating minor conflicts into dangerous situations. During his recent stay at a halfway house, Jeff admitted to planning how to harm an individual who allegedly stole his sandwich out of the refrigerator. He was subsequently referred to an inpatient facility by his parole officer for observation and assessment.

Jeff presented as arrogant and defiant in group work, but he also displayed a keen self-awareness when discussing resentment over perceived injustices during individual sessions. As seen in the following transcription, the individual therapist sought to validate the perceived injustice while tapping into Jeff's self-awareness to encourage a confrontation with how anger controls him:

THERAPIST: It seems like there are times when your anger gets out of control, like the incident with the sandwich. Can you tell me more about what happened there?

CLIENT: (shrugs) Look, I know it sounds stupid, but that sandwich was mine. And when I saw it was gone and found out who took it, something just snapped. I couldn't stop thinking about it.

THERAPIST: Your anger overtook you in that moment. Do you ever feel like your anger controls you, maybe more than you realize?

CLIENT: (pauses) Yeah . . . I mean, I don't know how to stop it. It's like, once something sets me off, I can't stop it. People just need to stop messin' with me.

THERAPIST: Your support team at the halfway house said they found you pacing back and forth in the alleyway with a knife in your hand, cussing and talking to yourself.

CLIENT: It wasn't a knife; it was a scrap of metal. But yeah. I blacked out, man.

THERAPIST: If only people would stop messing with you, then you could control yourself.

CLIENT: Nah, I mean, yeah. You know, I *can* control myself. But people do stupid shit like steal my sandwich, and then they need to learn a lesson.

THERAPIST: Ah, so you can control yourself *and* you wanted to teach that sandwich thief a lesson.

CLIENT: Yeah, something like that. I mean, people can't steal your shit and get away with it. Then they'll always just steal your shit.

THERAPIST: Makes sense. I do have a question for you, though, Jeff. Do you really think planning to shiv someone over a stolen sandwich makes sense? Like, is that really the best way for you to let people know not to steal from you?

CLIENT: (laughing) Uh, no, that's j-cat [crazy]. I'd go back to the clink.

THERAPIST: And you already said that you don't want to go back to prison. You mind if I share a thought with you?

CLIENT: Yeah, what is it?

THERAPIST: You're a smart guy who wants to be respected. You also don't want to go to prison for stabbing a guy over a sandwich. But the fact is, you were pacing around with a shiv and planning to do some j-cat shit because some dude stole your sandwich. Anger overpowers you sometimes, and you're going to land back in prison—over a sandwich—if you don't do something about it. I'd like to help you do something different so your anger isn't running the show ... so you can run the show. What do you think?

CLIENT: (nodding) Yeah, I get it. I've gotta stop raging out. You got ideas?

Isolating Automatic Thoughts

Automatic thoughts are often habitually repeated in a thought stream, with no conscious decision to actively form them (A. T. Beck, 1976; Vago et al., 2022). The phenomenon carried a slightly different connotation for Meichenbaum (1977), who emphasized its dialogical aspect and referred to it as negative self-talk, another commonly used term in therapy. For all practical purposes, automatic thoughts and self-talk are the same phenomenon and can be treated in the same fashion.

With regard to the awareness precursor, therapists should be alert to automatic thoughts that relate to denial of the severity or even the existence of a problem or issue. There are two primary methods of accessing and attending to automatic thoughts. The first is to have a client repeat a dysfunctional statement to the therapist, doing so out loud and with conviction. The second is for the therapist to repeat the statement directly to the client with the same level of conviction. In either case, the client then reports what thoughts arose when the statement was made, and the therapist writes the thoughts down so they may be examined and disputed later.

In the case of a client with a drinking problem, a therapist might choose the second method. The therapist can say to the client,

> I am going to make a statement to you. Please understand that I am not trying to convince you of anything; I just want to know what your immediate reaction is in your mind when I say it. The statement will be, "You have a drinking problem." Please remember that I am not trying to argue or force anything on you. Just tell me what thoughts go through your mind when I say it. Is this okay?

The client, presumably, gives consent. The therapist then repeats the statement with deep conviction and says, "Okay, what happened there?"

In many cases, the typical self-talk reported is something akin to: "My drinking is under control," "I don't have a problem," "I am doing just fine," "Nobody understands," or "I don't want to think about this." Defiant statements mixed with obscenities are also common. Such self-statements are designed to reduce awareness of the problem or issue. Awareness can be increased by focusing on these statements, examining their function in reducing awareness, and then disputing them in the cognitive behavioral fashion. Self-monitoring can be recommended so that clients can observe the operation of these thoughts in everyday life. If the therapist attempts to convince the client that there is a problem, however, the exercise can degrade into an argument.

Localizing Feelings as Sensations in the Body

Many difficult clients are unable to effectively communicate about, or differentiate among, nuanced feelings. Indeed, many such clients feel almost harassed by therapists who insist on working with feelings. According to Loevinger (1976) and other developmentalists, it is much easier for clients in lower developmental stages to identify sensations than emotions and feelings.

Often, clients can identify feelings by pointing to the location in the body where they feel tensions associated with the problem, as seen in Gendlin's (1981) focusing technique. It can be highly effective in enabling clients to eventually identify and verbalize feelings. For example, Joe, firefighter in his late 20s, sought therapy to resolve his shyness around women. When it became apparent that he had great difficulty identifying feelings, his therapist asked, "Joe, as you talk to me about your shyness, do you feel any sensations or tensions in your body?" Joe immediately pointed to his chest just under the sternum and said, "Yeah, I get really tight right here, and it's like I can't say anything."

Other clients report a feeling of tightness or heaviness in the pit of the stomach or in the chest or throat. Although a client may not be up to saying he is sad, he might have no trouble saying that he feels "choked up" in the throat. Once the area of the body is identified, the client can physically outline with the index finger how large the area is in which the sensation is felt. The sensation can then be used in self-monitoring rather than focusing on a feeling of sadness or fear that may still be too vague. From there, it is a natural progression to eventually identify actual emotions, although even then, they may only be felt and articulated in a primitive form.

Role-Plays of Others

Role-playing is part of many theories and schools of therapy and probably originated in the psychodrama school in the first decade of the 20th century (see Moreno, 1946). The adaptation of this old therapy technique is designed to facilitate the basic psychological act of stepping back or self-distancing. Role-plays can be highly effective with difficult clients when they are willing to engage in the procedure. However, many difficult clients dislike role-plays and will not engage until a relationship is well-established.

Awareness of a situation or problem can sometimes be enhanced by having a client do role-plays of people who know them well, such as current or former friends, current or former spouses or romantic partners, and family members. A trusted person is usually the best choice. Speaking as the person who cares about and knows the client well, the client becomes aware of how others view them. This technique can reveal aspects of self or behavior that have not been previously considered in therapy.

The role-play approach is particularly helpful for clients who demonstrate self-centeredness to the point of not knowing what effect they have on those around them. Following the role-play, the therapist can play the role of the trusted person and have the client respond to the earlier statements and observations given when the client played that same role.

The Empty Chair

The empty chair technique can be used as an alternative to role-plays. Typically listed in the category of experiential techniques, it is powerful in building awareness with some clients. It is another old technique that probably originated with psychodrama in the early part of the 20th century, although it is commonly associated with gestalt therapy. Unfortunately, many difficult clients will eschew this approach, steadfastly asserting that they will not talk to "a chair with nothing in it." Nevertheless, other clients will respond to this technique remarkably well in the context of a therapeutic relationship. It is one of those techniques that is best framed as an experiment (Polster & Polster, 1973).

Four modalities of roles qualify as candidates for "sitting" in the empty chair. The first is people currently available and accessible in a client's life. The second is people who have died or are otherwise no longer available. The third modality is subpersonalities or different parts of the self that have not been given a voice. The fourth modality is the client and consists of the current self, the past self (as in childhood), and the future self.

When doing the empty chair, it is important for the client to physically move into the other chair, assuming the identity that has been projected in it and communicating from that perspective. The interactive process continues until increased awareness results. The technique can result in the resolution of inner conflicts, reconciling unresolved issues with persons both dead and alive, and enhanced empathy for others.

Self-Monitoring

Behavioral and existential therapies use forms of self-monitoring to heighten awareness. Many difficult clients are not interested in homework assignments and will not do them. However, the therapist can issue a challenge to them to use awareness gained from a previous therapy session to apply in a self-monitoring context. When a client is only asked to observe themselves, the likelihood of compliance as a measured rate of homework completion increases (Kazantzis et al., 2014).

For example, Peter, a passive-aggressive client in his late 40s, was sarcastic, bitter, and defiant toward authority figures, especially in the workplace. He almost immediately displayed similar behaviors toward his therapist. Even though many of his complaints about his superiors were partially valid, his hostility made his interactions with them difficult.

During a role-play of a work situation, Peter was made aware of the intensity of his hostility toward authority. The therapist repeated the statement, "I need this by Friday," to him in the same condescending manner as his supervisor and processed it between repetitions. Before this role-play, Peter had firmly denied feeling any anger and resentment toward his supervisor, but he found it emerging through this process. Consequently, the therapist asked Peter to observe and monitor his feelings when in the general vicinity of his boss.

Peter soon reported that he had not suspected how much resentment he showed toward even the most benign directives from people in authority positions. Issues that arose between him and the therapist were used as opportunities to study his transference reactions based on his childhood spent with a verbally abusive father. As with many difficult clients, giving Peter homework tasks other than self-monitoring would have been premature and, quite likely, rebuffed.

Establishing the "Observer" or "Wise Mind"

Another tool involves helping a difficult client establish the "observer," or what Linehan (1993) called the "wise mind." Her use of the technique was effective with borderline clients and is a step beyond self-monitoring. This technique awakens the *observing self* (Deikman, 1982; Golubickis et al., 2016), that center of consciousness that is aware of and observes one's own emotions and behaviors even while acting out or giving into needs or demands. It often is present during or just after times of crisis or obsessive–compulsive behaviors (Zerubavel & Messman-Moore, 2015). It is also present when one sees the truth of a particular situation, "knows" the right thing to do despite temptations to the contrary, or when making mistakes in life.

The first step of this approach is to educate the client, letting them know that virtually every person has this capacity and that it is present and can be accessed at almost any moment. The second step is to point out a time when the client was cognizant or "used the observer" in various situations. The third step is to actively access the observer in the therapy session so that its manifestation is by choice, not circumstance. Homework assignments can then be issued in which the client uses the observer in various situations in their current environment. If a client learns this skill, the confronting and awareness precursors can be greatly enhanced.

The Use of Paradox

There are times when instructing a person to continue their lack of awareness can bring about awareness. For example, a 15-year-old girl in treatment for cocaine use was perceived as compulsively manufacturing lies by a treatment team and her fellow group members. She would not, however, admit to having such a problem and repeatedly and loudly declared that she was being unfairly accused. After some discussion, her therapist contracted with her the paradoxical directive that she was not to tell the truth in group. The therapist explained to her that she had to lie about her past, her beliefs, her behaviors, her family, and so forth. After she agreed to try this experiment, her fellow group members were also (necessarily) informed and were glad to help by going along with whatever she said, even asking her questions that required further lying on her part.

The intent of this intervention was for her to achieve an initial awareness of telling lies by consciously and constantly differentiating truth from falsehood, as required by the intervention. In other words, to be a successful liar, she also had to know what was true. Eventually, she asked the therapist if she could start telling the truth because she was "getting tired of lying." She articulated no specific insight, but after six of these group sessions, she reported that her lying was indeed a problem. Subsequently, it was much easier for her to address it as part of treatment, and her mother enthusiastically reported improvement in following up on the intervention's success. It was also a classic example of how insight can both follow and precede behavior change (see Arkowitz, 1989).

Setting and Contrasting Contexts

A problem will seem so only within particular contexts (Bateson, 1979; L. S. Brown, 2018). For instance, alcohol use does not seem a problem at all in the context of the bar and its inebriated inhabitants. In the context of family or workplace, however, alcohol use is usually seen as profoundly problematic. Similarly, some people may justify behaviors in a business context that are questionable in other contexts, such as friendships. If directed toward friends, businesslike behaviors and attitudes may be seen as cold or even cruel. Another obvious comparison arises when contrasting the killing of human beings in the context of war versus the context of murder. Both involve killing, yet the former is socially sanctioned (Shaw, 2015).

A client may come to understand that a problem, issue, or concern is acceptable in a particular context but does not transfer to others. The therapist's task in this strategy is to describe the contexts and then make the dialectical switch. By pointing out that the problem behavior is acceptable and fine in one context, the therapist can then switch the context so the negative aspects of that behavior stand in stark contrast. This approach often results in increased awareness, accompanied by a marked increase in cognitive dissonance.

One way to utilize contrasting contexts involves running groups with substance-using teenagers who insist they do not have a problem. For example, the group leader has members take turns naming the three most important things in their lives. Typically, the teens name friends, family, freedom, and love as the most important. The group leader then asks the group members to explain how using drugs and alcohol affects each one of those important aspects of their lives. This contrasting of contexts draws out many dysfunctional beliefs and behaviors and makes for spirited, animated group sessions and interesting insights. It can be done individually as well.

Acting-Out Theater

Acting-out theater is an awareness approach for groups and is not unlike what might be done in psychodrama. The technique involves designating a stage that the group can observe as an audience. One group member portrays another member's negative attitudes or acting out behaviors without saying who it is. The group's challenge is to guess which member is the one whose behavior is being portrayed. Group members must be reasonably familiar with the behaviors and attitudes of their fellows, or the approach is likely to fail. This method is especially effective with children and adolescents in both inpatient and outpatient settings where there is sufficient contact between members. One of the most interesting aspects of this technique is observing how the person being portrayed protests as the other group members loudly assert that the portrayal was indeed accurate. It has considerable potential to build awareness in clients.

Handling Intentional Unawareness

Some difficult clients deliberately reduce or diminish awareness in order to maintain a problem behavior. Such clients are unlikely to change without addressing this deliberate act. Clients whose presenting problems arise directly from treating others poorly or without compassion or empathy are often unwilling to admit they are unfeeling or insensitive. Many have deliberately dulled their awareness so they can remain convinced that they are good, decent people. These clients directly cause their presenting problems but typically blame them on others. In many clients who are harming those around them, a carefully considered choice seems to take place just outside of their awareness: Never inspect or evaluate harmful behaviors or attitudes. People who make this choice can thus be antisocial without the criminal elements that would normally qualify them for a personality disorder diagnosis.

People who have made this choice believe, usually without being aware of the belief, that if they are not aware of harming others, their selfish behaviors can continue unchecked. They deliberately and even passionately keep their continuous harmful behaviors out of awareness. We have found that many such clients will not improve until this mechanism is exposed or until they admit that their selfishness or self-centeredness is harming others. The executive who believes that he does what is best for the business or company and that he has to be "brutally honest" and direct with employees is one such example. Another is the harsh father who believes that he must be "stern" with his children and wife so he can keep them "disciplined," when in fact he is abusing them. The employee who embezzles property or company funds while rationalizing and complaining about management and how badly she is mistreated at work is another. Such people are unlikely to change because they are dedicated to remaining oblivious.

Getting a client to be cognizant of harmful or insensitive behaviors is difficult in part because the ignorance is intentional. A helpful strategy for reversing this ignorance is the use of paradox. A client can be told that, for 1 day per week, they are to be "deliberately unaware." The therapeutic task is to avoid observing how those around them—family, friends, coworkers, employers, and employees—respond to their actions. In some cases, this will actually spark awareness. In other cases, the therapist can directly challenge the client, courteously and empathically, about whether or not they really want to know how they are affecting others.

Empathy as a Remedy

In the CHANGES model, empathy is viewed not only as a necessary characteristic for an effective therapist but as a mark of mental health and well-being for people in general. Cautela (1996) observed that developing empathy in clients should be a routine aspect of therapy. He referred to the insensitivity of people who regularly harm others as *empathy immunization* (p. 341). Perhaps *empathy inurement* is more apt. Many clients who harm others lack empathy. These clients do not necessarily meet the criteria for antisocial or narcissistic personality disorders, and many have no criminal involvement, but they are similarly cold, callous, and unempathic. They are found in many walks of life, but that empathy deficit is a shared characteristic.

The idea of social interest or community feeling, Adler's most important concept (Adler, 1979; Watts & Bluvshtein, 2020), is based on the idea that a healthy person is one who is sensitive to and identifies with others, extending ultimately to all of humanity. The concept is built around empathy (see Adler, 1956; Buechner, 2023). Because they do not feel the pain of others, people lacking in empathy find it easier to perpetrate selfish acts on those around them. Nonempathic people may be so self-absorbed and limited in the awareness of others that they may be deeply and continuously upsetting those around them without a clue they are doing so. Such a person's road to change is built on becoming aware of how they contribute to so many of their problems. Rehabilitating empathy is one approach to such clients (Hart et al., 2018; Simard et al., 2023).

Perhaps the best approach for helping a person see how they affect others is the feedback received in group therapy. Yalom (1995) indicated that of all the therapeutic factors provided by group therapy, *interpersonal input* was the most important and provided the greatest therapeutic benefit. The feedback provided by group members concerning how unempathetic members affect them is invaluable, and the effective group therapist will take advantage of this at every opportunity. Such feedback can also occur in family and couples therapy. Although it can also be done in individual therapy, it lacks the dramatic impact of having one's insensitivity exposed by a group of peers.

Building Tolerance of Confusion

Another clinical observation of difficult clients lacking in the awareness precursor is that some seem to have a low tolerance for confusion, which inhibits or shuts down the development of awareness. These people can withdraw from or avoid any sort of complexity, disarray, or disorganization, and in that act of withdrawal, they abandon awareness. When the thoughts and events in their minds are in disarray, they are unwilling to explore the mess. For these people, it is helpful to outline, diagram, or draw problem situations, reactions, impulses, and anything else that may be excessively complex or confusing.

Examining Maladaptive Beliefs About Awareness

Several maladaptive beliefs interfere with cultivating awareness. If these can be isolated, identified, and successfully disputed, there may be fewer obstacles that prevent the client from developing awareness. These beliefs are so fundamental that clients cannot be expected to mention them, except perhaps in blurting one out unwittingly. The therapist would need to present each belief to a client for comment to see if they agree with the statement. In addition, the belief should be stated using language familiar to the client. Disputation and replacement are the best means for handling these beliefs. The following statements indicate maladaptive beliefs about awareness:

- "To be aware is to feel pain."
- "Awareness only makes pain more intense."
- "If I am not aware of something, it can't hurt me."
- "Awareness interferes with my fun."
- "If there is no awareness, there is no suffering."
- "Awareness is a risk and a liability."
- "Awareness only reveals how bad I am."
- "Awareness reveals how empty I am."
- "To be aware is to know that I am unlovable."
- "Awareness exposes all my faults."
- "Awareness will make me feel ashamed or guilty."

A CLOSING NOTE ON AWARENESS

Psychotherapy hardly functions without client awareness. Awareness is at the heart of all effective therapies that facilitate therapeutic change, even serving to bind theoretically opposed practices such as existential and behavioral therapies. Therapists can build awareness by calling almost any phenomenon into play for a client to consider or contemplate. At the same time, one of the hallmark challenges of therapy with difficult clients is figuring out how to activate client awareness of a problem when that client has no idea the problem exists. Upon realizing a problem, most clients become motivated to address it. However, if you don't see a problem, what is there to fix?

This is nowhere truer than in the field of drug and alcohol rehabilitation counseling. The techniques in this chapter and the addiction-focused Chapter 15 in this volume are designed to help therapists grow client awareness of problems, as well as the various thoughts, feelings, and beliefs that connect to those problems. The road from obliviousness to awareness is challenging for therapists and clients alike. However, when a client realizes they can actively contemplate nearly anything without having to be ensnared or entrapped by it, a feeling of empowerment emerges that can truly be powerful.

7

N: Necessity for Change

If necessity is the mother of invention, then it is also surely the mother of therapeutic change. A sense of necessity arises when one assesses a situation and determines that certain circumstances, feelings, thoughts, or conditions should not or must not be allowed to proceed further. When a client reaches this point, necessity for change can become a driving force in their life. Although therapeutic change can be a wonderful experience, it is not something that people seek out as a recreational diversion as they do with shopping or sporting events. Therapeutic change almost invariably requires some degree of sacrifice and disruption of routine, as well as an often uncomfortable degree of uncertainty and mystery. Before contemplating therapy, many clients must first consider it necessary. Without that motivational force, difficult clients are particularly unlikely to seriously pursue meaningful change beyond trifling with it or giving it mere lip service.

However, it must be noted that this precursor alone does not have sufficient power to produce change (Hanna, 1995). As a necessary yet insufficient condition for change, a client can have an abundance of necessity and a tremendous amount of urgency yet never accomplish therapeutic change when the other precursors are missing. A recognition of necessity may bring a person to seek therapy and indeed to seek help of almost any type, from fixing a muffler to getting one's computer repaired. However, simply having or recognizing the need is no guarantee that change will occur. It is only an admission or recognition that change is needed. There are six other precursors that work interdependently and in conjunction with necessity. At the same time, therapeutic change is unlikely without the presence of this precursor.

Case examples have been disguised to protect client confidentiality.

https://doi.org/10.1037/0000451-008
Therapeutic Change With Difficult Clients: Precursors and Techniques in the CHANGES Model, Second Edition, by B. D. Wilkinson and F. J. Hanna

HOW A SENSE OF NECESSITY LEADS TO THERAPEUTIC CHANGE

The emergence of a sense of necessity in life is a remarkable phenomenon. People often recall its precise moment of emergence with great clarity, including where it arose and how they responded to it, many years after an incident (Bellaert et al., 2022; Kemp, 2013). A recognition of necessity is more than mere motivation: It is a driving force for personal change and a major factor in many of the great political reforms throughout history (Goshe, 2019; Townshend, 1987).

A sense of necessity is thus not merely an intellectual matter of priorities or values. It has a definite affective component in the form of a felt urgency about the need for change. It also involves recognizing the importance of replacing a set of circumstances with different ones. There is a palpable feeling that the current conditions in one's inner life (e.g., sadness or anxiety) or outer life (e.g., intimidation or threat) are in some way unacceptable if left unattended. The awareness of necessity comes from assessing a situation, whether of emotion, mind, or environment, and declaring that it must be altered or ceased. A feeling of necessity is observed in the pressure felt by a child to get acceptable grades at school or a man reckoning with the need to change his behaviors to save his marriage. Other obvious examples are the desire to rid oneself of depression or the recognition that one's binge drinking must cease.

When someone recognizes that change is necessary and that things must not continue as they are, a remarkable reorganization of the person's perceptions and values takes place. This realignment orients priorities and actions toward the immediate activation of change. The way to change may not be clear, and the means of change may be obscure, but the person is indelibly certain that change must take place sooner rather than later.

When a sense of necessity arises in such a moment, something profound and deeply meaningful takes place. A cognizance dawns that something is desperately wrong and needs to be set to right. This is accompanied by a feeling of urgency, and sometimes even desperation, that indicates change must occur. Part of this feeling of "wrongness" can be viewed as a response to *cognitive dissonance* (Festinger, 1957), which arises as a feeling of discomfort or anxiety when two mutually exclusive or incompatible beliefs or perceptions are present in a person. The discomfort motivates the person to seek a reduction in the level of discomfort. Although such seeking behaviors do not always lead to a sustained change orientation, cognitive dissonance is closely tied to the complex process of therapeutic change (Axsom, 1989; Harmon-Jones, 2019).

When a client displays little or no awareness of necessity, cognitive dissonance is the therapist's best friend. As seen later in this chapter, there are ways to activate cognitive dissonance and, with it, a carefully nurtured sense of necessity. When dissonance arises, the goal is to inspire clients to actively reduce the discomfort via attitude or behavioral change rather than engage in dissonance reduction strategies, such as distraction, forgetting, trivialization, rationalization, or outright denial (McGrath, 2017). Some people deal with cognitive dissonance by seeking to resolve it, whereas others find creative ways to avoid it, particularly when the level of discomfort is insufficient to disturb maladaptive beliefs or attitudes about how life, relationships, and oneself ought to be.

SIGNIFICANT MOMENTS OF CHANGE INVOLVING A SENSE OF NECESSITY

A sense of necessity helps people act. For instance, a 33-year-old battered spouse initially questioned whether leaving her husband was the best decision. She felt there was a necessity for change, but it was not particularly strong. However, when she grasped the damage caused to her children by witnessing their mother being assaulted, her recognition of the necessity for change was sufficiently strengthened, such that she sought out a women's shelter. The woman did not yet value her own well-being enough to consider change necessary, but she considered the welfare of her children to be of the utmost importance. Preserving their safety made all risks tolerable, the threat of suffering notwithstanding.

In another example, a 41-year-old recovering alcoholic informed his therapist that a sense of necessity was crucial on the most important day of his life: The day he decided to seek recovery. Having used drugs and alcohol almost daily for 15 years, he awoke one morning with a broken tooth, stitches in his face, and no recollection of the previous night. His son recounted how he had chased some neighborhood kids who called him "a drunk," slipped while climbing a fence, and fell face-first onto the concrete. As his sons spoke, he felt deep shame and did not want to believe it. He even poured a drink to try to curb the shameful feelings, but something was different. For the first time in his life, he said, "I thought about killing myself and not hurting these people [his family] anymore." The realization that he was hurting his family activated a sense that his belligerent behaviors must end. He reoriented his life toward avoiding alcohol and drugs, put himself into treatment, and regularly attended Alcoholics Anonymous.

CLINICAL MARKERS: HOW TO RECOGNIZE A SENSE OF NECESSITY

Sometimes the easiest way to gauge the presence or absence of necessity is simply to ask the client. Some difficult clients are willing to answer this question honestly and forthrightly, but others obviously may not. If the direct question does not produce satisfactory results or if the client is manipulative or dishonest, therapists can look for certain behavioral markers in clients' nonverbal and verbal communications.

Nonverbal cues can be apparent from the very beginning of therapy, even during an initial telephone conversation. A client who feels a sense of necessity may have a tone of urgency in their voice or demeanor regardless of the problem described. This urgency might take the form of an intense determination or a plea for help. Certain emotions and feelings can also be clinical indicators. The most obvious is a fear that if change does not occur, some tragedy will ensue. In some people, the fear manifests as a frantic preoccupation with the need for change to take place. In other cases, the person broods or mopes. Apprehension, foreboding, or agitation may also manifest with the awareness of necessity for change. Anxiety, of course, is typical, especially when accompanied by cognitive dissonance.

A client with the necessity precursor tends to be alert and attentive during sessions and actively listens to what the therapist says. Such a client may seek clarification of a question or statement by the therapist and often makes intense eye contact, although people of some cultures avoid eye contact, as do many motivated children and adolescents. Posture can be indicative but is unreliable, as some clients have a high level of the necessity precursor but are also suffering from depression or fatigue that delimits their engagement. Additionally, clients with a high sense of necessity will generally attend sessions on time, but therapists must remain cognizant of the fact that childcare needs can be a hindrance for even the most motivated clients.

The clearest indicator of a client's acceptance of necessity is the importance they ascribe, either directly or indirectly, to change. Of course, this assumes that the client is being honest and is not scheming or being manipulative, as in the case of clients attempting to graduate (i.e., escape) from substance use treatment programs. Dishonesty aside, clients are often quite happy to inform a curious therapist about their awareness of necessity for change, despite being difficult in other ways.

If a person believes that change is important or that action needs to be taken when things go awry, this is an indicator that the person will likely assess a specific situation as in need of correction or change. Whether the client's locus of control is internal or external does not affect this precursor; it matters not whether the person considers the problem as being caused by the environment or by their own doing. Thus, a client with a high sense of necessity may still blame another person or institution for their problems. While such a belief may indicate a lack of other precursors, such as awareness and the readiness for anxiety or difficulty, it does not impact their necessity precursor. In fact, knowing what triggered a sense of necessity is technically irrelevant for therapists and clients alike. What matters is that it has been activated.

HOW TO DETECT THE ABSENCE OF A SENSE OF NECESSITY

When a sense of necessity is fully absent, change appears to be entirely unnecessary or a waste of effort and time. Some clients who adamantly deny a need to change have worked hard to get things the way they are in their lives and are not about to allow some therapist to wreak havoc with their carefully ordered, tightly knit system of thought and behavior. This stance can be particularly true for both criminals and executives. Some clients may appear listless, inattentive, or lackadaisical and will be late for or skip sessions altogether. Even if the client's living conditions or problems seem intolerable to the therapist, the client will appear resigned and apathetic, perhaps stating, "It's no big deal," or "There's no need to get excited." The client may make a statement that indicates a core belief is inhibiting this precursor, for instance, "Nothing really matters anyway," and will find little need to form the all-important emotionally charged bond with a therapist (Laska et al., 2014).

Alternatively, some clients low in necessity may vacillate between engagement, ambivalence, and disengagement. A lack of sustained and focused motivation to change is the most obvious indicator of a low awareness of necessity. Although such clients will usually show up for therapy, there is a feeling of tedium in session as they wait for the therapist to change their lives as though with a magic wand. They may put off or avoid completing homework assignments, even if they suggested some degree of interest during the session. In a similar vein, some court-ordered clients will appear engaged and compliant, but their sense of necessity is directed toward graduating from treatment programs and reclaiming their freedom rather than changing their problematic behaviors.

People who have experienced abuse and suffering often equate it with change that is negative and destructive. Consequently, they may unconsciously associate the prospect of any kind of change, positive or negative, with that same pain or historical trauma. Having had enough of change, they will not consider it a worthy pursuit. Others feel undeserving of anything better in their lives and may feel guilt or regret about things they have done. They may erroneously believe that getting better could be dangerous to the people around them, and the only solution is to remain publicly innocuous yet personally miserable—a twisted proof of their goodness. Finally, other clients appear uninterested in change because they have not envisioned its benefits. Such clients often battle hopelessness, becoming so discouraged with life that they believe enacting positive changes will set them up for more disappointment. They have learned that change is threatening and that the familiarity of misery is far better than the uncertainty of change.

TECHNIQUES: INSTILLING NECESSITY

Some therapists like to set goals for therapy with a client in the first or second session. The general wisdom for setting these goals is that the client and the therapist agree on their worthiness and importance. Unfortunately, without a sense of necessity, goal setting is often premature and unlikely to have its desired effect. Any goals initially set with a difficult client may not be genuine or realistic for that client, as they may be going through the motions. While there are tremendous variations in presentation, pattern, and temperament among difficult clients, a fairly reliable commonality is that difficult clients are typically low in this precursor. As a result, a more fruitful approach in the beginning stages of therapy with such clients is to instill an awareness of necessity, increase motivation, and get the client oriented in the general direction of change.

Rate the Necessity for Change From 1 to 10

Once a behavior or feeling that needs change is decided upon, the therapist can ask where, on a scale from 1 to 10 (with 10 representing the most important and urgent thing in one's life and 1 the least important), the client would rate the importance of personal change or solving the presenting problem. The response will indicate the level of urgency the client feels about the problem. This quick, simple rating helps determine how important therapy is to the client and provides a clue to their degree of motivation. If someone else says the client has a problem—for example, a judge or a spouse—they should rate the problem in the same way.

If the client gives a low rating, which is to be expected, the therapist can ask what it would take to get the rating up to, say, 7 on the scale. Do not imply that it should be 7; only ask what would have to happen to get it there. Answers to this question might be, "If I knew I was hurting someone," or "If I thought I would get in trouble." Such answers provide insights into what motivates the client and what they consider important. The next step is to connect the problem to those evaluations of importance, so as to increase the level of necessity.

Find Out if Change Is Important to the Client

For some clients, rating something on a scale of 1–10 is not particularly interesting or appealing. The problem may be too vague, or the person may blame someone else for causing the problem or issue and want the blamed person to solve it. Rating may also seem too much like a game the therapist is making them play, and the client may not be interested in what looks like "jumping through a hoop."

If the client is involuntary or has no presenting problem or issue, it can help if the therapist suggests a few possible examples to determine the kinds of change the client may be interested in. These suggestions can be as mundane as getting in better physical shape or as abstract as being better able to solve problems or make friends. Once a client hints at a desirable change, the therapist can ask if it would be nice to have that change actually happen. If a change is regarded as desirable, the therapist can ask if it is a little important, highly important, or urgent. This is a more casual alternative to the 10-point scale and takes a little more time. In any case, the desired change should not be dismissed, as it can serve as a stepping stone in the direction of change activation. Thus, therapy should be reoriented to help bring about the desired change.

Find Something Important to the Client and Align It With Therapy

In cases of a trace or nonexistent level of necessity, a client might deny any need to work on a problem and display outright apathy about resolving it. If a therapist were to point out a particularly negative problem or situation, the client might shrug and say, "So what?" A classic problem in psychotherapy is how to motivate such a client. For such clients, going directly after change is asking too much. It is much better to find out what is important to the client in their overall worldview. Reflecting meanings can help here (see Chapter 3, this volume). The therapist can then instill a sense of necessity by aligning therapy to whatever is most important to the client.

The Story of Carl

For example, Carl was a difficult client who was suspicious of therapy and lacking in verbal and articulation skills. He sat with his arms crossed and slouched far back in his chair. Carl was 36 years old and had two children with his wife, Ginger. He had worked in an automobile textile factory for 21 years. He claimed to have problems with his marriage. It was clear for the first half hour of the first session that Carl was rigid and unbending in his thinking and convinced he did not have a problem. He was especially low in the necessity and willingness precursors. The dialogue begins with addressing the metalogue:

THERAPIST: Tell me if I'm wrong, but we have been talking for a while now, and it seems you don't care much for any of this therapy stuff?

CLIENT: Yeah, you're right about that.

THERAPIST: I see. So here we are, doing something that you really don't care about, and it's all supposed to be about you anyway. That must be a pretty strange position to be in.

CLIENT: Right again. No offense, but this doesn't interest me much.

THERAPIST: Yeah ... Let me get this straight. You originally said you are here to save your marriage, but really you're just here because your wife demanded it. You really don't expect anything to actually happen.

CLIENT: Well, I guess. She said that I really needed it, so here I am talking to you.

THERAPIST: But you really don't expect much to change.

CLIENT: (looking at the therapist intently but briefly) To be honest, no, I don't.

THERAPIST: So, as far as you see it, you think you're pretty much okay.

CLIENT: Yeah. I mean, I ain't perfect, but she makes a big deal about lots of things that ain't important.

THERAPIST: Like what?

CLIENT: Like I said, it ain't important.

THERAPIST: You don't feel like talking about it.

CLIENT: You got it. (adds a smug smile)

THERAPIST: Okay. I get the message. But it's part of my job to ask questions. And we have to fill up the time somehow. Do you mind if I ask you a question now and then? You obviously know you don't have to answer anything if you don't feel like it.

CLIENT: (with a wave of his hand) Go ahead.

THERAPIST: You seem to be a guy who is pretty sure of himself. There must be something wrong in your life, or you wouldn't be here talking to somebody like me. I have to think that if you didn't care about your marriage, you wouldn't be here at all. Am I wrong?

CLIENT: No. You ain't wrong there.

THERAPIST: So, the marriage is important to you. (Client nods.) My guess is you think your wife is the one with the problems, and she's the one that should be in here.

CLIENT: Yep.

THERAPIST: Okay. But before we get into all that, can I ask you a question, Carl? (Client nods.) How important is it for you to save your marriage? A little? A lot? Give me an idea.

CLIENT: I guess it's pretty important. I don't want to get divorced. I don't want to go through all that bullsh— again.

THERAPIST: So, you've done it before, and it was pretty rough.

CLIENT: Yeah.

THERAPIST: Carl, I'll make you a deal. Tell me what you think. You and I will work on saving your marriage. I can see you went through a pretty rough time before, and you don't want to get divorced again. I might be able to help. Is it worth giving it a shot?

CLIENT: (nodding slowly and deliberately) Okay.

Carl's interest was sparked, and with it, an alignment of purposes arose between client and therapist. What was important to him was now connected to and aligned with therapy.

The Story of Nick

Another example of this approach was in the case of Nick, a 22-year-old client who said that having sex was the most important thing in his life. He had had a few relationships and dozens of one-night stands and said that sex made him feel "great." He was in therapy because of a problem with alcohol but was not interested in working on that at all, despite his employer recommending it. When asked if he ever felt empty after any of those one-night stands, Nick genuinely and openly laughed at the therapist. From his perspective, he had good reason. He was already so empty that sex was one of the few things that could occupy the void in his life. In his narcissism and lack of empathy, he was aware of little else other than his own needs.

Cautiously, the therapist advanced the idea that therapy could help him understand women better. For the first time, his interest was kindled. He had no idea, of course, that the therapist's intent was to help him acquire empathy for women as opposed to exploiting them. He eventually admitted that many women thought he was, in his words, "a jerk," and the therapist helped him to see why. This single inroad led to working on many of Nick's issues, including his alcohol use.

Addressing and Using Subpersonalities

An approach that is often successful in motivating indifferent and apathetic clients has to do with addressing subpersonalities. This approach is based on century-old ideas in psychotherapy that assume that the human personality comprises many separate parts, each containing different attitudes, purposes, and interests. It is not at all the same as multiple personality or dissociative identity disorder. This phenomenon has been described in various ways across the writings of Jung (1934/1969), Assagioli (1965), and James (1890/1981). While the concept of subpersonalities is central to internal family system therapy (Schwartz & Sweezy, 2019), it also plays a role in various transpersonal and humanistic therapies. One of the most effective ways of dealing with apathy is acknowledging and accepting its presence, building empathy, and avoiding power struggles. From there, using a subpersonality can be helpful. The following story illustrates the technique.

The Story of Tracey

Tracey was a 17-year-old high school senior who displayed a majority of the characteristics of borderline personality disorder. She engaged in self-mutilation (cutting and scratching designs into her wrists and forearms with various sharp implements) and reported that it made her feel better. She often stated that life was too depressing to be worth living. She had a history of sexual abuse perpetrated by an uncle and was promiscuous with boys who were clearly interested only in using her. She regularly abused drugs and alcohol and occasionally engaged in binging and purging. She was extremely vigilant for the slightest sign of a lack of caring or presence by the therapist and engaged in intense emotional outbursts at such times. A phrase she used several times to describe her perceived lack of affect was that she felt "empty inside," and it was clear she had little or no sense of self, being almost totally dependent on others for any sense of identity. At the time of the critical session, Tracey was failing three subjects in school.

There were a host of issues on which to focus with Tracey. Unfortunately, she was deeply apathetic about doing anything for herself. Whenever it was suggested to her that she do something for herself or that she might take better care of herself, she merely shrugged her shoulders and said, "Whatever." When asked if she was concerned that her boyfriends might be taking advantage of her desire to be liked, she admitted that she thought this was true and that she hated men, but she showed no awareness of the necessity for change. The "whatever" response appeared again when Tracey was asked if she cared about her failing grades and again when she was asked if she ever worried about what was going to happen to her. This did not appear to be a game or power struggle on her part. It was as though her necessity for change was frozen or paralyzed, as if something vital and dynamic in her personality was missing—a characteristic often seen in personality disorders.

When the critical session occurred, she had had 15 previous therapy sessions. By this time, a fairly strong relationship had been established with a reasonable amount of trust. After seeing no change but recognizing the pain this client routinely experienced, the therapist decided to use a different approach. Being met by indifference and apparent apathy at every turn, the therapist decided to believe this girl was not as dedicated to being as miserable as she appeared. Her wall of apathy was still imposed on the therapist, who struggled to find a way to get around, past, or through it. The therapist decided to try the subpersonality approach.

The therapist said, "Tracey, from what you have told me, it seems like you know things are not quite right in your life and could be much better, but you don't seem to really want to change anything. Is that right?" Tracey smiled and said, "Yeah, I guess so." "Well, can I ask you another question? And if it's okay, I would really like you to think about this one before you answer." She looked at the therapist a bit suspiciously but then agreed. The therapist continued. "Tell me, Tracey, is there a part of you, some tiny little part of you, that worries about what is going to happen to you?" Hesitantly, she asked, "What do you mean?" The therapist replied, "Maybe it's a little voice in the back of your mind or some part of you that you don't pay any attention to, but I wonder if there is that part of yourself that is worried about what is going on in your life."

She looked at the therapist, unsure of this new territory, and said, "Yeah, there are times when I hear a little voice that says things like 'You shouldn't do that' or 'You have to study,' things like that." The therapist was cautiously enthusiastic; it seemed some progress might be occurring. Even her tone was different. "How much of the total you is that part that cares?" Many clients can give percentages—for example, 20% or the like—but this was not her style. She was noncommittal. The therapist took out a piece of paper, drew a circle, and said, "Pretend this circle is like a pie that represents all of your personality. Can you draw a slice of it that shows me how big that part of you is that cares about what will happen to you?"

She drew a wedge that represented 25% of the total. This was, surprisingly, quite a bit, as many difficult clients draw 10% or less, although other clients judged as difficult can estimate up to 40%. "Can I talk to that part of you, Tracey?" She thought about it briefly and said, willingly, "Sure." Sensing that a therapeutic window was opening, the therapist immediately asked this newly identified part of Tracey, "I understand that you are worried about what's going to happen to you; is that true?" She looked at the therapist steadily, but her face seemed a touch softer than before. "I am very worried. I think I do a lot of things that are wrong and get myself into trouble. And like, I know better, but I don't stop it."

"Like, what do you do that's wrong?" Tracey paused and said, "Like what I do with boys, for one thing, and I drink too much and smoke too much weed." This was a brand-new side of Tracey coming out now. "Is it like your life is out of your control?" She looked down, seemingly sad as she agreed, "Yeah." Once again, the therapist tried to see if she was interested in change. "Would you be interested in changing the things that get you in trouble and get more control over your life?" Tracey paused for a moment, looked up at the therapist, and said, "Yeah."

"How about if we work together on this in counseling?" She nodded, and the therapist, gathering momentum, forged ahead. "Help me understand something, Tracey. Remember the pie?" She nodded again. "If the part of you that wants to change is 25% of the whole pie, is it fair to say that the other 75% of you is out of your control?" Tracey said softly, "I never thought about it like that, but yeah, I guess that's about right." The stage was finally set. "How much control would you like to have?" She thought about it for a while, then took the pen and paper and drew another pie circle. This time, she shaded in about 75%. "That's how much," she said, looking intently at the therapist. "So, you want 75% control instead of the 25% control over yourself that you have now?" She looked at the therapist almost longingly and asked, "Can I do that?" The therapist replied, "It is definitely possible, and we can give it our best shot. Are you willing to give it a try?" Tracey looked like she was preparing for a fight. Finally, she said, "Yes."

For the first time, Tracey was a motivated client with a sense of necessity for change. The apathetic Tracey would often emerge, but it was now identified as such. When it did happen, it was relatively easy to access the part of her that she eventually came to call "the real me," which grew in size relative to the rest of the pie. As far as the precursors were concerned, she was now on the rails toward change and was responsive to established therapy approaches. Although she never achieved change in leaps and bounds, she did manage to make steady progress.

The Use of Cognitive Dissonance

Therapists have long known that cognitive dissonance can lead to change (Festinger, 1957; Harmon-Jones, 2019). It begins with the presentation of incongruent, inconsistent, or contradictory information concerning the self, beliefs, behaviors, or lifestyle. This produces anxiety, which is followed by a desire to resolve the incongruency and anxiety. This desire is closely related to a sense of necessity. There are many ways to produce cognitive dissonance; some are innocuous and simple, whereas others are provocative and require more skill and finesse.

Are You Getting What You Want?

An established method of producing cognitive dissonance can be borrowed from reality therapy (Glasser, 1965). It is a three-stage process that begins with asking a client the general question, "What do you want in your life?" This may or may not be related to the presenting problem, if there is one named at all. The answer to this question should not be superficial, such as getting a Ferrari or winning the lottery. For this approach to work, it is important to concentrate on getting a proper reply from the client, that is, one that is somehow related to cognition, affect, behavior, or relationships. A client might say they want happiness, love, work, or more friends.

Once this is established, the second step is to find out what the person is doing in their life relative to the want or desire. For example, a 15-year-old young man may say he wants more friends. Some exploration reveals that he is angry at many people, including teachers, family members, and peers. Further inquiry reveals he is suspicious, untrusting, critical, and abrasive toward those around him and given to temper outbursts as well. It is not surprising that he lacks friends when what he is doing is alienating and antagonizing people.

With this information, the final question is posed to the client in this way: "You have told me how you interact with people—saying mean things and losing your temper. Is what you are doing getting you more friends? Is it working for you?" When the client admits that it is not working, the therapist can ask if the client is willing to explore different ways of making friends that may be more efficient than what he has already tried. On the other hand, if the client is hesitant to say that it is not working, it can be effective to present the problem in a single sentence and wait for a response: "Explain to me how you can make friends by driving people away from you." Assuming that the client really does want to make friends, this approach can be effective in many cases. Once again, producing cognitive dissonance produces discomfort or anxiety, which allows the therapist to point out a well-established way to alleviate that anxiety: therapy.

What Would Happen if Nothing Changed?

Another way of producing cognitive dissonance is by asking a client what would happen if nothing changed, that is, if the client continued to do little or nothing about the issue or problem. This approach assumes the client has at least trace levels of awareness that there is an undesirable situation in their life. The question is whether the problem would stay the same or get worse. Typically, a difficult client will assert that a problem stays the same if unattended. However, if a therapist can establish a pattern of decline or progressive worsening of the problem over time, this can lead to an increase in dissonance and, thus, a sense of necessity for change. For example, if a client consistently blames others for problems that are mostly due to their own actions, ask if the blaming has served to resolve the problem. One can also ask if the blaming makes the client feel better, and if so, how much and for how long. The answer is usually something on the order of "Not so much, and not so long as I'd like."

Spitting in the Client's Soup

A technique called *spitting in the client's soup* is a way of increasing anxiety levels in the interest of change. Adler (see Dinkmeyer et al., 1987) originated this version of raising anxiety levels and gave it its name. This technique is especially useful for clients who boast about disorderly or destructive behavior. There are many situations worthy of this dynamic confrontation technique. Obvious examples include when clients report destroying a coworker's reputation, enticing a friend's spouse into having an affair, or loving competition to the point of always trying to outdo friends or family members.

Spitting in the client's soup requires a considerable amount of skill to be done smoothly, using a dispassionate, casual, and somewhat aloof demeanor while remaining empathic and compassionate. To be successful, the therapist's voice must be free of even the slightest inflections or intonations that hint at resentment, impatience, condemnation, or moral judgment. Because many difficult clients are extremely sensitive to being judged as a bad or wicked person, the therapist's voice must remain steady, curious, and caring. If the slightest trace of the therapist's personal issues becomes evident, many clients will pick up on it and immediately initiate a power struggle. It is also important to remember that the client is never discouraged from, or told not to engage in, the questionable behavior. The purpose is to make a behavior unattractive but not interfere with free choice. The following illustrates this technique.

The Story of Jason

Jason was a 27-year-old factory worker presenting with depression about his marriage and life in general. In the first session, it was clear that Jason was only toying with therapy and minimizing his problem. His rating for the necessity precursor was 1. In the third session, he disclosed that he had cheated on his wife with a number of women, both before and since he had been married. When asked to be more specific, he reported with great certainty that his many transgressions were necessary to his self-confidence and that it was the only thing that could get him out of his depression. He was convinced that his affairs had nothing to do with the difficulties in his marriage, and he maintained that the failing marriage was all his wife's fault.

CLIENT: Look, I'm good at hooking up. Women love me. My wife doesn't love me like that anymore, so I've got to get mine. If you're good at something, you got to get it. I mean, it makes me feel good, you know what I mean? I love sex more than anything.

THERAPIST: So, having these affairs is important to you, almost like a lifesaver for your ego.

CLIENT: (pausing) Exactly ... I guess.

THERAPIST: Just curious, Jason; do these women know you're married?

CLIENT: Of course not.

THERAPIST: What would they say if they knew?

CLIENT: They wouldn't like it, so I don't tell them. C'mon, man! How do you expect me to hook up if I say I'm married?

THERAPIST: So, all these women that you seduce are just like your wife. They want loyalty and faithfulness just like she does.

CLIENT: Yeah (smiling), they're all the same, you know. They want me to be faithful and take care of 'em and all that. What they don't know won't hurt 'em.

THERAPIST: I see. Am I right to say that the pleasures in your life depend on hurting people?

CLIENT: (visibly stiffens at the statement) Well, I wouldn't say it like that. I make women feel real good, you know.

THERAPIST: Jason, I believe that you're a good person who doesn't want to hurt anybody.

CLIENT: Yeah, that's right. I am.

THERAPIST: But I'm also tempted to believe that feeling good is so important to you that you're willing to hurt some nice, decent people just so you can feel good yourself.

CLIENT: (long pause) Nah, man.

THERAPIST: Can I say something else, and you tell me what you think?

CLIENT: Yeah, you're going to tell me anyway.

THERAPIST: If I was mean and uncaring toward the feelings of people who cared about me the way you are toward your wife and these other women, I'd be depressed, same as you.

CLIENT: (shaking head) But it's the only thing that makes me happy.

THERAPIST: I'm not telling you to stop, Jason. I'm wondering if the one thing that makes you happy in one way makes you really unhappy in a bigger way.

CLIENT: (mixed anger and sadness) What, so you're saying I have to stop having sex?

THERAPIST: No, not at all. It's kind of like a man who's stuck on a diet of beef jerky. He has a need for a good steak, but he doesn't know there's such a thing. He knows there has to be something better, but all he ever gets is beef jerky. That might be the case with you. Maybe you never learned how close relationships can make you even happier than sex and that sex in a close relationship can be particularly good.

CLIENT: But hooking up makes me feel better *now*. I don't know what to do about my wife.

THERAPIST: I understand what you're saying, but the question is whether the effort might be worthwhile. If all you know is beef jerky, then you've never had the real pleasure of trying a porterhouse or filet mignon. It could be worth the extra effort.

CLIENT: (with a tone of protest) So, you're saying I need to try to fix things with my wife.

THERAPIST: I'm only asking if any of this makes sense to you, and if so, whether you're willing to try something different and see if it helps with your depression. I'd also like your permission to find out some of your beliefs about women and relationships. These attitudes might be part of your depression. Are you willing to give it a try?

CLIENT: Yeah. I mean, I see what you're saying. We can try it.

With this approach, the therapist instilled a sense of necessity in Jason by spitting in his soup and using a metaphor relevant to his situation. He was more willing to discuss his depression and no longer minimized the affairs to the same degree. The remainder of that session, and the next several, were spent discussing the notion of viable and satisfying relationships and how not having them in one's life can lead to depression. Jason's beliefs about women were examined and linked to his depression and failing marriage. Eventually, he engaged in couples therapy with his wife.

Identify Secondary Gains or Cross-Purposes

For many clients, therapeutic change is in direct conflict and at cross-purposes with other goals. A sense of necessity for change will not emerge if there is a greater necessity to remain the same. The problem, disorder, or issue is retained because it serves a purpose that the client deems important, usually referred to as *secondary gain*. Take, for example, the case of a person who receives worker's compensation for a painful back injury. The person is referred to a clinic that includes therapy for the emotional and psychological aspects of managing the pain. Some of these clients are highly difficult to work with because any decrease in pain threatens their eventual financial settlement. Thus, the client will resist any decrease in perceived pain to eventually gain financial reward, usually through a lawsuit.

Another example of secondary gain is in the case of a client with explosive anger outbursts. Such a client typically recognizes how quickly others comply just to avoid confrontation. Thus, it pays to hold on to and use the angry attitude to maintain control over others. Malingering, or illness used to gain attention and care from others, is an established example of secondary gain.

It is often tricky to determine what purpose the problem, disorder, or issue serves, and many clients find it difficult to identify and discuss secondary gain. It is sometimes up to the therapist to suggest it as such. When identified, it is often helpful and rewarding to weigh the benefits of change against the benefits associated with the secondary gain. It also helps to rate the benefits of change and of remaining the same, and when possible, to find a solution that will bring the benefits of both. If a client sees the irrationality of maintaining a disorder or problem, an awareness of necessity will likely arise that can motivate the person toward change.

Recognizing the Client's Need for Power or Control

Some difficult clients believe that change is tantamount to an admission of defeat and a loss of pride. This belief interferes dramatically with the necessity precursor. Such clients may compete with the therapist to prove they are fundamentally smart and savvy.

There is not necessarily a power struggle here, but there is a double bind. On the one hand, to admit to the need for change brings about a feeling of being stupid and wrong. On the other, to deny the need for change leads to feelings of hopelessness and apathy. It is extremely important for such clients to see any move toward change as their own idea and as brought about under their own power. If a therapist has any need to be acknowledged for great therapeutic ideas or interventions, there will be a higher chance of failure with this type of client. Such clients often take pride in the way they have lived their lives and in the way they have survived hard times. It is difficult for them to admit to making mistakes. The bottom line is they must feel like the final decision to change was their own brilliant, self-determined choice—and so it is.

Giving Voice to Loneliness

Loneliness is a common feature of anxiety and depression that has become increasingly prevalent (Haidt, 2024). There is an aspect of loneliness that can be painful, and some clients are responsive to the idea of loneliness as a kind of hurt due to separation. At some stage, the client who admits to being lonely might be ready for this approach. Some are ready sooner, and some not at all, but it can be worth a try. The prospect of easing the pain of loneliness can be used to motivate some clients to become involved in therapy by increasing a sense of necessity.

Story of Jeannie

Jeannie was a 17-year-old girl in treatment for a meth addiction. She had few or no friends by her own admission, and fellow clients in her treatment program for teens had rejected her due to their belief that she was constantly lying and "fronting" (i.e., putting on a false veneer to impress people). Jeannie was hostile toward peers and defiant toward treatment program personnel.

Nonetheless, her therapist had established a reasonably close therapeutic relationship that allowed for some honest self-disclosure on Jeannie's part. A significant self-disclosure was reporting to the therapist that she was lonely and that although she wanted friends, she didn't know how to make them. She said meth took away all of her negative feelings and made her feel good and believe in herself. But, upon questioning, she admitted that after the meth wore off, her loneliness was much more intense, which in turn, increased her desire to use again at the earliest opportunity.

The therapist asked her if she felt the loneliness somewhere in her body, and Jeannie nodded. The therapist asked her to name the part of the body where she felt it and then point to it. Jeannie pointed at the center of her chest and said, "Here, in my chest." The therapist asked her to outline the area with her index finger and asked, "Does it hurt?" Jeannie nodded again. The therapist asked her to look inside her chest and report what was in there. Jeannie replied, "Nothing." Next, the therapist highlighted that Jeannie had said there was loneliness and hurt there, to which Jeannie reluctantly responded, "It's kind of nothing, but it also feels lonely and hurtful."

At this stage, the therapist, having set up the key question, asked, "If that feeling of loneliness weren't there, would you use meth as much as you do?" Jeannie shook her head. The therapist then stated, "If you could make almost all of that loneliness feeling go away without the drug, would you do it?" Jeannie paused for some time, staring at the floor, and said, "Well, yeah. Can you do that?" The therapist carefully explained that reducing that feeling is the purpose of counseling, and if she agreed, they could make that their goal. Jeannie agreed, and in the following twice-weekly sessions over 60 days, a variety of cognitive and experiential techniques were applied to the point where Jeannie reported a 70% reduction in the loneliness and hurt and a corresponding greatly reduced degree of craving for meth.

Use Feeling Good as a Motivating Force

It is amazing what people will do to reduce suffering. It is one of the major reinforcers in all of life. Going to therapy is just one of the things that people will do to reduce suffering and bring desirable feelings into their lives. Taking illegal drugs is another. One of the hidden promises of therapy has to do with holding out the possibility of feeling good. In itself, it is a rather crude enticement, but it can be persuasive to clients who think in hedonistic terms. If a client can be convinced that therapy may make them feel better, a sense of necessity may well emerge out of dark emotions and bitter attitudes. No promises should be made, of course, but it is certainly fair to mention feeling good as a possible outcome.

Examining Maladaptive Beliefs About Change Itself

Maladaptive beliefs inhibit the necessity precursor for most clients, particularly those who have never verbalized such beliefs. Knowing this, it is sometimes helpful to infer the belief and submit it to the client, awaiting their comment. For example, a therapist could say, "I sometimes get the sense from you that, deep down, you believe that nobody ever gets any better, so why bother. Is this true for you?" Once identified and disputed in classical cognitive therapy fashion, the doors to change, including second-order change, can be opened. Examples of maladaptive beliefs that can inhibit or prevent a sense of necessity are

- "I don't care about anything."
- "I'm fine the way I am."
- "I don't care what happens to me."
- "If I change, I'll feel terrible."
- "I don't deserve anything good, so why even think about it?"
- "Only fools think they can better themselves."
- "There is nothing worth trying for in this world."
- "Nobody ever really gets better, so why bother?"
- "If the problem were to change, it would only get worse."
- "If I changed, it would only mean that [a specific disliked person] was right."
- "Everybody is screwed up, so why should I be any different?"

A CLOSING NOTE ON A SENSE OF NECESSITY

The ancient philosopher Epictetus (ca. 130/1944) held that the recognition of necessity was a guide for human action and a key to human freedom. A sense of necessity may lead a person to seek freedom from many limiting circumstances, although a certain amount of inspiration is necessary. For example, no one in their right mind would put their hand in a fire if prompted. However, if an infant were seen to be wandering into flames, most people would consider it necessary to risk the flames to save the child. Because therapy often involves dealing with painful memories, confusing thoughts, and intimidating situations, people will not generally contemplate such unpleasant things without an awareness of necessity. People tend to avoid pain unless there is some perceived necessity to face it. Therefore, it is often incumbent upon the therapist to learn ways to identify, lean into, and grow a sense of necessity for change with difficult clients.

8

G: Grit, or Willingness to Experience Anxiety or Difficulty

Psychological grit as a personality tendency denotes "perseverance and passion for long-term goals" (Duckworth et al., 2007, p. 1087) and is generally associated with measures of goal-oriented achievement (Lam & Zhou, 2022). Within the precursors model, *grit,* or *the willingness to experience anxiety* or *difficulty,* can be defined as an openness to allowing change processes to occur, accompanied by a preparedness to feel the anxieties, emotions, and difficulties that routinely manifest as a result. This precursor therefore involves a meta-cognitive capacity to tolerate and even welcome emotional pain, confusion, or intimidation. It denotes surrendering to the change process, as grit metaphorically requires rolling up one's sleeves, gritting one's teeth, and preparing to get one's hands dirty. There is also an element of courage in facing the anxiety and difficulty that often comes with both the process and therapeutic results of change.

When this precursor is established, the client acknowledges anxiety rather than evading or shunning it. Further, the client acknowledges the difficulty that comes with change and is willing to tolerate it so that change can take place. Alternatively, grit is often a missing ingredient when therapists are puzzled by unchanging clients. Many difficult clients want change to occur without having to endure anxiety or difficulty along the way. As such, consideration of this factor can aid therapists in understanding why certain difficult clients appear unable to make any lasting progress toward change and in providing suitable approaches and techniques that can help.

Case examples have been disguised to protect client confidentiality.

https://doi.org/10.1037/0000451-009
Therapeutic Change With Difficult Clients: Precursors and Techniques in the CHANGES Model, Second Edition, by B. D. Wilkinson and F. J. Hanna

Grit is the diametric opposite of *defensiveness*, which is usually defined as the effort to avoid anxiety due to some kind of threat. Openness to experience, as opposed to defensiveness, has long been associated with successful outcomes in literature reviews (Hill et al., 2022; Lynch et al., 2015; Orlinsky et al., 1994). If the person is not ready to meet the difficulties inherent in the change process, they will likely balk at any movement in that direction. Frank (1961) originally pointed out that a common factor of any successful therapy involves the client becoming emotionally aroused. This precursor is also related to the "corrective emotional experience" that interpersonal therapists tend to emphasize as a central aspect of change (see Teyber & Teyber, 2014).

ANXIETY OFTEN ACCOMPANIES CHANGE

Introducing change into a rigid mindset often requires some confusion and disorder. The same is true for a mindset that is chaotic and disjointed. Both types of client are likely to resist change, even though they may be in desperate need. Therapeutic change appears to such clients as confusion and disorder, when it is actually a process of establishing a higher degree of order and function and is a natural part of the improvement process.

For example, when one does spring cleaning, disorder manifests in displacing furniture and household items while sweeping and dusting. If the disorder is resisted, the improvement cannot take place or is significantly hindered. Cleaning sometimes initially requires making a mess. Similarly, to an out-of-shape, unconditioned body, the early part of a physical exercise program looks more like pain and stress rather than muscle toning and cardiovascular health.

Likewise, when one's habitual thoughts and behaviors need altering, some mental confusion and disorder are inevitable and often produce anxiety. A mindset filled with dysfunctional beliefs and painful memories is in delicate balance. When one is not secure in the first place, the mere thought of altering one's beliefs or dredging up memories can produce a sense of unease or difficulty that reverberates throughout the entire mental framework. If one fights this process, change does not occur. Thus, in the real world, if a therapist, judge, spouse, probation officer, or teacher requires or recommends a change in thinking or behavior, some people perceive such a recommendation as an insult or attack. This anxiety can arise as a factor in clinical practice. If a client can be educated to recognize and be willing to experience such anxiety and disorder, they will likely be more amenable to the change process.

PHILOSOPHICAL ROOTS

This precursor has some interesting philosophical grounding. Heidegger (1927/1962) was probably the first to describe it and note its importance for change. One of the most important philosophers of the 20th century, Heidegger wrote the classic, deeply insightful book *Being and Time*. Much of this book was highly psychological in nature, and he defined and noted the importance of achieving authenticity as a human being and outlined how to achieve it. Heidegger (1927/1962) used the term *resoluteness* to describe being "ready for anxiety" (p. 343) and emerging from a sense of "lostness" (p. 345) toward authenticity. He said that to be resolute, one has to follow one's own growth path, which often requires going against the grain of empty habits as well as social pressures and demands. He used the term *fallen* to indicate a person who has given in to societal and personal inclinations toward avoidance and disingenuousness. Resoluteness involves recognizing those inclinations and pressures, refusing to be carried away by them, and choosing instead to move steadfastly toward being authentic. "Resoluteness" would be quite acceptable as the name for this precursor if it were slightly more descriptive.

OWNING THE PROBLEM

It has often been noted in the psychotherapy literature that a client will begin to make progress when they "own" the problem and stop complaining about it or blaming others for it. A person owns the problem when they acknowledge and accept that it is real, that it must be addressed, and that it is their own responsibility to do so. *Responsibility* can be defined as the willingness to take charge, control, or command of any memory, thought, image, problem, behavior, relationship, or feeling. When a person takes responsibility for a problem or issue, they are far more likely to see the issue through to its resolution Acceptance of a problem or issue is central to 12-step programs, with the Alcoholics Anonymous (2012) literature citing acceptance as a powerful influence on abstinence and recovery.

When a client has a sense of necessity but is not willing to experience anxiety or difficulty, the client will typically want the therapist to do most of the work. It is like knowing that the house is in dire need of cleaning but not being willing to go to all the trouble to clean it. Such a client will dump rather than describe their issues and problems. *Dumping* is characterized by exaggerating the problem's severity, a desire to be rescued from the problem, and very little willingness to own the problem or issue. Such clients are avoidant of important or key issues and prefer to keep things at a superficial level. They may claim that therapy is not helping and blame the therapist for the lack of change. Deep down, these clients are experiencing pain and discomfort and want therapy to provide the magic wand to make it all go away. They may even mistakenly believe that therapy is designed and intended to enable them to avoid or eschew pain and discomfort.

HOW GRIT LEADS TO CHANGE

As a willingness or readiness to experience anxiety or difficulty, grit involves surrendering to the process of change itself. The person lets go of established routines and habitual behaviors or thought patterns and is open to the possibility of forever altering them. But the key contribution to change is delving headlong into the unknown. With therapeutic change—and especially second-order change—awareness of the unknown causes no small amount of anxiety.

Thus, the willingness to admit oneself into the disarray that comes with the disruption of established and settled mindsets and behaviors is an essential act in the change process. For change to occur, a client must be willing to "give up or modify clinging to the past" (Strupp, 1988, p. 78) and welcome the new or unfamiliar. Negative feelings and anxieties must be acknowledged as they arise. Over time, the client learns to assimilate and integrate problem experiences under new headings or schema, so what was once disordered becomes understandable, no longer rejected but recognized, acknowledged, and accepted (Tedeschi & Moore, 2021).

The psychodynamic term "working through" has become a household expression. Many things can be worked through, and it is grit that facilitates the process. The presence of this precursor readies the person to reduce hesitance, break through self-imposed limits, and take command of the uncomfortable emotions that result. This precursor does not cause these acts, which will become clear when we examine the precursor of effort or will. However, just as will can lead one to take such "risky" actions as going white-water rafting or riding a giant roller-coaster, the willingness to experience anxiety or difficulty also involves engaging in risk.

When this precursor is abundant, it can produce a sense of adventure that makes the change endeavor not only possible but pleasurable. Even when a client is examining the effects of, for example, experiences of failure, physical abuse, or destructive behaviors, this precursor provides the opportunity for exploration and discovery that often marks the chief difference between co-operative and difficult clients. The latter have to be coaxed and persuaded to engage in the process. The former often have to be held back and paced.

For instance, a suicidal, highly rigid client was referred by a hospital after discharge from its psychiatric unit. When she arrived, she said to her therapist, "I know I need to change, so tell me what to do. I'll do it." Her necessity precursor was high, but her willingness to experience anxiety was low. She was rigid and avoidant, often telling the therapist that she did not want to discuss certain topics. At such times, the therapist had to remind the client of her initial willingness to "do anything." In effect, the therapist consistently asked her to surrender to the process, which brought the client back to the task at hand and eventually paved the way for systematic desensitization work related to her panic attacks. Such a case highlights how a client's awareness of necessity can be effectively used to establish a willingness or readiness for anxiety or difficulty.

The discussion so far on grit as a willingness to experience anxiety or difficulty has described the change process from the perspective of inner experience. Worldly or environmental perspectives must also be tolerated. For example, a passive, submissive client had a considerable degree of anxiety when she contemplated confronting a friend who was ridiculing her. The anxiety was so paralyzing that it completely inhibited the client from acting to stop the ridicule. Yet, if she did not manage her anxiety and cope with this problem, the situation might have continued indefinitely. The willingness to experience anxiety often opens the door for the confronting precursor.

Another example is a woman who was compulsively shopping, running up credit card bills to the point of bankruptcy. Since compulsive shopping is usually accompanied by the secondary gain of fun and anxiety reduction, the mere thought of imposing some discipline on spending was an action that flooded the client with mixed feelings. Those feelings included anxiety and shame about spending too much and an unwillingness to lose the fun and satisfaction of spending money. She began to feel emptiness and despair at the thought of no more shopping. For change to manifest, she needed to tolerate those feelings of anxiety, shame, and emptiness and to manage them so that self-regulation and self-management could take effect.

SIGNIFICANT MOMENTS OF CHANGE INVOLVING GRIT

Vanessa, a 32-year-old woman, and one of her two children were being severely emotionally and sometimes physically abused by her husband. In great fear for years, she finally decided that she must leave him and did so unannounced during the day while he was at work. She moved herself and her children to a town 600 miles away without his knowledge.

He traced her and followed her to that city, and he repeatedly threatened by telephone to shoot and kill her with one of his many firearms. He called her regularly to inform her of her daily activities and tell her that she was in the crosshairs of his rifle. He also followed her to her job and falsely reported to her employer by telephone that he was an FBI agent and that she was about to be arrested for selling drugs, and requested their cooperation. He had flowers sent to her place of employment charged to her own credit card. He also stole her car and vandalized the cars of two friends who were visiting her at the time. Finally, she contacted the FBI, who recorded some of his death threats, and he fled and had not been in contact with her for more than 10 years.

According to Vanessa, this man had her immobilized until she realized that death was the source of her fears. She realized that she had to be willing to face death itself to escape this man. As she put it, "I was pushed to the brink in facing my fears." She described herself as having been a person who worried a lot, and she reported that she had since become "very low key" and had learned to "live in the present." Through such processes as enduring the situation, not reacting in the way her husband expected, working with the police and the FBI, and informing her employer of the situation, she acquired "inner control" as she withstood the threats and phone calls. She said, "I'm not the same person. I learned to survive. [The experience] altered the way I am today. Because of that horribleness, I'm a better person." She reported that this was the most valuable experience of her life in terms of lessons learned and coming to terms with death.

CLINICAL MARKERS: HOW TO RECOGNIZE GRIT

Although grit is likely the most difficult of the precursors to gauge or clinically assess, there are some signs that can discern its presence. Like a sense of necessity, this precursor can be detected by asking clients to assess its presence in themselves. Once the client has described the nature of the desired change, the therapist can ask a simple question such as, "Are you willing to go through the difficulty necessary to make this change?" If the client is not being manipulative or attempting to deceive the therapist, chances are the therapist will get an answer that can help with the assessment. Many clients are able and willing to inform a therapist about their willingness to experience turbulent affect or painful memories.

One of this precursor's most obvious behavioral manifestations is an eagerness to "dive in" to the problem or issue. The client will display a markedly high level of tolerance of pain or discomfort as if they know that one "has to get through this" and that this is what has to be done. Such clients are active and cooperative during sessions and convey to the therapist that they are willing to suffer through tough issues without backing off. In many respects, these clients will appear courageous, their resoluteness often evoking the admiration and respect of their therapists.

Intrinsic to the mindset of this precursor is the belief that problems are solved by holding one's ground and not avoiding or running away. Clients with this precursor believe that positive change is worth the sacrifice and difficulty that comes with it. Conversely, the belief that anxiety and discomfort are always to be avoided directly negates the operation of this precursor. Another cognitive aspect of this precursor is the belief that opening oneself to the unknown can have positive results. A client who is willing to experience anxiety or difficulty is more likely to explore the unknown by probing the recesses of the mind or testing new behaviors in new environments.

A person high in this precursor generally believes that taking risks is a valuable and important part of living life and achieving one's goals. Such a person believes the old saying "nothing ventured, nothing gained" and lives by it from a psychological perspective. In terms of anxiety, a person with an ample amount of this precursor believes that it is not intimidating or intolerable but something that is a source of new lessons to be learned and new self-knowledge to be gained. At its most optimum levels, a person views anxiety not as a threat but as a challenge that holds the promise of a reward of insight and enhanced understanding.

One of the most important affective markers is a mastery of fear. The client may be intimidated by the prospect of change and may even express outright fear of it, but the fear is not so intimidating that it inhibits the act of enduring discomfort. Another way this precursor might manifest amid difficult therapeutic moments is as perseverance. The client will persist in the task at hand regardless of the anxiety aroused by the prospect of becoming aware of painful issues or problems.

Ideally, a person with a willingness or readiness to experience anxiety or difficulty is also able to express emotion in a way that results in catharsis. The person is aware of various feeling states and is close to the inner source of emotions and feelings. This precursor also allows a person to explore emotions directly as an activity in itself and not only as a means of isolating thoughts and behaviors. Some people may be more willing to explore certain emotions at the expense of others.

HOW TO DETECT THE ABSENCE OF GRIT

When the willingness or readiness to experience anxiety or difficulty is absent, the client may believe that therapy is too difficult or that the therapist, however reasonable, is too demanding. Such a client may be inclined to miss or be late for sessions, not so much because a sense of necessity is lacking but because they wish to avoid the inherent difficulties connected with change. Thus, avoidance is a prominent and observable feature when this precursor is lacking.

An absence of grit can drain change potential from the therapeutic encounter. Just because a client is willing to talk about a problem does not mean they are willing to experience the anxiety or difficulty associated with it. Some clients approach problems with an emotional detachment that shields them from, or filters out, the hurt and pain connected with an issue so they can talk about it with some composure. Specifically, when the talk is not "curing," the person often seems remote, even if they may be intellectually active and engaged in the analysis of the problem. The remoteness comes from a hesitance to experience the emotions or feelings aroused by the issue. Such emotional self-distancing from a problem is often accompanied by a distancing from the therapist (Kross & Ayduk, 2017). A solid relationship is crucial to closing this distance.

In the case of substance use treatment programs, some clients learn the proper treatment language and use it to tell therapists and counselors exactly what they want to hear. Such clients' ostensibly honest and heartfelt testimony is mere lip service designed to manipulate the staff into granting early, undeserved graduation. In such cases, a pivotal missing ingredient is assessment of the willingness to experience anxiety or difficulty.

Other difficult clients are willing to acknowledge that a so-called problem is there but insist that there is "no need to get into it." They will insist that the problem can be solved without much effort. A classic example is the alcoholic who says, "I can quit drinking any time I want." Such clients are not willing to admit how difficult abstinence is or to experience the anxieties and discomfort involved. Of course, it may also be true that the person does not know how difficult it may be, in which case the next precursor, awareness, comes into play.

When a willingness or readiness to experience anxiety or difficulty is missing, a client may become evasive with the therapist and may resist probing questions about troubled areas. The client may divert the therapist's questions toward "safer" topics on which to focus or elaborate. Often, the client is not aware of such diversions, but at other times, the client is aware, and questioning them about such diversions is perfectly acceptable and often productive. At other times, the client becomes annoyed or irritable and engages in rationalizations in the honest belief that the therapist is causing them needless pain and anxiety. This indicates a classic mindset or belief system that maintains that anything worth pursuing should not be uncomfortable or painful.

TECHNIQUES: ESTABLISHING GRIT

A client with a high sense of necessity but no willingness to experience anxiety or difficulty will probably want the therapist to do the work. This section describes how to help clients prepare for the work that only they can do. Some clients need preparation for therapy and change, much like athletes need to warm up and mentally prepare for meets, tournaments, and games. Therapy can involve something akin to hardship, and working through issues requires some degree of "getting in shape" to deal with it. Dealing with difficult people, marital turmoil, and emotional pain requires determination. In promoting and building grit, therapy comes close to resembling active coaching.

An effective therapist can inspire a client, through persuasion, to work through difficult, painful, anxiety-ridden problems. Through the empathic connection, therapists can promote this precursor without being coercive; surpassing the tolerance limits or threshold of anxiety can be harmful and discourages a person from further work in therapy. This chapter catalogs techniques and strategies that can inspire a person to recognize and be willing to tolerate and experience the anxiety and difficulty that come with therapeutic change. The use of metaphors is a major approach to increasing this precursor.

Metaphors That Illustrate Grit

Many possible metaphors or reframes can help point out the importance of grit as a willingness to experience anxiety or difficulty in the therapeutic change process. All too often, difficult clients simply do not know that this precursor is helpful and can be enhanced. Some may see this precursor as threatening. Thus, education is often essential when seeking to activate grit as a willingness or readiness to experience anxiety or difficulty.

No Pain, No Gain: The Workout Metaphor

The workout metaphor essentially states that one gets out of therapy what one puts into it. In the world of physical conditioning, weight lifting, and bodybuilding, a common phrase is, "No pain, no gain." It is often advised that the muscles have to "burn" or be sufficiently stressed to grow and develop. Many clients can readily understand and relate to this perspective. One example was a 20-year-old bulimic woman who was regularly binging and purging. Like many people with this problem, she worked out furiously for 2–4 hours a day in an effort to lose weight. She knew and operated on the "no pain, no gain" principle.

However, when it came to looking at psychological issues, she was unwilling to tolerate anxiety to any marked degree. In fact, it was soon established that whenever she felt the anxiety or edge of her emotional pain, she was off to the gym or binging on junk food. She was able to recognize this pattern with relative ease. Her understanding of therapy and the therapeutic process shifted dramatically when it was explained to her that "no pain, no gain" also applied to therapy. The muscles burn in physical exercise, she was told, just like anxiety "burns" in mental therapy. In either case, the burn is important and necessary for growth, but it should not be overdone. In her case, the therapist promised not to deliberately ask her to confront painful memories or feelings unless she was willing and ready.

Workout Machines

Along these same lines, many people buy abdominal exercisers or various other workout machines, knowing they have to change some part of their body or health. Unfortunately, in so many cases, those machines are not used because many people are not willing to experience the difficulty that comes with actually exercising. As a result, the machines are bought with the best intentions but collect dust because of the person's lack of willingness to experience the discomfort that accompanies exercise. Similarly, people can pay for therapy and "have it around" for years, but they may not ever really use it to make changes in their lives. Thus, therapy gets wasted and "collects dust," just like those workout machines.

The Stuffed Closet

The stuffed closet is akin to the cleaning house metaphor. Clients who report that they do not like to think about their problems lack grit as a willingness to experience anxiety or difficulty. The problem can be likened to a closet that is packed full of "stuff": Every time the client wants to avoid something, they throw it into a mental closet. Eventually, the closet gets so full that the door will not close properly, and the client is constantly trying to cram the door shut. Eventually, the closet overflows and spills painful material into the rest of the house.

Alternately, the mind can be likened to a house with several rooms and closets. Some people stuff one mental room after another full of painful issues and difficult problems. They continue to ignore them until there are so many rooms full that the person has only a fraction of their mind left in which it is comfortable to live. Guests are also kept out of those rooms. Some people spend the bulk of their mental life keeping the doors to those rooms closed so they do not have to think about what is in there. Anxiety builds, and whatever peace of mind or contentedness the person may have had begins to fade progressively. As the metaphor goes, if a person were to take the time to sort out all the stuff, they would eventually be able to breathe easier and be more comfortable. The willingness to experience anxiety or difficulty has to do with being willing to go through and clean out all the junk and get the house in order.

Driving With the Brakes On

Another metaphor illustrates how some people want change to occur but inhibit themselves in various ways. A therapist can inform a client how research shows that experiencing emotions and anxiety is conducive to the change process. But being willing only to talk about things and not to actually experience the anxiety connected with those issues is rather like driving with the brakes on. The car might move along in jerky motions, or it might not move at all. Furthermore, after a while, the brakes begin to wear out from overuse. Similarly, a person who resists the pain while talking about it might go through years of therapy but may not ever arrive at any viable or lasting change, ending up exhausted after fighting the anxiety over the years.

Washing Hands With Gloves On

Talking about issues in therapy without ever getting into the experiential aspects is like washing one's hands while wearing gloves. The washing is good, but the hands never get clean. The problem never gets contacted or worked through because it is never really experienced. Eventually, it becomes time to take the gloves off.

To Properly Clean, You Have to Get Dirty

When one cleans, if the job is challenging, one is bound to come into contact with dirt and grime. The paradox is that cleaning up requires looking for, exposing, stirring up, and scrubbing away dirt and dust. To clean, you have to get dirty, and the same is often true in therapy. Therapy can be viewed as a cleansing of the mind, feelings, and behaviors. Things can get messy with emotions and beliefs, but that is part of any cleaning process. One can hire someone to clean the house and yard, but in therapy, only the person can do the work; no one else has direct access to another's mind, feelings, and behaviors.

Old Pipes and Dirty Water

For a person who is hesitant to touch or come into contact with what phenomenologically appears as messy—painful emotions and feelings—a metaphor of old, rusty pipes can be used. When the pipes have not been used for a long time and the faucet is turned on, the water will flow a dirty brown color, but if the water is allowed to flow, it will eventually turn clear and clean. The same is true for those "messy" emotions and feelings once they begin to be experienced, processed, and discussed.

It Takes Guts

The guts metaphor is ideally suited for many defiant adolescents who are suspicious of and ridicule the idea of counseling and therapy. With this metaphor, therapy is framed as an exercise in courage. Many difficult clients might respond to a statement like, "Anyone can ignore their thoughts and feelings, but it takes guts to be honest about them and not back off from what you are really about." A variation designed to increase this precursor with defiant, aggressive adolescents might be, "Anybody can go and punch someone that pisses them off, but it takes real guts to face up to your feelings of anger and hurt."

Identify the Internal Dialogue That Avoids Anxiety

This cognitive-behavioral strategy can effectively remove obstacles to being open to experience. If the therapist knows a possible issue for change, they can ask the client to just consider making the change in question, and then to describe any automatic thoughts or self-talk associated with the change. If the client is especially difficult, they can first be told, "Please do not make the change; I just want to ask you something related to it." Usually, the client will agree to this quasiparadoxical approach. Also, when clients are hesitant to experience anxiety or difficulty, they are prone to deflect questions and change topics. It is often essential to repeat questions and, if necessary, get prior permission to interrupt a client's rambling on other less disturbing topics.

For example, Tony, a man in his late 40s, admitted that he liked to fight with his wife, Sara, to get her agitated and upset. He knew this was mean and even cruel, but he said he enjoyed it and did not know why. Sara, meanwhile, was extremely angry with him and knew all too well that he enjoyed it. She told him she would not take this kind of treatment any longer. This client had not responded to any approach to changing this behavior, and the CHANGES assessment indicated a rating of 0 or 1 in the willingness precursor:

THERAPIST: I'm going to ask you only to think about changing your "agitating behavior" with your wife. Please don't change it or stop it. Once you are considering the change, I'll ask you a question or two about it. Is this okay with you?

CLIENT: Sure.

THERAPIST: Get the idea of supporting your wife's peace of mind and happiness, and tell me when you have it. (Client nods.)

THERAPIST: What do you feel like at this moment as you think about it?

CLIENT: It's okay with me.

THERAPIST: Didn't you just tell me that you like to get her agitated and upset?

CLIENT: Yeah, I guess I did, but it's not like I don't want her to be happy.

THERAPIST: Of course. Do you ever start a fight with her when she's sad?

CLIENT: No.

THERAPIST: Do you ever start a fight with her when she is angry about something?

CLIENT: No.

THERAPIST: Let me bounce something off you, and tell me if I am wrong. When Sara is happy, does it kind of bug you in a way? Does it sort of get under your skin, and you have this urge to mess with it?

CLIENT: Well, yeah. It's like I get kind of mischievous. (smiling and looking for agreement)

THERAPIST: I'm going to say something very direct to you, and I want your permission to do so. If you want me to sugarcoat it, I will.

CLIENT: Go on, tell me straight.

THERAPIST: It sounds like one of the roles you play in your marriage is to keep your wife from ever knowing happiness and joy. What do you think?

CLIENT: You make me sound like a real ass.

THERAPIST: I'm not saying that at all. I'm trying to understand your role in the marriage.

CLIENT: (irritated) I work hard, long hours and do lots of overtime. I keep the house and yard together.

THERAPIST: No doubt, and I bet you're good at what you do.

CLIENT: Damn right.

THERAPIST: Tony, I'm just trying to understand what's going on here between you and Sara, and pardon me for saying so, but it sounds like you don't want to answer me about not wanting her to be happy. Do you want me to repeat it?

CLIENT: I know what you said. (pauses ... silence) I feel kind of ashamed about it, you know.

THERAPIST: Did that feeling of shame get in the way when I asked you earlier to tell me your thoughts about your wife and your marriage?

CLIENT: But you don't know how much she bugs me. She's all the time acting like she knows everything and it drives me nuts. (Client continues in this vein for 2 or 3 minutes.)

THERAPIST: Can this be one of those times when I can interrupt you?

CLIENT: (mildly surprised) Yeah.

THERAPIST: I noticed that when I brought up that feeling of shame, you started talking about Sara instead of your feelings. Do you find it uncomfortable to talk about that feeling of shame?

CLIENT: (shifting in the chair) I don't know, whatever. Can we talk about something else?

THERAPIST: Sure. For now, though, if you can give me a yes or no, it would help me understand where you're coming from. Do you not like to talk about those kinds of feelings?

CLIENT: Yeah, I don't. It's like I just don't want to think about it.

THERAPIST: And it's easier to complain about her?

CLIENT: Maybe.

THERAPIST: What do you think would happen if you got into that feeling of shame?

CLIENT: I'd end up feeling down and depressed, and I don't feel like getting into all that.

THERAPIST: So, you believe that if you ever come into contact with or get into your feelings, you'll end up getting depressed?

CLIENT: Well, yeah, wouldn't you?

At this point, a dysfunctional core belief was isolated that obviously interfered with the therapeutic process and specifically the willingness precursor. Tony believed that contacting or experiencing his feelings would result in depression. The therapist disputed the belief and used the "old pipes and dirty water" metaphor to help Tony understand that contacting feelings is not a formula for depression but can be part of a way to alleviate his depression.

In addition, the therapist asked Tony if he had a habitual pattern of avoiding difficult feelings, as he had just displayed in therapy. He said that he did, indeed, all his life, and when asked, he said he thought this was what one "was supposed to do" with such feelings. The stuffed closet metaphor was then used to advantage with Tony to illustrate how he had stuffed so many aspects of his life into a mental closet that the door could no longer be locked shut, and "material" was seeping out from the cracks around the door. It took one session to establish this precursor with Tony, moving him from a rating of 0 to 2 on the scale. He was more open to emotional experiences after this and much less avoidant.

Typically, when clients are ready to experience anxiety or difficulty, they find therapy to be a much more tolerable and perhaps even rewarding endeavor. In Tony's case, he eventually explored his desire to prevent Sara from being happy. Therapy went relatively smoothly once he was willing to experience the anxiety that came with facing the incongruence between his self-image as a fun-loving, nice guy and his behavior of interfering with the happiness of the woman he married.

The Use of Paradox

Research findings on paradoxical techniques have shown them to be highly effective (Browning & Hull, 2021; Peluso & Freund, 2023). Paradox can help develop several of the precursors, including the willingness to experience anxiety or difficulty. For example, Cindy was a woman in her mid-40s who was extremely sensitive to emotional pain. She would consistently deflect attempts to discuss anything unpleasant or anxiety provoking. Her conversation seemed geared more toward making the therapist like her and think that she was acceptable as a human being than it was toward alleviating her presenting problem of depression.

The therapist attempted to validate and reassure her of her value and intelligence in many ways. After the initial relationship was formed, the therapist explained to her that therapy sometimes required difficult topics to be addressed and discussed. She said she knew this but that it was "really hard" for her. The therapist asked her what was so hard about doing so. She said, "I don't want to hurt anymore." Using the old pipes and dirty water metaphor helped a bit, but she was still extremely sensitive to any further pain, even though she was intellectually aware that she would eventually need to deal with it. Sensing her fragility in this regard, it was clear that to expose her to painful emotions at this point in therapy would be premature. The therapist decided to introduce a paradoxical technique:

THERAPIST: Cindy, I understand that you don't want to hurt anymore, and that's why you don't want to talk about anything uncomfortable.

CLIENT: It's so hard, and I just don't have the energy anymore.

THERAPIST: What would you prefer to talk about?

CLIENT: You mean, anything?

THERAPIST: Yes.

CLIENT: I like to talk about the novels I read. I also like to talk about stores and sales.

THERAPIST: Would you like to try an experiment?

CLIENT: What is it?

THERAPIST: Let's talk about any of the things you mentioned but with one added ingredient.

CLIENT: What's that?

THERAPIST: How about if you talk about what you find enjoyable and, during the entire time, deliberately avoid discussing anything painful or uncomfortable. Are you willing to try it?

CLIENT: Sure.

THERAPIST: Please remember, you can stop this at any time if you want.

Cindy was soon talking about books, and then about friends and neighbors and a variety of other topics. This went on for 10 minutes, with the therapist acknowledging, reflecting, and asking an occasional question. At two points in the conversation, the therapist interjected a question verifying that Cindy was still deliberately avoiding anything painful or uncomfortable. She said that she was.

THERAPIST: Okay, Cindy, you have been telling me about all kinds of things for about 10 minutes now. Did you notice anything while you were deliberately not talking about anything painful or uncomfortable?

CLIENT: Yes (speaking softly). It's the same thing that I do all the time anyway. I've been through so much in my life that it seems whenever I talk about anything, I'm trying to not think about the bad things that have happened to me. (begins to cry) I don't know if I can ever feel better.

THERAPIST: So, what you are saying is that your emotional hurt is so close to the surface that it's always there to some degree, even when you try to forget about it.

CLIENT: (softly crying) Yes. It's always been like that. A lot of the things I enjoy doing I'm really doing because I'm trying to forget about the bad things that have happened to me.

THERAPIST: Has it worked for you?

CLIENT: What do you mean?

THERAPIST: Have you been able to forget about the bad things by avoiding them in your actions and behaviors?

CLIENT: Not really.

THERAPIST: Would you be willing to try a different approach? Therapy exists so that feelings of hurt and pain can be diminished to the point of not having them on your mind all the time.

CLIENT: I guess I knew that.

THERAPIST: True, but it didn't seem you were willing to give it a try. Am I wrong here?

CLIENT: No. It's just that it's so hard.

THERAPIST: What if we only go a little bit at a time, and you tell me when it's getting too intense and then I'll back off as soon as you tell me.

CLIENT: That would be okay. I just don't want to cry all the time. I've done so much of that, and I'm afraid it's what you want me to do.

THERAPIST: I think it's cruel to ask a person to go swimming in all their pain if they can end up drowning in it. What I'm asking you to do is to go wading little by little, toes and ankles first, until you feel comfortable in the water.

CLIENT: You know there's a lot that has happened to me ...

Therapy was conducted very carefully at first, checking with Cindy to see if she was feeling overwhelmed at any time. Eventually, she disclosed several traumatic experiences and was able to work through these to her satisfaction. In this case, the paradoxical injunction to avoid anything painful showed her that it was what she was already doing. Cindy further realized that she had been fighting and resisting painful memories and emotions throughout her life. She eventually reported that she had become so accustomed to keeping those feelings at bay that the thought of not doing so was intimidating and disorienting.

Taking Ownership and Stopping Blame

In many ways, owning the problem is the opposite of blaming, or compulsively attributing responsibility for a problem to another or others. Difficult clients typically blame others for their problems and issues. Getting a difficult client to take ownership of a problem can be demanding work, but when successful, it is rewarding for all concerned.

A key to bringing about a sense of ownership is, of course, to help a person see that they have some influence over the problem. Even if the person is truly blameless or not at all responsible for what happened, as in cases of sexual abuse, the person is still responsible for how they think about that problem (see Frankl, 1992). A person's attitude toward a situation or problem can still come under their control. Blaming, and its close associate, hating, is often the result of a decision.

Following his experience at Auschwitz, Frankl (1992) learned a lesson that therapists can present to clients. He said that although he had no control over what the Nazis did to him and the other people there, he did have control over his attitude toward what they did to him and others. This story and lesson can be used to advantage with people who blame, complain, or detach from events that have affected them. Recommending his book can be helpful, if the client will read it.

The following list of questions and statements may be useful to those who habitually complain or blame others for a problem or a particular issue:

- How is this problem controlling you?
- Has complaining about the problem helped any?
- Do you feel better when blaming [a person or situation] for the problem?
- Has blaming [the person or situation] helped to ease the difficulty?
- Perhaps you don't give yourself enough credit for your own influence.
- Have you set this problem up so that you would be helpless over it?
- Have you set yourself up to be weighed down by this problem?

- Is having someone to blame important to you?
- How has this problem robbed you of your freedom?
- Is this problem worth giving up your freedom for?
- Sure, you didn't cause the problem, but can you control your reaction to it?

Some of these questions are highly confrontative, whereas others are designed to induce processing. Each is meant to be used appropriately to match different clients' situations at different times. All are questions that many difficult clients will seek to avoid, and therefore, they need to be pursued. If asked curiously and openly, each has the potential to open a therapeutic window. The questions or statements, and variations of the same, will need to be repeated in many instances to achieve some depth of processing. Processing these questions and statements can help a client understand that blaming involves the avoidance of anxiety that comes with responsibility (see Yalom, 1980).

Blaming inhibits the ability to respond by placing all control in the hands of the blamed person. This lack of ownership of a problem or issue often comes from never having accepted the problem and its associated thoughts and feelings. As such, acceptance of undesirable thoughts and feelings can be a powerful change factor (Hofmann & Asmundson, 2008). The goal is for a client to learn that blame brings neither solutions nor joy, whereas responsibility can bring both. Generally, *responsibility* is the willingness to take charge of the circumstances that make up the problem. Clients can be taught that "responsibility" is not a word that weighs 500 pounds and sits on one's shoulders. It can be described as the ability to respond, or "response-ability."

The Control or Freedom Challenge

Therapists can use a specific technique to help clients stop blaming others as a means of avoiding anxiety or difficulty. The technique focuses on a mechanism used by difficult clients of all ages. An example might be the sixth grader who says, "He made me do it," and another the domestic violence perpetrator who says, "If she didn't make me angry, I wouldn't have hit her." The proper response to these statements is to indicate somehow that the client themselves has given up control of their life to the person being blamed. This is not at all the same as indicating to a client that they lack self-control, which is a familiar song to these clients that will usually end up being uneventful.

The therapist must point out that the person has given up control over behaviors and thoughts to the person who is being blamed, as this is, in most cases, a more accurate representation (see Hanna & Hunt, 1999). For example, when a sophomore in high school says, "Joey made me do it," the response could be, "So Joey now controls your life?" Similarly, when an abusive husband says, "She nagged and nagged and got me so mad I just hit her," a proper response could be, "So you have no control over your emotions and behaviors?"

Such statements can be followed up by saying, "It's too bad you don't have any freedom." This is a provocative statement, and therapists must recognize it as such and be aware of the consequence of increased anxiety. Another statement might be, "It's too bad you no longer have any say over your life." The therapist can then offer to help the client get their life back.

This approach is especially effective with defiant and conduct-disordered adolescents (Hanna & Hunt, 1999). Because it is provocative, this technique usually evokes self-righteous protest and even defiance by clients prone to blaming others. The therapist responds to the increased anxiety levels by saying, "I mean you no disrespect. I'm only wondering if you know that you give up your life and your freedom to other people every time you blame them." After the anxiety level reduces a bit, the therapist can say something on the order of, "Tell me if I'm wrong, but it seems like [the person] is controlling your emotions and behaviors with their words and attitudes. Do you have any freedom left, or have you given it all away?" At this point, the therapist can again offer to help the person take back that freedom. This is initially done by having the client observe and reflect on times when they blindly acted out in response to stressful situations. For this technique to work, the therapist must be compassionate as well as dispassionate in using it.

Once again, the therapeutic strategy is to show that whenever people blame others for their own actions, they have given their self-control or freedom away to the person blamed, and even their thoughts are now controlled by the blamed person. This approach is especially effective if the person blamed is someone the client dislikes, does not respect, or is competing against in some way. It brings about considerable cognitive dissonance and will often influence a client to be willing to experience anxiety or difficulty in seeking to regain freedom or control over behaviors. The technique is a classic example of a difficult situation that produces a moment of great therapeutic potential. It is unwise to use the term "freedom" with children under 14 years of age, as they do not seem to respond to it. Using the term "control" is much better with that age group.

Self-Monitoring the Waxing and Waning of Anxiety

The behavioral technique of self-monitoring (Korotitsch & Nelson-Gray, 1999; Meichenbaum, 1977) can be used to show some difficult clients that they have a low tolerance for anxiety and resort to avoidant behaviors rather than face it. The technique involves pointing out moments in which a client shrinks from, deflects, or otherwise avoids anxiety. Many defiant adolescents have a low tolerance for anxiety, even though they may assert that they can handle any kind of confrontation or difficult situation. Highlighting the low tolerance is best done in vivo, during the session, with the therapist asking permission to do so.

For example, Marie was a teenager who seldom studied, being more interested in entertainment of various kinds. She was, nevertheless, highly sensitive about her low grades and would get irritated or annoyed at the mere mention of them. Her therapist used this, in combination with a strategy for increasing the client's necessity precursor, as an opportunity for her to learn about anxiety:

THERAPIST: Have you noticed that some people can deal with things without getting upset, and they just seem to be very cool and not let things get to them?

CLIENT: Yeah.

THERAPIST: Would you want to be more like that?

CLIENT: (protesting) I am like that. Don't you think I am?

THERAPIST: In many ways, you are. But you also know that many people know exactly what to say to get you annoyed and irritated, right?

CLIENT: Yeah, I guess.

THERAPIST: How would you like to get less annoyed and irritated and be more on top of things?

CLIENT: I can do that?

THERAPIST: Yes. Can I explain to you about a thing called anxiety?

CLIENT: I guess so.

THERAPIST: Anxiety is an agitated feeling that we all get when something happens that we don't want to happen or we feel sensitive about. It's like an itch you can't scratch.

CLIENT: Yeah, so?

THERAPIST: Well, I'm going to deliberately say something about your grades, but I don't mean what I say; I just want you to watch the feeling you get. Is that okay with you?

CLIENT: So now you are going to tell me to get good grades, too?

THERAPIST: Do you have that agitated feeling right now, Marie? The itch you can't scratch?

CLIENT: (protesting) I thought you said that you didn't care about my grades.

THERAPIST: I said that I cared more about you than your grades. But this is not about grades. Can I go on?

CLIENT: Whatever. I'm just so tired of people bothering me about my grades.

THERAPIST: I understand. Marie, have you noticed that any time someone mentions your grades, like right now, you get irritated with them and say things like "Whatever," or "I don't care," or "Leave me alone"?

CLIENT: Yeah.

THERAPIST: Well, I'm going to say it deliberately again so that you can watch the feeling rather than give in to it and get all irritated and angry. Is it okay if I do that?

CLIENT: Okay.

THERAPIST: (playing the role with a stern voice) Marie, your grades are terrible, and you can do a lot better. What happened when I said that?

CLIENT: I wanted to strangle you.

THERAPIST: Fair enough. But what did you feel the moment I said that?

CLIENT: I felt like I was going crazy ... like I can't stand it.

THERAPIST: Right. That "can't stand it" feeling is the anxiety I'm talking about. How often do you feel that feeling?

CLIENT: A lot.

THERAPIST: If that didn't bother you so much, would your life be different?

CLIENT: Well, yeah, a lot different.

THERAPIST: How?

CLIENT: I wouldn't, like, be going off on people and yelling at them to leave me alone all the time.

THERAPIST: Would you like to learn how to handle that anxiety feeling so it doesn't bother you as much? You can, you know. And it might help you to feel better.

CLIENT: Uh, yeah. What do I do?

The therapist then explained to Marie the idea of self-monitoring. In this case, it involved watching her anxiety levels at school, at home, and with her friends. She was told to watch what she did whenever she felt it and to "look, don't think." She did this quite well and soon was paying attention to her anxiety responses with relatively little effort. Once she was willing to experience the anxiety, she began to use the stress management techniques she had previously ignored. Her progress in therapy was enhanced as well. With difficult clients, merely speaking of anxiety usually will not communicate the point, but demonstrating it to a client can do so effectively. Unfortunately, the anxiety sometimes needs to be induced in the session.

Examining Maladaptive Beliefs About Anxiety, Difficulty, and Pain

There are several maladaptive beliefs that can interfere with establishing grit as a willingness to experience anxiety or difficulty. If these beliefs can be identified and disputed, therapy is likely to progress more quickly. It may be helpful to recall that many maladaptive beliefs related to anxiety and pain avoidance involve implicit learning (Cleeremans, 2019; Hartmann, 2019) and may have been formed preverbally. Thus, when seeking to isolate such a belief, the exact wording is not as important as the concept underlying the belief itself. Some beliefs related to this precursor include

- Experiencing emotional pain is self-torture.
- Experiencing emotional pain makes me weak.
- Experiencing anxiety drains my energy.
- Anything unpleasant is a sign of impending pain.
- Only fools immerse themselves in emotions.
- To open oneself up to emotional pain threatens the integrity of the self.
- If I begin to feel, I will begin to fall apart.
- There is always a way to avoid anything difficult.
- I can avoid anything.
- If I have no feelings, I have no pain.

A CLOSING NOTE ON GRIT

It is important to recognize that the grit precursor only manifests when the client willingly and intentionally experiences anxiety and difficulty. Simply having anxiety or finding oneself in difficult circumstances is insufficient. As mentioned earlier in this chapter, grit aligns with the workout motto of "no pain, no gain." When a person engages in weightlifting, running, or other forms of exercise, they willingly endure strain on the body because doing so increases strength, endurance, and resiliency. Transformative change demands a willingness to endure hardship and discomfort, whether in the gym or in a therapy session. People vary widely in their capacity for grit, but those clients who demonstrate the greatest capacity for psychological discomfort are often well suited to the challenge of therapeutic change. Helping clients grow this precursor can also have a far-reaching effect in life, as clients learn that anxiety and negative feelings will reduce in intensity, or even subside altogether, when willingly experienced. The strategies and techniques in this chapter provide a solid foundation for growing grit as a necessary internal capacity for change.

9

E: Effort, or Will to Change

Effort or will toward change seems quite simple on the surface, but its implications run deep to illuminate the complex relationship between perceived engagement in therapy and therapeutic change. Talk therapy can often be a verbal, passive, cerebral experience in which too little action takes place. It is common for the therapeutic relationship to become exceedingly comfortable, with conversations flowing freely in a pleasurable and meaningful way. It is extremely easy to talk to most therapists and counselors, so much so that the talking and sharing itself displaces the overarching goal of therapeutic change. Such therapy becomes a soothing opiate whereby a client engages in a close relationship with a therapist who seeks to nurture, reassure, and console.

Of course, some benefits can certainly arise from the pleasure of a listening ear and the catharsis that accompanies it. But talking is usually not enough to instantiate meaningful change in life. For most clients to stabilize and improve, they must actively and effortfully change their conditions of mind, behavior, and environment. A client may be aware that their house is a mess and actively talk about their frustrations with a dirty home in session, but such discussions are pointless if they do not exert the effort to physically clean their house. Thus, it is usually not enough to be aware of, open to, or to confront a problem. In so many cases, there is no substitute for effortful action, and therapeutic change may not occur without this crucial ingredient.

According to the CHANGES model, *effort* is the deliberate and self-determined exertion of physical or mental energy or resources toward therapeutic change in the form of self-improvement or positive alterations to one's environment. It is observed when one executes the psychological and behavioral tasks necessary to reach the goal of change and marks the transition between contemplation and

---

Case examples have been disguised to protect client confidentiality.

https://doi.org/10.1037/0000451-010
Therapeutic Change With Difficult Clients: Precursors and Techniques in the CHANGES Model, Second Edition, by B. D. Wilkinson and F. J. Hanna

action (Krebs et al., 2019). In the CHANGES model, effort or will is how the hard work of change is accomplished.

How much effort a person expends seems to depend on the presence of other precursors, even as effort itself can sometimes activate other precursors. Confronting is often enhanced when exerting effort toward resolving a problem or situation, but without confronting, effort might never be engaged at all. Hope also has a strong influence on effort. If a person can see no realistic, desirable outcome for an expenditure of effort, they are not likely to attempt it (Bandura, 1977). A therapist attempting to increase the effort precursor cannot lose sight of the importance of hope. Alternatively, a sufficient degree of a sense of necessity can be enough to spring a person into effort and action, even when neither hope nor self-efficacy is present.

THE NATURE OF THE WILL

When a difficult client who makes little or no progress toward change seems unable to act, it may be wise to recognize this as an affliction of the will, just as Rank (1936) noted. In fact, Rank recommended that clients strengthen their will as part of therapy and said in no uncertain terms that this personal quality was a core capacity of healthy human beings. Low (1952) also observed that the will, as the human quality that moves beyond instincts and drives, is the key to mental health. Similarly, Assagioli (1973) saw the will as the primary, active element in reaching higher stages of development. These early scholars were convinced that the will should be rehabilitated and restored in clients as a routine part of therapy. We do not believe, as they did, that an entire therapy should be built around the will. Nevertheless, it should not be overlooked as a point of rehabilitation in certain difficult clients, especially when a client simply won't act.

The idea of the will is a crucial yet seldom discussed and rarely appreciated aspect of effort (Robinson, 1990). It is conceptually ambiguous, to be sure, but just because a concept is ambiguous does not mean it should be ignored, especially if it has utility. Although the will is complex and even confounding, this is no excuse to avoid it. Effort toward change is seldom, if ever, engaged without some act of will. Commitment is a good example of the will in action, which serves as a major behavioral process in the transtheoretical mode of change (Prochaska et al., 1994). Glasser (1965) was one of the first to point out the importance of commitment to change on the part of a client as crucial to bringing about the effort needed to carry a plan through to its fruition.

Many clients are aware of a specific problem and the need to find a solution, and they can also be confronting and willing to experience anxiety or difficulty. But in spite of the presence of those precursors, they seem unable to garner sufficient momentum toward action, as though they are metaphorically chained or harnessed. Explorations into the phenomenology of agency have surged in the field of consciousness studies (see Bayne, 2008; Mylopoulos & Shepherd, 2020) and may provide valuable insight into how reticence arises as a pathology of agentive experience, perhaps in terms of the relation between directive mental causation and intentionality. It may be simpler to frame deficits in effort to more pragmatic neuroscientific causes, such as insufficient effort justification and the relative neurobiological cost of energy expenditure (Inzlicht et al., 2018). In any case, the issue of sustained effort or will is of central concern to therapeutic change practices, and understanding the complex nature of agentic effort should be prioritized in the field.

HOW EFFORT OR WILL LEADS TO THERAPEUTIC CHANGE

How does change come about through effort or will? To answer this question, it is helpful to first outline two essential modalities in which effort or will initiates change: control and maneuvering. In both modalities, change is implemented directly and with immediacy. Each takes place in three distinct though related contexts—the mind, the body, and the environment. Finally, we look briefly at how the efficient use of in-session time reflects a therapist's prioritization of effort in the therapeutic change process.

Will as Effortful Control of Thoughts and Images

Researchers generally tend to discuss will in terms of self-control (Duckworth et al., 2018), with some distinguishing between forms of self-control, such as resolve and suppression (Ainslie, 2021). However, will can also be understood as a sense of ownership over one's actions. This aligns with the classic definition set forth by the 18th-century English philosopher Hume (1739/1978), who said that "[will is] *the internal impression we feel and are conscious of,* when we knowingly give rise to any new motion of our body, or new perception of our mind" (p. 399, italics in original). If we substitute effort in this definition, it does not lose its descriptive power. That is, perhaps, why James (1890/1981) noted how effort and will are intimately related—so much so that he viewed them as essentially equivalent.

Control in this context refers to the psychological act of directly influencing the content and operations of the mind. Maladaptive beliefs are discarded and replaced with new, more reasonable beliefs. This can take place immediately, as in some cases, or it can be a psychological act that is repeated continuously until the original maladaptive belief is pushed to the background of the mind or deprived of its power. For example, if a client has the belief "I will not trust anyone," the act of will or effort replaces this notion with a more rational "I can trust some people but not just anyone."

The act of replacing the old belief may have a larger role to play in bringing about change than the content of the new belief. For example, a client learned that his idea of the perfect woman was one who is beautiful, intelligent, self-willed, and proud, and yet he also expected this imagined woman to be submissive, needful of direction, and willing to sacrifice her desires and goals for his own. Such a bundle of self-evidently contradictory traits cannot exist except within one's imagination. In the act of change, whereby effort and will were engaged, the client directly altered this flawed image in his mind, and this act was just as important as the image that was put in its place.

In the case of the will directly affecting the body, an obvious example is a client who is unhealthy and overweight and finds it necessary to establish a workout routine. If this client actively engages the will to get up and jog, do aerobics, climb stairs, or whatever the exercise may be, then change will surely result from such effortful engagement of the will.

In the case of directly affecting the world or environment, an example is a client who is troubled by the unethical actions of the company she works for. The act of will is engaged when she commits to finding a new job and exerts the effort toward quitting the old one. If the will is not engaged, the person can be aware of the need and confront the problem itself but will talk about the need to quit for years with no result.

Maneuvering Around Problem Areas

Maneuvering involves therapeutic action around or away from issues or problems that are beyond one's control. For example, a therapist may ask a depressed client if they "know where to find an area of pain" in their minds. The therapist then asks with curiosity whether the client can "go there" if they so wish. We have found that many people can do this with little trouble. If they know how to "go there," they also know precisely how to maneuver away from troubling regions of pain.

The body can also be maneuvered to avoid problem areas and difficulties. For example, a person insulted in a public meeting may need to engage the will to "hold their tongue," which is one way of maneuvering around a potentially difficult situation in which saying something at that moment may be ill advised. This simple lesson is often helpful for defiant teenagers.

Similarly, in maneuvering around worldly or environmental dangers, a recovering alcohol abuser who is driving by one of their old bar hangouts and experiences a powerful urge to stop for a drink may need a deliberate act of will to keep the car going in a direction away from the bar and the possibility of relapse.

Effort and the Efficient Use of Time

Effort must be integrated into a client's everyday life before substantial change can manifest. One way of looking at this is to consider that there are approximately 112 waking hours in a week. Depending on the treatment arrangement, a client sees a therapist for only one of those hours, or perhaps more if the person is hospitalized or in an outpatient format where therapy is conducted several times a week. The small number of therapy hours is often insufficient if the lessons from therapy are not converted or integrated into the client's everyday living. Effort toward change needs to be implemented by the client throughout the week for in-session lessons to merge with worldly experience. When effort is directed toward therapeutic tasks throughout the week, therapy time becomes more meaningful, productive, and valued by the client.

Therapists should also be cognizant of how many minutes per session are intentionally and meaningfully devoted to change. Session time is wasted if the precursors are not actively engaged, and client progress can sometimes be gauged by this measure. A 50-minute session with a difficult client may have only five truly productive minutes in which the therapeutic change process is actually engaged. The other 45 minutes of session time might be spent in denial, avoidance, manipulation, or any form of circuitous communication and evasive actions. However, when a client puts in substantial effort with a therapist who emphasizes change, in-session productivity radically increases as a result of efficient focus on important therapeutic change factors.

SIGNIFICANT MOMENTS OF CHANGE INVOLVING EFFORT OR WILL

Mary's experience of significant change occurred when she found out she was pregnant. The man had told her that he had had a vasectomy, and thus the pregnancy was a shock. Putting that and other information together, Mary recognized that the man she was involved with was a pathological liar. Her first duty was to attend to the pregnancy, as she had decided that she did not want the baby. The doctor did an abortion under a medical D&C and performed a tubal ligation at her request. It was while she was being taken to the operating room that she realized, "I was on my way to taking control of my life." She also realized that "fate wasn't dealing me a hand. I was responsible for my life—I did have control ... I have choices." Having attended to these matters, she faced him at home, spoke her mind without reservation, and removed him from her life. After telling her story, she looked at the therapist and said that before that experience, "I always felt things happened *to* me." That insight was a major factor in the long-term stability of her change.

Another story of change happened to a social worker named Ellen. She recounted considerable test anxiety as an undergraduate, with symptoms such as stomachaches and even vomiting before exams. She worked on her test anxiety with a therapist, hoping the results would influence what she called her "general anxiety." She engaged in cognitive-behavioral therapy that included desensitization, meditation twice a day, and countering each arising "negative self-statement with two positive ones." After 12 weeks of treatment, she was relaxing on the deck of her apartment and suddenly experienced a flood of memories and repressed feelings and emotions. She said, "I think of it as a dam bursting or a wall crumbling," as memories of childhood sexual abuse rose into her immediate awareness. These memories were associated with "anger, guilt, hurt, and pain," and she described the experience as simultaneously "exciting and scary."

Looking back on her realization, she said, "Up until that point, I thought people could do things to me and that I had no control over it." She also thought that "people and situations just happened to me, that I couldn't make things happen myself ... everything felt accidental." Later, she was reading about locus of control and realized, "I could wake up tomorrow and my life didn't have to be accidental; it could be intentional." Ellen's anxiety dramatically improved thereafter, and she reported knowing that she had handled the problem after successfully passing comprehensive exams for her master's degree.

CLINICAL MARKERS: HOW TO RECOGNIZE EFFORT OR WILL TOWARD CHANGE

The client with a high level of effort or will is relatively easy to identify, although one should avoid mistaking a manic or hyperactive client for one exhibiting effort to change. Clients with this precursor tend to display high energy and in some cases even eagerness to complete therapeutic tasks and homework assignments. Such clients seek out and ask for therapeutic exercises and activities they can perform on their own and then proceed to carry them out to completion with attention to beneficial change. They are eager to try new techniques, from role-playing to the empty chair, from paradox to journaling negative self-talk and disputing dysfunctional cognitions. They impress their therapists with their willingness to actually "do therapy" in the purest sense. In short, they take action to change.

In therapy sessions, these clients exert effort to actively help the therapist do a more thorough job. Such clients are movers and shakers where therapeutic change is concerned. They display an experimental attitude toward change with the stated intent of improving their lives, reaching their goals, or being of benefit to others. They are not hesitant to take action to correct difficult situations in their lives, such as getting out of debt, improving a relationship with a spouse, or disciplining a defiant teenager. Energy and actions taken toward change are chief markers, but the overlap with confronting should be obvious as well, although the two should not be confused.

When a client is endowed with this precursor, the therapist will sense an abundant determination to solve problems, accompanied by engagement of the body in action on the environment. Thus, such a client will actively transfer in-session learning experiences to understanding outside activities and situations. These clients may be interested in learning not only about their own issues but also about the process and nature of therapy itself. When they are hesitant to act, they will be curious about the nature of the hesitance, realizing that this is an issue in itself.

These clients can easily alter thoughts, change beliefs, and reorganize entire patterns of attitudes. They are also able to construct or deconstruct images or illusions to accomplish therapeutic tasks. For instance, a 30-year-old client named Marcus was puzzled by difficulties in his relationship with his fiancée. He realized after six intense, active sessions that he had harbored a false image of his fiancée functioning as his therapist. When he saw that he had a faulty or misguided image of a nurturing woman, in a matter of minutes, he reorganized his beliefs about women in this regard and told his therapist that he dissolved the image on the spot. In doing so, he realized that he had been difficult to get along with and that he had been demanding too much from women in his life. He reported that he had somehow grown up thinking that a woman should be his "emotional servant." In a whirlwind of insight and activity, he changed his mental attitude and patched up the relationship with his now overjoyed fiancée. This is effort at its height, attaining change remarkably quickly, deeply, and with great facility.

As one client who had a high level of this precursor once said, "I don't need a reason to be happy. I can just go ahead and force myself to feel that way whenever I want to." The apparent fact is that feelings do not have to be binding, and one can act to directly alter feelings (see James, 1890/1981) as well as behavior and cognitions.

HOW TO DETECT THE ABSENCE OF EFFORT OR WILL TOWARD CHANGE

A client lacking in effort or will consistently takes the path of least resistance and gives up easily on projects. Such a client seems to lack energy for therapeutic tasks, procrastinating and making excuses. Such clients may talk about procrastination as being a problem in itself, perhaps openly shrugging their shoulders and saying something like, "I know it is something I need to do, but I just don't seem to get around to doing it." Other clients will devote only a limited amount of energy to working in therapy and may make it appear like they are expending great amounts. This can be the case even though their abilities are considerable. Therapists may consider three particular styles of difficulty in this regard: the passive–compliant client, the falsely compliant client, and the ambivalent and conflicted client.

The Passive–Compliant Client

Some clients display a puzzling style to which therapists should be alert. The "passive–compliant" client is ostensibly compliant but is so passive that real action is outside their current capacity. This client initially seems to be anything but difficult, coming across as admirable and compliant. Such clients are highly cooperative in session and willing to discuss and disclose even the most intimate details about their lives. They appear to have many of the other precursors, such as the willingness to experience anxiety or difficulty, high awareness, a sense of necessity, and are even confronting the problem. They are willing to examine thoughts and beliefs. However, when the client is asked to carry out a therapeutic task or actually change anything, they will find excuses to avoid direct manipulation of mind or world.

Such clients may flatter the therapist or come up with amazingly creative diversions to avoid effort. A client once told her therapist that she could not do her homework assignment because the therapist was "so great that all she could do was think about our sessions and how much she learned from them." She said this during their 20th session while her life was coming down around her. Another client told his therapist that he was so fond of his old, maladaptive beliefs that "it would be a shame to lose them." This admission came following weeks of in-session work to identify, modify, and replace certain maladaptive beliefs. The client had been so compliant in sessions that the therapist genuinely believed the client had effectively replaced the beliefs many weeks earlier.

The Falsely Compliant Client

The falsely compliant client will return to the therapist and report with great authority, "I did what you said, and it didn't work." Usually, the therapist never gave direct advice and never told the client to do anything that would make anything "work." Such clients are attempting to manipulate the therapist, avoid actual effort, and prove that their situation is impossible. They rarely do the tasks assigned, and what they do usually leads to negative consequences. For such clients, effort expended has a way of making the problem even worse.

For example, a male client was given a homework assignment designed to deal with his tendency to get angry at his children. At the next session, he tells the therapist that it did not work and that he lost his temper, but he confessed several sessions later that he never attempted the assignment. In another case, a particularly difficult client in his mid-30s was convinced that his father hated him and was seeking to ruin his life. The therapist asked him, while gathering information, if he had ever talked to his mother or siblings about it. Rather than seeing this as a question, he took it as a "direct order." He came back and told the therapist, "I did what you said, and it was a waste of time." Eventually, he confessed that he had talked to his mother but only about whether she loved him. Then he told his father that the therapist said he was "neurotic," a term the therapist never used. His efforts were not consistent with the precursor of effort toward change.

Such falsely compliant clients may not be ready for homework assignments that require interpersonal skills. It may be more important to first establish other precursors before sending the client off to do tasks in their environment. For these clients, tasks should be confined to the realm of mind in the form of cognitive rehearsals and role-plays until the precursors increase and stabilize. Tasks assigned should be realistic, concrete, and clear.

The Ambivalent and Conflicted Client

Another client who is unlikely to be able to effect change experiences a massive infiltration of counter-intentions in the mind. An example is the drug user desperately seeking a way to get clean but is so in love with the drug that they cannot act on any intention to abstain. This person wants to keep using far more than they want to quit, but both intentions are there, nevertheless. If this is the case, the person is stymied, and real effort toward change will not be expended.

TECHNIQUES: INCREASING EFFORT

Effort is fundamental and essential for change. Even when it spontaneously arises out of insight and the resulting change appears "effortless," the client may have expended great amounts of energy. Where there is great interest, great effort appears effortless. Whenever possible, a sound approach is to increase the client's interest in the task, thereby increasing the degree of effort expended toward change. This section lists ways to encourage the client to expend effort in doing homework assignments and to increase the client's interest in exerting effort. These are methods to inspire a person to get off their fence, couch, or behind, as the case may be.

Metaphors That Illustrate Effort

Like so many other precursors, the idea of exerting effort lends itself well to the use of apt therapeutic metaphors. Many clichés also apply, such as "Nothing ventured, nothing gained" and "You don't know unless you try." The following additional metaphors may help clients understand the importance of effort or will.

The Finger Technique

We have found that some clients who do not understand that the act of will moves a person into action can be helped with a bit of psychoeducation. The finger technique is a thought experiment that illustrates the act of will through a simple demonstration. It is especially helpful with procrastinating clients, and the therapist should demonstrate it first and then ask the client to try it. Ideally, the client does the exercise along with the therapist.

The technique begins with the therapist holding up an index finger. The therapist explains to the client that exerting effort begins with an act of will. Holding the finger up, the therapist shows different ways of bending it. The therapist first explains that they are thinking about bending the finger. Of course, the finger does not move. Next, the therapist shouts at the finger to bend, and still, it does not move. The therapist begs and pleads with the finger to move, with no effect. The therapist then says in a procrastinating way, "Don't worry, I am going to bend it tomorrow." The finger does not move. The therapist says they are forming a mental image of bending it, and the finger almost moves, but not really. Finally, the therapist actually exerts the will to move the finger, explaining that this is the only thing that will get it to move and that the act of will accomplishes the task. It can be explained that virtually all human actions that are not on some automatic or autonomic routine are executed in this way.

The therapist asks the client to bend their finger several times and to closely observe the act of will that brings the motion about. Performing this simple act of willing, a person can often come to understand what procrastination is and what it takes to engage effort toward a task. If the will is not exerted, effort is not expended. Once the exercise has been done, if a client does not do a homework assignment, the therapist can use the metaphor by saying, "So you didn't move the finger?" In this way, the client will readily understand the metaphor as it relates to therapeutic goals and tasks and to the fundamental act of achievement and accomplishment.

The House Cleaning Metaphor

The house cleaning metaphor is appropriate for some clients in communicating the importance of effort. The person's mind or life can be likened to a house in need of cleaning. One can be acutely aware of how dirty the house is and even confront every detail of the dirt and grime, but it will not get cleaned by awareness or confronting. It will also not get cleaned by talking about how much it needs to be cleaned. At some point, a person has to sweep the carpets, do the dishes, scrub the floors, and disinfect the bathroom. That is the effort precursor, in essence. If a person is not willing to go through the actions, the house accumulates more and more dirt until it becomes unhealthy. The same is true for the effort needed to change the conditions of one's mind, behavior, or environment.

Acknowledging Freedom of Choice

From a slightly different perspective on will, studies have demonstrated that many selection tasks are more efficiently completed by those who perceive they have freedom of choice in approaching those tasks (Barlas et al., 2017; Barlas & Obhi, 2013). Therefore, it is unsurprising that clients are more likely to complete homework assignments, for instance, when provided options and their power of choice is validated (Kazantzis & Lampropoulos, 2002). In cases of antisocial personality disorder, therapy is more effective when the free choice of these clients is emphasized (Kierulff, 1988; McRae, 2013). Many young people with a criminal record will do nearly anything to gratify their desires and passions, regardless of law or convention (DeLisi et al., 2018; Shoemaker, 2018). In fact, the goal of "getting one over on" the law or societal conventions is the challenge that often guides their choices. Rather than attempt to inhibit this distorted way of being, it occasionally helps to demonstrate the paradox that while freedom is indeed precious, exercising it unwisely can lead to a loss of that same freedom.

Clarifying the Goal

Many clients display a lack of effort when it is not clear to them what exactly it is they are trying to achieve. Often, this occurs due to busy therapists inadvertently assuming that a client understands why they have recommended a particular task. To a lesser extent, there are times when a client knows the goal but does not have a sufficiently detailed image or conception of it. People tend to demonstrate increased effort to change when they believe that the task has value (Klug & Maier, 2015) and, as previously noted, the expenditure of effort is of their own choice (Barlas et al., 2017). When difficult clients clearly see how the task relates to the "What's-in-it-for-me" question, compliance is likely to result. Thus, the client must believe they are exerting the effort by free choice toward a realistic, clear, and desirable outcome or result.

Self-Observation

Some difficult clients simply will not complete homework assignments. For these clients, the most appropriate form of assigned work is self-observation or self-monitoring, which can be done in and out of therapy. Rather than issuing a homework assignment, the therapist can ask the client to not actually change anything but only to watch or be aware of a particular behavior or situation in everyday life and report on it in the next session. While this paradoxical approach is suitable for building awareness, it is also a good preparation for exerting effort and energy toward change. A client can also do role-plays in which they are not expected to do anything other than observe what happens in terms of behavior, self-talk, or emotions. Once the client has done some self-observation, it is possible for them to engage in tasks requiring specific acts. Resistance to self-observation could mean the client decided to dull or diminish their awareness, and that is the issue that needs to be addressed.

Assigning Graduated Tasks

Some clients are not up to performing complex tasks, and it is easy to overestimate what they are capable of doing. In such cases, graduated tasks increase the likelihood that a client will follow through with a homework assignment. There is a threefold requisite to assigning graduated tasks. First, the therapist should check with the client to establish that the task is realistic. Second, the therapist should ensure the client is interested, understands the rationale, and agrees it is necessary. Finally, the therapist should assign only those tasks that the client is capable of performing successfully because to do otherwise is to set the client up for failure, which could lead to premature termination. Once simple tasks are done successfully, the client can move on to those that are more difficult or challenging.

Empathic Validation

Clients who are not in therapy of their own volition are unlikely to exert effort toward change. They may even string the therapist along, making excuses or only pretending to be cooperative as a means to graduate from a treatment program. Of course, these clients are likely to lack a sense of necessity.

Empathic validation communicates a therapist's complete acceptance and understanding of the client's current goals or attitudes toward therapy. Rather than push a therapeutic agenda and a series of corresponding activities, it is more important for the therapist to deal with clients exactly where they are at in terms of how they are feeling and thinking about treatment itself. These clients need the freedom to admit (and the therapist to acknowledge) that they do not care about therapy, the therapist, or any related program. The therapist must make it clear that they completely understand and acknowledge that the client probably feels that their rights to freedom as a citizen are being violated. This acknowledgment will bring involuntary clients closer to a viable relationship and encourage a willingness to self-disclose to a therapist. No one will exert much effort toward a goal without it resonating with some sense of meaning or purpose.

Subpersonality Approach to Effort: Mode 1

Once empathic validation has helped establish a therapeutic relationship, a subpersonality approach (see Chapter 9, this volume) may be used to address a part of the unwilling client that does want to work toward change but has yet gone unacknowledged. The therapist can ask the client if there is a part of themselves that does indeed want to work toward change. Although this might take some discussion, most clients will eventually admit they do. The therapist can ask to talk to that part. Once a subpersonality has been identified, it can be directly addressed, with the permission of the client, so as to give it voice. Thus, from their own mouth, the client "hears" their own inclinations toward change. The subpersonality of the person that was uninterested in change can eventually be disidentified with and treated as being well intended but ultimately detrimental to the person's well-being.

Subpersonality Approach to Effort: Mode 2

Some clients say they do want to exert effort toward change, but they never get around to it. Many of these difficult clients appear to have a strong sense of lethargy. They might report that they wanted to change, but it seemed too difficult. Sometimes, this is related to a lack of grit, or willingness to experience anxiety or difficulty, but in other cases, the willingness precursor may be adequately present, yet effort is still rated at trace or nonexistent. For such clients, the subpersonality approach can be of occasional service. In this application, a subpersonality may be actively but silently opposing intentions or goals.

To address this brand of lethargy, the therapist can ask whether the client wants to complete the task. If the client replies in the affirmative, the therapist can validate the client and suggest that there is probably a good reason why they did not carry out the homework assignment. The therapist can ask if the client is willing to explore this possibility and ask any of the following subpersonality questions as appropriate:

Is there a part of you that ...

- ... does not want to do anything to change?
- ... does not want to do anything in therapy?

- ... does not want to do the homework assignment?
- ... is stopping you from completing the task?
- ... believes something bad will happen if you do this?
- ... feels nervous or hesitant about the task?

Next, the therapist asks to talk to that part of the client. In carrying on an empathic dialogue with the subpersonality, much information is gained that can then be addressed using other techniques offered in this chapter. Several approaches can follow. For example, the empty chair can be used, or perhaps a role-play in which the therapist plays the client or the subpersonality so that an ensuing dialogue leads to some integration of the lethargic self. The technique aims to increase expended effort by reducing the lethargy produced by being at cross-purposes with different aspects of the self. When the client's counter-effort reduces, more energy is available to devote to effort toward therapy goals without the previous inhibition or hesitancy.

Mapping Intentions

Once again, if the client genuinely states a desire for change but is not doing anything to achieve it, it is often helpful to recognize that there might be a good reason. At one level, the client may truly want change, but at another, less conscious level, change is threatening. It is often the case that a client "knows" change to be a liability and unconsciously works to sabotage all efforts even while consciously attempting it. A diagram of intentions oriented around and in relation to a therapeutic goal can reveal why effort is stalled.

The therapeutic task with such clients is to understand what the intentions and purposes are in the client's mind and to map or diagram them in terms of intentions and counter-intentions, or purposes and counter-purposes. The therapist records on the map both intentions that are conducive to achieving the goal and intentions that directly counter it.

This technique is not the same as listing the pros and cons of a particular activity or goal, which, in our experience, is largely unproductive. The mapping intentions approach is different because it lists active intentions rather than reasons. A counter-intention is not listed in negative terms and judged as a "con" but is recorded as a purpose in and of itself that simply runs in a different direction. In the mapping process, it is important to state each intention in an active context, avoiding terms such as "not." "don't," "should," or "must." After all, a counter-intention is nothing more than a conflicting intention. The intention "to travel the world" is an intention, but it can be a counter-intention when juxtaposed with "to get a doctorate." Each is active and each typically excludes or denies the other, at least in terms of time periods.

Naming each intention properly is also important, and the client must agree that the wording is accurate. After the intentions and counter-intentions are mapped, the therapist asks the client to examine each intention to determine if it should be supported, altered, suspended, or eliminated. The goal is to allow the person to see the patterns of intention and to be released from them so that effort can be freed up to pursue change.

For example, Carla, a woman in her mid-30s, was in great distress over the verbal abuse her husband heaped on her and her two children. He was a charming fellow who was often apologetic after a particularly cruel outburst. Carla had decided to leave him. She set a task for herself of calling a lawyer to initiate divorce proceedings, but she could not bring herself to act. Her progress in therapy was stalled, and she was low in the precursors of effort and confronting.

With the help of the therapist, she mapped her intentions and counter-intentions. Her intentions were to be happy, protect her children, be free, and grow as a person. Her counter-intentions to the goal of seeking a divorce were to have financial security, help her husband, be true to her marriage, show her children that marriage can work, and prove to her parents and friends that she had good judgment in men.

In therapy, each intention and counter-intention, as revealed and identified, was newly evaluated, with Carla deciding whether to support, alter, suspend, or eliminate each. She supported all of her intentions. With her counter-intentions, she chose to temporarily suspend her goal of having financial security, knowing that she would be facing hardship through the divorce. She also suspended (with considerable relief) her goal to be true to her marriage and eliminated her intention to prove to her parents and friends that she had good judgment in men. As a result, she finally went into motion and acted by filing the divorce proceedings.

Facilitating the Decision to Change

Therapeutic change often comes from a specific decision and manifests as a firm commitment. Many factors combine to culminate in a pervasive, integral decision. Unfortunately, we know no easy way to facilitate such a stalwart decision; many precursors seem to weave together in such cases. It seems that confronting and a sense of necessity are especially involved. At times, a client can be asked, straightforwardly, whether they have made an active decision to change. If no change has been forthcoming, the client will likely admit that they have not made such a decision. Last, the therapist can ask the client what it would take to make such a decision. They can also ask the client, in the tradition of the Adlerian "as if," to go ahead and pretend to make the decision, to try it on for size and see if it "feels right." This can sometimes influence a decision and a commitment.

Identifying and Disputing Automatic Thoughts or Negative Self-Talk

This approach consists of asking a person to perform a therapeutic task, preferably in the session, and then having them immediately watch for and report any opposing, contrary, or negating thoughts that arise. Such automatic thoughts can sabotage effort. For example, Michael, a client in his early 20s, had a female employer who was verbally abusive but veiled her hostility with biting jokes and laughter. Michael found it extremely difficult to ask his employer if she would stop and could not bring himself to do even graduated tasks. However, he was willing to do role-plays.

To discover possible automatic thoughts, the therapist and client agreed to set up a role-play in which Michael was to address the therapist, who was playing the role of his verbally abusive employer. The variation was that instead of going through with the role-play and being concerned with what to say to his boss, Michael was to report any opposing thoughts in the way of doubts, criticisms of self, negative expectations, and distorted images associated with speaking to her boss. Once these were addressed and disputed, Michael found it much easier to do realistic role-plays and eventually confronted his employer in an assertive and nonaccusatory manner. In their meeting, he informed her that even though she meant no harm, telling him that he was "dumb" and "silly" was hurtful and did not help him perform better on the job. What was more helpful, he told her, was when he was given details and instructions on how to do better. He was pleasantly surprised when his employer quietly but deliberately agreed to his request.

Attending to the Metalogue

It is easy to hand out homework assignments to difficult clients, and it is just as easy for clients to listen and show (or feign) interest. The question is whether a client will execute the assignments and make self-initiated actions toward change. In many of these cases, a cognitive approach may be enhanced by attending to the metalogue between client and therapist. The client may be engaged in a metalogue that is inhibiting therapeutic progress. For example, many difficult clients will say that as soon as their therapist suggested a task to be completed outside of therapy, they immediately responded with an unvoiced statement such as, "Yeah, right," "I don't care about this," or simply, "No way." Clients may be only dimly aware of this response until asked. Curiously, a client can simultaneously engage in such self-talk while externally agreeing to do the assigned tasks.

Addressing and vocalizing the metalogue can reduce interference with the effort precursor. Once the metalogue is identified and articulated, it can be handily treated by using the subpersonality approach, mapping intentions, or finding maladaptive beliefs. As long as the metalogue is not addressed, however, the therapist runs the risk of the client's continued ambivalence and lack of cooperation, both within and outside of therapy sessions.

Examining Maladaptive Beliefs About the Effort to Change

When effort is not occurring, a host of maladaptive beliefs may be inhibiting a person from moving toward change. When maladaptive beliefs are disputed, the person may spring into motion and make an effort toward therapeutic tasks and goals. While some maladaptive beliefs emphasize a tendency toward dispositional passivity, others indicate fears about the consequences of, or inability to enact, change. The following are possible maladaptive beliefs that may inhibit effort and exertion of the will:

- Being happy means doing as little as possible and just enjoying myself.
- I don't deserve to get better and won't do anything toward it.
- I've hurt too many people to deserve to devote effort to myself.
- I screw everything up anyway, so why try?
- I can't be trusted to succeed.
- Becoming effective will only make me capable of hurting people more.
- I know that if I try, I'll fail.
- I'm afraid of making more mistakes.
- I feel guilty if I try to help myself.
- Someone else should do the work for me.
- I'm waiting for someone to rescue me from all this.
- God will take care of me.

A CLOSING NOTE ON EFFORT

Effort or will toward change is a precursor that leads to multifaceted, multilevel transformation and ties together many of the other precursors. When effortful control is operative, there is a sense in which one's thinking, feeling, and behavior are seemingly driven by a deep wellspring of desire and determination. Some clients find such effortful control simple to tap into, but many difficult clients struggle in this area of action as if the will itself has been rendered immobile. When effort or will to change is inactive in therapy, it becomes the responsibility of the therapist to implement basic strategies and techniques that might help clients move toward self-directed action to change.

10

S: Social Support for Change

Social support is the condition in which a person receives physical, emotional, attentional, and other resources from other human beings. A person has social support when they are surrounded by a network of friends or family who feel positively about them, are empathic, and are willing to help. The ideal social support network consists of people who are available and willing to invite, receive, and accept emotional disclosures. Social support also includes a person's perception that the community or environment is conducive and actively contributing to their well-being. It gratifies the primary needs of acceptance and belonging in human relationships.

Social support is a powerful reinforcer in both positive and negative ways. In effect, this precursor is like fire in that it can be used positively or negatively to produce constructive or destructive results. It is important to consult with the client to determine whether a social support is beneficial and constructive or deleterious and dysfunctional. For example, many adolescent offenders will note that their peers have supported them in committing criminal acts. The supportive statements sound something like this: "It's okay, you can do it. We were scared too at first, but don't worry. You get used to it and don't even think about it after a while." This message could as easily be about getting good grades as committing a crime. In the armed forces, the same message is given to soldiers coming to grips with the fact that they may be called on to kill.

Likewise, social support can exist in varying grades of quality. Even in its lowest grades, it may be seen as precious, and its influence should not be underestimated. Probably the best perspective on understanding this phenomenon is Yalom's (1995) therapeutic factors of group therapy, which include universality, group cohesiveness, and catharsis. Such therapeutic factors can

---

Case examples have been disguised to protect client confidentiality.

https://doi.org/10.1037/0000451-011
Therapeutic Change With Difficult Clients: Precursors and Techniques in the CHANGES Model, Second Edition, by B. D. Wilkinson and F. J. Hanna

occasionally be found in street gangs, however twisted these may be. For instance, a gang member named Lollo once showed up for a session at an outpatient substance use treatment center with a severe hangover and a large bruise on his forehead. He had a devastatingly sad family history, having long witnessed his mother being sexually abused by his uncle, who also physically abused him.

The therapist inquired about the injury, but Lollo defiantly replied, "Yeah, I was drunk, so what?" and insisted the therapist would "never get it." The therapist said that although he might not understand, perhaps Lollo could teach him something. Lollo idly shrugged but explained how he had been drinking with fellow gang members the previous night when the gang leader, whom Lollo deeply admired, told him in an emotional outburst that they would always be friends. He then threw an empty beer bottle, hitting Lollo in the forehead and leaving the bruise while yelling that he loved him like a little brother. For Lollo, this distorted declaration of affection and care was precious and gave him an immense sense of belonging and acceptance. It is not surprising that he thought the therapist would not understand, and it is also evident that even low-grade social supports that prevent positive change can invoke a feeling of belonging and care.

A multitude of theorists, ranging from Adler and Maslow, to Bowen and Satir, to Johnson and Hayes, have noted the need for belonging and acceptance by fellow human beings. Social support is just another term for this primary human need. Recommending it is stating the blatantly obvious, rather like saying that eating nutritious foods is a good idea. In the same sense that certain vitamins can cure rickets and scurvy, social support can alleviate loneliness and a sense of abandonment. It meets the need for companionship, warmth, intimacy, and communication, all of which give meaning to life (Yalom, 1980, 1989). It may be that the service provided by support groups such as Alcoholics Anonymous (2012) is valuable because it replaces negative social support with a social support system oriented toward positive change.

HOW SOCIAL SUPPORT LEADS TO THERAPEUTIC CHANGE

Common factors researchers have long identified perceived social support as an active ingredient in therapeutic change (Wampold & Ulvenes, 2019), and it has also been identified as a source of spontaneous self-improvement (Lambert, 1992). It can be a major influence on recovery from depression (Woods et al., 2021) and a protective factor against suicidal ideation (Arenson et al., 2021). Various meta-analyses identify social support interventions as critical in improving mental health outcomes among survivors of intimate partner violence (Ogbe et al., 2020) as well as posttraumatic stress disorder among both civilian (Wang et al., 2021) and military (Blais et al., 2021) populations. It has been positively correlated with effective pain management among cancer patients (Warth et al., 2020). Preliminary research on the everyday use of chatbots indicates that artificial agents may be perceived as warmly supportive companions that reduce loneliness (Bendig et al., 2022; Ta et al., 2020), while companion animals produce even more pronounced results (Brooks et al., 2018).

According to the CHANGES model, the primary contribution of social support for change seems to be its interaction with the other precursors as a potentiating agent. Its presence makes the other precursors more powerful. The social support precursor affects nearly every aspect of a person's life, including physiological health. Like hope and a sense of necessity, it is seldom, if ever, a sufficient condition for change by itself, but it enhances the power of all the other precursors in both subtle and direct ways. It can be a source that inspires motivation and engagement, although we have found that in some cases of extraordinary human resourcefulness, it did not seem to be necessary at all. Even in those same cases, however, it probably would have been helpful.

Social support has a resounding influence on hope (Martínez-Martí & Ruch, 2017). Change seems easier to envision when the encouragement of loved ones and friends is present to convince a person to contemplate new possibilities. When one has little hope for oneself, knowing that trusted family and friends are convinced of a bright future can bring about a contagion of hope. Effort can be enhanced in kind, as one approaches tasks and goals with more conviction. When a person is filled with self-doubt, the confidence of family and friends can induce the exertion of effort. Similarly, social support makes anxiety or difficulty easier to tolerate. When one knows that people are there to help one get through the ordeals of life, the sheer difficulty of it can be much reduced.

Social support can also help establish a sense of necessity by informing a person of the importance of a particular change. For example, if a client expresses little or no interest in change, the advice, counsel, or concern of a friend or friends may convince them that things are not acceptable as they are and that a change is needed. Awareness may also become enhanced in this way. A verbally abusive husband may listen to a friend saying, "You are being too rough with your wife," but resist the same statement by a mother-in-law or therapist. Finally, in terms of confronting the problem, a safe and socially supportive environment is needed to look at threatening beliefs, mistakes, memories, or feelings. A person who feels secure is more likely to steadily and deeply confront problems (see Hanna & Puhakka, 1991).

SIGNIFICANT MOMENTS OF CHANGE INVOLVING SOCIAL SUPPORT

Alicia was a licensed psychologist in her mid-30s, and her second-order change experience took place 8 years prior to the following therapeutic encounter. She explained to her therapist how, 8 years earlier, she was planning to get married when her fiancée died from a chronic heart ailment. "I went nuts," she said, "and I started cutting myself." She revealed her arms, which still bore the extensive scars from her self-mutilation. She explained that she became "really depressed" and started to lose control of her life. "This stuff was always there" in the background, she reported; this tragedy had just brought it to the foreground.

She went for therapy at her university counseling center, but it was ineffective, and she eventually dropped out of school altogether.

She tried to overdose on sleeping pills soon thereafter. She was hospitalized in critical condition. As a psychiatric ward inpatient, she eventually found a particularly helpful therapist. With this therapist, she said, "There was some hope," and the road to her recovery began. Long after her release from the hospital, however, she was still very angry and resentful. That summer, Alicia worked with kids at a Christian camp "just to give something" to people. She was also in a therapy group and was getting a lot of "positive affirmations" and support to return to school. However, her therapist moved out of town, and after about 2 months, Alicia "started getting scared again."

By this time, she had entered another university and found her way to its counseling center. The therapist there proved to be extremely helpful and disclosed that she had once had similar problems. In Alicia's work with the new therapist, she learned more about her relationship with her mother and her mother's influence on her life. "I learned that I really did need her love and that she really did love me." This was an important insight for Alicia. At one stage, all of this had built to a climax, and the "therapist let me cry." This experience was the crux of change for her. It was a "letting go of anger, frustration, hurt, watching it all wash away. I learned to listen with a different heart. That was a major changing point," she said. She came to see that she could "get the things [she] needed from other people, and it's okay."

After this experience, she "became less angry" and "related to people differently." She learned to "phrase things differently." She added that she learned "that you have to take control of your life and not let life control you." She also said, "The experience reaffirmed my faith in God." During the ensuing years, she reported no acts of self-harm. This example is typical of the excellent social support for change available through effective therapy.

CLINICAL MARKERS: HOW TO RECOGNIZE SOCIAL SUPPORT FOR CHANGE

The therapist can assess the presence of social support from information provided by the client and the nature and tone of the conversation about it. There are many signs and indicators of the presence of social support. Perhaps the easiest and most direct method is to ask the client if they feel supported by friends and family and to what degree. Therapists can also ask if clients know of a network of friends who are supportive and helpful to each other, whether they are a part of that network, and, if they are, how much. It also helps to examine the client's conversation for descriptions of relationships and interactions with others.

To determine the quality of support in those relationships, one can gauge the degree of perceived empathy by asking if the client feels those people truly understand them and, if so, how much. Trust is another element that can be assessed when gauging the quality of those relationships. It helps to ask a client if they actually trust anyone and, if they do, to name those people. When trust and empathy are present, the relationship will tend to be especially valued by a client, and levels of ambivalence or hostility toward the person will be low. In addition, those people described by the client should be spoken of as accessible and available in times of need.

Common references to the support of family and friends are an obvious indicator, especially when the person takes comfort in and works to maintain and develop those relationships. It is also helpful to assess religious or community involvements in terms of social support to see if they are characterized by empathy and encouragement and are not critical, demeaning, or disempowering. Some difficult clients may be members of religious groups with cultlike features that invoke deep suspicion of therapy and try to sabotage any apparent progress (Goldberg, 2017).

A client rich in social support may also demonstrate relationship skills with the therapist; this is a vital sign of the client's interaction skills. A client with this kind of skill will make the therapist feel supported in attempting to help the client. Needless to say, this type of client is especially appreciated by a therapist. In terms of the therapeutic relationship, however, it is important to ask the client for their assessment of how empathic or trustworthy the therapist may be, as research clearly shows that client assessments of therapist empathy tend to be more accurate than self-assessments by therapists themselves.

HOW TO DETECT THE ABSENCE OF SOCIAL SUPPORT FOR CHANGE

A primary problem with difficult clients is often a profound lack of social support, especially for positive change. Many have destroyed the possibility of positive social support through manipulation and harmful acts toward others. We have known some difficult clients who did not change even with the help of the most compassionate and skilled therapists and well-meaning family and friends. A common example is the case of some chronic alcohol users who continue their self-destructive behaviors despite a network of caring friends, family, and therapists. By the time some clients have discovered a sense of necessity for change, they are profoundly alone.

A lack of social support is frequently quite obvious in therapy. In the absence of social support, a client may complain about feeling "all alone" or having "no one to talk to." There may be direct references to "not having anyone to rely upon," such as the single mother working multiple jobs without family or friends who can lend a hand for childcare needs. Depressed clients are also likely to report that they are without close friends or family who understand them (Arslan, 2019; Franck et al., 2016). When stated as a global declaration, the complaint that "there are so few understanding people in the world," or the related belief that life itself is unjust, should be explored in the context of lifelong changes and patterns of social support (Stauffer, 2015; Ucar et al., 2019).

Many people lacking social support are likely to perceive their families with either open hostility or deep ambivalence. They are likely to see their families as a source of anxiety, emotional hurt, depression, abuse, or ridicule. They may accuse their mother or father of having ruined their lives. They describe brothers and sisters as distant or having so little in common with the client that there is no point in maintaining the relationship. All this may be true, but this does not diminish the glaring fact that the person sadly lacks a vital source of social support.

Assessing social support in people with personality disorders is more complex. Many narcissistic clients will boast about possessing a wide network of friends and admirers. They may boldly assert their popularity and appeal to others. The need for admiration on the part of this client is a clue to the nature of their relationships. Such clients do not describe so-called supportive people as those to whom they might make vulnerable self-disclosures about shame, guilt, or mistakes made in life and will abruptly cut off relationships with critical people. With gentle but pointed questioning, such a client may admit that no one "really cares" and claim all people are out for their own gain.

Other difficult clients will see interpersonal support as a liability and will avoid it as a potential source of pain. This is often true of borderline clients, who may interpret any gesture of help as a cause for suspicion of eventual betrayal. Many difficult clients have issues of abandonment and engulfment. Gaining social support is both a need and an admission of weakness and while they actively seek it, they may simultaneously despise the fact that they need it. Many borderline clients are unconsciously convinced that they are so worthless and undeserving that anyone who likes them must be a fool or stupid or both, and they treat potential friends with mixed neediness and contempt. Thus, to help, a therapist must be willing to be seen as an object of contempt and derision, at least for a while. The end result, of course, is that a person with borderline traits may tolerate the idea of improvement. In addition, they often punish those who try to help.

Antisocial clients also tend to believe that needing the help of others is a sign of weakness and will manipulate those who try to help, perhaps by taking advantage of them in terms of money, sexual favors, gaining status or connections, or the like. Social support systems for these clients, if they can be considered that at all, are usually extremely low grade and are characterized by internal dishonesty, cheating, and betrayal. The underlying belief is that one can expect others to help only when they are ultimately manipulating the situation for their own gain. Alternatively, a client with schizoid personality disorder will be unlikely to complain about a lack of friends and will probably prefer to be alone in their private world.

In conclusion, when a person lacks social support, it is easier to minimize or explain away problems. Change will be viewed as an individual struggle requiring more effort than necessary, and experiencing anxiety or difficulty will look rather like torture, to be endured all alone. Further, the awareness of a problem may not crystallize with the clarity that comes with trusted feedback, and the problem may not seem quite so imminent or real to the point where confronting may actually be necessary or helpful. Clients without social support will also be inclined to give up hope more easily, lamenting that the future is an essentially lonely undertaking. Finally, clients will respond to therapy when they experience their therapists as having positive expectations for them (W. R. Miller & Rollnick, 2012), which is a vital aspect of social support generated by the therapist.

TECHNIQUES: GROWING SOCIAL SUPPORT

Helping clients develop social support consists of three separate but related modalities. The first modality is to minimize contact with people who are harmful, abusive, negative, exploitative, or otherwise oppressive to the client. If uncorrected or undeterred, such people can discourage the presence of the precursors of change just as surely as a good therapist can encourage them. In general, of course, it is usually futile to attempt to change people who are not supportive.

The second modality for increasing social support is to increase the presence of people who encourage the precursors of change. These people are generally empathic and helpful and serve as sources of understanding, trust, and active support. Increasing this kind of social support involves teaching a client social and communication skills.

The third modality is important for some clients who have lost valuable friendships. Therapy can help clients rebuild once-supportive relationships that have deteriorated. This goal is often unrealistic, of course. Many clients with personality disorders or alcohol or drug use problems tend to exploit people who trust them and make a habit of wasting relationships. It is common for them to regret having done so, and therapy can sometimes help clients recover those friendships. However, it can be delicate work because of the intense emotions and feelings, including betrayal, on both sides.

A Cautionary Note

Therapists should use caution in teaching skills that facilitate social support, such as reflection of feelings and other communication microskills, to clients who will use them in harmful ways. For example, sex offenders and criminal offenders may use advanced social support skills to victimize and exploit others, taking targets into their confidence only to betray them. The clinician must use careful judgment to ensure irresponsible clients do not use newly learned social skills to deceive, betray, or hurt the people around them. In our experience, empathy and responsibility need to be rehabilitated in such clients before they can be trusted with skills that engender the trust of others.

Teaching Clients to Recognize Empathy in Others

Many clients get themselves into trouble by trusting or becoming attached to people who are harmful and exploitative. This often happens in romantic relationships; a person marries or otherwise becomes attached to a partner who is not empathic. A battering husband may indeed love his wife, just as an abusive father may care for his children. However, these men have little or no intact empathy for the people they love. Love and empathy are not the same.

Clients can benefit from learning to recognize the presence of empathy in others. The telltale sign is whether the client feels understood by the other person, as if the person understands the client "from the inside out." Many clients need to be taught to recognize this fundamental phenomenon because they are simply not familiar with it.

Recognizing empathy is a skill that can be used throughout one's life. Teaching a client to recognize empathy or the lack of it in another person and to avoid confiding in nonempathic people can help them avoid many difficult situations and be able to recognize a vital aspect of healthy friendships. Perhaps the best standard of measurement for empathy is the therapeutic relationship. Clients can use it as a model, especially those who lack empathic relationships in other areas of their lives and have no way to compare the quality of relationships.

Identifying Sources of Social Support for Change

Once able to recognize empathy, the therapist should help the client identify sources of social support in their environment. In determining whether to discourage or increase a continued association, the therapist can explore with the client the value of reciprocal relationships and the presence of empathy in the person or group. It also helps to assess whether the person or group contributes to the presence of the precursors of change. Social skills training can be used to strengthen and reinforce positive relationships.

The Concentric-Circle Technique

Lazarus's (1989b) concentric-circle technique is an excellent means of gauging the closeness of the therapeutic relationship and the level of the client's self-disclosure to the therapist. It involves drawing five embedded, or concentric, circles on a full piece of paper. The inner circle represents closeness and trustworthiness, with each outer circle representing a degree of distance. The client is asked to write down the names of important people in their lives within an appropriate concentric circle, according to how much they trust that person. As part of developing social support, this technique can be used to gauge client closeness to people in their environment. Once this is done, the client can discuss whether the person in question contributes to their social support toward change, or whether the person is harmful to it, or even merely tangential.

The technique is useful in several ways. If a client has a habit of taking people into their confidence who then betray or abandon the client, the technique can be used as a tool to help delineate and establish boundaries. In other words, clients who have such boundary difficulties can begin to establish criteria for allowing a person entrance to each inner circle. For clients who have experienced pain from unwisely disclosing personal information to inappropriate people, this technique helps to determine who is a worthy candidate for this precious trust. If a client associates pain with allowing a person into the two innermost circles, this indicates that they need to be more perceptive and discreet. Another use for this technique is to count the number of people the client has allowed into the two inner circles. The therapist can ask the client how it feels to have so many people with intimate knowledge of them.

At the opposite polarity, some clients have allowed and plan to allow no one into the two innermost circles. This has consequences and a set of corresponding feelings. People who have shut out virtually everyone from emotional intimacy may feel alienated, lonely, sad, or resentful. The approach would be, once again, to determine what it would take to allow someone into those two inner circles. A client's criteria may be quite irrational, from too rigid on the one hand to too demanding on the other. A therapist can help clients formulate more realistic beliefs to ease the negative feelings associated with being disengaged from one's fellow human beings.

Identifying People Who Are Opposed to Therapeutic Change

Curiously, in almost every difficult client's life, there are people who oppose the client's therapeutic change. In such cases, beneficial change runs at cross-purposes to that person's intentions for the client. Some people need the client to remain exactly the way they are. Dysfunctional families tend to resist the improvement of one of its members because family systems naturally seek consistency, or homeostasis. Virtually all therapists have encountered these challenging situations. Many difficult clients swear that the people who are most harmful to them are also the most important people in their lives. Thus, helping clients identify these people's harmful or negative influences is often a major therapeutic task. This can include addressing the impact of influencers across social media platforms who may have an outsized influence on client's lives.

The therapist should assess, perhaps directly by asking the client, whether people in the client's life encourage therapeutic change. The client must be able to answer specific questions. For example, do the people in question tend to quell the sense of necessity, discourage confronting, or destroy hope? Do they seek to undermine efforts toward change? In the case of enabling, does a caring family member or spouse actively discourage the willingness to experience anxiety? Such questions can be formulated easily enough, and the therapist can present the evidence to the client for their consideration. The therapist can also ask some clients how they feel about themselves or their lives in the presence of the person to get a kind of visceral assessment. Another approach to facilitate insight is to ask what the person wants the client to believe about them. The overall goal is to help a client be able to gauge who is not supportive of their growth; in this way, a difficult client can recognize the consequences of associating with specific people and either avoid or directly address this source of needless anxiety and trouble.

Recovering Lost Sources of Social Support

Many clients live with the hidden regret of having lost relationships with close friends or family members who once provided social support. The relationship may have been lost due to the client's acting out, betrayal, manipulations, or dishonesty. Clients can feel considerable shame and guilt at the mere thought of these lost relationships. The client may try to convince themselves that the relationship was unimportant or that the other person was at fault. Many clients also use the excuse that the person in question "never really understood" them, so there is no loss when, in fact, such people may have understood a bit too well.

In some cases, social support can be rehabilitated with those relationships. In many other instances, the residual distrust is so great that the relationship is, for all practical purposes, beyond repair. Nevertheless, therapy can help determine how and why such relationships fell apart to begin with so the client can avoid making the same mistakes with new relationships. Help in recovering relationships is not recommended for clients who are criminal or extremely manipulative. In such cases, they may only again victimize people who were happy to be free of them. Protecting others from such clients is often the primary concern until they are no longer destructive in relationships.

Exploring Trust

For many difficult clients, the issue of trust is surrounded by irrational beliefs and righteous indignation. These clients must learn that trust has to be earned and that it is not a birthright. Many clients exploit and betray others, only to later complain that no one trusts them, as if they have been deeply wronged. Such blaming behavior can be addressed by setting up scenarios, role-plays, role reversals, and dialogue in which such clients can experience the effects of their manipulations. The purpose is to produce empathy in the client toward those who no longer trust them. In a role-play, the therapist can assume the role of the client, displaying the same behaviors that led to a loss of trust, followed by a role reversal. Responses and feelings can then be explored, which may lead to understanding how the relationship broke down.

Standard Social Skills Techniques

Standard therapy techniques and approaches can help clients develop social support networks. These are described in introductory textbooks on counseling and psychotherapy and are not described in detail here. Each is valuable in teaching clients how to find, build, and maintain supportive relationships based on empathy and caring.

Group Therapy

Group therapy provides a context and format conducive to developing social support. A well-run group can be tremendously powerful in teaching clients the power of empathy and compassion. The amount of empathy generated in a group that has achieved the working stage is exponentially more powerful than that of a single therapist. In a group, a person can come to experience empathic, supportive friendships that if provided by an individual therapist would violate ethical codes on dual relationships.

For a group to achieve the working stage, its members need to have progressed through the early-stage disagreements, personality conflicts, and the like and proceed to provide high-quality help to each other (see Corey, 2017). This is a standard aspect of the group process. The group therapist is only minimally active at the working stage unless carrying out a specific set of therapeutic techniques. In this regard, group therapy has far more utility than individual therapy. Ideally, group and individual therapy are done concurrently. They need not interfere with each other, and each can enhance the progress of the other. Done well, it is a model of empathic relationships. An additional benefit of group therapy that is not found in individual therapy is the opportunity the client has to learn helping skills and apply them in a therapeutic setting (Yalom, 1995). Unfortunately, we have seen many group therapists who have not experienced a group at the working stage due to limited training or poor treatment planning by agencies.

Examining Maladaptive Beliefs About People and Relationships

Difficult clients can possess an amazing array of dysfunctional, maladaptive beliefs about people and relationships. These can profoundly influence the kinds of people with whom they associate and the relationships they form and keep. When firmly held, maladaptive beliefs about relationships and people can influence a client to become attached to exploitative and self-serving people. Such beliefs also influence clients to develop behaviors that will subvert, undermine, and eventually damage potentially valuable relationships.

Many dysfunctional, maladaptive beliefs seem sound at first glance but have a logical flaw in them that, when acted on, can sabotage relationships. Identifying, disputing, and changing maladaptive beliefs can have an amazing effect on a person's willingness to engage in intimate relationships and choose to enter healthy, stable relationships. As with other maladaptive beliefs, clients require therapist support in examining personal belief statements, such as the following:

- People are sources of anxiety and pain.
- I can't live without love.
- I'm only alive when others are paying attention to me.
- If I can please people, they'll like me.
- I need other people to tell me how to be and who I am.
- I must have a person to whom I can give up control of myself.
- A friend is someone who will rescue me from my problems.
- Even cruel attention from others is better than no attention at all.
- I'm secure only when others are around.
- Only fools would be interested in supporting me.
- The people stupid enough to support me cannot be trusted for that reason.
- Nobody ever really cares about anybody.

- People only care about themselves.
- The presence of others threatens my survival.
- People are tolerable only when they are under my control.
- If I stay away from the company of others, they cannot hurt me.
- I'm not as good as anyone else and thus deserve to be treated badly.
- If I'm in conflict with someone, it's better to end the relationship.
- If a relationship requires work, it's not worth keeping.
- If I run into difficulty with one person, I can always find someone else.

A CLOSING NOTE ON SOCIAL SUPPORT

Contributions of social support to therapeutic change have long been discussed in psychotherapy, although its centrality as a client-specific change factor varies across theoretical perspectives. In terms of the CHANGES model, social support seems to enhance the potency of other precursors insofar as support by others can serve as a reminder to confront an issue, increase necessity, sustain hope, and take consistent action. When social supports are not in place, it is incumbent upon the therapist to determine the nature of relationships in a client's life so as to facilitate engagement of high-quality supports and confront the deleterious effects of negative or interfering persons.

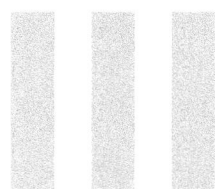

THE CHANGES ASSESSMENT

11

Rating Client Potential for Therapeutic Change

There are many active ingredients of change in addition to the seven precursors. Many therapy outcomes, such as finding meaning in life or a cathartic release of emotion, are also responsible for changes (Hanna & Ritchie, 1995). In other words, what looks like an effect of change processes can also become a cause in its own right, creating a snowball effect of more change. Kant (1787/1929) noted that each effect contains within it a seed of causality. This seems to be true in cases of therapeutic change. Certain variables can be both cause and effect. A profound insight, for example, is an effect brought about by multiple precursors, but the insight can have enormous additional effects on behaviors, feelings, and cognitions. However, insight and other variables are not listed as precursors because they are more of an effect than a cause.

In the case of second-order change, each change factor plays a role in combining, culminating, recombining, and subsiding in a kind of global chain reaction. There are so many cognitive, metacognitive, physiological, affective, behavioral, interpersonal, and environmental factors operative in therapeutic change that studying it is like tracking and gauging the motion and influence of all the emerging bubbles in a cauldron of boiling water. Nevertheless, the seven precursors do explain a significant portion of therapeutic change. Not completely, of course, but enough to provide a manageable and workable explanation that informs the work of therapy.

Thus, the CHANGES assessment (see Table 11.1) is not offered as some unequivocal representation of the change process. It is a pragmatic clinical tool for understanding and implementing change with difficult clients. The model is

Case examples have been disguised to protect client confidentiality.

https://doi.org/10.1037/0000451-012
Therapeutic Change With Difficult Clients: Precursors and Techniques in the CHANGES Model, Second Edition, by B. D. Wilkinson and F. J. Hanna

TABLE 11.1. Assessing for Active Client Precursors: The CHANGES Assessment Form

Precursor and relevant marker	None (0)	Trace (1)	Small (2)	Adequate (3)	Abundant (4)
C: Confronting the problem					
• Steadfastly faces problems					
• Sustained attention to issue					
H: Hope for change					
• Positive future outlook					
• Coping/humor intact					
A: Awareness					
• Able to identify problems					
• Identifies thoughts/feelings					
N: Sense of necessity					
• Expresses desire for change					
• Feels a sense of urgency					
G: Grit					
• Able to manage anxiety					
• Willing to take risks					
E: Effort or will to change					
• Eagerly practices skills					
• Cooperative; high energy					
S: Social support for change					
• Network of friends/family					
• Many confiding relations					

Total score: _____/28

Scoring guide[a]

0–6: Change unlikely. Educate client on change. Focus on precursors with lowest ratings.
7–14: Change limited or erratic. Educate client and focus on precursors with lowest ratings.
15–21: Change steady or noticeable. Use lowest rated precursors to stay on track.
22–28: Change occurs easily. Highly motivated client. Standard approaches work well.

[a] Scoring intended only as a general guide to a complex process. Some precursors may be more potent than others for individual clients and/or individual contexts.

insufficient to explain the vastly complex change process itself. However, the seven precursors of the CHANGES model, in their nonlinear, interactive interdependence, can certainly provide a skeletal outline that indicates the bones of change, even if it does not fully capture the finer points of flesh, fluids, tendons, and tissues.

EMPIRICAL VALIDATION OF THE ASSESSMENT

Each of the seven precursors enjoys empirical validation in the psychotherapy literature as a catalyst of change. The assessment form itself, although not empirically validated, is empirically informed. It may be a mistake to exclusively insist on empirically validated treatment approaches (Beutler, 2000; Beutler & Forrester, 2014). We might alternatively insist that therapy practices be empirically effective and clinically sensible. The CHANGES assessment is neither a psychometric evaluation nor a diagnostic instrument. Rather, it is a clinical guide for identifying the presence or absence of each of the seven change factors. It is a tool, a clinical heuristic, that reduces some of the mystery surrounding difficult clients and simultaneously indicates a path toward change.

The CHANGES assessment has been particularly useful in clinical practice in charting a course of approach to difficult clients. It has been used in various forms with many populations having difficulty with the change process or not responding to treatment. We have given the assessment form to more than 1,000 therapists and trainees to rate their most difficult clients. Almost without exception, those clinicians have reported that their clients were indeed lacking in the precursors of change, and they routinely reported that the assessment was helpful in the treatment planning process.

The CHANGES assessment has helped put together treatment programs for antisocial, borderline, and narcissistic clients, as well as adolescent criminal offenders, adolescent sex offenders, and alcohol and substance users in general. The assessment was also found to be helpful with adolescent female and male victims of sexual abuse and adults in general who were perceived or reported to be resistant to therapy. It appears to be transcultural in its capacity and relevance to describe human change across cultures.

USE OF THE CHANGES ASSESSMENT FORM

Fortunately, this assessment is among the most simple and user-friendly that a clinician is likely to learn or use. The form is helpful for individual practice, individual or group supervision, and consultations in which difficult clients are the focus of attention. The form can reveal a configuration or pattern among difficult clients in a way that also makes its use in group contexts valuable. As seen in Chapter 12 in this volume, precursor activation among difficult therapists can be rated, too.

The assessment need not be applied to clients already doing well in therapy because the precursors are already amply present in such clients. Applying the CHANGES model with these clients might deter progress away from change. The assessment is of greatest help in identifying precursors that need attention when a client makes slow or inconsistent progress toward change.

Rating the Precursors

The ratings for the assessment form are based on the clinical marker sections for each precursor in their respective chapters. These sections provide a guide to determining the presence or absence of a precursor for the purpose of rating it on this form.

Each precursor is rated from 0 to 4. At 0, there is no evidence of that particular precursor with the client in question. At 1, just a trace of the precursor shows itself only rarely, or perhaps a hint is given every now and then. A rating of 2 indicates an identifiable but small degree of presence of that precursor; it is not much, but it is also not as elusive as in the trace rating. With a precursor rated 2, the clinician knows there is at least something with which to work. At a rating of 3, the precursor is evident and readily apparent and, therefore, active in a person's approach to therapy, change, and resolution of problems. Precursors rated 4 are extremely evident, and therapists often find themselves spontaneously admiring these qualities in a client. Ratings of 4 are seldom found in difficult clients. Total scores range from 0 to 28, with higher scores representing greater levels of precursors for change.

We strongly recommend consulting with clients to ascertain the presence or absence of precursors. Clients seem to have no difficulty understanding the precursors. Difficult clients can be willing to assess the presence of their precursors and their own potential and ability to increase them. Of course, with highly manipulative or deceptive clients, caution is needed because the intentions of these clients are not always honest. Nevertheless, the client should be consulted whenever possible and appropriate. Consulting with clients has worked well in clinical settings and is often therapeutic in itself in terms of its educational function. Additionally, both trainees and experienced therapists seem to be able to rate difficult clients easily and accurately.

Do Not Include the Therapist in Rating Social Support for Change

The therapist is expected to provide social support for change, but including the therapist in the rating of this precursor will confound the actual score. The focus of the assessment in this context is the amount of support in the client's environment. The therapist is viewed more as an agent of social support than a source of it in the client's living environment. This point is debatable, of course, but social support scores seem more accurate when the therapist is not included.

Rate the Actual Presence of Precursors

Rate only the actual presence of precursors as manifested in the here and now and according to the current problem or issue. One could rate the person according to their general ability or potential for change, but this is misleading and should be avoided.

For instance, Darnell was a 32-year-old bisexual African American who identified as genderfluid. The therapist used the CHANGES assessment to rate their overall capacity for change, based upon the evidence that Darnell had instituted many changes in their life. As a result, it was assumed that Darnell had high levels of precursors, and Darnell confirmed as much when speaking in general terms. When the counselor asked, "How willing are you to experience anxiety or difficulty, or to put forth effort?" Darnell said, "Super willing. I face it every day and work hard every single day." Of course, this was true in many instances since Darnell was avidly working to support lesbian, gay, bisexual, transgender, queer/questioning, plus (others) community members as a resource coordinator for a nonprofit advocacy group.

Yet a capacity for grit, effort, awareness, or any other precursor says nothing about particular instances in which change is needed. While Darnell exhibited tremendous grit and effort in many areas of life, there were other areas in which it was absent. For one, Darnell sought out therapy because they had recently been diagnosed with Type 2 diabetes and a related lung dysfunction. They needed to build an exercise routine but had tremendous difficulty with motivation in this area. Their CHANGES assessment score for the issue of maintaining an exercise routine was

1. Confronting the problem 2
2. Hope for change 1
3. Awareness 3
4. Necessity for change 3
5. Grit, or readiness for difficulty 1
6. Effort or will toward change 0
7. Social support for change 2

Total score: 12

The score of 12 was low but also accurately indicated that change was limited or erratic, as it had proved to be in therapy. Darnell was aware of the problem and felt enough necessity to seek therapeutic support. They would also discuss the issue in therapy but would change the subject after a few minutes to discuss other topics. The therapist had noticed this avoidance pattern but followed Darnell into new areas of conversation based on the belief that they had the capacity for change but just needed validation and consistent support.

Through consultation, the therapist was encouraged to review the CHANGES assessment scores in session. It became apparent through that conversation that Darnell had lied to the therapist about doing any exercises because they were embarrassed and wanted the therapist to be proud of them. Throughout his entire life, Darnell had never exercised and did not have any friends or family who exercised. It was a very unfamiliar activity that also connected to childhood traumas in which Darnell had been made fun of by peers due to their weight. The avoidance of exercise was an avoidance of shame, and Darnell had to work with the therapist to disconnect these fused issues.

If you are interested in rating the client according to their overall ability for change, then, by all means, do so. This might indicate what expectations to have or what to aim for in a client's overall growth potential. However, when a client comes into therapy, it is important to rate the presenting problem and have the client help you complete the rating if appropriate.

With clients who are court referred to therapy against their wishes, therapists should rate each person not only according to the issue at hand but according to the perceived change ability as well. When rating perceived ability to change, it might help to ask if the person has ever made changes in their life relative to, for example, diet or exercise, anxiety, depression, relationship problems, or work issues.

SCORING: DETERMINING THE LIKELIHOOD OF CHANGE

The scoring of the CHANGES assessment is a general guide. Nothing is carved in stone. Since the assessment is a clinical guide and not a product of psychometrics, no claims can be made as to the precise accuracy of scoring. However, scoring in clinical practice has been remarkably sufficient as an indicator or predictor of therapeutic change. In our experience doing supervision, trainings, consultations, workshops, and lectures on this topic, there is remarkably high agreement among raters on the presence or absence of various precursors. In any case, being off by a point here or there on a precursor is insignificant since the real purpose of the scoring is to identify the configuration of precursors for each person and find which precursors need attention.

Another source of scoring ambiguity is that precursors can wax and wane from day to day, hour to hour, or session to session. If we were to determine that the precursors could be consistently ranked according to potency, the scoring would be weighted, with the more potent precursors getting a higher percentage of the overall score. However, the variable nature of presenting concerns and the overarching complexity of human beings makes this possibility unlikely.

The precursors can also be far more present with one therapist than with another, as seen in Chapter 12 in this volume. Nevertheless, with difficult clients—who typically receive very low ratings—the precursors are likely to be far less variable than with clients whose ratings are in the middle range and above. It is sometimes helpful to do a quick rating on the precursors with the same client from week to week or even from session to session. A client can be consistently low for the first five sessions and then go consistently higher as therapy continues. Change is a skill, and it is born of knowledge and work. The precursors will naturally emerge in greater magnitude as clients learn *how* to change, especially when its suits their purpose.

Change Unlikely

When a client has a total score of 0–6, change is unlikely any time in the immediate future. As long as the score remains this low, standard therapy approaches may be inappropriate with this client, who might be better served by focusing on growing the individual precursors. Clients who score this low should be informed—in an empowering way—about what it takes to change.

Insofar as no one completely understands the therapeutic change process, clients should not be expected to either. In some cases, psychoeducation on the precursors themselves can be helpful and important. Therapeutic change can also be presented as a skill to be learned. Education on the CHANGES model and the change process itself is generally more effective when a client demonstrates some sense of necessity for change. The sense of necessity can drive a client to learn more about what it would take to make things better. However, education about change and the precursors may be inappropriate for other clients, such as those with paranoid or antisocial personality disorders. For the latter, it is more important, initially at least, to go straight to work on implementing the absent precursors.

Change Limited or Erratic

A score of 7–14 indicates that change will take place at a rate that is slow and unnoticeable. The process will be limited, erratic, and fraught with setbacks but may also yield triumphs so small they might not be noticed at all by outside observers or even the therapist. However, these small changes can seem large to the client. Educating the client is important at this level as well.

It may be of great help for a client to understand why progress is not happening as quickly as they might like. After six sessions of working with a client who had been diagnosed with obsessive-compulsive personality disorder, the client complained to the therapist that he was not making progress. The client was relieved when the therapist explained the precursor of grit and noted how the client sometimes deftly avoided or redirected conversations in therapy when he became uncomfortable. Of course, it took a while for the client to realize the ramifications of this issue, but it was an important educational experience that he did confront after a couple more sessions.

It can also be helpful for clients to understand why changes are being made. Clear communication between therapist and client is paramount in specifying such intent. In addition, for clients who had lower scores earlier in the therapeutic process, increased scores may indicate it is time for further education on change. The idea of change as a skill may be more readily grasped when scores reach a more advanced level, especially when a high sense of necessity is present.

Change Steady and Noticeable

A score of 15–21 indicates the person will make changes at a rate that is steady and noticeable, both to the therapist and to others. There is little need for the precursors approach with many of these clients. Nevertheless, scoring can be used to determine which precursors are lowest. A person's rate of change can still be increased by focusing on the precursors with the lowest scores.

Change Highly Likely

When the total score is above 21, the client will likely achieve change with relative ease. Using the CHANGES model is unnecessary and perhaps a waste of time since the client may respond to almost any appropriate therapeutic tool, technique, or procedure. Similarly, any theoretical approach will tend to be effective as long as the therapeutic relationship is established and the interventions and in-session rituals used match the client's problem and needs. Client-centered, psychodynamic, or cognitive therapy, for example, can work equally well with a client at this level.

Clients with high scores can be very insightful and surprisingly sensitive to the therapist's approach. Therapists do not become preoccupied with or worry about these clients, simply because these people are likely to do well even on their own. In many cases, these people do not need therapy and are likely to improve with the right book or with a friend who listens. But when they do roll up their sleeves with an intention to change unwanted conditions, the therapeutic sparks fly.

EXAMPLES OF ASSESSMENTS

In Chapter 2 in this volume, we describe three difficult clients, Tommy, Joy, and Ricky, who did not achieve success in therapy. To illustrate the CHANGES assessment, these clients are revisited here, and their configuration of precursors is charted. For convenience, we briefly summarize each case before reviewing their ratings. Again, no diagnoses are mentioned in these examples, although therapists will recognize many of the symptoms. Diagnoses are omitted to focus on the precursors and avoid the stereotypes that can occur with dependence upon diagnoses for client characterizations. In addition to the case descriptions, further details that were gathered by the CHANGES assessment are provided.

Case Example 1: Tommy

Tommy was the charming and thoroughly self-centered man who claimed he could not stop his promiscuous behavior and who was both proud of it and ashamed about hurting his wife, Julie. He had been to three previous therapists and had fought with them all. His chief anxiety was the fear of going to hell for his immoral behavior. He was in therapy at the insistence of Julie, who had no idea of his sexual improprieties. His apparent interest in therapy amounted to an attempt to reduce his anxiety without changing his behaviors. He wanted a therapist who would tell him that he was fine, that he was a good person, and that he would not go to hell.

Therapy was at an impasse for Tommy. No progress was made after three sessions, nor did it seem likely. The therapist soon realized that Tommy came to therapy with a purpose quite contrary to anything that could be regarded as therapeutic. His precursor profile looked like this:

1. Sense of necessity (rating: 2): When asked directly, Tommy readily admitted that something had to change. He said he could not "go on like this" for much longer without "burning out." On the other hand, he continued all his problem behaviors with the same intensity.

2. Readiness for anxiety (rating: 0): Tommy was unwilling to feel the anxiety associated with his behaviors and attitudes. He evaded or deflected any questions that led to exploring fears, feelings, or difficulty. He insisted the therapist help him feel better.

3. Awareness (rating: 1): Despite all his contradictory statements, Tommy seemed to know he was feeling bad and that something was wrong but saw his behaviors as a problem only from a religious perspective. He insisted that his infidelity was natural for a man and unrelated to the tension between him and his wife. It was fine as long as she did not find out. He seemed to have little awareness that his behavior was out of his control. In addition, he had trouble identifying specific beliefs and feelings.
4. Confronting the problem (rating: 0): Tommy was doing nothing in terms of studying or contemplating the issues that brought him to therapy. He expected the therapist to tell him what he needed to know.
5. Effort or will toward change (rating: 1): Tommy was also doing nothing to resolve his problem other than going to therapy, which was a kind of effort in itself. Once in therapy, he showed little actual cooperation.
6. Hope for change (rating: 1): The therapist asked if Tommy ever had thoughts of "giving up" or if "life wasn't worth living." He said no but that sometimes he did not think he could ever be happy. He could not envision a positive outcome.
7. Social support for change (rating: 2): Tommy reported that his wife was supportive of him to get through his therapy, although she had no idea he was cheating on her. He also said that although most of his friends admired and encouraged his promiscuous behaviors, two had told him that he was making a mistake.

Tommy's CHANGES profile, which yielded a total score of 7, accurately showed that he was not likely to change, just as his therapy history had indicated. He was lowest on precursors 2 and 4, and 3, 5, and 6 were also low.

Case Example 2: Joy

Joy was the former showgirl from the Lake Tahoe region in Nevada. She was so involved with new age pursuits that she insisted on avoiding anything that was not "positive," which meant anything associated with the slightest pain or anxiety. She also claimed that she was clairvoyant and psychically gifted and that therapy might help her to clear the blocks that stood in the way of her becoming an occult "adept." She also believed that therapy could help her forget and cut herself off from any negative emotions and relationships in her past. Joy reached an impasse in her therapy after eight sessions, impatiently claiming to have derived no benefit, and she often lectured to her therapist on several new age principles.

In Joy's mindset, change was clearly desired but, because of the conditions she set, almost impossible to attain. For her, achieving change was like insisting that a fire be started with wet logs and then becoming upset because there was no combustion.

1. Sense of necessity (rating: 3): Joy made it clear that change was important and necessary for her to reach her goal of attaining a high state of being.

2. Readiness for anxiety (rating: 0): Joy would not consider or allow any anxiety into her consciousness and became visibly upset at the prospect of it. Although it was clear that she was consumed with anxiety, she insisted on staying "strong." She believed she had to remain "above" all negativity.

3. Awareness (rating: 1): Joy was not aware of the obvious deeper issues within which she was immersed, nor was she particularly adept at identifying beliefs and feelings. Any awareness of negativity was shunned.

4. Confronting the problem (rating: 0): Caught in a fixed belief system, she was not willing to explore the painful or confusing aspects of her life and clung instead to her belief in the self she would become. She believed that addressing and attending to difficult issues was a grave mistake.

5. Effort or will toward change (rating: 1): Joy meditated regularly and used it as a means to ease her anxiety, but as an escape rather than as a means of dealing with it. She showed little cooperation once in therapy itself.

6. Hope for change (rating: 2): Joy was expecting a future state of peace and attainment, but not without difficulty. That was realistic to her, but how to proceed was not.

7. Social support for change (rating: 1): Joy reported many superficial friendships but only one friend in whom she could confide. However, she often argued with her friend, who she felt did not really understand her potential.

Joy's CHANGES profile, with a total score of 8, presents an interesting configuration. She was relatively high in necessity for a difficult client. She also made some effort, but in this regard, her effort was not focused on psychotherapeutic change. She showed little or no cooperation in the therapy sessions themselves. That she had reached a therapeutic impasse is no surprise.

Case Example 3: Ricky

Ricky was the street gang member, a drug dealer to many children, and he had fathered at least three children with three different adolescent girls. Although he never exhibited any overt violence, he was adept at displaying nonverbal threats of intimidation. Ricky was also highly intelligent, manipulative, and spiteful toward the world and people in general.

Ricky did not disclose anything other than superficial information in 6 months of treatment. He viewed therapy as an intrusion into his personal life and therapists as tools of the police. For him, therapy was an attempt to make him weak and rob him of his freedom and virility. Therapists were avowed enemies, and therapy was a war that he could easily "win." Unfortunately, he was correct.

1. Sense of necessity (rating: 0): Ricky claimed to have had a great life until the police interfered. There was nothing wrong with him, and nothing needed to be changed. If anything needed to be changed, it was to get out of the treatment program.

2. Readiness for anxiety (rating: 0): Ricky was not willing to experience anxiety or difficulty in any way. He believed that drugs existed to relieve that sort of thing.

3. Awareness (rating: 0): Ricky gave no indication that he was aware of any problems. He was certainly intelligent, but not in identifying problems, feelings, or issues. He did not see his drug and alcohol use as a problem.

4. Confronting the problem (rating: 0): Ricky was not confronting any aspect of his life other than sex, drugs, and criminal activity.

5. Effort or will toward change (rating: 0): Ricky was not putting any energy into changing.

6. Hope for change (rating: 0): Ricky did not believe that he would live past the age of 21. He was certain he was going to die, having seen several members of his own and other gangs shot and killed. As far as he was concerned, the future held little or nothing for him.

7. Social support for change (rating: 2): Ricky reported that he cared for his mother and his 14-year-old sister and was extremely protective of them. Both loved him dearly and came to family counseling sessions, wanting to help. His gang members were also supportive.

Ricky's CHANGES profile presents an extraordinary therapeutic challenge. Therapeutic change is nowhere in sight with that configuration of precursors and a score of 2. Confronted by such difficult clients, therapists are likely to give up on the prospect of change, and a caseload of such clients is a recipe for burnout (J. J. Kim et al., 2018). This challenge clarifies how therapists have a unique configuration of precursors for each client, as we examine in Chapter 12 in this volume.

USING THE CHANGES ASSESSMENT WITH GROUPS AND FAMILIES

The CHANGES assessment is easily used in therapy with groups or families that are considered difficult or not exhibiting therapeutic change. In groups, it is helpful to rate each group member on the CHANGES assessment first and then to rate the group as a whole. Clinical impressions of groups indicate that many groups have a unique character of their own. Rating a group as an entity provides insights into the precursors needed to enhance the group process toward change.

Similarly, in family therapy, each family member can be rated to determine their configuration of precursors (Wilkinson & Hanna, 2018). As in the case of groups, Whitaker (see Simon, 1985) pointed out that a family system has a character and personality of its own. A difficult family can also be rated as a whole to determine which precursors are missing in the family system and need attention. Many of the techniques given in later chapters for implementing the precursors can be adapted for therapy with groups and families.

Even when not mentioned specifically, many of the techniques described have been adapted as appropriate for use in group and family settings.

USING THE CHANGES ASSESSMENT ACROSS VARIOUS DIAGNOSES

The CHANGES assessment outlines and elucidates the potential for therapeutic change across the wide range of disorders classified in the various incarnations of the *Diagnostic and Statistical Manual of Mental Disorders* (5th ed., text rev.; *DSM-5-TR*; American Psychiatric Association, 2022). It is applicable to people with depression, anxiety, and personality disorders, as well as substance use disorders. In terms of disorders with organic or congenital components, such as bipolar disorders, traumatic brain injury, or schizophrenia, the CHANGES model may still be applicable, but further evidence is needed to determine its efficacy with such populations.

The *DSM-5* defines a *personality disorder* as consisting, in part, of unchanging, dysfunctional personality traits (American Psychiatric Association, 2013). Thus, part of the essential definition of a personality disorder in general is precisely the lack of qualities and conditions described by the precursors. In one important sense, a personality disorder is a lack of therapeutic change characteristics. The CHANGES model seems to apply quite appropriately in work with clients who have received various personality disorder diagnoses, but each personality disorder has its own style. Although that style may differ from disorder to disorder and person to person, the different precursors apply to all at a fundamental level that lies well beyond personality traits and styles. For example, an antisocial client and a borderline client may both be low in the confronting precursor, and each will have their own style of avoiding confronting, but both are likely to improve if confronting is activated. A person with a personality disorder who also has ample presence of the precursors will tend to change faster than the person with the same personality disorder who is lacking in precursors. The therapist's task is to use strategies and techniques that will activate the precursors.

A FINAL NOTE ON RATING CLIENT POTENTIAL FOR CHANGE

It seems to be the case that clients with a high degree of precursor activation will find therapeutic change likely and welcome. Conversely, a client unlikely to move toward beneficial change lacks the critical mass that an adequate presence of precursors provides. The CHANGES assessment is a relatively simple clinical tool to determine precursor presence and level of activation, so its implementation has considerable flexibility. Therapists are encouraged to use it in session with clients to facilitate dialogue about therapeutic change, between sessions in the process of case conceptualization and treatment planning, as well as in training, supervision, and consultation to better understand why client change is not taking place. In the next chapter, we explore its use as a supervision tool for identifying when therapists are hindering the change process.

12

Rating Therapist Interference in the Change Process

This chapter deals with what is probably the most underestimated aspect of therapy with difficult clients: therapist disposition and reactivity. It is usually referred to as countertransference and is often given lip service but is seldom given the attention it deserves. The idea of countertransference is one of the most valuable contributions made by psychodynamic schools of therapy. Unfortunately, the intricate phenomenon itself has been obscured by terminology that is not readily translatable into popular cognitive and behavioral languages (Holmes, 2017; Mills, 2004).

Transference is when a patient treats the therapist as they would a significant other, such as a parent. *Countertransference* is defined as a therapist's reactions to a client's transference. Freud (1910) originally described countertransference as the therapist's unconscious emotional response to the patient. Freud recognized the potential adverse effects that a therapist's emotional reactions can have on a client. Psychodynamic scholars eventually broadened the concept to include conscious and unconscious reactions to a client based on a therapist's own past relationships (e.g., Kernberg, 1967; Reich, 1951; Segal, 1977).

Modern psychodynamic therapists and scholars tend to agree that therapists' reactions to clients' provocative and evocative behaviors and statements must remain under conscious control (Bager-Charleson, 2010). It is further agreed that the therapist must develop sufficient maturity to avoid seeking to meet their own needs with clients in therapy sessions. When therapy has stalled, countertransference

---

Case examples have been disguised to protect client confidentiality.

https://doi.org/10.1037/0000451-013
Therapeutic Change With Difficult Clients: Precursors and Techniques in the CHANGES Model, Second Edition, by B. D. Wilkinson and F. J. Hanna

is often the cause (Holmes, 2017; Weiner, 1982). Blaming treatment failure on the client's difficulty or resistance is often a rationalization—and a poor one.

Unfortunately, this aspect of therapy is discussed almost exclusively by the psychodynamic schools and is too often ignored or mentioned only in passing by other approaches. When it comes to working with a large portion of difficult clients, a therapist of any theoretical persuasion can get frustrated, irritated, sad, anxious, angry, or bored and can have their own issues, biases, or sensitivities inflamed. Fromm-Reichmann (1950) gave the example of a therapist who, as a child, was forced to listen to his elderly grandmother drone on in a seemingly endless chatter. As a result, he automatically detached from any long-winded communication with anyone, including clients.

Countertransference reactions are not limited to particular diagnoses. A depressed person can provoke anger in a therapist, and so can an antisocial client (Giovacchini, 1989). As a consequence, the relationship degrades; the therapist is no longer an empathic helper and the effectiveness of therapy declines. Whenever a therapist experiences frustration, anger, resentment, hurt, or loss of confidence with a client, the potential is there for these feelings to interfere with client change. When the therapist acts on those feelings, therapy deteriorates into a power struggle or inquiry into "who wronged whom." Therapeutic change, as a goal of therapy, is then disregarded.

DIFFICULT CLIENTS OR DIFFICULT THERAPISTS?

For over 100 years, case studies and research have shown how countertransference can interfere with the change process in even the most well-intentioned therapists (Stefana, 2017). The client influences the therapist far more than is generally acknowledged in the psychotherapy literature (Holmes, 2017; Singer & Luborsky, 1977; Valerio, 2017). Outside of psychodynamic programs, countertransference is rarely prioritized in therapist training. It is also seldom controlled for in research since the phenomenon is difficult to study due to its subjective nature, and the methodological question persists as to how to reliably identify its effects. Countertransference could be confounding a wide range of therapeutic outcome and process studies without our knowledge (Persons, 1991).

In the CHANGES model, the key is understanding that there is nothing pathological or disordered about a therapist experiencing countertransference feelings and attitudes. Quite the contrary, it is a completely natural and routine aspect of doing therapy. What is vitally important to understand is that there is a difference between merely experiencing countertransference and acting on those feelings and attitudes. The feelings evoked and provoked can be used as tools to further understand a client, as psychodynamic therapists know so well (Stefana, 2017). The lesson highlighted across a long history of psychotherapy outcome and process research is that the difference between average therapists and excellent therapists seems to be how well they can manage their countertransference reactions (Hayes et al., 2018; Van Wagoner et al., 1991).

Many difficult clients are experts at thwarting well-intended attempts to establish a therapeutic relationship. They perceive help as a threat and can undermine it with great finesse, which is often one of the reasons they are seen as difficult. Some therapists react with frustration, such as one who said in supervision, "Here I am trying to help this woman, and all she does is block my work at every turn! Frankly, I'm sick of it." This chapter helps therapists make sense of countertransference reactions and uses the CHANGES assessment to show how change processes degrade in the therapist.

THERAPIST INTERFERENCE: A NEW TERM FOR A VENERABLE CONCEPT

The term countertransference represents a vitally important concept with a considerable amount of historical and theoretical baggage that needs repackaging. As an alternative, the term *therapist interference* directly conveys the idea of a therapist's hindrance of the change process, in or out of the context of countertransference. Therapist interference has the advantage of describing the detrimental effects of a therapist's reactions toward a client purely in terms of interpersonal interaction. It can include countertransference issues as well as other factors, ranging from lack of skill to being overwhelmed by a client's complexity, difficulty, or degree of suffering and misfortune. The benefit of this perspective lies in the fact that interference can be viewed in the context of each of the precursors. A therapist's set of precursors toward a defiant, insulting, and arrogant client can be far lower than toward a cooperative, willing, and respectful client. This difference within and between therapists can profoundly affect the change process.

When the therapist acts on negative feelings by criticizing the client under the guise of helping, making covertly hostile comments, or prematurely terminating therapy, a valuable opportunity to better know the client is lost. This is often the hidden benefit of understanding therapist interference. Although many therapists know of it, few develop this essential skill.

ASSESSING THERAPIST INTERFERENCE VIA THE CHANGES ASSESSMENT

The CHANGES assessment can be used to rate the potential of a therapist to inhibit the therapeutic change process. In this section, we list each precursor and show how a therapist with a deficit in that category can interfere with client progress in therapy. This tool can be helpful in supervision, and therapists working with difficult clients can use it to draw a more complete picture of their own possible contribution to an impasse in therapy.

Deficit in Confronting the Client's Problems

The primary clue to the lack of this precursor lies in the disposition of the therapist toward the client's issues. Just as a client can be aware of issues and not confront them, a therapist can also be daunted by a client's problems and choose to avoid them. A therapist must be able to directly identify, address, and work toward the resolution of the client's issues and persevere in a sustained, steadfast manner. If not, the client may not be inclined to confront those issues either, and change will be mysteriously elusive.

For example, Colleen worked in a university counseling center for 7 years before returning to school to earn a doctorate. Colleen was always well dressed, energetic, and likable. George, her client, was a graduate student in history who was in his late 20s and working on his master's thesis. He was low-key, unexpressive, and interpersonally awkward, paying little attention to his dress or appearance. He sought therapy at the university because his wife was threatening to leave him, and he was deeply troubled by the possibility. After 4 months, no change had occurred.

Supervision eventually revealed that Colleen had focused primarily on developing George's social skills but, for some reason, was not addressing the marital relationship itself. Colleen told her supervisor that she thought if George were more "appealing," his wife might not want to leave him. Meanwhile, it was apparent that George had become romantically attracted to Colleen. It was a good way to forget about his troubles with his wife. He interpreted her interventions as a means for him to become more appealing to her instead of his wife. This entanglement became clear when Colleen revealed that, in a way, George reminded her of her husband, whom she wanted to change to become more "sensitive." Not surprisingly, she also had avoided dealing with the relationship issues in her marriage. Colleen eventually confronted this issue in her own therapy.

Sometimes, therapists avoid confronting a client's problem when it seems foreign or bizarre. In one case, a married woman disclosed that her job was in danger and that she needed help processing how to manage interpersonal issues at the office. After a few weeks, she revealed that she was involved in encounters involving sexual dominance with random men she had met online. She met these men first in chat rooms and then in person, where she was having them tie her up in ropes and chains for sexual interludes. Although her therapist admitted concern in consultation, she had not discussed the issue in any depth with her client. Despite the potential danger, the therapist avoided the topic because it was unfamiliar and, per her own admission, frightening. For some therapists with a deficit in confronting, staying in familiar territory feels more comfortable.

Other potentially important issues that can be ignored are unethical business practices, thievery, cheating on exams, and insensitivity to others. If a therapist is uncomfortable with and thus avoids a client's problem, that problem is unlikely to be resolved.

Deficit in Hope for Client Change

If a therapist has little hope that a client will change, it is likely to be contagious through their nonverbal signals and cues to the client. It can show up in vocal intonations and inflections as well as physical gestures and postures. The expectations of a therapist for a client's success indicate how much hope the therapist has for a client. Just as a longstanding body of research substantiates the power of expectancy effects for clients (Constantino et al., 2023; Weinberger & Eig, 1999), a growing body of contemporary research shows that a therapist's positive expectations for client change are profoundly impactful in terms of client outcomes (Bartholomew et al., 2020). The degree to which a therapist maintains hope for client change may even explain 7.3% of the variance in positive change outcomes, as determined by the client (Connor & Callahan, 2015).

Unreasonably positive expectations for a client, however, can amount to fantasy and are not helpful at all. Just as the precursor of hope must be realistic in a client, it helps for a therapist to be able to envision the realistic possibility of a client making changes. Hope is also encouraged by the knowledge that in work with difficult clients, a therapeutic window can open at almost any time, and change, although not evident at the moment, can nevertheless happen as a result. This is a realistic expectation born of actual experience for many therapists. This contagion of hope communicates to and affects a client in various ways, one of which is showing a client that improvement and a better future is possible.

Deficit in Awareness of Client Issues or Own Corresponding Issues

The lack of awareness in a therapist primarily manifests as a lack of empathy and a preoccupation with one's own issues, agenda, or needs. For example, some therapists resent difficult clients who do not change. They may not readily admit to this, of course, but it often results from a lack of self-esteem in the therapist. Being unaware of their own self-esteem needs, a therapist might be using therapy to meet these needs. Such a person needs the belief and assurance that they are competent and successful at doing therapy, as a source of self-esteem. A difficult client can cast doubt on that belief, making a therapist feel awkward, inept, or incompetent. The client becomes a threat to the therapist's self-esteem, and the therapist then detests and resents the client precisely because the therapist detests and resents their own lack of self-esteem. In supervision, this usually comes as a great insight to therapists, who quickly see the new awareness as beneficial.

Another manifestation of the lack of awareness is when a therapist does not know that a client is difficult. In some cases, this is because the level of dysfunction is not apparent. For example, a supervisee worked with a client who was likely to be misusing alcohol but had not disclosed this to the therapist. Since the client did not mention it, the therapist overlooked warning signs and failed to address the issue. When asked if he had suspected alcohol use, it came as a surprise, and he responded, "I wondered what was going on, but I wasn't sure and didn't know how to ask." While the symptoms would have been clear to any advanced practitioner who has studied addictive behaviors, it constituted a fundamental blind spot to this early career therapist.

Other therapist issues can be evoked when doing therapy, and odd consequences can occur if they are not managed appropriately. A male and female therapist co-led a group of substance-abusing teens aged 14–18. During one group session, a female client disclosed that cocaine made her feel sexy and that she would flirt with men of all ages when she was high. At this point, the female cotherapist disclosed that while using cocaine she had had sex with many men and never found fulfillment. She ended by saying that she still had not found fulfillment. The entire group was intensely interested and focused on her startling self-disclosure.

A few weeks later, one of the boys from the group, Timmy, spread the rumor that he had had sex with the female cotherapist over the weekend. It was not true, of course, but the female cotherapist was obviously horrified. In consultation, it was determined that nearly a third of the client population in that agency believed the story. Tina could not get Timmy to confess his deception and asked the male cotherapist to intervene. After considerable effort, Timmy finally admitted his ruse. However, the damage had already been done, as the group dynamic was negatively impacted and the female cotherapist had to deal with the ongoing embarrassment of the issue within the agency setting.

The change process becomes derailed when therapists have so little awareness of their issues that they have difficulty identifying when client issues resonate with their own. A therapist will be inclined to treat such an issue in a client the same way they treat it in themselves or may treat the client in the exact opposite way, perhaps out of guilt. For example, if a therapist has a problem communicating with a partner, they might ignore the same issue in a client. Or they might overemphasize it and coerce a client to deal with it above and before all else. When seen through the filter of the therapist's own issues, a client's problems can be obscured. Empathy becomes impaired. In short, a lack of awareness of one's issues can ruin clinical perception and empathy.

Deficit in Necessity to Help the Client

A therapist can lack a sense of necessity to help a client, which can manifest as indifference or lack of interest. It can result from "giving up" on a client and "giving in" to the unconscious conclusion that change is probably never going to happen. In such instances, a therapist is merely going through the motions of therapy, engaging in conversation without a true intent to help.

The most important characteristic for identifying a lack of this precursor is a distinct absence of the therapist's felt sense of a client's need for change. In other words, the helping instinct that may have been strong at one time in that therapist has diminished or is no longer present. The therapist may feel defeated, apathetic, or bored; they may daydream during sessions with this client or shift the conversation into areas more of interest to the therapist and not of any particular therapeutic value.

Another possibility contributing to a lack of a sense of necessity is if the therapist has the same or similar issue as a client that also has not been addressed, such as drinking or a failing relationship. Self-awareness is required on the part of a therapist to recognize and admit that one's apathy about, or lack of attention to, an issue may be stirred up by a client.

Deficit in Grit, or the Willingness to Experience Anxiety or Difficulty With a Client

Sometimes, a therapist can be overwhelmed by the disrespectful treatment from clients. Such treatment can be discouraging and daunting, especially if the client is overtly or covertly critical of the therapist. For example, some narcissistic clients find it necessary to criticize the therapist to keep their own envy and jealousy under control (Adler, 1992). The thought of wading through this kind of treatment can be disconcerting to a therapist who is not aware that this behavior is to be expected. The end result is that the therapist becomes unwilling to experience the anxiety or difficulty of being with the client. The same is also true of work with antisocial or borderline clients, who can be a source of considerable discomfort to a therapist.

On a different front, some therapists can become overwhelmed by the vast amount of raw, painful emotion of a client. Similarly, when therapists are continuously exposed to the pain of client after client, they might begin to avoid any further exposure to this type of painful experience due to vicarious traumatization (Aafjes-van Doorn et al., 2020; McNeillie & Rose, 2021). As a result, a therapist will avoid any inroads into a client's case that lead to more or similar painful emotions. The therapist may have also reached their threshold of tolerance for the pain of others. If this occurs, the therapist and client may end up with an unconscious collusion that is designed to spare the client any more painful experience, even when appropriate. That willingness or readiness to undergo the anxiety of helping a client with this process is the chief aspect of this precursor.

In other cases, a therapist can be discouraged by the sheer amount of work a client has to do to change, especially when the therapist's assessment of the amount of change needed is vastly greater than what the client believes. Normally, an experienced therapist patiently prepares for the long haul, but less experienced or less patient therapists can become discouraged. In supervision, a supervisee once said, "There is so much work to do here, and the client is so difficult to deal with. She has so little awareness and insight. I don't know if I have the energy to go through it all." When the therapist no longer wants to "go through it all," this precursor is nearly gone.

Deficit in Effort and Will to Work Through Issues With a Client

There is a myth that a therapist should kick back and let the therapeutic process take care of itself. Some therapists believe they can just reflect the feelings and statements of the client without getting involved, which is justified as "maintaining professional distance." Although important, the concept of distance is often misunderstood. Some therapists believe that to roll up one's sleeves and get involved with a client's problem is to risk becoming enmeshed. This is also a mistake.

Lazarus (1989a) noted that success in therapy requires effort on the part of the therapist as well as the client. Simply stated, a therapist has to work hard not only at confronting the problem with the client but also at persuading the client to work hard. Being persuasive is part of working with difficult clients, and so is modeling for a client how to exert energy toward change.

However, when a therapist is doing all the work and the therapy itself is going nowhere, it may be time for a slight withdrawal. If all that effort is not leading to change, it may be time for confrontation or even limited provocation (see Chapter 7, this volume) to activate precursors. A therapist may be helped by the Zen state of being fully engaged in actions while remaining unattached to the rewards, as recommended by Horney (1952/1987). Exerting effort toward helping a client is not the same as becoming enmeshed. It is the therapist's responsibility to model effortful action.

Deficit in Social Support for Facilitating Change

Some therapists are in desperate need of social support. If social support is lacking in the therapist's personal life, therapist and client can become so friendly and close that the therapist may inadvertently seek to keep that fuzzy feeling. Consequently, the therapist may find confronting the client on sensitive issues threatening to the delicate security of the relationship. This most often occurs when a therapist is lonely and wants to be friends with the client or desperately needs empathic supervision or collegial support. To avoid this scenario, therapists must ensure that their social support systems are adequate and fulfilling.

Another social support problem arises when the therapist lacks a professional support system. Although isolation is obviously a unique challenge for therapists in private practice, professional burnout due to a lack of social support is most often observed in community agency settings despite ready access to supervision support (Yang & Hayes, 2020). A therapist should feel, as much as possible, part of a therapeutic team. This can be accomplished by attending conferences and workshops, but it is much more fruitful to engage in positive and productive group, individual, or peer supervision. These latter sources of social support can alleviate feelings of isolation and bolster the presence of a therapist's other precursors. The unique benefits of clinical and reflective supervision to prevent burnout and bolster the precursors will be discussed further in Chapter 16 in this volume.

CASE EXAMPLES

The CHANGES assessment form can be helpful to therapists not only in rating clients but in rating themselves. It can also be effectively applied in supervision or consultation. In Chapter 2 in this volume, we review examples of therapists who had "trained" their clients to be difficult. Such unnecessary difficulties arise when therapist interference goes unrestrained and uncorrected, thereby diluting the change process. We now examine two of those examples in further detail and provide an additional example involving a severe ethical violation. Following each example, the case study therapist is rated according to their own precursors.

Rating therapists is similar to rating clients. The therapist's potential is not the focus of the rating; what is rated is their current attitude and actions with a particular client. Just as client precursors can vary on different issues, therapist precursors can vary widely from client to client. The focus is on the therapist's actions with a particular client in a particular moment or with a particular client over several sessions. The score serves as a general indicator of a therapist's potency as a change agent for that particular client at that particular time.

Rating Kurt's Precursors

Kurt was a 39-year-old therapist working with Janey, a female university senior in her mid-20s. Janey's major complaint was that the men in her life verbally mistreated her. Although she was an engaging person in therapy, there was a certain naivete about her and a tendency to smile even while in pain and to laugh with a shrill, almost tinny tone. After 20 therapy sessions, her current male partner continued to verbally abuse her, and she stopped showing up for sessions and then terminated by telephone.

In reflective supervision, Kurt discovered that he harbored a hidden belief formed in high school that "girls with fake laughter" and who "smiled for no reason" were "stupid" and not likely to ever achieve any significance in their lives. At a deeper level, he never took Janey seriously as a person. When confronted further, he also realized that he was attracted to women like Janey in high school but that he also "resented the popular girls" because they wouldn't go out with him. He expressed genuine surprise that these "old attitudes can creep into a counseling session."

Kurt's CHANGES assessment could be rated as follows:

1. Confronting (rating: 1): Kurt only superficially attempted to get Janey to confront her problem with verbally abusive men. Although he did get her to try to escape from them, part of his approach was to subtly, almost imperceptibly blame her for causing her problems.

2. Hope (rating: 2): Due to his maladaptive beliefs about women like Janey, he had little hope for her improvement and did not envision her empowerment. He did, however, believe in the therapy process and spoke with her as though she could improve.

3. Awareness (rating: 1): Kurt was unaware of gender issues affecting his performance as a therapist. He was also unaware of the gender issues affecting Janey's problems with her current and past partners.

4. Necessity (rating: 1): Never taking her seriously, Kurt did not see her problems as worthy of concern and solution, other than just doing his job.

5. Grit (rating: 1): As a part of considering her at some level "stupid" and insignificant, Kurt was not willing to experience her pain as part of an empathic process beyond a superficial devotion to duty.

6. Effort or will (rating: 1): Kurt's exertion of effort amounted to going through the motions with no real dedication or focus.

7. Social support (rating: 3): Kurt did not provide a therapeutic relationship of warmth and support for Janey, but he did have the support of a supervisor and other therapists in the counseling center.

With this small presence of precursors and the corresponding sum score of 10, it is not surprising that Kurt did not act as an agent of change for Janey nor that she terminated therapy prematurely. To his credit, he readily admitted that he understood why his effort was unsuccessful and sought his own therapy to work through some of the underlying issues that arose in reflective supervision.

Rating Nancy's Precursors

Nancy was a 38-year-old therapist working toward her doctoral degree. Her client was Jerry, a pleasant 30-year-old man who reported feeling "lost" in his life. Although intelligent, he had no career plans, and nothing seemed to inspire him to want to do anything with his life. However, he was aware enough to worry about it and sought counseling for help. Jerry was courteous and friendly and genuinely tried to cooperate with Nancy's probes and questions. He would even sometimes apologize for being so noncommittal about any goals or future direction. After 16 sessions, little progress was made, and Jerry was still worried about "wasting" his life.

In supervision, Nancy reported that Jerry was so "totally passive" and so "resistant" that she could not get him to do anything about his problem. She also named a variety of personality disorder traits that Jerry was exhibiting, such as those of avoidant and obsessive–compulsive personality disorders. When asked to report her feelings when in the presence of Jerry, she said she found him "irritating" and "maddening" to work with. She felt bad about feeling this way about him but also said she was not able to help and requested that he be transferred to another therapist. Before doing so, she agreed to have the supervisor explore the dynamics between the two of them.

Nancy's CHANGES assessment could be rated as follows:

1. Confronting (rating: 0): Nancy was unsuccessful in getting Jerry to confront anything of magnitude or substance, and she did not do so with herself in relation to him.

2. Hope (rating: 1): Nancy had not envisioned a future for Jerry that had a realistically positive outcome. Such was her frustration that the future she did

envision for Jerry was referring him to another therapist, although she did believe another therapist might be able to help.

3. Awareness (rating: 0): Nancy spoke of Jerry's issues superficially, occasionally speaking of him as though he were lazy and docile, which was inaccurate. She was not aware of her own issues being stirred up by his behaviors.
4. Necessity (rating: 2): Nancy did recognize Jerry's predicament, in spite of her feeling irritated.
5. Grit (rating: 0): Nancy found it difficult to empathize with Jerry and was hesitant to form a close relationship with him. She never really engaged him in a warm and genuine manner as she did her other clients. It was clear that rather than being ready for the difficulty of working with him, she was waiting to terminate.
6. Effort or will (rating: 2): Although Jerry seemed cooperative and willing to work, Nancy was hesitant to apply herself to the task of working with him. Most of her session time was spent getting him to talk about various aspects of his life while she reflected and summarized. Her effort was in "doing her job."
7. Social support (rating: 3): Nancy did enjoy a fairly strong social support system. She also had several supportive friendships with other therapists and an empathic supervisor.

With a sum score of 8, the lack of change was unsurprising. As noted in Chapter 2 in this volume, the supervisor asked Nancy to write down all the people in her life whom Jerry reminded her of. She later reported that Jerry's mannerisms were similar to her sexually abusive brother, whom she had not spoken to in 20 years. With continued reflective supervision support, Nancy actively worked through the countertransference issue with Jerry, and therapy proceeded in a much more effective way. Nancy also sought personal therapy to work through the issue further on her own terms. If the CHANGES assessment had been done again following Nancy's insight and subsequent change of behavior toward Jerry, her precursors rating would surely have been much higher.

Rating James's Precursors

Any therapist who would engage in a sexual relationship with a client would, of course, be severely lacking in precursors of change. James was a 35-year-old man who took great pride in his ability as a therapist. Psychodynamic by theoretical orientation, he had 10 years of experience working as a counselor in a small East Coast business college. He had been married for 17 years and had three children, all under age 15. He arranged to see a fellow therapist one day for peer consultation and, almost as an aside, told the therapist that he had just met "the woman of his dreams."

Her name was Rachel. He said she was 24 years old, recently divorced, and "incredibly understanding and loving." He made it a point to tell the peer consultant several times how attractive she was, how "perfect" they were for each other, and that he was going to leave his wife and children to be with his new love.

"Unfortunately," he added hesitantly, "she was my client." He immediately emphasized with great sincerity, "But that doesn't matter." He explained that they had, just a week earlier, mutually decided to end therapy because they "knew" that their relationship was "meant to happen," and it would have happened after termination anyway. Fate had just so arranged it, he claimed, that they were to meet in the therapeutic encounter. He explained that their mutual decision was made during their eighth session together, and he asked the peer consultant for his view of the situation.

The consultant decided to be as direct as possible. "James, she doesn't know you." James looked incredulous and, with great frustration in his voice, said, "She knows me better than anyone I've ever met!" When the consultant reminded James that such behavior was unethical and harmful, James pleaded for understanding. The consultant paused, leaned forward, and said, "James, please listen. She only knows the idealizations she projected onto you. You know this. You've thoroughly studied this aspect of therapy. She was your client, James. The sessions were all about her, right?" James nodded. The consultant continued. "She doesn't know you, James. She has obviously filled in all the mysteries about you with her fantasies and desires. You were her understanding, caring therapist. It's natural she would feel this way. You're deceiving yourself and her." The consultant then reminded James of the American Psychological Association ethical codes and implored him to consider the potential harm.

Unfortunately, James responded that "rules don't apply 100% of the time" and that anyone who saw him and Rachel together would understand. James was entrenched, so the consultant acknowledged the impasse and said, "Please do one thing regardless of what you think about me and the rules. Answer this question for your sake and hers: What unmet need is this client fulfilling for you?" As might have been predicted, James ignored the peer consultant as well as numerous other psychotherapists whom he had told about the affair. He promptly left his family for Rachel.

James's CHANGES assessment could be rated as follows:

1. Confronting (rating: 0): James carefully avoided direct examination of Rachel's or his own issues.

2. Hope (rating: 1): James envisioned a bright future for the two of them. Was it realistic? It seemed so to him, but it seemed like fantasy to outside observers.

3. Awareness (rating: 0): It is clear that James had little or no idea of what actual issues Rachel faced. In addition, he had no apparent awareness of his own unmet needs or unresolved marital problems.

4. Necessity (rating: 4): James had an apparently sincere, heartfelt sense of necessity to help Rachel with her emotional issues and problems.

5. Grit (rating: 0): James, being in love, was probably more interested in helping Rachel escape from her anxiety rather than experience it. He was obviously unwilling to experience his own anxiety in terms of issues brought up in his interactions with her as a client.

6. Effort or will (rating: 0): Curiously, James was going to great effort to help Rachel gain what he thought was happiness and growth. But the effort was not toward therapeutic change so much as it was toward the gratification of needs.

7. Social support (rating: 3): James had many friends and a good deal of social support, although no one that I knew of approved of his actions. He ignored his professional and personal social support systems.

This scenario involves the harmful exploitation of a client, ethical violations, and a lack of concern for a client's therapeutic change. With a sum score of 8, James was unlikely to ever be an effective agent of change. There were ongoing ramifications for his actions, as well. His social support system collapsed when he was reported and disciplined for unethical behavior. His wife divorced him, and he lost his job. After a year and a half, his relationship with Rachel came to an end, leaving James severely depressed and alone.

After seeking out therapy, James told the peer consultant that if he had been aware of his issues and needs, "it never would have happened." In spite of his training and intellectual understanding of this cardinal rule of therapy, James did not have sufficient self-awareness to recognize the harm he was doing to his client, as well as himself, until it was too late. The lesson learned from this tragic incident is that a helper can exert great effort, feel tremendous urgency for change, possess great affection for a client, and still utterly fail in helping that client to change. This case demonstrates that change does not arise out of only two precursors on the part of the therapist. A dynamic interplay of precursors seems to be necessary for therapists as well as clients.

A FINAL NOTE ON RATING THERAPIST INTERFERENCE

Just as the CHANGES assessment can be used to measure the precursor activation of clients, it is also an effective measure of the same among therapists. Therapist interference is akin to countertransference but casts a wider net insofar as it includes any behavior that impedes the therapeutic relationship and/or client change efforts. It is important for therapists who use the CHANGES model to consider how their own issues and histories impact client outcomes.

A therapist who turns the light of awareness on themselves to examine deficits in precursors with a client or issue can dramatically alter the course of treatment. With this in mind, we suggest that therapists consider the value of supervision, consultation, and personal therapy at any career stage. Therapists have an ethical duty to examine and work through barriers to empathy, not only at the prelicensure training levels but throughout their many years of practice. Additionally, it is important for therapists to honestly assess their limits. We cannot work with every client, and each of us has limits. Owning such limits requires self-honesty and, at times, the candid feedback of supervisors or consultants to help a therapist overcome blind spots or even a lack of humility.

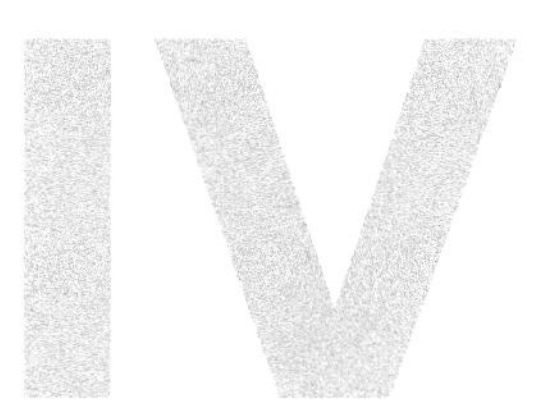

SPECIALIZED APPLICATIONS

13

Guidance for Advanced Training and Supervision

A unique aspect of the CHANGES model is that therapists can be rated according to their own precursors. The previous chapter looks at rating therapists according to their precursors, which can enhance an understanding of how therapists contribute, or not, to the change process in that client (see Chapter 12, this volume). Since the configuration of precursors can vary from problem to problem or even from day to day, a therapist can be more oriented toward change with one client and less so with another. It is crucial for therapists to self-monitor the precursors in practice.

However, supervision during clinical training is an equally important avenue by which to monitor the precursors. The CHANGES model has clear value in supervision and training, not just in helping supervisees learn ways to foster client change in therapy, but because it highlights the central and deeply relational role played by therapists in the change process. The obvious way to incorporate precursors into supervision involves using the CHANGES assessment to identify whether client progress is hindered or facilitated by unidentified therapist effects.

It is important for supervisors, consultants, and trainers to be able to identify the observable source points of therapist interference, as well as strategies to help therapists monitor and deal with their own interference in the change process with difficult clients. Rating therapist interference can lead to thoughtful dialogues in supervision as to what is happening for the trainee or supervisee while working with a difficult client. Yet there are also notable markers of therapist interference that should be actively identified and targeted for reduction or elimination in training to improve therapist skill and increase the probability of therapeutic change outcomes for clients.

https://doi.org/10.1037/0000451-014
Therapeutic Change With Difficult Clients: Precursors and Techniques in the CHANGES Model, Second Edition, by B. D. Wilkinson and F. J. Hanna

In kind, supervisors and trainers who recognize the signs of wisdom in therapy can highlight the path toward greater efficacy in facilitating therapeutic change for difficult clients. As such, another consequence of understanding the CHANGES model is that the knowledge gained sheds indirect light on the central characteristics of the most effective "master" therapists. A complete model of therapeutic change must define the kind of therapist most adept at facilitating it. As seen later in this chapter, the CHANGES model makes the characteristics of effective therapists stand out in bold relief and helps us better understand exactly what they do.

MARKERS OF THERAPIST INTERFERENCE

It is important for therapists to recognize when they play a role in preventing therapeutic change. We have identified 13 ways in which therapists inadvertently undermine client outcomes.

Wanting to Be Liked

Many therapists believe that difficult clients will be more likely to work in therapy if they like their therapists. While there is evidence to support this assertion (Fletcher & Delgadillo, 2022; Orlinsky et al., 1994; Russell et al., 2022), placing too much emphasis on it can be a mistake. If a therapist has a neurotic craving to be liked by clients, the client may only validate and reinforce certain therapist actions that lead to being liked but do not lead to client change. It is more important to be respected than liked. If a client respects a therapist, this is a tacit permission for the therapist to operate as an active change agent rather than a passive supporter of the client.

Engaging in Long-Winded Explanations

Packing a lot of meaning into a single sentence is an art form that is seldom appreciated. It gets easier with practice. If a therapist fumbles for words in an explanation, the verbiage can hinder the therapeutic process. For example, instead of a long-winded explanation about drug use to an addicted client, one could say instead, "So you figure you are your own physician, and you are prescribing illegal medicine to heal your pain. Am I getting it?" The client will usually correct the therapist with a better description or approximation, leading the conversation in the direction of further awareness and confronting.

Prioritizing Care Over Empathy

Caring and empathy are not the same. In fact, a therapist can show caring toward a client and still cause the client to deteriorate (Hardy, 2019; Lambert et al., 1977). For example, when a therapist genuinely cares for the client but lacks empathy and is authoritarian and intrusive, not only is change unlikely, but such attitudes disempower a client. Empathy can also degrade when a therapist is deeply affected by a client's stories of abuse and tragedy. Therapists can get caught up in feelings of protest, sympathy, protectiveness, and righteous anger toward those who hurt the client, which can interfere with the therapist's ability to empathically see the world from the client's perspective. The therapeutic process can stop at this point if one is not aware.

Losing Focus in Session

When the therapist's thoughts are swirling and feelings and emotions are in turmoil in the presence of a client, the therapist has likely lost the sense of being centered and should seek consultation. In some cases, the depersonalization and emotional exhaustion associated with therapist burnout serve as a means to cope when working with many difficult clients during the same day or week. In a closely related vein, difficulty focusing on a client may also be a sign that one has taken on the emotional baggage of clients' unresolved issues, as with compassion fatigue (Figley, 2013).

When one loses focus, it is time to get centered and proactive. Therapists are encouraged to identify client behaviors and attitudes that are stressful or curry one's attention and try to consciously let them go, one by one (Teater & Ludgate, 2014). There can also be value in reflecting upon the rewarding aspects of work as a therapist to help maintain perspective and sense of purpose. Kottler (2022) developed a helpful, six-step self-reflection exercise for identifying and confronting points of therapist interference that stymie client progress and lead to compassion fatigue and burnout:

1. List what one is doing to exacerbate the situation.
2. List the therapist's "buttons" that are being pushed by the client.
3. List the people from the therapist's past with behaviors similar to the client.
4. List the ways the therapist acts out their impatience.
5. List the therapist's expectations of the client and if they are being met.
6. List the therapist's needs that are not being met by the client.

Challenging Clients Too Soon

When working with difficult clients, the temptation to confront and challenge can come too soon and can actually contribute to uncooperativeness. Before engaging in confrontation, the therapist must thoroughly convince the client that they are understood through empathic reflections of meaning and feelings to the point where the client has little choice but to believe the therapist genuinely understands. Only then can challenges and confrontations be made that will be respected and not dismissed out of hand. Kiesler (1988) called this approach "hooking and unhooking," and it is quite effective with difficult clients.

Asserting One's Credentials or Degrees

When a client is particularly difficult, it is easy for therapists to "pull rank" and try to assert one's college degrees, license, credentials, or years of experience as proof of value and competency. This is seldom effective and typically serves as a symptom of the therapist's frustration and a sign of burnout. It is also a possible indicator of a power struggle the therapist is losing. Any attempt to one-up clients will create barriers that prevent genuine interaction and hinder the therapeutic process. It is often a last-ditch attempt by the therapist to gain respect, but unfortunately, respect usually has to be earned in other ways.

Many difficult clients will take an adolescent "whatever" attitude toward this display and become resentful and rebellious. As part of their metalogue, some privately ridicule people with degrees and see the assertion of one's degrees or education as similar to boasting about one's wealth or property. The therapist should be aware of any inclinations to do so and note carefully when the urge arises, which often signals the arrival of an impasse.

Engaging in Power Struggles

Many difficult clients expect a power struggle with a therapist, just as they would with anyone who has a potential influence on their lives. One way to avoid power struggles is to call out or identify any agenda or "game" the client is playing, whether it involves sex, intimidation, or simple evasiveness. To win the struggle, the client must keep their game a secret. For the change process to occur, the therapist must bring the game into the open and call it faster than the client can reset it. Empathy is key and is indispensable in identifying the client's power agenda. Calling the game is superior to winning the struggle, especially when the client admits to the game itself.

Take, for instance, the 16-year-old girl who had steadfastly refused to disclose anything for more than 90 days in a unit for adolescent criminal offenders. The therapy team described her as a "classic borderline," and the agency director decided to call in a consultant. Upon meeting the girl, the consultant told her that she must be remarkably intelligent and powerful to be able to frustrate and resist all the highly educated people in that agency. It was about time someone gave her credit for being so savvy. "You've won," the consultant said. "You've proved that you can stop these people from ever affecting your life." The consultant then told her she was so smart that she might be outsmarting herself by not letting these people help her. That acknowledgment was all she needed, and her change process began with the knowing smile she radiated in spite of herself.

Some therapists feel a need to be in continuous control of a session and feel threatened by a client who rebels against that control (Shamoon et al., 2017). In many cases, if the therapist was not enforcing the control, the client would not be so likely to rebel against it, creating a power struggle. It is sometimes helpful to reverse the interaction by getting a rebellious client to admit that they hate being told what to do, even by a therapist. Role reversals can be used with defiant teenagers to great effect, whereby a therapist asks a client to "boss me around for a while," within reason, of course. If done in a group setting, the process can be entertaining.

Confronting With Impatience or Rancor

Some therapists, counselors, and social workers in community-based treatment programs confront clients in a hostile way. Such confrontation, replete with accusations of denial and dishonesty, are still seen in the addictions field. Defenders of this approach have described it to us as "tear 'em down and build 'em up." Unfortunately, these people are ignoring the basic tenets of motivational interviewing (W. R. Miller & Moyers, 2021; W. R. Miller & Rollnick, 2012). Heavy confrontation tactics are a mistake of considerable magnitude made by people who have neglected to resolve their own hostilities and frustrations or who are tormented by their own unresolved issues. Being so tormented, they also torment clients.

Confrontation can be an extremely effective tool when it is done with curiosity, empathy, compassion, serenity, and the intention to help. When done in anger, it simply activates a power struggle with the client, who may become defiant or hostile or, conversely, submissive and docile, or it may only produce premature termination. Either way, change is unlikely. Therapy with difficult clients is not the same as training soldiers or disciplining inmates. It requires finesse.

Avoiding Confrontation

Some therapists eschew confrontation altogether, instead prioritizing unconditional positive regard to such an extent that confrontation becomes a dirty word. From this perspective, the warm feeling that accompanies a therapeutic relationship must be preserved at any cost, and confrontation is framed as a dire threat to the relationship. This threat can cause a therapist to retreat from sharing even basic observations that might upset the client or otherwise "rock the boat." Many clients happily engage in such warm, supportive therapy for years without making therapeutic progress.

Becoming Either Too Rigid or Too Flexible

A therapist who has been adversely affected by working with difficult clients is at risk of becoming rigid, which serves as a last-ditch effort to gain a sense of control. When one notices this tendency in oneself, it is important to take a step back to determine when it began and what was said or done that led to the rigidity. On the other hand, most therapists are nice people rather than taskmasters and do not want to be controlling or dominating. Some difficult clients are adept at persuading such exceedingly flexible therapists to take a break from all the "hard work," saying something like, "Do we have to do this now? I'm not really in the mood." Once again, when one becomes lackadaisical in terms of sustaining efforts toward client change, the tendency should be explored with attention to when and how it started. Chances are there is a lesson to be learned in reflection upon that moment. Either extreme is a potential move away from therapeutic change.

Defending Oneself Against Clients

There are times when an overworked therapist is so weary from seeing difficult clients that they might create artificial distance, unconsciously perceiving clients as a threat to their sanity or peace of mind. Again, this can be a sign of burnout, but it can also result from the fear of becoming somehow "polluted" by the negative attitudes and problems of unsavory clients. An indicator that this may be happening is a reluctance to form a relationship, perhaps coupled with bitterness, sadness, or helplessness. Numerous therapists have shared with us how they erect mental walls to maintain a personal space that difficult clients cannot touch or violate. The problem with walls is that they also incarcerate therapists, keeping them from making empathic contact with clients. When this kind of distancing occurs, it is time to seek help.

Accusing Clients of Defensiveness

Some therapists will directly say to a client, "You're being defensive" or "You're in denial." This might work on some occasions but more often makes clients resentful and recalcitrant. To many difficult clients, such a statement appears to be an accusation. Such "in-your-face" confrontation tactics are outmoded and have been shown to be ineffective (W. R. Miller & Rollnick, 2012). Rather than accuse the client of being defensive, ask whether the client dislikes someone trying to get inside their head. Also, using the term "protective" can be helpful. Saying, "Are you trying to protect yourself by not answering my question?" may be far more to the point. One might also say, "Did you notice how you did not answer my question just then? Do you think it is too much of a hassle to deal with?" This latter question addresses grit as being willing to experience difficulty.

Disliking a Client Because They Do Not Change

In consulting and supervision, many therapists will reveal that they dislike certain clients. To be clear, we do not maintain that a therapist has to like their client to facilitate change. However, further exploration of this issue with some therapists reveals that the dislike arises when a client is not responding to therapist interventions and strategies or attempts to form a relationship. In other words, dislike can arise when a client threatens a therapist's self-esteem. The same can occur when a client does not change, as the therapist begins to question their own competence, and self-esteem is diminished. From that point forward, the client becomes a symbol of the therapist's low self-esteem, and the therapist is likely to respond in ways that further interfere with client progress.

GUIDEPOSTS TO WISDOM IN THERAPY

Standardized and formulaic approaches to therapy tend to be of limited value when working with difficult clients. When a client has refined the skills of averting and avoiding positive change, standard treatment methods tend to fall short. Many difficult clients are more adept at undermining the therapeutic process than therapists are at facilitating therapeutic engagement toward change. Remarkably, it is in work with difficult clients that a therapist's skills become honed and refined, and a therapist's knowledge of human beings becomes deepened and seasoned. Mere theories, techniques, and even empathy are often not enough. Influencing difficult clients to change often calls for therapeutic wisdom, a set of abilities that transcend formal therapy training (Karasu, 1992).

Wisdom is not the same as intelligence (Grossmann et al., 2013; Sternberg & Glück, 2021). It involves abilities and kinds of understanding that are not assessed by an IQ test. Such abilities include dialectical thinking, self-awareness, self-transcendence, impeccable timing, and the ability to formulate reframes and metaphors that can easily transfer into and out of a client's worldview (Hanna & Ottens, 1995). Therefore, the wise therapist is centered, genuine, engaged, empathic, and committed to helping clients activate the precursors so as to facilitate therapeutic change. The therapist is intellectually and intuitively present in awareness to all verbal and nonverbal client offerings and remains open to facing any difficulties that arise in the therapeutic relationship.

Difficulty for the therapist surfaces when a client engages in behaviors that compromise or undermine these foundational conditions of therapeutic wisdom. A client may do or say something that introverts and collapses the therapist into their own issues, knocking them off center, breaking contact between client and therapist, and compromising the therapist's empathy. Wisdom helps a therapist to maintain empathy and poise despite inclinations to the contrary. Therapists should remain alert to the breakdown of this essential state and learn to recognize certain behaviors and states of mind that tend to indicate one is operating from a place of wisdom. In what follows, we explore a subset of such healthy behaviors and states of mind.

Seeing Empathy as an Act of Will

Empathy does not always occur naturally and automatically. If it did, the entire human race and the living of life would be qualitatively and radically different. A rule of thumb: Empathy will tend to break down to the degree that one's unresolved issues begin to emerge. Another rule is that empathy breaks down to the degree that the other person presents with viewpoints or perspectives that seem strange or foreign. Thus, when these conditions manifest, empathy requires other-directed intentionality as an act of will to be maintained (Margulies, 1989; Zahavi, 2014). In addition, some clients are uncomfortable with empathy and prove adept at discouraging it in a therapist. Mindfulness of this phenomenon in the moment that the client is making the attempt is extraordinarily helpful in recognizing how and why one's empathy begins to dissipate. At that point, empathy must be intentionally recalled, reestablished, and maintained as an act of will.

Cultivating Nonattachment to the Fruits of One's Actions

A personal state of equanimity is invaluable when working with difficult clients. To learn how to develop it, we highly recommend *Everyday Zen: Love and Work* by C. J. Beck (1989), which provides invaluable mindfulness guidance with remarkable implications for therapists. Horney (1952/1987) noted that Zen can teach therapists to be wholeheartedly and fully present, paying full attention to the client without any involvement in self. Of course, this is not a matter of being a cold observer, and the therapist should not have any "personal axe to grind and no neurotic craving" (p. 31) attached to therapeutic outcomes. This means having, as Horney put it, both "the highest presence and the highest absence" (p. 34). She also recommended being so totally absorbed in the practice of therapy that one is putting "all of yourself in what you do" (p. 31).

Some difficult clients are extremely alert to a therapist's desire that they improve and will react to that desire by attempting to frustrate the therapist and sabotage therapy in a variety of ways. Such clients learned these skills long before meeting the therapist, perhaps as a result of their interactions with those who oppressed them or denied them in their families, schools, neighborhoods, and other environments. If a therapist is open and not egotistically craving a particular outcome, or if they are not defending their ego or sense of competence, a difficult client has no "hook" or "tug" that undermines the change process. Remaining unattached to our ego-driven need for control and admiration helps therapists maintain a calmness that fosters creativity and tolerance for ambiguity.

Maintaining Self-Compassion

Working with difficult clients can make one feel foolish and inept. The benefits of maintaining a positive and empathic attitude toward oneself are wide reaching, with extensive research supporting the value of self-compassion as a means of coping with stress and hardship (Ewert et al., 2021; Muris & Otgaar, 2020; Neff, 2023). If one is understanding toward oneself, it is easier to recover from the inevitable mistakes one makes with difficult clients. This has the added benefit of indirectly modeling self-compassion for clients. No therapist needs a critical, invalidating stream of negative self-talk sabotaging awareness and spontaneity in therapy; this can be handled through standard cognitive and mindfulness-based methods (see Hayes, 2005; Leahy, 2017).

Attending to One's Own Metalogue

Both clients and therapists alike have a metalogue, or unspoken streams of thoughts, whether we attend to them or not. Well-trained therapists naturally attend to the metalogue of clients. It can be extremely revealing when therapists listen attentively to their own thought streams during sessions, especially when listening to a difficult client tell a story or describe a problem. One thought stream might think the client is being deceptive, while another wants to believe the client. Simultaneously, another reaction could be disgust, yet another sympathy, and still another a desire to console the client. Accompanying all this may be the thought that it may take years for this client to change, while a concurrent stream is weighing the value of various approaches and strategies that may help. Still another stream may wonder if therapy is having any beneficial effect.

These reactions can be extremely revealing to a therapist who is unsure where to go with a client or what to believe about what a client says. The more these parallel processes are present in one's mind at a given time, the more concerned the therapist should be about the success of the therapy. Noting all these thought streams may be helpful in the case of difficult clients who are confusing, misleading, or extraordinarily complex. Of course, one must be able to do this while still maintaining active contact in the relationship. It takes practice, but attending to one's metalogue can be extraordinarily helpful as a barometer or centeredness, as well as insight, in session.

Monitoring One's Own Visceral Reactions

That queasy feeling in the pit of one's stomach may be communicating something of inestimable value. One may be going along listening and reflecting a client's statements, not suspecting anything amiss at the intellectual level, yet one's "gut" may be churning in a way that says something is wrong. The gut feeling may be a nearly imperceptible tightness in the stomach, an empty feeling, or some other indication of anxiety. It is all too easy to ignore.

Gut feelings are often a sign that something is going on at the level of one's own metalogue that needs to be examined. Therapists who recognize these feelings can then suspend them and explore how and why they arose. Often, there is much more to such a client than meets the intellectual eye. Perhaps this client is intimidating or threatening in a way that is easy to ignore, and the therapist wants to give them the benefit of the doubt. On the other hand, perhaps the therapist is feeling nervous, self-conscious, or worried and wants to escape. Or it could be a projective identification on the part of the client. Whatever the case, monitoring and attending to gut reactions reap rewards in terms of greater learning and insight.

Confronting With Compassion and Empathy

It can be very difficult to care about some difficult clients, such as those who are cruel, demeaning, or insulting or who make sexist, racist, or other prejudicial and harmful comments. Some difficult clients blatantly lie, cajole, and deny responsibility regardless of the evidence at hand. It can be challenging to meet such clients with compassion and empathy, much less confront the client with an open heart. Then, when it would be most useful to confront such a client, the agitated therapist may feel so much hostility or disgust that they become afraid it will be revealed and therefore avoid confronting altogether, or otherwise does so ineffectively, as a sort of half measure.

Confrontation can be an effective tool if done with care and respect. If a therapist does not honestly like a client and a confrontation is necessary, the therapist should be respectful and courteous to that individual despite their own feelings. Ideally, a therapist would work through any negative feelings in therapy or peer supervision. If this is not possible, a therapist can suspend or bracket those feelings, leaving the mind clear to go ahead with the confrontation, knowing that their negative reactions to a client are natural and, to some degree, expected. Suspending or bracketing negative feelings is a skill that one is well advised to develop. It allows one's natural helping instincts to come to the fore and take over the process. One should never ignore such feelings but rather suspend them with full awareness of what one is doing.

Meeting Hostility With Equanimity and Humor

If a client knows that a therapist shrinks away from open hostility, the client may use it as a control tactic. Clients tend to respect therapists who genuinely show no negative or fearful reaction in the face of hostility. Equanimity serves a dual purpose. It shows the client that the therapist will not act out, and it models how to respond to hostility. If the hostility continues, it needs to be discussed empathically with the client. If it continues, termination may be necessary, but this is rare.

Defiance and hostility may also present an opportunity to use humor. Therapists can occasionally use self-deprecating humor to place themselves on an even status with the client. This is especially helpful with aggressive adolescents who use obscene language. If a therapist can insult themselves more severely than the client can, the act not only diminishes their momentum but can paradoxically garner respect. Of course, the use of humor must be natural and not strained.

For instance, if a client calls the therapist an "ass," one reply might be "Have you been talking to my friends?" If a client calls the therapist stupid, one response might be, "I wish I was smart enough to see that." These cannot be canned lines, however. They must be spontaneously delivered in the moment to be fully effective. The purpose is to lessen the client's impression of the therapist as a symbol of authority or a bundle of predictable responses. Role-playing such situations with a supervisor or with another therapist can be extremely helpful along these lines.

The value of humor in therapy cannot be overestimated. If a difficult client can be helped to see the humor in a situation or a behavior, the aura of seriousness and tragedy can diminish and make it easier to confront. It is vital, of course, to remain sensitive to the needs and condition of the client in the moment. Empathy rules. The therapist should avoid any humor that has traces of residual anger or resentment toward a client. For some therapists, healthy humor tends toward tongue-in-cheek situational quips or observations laden with existential irony, whereas sarcasm or dark cynicism tends to indicate unresolved issues being stirred up by a client. The therapist's duty is to resolve personal issues that a client evokes, and healthy humor tends to be a good barometer of effective boundaries and a lack of countertransference in the room.

Recognizing Projective Identification in Action

Projective identification involves feelings produced in a therapist by a client that do not originate with the therapist (Ogden, 2018). It is the responsibility of therapists to thoughtfully assess whether such feelings originated within oneself or are a projection of the client. For example, if a therapist feels intimidated by a client, it could be the client's projection of power. If a therapist experiences sexual feelings for no apparent reason, it could be the client's projection of sexuality. If a therapist feels indebted to a client, or as though they owe something to a client, it could be a projection of ingratiation. If a therapist is disgusted by a client, it could be a projection of repugnance. All of these reactions can be used in therapy to deepen empathy with and insight into the client through the recognition that this may be how people in the client's environment feel in their presence.

Projective identifications can be approached passively or actively. For example, in the case of an intimidating client, one may feel afraid, powerless, or weak. Although it is not a good idea to admit when one is afraid of a client, one can ask, "Do you find that people in your life are scared of you?" A client will often respond to this in the affirmative in some manner. The therapist can then ask how the client came to be that way. One can also ask if it works for the client's life in terms of advantages and disadvantages. In addition, one can find out who the client's intimidating models have been. If the client can identify and describe the models, the therapist can work with the client to reconsider the value of those models. Each of these approaches can be fruitful.

Knowing When Discouragement or Disgust Is Not Your Own

When working with difficult clients, it is common to feel discouraged, disgusted, or hopeless. Such feelings may actually be an empathic "borrowing" of how that client habitually feels toward themselves. To test this, it is sometimes possible to describe to the client the feeling of disgust, discouragement, or hopelessness that one is experiencing as a therapist and to inquire whether this is, in fact, what the client is experiencing. It is often accurate to some degree. Clients can sometimes be so surprised by this revelation that trust is deepened. This is also true of anger and resentment. Phrasing is important; one would never tell a client that they are disgusting or hopeless.

Drawing Clear Boundaries Around What Is Tolerable and Acceptable

Some clients have a way of making a therapist feel silly about not being willing to discuss their personal lives or fantasies, not being willing to meet outside the office, or not being able to "take a joke." Too many of these clients on one's caseload can eventually cause a therapist to doubt their sanity. The only course of action is to be sensitive to one's sense of propriety, personal space, privacy, and limits. These must be communicated to the client clearly and unhesitatingly for the sake of the therapeutic change process. It is also important to stay in contact with colleagues or a supervisor to maintain one's social support system.

Asking "Therapy Veterans" How They Managed to Defeat Previous Therapists

Occasionally, a therapist will run into a client with a long history of confounding and defeating therapists. Some even seem to have made it a hobby. Many enjoy being so complex and mysterious that no one can understand or reach them. In a convoluted twist, many clients believe that to be understood is to be humiliated. It is advisable to ask such "therapy veterans" how they managed to foil the attempts of so many intelligent people, as well as to praise their intelligence and credit them for the accomplishment. It can be helpful to reframe their "failure" as a client as a "success" in defeating a therapist, which can be noted as a skill born of intelligence, craftiness, or shrewdness.

Therapists can also say that the client will, no doubt, defeat them well. Therapists should make no pretense about being better than other therapists or more capable of handling the client's difficulties. Instead, the therapist can convey curiosity and a desire to learn from the client about how to defeat therapists. One might even suggest the client could probably give lectures on therapy if they were so inclined. Eventually, it is important to point out that the client has not really tried therapy yet, and that maybe they could try it for the first time. If the client can be persuaded to teach their skills, a therapeutic window may open. Many such clients are secretly convinced that they are empty inside, devoid of soul or self. If a therapist, or anyone, were to discover their true nature, it would result in a kind of mortal disintegration.

Understanding That Difficult Clients Function as Teachers

Here is a simple rule: Treat a client who does not change as an opportunity for learning rather than a professional failure. There are times when it is helpful to tell a client that they are functioning as a teacher. This approach can produce a shift in how the client regards the therapist in a way that reduces power struggles and promotes the genuineness of the relationship. For instance, a power-seeking client will sometimes be responsive to admiration. Thus, the therapist might say, "From what you've told me, I'm willing to bet you've learned some pretty remarkable lessons about life and people. Do you ever share any of that knowledge? I'll bet there's a lot I could learn from you."

A FINAL NOTE ON GUIDANCE FOR ADVANCED TRAINING AND SUPERVISION

Jung (1934/1969) suggested that we may better understand ourselves by examining what we dislike about others. The client who evokes graded reactions of irritation, annoyance, frustration, or even utter disgust provides a wonderful opportunity for therapists to learn more about their own values and biases. Recognizing the various markers of therapist interference can lead to personal and professional growth, but such recognition may require the observation or input of supervisors, consultants, or other therapists. The path to wisdom is seldom an isolated affair. We learn from our interactions with, and the supportive insight provided by, others. There are few careers in which one's professional talents may grow in close tandem with one's evolutions in personal wisdom. It would surely be wise for all of us to take this unique opportunity seriously.

14

Oppression, Perspicacity, and Liberation

The prioritization of multicultural awareness has become a pillar of knowledge and practice in the fields of counseling and psychotherapy during the 21st century. After years of neglect, it has finally been widely acknowledged that clients from cultures different from those of a therapist need to be understood from within their own unique sociocultural perspectives. Significant emphasis has been placed on studying various customs and beliefs to develop cultural competency in tandem with better communication of empathic understanding in therapy (Fuertes et al., 2006; Sue et al., 2022).

At the same time, there is a fundamental and profound limit to a therapist's capacity to understand the lived experience of clients from different cultural backgrounds. White therapists are unable to fully grasp the trauma of racism due to cultural privilege, for example, and the same can be applied beyond race or ethnicity to areas such as gender, sexuality, age, disability, religion, and so forth. It is vital to acknowledge such limits while still seeking to build sociocultural knowledge and understanding that grows empathy. Effective multicultural practice therefore requires cultural humility, by which "therapists are able to have an accurate perception of their own cultural values as well as maintain an other-oriented perspective that involves respect, lack of superiority and attunement regarding their own cultural beliefs and values" (Hook et al., 2017, p. 29).

Another dimension of multiculturalism that must be acknowledged in practice is the phenomenon of oppression, which clarifies the context or situatedness of many difficult clients. If one examines the wide range of problems

Case examples have been disguised to protect client confidentiality.

https://doi.org/10.1037/0000451-015
Therapeutic Change With Difficult Clients: Precursors and Techniques in the CHANGES Model, Second Edition, by B. D. Wilkinson and F. J. Hanna

encountered in psychotherapy from this perspective, it is readily apparent that oppression is an active ingredient in the formation of psychological and emotional problems. Yet there is also a curious and widely overlooked perceptual skill that arises from oppression that has implications for empowerment with difficult clients: *perspicacity*, a characteristic of wisdom that enables one to see beyond appearances (Sternberg, 1990) and "intuitively understand, read, and accurately interpret the environment" (Hanna et al., 2000; p. 433). In this chapter, we examine the nature and trauma of oppression, the power of perception, and how an integrative existential-cognitive therapy can support psychological liberation for oppressed clients.

THE NATURE OF OPPRESSION

Oppression may be defined as an abuse of power at the expense of the well-being of others. As a phenomenon, oppression is the central concern guiding both professional and broader sociocultural discourses on multiculturalism and social justice. It manifests in our relationships as well as through institutional and systemic influences that, while pervasive, are sometimes more difficult to identify, diagnose, and modify than interpersonal abuses of power. However, in both interpersonal and systemic cases, oppression involves two modalities: force and deprivation.

Oppression by force, coercion, or *duress* is the act of imposing an object, label, role, experience, or set of living conditions on another or others that is unwanted, brings needless pain, or detracts from physical or psychological well-being (David & Derthick, 2018; Hanna et al., 2000). An imposed object, in the context of oppression, can be a bullet, a bomb, shackles, a bludgeon, a fist, a penis, unhealthy food, or abusive messages designed to degrade or reduce the self-determination of the person in question. Other examples of oppression by force can be coerced labor, enforced religion, degrading jobs, and negative media images and messages that foster distorted, negative beliefs. Oppression by force is directly tied to physical harm in interpersonal relationships, as seen in cases of intimate partner violence or child abuse.

Oppression by deprivation is the act of denying to another an object, role, experience, or set of living conditions that is desirable and conducive to physical or psychological well-being (Hanna et al., 2000; Sue, 2010). Neglect is a central form of oppression by deprivation, whether in terms of neglecting to provide basic needs such as food, shelter, or clothing or withholding relational needs such as respect, dignity, or love. Objects deprived can be a house, a plot of land in a desirable neighborhood, various forms of wealth, or gainful employment. Oppression can deprive one of one's children, parents, friends, freedom, or even one's childhood. Religious practice can also be deprived, as was the case from 1890 to 1940 when the United States banned certain practices of the Sioux tribes (D. Brown, 1970).

Implicit Oppression

There has been a steady, multigenerational transition in American culture from explicit and thus wholly undeniable acts of oppression to implicit, easily deniable acts of oppression grounded in phenomena such as color blindness and microaggressions. By no means are we suggesting that oppression by force is not a major contemporary cultural issue. The Black Lives Matter movement is a clear example of protests designed to address explicit and ongoing oppression by force in American culture. However, the blatant nature of daily prejudicial acts that buoyed the civil rights movement has largely been replaced with subtler, more easily deniable acts of discrimination.

As proposed by the cultural philosopher Harvey (2015), the transition from explicit acts of oppression that are sanctioned and normalized by a dominant cultural group to the outright denial of oppression as a phenomenon of sociocultural interest or concern by many individuals and organizations within the same dominant group marks the sociocultural onset of

> *Civilized oppression* ... [which] refer[s] to oppression that involves neither violence, nor the use of law. It is systematic and disadvantages and demeans members of certain groups and in Western society it is pervasive. The phenomena involved are routinely trivialized, given their subtle nature, yet both their effects and their nonconsequentialist implications are highly significant. (p. 1; italics added)

The visceral nature of abuse and violence in oppression by force tends to take center stage in our collective consciousness. When one thinks of oppression, it is quite natural to first think of harms that are obviously imposing or abusive. At a systemic level of analysis, we reference examples of enslavement, genocide, or other violent acts in which a group of people are disproportionally and deleteriously impacted. At an interpersonal level, we may identify specific instances of physical or verbal violence in the home, such as an alcoholic parent belittling their child or a physically abusive partner striking their spouse. In systemic and interpersonal conditions, oppression by force is often the default reference when wrangling with the complexities of harmful acts.

In contrast, the notion of implicit oppression by deprivation, or civilized oppression, can be more difficult to grasp while simultaneously being central to contemporary discussions on the role of social justice and multicultural awareness in therapy. At the interpersonal level, we can easily cite issues of child neglect or a partner unwilling to apologize for abusive actions as oppression by deprivation. Yet deprivation also includes interpersonal acts that are not so obviously identifiable, such as microinvalidations that arise from implicit biases, or denials of difference that arise from "color blindness," or a disavowal of oppression as a concern for some group of people altogether.

For instance, the White therapist who decides not to broach the topic of race with a client of color is engaging, inadvertently or not, in oppression by deprivation because avoiding the discussion amounts to a denial of difference. Such a choice also runs counter to research that has established the value of broaching, particularly since therapists occupy a position of privilege (Day-Vines et al., 2020; Lee et al., 2022). It has also been shown that unresolved countertransference issues and self-ascribed color blindness increase the likelihood that a therapist will either microinvalidate or overpathologize racial and sexual minority clients (Dictado & Torres-Harding, 2023).

Epistemic Injustices

Invaluable scholarly work has been done on epistemic injustice in therapy. It has been argued that therapists should avoid epistemic injustice in two forms: testimonial injustice and hermeneutical injustice (Lee et al., 2022). Testimonial injustice clearly serves as a form of interpersonal oppression by deprivation. For instance, a therapist might explicitly (and without malintent) deny the validity of an experience by casually reframing a client's report of racial profiling at work as "a misunderstanding." Such a denial of lived experience amounts to "an ontological violation" that perpetuates an erroneous yet oft-perceived "credibility deficit" among marginalized clients in therapeutic settings (Fricker, 2007, p. 137).

While hermeneutical injustice is also enacted interpersonally, it more directly speaks to systemic oppression by deprivation. For instance, even if the therapist validates a testimonial provided by a client who reports racial profiling at work, the therapist may have little to no grasp of the deep significance of this event for the client due to the therapist's own privilege. In such an event, the therapist actively reflects the perspective of the client, yet the client feels viscerally unheard anyway. Such instances speak to structural inequities in larger social systems, whereby a client experiences systemic oppression by deprivation in an encounter with a therapist who, despite good intentions, has not done the self-work necessary to empathize with that client.

In either case, therapists should obviously avoid perpetuating such epistemic injustices. There is strong evidence that therapy can be traumatizing for minoritized clients when therapists, however well-intended, perpetuate oppression (Bennett-Leighton, 2018). Therapists are ethically bound to communicate with deep empathy to avoid testimonial injustices, as well as to engage in the self-work required to avoid hermeneutical injustices. The self-reflection and personal work required is perpetual, in the sense that one must always work to deepen empathy and awareness; it is an ongoing journey without a final destination or state of being.

THE TRAUMA OF OPPRESSION

Our discussion thus far primarily focuses on the duty of therapists to avoid oppressive acts. From the client perspective, oppression can be a stalwart barrier to therapeutic change, whether it is interpersonal or systemic in nature. It can feel impossible for many clients to imagine positive change when one has experienced a lifetime of relational and systemic disempowerment. Most difficult clients encountered in therapy have had tough lives, often filled with pain and suffering inflicted upon them by others. Many difficult clients cause similar pain and suffering for others, perpetuating a traumatic cycle of anguish and misfortune. Arguably, various forms of oppression are at the root of all traumatic coping (Bennett-Leighton, 2018; Jacobs, 1994; Watson et al., 2016).

The therapist's duty is to validate feelings and perceptions and not discourage anger. Anger management strategies, in our estimation, risk failure with oppressed groups when they do not acknowledge that oppression causes justifiable anger (Archer & Mills, 2019). Anger often derives from hurt and serves as a coping response to protect against further hurt. Anger is also a much safer emotion to display than hurt, which demands some vulnerability and, thus, perceived risk. At the same time, we maintain that hurt is more of a sensation than an emotion, making it easier for clients to identify than contextual feelings that can be difficult to grasp. The following story is an actual occurrence showing how this approach can be used with traumatized clients.

The Story of Carlos

A consultant was providing training to staff in an adolescent runaway shelter at the time of this case example. The shelter was for kids living on the streets with no place to sleep. There was no court involvement here, and kids could stay to get possible placement in a foster home or stay one night and move on. There was no leverage to get kids to engage in therapy, but all adolescents who stayed there were encouraged to seek on-site counseling services. The staff ranged from licensed therapists to social workers with bachelor's degrees to volunteers without degrees. The consultant provided training days once every 2 weeks and also worked with some of the teens on-site.

Carlos was a 17-year-old who had been at the shelter for only 24 hours. The consultant entered the building that morning to find Carlos yelling and cussing at staff and other kids. Apparently, he had been throwing furniture. The consultant walked directly up to Carlos in a nonthreatening manner and spoke to him loudly. People do not listen well when angry, so the standard therapy statements can be made, but the volume often needs to be turned up without any hint of hostility or disrespect:

THERAPIST: (loudly, but respectful and unimposing) Wow! Man, are you angry! Wow! Look at all of that anger!

CARLOS: (turning around, breathing hard) You're goddam right! I'm pissed off!

THERAPIST: (loudly) I can see that! What's got you so pissed off?

CARLOS: I'm sick and tired of this shit!

THERAPIST: (more softly now) I hear you, Carlos. I am sorry you are taking a lot of shit. Come on, man. Let's get away from all this and go talk somewhere. (shakes hands with Carlos and leads him to a group room followed by three trainees from the program designated to sit in to observe that day)

THERAPIST: Can we talk, Carlos? If I make you angry, you just tell me to back off.

CARLOS: (sits down with arms crossed) Whatever.

THERAPIST: What's got you so angry?

CARLOS: I'm sick and tired of people all the time telling me what to do.

THERAPIST: Yeah. I get it. That shit gets old fast.

CARLOS: (a bit calmer) Yep.

THERAPIST: Do you get angry a lot, Carlos?

CARLOS: Yeah. Like, all the time, almost.

THERAPIST: Can I ask you a tough question, Carlos? You don't have to answer.

CARLOS: Yeah.

THERAPIST: Have you been through a lot of shit in your life? Don't tell me what it was.

CARLOS: (calmly) Hell yeah.

THERAPIST: Have you ever been hurt?

CARLOS: (stiffens up a bit) Hell no. (pausing) What do you mean?

THERAPIST: Like somebody cut you deep, but in your feelings where no knife or bullet could ever go.

CARLOS: (hesitantly, turning away slightly) Yeah.

THERAPIST: Thank you for telling me. Do you think about it much?

CARLOS: I think about it a lot. Yeah, and f*** him.

THERAPIST: Don't tell me who it was, but is there a place in your body where you feel the hurt?

CARLOS: (points at his chest area) Here.

THERAPIST: That's where you feel the hurt?

CARLOS: (nodding)

THERAPIST: Thank you, Carlos. Can I ask you another question? If all that hurt inside of you went away, what would happen to your anger?

CARLOS: (thinks about it, then laughs) If it went away? Man, I wouldn't be angry.

THERAPIST: Okay. Have you ever been told you have an anger problem?

CARLOS: (still laughing) All my life.

THERAPIST: Maybe you don't have an anger problem. Maybe what you really have is a hurt problem. Do you like having all that hurt inside you?

CARLOS: (quietly, looking down) No.

THERAPIST: What do you do to get rid of the hurt?

CARLOS: I don't know. I get high. If people mess with me, I mess with them back.

THERAPIST: Does any of that actually make the hurt go away?

CARLOS: (smiling) Yep, sure does.

THERAPIST: Okay, but does the hurt go away for good, or does it come back?

CARLOS: (looking down, not smiling) Nah, it comes back.

THERAPIST: So, hurting others doesn't *really* make the hurt go away.

CARLOS: Not really. No.

THERAPIST: Would you like to try something that actually works?

CARLOS: Yeah, what's that?

THERAPIST: It's called counseling. (smile) It can make the hurt go away, Carlos. Not like getting high or fighting people, but it works, and it makes you feel a lot better. You want to try it? You can get it right here. You don't have to do it forever. But maybe you can give it a try and see if it helps.

CARLOS: (hesitating)

THERAPIST: You've already given drugs and (smiling) throwing furniture around your best shot, and I give you credit for trying to feel better. But maybe getting high and breaking stuff and yelling at people isn't really the answer.

CARLOS: Yeah, okay. But I'll only talk to you.

This session was done with a client who was never introduced to the therapist and the two had never seen each other previously. The therapist identified the underlying issue behind the anger that seemed to disturb Carlos the most, which was hurt. By offering Carlos the option of reducing the hurt, Carlos became motivated to engage in therapy, and he did, although with limited success.

Activating Anger Related to Systemic Oppression

Effective multicultural therapists recognize the depth of traumatic tension, anger, and resentment felt by marginalized groups toward dominant groups (L. S. Brown, 2017). In regard to systemic forms of oppression, anger can be an effective catalyst of action. The key to validating anger is to help clients express, manage, and redirect it toward worthwhile goals such as community engagement and activism. While it may be helpful in some cases to suggest that clients make anger a "friend" so as to become acquainted with it, in many other cases, the anger is so destructive that only acknowledging it, validating it, and reflecting its intensity and meaningfulness can take oppressed clients beyond it. On the other side of destructive anger lies the potential for healing, personal transformation, and constructive sociopolitical action (Comas-Díaz, 2016).

In most cases, detailed discussions should move into the relationship between anger and cultural positioning. The following questions can foster discussion on the impact of oppression in the lives of minoritized clients. Such questions, even if pursued briefly, can be helpful in relationship building:

- What beliefs does the dominant culture have about the client's cultural group?
- How does the dominant culture hold control over the client's cultural group?
- How does the dominant culture want marginalized persons to be, think, and behave?
- What messages about the client's cultural group are seen in movies and television?
- What behaviors by the client's cultural group are rewarded in this society?
- What attitudes by the client's cultural group are rewarded in this society?

The purpose of this approach is not to build resentment or fan the flames of anger but to produce a sense of liberation from negative beliefs about oneself and one's cultural group (David et al., 2019; Vickery et al., 2023). The minority stress model (Meyer, 2003) highlights the mental health strain that arises for clients who identify with historically marginalized groups. While originally developed to address the challenges faced by gender and sexual minorities, it has been expanded to examine the intersectional identities among racial and ethnic minoritized groups (Cyrus, 2017) and people with disabilities (Botha & Frost, 2020). It emphasizes the complex relationship between external, sociocultural experiences of discrimination and the internalization of oppressive ideals.

When clients from historically marginalized communities internalize stigmatizing ideas and beliefs held by dominant cultural groups, there is an increased likelihood of negative physical and mental health outcomes (Boykin et al., 2016; Hendricks & Testa, 2012). Ever since Crenshaw's (1989) examination of the sociopolitical ramifications of ignoring how overlapping axes of gender and race "reveal how Black women are theoretically erased" (p. 139), research has made it increasingly clear that such outcomes are even more deleterious for minoritized clients with intersectional identities (Vargas et al., 2020).

Of course, there is also considerable variation in the attitudes of members of oppressed groups toward oppressors. Some may not have experienced the intensity of oppression that other members of their groups have. Assuming that all

Latino Americans, for example, have had the same experiences or attitudes is a mistake and could be called "benign stereotyping." As a presenting issue, discrimination may not be of the same magnitude for all clients from oppressed groups.

LIBERATION AND THE POWER OF PERCEPTION

As that which arises by means of dismantling oppression, *liberation* may be defined as the sociopolitical freedom to live under self-determined conditions, unconstrained or unaffected by oppressive forces (Martín-Baró, 1996; Watkins, 2002). According to the CHANGES model, psychological liberation may occur when a person engages in practices that facilitate therapeutic change via the activation of precursors, the bulk of which are inhibited by harmful messages imparted by means of racism and discrimination. Psychotherapists are in a particularly good position to support clients in the process of confronting the impact of internalized racism, misogyny, homophobia, ableism, and other forms of discrimination (David & Derthick, 2018).

The path to psychological liberation partly involves identifying, challenging, and altering one's relation to negative internalized beliefs imposed by individuals, institutions, and systems. Asking an individual from an underrepresented group to adjust or adapt to an oppressive society that does not work toward the best interests of some of its members clearly reduces the integrity of that client. However, working with minoritized clients to identify the traumatic impact of dominant narratives on self-concept can be a path to psychological liberation, particularly if we view trauma as "any experience that is subjectively unbearable" (Greenberg, 2019, p. 1144) and which subsequently results in some form of psychological and behavioral adaptation to environmental conditions.

Since oppression arises from power imbalances, those who are denied power must cope somehow to survive under threat. As noted by J. B. Miller (1986), a primary mechanism for coping with the threat of harm arises through an enhanced perception of the oppressor, be it an individual or a group. In this context, perception has to do with cognizance, recognition, or noticing and is thus related to the awareness precursor. It also relates to what Sternberg (1990) called *perspicacity*, an aspect of wisdom that involves the ability to see beyond appearances, to "see through" situations, or to "read between the lines."

For example, when a battered spouse becomes increasingly aware of how her abusive partner reacts to certain situations, she learns to deftly maneuver around sensitive topics. Her enhanced empathic attunement to the oppressor allows her to read into subtle verbal and nonverbal cues so as to better predict danger and ensure self-preservation. As her awareness becomes more acute, she may also come to see through and understand the mechanisms used by her partner to entrap her and how her life has become limited by abusive actions. Eventually, as awareness increases in therapy, she may learn how to maneuver herself safely out of the relationship. Perspicacity, born to ensure safety in the face of suffering, may yet serve as the catalyst for freedom and well-being.

Deprivation of power can awaken perception (J. B. Miller, 1986). Being rejected from a group or excluded from its benefits can inspire one to notice and study the oppressive group. Whether within or between cultures or among individuals of the same or differing cultures, this phenomenon of heightened perception has the paradoxical yet remarkable advantage of keeping the oppressed person or group alive and aware. Although an oppressed client's life may be ruled by harsh realities, their therapeutic ticket is often the raw and penetrating perception that develops out of those painful experiences. In this light, their accurate perception can be reframed in therapy as a strength, not only in perceiving their current situation but as a way out of it.

Leveraging Perspicacity

Although oppressed people are largely unaware of any hidden benefit, it is within this dynamic that we find a critical mechanism of empowerment that can stimulate the precursor of hope. This mechanism operates in members of systemically oppressed groups and among oppressed persons who have been relationally abused. It can be useful to empower certain clients by pointing out, or helping them realize, that they possess the hard-won, valuable perceptual skill of perspicacity. The therapeutic aim is to get a client to describe the oppressor in as much detail as possible.

Victims of intimate partner violence and domestic abuse, for example, can typically provide very descriptive accounts of the batterer's triggers, behaviors, attitudes, and emotions, complete with predictions of when and under what circumstances the abuse will occur. They seldom see evidence of this being a skill, however. When told that this same perceptive ability can be used to read people in a variety of settings, from the workplace to romantic relationships, they are often surprised to learn that they possess a valuable ability. For instance, group work with heterosexual women and teenager girls can be guided by questions to refine, clarify, and heighten perception:

- How does this society expect women to be, think, or behave?
- What kinds of behavior by women are rewarded in this society?
- How does society treat women who are considered overweight?
- How are women's bodies typically portrayed in popular culture?
- What does a man gain from having a woman with low self-esteem?
- What kind of partner is a man with little or no empathy?
- What kind of man is likely to physically abuse women or girls?
- How do women benefit, or suffer from, being dependent on men?
- Why are women more often depressed than men?

The phenomenon of perspicacity is also transcultural. It is observable among members of many culturally oppressed groups, such as the Tamils of Sri Lanka, the Uighurs of Northwest China, the untouchables of India, and the Bataks of Sumatra in Indonesia (see Hanna, 1998). In nearly all cases, members of the oppressed groups make accurate, detailed observations about the behavior and attitudes of their oppressors. Even a cursory examination of descriptions of oppressors by members of oppressed groups reveals remarkable insights. For example, the writings of Douglass (1855) are far more detailed and accurate than any of the bankrupt descriptions by enslavers. Northup (1853/1968) was a free African American who was kidnapped from New York State and forced into enslavement in Louisiana. He had no education and no status but was intelligent and wise, as seen in the following systemic clinical analysis of enslavers:

> It is not the fault of the slaveholder that he is cruel, so much as it is the fault of the system under which he lives. He cannot withstand the influence of habit and associations that surround him. Taught from earliest childhood, by all that he sees and hears, that the rod is for the slave's back, he will not be apt to change his opinions in maturer years. (pp. 157–158)

During the years of Northup's bondage, he observed how being an oppressor hardens a person, turning men and women alike into callous and cold beings:

> The existence of Slavery in its most cruel form ... has a tendency to brutalize the humane and finer feelings of their [the slaveholders'] nature. Daily witnesses of human suffering—listening to the agonizing screeches of the slave—beholding him writhing beneath the merciless lash—bitten and torn by dogs—dying without attention, and buried without shroud or coffin—it cannot otherwise be expected, than that they should become brutified and reckless of human life. (p. 157)

Acton (1887) offered an oft-quoted observation: "Power tends to corrupt, and absolute power corrupts absolutely" (p. 3). Members of dominant groups tend to describe members of subordinate groups in ways that justify their own possession and superiority (Gevisser, 2020). In both remote and well-traveled parts of the world, one may hear standard, stereotypical accusations against members of subordinate groups as "lazy," "dirty," and "stupid" and accused of not caring about or being harmful for or toward children. In this same way, boys come to believe that women are sex objects, heterosexuals frame sexual orientation as a moral quandary, and whole societies become convinced that underrepresented groups are somehow "less than." To be immersed in a culture is to be immersed in its context, and context, at its core, is made up of beliefs (David & Derthick, 2018).

Power corrupts, whether it is abused in the context of a country, culture, workplace, family, or marriage. The road to healing for oppressors is through recovering empathy and awareness lost due to the effects of excessive power on accurate perception. The goal is to bring them out of a state of obliviousness that prevents empathy for the oppressed and to increase awareness of their own psychological states. There are also mixed cases in which a client is oppressed in one context but an oppressor in another. For example, a woman may be oppressed at the workplace but be abusing her children at home. In these not-uncommon cases, the client needs to be helped to restore empathy for anyone they oppress and be liberated from their own oppressors in kind.

INTEGRATIVE EXISTENTIAL-COGNITIVE THERAPY FOR OPPRESSED CLIENTS

Liberation is, at the most fundamental level, a process of being freed from limiting beliefs. Many difficult clients have trouble isolating and identifying distorted cognitions or beliefs. However, there is an alternative approach that may make it easier for some clients who have been oppressed. The freedom-from-oppression model (FOM) promotes the use of clients' advanced perception to identify the beliefs of those who have hurt and forced dysfunctional beliefs upon them (Vickery et al., 2023). Many oppressed clients with heightened perceptive capabilities can identify the irrational beliefs and messages of oppressors, which they consequently absorb.

An oppressed person's negative beliefs are heavily influenced by oppressors, whether a group or an individual. Part of this oppression is a series of beliefs that the oppressor "inflicts" on the victim. As noted by Vickery et al. (2023), "Repeated exposure to oppressive messages or actions is coercive, resulting in the internalization of an oppressive and undermining self-structure" (p. 175). For example, a belief that one is stupid, worthless, incompetent, or unlovable can be seen as inflicted rather than self-determined. The standard cognitive approach would be to help a client identify irrational or dysfunctional beliefs as though they generated them. Yet a client is often not the origin of such beliefs; the oppressor is. The client's mistake is in agreeing with the beliefs inflicted or forced upon them, creating a negative array of feelings and behaviors.

The therapeutic goal is to end the agreement and replace the dysfunctional beliefs with those that are accurate and self-determined. To this end, the FOM includes three existential stages and 12 cognitive steps that support clients in relinquishing internalized, oppressive belief structures. Moving across stages that emphasize awareness, empowerment, and liberation, clients are supported in discovering their perspicacity and applying it with resoluteness to discover liberation from internalized oppression (Vickery et al., 2023). In every stage of the FOM process, it is incumbent upon the therapist to maintain empathy and compassion for the client. The approach is based on Heidegger's (1927/1962) descriptions of how a person may lose authenticity.

Stage 1: Awareness

In the first stage, the therapist must help the client gain awareness by identifying the oppressive persons or groups that impact their life. Following identification and description of the oppressor and oppressive beliefs, the therapist asks the client, "What did [the oppressive group or person] want you to believe about yourself?" For example, if a man was oppressed by his father, his father may have given him beliefs that he was a mistake or that he was worthless or ugly. African Americans may identify the dominant group as White society that conveys messages that Black people are lazy or incompetent. Specific people, such as teachers, neighbors, or employers, can be cited as oppressors as well.

Women have long been inundated with messages of inferiority (Saini, 2017). A therapeutic focus therein might be on listing dysfunctional beliefs imparted by messaging from popular culture, as well as men or boys in the client's life. It is important to cite actual sources when possible, such as movies, musicians, social media influencers, or family members. For example, when asked, a severely battered woman once listed the following beliefs forced upon her by her abusive husband:

- I'm stupid.
- I'm immoral.
- I'm a terrible mother.
- I'm fat and ugly.
- I can't make it without him.
- I'm a chronic liar.

She could easily recognize how much she had agreed with these beliefs and how he not only wanted but needed her to believe all these things so that she would not leave him. Once again, client perception needs to be validated to highlight that the belief was not created by the client but rather imposed by the oppressive source. Such perception is not perfectly accurate, of course, and like any skill, it often needs to be honed. In this instance, the approach may be framed in terms of the idea that women should not be blamed for having negative self-beliefs when those beliefs are so often a direct consequence of patriarchal culture.

It can be relatively easy for perceptive clients to spot the beliefs and tactics of oppressive individuals, but it is more difficult when those beliefs come from institutions or systems. In such cases, the therapist should help clients identify those oppressive beliefs. For example, if a lesbian were asked to list beliefs forced upon her by a rigidly heterosexual society, it is possible that she might omit certain oppressive beliefs, such as "All gays are immoral." The therapist could help by submitting that belief for the client's verification and then determine if she agrees with it and, if so, to what extent.

Alternatively, if a teenage girl reports a negative self-belief but has difficulty grappling with the possibility that her family plays some role in imparting that belief, she can be asked if there is any group in the world that might like a young woman such as herself to believe it. Helping the client address that larger group may provide the distance needed to directly challenge the belief and identify that the client's only mistake was in agreeing with something false. Sometimes, broader sources of oppression can be easier to discredit than more specific and proximal ones, and it is important to remember that challenging the internalized belief remains the therapeutic goal.

After isolating beliefs that were inflicted or otherwise fed to the client, the therapist must ask how much the client agreed with the message. A little? A lot? It might be rated on a scale from 1 to 10, with 10 being the most that one could ever believe anything and 1 being just a tiny bit. The belief will have its most damaging effect, of course, if the client still heavily agrees with it.

Stage 2: Empowerment

In the second stage, a therapist helps the client examine the consequences of agreeing with the negative internalized belief and any reasons the oppressor wanted or demanded agreement. For example, a client's agreement with the negative belief, "I'm not attractive because I'm not skinny enough," serves the purpose of always having the client doubt themself and never feeling worthy enough to challenge their partner on possible sexual manipulations or power schemes. The therapist must then work with the client to confront the pain associated with the belief in an effort to build tolerance as grit, or the willingness to experience anxiety and difficulty. After closely examining how one feels and acts as a result of agreeing with the belief, the client is then supported to dispute the belief to reclaim power and self-determination from the oppressor.

This process involves identifying evidence against an imposed belief and identifying evidence that discredits the standing of the oppressor. Sometimes, these processes merge in a broader effort of grasping intergenerational transmission patterns. For example, a young bisexual Latina woman named Maria once shared how her family's enactment of marianismo culture derived primarily from her grandfather's values. She came to believe that her mother and grandmother had long been depressed because they had submitted to familial expectations that ran counter to their nature. Her grandmother was raised in Guadalajara by an artistic, willful mother and progressive father, both of whom died in a tragic accident when her grandmother was just a teenager. Her paternal aunt soon pressed her to marry into a wealthy family with rigid views on traditional gender roles.

At the intersection of race, gender, sexual orientation, and first-generation immigrant status, Maria was confronting a highly complex, multigenerational family pattern. In observing how her grandmother and her mother would quietly shift from expressiveness to submissiveness around the men in her family, she had come to see her depression as a result of internalizing the belief that "women only have value as quiet homemakers, wives, and mothers." Determined to break the pattern, she not only identified and disputed the belief but discredited the oppressive message not by attacking her grandfather (although this was her initial path, she found it difficult to maintain focus on the issue due to deep, cultural guilt) but confronting embedded cultural patriarchy.

Finally, clients consciously and actively terminate agreement with the belief. If the client does not readily see the benefit of terminating agreement, the therapist can ask that they try it just as an experiment to see how the act of disagreeing with the belief feels. The Jamesian device technique (see Chapter 5, this volume) can also be used here to determine the truth or falsehood of an enforced belief. This approach is not about clients cutting themselves off from or ending all ties with the oppressor; that is another decision entirely, and the therapist should explicitly discuss the distinction.

Stage 3: Liberation

In the third stage, clients work toward liberation. Once harmful or dysfunctional aspects of a belief are revealed through disputation, the client can proceed to disagree with it and then replace it with more accurate, self-determined beliefs. This approach can be easily adapted to group therapy. Finding hope and social support for clients in this stage is also important. Therapists can tell stories of, and help clients connect with, people who have escaped difficult conditions, such as gang involvement or abusive relationships. Such exemplars demonstrate a path to establishing self-determined and empowering beliefs and show that achieving a degree of happiness or fulfillment is possible. Therapists can also link clients directly with community resources or organizations that offer help to people in need.

A FINAL NOTE ON OPPRESSION, PERSPICACITY, AND LIBERATION

For people who have faced a lifetime of disempowerment by others and by society, it can be hard to imagine life getting better. The various traumas of interpersonal and systemic oppression often underlie the suffering and ineffective coping mechanisms of difficult clients. At the same time, a significant increase in perspicacity can arise among oppressed persons and groups, as the trauma of oppression tends to enhance perceptive awareness and empathic attunement. A therapeutic path to liberation for oppressed clients involves recognizing how oppressors maintain coercive power by introjecting destructive beliefs at both individual and sociocultural narrative levels. Helping clients see how dominant narratives have shaped their self-image can be a powerful way to heal.

15

Addiction and Substance Use

Therapy with addicted clients is one of the most challenging and difficult kinds of therapy, especially if considered in the context of achieving successful outcomes. This kind of therapy can be plodding, stressful, disappointing, and discouraging to therapists. Burnout is a common side effect of doing therapy with these clients, especially if the agency or setting does not provide support, encouragement, and high-quality additional training (J. J. Kim et al., 2018; Yang & Hayes, 2020). Such additional training should not only be on the nature of drugs, alcohol, and addictions but also on how to work with difficult clients. Burnout is often due to empathy degradation and other countertransference phenomena, such as resisting resistance, as well as inadequate training.

This chapter demonstrates how the activation of precursors helps clients enter the change mindset needed for successful treatment. A combination of practical strategies and techniques are provided to illustrate the ideas behind the CHANGES model approach to drug and alcohol treatment. There is no single strategy or technique that will lead to sobriety, so a competent addictions therapist needs a well-assembled toolbox of techniques. We will begin with a reframe of addiction itself.

DRUG USE VERSUS THERAPY: THE GRAND REFRAME

Some therapeutic approaches to addictions therapy suggest that clients are wrong in their use of drugs or alcohol, gambling, or whatever the addiction happens to be. It is common in psychoeducation-focused addictions groups for clients to

---

Case examples have been disguised to protect client confidentiality.

https://doi.org/10.1037/0000451-016
Therapeutic Change With Difficult Clients: Precursors and Techniques in the CHANGES Model, Second Edition, by B. D. Wilkinson and F. J. Hanna

learn that their beliefs, attitudes, and behaviors are simply wrong. The moral implication is clear: They would not be in active treatment for drug and alcohol use if such uses were the "right" thing to do.

The grand reframe takes a different tack by establishing why the client is using in the first place and how addictive drug use in general is often pursued as a solution to the client's problems in life. Eschewing the impulse to repudiate addiction as immoral, therapists can demonstrate how drug or alcohol use can be understood as an attempt, at least initially, to improve one's life experience.

Many reasons explain why a person uses substances, becomes addicted, or becomes chemically dependent, including genetic makeup, family dynamics, social pressure, reducing stress, self-medication, thrill seeking, or sheer boredom. This technique works best, in our experience, with the self-medicating user. Clients have reported self-medicating to address a wide variety of symptoms, including physical pain, emotional pain, anxiety, depression, emptiness, loneliness, relationship problems, low self-esteem, self-loathing, bitterness, and apathy.

The grand reframe is that drug use and psychotherapy have the same purpose: to produce positive changes in thinking, feeling, behaving, and relating. Thus, the grand reframe presents drug use as purposeful in a manner similar to psychotherapy and counseling. When clients are presented with this reframe, they tend to be surprised, if not shocked, to hear a mental health professional say such a thing. It can also increase interest in therapy and bolster the motivation to do the work. In essence, clients can learn that their behavior has been purposeful, even if the use of drugs is not the optimal solution to attain their goal. Rather, psychotherapy is.

Step 1: The Backdoor Question—Finding Out Why the Client Is Using

This technique aims to discover why a client is using and reframe the entire endeavor. Of course, one cannot ask a client, "Why do you use?" We have tried variations of this many, many times, but the typical responses are "to feel good" or "to get high." However, a strategy that we refer to as "the backdoor" approach often renders clinically useful responses.

The reframe can be initiated in the first session after intake. It is important to mention that the technique may need to be repeated several times, depending on the client's awareness level or cognitive condition, before it takes hold, or "bites." Once the client has disclosed their drug of choice, it is appropriate to inquire as to why the client uses at all. This is done in a context that reveals their purpose through inference, as a direct answer is unlikely. The backdoor question is presented next with a typical dialogue with an opiate user that includes follow-up questions:

THERAPIST: (backdoor question) What do you get out of being high?

CLIENT: It makes me feel relaxed.

THERAPIST: Does that mean that when you're not high, you're not relaxed?

CLIENT: Yep. Getting high relaxes me.

THERAPIST: What are you trying to get relaxed about?

CLIENT: I just feel uptight a lot, like everything is all messed up.

At this point, a dialogue may ensue that invites self-disclosure from the client. The therapist can ask questions that do not demand specific information but rather emphasize what burdens the client is experiencing and the purpose that using a substance serves in that context.

THERAPIST: Don't tell me what it is, but are there some things in life that are keeping you from feeling relaxed?

CLIENT: Yeah. (looking down) I don't want to talk about it.

THERAPIST: Fair enough. Don't tell me what it is, but can I ask how it's stopping you? And listen, if I start to bother you, just tell me to back off and I will.

CLIENT: Okay.

THERAPIST: I'm going to guess that you've been through a lot in life. Is that fair to say?

CLIENT: Yeah. I've been through more shit than anybody knows.

THERAPIST: You mean you've never told anyone about it?

CLIENT: Nope. (shifting in seat) Nah ... what's the point? Nobody cares anyway.

THERAPIST: What's it like to be carrying around all of that bad stuff inside of you 24/7?

CLIENT: That's why I use. It helps me forget.

Step 2: The Grand Reframe: Admiring the Intention

THERAPIST: You know, I've gotta give you credit for finding a way to forget and relax.

CLIENT: (surprised) Really?

THERAPIST: Yeah. You're trying to fix the problem. Are the painkillers working for you?

CLIENT: Yeah, a little.

THERAPIST: How much?

CLIENT: What do you mean, how much? Like, enough, I guess.

THERAPIST: As a percentage, how much is the codeine working to fix your problem and help you relax?

CLIENT: Maybe, like, half the time it works, so 50%. Used to be a lot more.

THERAPIST: Pardon me, but 50% isn't much. It sounds like you've noticed the drug is less and less effective in helping you relax and forget.

CLIENT: Yeah, my tolerance is up, so I take more, but it's never enough.

THERAPIST: You know, counseling and drug use are very similar. Both aim to help you find some relief. The big difference is that counseling works over the long haul. Codeine is not a method that works forever, you know?

CLIENT: Yeah, I get that, but it works right now, and that feels like enough sometimes.

THERAPIST: Right, but now you're addicted to a pain pill that controls your life through addiction, and you're stuck in this treatment center. Here you are, having to talk to someone like me for something that only helps half the time, at best?

CLIENT: Well, it's better than nothing.

Step 3: The Redirect

THERAPIST: Fair enough, it is better than nothing. Nothing would be no help at all. Can I say something else, and you tell me what you think?

CLIENT: Sure.

THERAPIST: The goals of counseling and drug use are the same, but the methods are way different. Counseling has the same goal to help you relax, but unlike drugs, it actually works. You've given oxy and other drugs a shot, and you deserve credit for trying to fix problems. But it hasn't helped, 'cause here you are.

CLIENT: Yeah, okay. So what?

THERAPIST: So, you gave drugs your best shot. Are you willing to try something that really works? The good thing about counseling is that you can easily quit if you don't like it (laughing). You're in treatment now, so it might be a good time to give it a try. And if you want to get out of this program, working in counseling is the best way to do it.

CLIENT: Well, I'm here, so why not. I'll give it a shot.

We have found that this reframe can be remarkably motivating, activating the precursors of grit, awareness, confronting, and necessity. Although the use of affirmations in motivational interviewing involves praising positive behavior (W. R. Miller & Rollnick, 2012), few addiction therapists would naturally consider affirming or admiring drug use itself as an attempt to fix a problem. As such, admiring clients for their intention to feel better is a big part of the effectiveness of the grand reframe. When used early in treatment, it can lay the groundwork for treatment success and eventual sobriety. If it doesn't work for a client early in treatment, it can always be attempted later in the therapy process.

CLIENTS UNDERGOING PHARMACOTHERAPY

It is widely acknowledged that medication is best combined with psychotherapy to facilitate change. For most people, medication alone is an insufficient condition for change. In terms of the CHANGES model, medications may make it easier for some clients to activate the precursors and achieve lasting changes. Many clients who report even minor improvements in daily functioning due to medication use are better able to work toward change in therapy. A good metaphor might be weightlifting: Accomplishing therapeutic change is the equivalent of lifting a 500-pound weight. An appropriately prescribed medication can reduce that weight to something more manageable. The effects of the medications vary from person to person, of course; a medication might reduce that 500-pound burden for some people to 100 pounds but for others, only to 400 pounds.

FIRST HIGH, BEST HIGH

This strategy is best done in group therapy but can be done individually as well. After years of work with drug-using clients, we found that there is a large percentage of clients who had their most euphoric and impacting experience the very first time they became intoxicated on their drug of choice. This can be a powerful topic to pursue, demonstrating that drug use is so insidious that it holds out a promise for happiness on the first high and then yanks it away by never again delivering a high as good as the first one or, in some few cases, a second. Meanwhile, the user tries over and over to achieve the intensity and wonder of that first high. It usually doesn't happen again.

The drug experience can be reframed to show that

> Drugs are like con artists or swindlers. They suck us in, getting us to buy some line of crap about how great it all is, making us think that we just discovered the secret of happiness. But it's seldom, if ever, that good again. And we are left holding an addiction as a reward for our efforts, along with a life in ruins.

This type of phrasing gets to the essence of the issue but can be varied to fit the situation at hand.

This technique can be done in groups, where each group member recounts their first high and talks about how great it was. After recounting the experience, the therapist or the other group members ask, "Was it ever that good again?" The therapist or group members then go further by comparing drug use to "buying a line of crap" or something to that effect.

Clients typically report that it was never that good again, and indeed, drug users begin to see that it is even harder to get high with continued use as tolerance of the drug builds. Presenting therapy as an alternative has the advantage of providing hope for the client rather than their compulsively pursuing drugs to squeeze every last bit of joy out of them, as often happens. This technique can increase a sense of necessity, grit, awareness, confronting, and hope.

SUBPERSONALITY TECHNIQUE FOR ADDICTION

The subpersonality technique was explored in relation to the sense of necessity precursor (see Chapter 7, this volume). The term itself refers to different parts of the personality (Assagioli, 1965, 1973; Rowan, 1990), and it has an important role in addictions counseling as a way of using the person's inner resources to work toward change. It can be done relatively quickly when a client expresses apathy or defiance about stopping their drug or alcohol use. Listen for statements about their drug use or regarding quitting drug use, such as "I don't care" or "I'm never going to quit using."

Therapists must first acknowledge, fully accept, and show empathy for the defiance, apathy, and resentment. Defiance can often be reframed as the love of freedom. Make sure that the client feels understood. In our experience, ignoring this first step will lead to technique failure. Then, ask the client, "I know you said you really don't care, but is there a small, perhaps 'little teeny' part of you that worries about what's going to happen to you—a little part of you that does care?" Often, the addicted client will nod or give a simple "yeah." Other possible subpersonality questions are provided next, but do not utilize more than one, as it will disperse the focus of the technique:

- Is there a part of you that thinks you use too much?
- Is there a part of you that wants to quit using?
- Is there a part of you that is tired of living like this?
- Is there a part of you that knows you can't go on like this for much longer?
- Is there a part of you that knows you need to make a change?

If the answer is positive, the therapist can then ask what percentage of their mind is that part, from 0% to 100%. Answers tend to range from 3% to 40%, and any client-identified percentage can be addressed. It is important to note that this percentage step can also be done later in the process if the therapist thinks it better to wait. However, getting a percentage estimate will give the therapist an indication of how much of the client wants to improve. The next step is crucial and involves directly addressing the newly discovered subpersonality. The following question should be asked:

- Can I talk to that part of you?

If the client can identify that there is a part of them that runs contrary to what they have been saying, and if they feel understood by the therapist, the client will almost always let the therapist talk to that positive part of them. However, if the client does not experience the empathy of the therapist, the technique is unlikely to work. This approach can change the entire tone of the conversation and the counseling process itself. It is common to see a completely new part of the client that had not previously been present. The goal is to strengthen the positive part and make it bigger and stronger. In theory, at least, this could be the real client. Other questions to identify the positive subpersonality, according to the issue at hand, could be

- Is there a part of you that wants to change?
- Is there a part of you that wants to quit using drugs?
- Is there a part of you that wants to stay alive? (in the case of suicidal ideation)
- Is there a part of you that is tired of living this way?
- Is there a part of you that thinks you use too much?

When the client acknowledges that there is indeed a part of them that wants to change, and the therapist is given permission to talk directly to that part of them that is interested in positive change, the clinical adventure begins. This technique can be done in the first session, and we have demonstrated this in live consulting practice with actual clients in treatment programs before a small audience, without ever meeting the client previously.

The next step involves speaking directly to the positive part. The therapist can say to the positive part of the client, "What is it like to watch yourself using and taking risks with drug use?" Or, the client could be asked, "What are you thinking when you watch yourself using and getting wasted day after day?" This can sometimes lead to some very positive and encouraging conversations.

Finally, the procedure calls for an assessment on that part of the client. The client subpersonality is asked something on the order of, "You said that you're 5% of the total you, but I wonder how big you'd like to be?" At this point, the client may answer with a much larger percentage. The response from the therapist might then be, "How about we work on making you bigger and stronger so that you are no longer controlled by your using self?" If they get this far, the client is usually interested in pursuing this line of therapy. Validate the client for their good intentions and ask to make this a goal of future sessions. If the client moves back into defiance or apathy in the following sessions, the therapist can ask politely to again speak to the subpersonality that wants to change.

EMPTY CHAIR WORK FOR ADDICTION

It should be mentioned at the outset that in order to use the empty chair technique from gestalt therapy (Mann, 2010; Perls, 1969, 1973; Polster & Polster, 1973), it is helpful to have studied the technique itself, as it is commonly misapplied. Done correctly, this technique can be remarkably powerful as evidenced by our own experience both as therapists and as clients in therapy. It is not within the purview of this book to teach the use of the empty chair technique, as this can be obtained elsewhere easily enough. However, we feel it important to mention that maintaining the dialogue between the entities in the chairs is what brings about resolution and closure. A common mistake, seen even in online training videos, is that the dialogue is compromised by too much conversation between therapist and client and there is not enough focus on continuing the interchange carried by the dialogue itself. It is in that dialogue that the personality integration takes place.

Empty Chair With Subpersonalities

Once the subpersonality is identified, the empty chair technique can help resolve the split between personality parts by bringing about a personality integration. In one chair is placed the part that wants to quit using and in the other chair is placed the subpersonality that wants to continue using. The empty chair procedure is conducted, continuing the interaction until the aforementioned personality integration takes place. Addressing subpersonalities, or disowned parts of the personality, as they are called in gestalt therapy (Mann, 2010; Perls, 1969, 1973; Polster & Polster, 1973), is a standard utilization of the empty chair. The results can be remarkable. We have seen this technique occasionally turn the tide of drug use in some clients.

Other Applications of the Empty Chair Technique in Substance Use Therapy

In addition to use with subpersonalities, there are other applications of the empty chair technique that can be effective in addictions work. The following list includes various applications of the empty chair in substance use that we have found helpful. However, do not attempt the technique unless the client can actively visualize something or someone in the empty chair. If they cannot visualize it, the technique will likely fail, or the client will be unwilling to enact it.

- Place a supportive deceased person in the empty chair.
- Place an oppressive deceased person in the empty chair.
- Place the addiction itself in the empty chair.
- Place the drug of choice or the craving in the empty chair.
- Place an influential drug-using person in the empty chair.

That said, some clients find this technique unappealing and will not attempt it at all, although they might do so later in the therapy process. Unfortunately, we have seen several occasions where a client declares, "I'm not talking to an empty chair," and that's that. Confronting the problem, grit, and the awareness precursor are most operative herein.

THE DRUG HERO TECHNIQUE

Clients occasionally disclose that they have a friend who can consume any amount of alcohol or any quantity of a particular drug with no ill effects at all. When asked, clients in such a situation will typically say that they think about this "drug hero" during psychoeducation sessions or videos illustrating the dangers and destructiveness of drugs. The memory or image of the drug hero will invalidate any therapeutic message or information about the problems that come with drug use.

Clients have reported that they often see these videos in treatment programs. However, many of them listen to and then dismiss the messages because they think, "Yeah, but that doesn't apply to Joey." In this instance, the client will keep their drug hero, Joey, alive in their mind as a means to hold onto the defiant or desperate wish that drug use really is not all that bad.

Although this technique can be done individually, it is perhaps best in group therapy by having each client tell a story about how their drug hero (if they have one) was virtually immune to the negative effects of alcohol and/or drugs. The group therapist can also share an example of a drug hero from their life, if applicable. The next step consists of having clients talk about what the hero is like in the present, presumably after years and years of drug use. Clients will often report that their hero is now in jail, on the street, lost their job, rejected by their family, selling drugs, or committing other crimes to support their addiction. If not, the hero is the exception, not the rule.

Either way, it is better to bring the hero out of the client's mind and into therapy than allow them to take up space in the mind playing an oppositional role that is illusory if not delusional. Clients tend to enjoy this technique when done in groups and can tell interesting stories. The point of this technique is to show that some people indeed have higher resistance to drug effects than others, but eventually, the drug wins, and it destroys the hero's life just like it does for everyone else.

Finally, it is a good idea to ask clients in group if they were ever considered a drug hero, capable of consuming huge doses with minimal negative effects. This is particularly helpful when clients hear that a fellow group member has been so badly affected that they are here, in the same treatment program as the rest of the group. When the heroes can self-disclose in group, it can be beneficial to the group as a whole to see how drugs can take down the hardiest among us. The awareness, necessity, and hope precursors are primary here, but when used in a group dynamic, this technique can obviously influence social support, as well.

VALUE DISSONANCE AND REALIGNMENT

This is another technique that can be done individually and in group therapy. It is a cognitive approach that is combined with an existential component. It is a way of restoring value systems among users who have had their values devastated by drug and alcohol addiction. Like most of the techniques presented in this chapter, this technique is not powerful enough to get the user to quit using altogether, but it may be powerful enough to put a dent in the user's drug mindset.

The technique begins with asking group members, in a round, to name the three most important things in their lives. They can briefly describe each and explain why each is important. Typical values can include family, friends, love, jobs, money, romantic partners, and freedom. The second step is to have each group member discuss how drug use has impacted each of these values. For example, using can destroy families, lead to loss of freedom, break up marriages, and so on.

The third step consists of asking, still using rounds, how much the drug is valued compared with the three major values. In the fourth step, the therapist asks the client to insert the drug into the list of values. This begins to reveal cognitive dissonance in the client's mindset in terms of how the pursuit of the drug has degraded the precious things in life that the client has held dear. Finally, the group is asked, one client at a time and still in the rounds format, how they would or could realign their values without the drug taking precedence and what this might look like.

This technique can help a client build a clearer picture of what they want their life to look like. It makes goal setting easier due to removing a good portion of the conflicting value that arises between drug use and the living of life itself. It is also valuable because it presents a very stark image of how drug use is destroying the important things in their lives and yet how there remains hope in recovery. Awareness, confronting, necessity, and hope are the operative precursors in this technique.

EXISTENTIAL-COGNITIVE THERAPY OF OPPRESSION FOR DRUG USE

This technique has been described elsewhere (Hanna & Cardona, 2013; Hanna et al., 2000; Vickery et al., 2023) and can be used in the drug and alcohol context simply by identifying the specific people in the client's life who have been harmful to them. Make a list of these people and ask who among them seems to be related to their drug or alcohol use. With each of these people, list the beliefs that the oppressive person forced or imposed upon the client and how much the client agreed with each of the beliefs on a scale from 1 to 10.

By getting the client to disagree with and let go of the enforced harmful beliefs, the client may be able to free themselves of the psychological attachment to these oppressors and become empowered to stop drug use and aim for sobriety. Another way of addressing this is by asking the client if they are hard on themselves, hate themselves, or "give themselves a lot of shit." All that is necessary for therapeutic benefit is for the client to disagree with the imposed messages.

If this is insufficient to get a client to disagree, the therapist can ask what they say to themselves in classic cognitive behavior style (Meichenbaum & Cameron, 1974; Meichenbaum, 1977) and then restructure the self-talk. Sometimes, discrediting the source of the messages can help dismiss the harmful belief. This technique involves confronting the problem and liberating the client from the oppressive influence rather than merely adjusting to the situation.

ROLE-PLAY: THE DEVELOPMENT OF REFUSAL SKILLS

Refusal skills are an important strategy in the establishment and maintenance of sobriety. Role-playing situations in which the client has been, or is likely to be, offered drugs and/or alcohol can be empowering. This role-play is best done in group therapy, with each group member getting a turn to play the client. The technique is carried out in stages. The emphasis is on situations where the offer is strong and a simple "no" is an insufficient response due to active social pressure.

The first stage consists of the therapist and client sitting in the middle of the group. A marijuana blunt, for instance, can be easily simulated by rolling up a piece of paper. The therapist mimics smoking from the blunt before offering it to the client, saying with insistence, "Hit it." as if the therapist's lungs are filled with smoke. The therapist, as in real life, pushes the blunt toward the face of the client, insisting they take it. Meanwhile, the client is informed only that they should mindfully observe the process, watching for thoughts or feelings that arise in role-play. The client is then asked if any memories are triggered. Thoughts and feelings that arise during the role-play are actively processed in the group setting.

The second stage of the role-play involves having a group member sit in for the therapist and offer the blunt to add an element of reality to the scene. The group member foists the blunt on the client, using condescending or ridiculing statements to get them to use the drug. Once again, the client does not answer but mindfully observes any feelings and thoughts that arise as well as any memories that might be triggered. These are also processed in the group.

The third stage of the role-play involves an element of psychodrama called doubling (Goldman & Morrison, 1984). The client is again offered the blunt by a group member while being mindful, as in the earlier stages. However, in this stage, two group members are asked to provide verbal, audible "self-talk" by standing alongside the seated client and speaking into the client's ear. When the client is offered the blunt, the two standing "doubles" actively participate. The double on the client's right side speaks positive statements into the client's right ear, saying, "You know you shouldn't hit it," or "You've been clean; don't wreck it now." Meanwhile, the double on the left side says in the client's left ear, "You know you want to hit it," or "Go ahead, get high." Clients consistently report that this method is close to their lived experience and reproduces their mindset in the real world. They also tend to report how helpful it is to role-play such situations.

The fourth stage is called refusal dialogue, and this is the stage in which the refusal itself is role-played. With the continued engagement of mindfulness, the third stage scenario is duplicated with an additional step of speaking actual refusal lines that convince the person foisting the drug or alcohol to back off or give up. Done properly, this stage can be remarkably empowering and encouraging for clients, if not a bit humorous as well.

In the midst of being offered the drug, and with the doubles continuing their messages in the ears of the client, a list of refusal lines is provided. Refusal lines go beyond the simple "no" for an answer in a way that discourages the person from continuing to offer the drug. Some of these responses are intentionally sarcastic or absurd so as to activate cognitive dissonance. Please note that refusals should be spoken respectfully, as rude as they may appear in writing. There is no point in antagonizing others, and it is preferable for the lines to be spoken with a sense of irony, sarcasm, or humor. Here are some refusal lines we have found to be effective in group therapy:

- I appreciate your wanting to get me high, but sorry, I'm into depression!
- I don't do that anymore. Drugs make me too happy. I hate being happy.
- Nah, I like feeling like shit. Can't be getting high and messing that up.
- Thanks, but I got better things to do than to feel good.

Of course, group members can work to come up with their own refusal lines. Clients in outpatient programs often come back to the treatment program laughing at the effects these kinds of lines have on sellers or people who try to get them high. The confronting precursor as well as grit and awareness precursors are clearly at work here.

ALCOHOL AND DRUG USE IN THE SERVICE OF THE FAMILY

There is an aspect of family systems that is often overlooked by therapists, which may be because the *Diagnostic and Statistical Manual of Mental Disorders*, in its various iterations, does not take systems theory into account (Bonino & Hanna, 2018). There are some instances we have found where the only way to reach the client was by treating the client as the family healer (Whitaker, 1989). As is well known, in systems theory, the problematic person in a family is often referred to as the scapegoat or symptom bearer (Bowen, 1976). Whitaker (1989) added the concept that the symptom bearer or scapegoat can also serve as the family healer.

A drug or alcohol user may be engaged in addictive behavior not only because of addiction but to actually serve the family. Hanna (2004) told the story of how even a 2-year-old boy with serious symptoms could function as the family healer. Whitaker (1989) informed us that we should not be too quick to label maladaptive behavior as psychopathology. What follows is the story of Jimmy, an 18-year-old boy with a serious alcohol problem. It illustrates how negative behavior can be well-intended and purposive. But as the saying goes, "The road to hell is paved with good intentions."

The Story of Marco

At 18 years old, Marco had been in an inpatient facility for 2 months. He was reported to the consultant by the clinical supervisor of the agency as being unmotivated to quit drinking and disinterested in sobriety. The consultant was informed that he had made no progress at all. Marco openly admitted that he was going to drink as soon as he was released from the facility, and he was highly defiant toward therapists and staff members, as well as showing a certain amount of contempt for therapy itself. The consultant was also informed that Marco's father was a severe alcoholic who had been hospitalized several times because of it and had developed a serious liver disease.

The consultant was asked if he would see Marco with the goal of getting him motivated to participate in counseling and make positive changes. The consultant met with the treatment team to learn what had been previously done with Marco to help with his excessive alcohol use. They reported that no discernible progress had been made during Marco's time in treatment. The treatment team reported using cognitive behavioral therapy and motivational interviewing to talk with Marco about his father, with particular emphasis on building some dissonance around ending up sick like his father had been.

The consultant sat down with Marco, who had been informed of this session. Initial probing and precursors techniques such as admiration of his defiance as an expression of freedom were just effective enough to establish some small amount of rapport. After the consultant asked Marco if he was worried that he might end up like his father, Marco, as predicted, replied that he was not worried at all. He also said that he did not care if he became addicted and was not concerned that alcohol could ruin his life. He said that he did not know why he loved to drink and did not care to find out.

The consultant attempted the subpersonalities technique, addressing the hurt, freedom challenge, admiring negatives, and others, but none were effective. The only reason Marco was talking to the consultant was due to a few effective expressions of empathy. Seeing no pathway into his mindset, it dawned on the consultant that Marco might be an alcoholic in service of the family when the following sequence occurred in the conversation:

CONSULTANT: Do you like your father?

MARCO: Nope.

CONSULTANT: How well do you know your father?

MARCO: I mean, I don't really know him. Nobody does.

CONSULTANT: Did you ever try to get close to him, talk with him, anything like that?

MARCO: Yeah, he didn't want to know me.

CONSULTANT: I see. So, you somehow became much like the guy you don't like or know?

MARCO: (hesitantly) I guess.

At this point, the consultant sensed an inroad. Through a series of questions and reflections, the consultant highlighted how Marco's father chose alcohol as his only meaningful relationship, over and above that of Marco and his siblings. The consultant validated the difficulty of knowing an alcoholic and suggested that Marco was trying to not only understand his father but to save him:

CONSULTANT: Is it possible that you're drinking to learn all that you can about alcohol ... as a way to better understand your father?

MARCO: Uh, I don't know. Maybe? Never thought about it like that ...

CONSULTANT: Think about it ... take your time ...

MARCO: Maybe, yeah. I just assumed we were both addicts.

CONSULTANT: Maybe. But maybe it's different for you. Maybe you started drinking because you wanted to learn about it so that you might be able to save him. You started drinking to feel close to your father, to save your father.

MARCO: (staring at the ground) Yeah, I mean, I was just curious, when I started, about why he loved it so much. I just wanted to know what it was like ...

CONSULTANT: Jimmy, you're a hero.

MARCO: (looking up with a scowl) Yeah, right. What are you talking about?

CONSULTANT: (reframes) You could have had a very different life, but you're such a good person—such a good son—that you decided to help your father with his drinking regardless of what it does to you. That's what heroes do; they help others despite the risk to themselves. And that's what you're doing.

MARCO: (with tears in his eyes but shrugging his shoulders) If you say so.

CONSULTANT: Well, to your credit, you've become very familiar with drinking. Maybe it's time to learn about sobriety and how to stop drinking. What do you think?

MARCO: (wiping tears) Yeah, maybe it's time to do that.

From that session forward, Marco responded to most therapy techniques and, after a few weeks, had become active as a leader in group therapy and strongly involved in individual therapy as well. The consultant informed the treatment team, the clinical supervisor, and the director of the agency that Marco was not a typical kid with an alcohol problem. He was a family healer attempting to acquire knowledge of alcohol use through experiential learning and that he was now ready to work toward sobriety. It was a classic example of how systems can play a major role in addictions.

TECHNIQUES FOR REDUCING CRAVINGS

According to research, cravings are among the most important phenomena to address in drug and alcohol treatment due to evidence showing that once the cravings are greatly reduced, the urge or need to use is also greatly reduced (Tapper, 2018). In addition, the presence of cravings is related to the urge to relapse, and conversely, without cravings, the user will be much more likely to achieve stable sobriety (Vallejo & Amara, 2009). Mindfulness practice has been shown to be consistently effective in reducing cravings (Tapper, 2018). What follows in this section are various techniques and approaches to reducing cravings through the application of mindfulness.

Craving Management

Ask the client to locate in the body where exactly they feel the craving. This is done so that the client can "anchor" the craving and reduce its vagueness so they can identify some specific details about how they experience it. Typical responses are that the craving is in the area of the chest, solar plexus, or throat. Teach the client how to be mindful of the craving and whether the intensity varies, as well as when and in what situations the craving becomes most and least intense. Ask the client to set their phone to sound an alarm every 30 minutes (or whatever is most appropriate) so they can then enter a craving intensity rating into a tracking app. This will give the client and therapist a good indication of what kind of state the client is in relative to the nature of the cravings on a daily basis. Part of craving management also involves tracking the craving itself, noting if it grows or shrinks in the body at various times and across various situations.

Mindfully "Riding" the Craving

Also known as "urge surfing" (Shonin & Van Gordon, 2016), this technique involves deliberately confronting a craving but not reacting to it. This technique is best done in the beginning with the therapist present. Inform the client that cravings do not have to be the most controlling phenomenon in their life. The therapist then asks the client to establish a mindful state, enter the craving itself, and describe the sensations, feelings, and temptations associated with it.

The key is to fully concentrate on the craving and stay with it until the client achieves some stability in the sense that they can accept its presence and not feel overpowered by it. If a client can immerse oneself in the craving without reacting, the craving has been observed to reduce in some cases. But at first, the riding of the craving should be done with the therapist present and guiding. The "riding" is essentially the capacity to experience the craving itself and watch how one is "drawn into" or "sucked into" the craving, and noticing where and how one loses control to it. Once again, mindfulness is the key practice in this technique, which requires deliberate client practice.

Influencing the Craving

Once the client has shown some skill in riding the craving, they can begin to influence it. The key here is to consciously increase the intensity of the craving while continuing to ride it. Once the craving has increased, it can be somewhat controlled by intentionally reducing its intensity. Mindfulness reveals that some of the intensity of a craving is due to resisting it, whereas accepting it (as is done by riding the craving) removes some of the power of the craving.

Working with the therapist in session, some clients start to show self-determined degrees of direct influence by taking a measure of control of cravings.. The effect of this technique for some clients is significant in that the client is no longer helpless before the craving and not controlled by it to the degree they were previously. The result, when successful, gives a client a sense of hope for the possibility of achieving sobriety. But as we said, one cannot become overconfident here.

A FINAL NOTE ON ADDICTION AND SUBSTANCE USE

In important ways, this technique-focused chapter serves as an appropriate endcap to the book due to its fully pragmatic bent. Since the purpose of the CHANGES model is to help therapists facilitate change with difficult clients, it is reasonable to conclude with the immense challenge faced in the battle against addiction and substance use. In the last chapter, we review the advantages of the CHANGES model, highlight potential applications of the CHANGES model in the business sector, and discuss how the CHANGES model informs the development of a metatheoretical framework on teleological freedom that uniquely contributes to the psychotherapy integration movement.

CONCLUSION

16

At the Horizon of Change

The CHANGES model may be a valuable step in isolating change principles that are not bound by jargon, techniques, theory, stages, or personality traits. The model moves in a direction that cuts to the essence of therapy in attempting to isolate the necessary conditions or common factors of therapeutic change itself. Such an approach has several practical advantages.

The first advantage arises when a difficult, perplexing client seems unlikely to change. Rating difficult clients using the CHANGES assessment may help formulate new approaches to change, and using the techniques provided in this book could assist in implementing the missing precursors. There are, of course, many techniques and strategies that can be developed for each precursor beyond what is listed herein. We leave it to therapists and researchers to add new approaches and techniques. When one understands the nature of the precursors, suitable techniques for each tend to spontaneously develop. Indeed, many therapists do this already.

A second advantage is gained through rating the precursors of a therapist. Possibly for the first time, a gauge is provided that supervisors and trainers can use to take some measure of a therapist's potential as a change agent with specific clients. Similarly, when having difficulty with a client, the CHANGES assessment can help practitioners rate themselves and pinpoint missing precursors. The aim in this context is to bolster missing precursors to help therapists maximize helping efforts. It can also be of assistance in rehabilitating impaired therapists or those suffering from burnout.

A third advantage is that the CHANGES model can help a therapist better understand therapist interference. When therapists recognize that their actions or attitudes are diminishing the presence of specific precursors in clients, corrections can be made to increase, rather than reduce, those precursors. For example,

https://doi.org/10.1037/0000451-017
Therapeutic Change With Difficult Clients: Precursors and Techniques in the CHANGES Model, Second Edition, by B. D. Wilkinson and F. J. Hanna

sometimes therapists make the mistake of thinking that certain actions or attitudes that were helpful with one client will be helpful with others. Unfortunately, this can impede the emergence of precursors in some clients. Awareness of therapist interference is important for developing self-awareness, facilitating self-correction, and growing wisdom. It also helps therapists avoid automatizing their approach with difficult clients.

A fourth advantage is that focusing on precursors may help train therapists and counselors in client-specific change processes and variables. When trainees are taught to think in terms of therapeutic change, including what causes it and what discourages it, the psychotherapy enterprise is seen as a process of removing the obstacles to change and implementing elements that expedite its progress. Manuals and treatment procedures should be a secondary priority after change itself.

When trained in the CHANGES model, students can recognize that resistance and difficulty are spontaneously removed by increasing the precursors. A resistant or difficult client, then, is not to be thought of in rigidly stereotypical or diagnostic terms. A person who is difficult to work with may simply be a person who is missing precursors at that time. Because the precursors wax and wane from day to day or week to week, almost everyone is bound to be difficult now and then.

Finally, research seems to indicate that the production and enhancement of client change in psychotherapy lies in four areas: (a) intensifying the client's motivation to change, (b) augmenting involvement in the change process, (c) enhancing the feasibility of change, and (d) removing obstacles to change. If each of these four areas of change is addressed, the rate and magnitude of client change in psychotherapy will likely increase. If the field had a variety of procedures designed to increase each of these areas for clients, the efficiency of therapy might be enhanced, and psychotherapy may become far more reliable in terms of expected outcomes.

Figure 16.1 outlines precursors that play a role in enhancing each of these four areas. Of course, the therapeutic relationship is assumed to be operational in each area. The CHANGES model lends itself to these tasks by pinpointing specific avenues of approach with difficult clients.

POTENTIAL FUTURE APPLICATIONS OF THE CHANGES MODEL

The CHANGES model can support new frameworks for enhancing clinical outcomes by fully centralizing the potency of client-specific change factors. We explore two such prospective frameworks in the following two sections. First, we briefly demonstrate how the CHANGES model might be applied as a client-specific change factor framework in organizational leadership, emphasizing identifying barriers to business and corporate reorganizational change efforts.

Second, and more importantly, we examine how the concept of freedom may serve as a teleological approach to psychotherapy integration. Our proposed

FIGURE 16.1. Increasing the Rate and Magnitude of Change

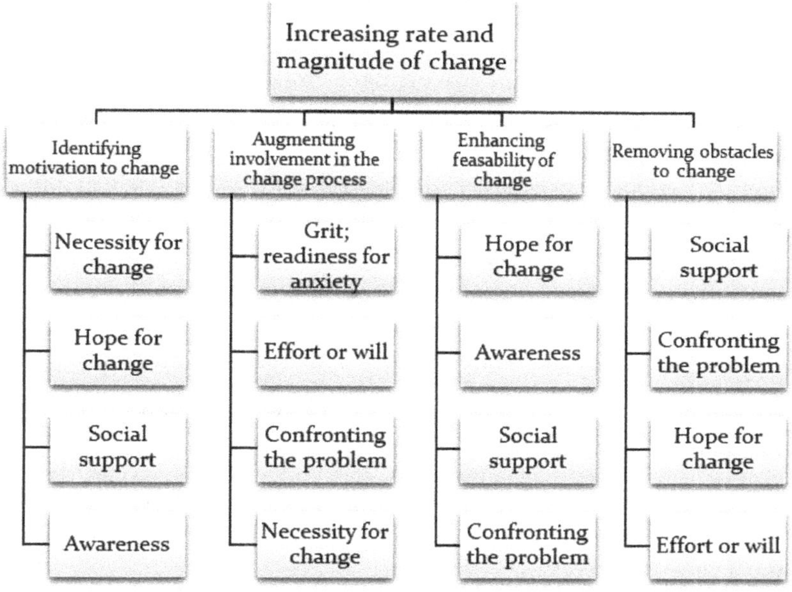

metatheoretical framework suggests that the purpose of psychotherapy is to facilitate psychological freedom, which occurs via activation of the precursors as client-specific factors in the CHANGES model. Such a teleological integration preserves the value of different theories, techniques, and even the common factors but uniquely positions psychotherapy as an integrative endeavor with an overarching purpose. By acknowledging that freedom is the guiding purpose of therapeutic change efforts, we can begin to collectively focus on the pivotal nature of client-specific factors in counseling and psychotherapy.

ASSESSING INDIVIDUAL, TEAM, AND ORGANIZATIONAL CHANGE

Just as the CHANGES model is used to determine the change potential of groups and families, it might also be used for the same with employees, departments, or entire businesses. A particularly rigid, compulsive, or driven executive, for example, may be lacking in certain precursors and thus negatively impact the precursors of employees during a corporate transition. A modified version of the CHANGES assessment form has the potential to help business leaders and employees think in terms of generating, tolerating, and working toward constructive organizational change.

The CHANGES Assessment for Businesses and Corporations (CHANGES-ABC) is a prospective tool for understanding and supporting employees, management teams, and other units during organizational transitions. The proposed assessment form is not yet empirically validated, but it is empirically informed and may serve as a guide for change agents in the business sector to identify the presence or absence of the seven change factors from the

CHANGES model. CHANGES-ABC can identify where a team is in the change process, which deficits may be preventing development along the change process continuum, and whether specified change factors need to be activated in order to better facilitate the leadership or organizational transition.

Scoring the CHANGES-ABC allows assessors to determine the likelihood that change will occur based on individual scores, whether for project coordinators, midlevel managers, or executives. The assessment could be taken at a number of points throughout the change initiative process, letting assessors track the probability of project success based on normed metrics and within-group comparisons across project teams, units, departments, or full organizations. The CHANGES-ABC would support business leaders across various sectors to ensure the efficient facilitation of change initiatives and transitions by

- Assessing a business's overall readiness for change when preparing for a transition of any kind.
- Supporting new team leaders and/or managers who want to understand their teams' readiness for change and to identify gaps that need to be filled in order for the change process to be realized.
- Onboarding executives placed directly in leadership roles who must join with and adapt to a new organizational culture. Research indicates that a significant percentage of business executives derail in the first 18 months due to an inability to join with company culture.
- Training leaders and executives responsible for organizational change, providing a framework for understanding how to increase change-related opportunities and reduce barriers to change.
- Monitoring executives perceived to be in "derailment" who need to make immediate and sustainable positive changes in their role to get on track and maintain their current position.
- Helping leaders determine the probability of success during major business transitions by measuring how responsive employees/teams are to specific change initiatives.
- Being integrated into company culture as part of a 360° feedback assessment. Human resources departments can use it as an educational/training tool on adaptability to change and what is required to facilitate new initiatives. Managers can be trained to understand and grow change readiness—a problem that can be fixed via training—and to support new talent acquisitions.

To support CHANGES-ABC, each of the precursors in the CHANGES model can be updated to capture their applicability to the business sector, as seen in the following expanded definitions:

1. *Confronting the problem:* This is the culmination of awareness but is not the same. This is the steady and deliberate attending to and observing of anything intimidating, painful, or confusing in the change process. It is looking at the

problem dead in the face and continuing to look in spite of the tendency to avoid it. *Expanded definition:* This precursor involves the individual's, team's, or organization's willingness and perceived ability to constructively confront the problem(s) the organization is presently seeking/struggling to overcome through initiating the change process. The ability to constructively confront the problem also involves the willingness of the individual, team, or organization to learn new skill sets that they may not possess but are necessary to learn in order to confront the problem areas and develop a sustainable solution. At the organizational level, this precursor is indicated through one's willingness to accept and offer constructive feedback to or from fellow employees regarding their role in creating and sustaining the problem and/or the change process. This precursor is most closely interrelated to the grit precursor in that the individual, team, or organization must be willing to experience discomfort/anxiety so they can effectively and constructively confront the problem they are facing. Often, fear or anxiety are the underlying psychological mechanisms that prevent obstacles from being overcome and goals achieved.

2. *Hope for change:* This is the realistic expectation that change can and will occur. It is not wishing, longing, desiring, or yearning. Hope sees the possibility of change and motivates a person, knowing that change can be accomplished. *Expanded definition:* When an employee has hope, they can envision a positive outcome and how it can be achieved. This precursor is intimately linked to the effort precursor, as hope, more often than not, precedes will, and effort and is needed to stimulate their development. If one has a low hope score, it is likely that effort will also be low. This is important because motivation has no force in and of itself without a strong and emerging foundation of hope guiding it forward.

3. *Awareness of the problem:* This is knowing that a problem exists and having a good sense of what the problem or issue is. Awareness is the opposite of denial. Without it, a person has no idea where to direct their resources toward change. Awareness has to do with the ability to identify thoughts, feelings, and perceptions that may be obstructing change processes and those needed to embrace it. *Expanded definition:* Awareness can also be defined as the employee, team, or organization understanding of their role in creating, developing, and maintaining the present issue and/or problem that needs to be changed; the role they play in obstructing change processes; and the role they can play in affecting positive change to procure a future solution. Once an awareness precursor score has been determined, the change agent can determine the individual's underlying assumptions and limiting and mistaken beliefs that hinder them from developing the insight necessary to effectively participate in and influence the change process. This precursor helps to measure the level of leadership interference that may be present in the organization.

4. *Necessity for change:* This is the recognized urgency or need that change take place. It considers that change is important and that current conditions are

not satisfactory and must give way to a different set of circumstances. *Expanded definition:* This precursor indicates the employee's, team's, or organization's level of understanding that change is essential for problems to be resolved and a solution developed. It indicates that current dispositional, environmental, and situational conditions are not satisfactory and must give way to a different set of circumstances if the change process is to be initiated and change outcomes are to be achieved.

5. *Grit, or the willingness to experience anxiety/difficulty:* This is the simple surrender to the change process. It is the recognition that one is willing to feel the discomfort that comes with change. Defensiveness is usually defined as an attempt to avoid anxiety. This precursor is the diametric opposite. The person is open to and allows for the presence of any anxiety or difficulty in order to bring about change. The willingness to take risks and be open to experience is also an important aspect of this precursor. *Expanded definition:* This precursor indicates the employee's, team's, or organization's willingness to experience the anxiety, discomfort, uncertainty, and fear that originate from not understanding what to do and how to effectively initiate and/or engage in the transitional process. This precursor involves employee, team, or organizational acceptance of their own limitations and creates a platform for learning new skillsets needed to initiate and participate in the change process. A fundamental part of this is being willing to confront the very things one has been actively avoiding.

6. *Effort toward change:* This is the precursor that indicates action engaged and taken to solve the problem. It is the actual expending of energy as well as movement taken. It also involves the will, in the sense of commitment and a decision to change. Effort takes place in changing the mind, behavior, feelings, or environment. *Expanded definition:* This precursor measures the will to action that is essential to effectively achieve the change being targeted through an initiative. It equally measures the level of anabolic and/or catabolic energy the individual puts forth to resolve the issue. From the score, the change manager can then process with the employee the factors influencing their levels and/or type of energy and how those influential factors are preventing them from overcoming the obstacles to participating in and fulfilling change initiatives. Such obstacles may be motivational factors that can be questioned and addressed once the individual's level of energy and effort has been determined. The level and type of energy indicate the obstacle the client is not willing to confront.

7. *Social support for change:* This consists of confiding, supportive relationships that are dedicated to the well-being of the person. Such relationships make the change process more tolerable and can inspire each of the other precursors. Conversely, relationships not dedicated to the well-being of the person negatively impact a person's ability to change. *Expanded definition:* Supportive relationships in an organization make overall change processes far more tolerable and achievable and can motivate the formation and

development of each of the other precursors. Without high scores in social support, all other precursors can be extremely difficult to inspire or motivate. This precursor has a secondary gain in that it helps to identify relational problem areas in departments, teams, and between individuals within the organization. Conversely, professional relationships that are not dedicated to the well-being and future growth of employees, teams, or the organization as a whole negatively impact change initiatives. Once the score of this precursor has been determined, change agents can explore the depth of the problem and how to establish and/or reestablish relationships that have been damaged or severed between departments or among employees. The purpose of this is to affect substantive change for team building processes and motivate team members and departments to work more constructively toward the realization of an initiative.

FROM MODEL TO METATHEORETICAL FRAMEWORK: THE FREEDOM PARADIGM

The driving ideological force behind the psychotherapy integration movement has been the pursuit of consensus, without which we are ostensibly left stranded within the much-frowned-upon realm of preparadigmatic scientific status. The CHANGES model finds its footing in the ground between the common factors model and longstanding calls from integrative researchers (see Goldfried, 1980, 2019) to closely examine principles of change that arise "at an intermediate level of abstraction between theoretical frameworks ... and specific techniques or clinical procedures" (Gaines & Goldfried, 2021, p. 268). Based on such relative positioning of the CHANGES model, there is the possibility that it can support conceptual extrapolations into metatheoretical territory.

We believe that all effective psychotherapists have a distinct capacity for phenomenological engagement; that is to say, being predisposed to examine the structures of consciousness both as a general form of experiential-psychological inquiry (i.e., psychological mindedness) and an ability to attune to the inner universe of others by means of deep relational contact (i.e., presence). When such therapists engage clients in a second-order change process, they are not just trying to reduce symptoms. There is a deeper, more meaningful transformation taking place to which the savvy therapist is distinctly attuned. This transformative process transcends the language of particular theories and thus may only be conceptualized in terms of a metatheory or paradigm.

The metatheoretical framework we propose as an extension of the CHANGES model is rooted in the concept of freedom. Simply put, the purpose of nearly all psychotherapy endeavors, at some level and to some degree, is to set people free (Hanna, 2011). This deceivingly simple statement, hidden in plain sight, reflects what might be considered the preeminent goal of therapy for individuals, families, groups, and communities. We propose that coordinating therapeutic practices under the metatheoretical framework of freedom provides a common teleology, or purpose, to the act of therapeutic engagement that fosters a client-specific change emphasis (Hanna & Black, 2007).

Controversy and suspicion have long surrounded the notion of freedom, particularly among traditional behavioral scientists inclined to doubt its existence. Whereas James (1981/1890) held the subject of freedom in high regard, Skinner (1971) regarded it as an illusion (see Korn et al., 1991). The free will–determinism debate continues today, as seen between various philosophical heavyweights such as Daniel Dennett and Robert Sapolosky (D. Miller, 2024). Clarifying the concept of freedom in its present use is thus necessary to skirt controversy.

To begin, freedom as discussed herein is not the same as free will, which is a metaphysical perspective on freedom that denotes that the human condition is one of rational, freely chosen actions, behaviors, and decisions. Freedom in its use here does not relate to free will. The debate between determinism and free will cannot be empirically validated because it is an antinomy (see Kant, 1787/1929), and the freedom paradigm is not designed to address, much less resolve, that philosophical dispute.

Furthermore, freedom in its present use is related to political freedom only indirectly and should not be considered in a political or liberty-related context. Within our purview, freedom is considered through a psychological rather than a physical lens. As such, the 27-year imprisonment of Nelson Mandela by the South African Apartheid government signifies the difference between political freedom or liberty and psychological freedom. Reflecting on his experience, Mandela (1995) said, "I learned that courage was not the absence of fear, but the triumph over it. The brave man is not he who does not feel afraid, but he who conquers that fear" (p. 622). A similar perspective is expressed by Frankl (1992), who wrote, "Everything can be taken from a man but one thing: the last of the human freedoms—to choose one's attitude in any given set of circumstances" (p. 86). Such statements appropriately convey the disparity between physical and psychological freedom.

With these parameters in mind, freedom may be defined in terms of a psychological state absent those psychological restrictions and inhibitions that result in symptoms of depression, anxiety, compulsions, obsessive thoughts, or emotional pain. This definition of freedom includes the active capacity to ameliorate or alter undesired thoughts, emotions, and behaviors so as to enhance positive conditions and increase the range of available options for consideration. This also corresponds with a mastery of or the ability to navigate conflictual conditions by acquiring whatever skills necessary to handle mental, interpersonal, systemic, or societal challenges. Responsibility is another key component of such freedom (Sartre, 1943/2003) insofar as the individual is willing to proactively manage, influence, or otherwise respond to such conditions without succumbing to a sense of burden, duty, or blame. In this manner, empathy, understanding, and tolerance supersede forms of prejudice, bias, and obliviousness when the individual chooses responsibility and freedom.

Maintaining a permanent state of freedom is clearly not possible. However, it is assumed that greater degrees of freedom can be attained through intentional and constructive change processes. Influenced by myriad internal/psychological and external/environmental factors, freedom involves a capacity to alter or amend our response to influences. As such, the metatheoretical freedom framework highlights the importance of understanding the process of becoming free such that one may adaptively respond to a variety of influences with empathy, compassion, and common sense.

This conception of freedom is closely related to previous discussions in this book on active agency, although it represents a more fluid take on the concept. In many ways, the metatheoretical freedom framework is more closely akin to liberation as conceptualized in certain ancient Indian philosophies (Pereira, 1976), particularly regarding the attainment of *mukti* or *moksha* as rendered in Sanskrit. The ancient Yoga concept of *kaivalya* is also similar (Aranya, 1983; Srinivasan, 2021), by which freedom is an attainable state of being. The foundational concept of *nirvana* in Buddhism also relates to freedom as an attained state corresponding to the cessation of suffering (Rahula, 1978).

The similarity between these ancient philosophical paradigms and the proposed metatheoretical freedom framework rests in the shared idea that freedom can be attained only by degrees, is never complete, and is neither intrinsic to human action nor divinely given. To be absolutely clear, we are not recommending *moksha, kaivalya,* or *nirvana* as a psychotherapeutic goal or practice. It cannot be disputed, however, that ideas originating from such Asian philosophies have begun to deeply influence psychotherapy research and practice in the Western world (see Moodley et al., 2017). Mindfulness and meditation practices taken from the ancient Buddhist and Hindu traditions have been adopted and adapted, rapidly becoming a central tenet in many psychotherapy practices.

All this aside, the idea is that clients benefit when therapists trade a restrictive mindset imposed by singular theories for a metatheoretical framework that emphasizes the value of increasing a client's degree of freedom. Such teleological integration signifies an attempt to align classical psychotherapy theories and subspecialties under a common conceptual heading that may result in greater cohesion, collaboration, and innovation across helping professions. In an effort to support and unify the field, the teleological approach aims to enhance distinct subspecialty areas through the integration process rather than attenuate them. Freedom-based teleological integration purposefully aligns theories and subspecialties so that they may retain their tradition, authority, and power (Hanna & Black, 2007).

Defining Four Modes of Freedoms

According to Weiss (1958), we can conceptualize freedom via four different modes: freedom-from, freedom-to, freedom-with, and freedom-for. These modes are interdependent and interactive, and they align with majortheories and practices in psychotherapy. For each freedom mode, corresponding knowledge and practices must be applied to increase them in accordance with the other modes.

Freedom-From

This mode of freedom relates to the alleviation of symptoms such as depression, anxiety, obsessions, compulsions, and emotional pain, including freedom from addictions. *Freedom-from* is a psychological state absent those psychological restrictions and inhibitions resulting in symptomatology. Freedom-from relates to internally oppressive conditions seen in psychotherapy, such as negative self-talk, distorted beliefs, disturbing images, and otherwise harmful, violent, and self-defeating impulses and behaviors. It also refers to a reprieve or escape from oppressive external conditions that may include abuse, bullying, racism, sexism, and homophobia. Metacognitive skills are a key component of freedom-from insofar as the process of increasing freedom necessitates an ability to disidentify with issues, choose how to conceptualize particular problems or psychological phenomena, and cultivate an understanding of both one's awareness and the limits of that awareness.

Freedom-To

This mode of freedom involves personal growth and development toward self-determined goals that enhance well-being and self-efficacy. *Freedom-to* includes the active capacity to ameliorate undesired thoughts, emotions, and behaviors so as to enhance adaptiveness, improve decision-making and self-regulation, and expand behavioral options. It thus includes the ability to actively change aspects of psychological experience, such as fluidly replacing maladaptive beliefs or directly altering, controlling, or dissolving mental phenomena such as memories, images, thoughts, and emotions (see Puhakka & Hanna, 1988). Freedom-to also suggests an enhanced ability to navigate conflictual conditions by acquiring the skills to handle mental, interpersonal, systemic, or societal challenges. Perspicacity and wisdom are relevant considerations here such that adaptability in any given situation reflects a heightened capacity to perceive reasonable courses of action from a broad range of possibilities. Furthermore, freedom-to involves a sense of autonomy and personal responsibility whereby thoughts and behaviors remain independent from internal and external pressures that can result in abandoning one's integrity or rationality.

Freedom-With

This mode of freedom involves respecting the agency of others and recognizing the interpersonal or systemic nature of the human condition (Hanna et al., 1999). *Freedom-with* highlights both that personal development is inextricably linked to social relationships and denials of others' freedom diminishes our own freedom in kind (Sartre, 1943/2003). Individuals typically do not live in isolation but rather in a complex web of social relationships that thrive when imbued with empathy, compassion, tolerance, and a desire for the freedom of others. The bifurcation of competitive societies and the accompanying issues of privilege, status, and exclusivity result from a lack of freedom-with. Enhancing receptivity and respect toward others makes communicating differences in opinion or perspective possible without harboring ill will. At the same time, freedom-with demands a capacity to maintain boundaries such that the individual self-protects against verbal, emotional, or physical harm. Balancing autonomy and interdependence is, at its foundation, an attempt to establish freedom-with while sustaining one's freedom-to.

Freedom-For

This mode of freedom involves proactively responding to, assessing, managing, or otherwise influencing sociocultural conditions without succumbing to a sense of burden, duty, or blame. *Freedom-for* is the grounds for advocacy, a central goal in the fields of psychotherapy and counseling that both includes and transcends the provision of services that lead to the other three freedoms. Those who are actively denied freedom due to oppressive conditions or circumstances need liberation, as is so often seen among clients (Hanna et al., 2000; Ivey, 1995). As such, the therapists actively seek to combat social oppressions such as sexism, racism, and homophobia by supporting the development of freedom-with. However, it is not restricted to therapists or other health and human service providers. Insofar as clients can serve as advocates for others, freedom-for can be an active client goal in psychotherapy. Positively transforming the world necessitates all people providing freedom for others. By seeking freedom-for, counseling and psychotherapy become instruments of transformation across interpersonal, institutional, and systemic levels.

Freedom as a Teleological Approach to Integration

According to teleological integration, *freedom* is a state that includes the absence of restrictive or inhibitive psychological symptoms, the ability to enhance positive psychological conditions, the experience of respect for the agency of others, and the sense of responsibility for supporting or otherwise enhancing the agency of others. Such a broad conception of freedom is notably related to the concept of human agency outlined in social-cognitive theory, insofar as "freedom is conceived not just passively as the absence of constraints, but also proactively as the exercise of self-influence in the service of selected goals and desired outcomes" (Bandura, 2006, p. 165). Furthermore, freedom is not an isolated activity, as "in thus willing freedom, we discover that it depends entirely upon the freedom of others and that the freedom of others depends upon our own" (Sartre, 1943/2003, p. 48). Freedom is an expression of agency at both personal and communal levels (Bandura, 2006).

Rather than alter, combine, or otherwise integrate psychotherapy theories and practices, the proposed approach is meant to focus and coordinate them within the framework of freedom as a common teleology, or purpose (Hanna & Black, 2007). The traditionally espoused purpose of theory has been to provide a framework for conceptualization, a coherent explanation for presenting concerns, and a general guide for clinical interventions (Fall et al., 2017). A freedom-based teleological approach to integration supports each of these goals, whereas selectivity is paramount across the four standard approaches to integration. Technical eclecticism focuses on techniques, common factors identifies nonspecific change principles, assimilative integration selects a single theoretical basis, and theoretical integration seeks a unified theory (Zvi-Beiman & Shahar, 2015).

The teleological approach serves as a lens through which to view theories as unique approaches to enabling client freedom. Such a perspective circumvents the competitive nature of divergent theoretical perspectives and replaces it with the remedial view that each theory or approach provides insight into different facets of lived experience. The purposes of the four freedoms align with four major models in counseling and psychotherapy: the medical model, the wellness model, the systems model, and social justice and advocacy models (see Table 16.1). This conceptual alignment indicates that each model lends a unique perspective to the therapeutic process as it relates to freedom. If the purpose of counseling is freedom, then all theories, subspecialties, and therapeutic change factors serve a role for clients in attaining freedom. The proposed teleological approach to integration suggests that each model centralizes and restores a particular form of freedom.

Freedom-From: Symptoms, as a Medical Model Perspective

The medical model arises from "a scientific process involving observation, description and differentiation which moves from recognizing and treating symptoms to identifying disease etiologies and developing specific treatments" (Shah & Mountain, 2007, p. 375). From the perspective of the medical model, psychotherapy denotes a process of alleviating symptoms of psychological distress, which, in turn, may enhance mental health. Teleological integration in freedom-from aligns with efforts to reduce maladaptive symptomatology, highlighting an inverse relationship between negative symptoms and client freedom. In effect, the medical model approach to psychotherapy aims to promote freedom-from symptoms within a disease-oriented framework of mental health.

Freedom-To: Grow, as a Wellness Perspective

The wellness model emphasizes holistic and strength-based mental health along with practices that increase "a positive state of well-being, through developmental, preventive, and wellness-enhancing interventions" (Myers & Sweeney,

TABLE 16.1. Johari's Window of the Four Freedoms

	Being stance	Becoming stance
Individual lens	Freedom-from: symptoms *Medical model*	Freedom-to: grow *Wellness model*
Relational lens	Freedom-with: relationships *Systems model*	Freedom-for: engagement *Social justice/advocacy model*

2008, p. 482). The wellness model prioritizes developmental, humanistic, and other growth-oriented perspectives on mental health. Teleological integration in freedom-toward aligns with efforts to grow adaptive psychological functioning, asserting that client freedom is enhanced when capacities for personal growth are promoted and enhanced. In effect, the wellness model approach to psychotherapy promotes freedom-toward growth by encouraging personal development within a holistic-oriented framework of mental health.

Freedom-With: Relationships, as a Systems Perspective

The systems model emphasizes the influence of interpersonal relationships within and across interconnected social structures (McDowell et al., 2022). The systems model examines relationships through the lens of nonlinear dynamics, with an eye toward disentangling how various intersecting points of complex interaction influence individual development and system functioning. Teleological integration in freedom-with aligns with efforts to support effective interpersonal dynamics, maintaining that client freedom is enhanced by forming meaningful, congruent, and authentic relationships with others. In effect, the systems model approach to psychotherapy promotes freedom-with by improving the quality of interpersonal relationships within and across various ecological systems.

Freedom-For: Engagement, as a Social Justice/Advocacy Perspective

Social justice advocacy has been defined as "intentional and sustained action intended to influence public policy outcomes, with and/or on behalf of a vulnerable individual, group, community, or the public at large" (Marshall-Lee et al., 2020, p. 12). Taken together, social justice and advocacy models in psychotherapy encourage political, economic, and sociocultural engagement as a professional practice to address metasystemic issues of oppression, discrimination, and inequality. Teleological integration in freedom-for upholds the idea that denying others freedom diminishes our own freedom in kind, such that no one can truly be free until everyone is free. In effect, social justice and advocacy models promote freedom-for by motivating psychotherapists to acknowledge and confront oppressive sociocultural forces that impede mental health and wellness.

Teleological Freedom and the CHANGES Model

The CHANGES model identifies the necessary preconditions for any effective implementation of theory-based therapeutic interventions. As such, precursors are the conditions upon which freedom manifests, particularly in relation to freedom-from and freedom-to. Freedom-with is incorporated into the CHANGES model vis-à-vis the development of client empathy through the awareness and social support precursors. The metatheoretical freedom framework suggests that change precursors are the necessary prerequisites of freedom. By means of activating and maintaining precursors, the path is cleared to enable the exploration of the individual's four freedoms. In sum, we maintain that the CHANGES model is the fundamental catalyst of an individual's freedom. Therapeutic change involves distinctive processes, stages, and precursors which have the capacity to set people free.

FINAL THOUGHTS

As coined by the sociologist Merton (1968), the Matthew effect serves as a foundational principle of inequality in Western capitalist societies, aligning with the old saying that the rich get richer and the poor get poorer. The underlying notion of accumulated advantage indicates that inequality grows over time because those who start with sufficient resources garner compounded resource advantages over time, whereas those who start without resources remain at a perpetual disadvantage (Rigney, 2010). It could be easily argued that such advantages derive not only from wealth, power, prestige, and social status but also from knowledge or education.

As such, a similar phenomenon can be seen in the realm of psychotherapy: Clients who are already in a good state of mental health seem to make the greatest gains in psychotherapy. Alternatively, clients who need change the most tend to be the ones who are least capable of achieving it. So, what are the assets of the rich who get richer in therapy? It may well be that the precursors of change are the mental and emotional equivalent of wealth and resources needed to achieve and maintain mental health. Even a cursory analysis indicates that a person with an ample supply of mental health and wellness would also have an abundance of active precursors. If the rich really do get richer, time invested in establishing precursors among people who find change difficult may pay healthy dividends in terms of the relative degree of productivity per psychotherapy session.

Of course, this line of thought begets still larger questions: Will psychotherapy ever advance to the stage in which every single client, regardless of background or presenting concern, can achieve productive and lasting therapeutic change? What will it take to level the playing field, so to speak, and thereby produce a reliable and reproducible science of psychotherapeutic change? Do we have a professional responsibility to strive for such outcome effectiveness? Is it even possible to establish an efficacious science of change through the efforts of psychotherapy researchers and practitioners alone, or will it require coordinated efforts across multiple fields of study? Is it even ethical to propose such a possibility within our neoliberal, capitalist, sociocultural milieu as we struggle to reconcile mental health with historical economic advantages and social justice?

There are no easy answers to such questions, and we make no claim to the contrary. However, we do wish to acknowledge the value of shifting contemporary dialogues on mental health away from an emphasis on categorical and diagnostic features toward an emphasis on experiential features that may provide greater insight into therapeutic change mechanisms. In other words, we might do well to prioritize the phenomenology of lived experience over and above any intellectual fetish for conceptual abstraction. We are not suggesting that concepts are inherently problematic but rather that a preoccupation with abstract ideas can impede our ability to identify change processes as they unfold, in real time, with clients facing difficult problems while embedded in dynamic situations.

What moves the cynical, callous client from selfishness to relationality, from a sense of disdain to a sense of respect for others? What shifts the hopelessly depressed client from reactivity to proactivity, from a sense of helplessness to a sense of control? What nudges the defiant, mandated client from blaming and withdrawal to responsibility and engagement? There are subtle decision processes involved in replacing closed-mindedness with empathic openness, moving from a reactive to a proactive stance, and replacing blame with responsibility, or disengagement with engagement.

We know far too little about these subtleties. Our knowledge of change processes demarcated by the seven precursors is still remarkably incomplete. An activated precursor denotes a dynamic, contextually embedded psychological act. We may see the dawning of awareness, the shift into a sense of necessity, the onset of grit, or the leaning-into of social supports, but we are limited to observing these psychological acts via behavioral concomitants. What are these psychological acts, and how can they be better addressed in clients to enhance the likelihood of change?

All seven precursors would benefit from rigorous qualitative studies that are thoughtfully informed by advancements in various other academic areas, including the cognitive sciences and philosophy of mind. As noted in the introduction, change process research will also be key to unlocking these mysteries. Studies using neurophenomenology along with functional magnetic resonance imaging may lend insight into the therapeutic change transition as both an experiential and a neurobiological phenomenon. Conceptually driven research informed by works in philosophy of mind may support the phenomenological examination of how meaning is altered at the ontological level. In kind, microphenomenology reports might be conducted into the precise experiential concomitants of, for instance, the therapeutic activation of a sense of necessity for change. Newer methodologies of psychotherapy process analysis (Kiesler, 2017) could help us understand how unique therapist–client communication patterns contribute to the activation of specific precursors.

If the discipline of psychotherapy is to advance to the point of quickly and reliably producing lasting therapeutic change for clients, we must closely examine what it takes to activate each of the precursors. Again, the precursors are not our invention. They have been a part of the mythos of human knowledge and wisdom across the ages. They themselves are indeed broad categories that likely involve an extraordinarily complex array of neurocognitive subprocesses. Yet we are not calling for a complete understanding of the neurobiology of conscious awareness, which is perhaps an insurmountable task, although surely worth the effort (Chalmers, 2020; Gallagher & Zahavi, 2020).

Instead, we are calling for advanced research at the nexus of, for instance, insight, awareness, and change. There is perhaps as much, if not more, to glean from phenomenological investigations into these complex psychological acts than can be found via neuroscientific inquiries. The CHANGES model highlights seven well-established yet subtle change mechanisms that are extraordinarily pervasive and thus taken for granted, often to the point of disappearing from both the conceptual and practical view of therapists. Understanding the nature of within-individual precursors as well as pathways to their activation in the therapeutic encounter should be a top research priority.

REFERENCES

Aafjes-van Doorn, K., Békés, V., Prout, T. A., & Hoffman, L. (2020). Psychotherapists' vicarious traumatization during the COVID-19 pandemic. *Psychological Trauma: Theory, Research, Practice, and Policy, 12*(S1), S148–S150. https://doi.org/10.1037/tra0000868

Acton, L. (1887). *Letter to bishop Mandell Creighton: Historical essays and studies.* Macmillan.

Adler, A. (1927). *Understanding human nature* (W. B. Wolf, Trans.). Greenburg.

Adler, A. (1956). *The individual psychology of Alfred Adler: A systematic presentation in selections from his writings* (H. L. Ansbacher & R. R. Ansbacher, Eds.). Harper & Row.

Adler, A. (1979). *Superiority and social interest* (H. L. Ansbacher & R. R. Ansbacher, Eds.). W. W. Norton.

Adler, G. (1992). Psychotherapy of the narcissistic personality disorder patient: Two contrasting approaches. In N. G. Hamilton (Ed.), *From inner sources: New directions in object relations psychotherapy* (pp. 195–212). Jason Aronson.

Afonseca, M., Sousa, D., Vaz, A., Santos, J. M., & Batista, A. (2023). Psychotherapist's persuasiveness in anxiety: Scale development and relation to the working alliance. *Journal of Psychotherapy Integration, 33*(2), 169–184. https://doi.org/10.1037/int0000288

Ainslie, G. (2021). Willpower with and without effort. *Behavioral and Brain Sciences, 44,* Article e30. https://doi.org/10.1017/s0140525x20000357

Alcoholics Anonymous. (2012). *Alcoholics anonymous—Big book* (4th ed.). AA World Services.

Al-Yagon, M., & Margalit, M. (2017). Hope and coping in individuals with specific learning disorders. In M. W. Gallagher & S. J. Lopez (Eds.), *The Oxford handbook of hope* (pp. 1–21). Oxford University Press.

American Psychiatric Association. (2013). *Diagnostic and statistical manual of mental disorders* (5th ed.). https://doi.org/10.1176/appi.books.9780890425596

American Psychiatric Association. (2022). *Diagnostic and statistical manual of mental disorders* (5th ed., text rev.).

Anderson, T., Finkelstein, J. D., & Horvath, S. A. (2020). The facilitative interpersonal skills method: Difficult psychotherapy moments and appropriate therapist responsiveness. *Counselling & Psychotherapy Research, 20*(3), 463–469. https://doi.org/10.1002/capr.12302

Anderson, T., McClintock, A. S., Himawan, L., Song, X., & Patterson, C. L. (2016). A prospective study of therapist facilitative interpersonal skills as a predictor of treatment outcome. *Journal of Consulting and Clinical Psychology, 84*(1), 57–66. https://doi.org/10.1037/ccp0000060

Arango-Muñoz, S. (2019). Cognitive phenomenology and metacognitive feelings. *Mind & Language, 34*(2), 247–262. https://doi.org/10.1111/mila.12215

Aranya, H. (1983). *Yoga philosophy of Patanjali.* State University of New York Press.

Archer, A., & Mills, G. (2019). Anger, affective injustice, and emotion regulation. *Philosophical Topics, 47*(2), 75–94. https://doi.org/10.5840/philtopics201947216

Ardelt, M., & Ferrari, M. (2014). Wisdom and emotions. In P. Verhaeghen & C. Hertzog (Eds.), *The Oxford handbook of emotion, social cognition, and problem solving in adulthood* (pp. 256–272). Oxford University Press.

Arenson, M., Bernat, E., De Los Reyes, A., Neylan, T. C., & Cohen, B. E. (2021). Social support, social network size, and suicidal ideation: A nine-year longitudinal analysis from the Mind Your Heart Study. *Journal of Psychiatric Research, 135*, 318–324. https://doi.org/10.1016/j.jpsychires.2021.01.017

Arkowitz, H. (1989). From behavior change to insight. *Journal of Eclectic and Integrative Psychotherapy, 8*(3), 222–232.

Arslan, G. (2019). Mediating role of the self-esteem and resilience in the association between social exclusion and life satisfaction among adolescents. *Personality and Individual Differences, 151*, Article 109514. https://doi.org/10.1016/j.paid.2019.109514

Assagioli, R. (1965). *Psychosynthesis: A manual of principles and techniques.* Penguin.

Assagioli, R. (1973). *The act of will.* Penguin.

Aubuchon-Endsley, N. L., Callahan, J. L., González, D. A., Ruggero, C. J., & Abramson, C. I. (2015). The impact of hope in mediating psychotherapy expectations and outcomes: A study of Brazilian clients. *International Journal of Integrative Psychotherapy, 6*, 63–80.

Avdi, E., & Georgaca, E. (2007). Discourse analysis and psychotherapy: A critical review. *European Journal of Psychotherapy and Counselling, 9*(2), 157–176. https://doi.org/10.1080/13642530701363445

Avdi, E., Lerou, V., & Seikkula, J. (2015). Dialogical features, therapist responsiveness, and agency in a therapy for psychosis. *Journal of Constructivist Psychology, 28*(4), 329–341. https://doi.org/10.1080/10720537.2014.994692

Axsom, D. (1989). Cognitive dissonance and behavior change in psychotherapy. *Journal of Experimental Social Psychology, 25*(3), 234–252. https://doi.org/10.1016/0022-1031(89)90021-8

Bager-Charleson, S. (2010). *Reflective practice in counselling and psychotherapy*. Sage Publications.

Bandura, A. (1977). Self-efficacy: Toward a unifying theory of behavioral change. *Psychological Review, 84*(2), 191–215. https://doi.org/10.1037/0033-295X.84.2.191

Bandura, A. (2006). Toward a psychology of human agency. *Perspectives on Psychological Science, 1*(2), 164–180. https://doi.org/10.1111/j.1745-6916.2006.00011.x

Barker, P. (2013). Reframing: The essence of psychotherapy? In J. K. Zeig (Ed.), *Ericksonian methods: The essence of the story* (pp. 211–233). Routledge.

Barlas, Z., & Hockley, W. E., & Obhi, S. S. (2017). The effects of freedom of choice in action selection on perceived mental effort and the sense of agency. *Acta Psychologica, 180*, 122–129. https://doi.org/10.1016/j.actpsy.2017.09.004

Barlas, Z., & Obhi, S. S. (2013). Freedom, choice, and the sense of agency. *Frontiers in Human Neuroscience, 7*, Article 514. https://doi.org/10.3389/fnhum.2013.00514

Bartholomew, T. T., Gundel, B. E., Scheel, M. J., Kang, E., Joy, E. E., & Li, H. (2020). Development and initial validation of the Therapist Hope for Clients Scale. *The Counseling Psychologist, 48*(2), 191–222. https://doi.org/10.1177/0011000019886428

Bartlett, R. C. (2019). *Aristotle's art of rhetoric*. University of Chicago Press.

Bateson, G. (1979). *Mind and nature*. Bantam Books.

Bateson, G., & Bateson, C. G. (1987). *Angels fear: Toward an epistemology of the sacred*. Bantam Books.

Bayne, T. (2008). The phenomenology of agency. *Philosophy Compass, 3*(1), 182–202. https://doi.org/10.1111/j.1747-9991.2007.00122.x

Beck, A. T. (1976). *Cognitive therapy and the emotional disorders*. New American Library.

Beck, A. T., Rush, A. J., Shaw, B. F., & Emery, G. (1979). *Cognitive therapy of depression*. Guilford Press.

Beck, A. T., Steer, R. A., Kovacs, M., & Garrison, B. (1985). Hopelessness and eventual suicide: A 10-year prospective study of patients hospitalized with suicidal ideation. *The American Journal of Psychiatry, 142*(5), 559–563. https://doi.org/10.1176/ajp.142.5.559

Beck, A. T., Weissman, A., Lester, D., & Trexler, L. (1974). The measurement of pessimism: The Hopelessness Scale. *Journal of Consulting and Clinical Psychology, 42*(6), 861–865. https://doi.org/10.1037/h0037562

Beck, C. J. (1989). *Everyday Zen: Love and work*. Harper Collins.

Bellaert, L., Van Steenberghe, T., De Maeyer, J., Vander Laenen, F., & Vanderplasschen, W. (2022). Turning points toward drug addiction recovery: Contextualizing underlying dynamics of change. *Addiction Research and Theory, 30*(4), 294–303. https://doi.org/10.1080/16066359.2022.2026934

Bendig, E., Erb, B., Schulze-Thuesing, L., & Baumeister, H. (2022). The next generation: Chatbots in clinical psychology and psychotherapy to foster mental health—A scoping review. *Verhaltenstherapie, 32*(Suppl. 1), 64–76. https://doi.org/10.1159/000501812

Benn, P. (2021). Freedom, resentment and the psychopath. In C. Heginbotham (Ed.), *Philosophy, psychiatry and psychopathy* (pp. 29–46). Routledge.

Bennett-Leighton, L. (2018). The trauma of oppression: A somatic perspective. In C. Caldwell & L. Bennett-Leighton (Eds.), *Oppression and the body: Roots, resistance, and resolutions* (pp. 17–30). North Atlantic Books.

Bernstein, A., Hadash, Y., Lichtash, Y., Tanay, G., Shepherd, K., & Fresco, D. M. (2015). Decentering and related constructs: A critical review and metacognitive processes model. *Perspectives on Psychological Science, 10*(5), 599–617. https://doi.org/10.1177/1745691615594577

Bettelheim, B. (1960). *The informed heart*. Free Press.

Beutler, L. E. (2000). David and Goliath: When empirical and clinical standards of practice meet. *American Psychologist, 55*(9), 997–1007. https://doi.org/10.1037/0003-066X.55.9.997

Beutler, L. E., & Forrester, B. (2014). What needs to change: Moving from "research informed" practice to "empirically effective" practice. *Journal of Psychotherapy Integration, 24*(3), 168–177. https://doi.org/10.1037/a0037587

Birditt, K. S., Polenick, C. A., Luong, G., Charles, S. T., & Fingerman, K. L. (2020). Daily interpersonal tensions and well-being among older adults: The role of emotion regulation strategies. *Psychology and Aging, 35*(4), 578–590. https://doi.org/10.1037/pag0000416

Blais, R. K., Tirone, V., Orlowska, D., Lofgreen, A., Klassen, B., Held, P., Stevens, N., & Zalta, A. K. (2021). Self-reported PTSD symptoms and social support in U.S. military service members and veterans: A meta-analysis. *European Journal of Psychotraumatology, 12*(1), Article 1851078. https://doi.org/10.1080/20008198.2020.1851078

Bohart, A. C. (2000). The client is the most important common factor: Clients' self-healing capacities and psychotherapy. *Journal of Psychotherapy Integration, 10*(2), 127–149. https://doi.org/10.1023/A:1009444132104

Bohart, A. C., & Tallman, K. (1999). *How clients make therapy work: The process of active self-healing*. American Psychological Association.

Bonino, J. L., & Hanna, F. J. (2018). Who owns psychopathology? The *DSM*: Its flaws, its future, and the professional counselor. *The Journal of Humanistic Counseling, 57*(3), 118–137. https://doi.org/10.1002/johc.12071

BonJour, L. (2009). *Epistemology: Classic problems and contemporary responses*. Rowman & Littlefield.

Botha, M., & Frost, D. M. (2020). Extending the minority stress model to understand mental health problems experienced by the autistic population. *Society and Mental Health, 10*(1), 20–34. https://doi.org/10.1177/2156869318804297

Bowen, M. (1976). Theory in the practice of psychotherapy. In P. J. Guerin, Jr. (Ed.), *Family therapy: Theory and practice* (pp. 42–90). Gardner Press.

Boykin, A. W., Dixon, D., Mitchell, D. S. B., Bruce, A. W., Akinola, Y. O., & Holt, N. P. (2016). The intersection of racial and cultural identity for African Americans: Expanding the scope of black self-understanding. In J. M. Sullivan & W. E. Cross, Jr. (Eds.), *Meaning-making, internalized racism, and African American identity* (pp. 159–174). State University of New York Press.

Brodley, B. T. (2002). Client-centered: An expressive therapy. *The Person-Centered Journal, 9*(1), 59–70.

Brooks, H. L., Rushton, K., Lovell, K., Bee, P., Walker, L., Grant, L., & Rogers, A. (2018). The power of support from companion animals for people living with mental health problems: A systematic review and narrative synthesis of the evidence. *BMC Psychiatry, 18*(1), Article 31. https://doi.org/10.1186/s12888-018-1613-2

Brown, D. (1970). *Bury my heart at wounded knee*. Bantam Books.

Brown, L. A., Zandberg, L. J., & Foa, E. B. (2019). Mechanisms of change in prolonged exposure therapy for PTSD: Implications for clinical practice. *Journal of Psychotherapy Integration, 29*(1), 6–14. https://doi.org/10.1037/int0000109

Brown, L. S. (2017). Contributions of feminist and critical psychologies to trauma psychology. In S. N. Gold (Ed.), *APA handbook of trauma psychology: Foundations in knowledge* (pp. 501–526). American Psychological Association. https://doi.org/10.1037/0000019-025

Brown, L. S. (2018). *Feminist therapy* (2nd ed.). American Psychological Association. https://doi.org/10.1037/0000092-000

Browning, S., & Hull, R. (2021). Reframing paradox. *Professional Psychology: Research and Practice, 52*(4), 360–367. https://doi.org/10.1037/pro0000384

Buechner, B. D. (2023). Empathy versus tyranny: Witnessing moral conflict through Adlerian lenses. *Journal of Individual Psychology, 79*(4), 425–442. https://doi.org/10.1353/jip.2023.a915977

Cannon, W. B. (1942). "Voodoo" death. *American Anthropologist, 44*(2), 169–181. https://doi.org/10.1525/aa.1942.44.2.02a00010

Carcione, A., Riccardi, I., Bilotta, E., Leone, L., Pedone, R., Conti, L., Colle, L., Fiore, D., Nicolò, G., Pellecchia, G., Procacci, M., & Semerari, A. (2019). Metacognition as a predictor of improvements in personality disorders. *Frontiers in Psychology, 10*, Article 170. https://doi.org/10.3389/fpsyg.2019.00170

Castonguay, L. G., & Hill, C. E. (Eds.). (2017). *How and why are some therapists better than others? Understanding therapist effects*. American Psychological Association. https://doi.org/10.1037/0000034-000

Cautela, J. R. (1996). Training the client to be empathetic. In J. R. Cautela & W. Ishaq (Eds.), *Contemporary issues in behavior therapy* (pp. 337–353). Plenum.

Chalmers, D. J. (2020). How can we solve the meta-problem of consciousness? *Journal of Consciousness Studies, 27*(5–6), 201–226.

Chemero, A. (2013). Radical embodied cognitive science. *Review of General Psychology, 17*(2), 145–150. https://doi.org/10.1037/a0032923

Christensen, A., & Jacobson, N. S. (1994). Who (or what) can do psychotherapy: The status and challenge of nonprofessional therapies. *Psychological Science, 5*(1), 8–14. https://doi.org/10.1111/j.1467-9280.1994.tb00606.x

Chu, C., Walker, K. L., Stanley, I. H., Hirsch, J. K., Greenberg, J. H., Rudd, M. D., & Joiner, T. E. (2018). Perceived problem-solving deficits and suicidal ideation: Evidence for the explanatory roles of thwarted belongingness and perceived burdensomeness in five samples. *Journal of Personality and Social Psychology, 115*(1), 137–160. https://doi.org/10.1037/pspp0000152

Clark, A. J. (2023). *Empathy and mental health: An integral model for developing therapeutic skills in counseling and psychotherapy*. Routledge.

Cleeremans, A. (2019). The mind is deep. In A. Cleeremans, V. Allakhverdov, & M. Kuvaldina (Eds.), *Implicit learning: 50 years on* (pp. 38–70). Routledge.

Cochran, J. L., & Cochran, N. H. (2015). *The heart of counseling: Counseling skills through therapeutic relationships*. Routledge.

Coffey, K. A., Hartman, M., & Fredrickson, B. L. (2010). Deconstructing mindfulness and constructing mental health: Understanding mindfulness and its mechanisms of action. *Mindfulness, 1*(4), 235–253. https://doi.org/10.1007/s12671-010-0033-2

Comas-Díaz, L. (2016). Racial trauma recovery: A race-informed therapeutic approach to racial wounds. In A. N. Alvarez, C. T. H. Liang, & H. A. Neville (Eds.), *The cost of racism for people of color: Contextualizing experiences of discrimination* (pp. 249–272). American Psychological Association. https://doi.org/10.1037/14852-012

Connor, D. R., & Callahan, J. L. (2015). Impact of psychotherapist expectations on client outcomes. *Psychotherapy, 52*(3), 351–362. https://doi.org/10.1037/a0038890

Constantino, M. J., Goodwin, B. J., Muir, H. J., Coyne, A. E., & Boswell, J. F. (2021). Context-responsive psychotherapy integration applied to cognitive behavioral therapy. In J. C. Watson & H. Wiseman (Eds.), *The responsive psychotherapist: Attuning to clients in the moment* (pp. 151–169). American Psychological Association. https://doi.org/10.1037/0000240-008

Constantino, M. J., Muir, H. J., Gaines, A. N., & Ouimette, K. (2023). Hope and expectancy factors. In S. D. Miller, D. Chow, S. Malins, & M. A. Hubble (Eds.), *The field guide to better results: Evidence-based exercises to improve therapeutic effectiveness* (pp. 131–153). American Psychological Association. https://doi.org/10.1037/0000358-007

Corey, G. (2017). *Theory and practice of counseling and psychotherapy* (10th ed.). Cengage.

Cozolino, L. (2017). *The neuroscience of psychotherapy: Healing the social brain* [Norton Series on Interpersonal Neurobiology]. W. W. Norton.

Crenshaw, K. (1989). Demarginalizing the intersection of race and sex: A Black feminist critique of antidiscrimination doctrine, feminist theory and antiracist politics. *University of Chicago Legal Forum, 1989*(1), 139–167.

Cuijpers, P., Reijnders, M., & Huibers, M. J. H. (2019). The role of common factors in psychotherapy outcomes. *Annual Review of Clinical Psychology, 15*, 207–231. https://doi.org/10.1146/annurev-clinpsy-050718-095424

Cyrus, K. (2017). Multiple minorities as multiply marginalized: Applying the minority stress theory to LGBTQ people of color. *Journal of Gay & Lesbian Mental Health, 21*(3), 194–202. https://doi.org/10.1080/19359705.2017.1320739

Daly, M. (1978). *Gyn/ecology: The metaethics of radical feminism.* Beacon Press.

David, E. R., & Derthick, A. O. (2018). *The psychology of oppression.* Springer.

David, D., Lynn, S. J., & Ellis, A. (2009). *Rational and irrational beliefs: Research, theory, and clinical practice.* Oxford University Press.

David, E. R., Schroeder, T. M., & Fernandez, J. (2019). Internalized racism: A systematic review of the psychological literature on racism's most insidious consequence. *Journal of Social Issues, 75*(4), 1057–1086. https://doi.org/10.1111/josi.12350

Davidson, R. J. (2003). Seven sins in the study of emotion: Correctives from affective neuroscience. *Brain and Cognition, 52*(1), 129–132. https://doi.org/10.1016/s0278-2626(03)00015-0

Davis, D. E., DeBlaere, C., Owen, J., Hook, J. N., Rivera, D. P., Choe, E., Van Tongeren, D. R., Worthington, E. L., & Placeres, V. (2018). The multicultural orientation framework: A narrative review. *Psychotherapy, 55*(1), 89–100. https://doi.org/10.1037/pst0000160

Dawson, G. C. (2018). Years of clinical experience and therapist professional development: A literature review. *Journal of Contemporary Psychotherapy, 48*(2), 89–97. https://doi.org/10.1007/s10879-017-9373-8

Day-Vines, N. L., Cluxton-Keller, F., Agorsor, C., Gubara, S., & Otabil, N. A. A. (2020). The multidimensional model of broaching behavior. *Journal of Counseling and Development, 98*(1), 107–118. https://doi.org/10.1002/jcad.12304

Delgadillo, J., Branson, A., Kellett, S., Myles-Hooton, P., Hardy, G. E., & Shafran, R. (2020). Therapist personality traits as predictors of psychological treatment outcomes. *Psychotherapy Research, 30*(7), 857–870. https://doi.org/10.1080/10503307.2020.1731927

de Shazer, S. (1985). *Keys to solutions in brief therapy.* W. W. Norton.

de Silva, P. (1985). Early Buddhist and modern behavioral strategies for the control of unwanted intrusive cognitions. *The Psychological Record, 35*, 437–443.

De Vos, J., & Pluth, E. (2016). *Neuroscience and critique: Exploring the limits of the neurological turn.* Taylor & Francis.

Decety, J., & Jackson, P. L. (2004). The functional architecture of human empathy. *Behavioral and Cognitive Neuroscience Reviews, 3*(2), 71–100. https://doi.org/10.1177/1534582304267187

Deikman, A. (1982). *The observing self: Mysticism and psychotherapy.* Beacon Press.

DeLisi, M., Tostlebe, J., Burgason, K., Heirigs, M., & Vaughn, M. (2018). Self-control versus psychopathy: A head-to-head test of general theories of antisociality. *Youth Violence and Juvenile Justice, 16*(1), 53–76. https://doi.org/10.1177/1541204016682998

Dictado, J., & Torres-Harding, S. R. (2023). Predictors of therapy trainees' pathologizing and invalidating microaggressions with sexual and racial minority therapy clients. *Training and Education in Professional Psychology, 17*(3), 304–313. https://doi.org/10.1037/tep0000424

Di Giuseppe, M., Perry, J. C., Prout, T. A., & Conversano, C. (2021). Recent empirical research and methodologies in defense mechanisms: Defenses as fundamental contributors to adaptation. *Frontiers in Psychology, 12,* Article 802602. https://doi.org/10.3389/fpsyg.2021.802602

Dinkmeyer, D. C., Dinkmeyer, D. C., Jr., & Sperry, L. (1987). *Adlerian counseling and therapy.* Merrill.

Douglass, F. (1855). *My bondage and my freedom.* Penguin Books.

Drozd, J. F., & Goldfried, M. R. (1996). A critical evaluation of the state-of-the-art in psychotherapy outcome research. *Psychotherapy: Theory, Research, Practice, Training, 33*(2), 171–180. https://doi.org/10.1037/0033-3204.33.2.171

Duckworth, A. L., Milkman, K. L., & Laibson, D. (2018). Beyond willpower: Strategies for reducing failures of self-control. *Psychological Science in the Public Interest, 19*(3), 102–129. https://doi.org/10.1177/1529100618821893

Duckworth, A. L., Peterson, C., Matthews, M. D., & Kelly, D. R. (2007). Grit: Perseverance and passion for long-term goals. *Journal of Personality and Social Psychology, 92*(6), 1087–1101. https://doi.org/10.1037/0022-3514.92.6.1087

Dumont, F. (1991). Expertise in psychotherapy: Inherent liabilities of becoming experienced. *Psychotherapy: Theory, Research, Practice, Training, 28*(3), 422–428. https://doi.org/10.1037/0033-3204.28.3.422

Easwaran, E. (2007). *The upanishads* (Vol. 2). Nilgiri Press.

Eccles, D. W., & Feltovich, P. J. (2008). Implications of domain-general "psychological support skills" for transfer of skill and acquisition of expertise. *Performance Improvement Quarterly, 21*(1), 43–60. https://doi.org/10.1002/piq.20014

Edmondstone, C., Pascual-Leone, A., Soucie, K., & Kramer, U. (2023). Therapist effects on outcome: Meaningful differences exist early in training. *Training and Education in Professional Psychology, 17*(2), 149–157. https://doi.org/10.1037/tep0000402

Elliott, R. (2010). Psychotherapy change process research: Realizing the promise. *Psychotherapy Research, 20*(2), 123–135. https://doi.org/10.1080/10503300903470743

Elliott, R., Bohart, A. C., Watson, J. C., & Murphy, D. (2018). Therapist empathy and client outcome: An updated meta-analysis. *Psychotherapy, 55*(4), 399–410. https://doi.org/10.1037/pst0000175

Epictetus. (1944). *Epictetus: Discourses and enchiridion.* Walter J. Black. (Original work published circa 130)

Esterman, M., & Rothlein, D. (2019). Models of sustained attention. *Current Opinion in Psychology, 29,* 174–180. https://doi.org/10.1016/j.copsyc.2019.03.005

Eubanks, C. F., & Goldfried, M. R. (2019). A principle-based approach to psychotherapy integration. In J. C. Norcross & M. R. Goldfried (Eds.), *Handbook of psychotherapy integration* (pp. 88–104). Oxford University Press.

Ewert, C., Vater, A., & Schröder-Abé, M. (2021). Self-compassion and coping: A meta-analysis. *Mindfulness, 12*(5), 1063–1077. https://doi.org/10.1007/s12671-020-01563-8

Eysenck, H. J. (1952). The effects of psychotherapy: An evaluation. *Journal of Consulting Psychology, 16*(5), 319–324. https://doi.org/10.1037/h0063633

Fall, K. A., Holden, J. M., & Marquis, A. (2017). *Theoretical models of counseling and psychotherapy*. Routledge.

Farber, M. (1968). *Theory of suicide*. Funk & Wagnalls.

Fear, R. M. (2018). *Systematic desensitization for panic and phobia: An introduction for health professionals*. Routledge.

Feinstein, R. A. (2018). *When rape was legal: The untold history of sexual violence during slavery*. Routledge.

Festinger, L. (1957). *A theory of cognitive dissonance*. Stanford University Press.

Feuerstein, G. (1989). *The yoga-sutra of Patanjali*. Inner Traditions International.

Figley, C. R. (2013). *Compassion fatigue: Coping with secondary traumatic stress disorder in those who treat the traumatized*. Routledge.

Firth, N., Barkham, M., Kellett, S., & Saxon, D. (2015). Therapist effects and moderators of effectiveness and efficiency in psychological wellbeing practitioners: A multilevel modelling analysis. *Behaviour Research and Therapy, 69*, 54–62. https://doi.org/10.1016/j.brat.2015.04.001

Flavell, J. H. (1979). Metacognition and cognitive monitoring: A new area of cognitive–developmental inquiry. *American Psychologist, 34*(10), 906–911. https://doi.org/10.1037/0003-066X.34.10.906

Fletcher, A. C., & Delgadillo, J. (2022). Psychotherapists' personality traits and their influence on treatment processes and outcomes: A scoping review. *Journal of Clinical Psychology, 78*(7), 1267–1287. https://doi.org/10.1002/jclp.23310

Franck, L., Molyneux, N., & Parkinson, L. (2016). Systematic review of interventions addressing social isolation and depression in aged care clients. *Quality of Life Research, 25*(6), 1395–1407. https://doi.org/10.1007/s11136-015-1197-y

Frank, J. (1961). *Persuasion and healing*. Johns Hopkins University Press.

Frank, J. D. (1987). Psychotherapy, rhetoric, and hermeneutics: Implications for practice and research. *Psychotherapy: Theory, Research, Practice, Training, 24*(3), 293–302. https://doi.org/10.1037/h0085719

Frank, J. D., & Frank, J. B. (1991). *Persuasion and healing* (3rd ed.). Johns Hopkins University Press.

Frankfurt, H. G. (2005). *On bullshit*. Princeton University Press.

Frankl, V. E. (1992). *Man's search for meaning: An introduction to logotherapy* (I. Lasch, Trans.; 4th ed.). Beacon Press. (Original work published 1959)

Freud, S. (1910). The future prospects of psychoanalytic therapy. In J. Strachey (Ed.), *The complete psychological works of Sigmund Freud* (Standard ed., pp. 139–158). Hogarth.

Fricker, M. (2007). *Epistemic injustice: Power and the ethics of knowing*. Oxford University Press.

Friedlander, M. L. (2015). Use of relational strategies to repair alliance ruptures: How responsive supervisors train responsive psychotherapists. *Psychotherapy, 52*(2), 174–179. https://doi.org/10.1037/a0037044

Fromm-Reichmann, F. (1950). *Principles of intensive psychotherapy.* University of Chicago Press.

Fuertes, J. N., & Nutt Williams, E. (2017). Client-focused psychotherapy research. *Journal of Counseling Psychology, 64*(4), 369–375. https://doi.org/10.1037/cou0000214

Fuertes, J. N., Stracuzzi, T. I., Bennett, J., Scheinholtz, J., Mislowack, A., Hersh, M., & Cheng, D. (2006). Therapist multicultural competency: A study of therapy dyads. *Psychotherapy: Theory, Research, Practice, Training, 43*(4), 480–490. https://doi.org/10.1037/0033-3204.43.4.480

Gaines, A. N., & Goldfried, M. R. (2021). Consensus in psychotherapy: Are we there yet? *Clinical Psychology: Science and Practice, 28*(3), 267–276. https://doi.org/10.1037/cps0000026

Gallagher, S. (2017). Phenomenological approaches to consciousness. In M. Velmans & S. Schneider (Eds.), *The Blackwell companion to consciousness* (2nd ed., pp. 686–696). Blackwell.

Gallagher, S., & Zahavi, D. (2020). *The phenomenological mind.* Routledge.

Gazzillo, F. (2023). Toward a more comprehensive understanding of pathogenic beliefs: Theory and clinical implications. *Journal of Contemporary Psychotherapy, 53*(3), 227–234. https://doi.org/10.1007/s10879-022-09564-5

Gendlin, E. (1981). *Focusing.* Bantam Books.

Gendlin, E. (1992). Celebrations and problems of humanistic psychology. *The Humanistic Psychologist, 20*(2–3), 447–460. https://doi.org/10.1080/08873267.1992.9986809

Gendlin, E. T. (1986). What comes after traditional psychotherapy research? *American Psychologist, 41*(2), 131–136. https://doi.org/10.1037/0003-066X.41.2.131

Gevisser, M. (2020). *The pink line: Journeys across the world's queer frontiers.* Picador.

Gilbert, P. (2020). Compassion: From its evolution to a psychotherapy. *Frontiers in Psychology, 11,* Article 586161. https://doi.org/10.3389/fpsyg.2020.586161

Gilligan, C. (1982). *In a different voice: Psychological theory and women's development.* Harvard University Press.

Giovacchini, P. L. (1989). *Countertransference: Triumphs and catastrophes.* Jason Aronson.

Gladding, S. T., & Drake Wallace, M. J. (2016). Promoting beneficial humor in counseling: A way of helping counselors help clients. *Journal of Creativity in Mental Health, 11*(1), 2–11. https://doi.org/10.1080/15401383.2015.1133361

Glasser, W. (1965). *Reality therapy.* Harper & Row.

Golubickis, M., Tan, L. B., Falben, J. K., & Macrae, C. N. (2016). The observing self: Diminishing egocentrism through brief mindfulness meditation. *European Journal of Social Psychology, 46*(4), 521–527. https://doi.org/10.1002/ejsp.2186

Goldberg, L. (2017). Therapy with former members of destructive cults. In S. Harvey, S. Steidinger, & J. A. Beckford (Eds.), *New religious movements and counselling* (pp. 63–79). Routledge.

Goldfried, M. R. (1980). Toward the delineation of therapeutic change principles. *American Psychologist, 35*(11), 991–999. https://doi.org/10.1037/0003-066X.35.11.991

Goldfried, M. R. (1995). Toward a common language for case formulation. *Journal of Psychotherapy Integration, 5*(3), 221–244. https://doi.org/10.1037/h0101272

Goldfried, M. R. (2019). Obtaining consensus in psychotherapy: What holds us back? *American Psychologist, 74*(4), 484–496. https://doi.org/10.1037/amp0000365

Goldfried, M. R., Greenberg, L. S., & Marmar, C. (1990). Individual psychotherapy: Process and outcome. *Annual Review of Psychology, 41*, 659–688. https://doi.org/10.1146/annurev.ps.41.020190.003303

Goldman, E. E., & Morrison, D. S. (1984). *Psychodrama: Experience and process*. Kendall Hunt.

Goldstein, W. N. (2013). *A primer for beginning psychotherapy*. Routledge.

Goodwin, B. J., Coyne, A. E., & Constantino, M. J. (2018). Extending the context-responsive psychotherapy integration framework to cultural processes in psychotherapy. *Psychotherapy, 55*(1), 3–8. https://doi.org/10.1037/pst0000143

Gorlin, E. I., & Békés, V. (2021). Agency via awareness: A unifying meta-process in psychotherapy. *Frontiers in Psychology, 12*, Article 698655. https://doi.org/10.3389/fpsyg.2021.698655

Goshe, S. (2019). The lurking punitive threat: The philosophy of necessity and challenges for reform. *Theoretical Criminology, 23*(1), 25–42. https://doi.org/10.1177/1362480617719450

Greenberg, J. (2019). Trauma and the metaphor of oppression. *The International Journal of Psychoanalysis, 100*(6), 1144–1153. https://doi.org/10.1080/00207578.2019.1642760

Greenberg, L. S. (1986). Change process research. *Journal of Consulting and Clinical Psychology, 54*(1), 4–9. https://doi.org/10.1037/0022-006X.54.1.4

Grencavage, L. M., & Norcross, J. C. (1990). Where are the commonalities among the therapeutic common factors? *Professional Psychology: Research and Practice, 21*(5), 372–378. https://doi.org/10.1037/0735-7028.21.5.372

Griffith, J. L., & Dsouza, A. (2012). Demoralization and hope in clinical psychiatry and psychotherapy. In R. D. Alarcon & J. B. Frank (Eds.), *The psychotherapy of hope: The legacy of persuasion and healing* (pp. 158–177). Johns Hopkins University Press.

Grossmann, I., Na, J., Varnum, M. E. W., Kitayama, S., & Nisbett, R. E. (2013). A route to well-being: Intelligence versus wise reasoning. *Journal of Experimental Psychology: General, 142*(3), 944–953. https://doi.org/10.1037/a0029560

Gupta, B. (1998). *The disinterested witness: A fragment of Advaita Vedanta phenomenology*. Northwestern University Press.

Gutierrez, D., & Czerny, A. (2018). Transtheoretical model for change. In P. Lassiter & J. Culbreth (Eds.), *Theory and practice of addiction counseling* (pp. 199–216). Sage Publications.

Hackert, B., & Weger, U. (2018). Introspection and the Würzburg school. *European Psychologist, 23*(3), 217–232. https://doi.org/10.1027/1016-9040/a000329

Hahn, R. A., & Kleinman, A. (1983). Belief as pathogen, belief as medicine: "Voodoo death" and the "placebo phenomenon" in anthropological perspective. *Medical Anthropology Quarterly, 14*(4), 3–19. https://doi.org/10.1525/maq.1983.14.4.02a00030

Haidt, J. (2024). *The anxious generation: How the great rewiring of childhood is causing an epidemic of mental illness*. Penguin.

Hamonniere, T., & Varescon, I. (2018). Metacognitive beliefs in addictive behaviours: A systematic review. *Addictive Behaviors, 85*, 51–63. https://doi.org/10.1016/j.addbeh.2018.05.018

Hanna, F. J. (1993). The transpersonal consequences of Husserl's phenomenological method. *The Humanistic Psychologist, 21*(1), 41–57. https://doi.org/10.1080/08873267.1993.9976905

Hanna, F. J. (1994). A dialectic of experience: A radical empiricist approach to conflicting theories in psychotherapy. *Psychotherapy: Theory, Research, Practice, Training, 31*(1), 124–136. https://doi.org/10.1037/0033-3204.31.1.124

Hanna, F. J. (1995). Husserl on the teachings of the Buddha. *The Humanistic Psychologist, 23*(3), 365–372. https://doi.org/10.1080/08873267.1995.9986837

Hanna, F. J. (1996). Precursors of change: Pivotal points of involvement and resistance in psychotherapy. *Journal of Psychotherapy Integration, 6*(3), 227–264. https://doi.org/10.1037/h0101102

Hanna, F. J. (1998). A transcultural view of prejudice, racism, and community feeling: The desire and striving for status. *The Journal of Individual Psychology, 54*(3), 336–345.

Hanna, F. J. (2004). Holding the family together. In L. Golden (Ed.), *Case studies in marriage and family therapy* (pp. 91–98). Merrill; Prentice-Hall.

Hanna, F. J. (2011). Freedom: Toward an integration of the counseling profession. *Counselor Education and Supervision, 50*(6), 362–385. https://doi.org/10.1002/j.1556-6978.2011.tb01921.x

Hanna, F. J., Bemak, F., & Chung, R. C. (1999). Toward a new paradigm for multicultural counseling. *Journal of Counseling and Development, 77*(2), 125–134. https://doi.org/10.1002/j.1556-6676.1999.tb02432.x

Hanna, F. J., & Black, L. L. (2007, October). *Liberation and freedom: An integration of counseling theories, social justice, and multiculturalism* [Paper presentation]. Association for Counselor Education and Supervision, Columbus, OH, United States.

Hanna, F. J., & Cardona, B. (2013). Multicultural counseling beyond the relationship: Expanding the repertoire with techniques. *Journal of Counseling and Development, 91*(3), 349–357. https://doi.org/10.1002/j.1556-6676.2013.00104.x

Hanna, F. J., Giordano, F., Dupuy, P., & Puhakka, K. (1995). Agency and transcendence: The experience of therapeutic change. *The Humanistic Psychologist, 23*(2), 139–160. https://doi.org/10.1080/08873267.1995.9986822

Hanna, F. J., & Hunt, W. P. (1999). Techniques for psychotherapy with defiant, aggressive adolescents. *Psychotherapy: Theory, Research, Practice, Training, 36*(1), 56–68. https://doi.org/10.1037/h0087842

Hanna, F. J., & Ottens, A. J. (1995). The role of wisdom in psychotherapy. *Journal of Psychotherapy Integration, 5*(3), 195–219. https://doi.org/10.1037/h0101273

Hanna, F. J., & Puhakka, K. (1991). When psychotherapy works: Pinpointing an element of change. *Psychotherapy: Theory, Research, Practice, Training, 28*(4), 598–607. https://doi.org/10.1037/0033-3204.28.4.598

Hanna, F. J., & Ritchie, M. H. (1995). Seeking the active ingredients of psychotherapeutic change: Within and outside the context of therapy. *Professional Psychology: Research and Practice, 26*(2), 176–183. https://doi.org/10.1037/0735-7028.26.2.176

Hanna, F. J., & Shank, G. (1995). The specter of metaphysics in counseling research and practice: The qualitative challenge. *Journal of Counseling and Development, 74*(1), 53–59. https://doi.org/10.1002/j.1556-6676.1995.tb01822.x

Hanna, F. J., Talley, W. B., & Guindon, M. H. (2000). The power of perception: Toward a model of cultural oppression and liberation. *Journal of Counseling and Development, 78*(4), 430–441. https://doi.org/10.1002/j.1556-6676.2000.tb01926.x

Hardy, C. (2019). Clinical sympathy: The important role of affectivity in clinical practice. *Medicine, Health Care, and Philosophy, 22*(4), 499–513. https://doi.org/10.1007/s11019-018-9872-8

Harmon-Jones, E. (Ed.). (2019). *Cognitive dissonance: Reexamining a pivotal theory in psychology* (2nd ed.). American Psychological Association. https://doi.org/10.1037/0000135-000

Harré, R. (1984). *Personal being: A theory for individual psychology*. Harvard University Press.

Harris, S. J. (1986). *Clearing the ground*. Houghton Mifflin.

Hart, C. M., Hepper, E. G., & Sedikides, C. (2018). Understanding and mitigating narcissists' low empathy. In A. D. Hermann, A. B. Brunell, & J. D. Foster (Eds.) *Handbook of trait narcissism: Key advances, research methods, and controversies* (pp. 335–343). Springer.

Hartmann, I. C. (2019). Forms of expression of a preverbal reality in child psychotherapy. *Journal of Prenatal & Perinatal Psychology & Health, 33*(4), 259–281.

Harvey, J. (2015). *Civilized oppression and moral relations: Victims, fallibility, and the moral community*. Palgrave Macmillan.

Hattie, J. A., Sharpley, C. F., & Rogers, H. J. (1984). Comparative effectiveness of professional and paraprofessional helpers. *Psychological Bulletin, 95*(3), 534–541. https://doi.org/10.1037/0033-2909.95.3.534

Hayes, J. A., Gelso, C. J., Goldberg, S., & Kivlighan, D. M. (2018). Countertransference management and effective psychotherapy: Meta-analytic findings. *Psychotherapy, 55*(4), 496–507. https://doi.org/10.1037/pst0000189

Hayes, S. C. (2005). *Get out of your mind and into your life: The new acceptance and commitment therapy*. New Harbinger Publications.

Hayes, S. C., & Linehan, M. M. (2018). Third-wave therapies. In J. O. Prochaska & J. C. Norcross (Eds.), *Systems of psychotherapy: A transtheoretical analysis* (9th ed., pp. 291–311). Oxford University Press.

Hayes, S. C., Strosahl, K. D., & Wilson, K. G. (2011). *Acceptance and commitment therapy: The process and practice of mindful change*. Guilford Press.

Heidegger, M. (1962). *Being and time*. Harper and Row. (Original work published 1927)

Hendricks, M. L., & Testa, R. J. (2012). A conceptual framework for clinical work with transgender and gender nonconforming clients: An adaptation of the minority stress model. *Professional Psychology: Research and Practice, 43*(5), 460–467. https://doi.org/10.1037/a0029597

Henriques, G. R. (2019). Toward a metaphysical empirical psychology. In T. Teo (Ed.), *Re-envisioning theoretical and philosophical psychology* (pp. 209–237). Palgrave Macmillan.

Hergenhahn, B. R. (1996). *An introduction to the history of psychology*. Wadsworth.

Herrnstein, R. J., & Boring, E. G. (Eds.). (1965). *A sourcebook in the history of psychology*. Harvard University Press.

Hibberd, F. J. (2014). The metaphysical basis of a process psychology. *Journal of Theoretical and Philosophical Psychology, 34*(3), 161–186. https://doi.org/10.1037/a0036242

Hill, C. E., Morales, K., Gerstenblith, J. A., Bansal, P., An, M., Rim, K., & Kivlighan, D. M., Jr. (2022). Therapist challenges and client responses in psychodynamic psychotherapy: An empirically supported case study. *Psychotherapy, 59*(1), 74–83. https://doi.org/10.1037/pst0000424

Hipson, W. E., Coplan, R. J., & Séguin, D. G. (2019). Active emotion regulation mediates links between shyness and social adjustment in preschool. *Social Development, 28*(4), 893–907. https://doi.org/10.1111/sode.12372

Hoener, C., Stiles, W. B., Luka, B. J., & Gordon, R. A. (2012). Client experiences of agency in therapy. *Person-Centered & Experiential Psychotherapies, 11*(1), 64–82. https://doi.org/10.1080/14779757.2011.639460

Hofmann, S. G., & Asmundson, G. J. G. (2008). Acceptance and mindfulness-based therapy: New wave or old hat? *Clinical Psychology Review, 28*(1), 1–16. https://doi.org/10.1016/j.cpr.2007.09.003

Høglend, P., & Hagtvet, K. (2019). Change mechanisms in psychotherapy: Both improved insight and improved affective awareness are necessary. *Journal of Consulting and Clinical Psychology, 87*(4), 332–344. https://doi.org/10.1037/ccp0000381

Holmes, C. (2017). *The paradox of countertransference: You and me, here and now.* Bloomsbury.

Hook, J. N., Davis, D., Owen, J., & DeBlaere, C. (2017). *Cultural humility: Engaging diverse identities in therapy*. American Psychological Association. https://doi.org/10.1037/0000037-000

Horney, K. (1987). *Final lectures*. W. W. Norton. (Original lectures given in 1952)

Howard, K. I., Kopta, S. M., Krause, M. S., & Orlinsky, D. E. (1986). The dose–effect relationship in psychotherapy. *American Psychologist, 41*(2), 159–164. https://doi.org/10.1037/0003-066X.41.2.159

Hume, D. (1978). *A treatise of human nature*. Oxford University Press. (Original work published 1739)

Husserl, E. (1970). *The crisis of European sciences and transcendental phenomenology.* Northwestern University Press. (Original work published 1936)

Husserl, E. (1982). *Ideas pertaining to a pure phenomenology and to a phenomenological philosophy: First book*. Martinus Nijhoff. (Original work published 1913)

Insel, T. R. (2022). *Healing: Our path from mental illness to mental health*. Penguin Press.

Inzlicht, M., Shenhav, A., & Olivola, C. Y. (2018). The effort paradox: Effort is both costly and valued. *Trends in Cognitive Sciences, 22*(4), 337–349. https://doi.org/10.1016/j.tics.2018.01.007

Ivey, A. E. (1995). Psychotherapy as liberation: Toward specific skills and strategies in multicultural counseling and therapy. In J. G. Ponterotto, J. M. Casas, L. A. Suzuki, & C. M. Alexander (Eds.), *Handbook of multicultural counseling* (pp. 53–72). Sage Publications.

Jacobs, D. H. (1994). Environmental failure: Oppression is the only cause of psychopathology. *Journal of Mind and Behavior, 15*(1–2), 1–18.

James, W. (1965). *Pragmatism and four essays from the meaning of truth*. New American Library. (Original work published 1909)

James, W. (1977). Does consciousness exist? In J. J. McDermott (Ed.), *The writings of William James: A comprehensive edition* (pp. 169–183). University of Chicago Press. (Original work published 1904)

James, W. (1981). *The principles of psychology*. Harvard University Press. (Original work published 1890)

Jankowski, T., & Holas, P. (2014). Metacognitive model of mindfulness. *Consciousness and Cognition, 28*, 64–80. https://doi.org/10.1016/j.concog.2014.06.005

Janosik, E. H. (1986). *Crisis counseling: A contemporary approach*. Jones & Bartlett.

Jung, C. G. (1969). *Collected works: Vol. 8. The structure and dynamics of the psyche.* Princeton University Press. (Original work published 1934)

Kajonius, P. J., & Dåderman, A. M. (2017). Conceptualizations of personality disorders with the five factor model-count and empathy traits. *International Journal of Testing, 17*(2), 141–157. https://doi.org/10.1080/15305058.2017.1279164

Kanfer, F. H., & Grimm, L. G. (1978). Freedom of choice and behavioral change. *Journal of Consulting and Clinical Psychology, 46*(5), 873–878. https://doi.org/10.1037/0022-006X.46.5.873

Kant, I. (1929). *Critique of pure reason.* St. Martin's Press. (Original work published 1787)

Karasu, T. B. (1992). *Wisdom in the practice of psychotherapy.* Basic Books.

Kazantzis, N., Dattilio, F. M., Cummins, A., & Clayton, X. (2014). Homework assignments and self-monitoring. In S. G. Hofmann, D. J. A. Dozois, W. Rief, & J. A. J. Smits (Eds.), *The Wiley handbook of cognitive behavioral therapy* (pp. 311–330). Wiley.

Kazantzis, N., & Lampropoulos, G. K. (2002). Reflecting on homework in psychotherapy: What can we conclude from research and experience? *Journal of Clinical Psychology, 58*(5), 577–585. https://doi.org/10.1002/jclp.10034

Kemp, R. (2013). Rock-bottom as an event of truth. *Existential Analysis, 24*(1), 106–116.

Kensit, D. A. (2000). Rogerian theory: A critique of the effectiveness of pure client-centred therapy. *Counselling Psychology Quarterly, 13*(4), 345–351. https://doi.org/10.1080/713658499

Kernberg, O. (1967). Borderline personality organization. *Journal of the American Psychoanalytic Association, 15*(3), 641–685. https://doi.org/10.1177/000306516701500309

Kierulff, S. (1988). Sheep in the midst of wolves: Personal-responsibility therapy with criminal personalities. *Professional Psychology: Research and Practice, 19*(4), 436–440. https://doi.org/10.1037/0735-7028.19.4.436

Kiesler, D. J. (1988). *Therapeutic metacommunication.* Consulting Psychologists Press.

Kiesler, D. J. (2017). *The process of psychotherapy: Empirical foundations and systems of analysis.* Routledge.

Kim, B. S., Ng, G. F., & Ahn, A. J. (2005). Effects of client expectation for counseling success, client-counselor worldview match, and client adherence to Asian and European American cultural values on counseling process with Asian Americans. *Journal of Counseling Psychology, 52*(1), 67–76. https://doi.org/10.1037/0022-0167.52.1.67

Kim, J. J., Brookman-Frazee, L., Gellatly, R., Stadnick, N., Barnett, M. L., & Lau, A. S. (2018). Predictors of burnout among community therapists in the sustainment phase of a system-driven implementation of multiple evidence-based practices in children's mental health. *Professional Psychology: Research and Practice, 49*(2), 131–142. https://doi.org/10.1037/pro0000182

Klein, A. C., & Wangyal, T. (2006). *Unbounded wholeness: Dzogchen, bon, and the logic of the nonconceptual.* Oxford University Press.

Klug, H. J., & Maier, G. W. (2015). Linking goal progress and subjective well-being: A meta-analysis. *Journal of Happiness Studies, 16*(1), 37–65. https://doi.org/10.1007/s10902-013-9493-0

Kocalevent, R. D., Finck, C., Pérez-Trujillo, M., Sautier, L., Zill, J., & Hinz, A. (2017). Standardization of the Beck Hopelessness Scale in the general population. *Journal of Mental Health, 26*(6), 516–522. https://doi.org/10.1080/09638237.2016.1244717

Koch, S. (1981). The nature and limits of psychological knowledge: Lessons of a century qua "science." *American Psychologist, 36*(3), 257–269. https://doi.org/10.1037/0003-066X.36.3.257

Korn, J. H., Davis, R., & Davis, S. F. (1991). Historians' and chairpersons' judgments of eminence among psychologists. *American Psychologist, 46*(7), 789–792. https://doi.org/10.1037/0003-066X.46.7.789

Korotitsch, W. J., & Nelson-Gray, R. O. (1999). An overview of self-monitoring research in assessment and treatment. *Psychological Assessment, 11*(4), 415–425. https://doi.org/10.1037/1040-3590.11.4.415

Kottler, J. A. (2022). *On being a therapist* (6th ed.). Oxford University Press.

Kramer, U., Levy, K. N., & McMain, S. (2024). *Understanding mechanisms of change in psychotherapies for personality disorders*. American Psychological Association. https://doi.org/10.1037/0000388-000

Kraus, D. R., Bentley, J. H., Alexander, P. C., Boswell, J. F., Constantino, M. J., Baxter, E. E., & Castonguay, L. G. (2016). Predicting therapist effectiveness from their own practice-based evidence. *Journal of Consulting and Clinical Psychology, 84*(6), 473–483. https://doi.org/10.1037/ccp0000083

Krebs, P., Norcross, J. C., Nicholson, J. M., & Prochaska, J. O. (2019). Stages of change. In J. C. Norcross & B. E. Wampold (Eds.), *Psychotherapy relationships that work: Evidence-based therapist responsiveness* (pp. 296–328). Oxford University Press.

Kross, E., & Ayduk, O. (2017). Self-distancing: Theory, research, and current directions. In J. M. Olsen (Ed.), *Advances in experimental social psychology* (Vol. 55, pp. 81–136). Academic Press.

Kuhn, D. (2022). Metacognition matters in many ways. *Educational Psychologist, 57*(2), 73–86. https://doi.org/10.1080/00461520.2021.1988603

Lam, K. K. L., & Zhou, M. (2022). Grit and academic achievement: A comparative cross-cultural meta-analysis. *Journal of Educational Psychology, 114*(3), 597–621. https://doi.org/10.1037/edu0000699

Lambert, M. J. (1992). Psychotherapy outcome research: Implications for integrative and eclectic therapists. In J. C. Norcross & M. R. Goldfried (Eds.), *Handbook of psychotherapy integration* (pp. 94–129). Basic Books.

Lambert, M. J., Bergin, A. E., & Collins, J. L. (1977). Therapist-induced deterioration in psychotherapy. In A. S. Gurman & A. M. Razin (Eds.), *Effective psychotherapy: A handbook of research* (pp. 452–481). Pergamon.

Larsen, D. J., & Stege, R. (2010). Hope-focused practices during early psychotherapy sessions: Part I: Implicit approaches. *Journal of Psychotherapy Integration, 20*(3), 271–292. https://doi.org/10.1037/a0020820

Laska, K. M., Gurman, A. S., & Wampold, B. E. (2014). Expanding the lens of evidence-based practice in psychotherapy: A common factors perspective. *Psychotherapy, 51*(4), 467–481. https://doi.org/10.1037/a0034332

Lavik, K. O., Veseth, M., Frøysa, H., Binder, P. E., & Moltu, C. (2018). 'Nobody else can lead your life': What adolescents need from psychotherapists in change processes. *Counselling & Psychotherapy Research, 18*(3), 262–273. https://doi.org/10.1002/capr.12166

Lazarus, A. A. (1976). *Multimodal behavior therapy*. Springer.

Lazarus, A. A. (1989a). Multimodal therapy. In R. J. Corsini & D. Wedding (Eds.), *Current psychotherapies* (4th ed., pp. 503–544). F. E. Peacock.

Lazarus, A. A. (1989b). *The practice of multimodal therapy*. Johns Hopkins University Press.

Lazarus, A. A. (1990). Can psychotherapists transcend the shackles of their training and superstitions? *Journal of Clinical Psychology, 46*(3), 351–358. https://doi.org/10.1002/1097-4679(199005)46:3%3C351::aid-jclp2270460316%3E3.0.co;2-v

Lazarus, A. A. (1993). Tailoring the therapeutic relationship, or being an authentic chameleon. *Psychotherapy: Theory, Research, Practice, Training, 30*(3), 404–407. https://doi.org/10.1037/0033-3204.30.3.404

Lazarus, A. A. (1996). The utility and futility of combining treatments in psychotherapy. *Clinical Psychology: Science and Practice, 3*(1), 59–68. https://doi.org/10.1111/j.1468-2850.1996.tb00058.x

Leahy, R. L. (2017). *Cognitive therapy techniques: A practitioner's guide*. Guilford Press.

Lee, E., Greenblatt, A., Hu, R., Johnstone, M., & Kourgiantakis, T. (2022). Developing a model of broaching and bridging in cross-cultural psychotherapy: Toward fostering epistemic and social justice. *American Journal of Orthopsychiatry, 92*(3), 322–333. https://doi.org/10.1037/ort0000611

Leschziner, V., & Brett, G. (2019). Beyond two minds: Cognitive, embodied, and evaluative processes in creativity. *Social Psychology Quarterly, 82*(4), 340–366. https://doi.org/10.1177/0190272519851791

Lester, D. (1982). Astrologers and psychics as therapists. *American Journal of Psychotherapy, 36*(1), 56–66. https://doi.org/10.1176/appi.psychotherapy.1982.36.1.56

Levitt, H. M., & Piazza-Bonin, E. (2016). Wisdom and psychotherapy: Studying expert therapists' clinical wisdom to explicate common processes. *Psychotherapy Research, 26*(1), 31–47. https://doi.org/10.1080/10503307.2014.937470

Lewin, K. (1935). *A dynamic theory of personality*. McGraw-Hill.

Lewin, K. (1936). *Principles of topological psychology*. McGraw-Hill.

Linehan, M. M. (1993). *Cognitive-behavioral treatment of borderline personality*. Guilford Press.

Liu, H., Peng, H., Song, X., Xu, C., & Zhang, M. (2022). Using AI chatbots to provide self-help depression interventions for university students: A randomized trial of effectiveness. *Internet Interventions, 27*, Article 100495. https://doi.org/10.1016/j.invent.2022.100495

Loevinger, J. (1976). *Ego development*. Jossey-Bass.

Low, A. A. (1952). *Mental health through will-training*. Willett.

Lyddon, W. J. (1989). Personal epistemology and preference for counseling. *Journal of Counseling Psychology, 36*(4), 423–429. https://doi.org/10.1037/0022-0167.36.4.423

Lyddon, W. J. (1990). First- and second-order change: Implications for rationalist and constructivist cognitive therapies. *Journal of Counseling and Development, 69*(6), 122–127. https://doi.org/10.1002/j.1556-6676.1990.tb01472.x

Lynch, T. R., Hempel, R. J., & Dunkley, C. (2015). Radically open-dialectical behavior therapy for disorders of over-control: Signaling matters. *American Journal of Psychotherapy, 69*(2), 141–162. https://doi.org/10.1176/appi.psychotherapy.2015.69.2.141

MacFarlane, P., Anderson, T., & McClintock, A. S. (2017). Empathy from the client's perspective: A grounded theory analysis. *Psychotherapy Research, 27*(2), 227–238. https://doi.org/10.1080/10503307.2015.1090038

Machado, P. P., & Beutler, L. E. (2016). Research methods and randomized clinical trials in psychotherapy. In A. J. Consoli, L. E. Beutler, & B. Bongar (Eds.), *Comprehensive textbook of psychotherapy: Theory and practice* (pp. 445–461). Oxford University Press.

Magnavita, J. J. (2006). The centrality of emotion in unifying and accelerating psychotherapy. *Journal of Clinical Psychology, 62*(5), 585–596. https://doi.org/10.1002/jclp.20250

Mahoney, M. J. (1991). *Human change processes: The scientific foundations of psychotherapy*. Basic Books.

Maibom, H. (Ed.). (2017). *The Routledge handbook of philosophy of empathy*. Taylor & Francis.

Maibom, H. (2020). *Empathy*. Routledge.

Mandela, N. (1995). *Long walk to freedom*. Abacus.

Manganaro, P. (2017). The roots of intersubjectivity—Empathy and phenomenology according to Edith Stein. In V. Lux & S. Weigel (Eds.), *Empathy: Epistemic problems and cultural-historical perspectives of a cross-disciplinary concept* (pp. 271–286). Springer.

Mann, D. (2010). *Gestalt therapy: 100 key points and techniques*. Routledge.

Marchetti, I. (2019). Hopelessness: A network analysis. *Cognitive Therapy and Research, 43*(3), 611–619. https://doi.org/10.1007/s10608-018-9981-y

Margulies, A. (1989). *The empathic imagination*. W. W. Norton.

Marquis, A., Henriques, G., Anchin, J., Critchfield, K., Harris, J., Ingram, B., Magnavita, J., & Osborn, K. (2021). Unification: The fifth pathway to psychotherapy integration. *Journal of Contemporary Psychotherapy, 51*(4), 285–294. https://doi.org/10.1007/s10879-021-09506-7

Marsh, A. A. (2018). The neuroscience of empathy. *Current Opinion in Behavioral Sciences, 19*, 110–115. https://doi.org/10.1016/j.cobeha.2017.12.016

Marshall-Lee, E. D., Hinger, C., Popovic, R., Miller Roberts, T. C., & Prempeh, L. (2020). Social justice advocacy in mental health services: Consumer, community, training, and policy perspectives. *Psychological Services, 17*(S1), 12–21. https://doi.org/10.1037/ser0000349

Martín-Baró, I. (1996). *Writings for a liberation psychology*. Harvard University Press.

Martínez-Martí, M. L., & Ruch, W. (2017). Character strengths predict resilience over and above positive affect, self-efficacy, optimism, social support, self-esteem, and life satisfaction. *The Journal of Positive Psychology, 12*(2), 110–119. https://doi.org/10.1080/17439760.2016.1163403

Masterson, J. F. (1988). *The search for the real self: Unmasking the personality disorders of our time*. Free Press.

Mayrhofer, R., Kuhbandner, C., & Lindner, C. (2021). The practice of experimental psychology: An inevitably postmodern endeavor. *Frontiers in Psychology, 11*, Article 612805. https://doi.org/10.3389/fpsyg.2020.612805

McCloskey, K. D., Cox, D. W., Ogrodniczuk, J. S., Laverdière, O., Joyce, A. S., & Kealy, D. (2021). Interpersonal problems and social dysfunction: Examining patients with avoidant and borderline personality disorder symptoms. *Journal of Clinical Psychology, 77*(1), 329–339. https://doi.org/10.1002/jclp.23033

McDowell, T., Knudson-Martin, C., & Bermudez, J. M. (2022). *Socioculturally attuned family therapy: Guidelines for equitable theory and practice.* Routledge.

McGrath, A. (2017). Dealing with dissonance: A review of cognitive dissonance reduction. *Social and Personality Psychology Compass, 11*(12), Article e12362. https://doi.org/10.1111/spc3.12362

McKay, M., Davis, M., & Fanning, P. (2021). *Thoughts and feelings: Taking control of your moods and your life.* New Harbinger.

McMullin, R. E. (1986). *Handbook of cognitive therapy techniques.* W. W. Norton.

McNeillie, N., & Rose, J. (2021). Vicarious trauma in therapists: A meta-ethnographic review. *Behavioural and Cognitive Psychotherapy, 49*(4), 1–15. https://doi.org/10.1017/s1352465820000776

McRae, L. (2013). Rehabilitating antisocial personalities: Treatment through self-governance strategies. *The Journal of Forensic Psychiatry & Psychology, 24*(1), 48–70. https://doi.org/10.1080/14789949.2012.752517

Medawar, P. B. (1984). *The limits of science.* Harper & Row.

Meichenbaum, D. B. (1977). *Cognitive-behavior modification: An integrative approach.* Plenum.

Meichenbaum, D., & Asarnow, J. (1979). Cognitive-behavioral modification and metacognitive development: Implications for the classroom. In P. C. Kendall & S. D. Hollon (Eds.), *Cognitive-behavioral interventions: Theory, research and procedures* (pp. 11–35). Academic Press.

Meichenbaum, D., & Cameron, R. (1974). The clinical potential of modifying what clients say to themselves. *Psychotherapy: Theory, Research & Practice, 11*(2), 103–117. https://doi.org/10.1037/h0086326

Merleau-Ponty, M. (1962). *The phenomenology of perception.* Humanities Press.

Merton, R. K. (1968). The Matthew effect in science. The reward and communication systems of science are considered. *Science, 159*(3810), 56–63.

Metzinger, T. (2004). *Being no one: The self-model theory of subjectivity.* MIT Press.

Meyer, I. H. (2003). Prejudice, social stress, and mental health in lesbian, gay, and bisexual populations: Conceptual issues and research evidence. *Psychological Bulletin, 129*(5), 674–697. https://doi.org/10.1037/0033-2909.129.5.674

Miller, D. (2024, January 14). *How to academy mindset: Do we have freewill? Daniel Dennett vs. Robert Sapolsy* [Video]. Youtube. https://www.youtube.com/watch?v=aYzFH8xqhns

Miller, J. B. (1986). *Toward a new psychology of women.* Beacon Press.

Miller, W. R., & Moyers, T. B. (2021). *Effective psychotherapists.* Guilford Press.

Miller, W. R., & Rollnick, S. (2012). *Motivational interviewing: Helping people change* (2nd ed.). Guilford Press.

Mills, J. (2001). Philosophical counseling as psychotherapy: An eclectic approach. *International Journal of Philosophical Practice, 1*(1), 25–47. https://doi.org/10.5840/ijpp2001112

Mills, J. (2004). Countertransference revisited. *Psychoanalytic Review, 91*(4), 467–515. https://doi.org/10.1002/j.1556-6676.1985.tb02719.x

Moodley, R., Lo, T., & Zhu, N. (Eds.). (2017). *Asian healing traditions in counseling and psychotherapy.* Sage Publications.

Moreno, J. L. (1946). *Psychodrama: First volume.* Beacon House.

Muris, P., & Otgaar, H. (2020). The process of science: A critical evaluation of more than 15 years of research on self-compassion with the Self-Compassion Scale. *Mindfulness, 11*(6), 1469–1482. https://doi.org/10.1007/s12671-020-01363-0

Myers, J. E., & Sweeney, T. J. (2008). Wellness counseling: The evidence base for practice. *Journal of Counseling & Development, 86*(4), 482–493. https://doi.org/10.1002/j.1556-6678.2008.tb00536.x

Mylopoulos, M., & Shepherd, J. (2020). Agentive phenomenology. In U. Kriegel (Ed.), *The Oxford handbook of the philosophy of consciousness* (pp. 215–234). Oxford University Press.

Nardini, J. E. (1952). Survival factors in American prisoners of war of the Japanese. *The American Journal of Psychiatry, 109*(4), 241–248. https://doi.org/10.1176/ajp.109.4.241

Neff, K. D. (2023). Self-compassion: Theory, method, research, and intervention. *Annual Review of Psychology, 74*, 193–218. https://doi.org/10.1146/annurev-psych-032420-031047

Nelissen, R. M. (2017). The motivational properties of hope in goal striving. *Cognition and Emotion, 31*(2), 225–237. https://doi.org/10.1080/02699931.2015.1095165

Nguyen, T. (2016). *The Patanjali Yoga Sutras and its spiritual practice*. Balboa Press.

Nissen-Lie, H. A., Goldberg, S. B., Hoyt, W. T., Falkenström, F., Holmqvist, R., Nielsen, S. L., & Wampold, B. E. (2016). Are therapists uniformly effective across patient outcome domains? A study on therapist effectiveness in two different treatment contexts. *Journal of Counseling Psychology, 63*(4), 367–378. https://doi.org/10.1037/cou0000151

Norcross, J. C., & Prochaska, J. O. (1986a). Psychotherapist heal thyself: I. The psychological distress and self-change of psychologists, counselors, and laypersons. *Psychotherapy: Theory, Research, Practice, Training, 23*(1), 102–114. https://doi.org/10.1037/h0085577

Norcross, J. C., & Prochaska, J. O. (1986b). Psychotherapist heal thyself: II. The self-initiated and therapy-facilitated change of psychological distress. *Psychotherapy: Theory, Research, Practice, Training, 23*(3), 345–356. https://doi.org/10.1037/h0085622

Norman, E., & Furnes, B. (2016). The concept of "metaemotion": What is there to learn from research on metacognition? *Emotion Review, 8*(2), 187–193. https://doi.org/10.1177/1754073914552913

North, C. (1987). *Welcome, silence: My triumph over schizophrenia*. Simon & Schuster.

Northup, S. (1968). *Twelve years a slave*. Louisiana State University Press. (Original work published 1853)

Oberst, U. E., & Stewart, A. E. (2014). *Adlerian psychotherapy: An advanced approach to individual psychology*. Routledge.

O'Donohue, W. (1989). The (even) bolder model. The clinical psychologist as metaphysician–scientist–practitioner. *American Psychologist, 44*(12), 1460–1468. https://doi.org/10.1037//0003-066x.44.12.1460

Oettingen, G., & Chromik, M. P. (2017). How hope influences goal-directed behavior. In M. W. Gallagher & S. J. Lopez (Eds.), *The Oxford handbook of hope* (pp. 69–81). Oxford University Press.

Ogbe, E., Harmon, S., Van den Bergh, R., & Degomme, O. (2020). A systematic review of intimate partner violence interventions focused on improving social support and/mental health outcomes of survivors. *PLOS ONE, 15*(6), Article e0235177. https://doi.org/10.1371/journal.pone.0235177

Ogden, T. (2018). *Projective identification and psychotherapeutic technique*. Routledge.

O'Leary, E. (2021). The need for integration. In E. O'Leary & M. Murphy (Eds.), *New approaches to integration in psychotherapy* (pp. 3–11). Routledge.

Omer, H., & London, P. (1988). Metamorphosis in psychotherapy: End of the systems era. *Psychotherapy: Theory, Research, Practice, Training, 25*(2), 171–180. https://doi.org/10.1037/h0085329

Ong, A. D., Standiford, T., & Deshpande, S. (2017). Hope and stress resilience. In M. W. Gallagher & S. J. Lopez (Eds.), *The Oxford handbook of hope* (pp. 255–285). Oxford University Press.

Orlinsky, D. E., Grawe, K., & Parks, B. K. (1994). Process and outcome in psychotherapy: Noch einmal. In A. E. Bergin & S. L. Garfield (Eds.), *Handbook of psychotherapy and behavior change* (4th ed., pp. 270–376). Wiley.

Ornstein, R. (2003). *Multimind: A new way of looking at human behavior.* Macmillan.

Pagnini, F., Bercovitz, K., & Langer, E. (2016). Perceived control and mindfulness: Implications for clinical practice. *Journal of Psychotherapy Integration, 26*(2), 91–102. https://doi.org/10.1037/int0000035

Pandita, S. U. (1991). *In this very life: The liberation teachings of the Buddha.* Wisdom Publications.

Paris, J. (2017). *Psychotherapy in an age of neuroscience.* Oxford Academic Press.

Park, G., Chung, J., & Lee, S. (2023). Effect of AI chatbot emotional disclosure on user satisfaction and reuse intention for mental health counseling: A serial mediation model. *Current Psychology, 42*(32), 28663–28673. https://doi.org/10.1007/s12144-022-03932-z

Pascual-Leone, A., Greenberg, L. S., & Pascual-Leone, J. (2009). Developments in task analysis: New methods to study change. *Psychotherapy Research, 19*(4–5), 527–542. https://doi.org/10.1080/10503300902897797

Peluso, P. R., & Freund, R. (2023). Paradoxical interventions: A meta-analysis. *Psychotherapy, 60*(3), 283–294. https://doi.org/10.1037/pst0000481

Peräkylä, A. (2019). Conversation analysis and psychotherapy: Identifying transformative sequences. *Research on Language and Social Interaction, 52*(3), 257–280. https://doi.org/10.1080/08351813.2019.1631044

Pereira, J. (1976). *Hindu theology: A reader.* Image Books.

Perls, F. S. (1969). *Gestalt therapy verbatim.* Real People Press.

Perls, F. S. (1973). *The gestalt approach & eyewitness to therapy.* Science and Behavior Books.

Perls, F. S., Hefferline, R. F., & Goodman, P. (1951). *Gestalt therapy.* Dell.

Persons, J. B. (1991). Psychotherapy outcome studies do not accurately represent current models of psychotherapy: A proposed remedy. *American Psychologist, 46*(2), 99–106. https://doi.org/10.1037/0003-066X.46.2.99

Pessoa, L. (2023). The entangled brain. *Journal of Cognitive Neuroscience, 35*(3), 349–360. https://doi.org/10.1162/jocn_a_01908

Peterson, B. S. (2019). Editorial: Common factors in the art of healing. *Journal of Child Psychology and Psychiatry, 60*(9), 927–929. https://doi.org/10.1111/jcpp.13108

Peterson, D. R. (1995). The reflective educator. *American Psychologist, 50*(12), 975–983. https://doi.org/10.1037//0003-066x.50.12.975

Philips, B., & Falkenström, F. (2021). What research evidence is valid for psychotherapy research? *Frontiers in Psychiatry, 11*, Article 625380. https://doi.org/10.3389/fpsyt.2020.625380

Polster, I., & Polster, M. (1973). *Gestalt therapy integrated: Contours of theory and practice.* Vintage Books.

Pope, K. S., & Keith-Spiegel, P. (2008). A practical approach to boundaries in psychotherapy: Making decisions, bypassing blunders, and mending fences. *Journal of Clinical Psychology, 64*(5), 638–652. https://doi.org/10.1002/jclp.20477

Popper, K. (1963). *Conjectures and refutations.* Basic Books.

Powers, W. T. (1973). *Behavior: The control of perception.* Aldine.

Prochaska, J. O., DiClemente, C. C., & Norcross, J. C. (1992). In search of how people change: Applications to addictive behaviors. *American Psychologist, 47*(9), 1102–1114. https://doi.org/10.1037//0003-066x.47.9.1102

Prochaska, J. O., & Norcross, J. C. (2018). *Systems of psychotherapy: A transtheoretical analysis.* Oxford University Press.

Prochaska, J. O., Norcross, J. C., & DiClemente, C. C. (1994). *Changing for good.* Avon Books.

Puhakka, K., & Hanna, F. J. (1988). Opening the POD: A therapeutic application of Husserl's phenomenology. *Psychotherapy: Theory, Research, Practice, Training, 25*(4), 582–592. https://doi.org/10.1037/h0085385

Rahula, W. (1978). *What the Buddha taught.* Gordon-Fraser.

Rank, O. (1936). *Will therapy.* W. W. Norton.

Reber, P. J., Batterink, L. J., Thompson, K. R., & Reuveni, B. (2019). Implicit leanring: History and application. In A. Cleeremans, V. Allakhverdov, & M. Kuvaldina (Eds.), *Implicit learning: 50 years on* (pp. 16–37). Routledge.

Reggia, J. A., Katz, G., & Huang, D. W. (2016). What are the computational correlates of consciousness? *Biologically Inspired Cognitive Architectures, 17,* 101–113.

Reich, A. (1951). On countertransference. *The International Journal of Psychoanalysis, 32,* 25–31.

Reisenzein, R. (2019). Cognition and emotion: A plea for theory. *Cognition and Emotion, 33*(1), 109–118. https://doi.org/10.1080/02699931.2019.1568968

Rhodes, M. G. (2019). Metacognition. *Teaching of Psychology, 46*(2), 168–175. https://doi.org/10.1177/0098628319834381

Rholes, W. S., Michas, L., & Shroff, J. (1989). Action control as a vulnerability factor in dysphoria. *Cognitive Therapy and Research, 13*(3), 263–274. https://doi.org/10.1007/BF01173407

Richard, D. C., & Lauterbach, D. (Eds.). (2011). *Handbook of exposure therapies.* Elsevier.

Rigney, D. (2010). *The Matthew effect: How advantage begets further advantage.* Columbia University Press.

Robinson, D. N. (1990). Wisdom through the ages. In R. J. Sternberg (Ed.), *Wisdom: Its nature, origins, and development* (pp. 13–24). Cambridge University Press.

Rogers, C. R. (1951). *Client-centered therapy.* Houghton Mifflin.

Rogers, C. R. (1957). The necessary and sufficient conditions of therapeutic personality change. *Journal of Consulting Psychology, 21*(2), 95–103. https://doi.org/10.1037/h0045357

Rosengren, D. B. (2017). *Building motivational interviewing skills: A practitioner workbook.* Guilford Press.

Rowan, J. (1990). *Subpersonalities: The people inside us.* Routledge.

Ruimi, L., Hadash, Y., Zvielli, A., Amir, I., Goldstein, P., & Bernstein, A. (2018). Meta-awareness of dysregulated emotional attention. *Clinical Psychological Science, 6*(5), 658–670. https://doi.org/10.1177/2167702618776948

Russell, B. (1972). *A history of Western philosophy.* Simon & Schuster.

Russell, K. A., Swift, J. K., Penix, E. A., & Whipple, J. L. (2022). Client preferences for the personality characteristics of an ideal therapist. *Counselling Psychology Quarterly, 35*(2), 243–259. https://doi.org/10.1080/09515070.2020.1733492

Sahakian, W. S. (1976). Philosophical psychotherapy. In W. S. Sahakian (Ed.), *Psychotherapy and counseling: Techniques in intervention* (pp. 286–302). Rand McNally.

Saini, A. (2017). *Inferior: How science got women wrong—And the new research that's rewriting the story.* Beacon Press.

Sakaluk, J. K., Williams, A. J., Kilshaw, R. E., & Rhyner, K. T. (2019). Evaluating the evidential value of empirically supported psychological treatments (ESTs): A meta-scientific review. *Journal of Abnormal Psychology, 128*(6), 500–509. https://doi.org/10.1037/abn0000421

Samenow, S. E. (1998). *Straight talk about criminals.* Jason Aronson.

Samoilov, A., & Goldfried, M. R. (2000). Role of emotion in cognitive-behavior therapy. *Clinical Psychology: Science and Practice, 7*(4), 373–385. https://doi.org/10.1093/clipsy.7.4.373

Sartre, J. P. (2003). *Being and nothingness* (H. E. Barnes, Trans.). Routledge. (Original work published 1943)

Scalabrini, A., Mucci, C., Angeletti, L. L., & Northoff, G. (2020). The self and its world: A neuro-ecological and temporo-spatial account of existential fear. *Clinical Neuropsychiatry, 17*(2), 46–58. https://doi.org/10.36131/clinicalnpsych20200203

Schiepek, G., & Pincus, D. (2023). Complexity science: A framework for psychotherapy integration. *Counselling & Psychotherapy Research, 23*(4), 941–955. https://doi.org/10.1002/capr.12641

Schmid, P. F. (2019). The power of hope: Person-centered perspectives on contemporary personal and societal challenges. *Person-Centered and Experiential Psychotherapies, 18*(2), 121–138. https://doi.org/10.1080/14779757.2019.1618371

Schwartz, R. C., & Sweezy, M. (2019). *Internal family systems therapy* (2nd ed.). Guilford Press.

Segal, H. (1977). Countertransference. *International Journal of Psychoanalytic Psychotherapy, 6,* 31–37.

Seligman, M. E. P. (2006). *Learned optimism: How to change your mind and your life* (3rd ed.). Vintage Books.

Selmi, P. M., Klein, M. H., Greist, J. H., Sorrell, S. P., & Erdman, H. P. (1990). Computer-administered cognitive-behavioral therapy for depression. *The American Journal of Psychiatry, 147*(1), 51–56. https://doi.org/10.1176/ajp.147.1.51

Seow, T. X. F., Rouault, M., Gillan, C. M., & Fleming, S. M. (2021). How local and global metacognition shape mental health. *Biological Psychiatry, 90*(7), 436–446. https://doi.org/10.1016/j.biopsych.2021.05.013

Shah, P., & Mountain, D. (2007). The medical model is dead—Long live the medical model. *The British Journal of Psychiatry, 191*(5), 375–377. https://doi.org/10.1192/bjp.bp.107.037242

Shamoon, Z. A., Lappan, S., & Blow, A. J. (2017). Managing anxiety: A therapist common factor. *Contemporary Family Therapy, 39,* 43–53. https://doi.org/10.1007/s10591-016-9399-1

Shaw, M. (2015). *War and genocide: Organized killing in modern society.* Wiley.

Shoemaker, D. J. (2018). *Theories of delinquency: An examination of explanations of delinquent behavior.* Oxford University Press.

Shonin, E., & Van Gordon, W. (2016). The mechanisms of mindfulness in the treatment of mental illness and addiction. *International Journal of Mental Health and Addiction, 14*(5), 844–849. https://doi.org/10.1007/s11469-016-9653-7

Silberschatz, G. (2017). Improving the yield of psychotherapy research. *Psychotherapy Research, 27*(1), 1–13. https://doi.org/10.1080/10503307.2015.1076202

Silk, J. S., Shaw, D. S., Skuban, E. M., Oland, A. A., & Kovacs, M. (2006). Emotion regulation strategies in offspring of childhood-onset depressed mothers. *Journal of Child Psychology and Psychiatry, 47*(1), 69–78. https://doi.org/10.1111/j.1469-7610.2005.01440.x

Simard, P., Simard, V., Laverdière, O., & Descôteaux, J. (2023). The relationship between narcissism and empathy: A meta-analytic review. *Journal of Research in Personality, 102*, Article 104329. https://doi.org/10.1016/j.jrp.2022.104329

Simon, R. (1985). Take it or leave it: An interview with Carl Whitaker. *Family Therapy Networker, 9*(5), 27–37.

Simourd, D. J., Olver, M. E., & Brandenburg, B. (2016). Changing criminal attitudes among incarcerated offenders: Initial examination of a structured treatment program. *International Journal of Offender Therapy and Comparative Criminology, 60*(12), 1425–1445. https://doi.org/10.1177/0306624x15579257

Singer, B. A., & Luborsky, L. (1977). Countertransference: The status of clinical versus quantitative research. In A. S. Gurman & A. M. Razin (Eds.), *Effective psychotherapy: A handbook of research* (pp. 433–451). Pergamon.

Singer, M. T., & Lalich, J. (1996). *"Crazy" therapies: What are they? Do they work?* Jossey-Bass.

Singer, W. (2017). Conscious processing: Unity in time rather than in space. In S. Schneider & M. Velmans (Eds.), *The Blackwell companion to consciousness* (pp. 607–620). Wiley.

Skinner, B. F. (1971). *Beyond freedom and dignity*. Knopf.

Skovholt, T. (2017). *Master therapists: Exploring expertise in therapy and counseling*. Oxford University Press.

Slife, B. D. (1987). Can cognitive psychology account for metacognitive functions of mind? *Journal of Mind and Behavior, 8*(2), 195–208.

Smith, M. L., & Glass, G. V. (1977). Meta-analysis of psychotherapy outcome studies. *American Psychologist, 32*(9), 752–760. https://doi.org/10.1037/0003-066X.32.9.752

Smith, R., Lane, R. D., Nadel, L., & Moutoussis, M. (2020). A computational neuroscience perspective on the change process in psychotherapy. In R. D. Lane & L. Nadel (Eds.), *Neuroscience of enduring change* (pp. 395–432). Oxford University Press.

Smoliak, O., & Strong, T. (2018). *Therapy as discourse*. Palgrave Macmillan.

Snyder, C. R. (1994). *The psychology of hope*. Free Press.

Sperry, L. (2022). *Highly effective therapy: Effecting deep change in counseling and psychotherapy*. Routledge.

Srinivasan, T. M. (2021). Kaivalya: The ultimate freedom. *International Journal of Yoga, 14*(3), 173–174. https://doi.org/10.4103/ijoy.ijoy_123_21

Stauffer, J. (2015). *Ethical loneliness: The injustice of not being heard*. Columbia University Press.

Stefana, A. (2017). *History of countertransference: From Freud to the British object relations school*. Routledge.

Stein, D. M., & Lambert, M. J. (1995). Graduate training in psychotherapy: Are therapy outcomes enhanced? *Journal of Consulting and Clinical Psychology, 63*(2), 182–196. https://doi.org/10.1037//0022-006x.63.2.182

Sternberg, R. J. (Ed.). (1990). *Wisdom: Its nature, origins, and development.* Cambridge University Press.

Sternberg, R. J., & Glück, J. (2021). *Wisdom: The psychology of wise thoughts, words, and deeds.* Cambridge University Press.

Stevens, F. L. (2024). Revisiting the cognitive primacy hypothesis: Implications for psychotherapy. *Journal of Psychotherapy Integration, 34*(4), 450–462. https://doi.org/10.1037/int0000313

Stice, E., Shaw, H., Becker, C. B., & Rohde, P. (2008). Dissonance-based interventions for the prevention of eating disorders: Using persuasion principles to promote health. *Prevention Science, 9*(2), 114–128. https://doi.org/10.1007/s11121-008-0093-x

Strean, H. S. (1993). *Resolving counterresistances in psychotherapy.* Brunner/Maze.

Stricker, G., & Gold, J. R. (Eds.). (2013). *Comprehensive handbook of psychotherapy integration.* Springer.

Strupp, H. H. (1988). What is therapeutic change? *Journal of Cognitive Psychotherapy, 2*(2), 75–82.

Strupp, H. H. (1996). The tripartite model and the consumer reports study. *American Psychologist, 51*(10), 1017–1024. https://doi.org/10.1037//0003-066x.51.10.1017

Strupp, H. H., & Hadley, S. W. (1979). Specific versus nonspecific factors in psychotherapy. *Archives of General Psychiatry, 36,* 1125–1136.

Sue, D. W. (2010). Microaggressions, marginality, and oppression: An introduction. In D. W. Sue (Ed.), *Microaggressions and marginality: Manifestation, dynamics and impact* (pp. 3–22). Wiley.

Sue, D. W., Sue, D., Neville, H. A., & Smith, L. (2022). *Counseling the culturally diverse: Theory and practice.* Wiley.

Sultanoff, S. M. (2013). Integrating humor into psychotherapy: Research, theory, and the necessary conditions for the presence of therapeutic humor in helping relationships. *The Humanistic Psychologist, 41*(4), 388–399. https://doi.org/10.1080/08873267.2013.796953

Swift, J. K., Owen, J., & Miller, S. D. (2023). Client factors. In S. D. Miller, D. Chow, S. Malins, & M. A. Hubble (Eds.), *The field guide to better results: Evidence-based exercises to improve therapeutic effectiveness* (pp. 47–78). American Psychological Association. https://doi.org/10.1037/0000358-004

Ta, V., Griffith, C., Boatfield, C., Wang, X., Civitello, M., Bader, H., DeCero, C., & Loggarakis, A. (2020). User experiences of social support from companion chatbots in everyday contexts: Thematic analysis. *Journal of Medical Internet Research, 22*(3), Article e16235. https://doi.org/10.2196/16235

Tapper, K. (2018). Mindfulness and craving: Effects and mechanisms. *Clinical Psychology Review, 59,* 101–117. https://doi.org/10.1016/j.cpr.2017.11.003

Teater, M., & Ludgate, J. (2014). *Overcoming compassion fatigue: A practical resilience workbook.* PESI Publishing & Media.

Tedeschi, R. G., & Moore, B. A. (2021). Posttraumatic growth as an integrative therapeutic philosophy. *Journal of Psychotherapy Integration, 31*(2), 180–194. https://doi.org/10.1037/int0000250

Teyber, E., & Teyber, F. M. (2014). Working with the process dimension in relational therapies: Guidelines for clinical training. *Psychotherapy, 51*(3), 334–341. https://doi.org/10.1037/a0036579

Theise, N. D., & Kafatos, M. C. (2016). Fundamental awareness: A framework for integrating science, philosophy and metaphysics. *Communicative & Integrative Biology, 9*(3), Article e1155010. https://doi.org/10.1080/19420889.2016.1155010

Tirch, D., Silberstein, L. R., & Kolts, R. L. (2015). *Buddhist psychology and cognitive-behavioral therapy: A clinician's guide*. Guilford Press.

Totton, N. (2018). Power in the therapeutic relationship. In R. Tweedy (Ed.), *The political self* (pp. 29–42). Routledge.

Townshend, C. (1987). The necessity of political violence: A review article. *Comparative Studies in Society and History, 29*(2), 314–319. https://doi.org/10.1017/S0010417500014523

Travers-Hill, E., Dunn, B. D., Hoppitt, L., Hitchcock, C., & Dalgleish, T. (2017). Beneficial effects of training in self-distancing and perspective broadening for people with a history of recurrent depression. *Behaviour Research and Therapy, 95*, 19–28. https://doi.org/10.1016/j.brat.2017.05.008

Ucar, G. K., Hasta, D., & Malatyali, M. K. (2019). The mediating role of perceived control and hopelessness in the relation between personal belief in a just world and life satisfaction. *Personality and Individual Differences, 143*, 68–73. https://doi.org/10.1016/j.paid.2019.02.021

Vafaie, N., & Kober, H. (2022). Association of drug cues and craving with drug use and relapse: A systematic review and meta-analysis. *JAMA Psychiatry, 79*(7), 641–650. https://doi.org/10.1001/jamapsychiatry.2022.1240

Vago, D. R., Farb, N., & Spreng, R. N. (2022). Clarifying internally-directed cognition: A commentary on the attention to thoughts model. *Psychological Inquiry, 33*(4), 261–272. https://doi.org/10.1080/1047840X.2022.2141005

Vahdani, R., & Phillips, M. (2021). Existential–Jungian analysis: Reconciling the personal and archetypal realms in the consulting room. *Journal of Humanistic Psychology, 61*(5), 806–827. https://doi.org/10.1177/0022167819880039

Vallejo, Z., & Amara, H. (2009). Adaptation of mindfulness-based stress reduction program for addiction relapse prevention. *The Humanistic Psychologist, 37*(2), 192–206. https://doi.org/10.1080/08873260902892287

van der Kolk, B. A. (2014). *The body keeps the score: Brain, mind, and body in the healing of trauma*. Viking.

van Ede, F., & Nobre, A. C. (2021). Toward a neurobiology of internal selective attention. *Trends in Neurosciences, 44*(7), 513–515. https://doi.org/10.1016/j.tins.2021.04.010

van Inwagen, P. (2024). *The abstract and the concrete: Further essays in ontology*. Oxford University Press.

Van Kaam, A. (1966). *Existential foundations of psychology*. Duquesne University Press.

Van Wagoner, S. L., Gelso, C. L., Hayes, J. A., & Diemer, R. A. (1991). Countertransference and the reputedly excellent therapist. *Psychotherapy: Theory, Research, Practice, Training, 28*(3), 411–421. https://doi.org/10.1037/0033-3204.28.3.411

Vargas, S. M., Huey, S. J., Jr., & Miranda, J. (2020). A critical review of current evidence on multiple types of discrimination and mental health. *American Journal of Orthopsychiatry, 90*(3), 374–390. https://doi.org/10.1037/ort0000441

Vickery, P. J., Hanna, F. J., & Wilkinson, B. D. (2023). Techniques in the freedom-from oppression model: An integrative existential-cognitive therapy. *The Journal of Humanistic Counseling, 62*(3), 173–186. https://doi.org/10.1002/johc.12205

Wachtel, P. L. (1977). *Psychoanalysis and behavior therapy: Toward an integration*. Basic Books.

Wachtel, P. L. (2018). Pathways to progress for integrative psychotherapy: Perspectives on practice and research. *Journal of Psychotherapy Integration, 28*(2), 202–212. https://doi.org/10.1037/int0000089

Walrond-Skinner, S. (1986). *Dictionary of psychotherapy*. Routledge; Kegan Paul.

Walsh, L. M., Roddy, M. K., Scott, K., Lewis, C. C., & Jensen-Doss, A. (2019). A meta-analysis of the effect of therapist experience on outcomes for clients with internalizing disorders. *Psychotherapy Research, 29*(7), 846–859. https://doi.org/10.1080/10503307.2018.1469802

Wampold, B. E. (2001). *The great psychotherapy debate: Models, methods, and findings*. Routledge.

Wampold, B. E., & Flückiger, C. (2023). The alliance in mental health care: Conceptualization, evidence and clinical applications. *World Psychiatry, 22*(1), 25–41. https://doi.org/10.1002/wps.21035

Wampold, B. E. & Imel, Z. E. (2015). *The great psychotherapy debate: The evidence for what makes psychotherapy work* (2nd ed.). Routledge.

Wampold, B. E., & Ulvenes, P. G. (2019). Integration of common factors and specific ingredients. In J. C. Norcross & M. R. Goldfried (Eds.), *Handbook of psychotherapy integration* (pp. 69–87). Oxford University Press.

Wang, M., Zhang, L. J., & Hamilton, R. (2023). Developing the Metacognitive Awareness of Grit Scale for a better understanding of learners of English as a foreign language. *Frontiers in Psychology, 14*, Article 1141214. https://doi.org/10.3389/fpsyg.2023.1141214

Wang, Y., Chung, M. C., Wang, N., Yu, X., & Kenardy, J. (2021). Social support and posttraumatic stress disorder: A meta-analysis of longitudinal studies. *Clinical Psychology Review, 85*, Article 101998. https://doi.org/10.1016/j.cpr.2021.101998

Warth, M., Zöller, J., Köhler, F., Aguilar-Raab, C., Kessler, J., & Ditzen, B. (2020). Psychosocial interventions for pain management in advanced cancer patients: A systematic review and meta-analysis. *Current Oncology Reports, 22*(1), Article 3. https://doi.org/10.1007/s11912-020-0870-7

Watkins, M. (2002). Seeding liberation: A dialogue between depth psychology and liberation psychology. In D. Slattery & L. Corbett (Eds.), *Depth psychology: Meditations in the field* (pp. 204–225). Daimon Verlag.

Watson, L. B., DeBlaere, C., Langrehr, K. J., Zelaya, D. G., & Flores, M. J. (2016). The influence of multiple oppressions on women of color's experiences with insidious trauma. *Journal of Counseling Psychology, 63*(6), 656–667. https://doi.org/10.1037/cou0000165

Watts, R. E., & Bluvshtein, M. (2020). Adler's theory and therapy as a river: A brief discussion of the profound influence of Alfred Adler. *Journal of Individual Psychology, 76*(1), 99–109. https://doi.org/10.1353/jip.2020.0021

Watzlawick, P., Weakland, J. H., & Fisch, R. (1974). *Change: Principles of problem formation and problem resolution*. W. W. Norton.

Weinberger, J., & Eig, A. (1999). Expectancies: The ignored common factor in psychotherapy. In I. Kirsch (Ed.), *How expectancies shape experience* (pp. 357–382). American Psychological Association. https://doi.org/10.1037/10332-015

Weiner, M. F. (1982). *The therapeutic impasse*. Free Press.

Weintraub, J., Nolan, K. P., & Sachdev, A. R. (2023). The cognitive control model of work-related flow. *Frontiers in Psychology, 14*, Article 1174152. https://doi.org/10.3389/fpsyg.2023.1174152

Weiss, P. (1958). Common sense and beyond. In S. Hook (Ed.), *Determinism and freedom: In the age of modern science* (pp. 211–224). Collier Books.

Wells, A. (2007). Cognition about cognition: Metacognitive therapy and change in generalized anxiety disorder and social phobia. *Cognitive and Behavioral Practice, 14*(1), 18–25. https://doi.org/10.1016/j.cbpra.2006.01.005

Whitaker, C. (1976). The hindrance of theory in clinical work. In P. D. Guerin (Ed.), *Family therapy: Theory and practice* (pp. 154–164). Gardner Press.

Whitaker, C. (1989). *Midnight musings of a family therapist*. W. W. Norton.

Whitehead, A. N. (1925). *Science and the modern world*. Free Press.

Whitehead, A. N. (1938). *Modes of thought*. Capricorn Books.

Wilken, B., & Miyamoto, Y. (2018). Dialectical emotions. In J. Spencer-Rodgers & K. Peng (Eds.), *The psychological and cultural foundations of East Asian cognition: Contradiction, change, and holism* (pp. 509–546). Oxford University Press.

Wilkinson, B. D. (2019). A refined and further defined argument on the limits of neuroscience in counseling: Response to Field, Luke, and Beeson and Miller. *The Journal of Humanistic Counseling, 58*(2), 119–134. https://doi.org/10.1002/johc.12101

Wilkinson, B. D. (2023). Understanding experiential awareness in humanistic-phenomenological counseling. *The Journal of Humanistic Counseling, 62*(2), 145–159. https://doi.org/10.1002/johc.12196

Wilkinson, B. D., & Hanna, F. J. (2018). Using the precursors model of change to facilitate engagement practices in family counseling. *The Family Journal, 26*(3), 306–314. https://doi.org/10.1177/1066480718795502

Wilkinson, B. D., & Wilkinson, K. A. (2024). The ecological-enactive approach to embodiment in humanistic psychotherapy. *The Humanistic Psychologist*. Advance online publication. https://doi.org/10.1037/hum0000349

Winnicott, D. W. (2018). Ego distortion in terms of true and false self. In L. Caldwell (Ed.), *The person who is me* (pp. 7–22). Routledge. (Original work published 1960)

Wolpe, J. (1958). *Psychotherapy by reciprocal inhibition*. Stanford University Press.

Woods, A., Solomonov, N., Liles, B., Guillod, A., Kales, H. C., & Sirey, J. A. (2021). Perceived social support and interpersonal functioning as predictors of treatment response among depressed older adults. *The American Journal of Geriatric Psychiatry, 29*(8), 843–852. https://doi.org/10.1016/j.jagp.2020.12.021

Yaden, D. B., Johnson, M. W., Griffiths, R. R., Doss, M. K., Garcia-Romeu, A., Nayak, S., Gukasyan, N., Mathur, B. N., & Barrett, F. S. (2021). Psychedelics and consciousness: Distinctions, demarcations, and opportunities. *The International Journal of Neuropsychopharmacology, 24*(8), 615–623. https://doi.org/10.1093/ijnp/pyab026

Yadlin-Gadot, S. (2016). *Truth matters: Theory and practice in psychoanalysis*. Brill.

Yalom, I. D. (1980). *Existential psychotherapy*. Basic Books.

Yalom, I. D. (1989). *Love's executioner and other tales of psychotherapy*. HarperCollins.

Yalom, I. D. (1995). *Theory and practice of group psychotherapy*. Basic Books.

Yang, Y., & Hayes, J. A. (2020). Causes and consequences of burnout among mental health professionals: A practice-oriented review of recent empirical literature. *Psychotherapy, 57*(3), 426–436. https://doi.org/10.1037/pst0000317

Zahavi, D. (2014). Empathy and other-directed intentionality. *Topoi, 33*(1), 129–142. https://doi.org/10.1007/s11245-013-9197-4

Zahavi, D. (2017). *Husserl's legacy: Phenomenology, metaphysics, and transcendental philosophy*. Oxford University Press.

Zahavi, D. (2018). Brain, mind, world: Predictive coding, neo-Kantianism, and transcendental idealism. *Husserl Studies, 34*(1), 47–61. https://doi.org/10.1007/s10743-017-9218-z

Zerubavel, N., & Messman-Moore, T. L. (2015). Staying present: Incorporating mindfulness into therapy for dissociation. *Mindfulness, 6*(2), 303–314. https://doi.org/10.1007/s12671-013-0261-3

Zilcha-Mano, S. (2021). Toward personalized psychotherapy: The importance of the trait-like/state-like distinction for understanding therapeutic change. *American Psychologist, 76*(3), 516–528. https://doi.org/10.1037/amp0000629

Zvi-Beiman, S., & Shahar, G. (2015). Psychotherapy integration. In R. Cautin & S. Lilienfeld (Eds.), *The encyclopedia of clinical psychology* (pp. 1–6). Wiley.

INDEX

A

Accusing clients of defensiveness, 230
Acting-out theater, 123
Active agency, 8–9
Active–passive emotion regulation strategies, 30
Acton, L., 249
Acts of consciousness, 6
Addiction, 270
  empty chair work for, 262–263
  grand reframe of, 255–259
  subpersonality technique for, 260–261
Adler, A., 25, 42–43, 51, 88, 124, 139, 184
AI. *See* Artificial intelligence
Alcoholics Anonymous, 147, 184
Alcohol use, 95, 102, 122, 129, 213, 266–269. *See also* Addiction; Drug use
Aligning therapy, 133
  Carl case, 133–134
  Nick case, 135
Ambivalent client, 174–175
Antisocial clients, 62, 64, 67, 188, 208, 210
Anxiety. *See also* Grit, or the willingness to experience anxiety or difficulty
  accompanying change, 146
  internal dialogue avoiding, 155–157
  waxing and waning of, 162–165
Arkowitz, H., 72
Artificial intelligence (AI), 36
Asking permission, 54–55
Assagioli, R., 135, 168
Automatic thoughts
  identifying and disputing, 180–181
  isolating, 118–119
  reframing, 116
Autonomy, respecting and encouraging, 62–63
Awareness of the problem, 5, 18, 30, 126, 277
  absence of, 111–112
  acting-out theater, 123
  automatic thoughts, 116, 118–119
  and change, metacognitive aspects of, 106–107
  clinical markers of, 110–111
  confrontation, 116–118
  confusion, building tolerance of, 125
  conscious, degrees of, 106
  consciousness raising, 107–108
  consistency of, 108–109
  contexts, setting and contrasting, 122–123
  deficit in, 213–214
  defined, 105
  empathy as a form of, 109
  empathy as remedy, 124–125
  empty chair technique, 120–121
  feelings, identifying, 119–120
  intentional unawareness handling, 123–124
  lack of, metaphors for, 113–114
  leading to therapeutic change, 106–109
  maladaptive beliefs about, 125
  negative behaviors, reframing, 115
  observer or wise mind, 121–122
  oppressive clients, 250–251

paradox, use of, 122
role-plays of others, 120
as savoir faire, 115–116
self-monitoring, 121
significant moments of change involving, 109–110
situation, defined, 116
techniques for cultivating, 113–125
unaware person, example of, 114–115

B

Bad breath metaphor, 114
Bandura, A., 95
Bateson, G., 57
Beck, A. T., 107, 114
Beck Hopelessness Scale, 94
Being and Time (Heidegger), 146
Benign stereotyping, 247
Bettelheim, B., 90
Bibliotherapy, 31–32, 36
Black Lives Matter movement, 241
Blaming others, 160–161
Blind spots metaphor, 113
Body odor metaphor, 114
Boundaries setting, 65
Brodley, B. T., 117
Burnout, 216, 227–228, 230, 255

C

Cannon, W. B., 90
Caring, 63–64
prioritizing, 226–227
Cautela, J. R., 124
Cautionary note, 189
Change deficits, 20, 35
CHANGES assessment
case examples, 216–221
across diagnoses, 208
empirical validation of, 198–199
examples of, 204–207
form, 198, 199–202
with groups and families, 207–208
James case, 219–221
Joy case, 205–206
Kurt case, 217–218
Nancy case, 218–219
rating for, 197–208
Ricky case, 206–207
scoring of, 202–204
therapist interference, 211–216
Tommy case, 204–205
CHANGES Assessment for Businesses and Corporations, 275–279
CHANGES model, 4. *See also* Therapeutic change
advantages of, 273–274
assessing individual, team, and organizational change, 275–279
future applications of, 274–275
to metatheoretical framework, 279–286
technological freedom and, 285–286
unique aspects of, 6–9
Chatbots, 36, 184
Civilized oppression, 241
Cleansing metaphor, 154
Client-specific factors, 16–17
Clinical markers
awareness, 110–111
confronting, 76–77
effort, 172–173
grit, 150–151
hope, 92–93
necessity, 129–130
social support, 186–187
Cognitive behavioral therapy, 24, 26, 42, 107, 171, 267
Cognitive dissonance, 138–141
asking questions, 138–139
Jason case, 139–141
role of, 128–129
spitting in the client's soupis technique, 139, 141
Compartmentalizing, 112
Compassion, 63–64, 233–234
Compassion fatigue, 227
Concentration-based techniques, 87
Concentric-circle technique, 68, 190–191
Concretized confronting, 85–87
Confidence, 92
Conflicted client, 174–175
Confrontation, 116–118
avoiding, 229
with compassion and empathy, 233–234
with impatience, 229
Confronting the problem, 5, 18, 276–277
absence of, 77–78
ancient approaches to, 72–73
clinical markers of, 76–77
concentration-based techniques, 87
concretized, 85–87
cutting to essence of, 71–72
deficit in, 212
defined, 69
hesitancy to confront, 79
holding problem steady, 70
idea of change itself, 79
impatience, 229
leading to therapeutic change, 70–75
metaphor of strength, 78–79
miracle question, 88
mirroring, 85
onion peeler, 83–84
power of problem, removing, 74–75
reality testing, 73

recognizing, 76–77
representational, 85–87
significant moments of change involving, 75–76
simplicity and complexity, 73
smaller problem space, 74
and systematic desensitization, 71
techniques for enhancing, 78–88
vaccine effect of, 75
in vivo, 80–83
Confusion, building tolerance of, 125
Consciousness-raising strategy, 31–32, 107–108
Contagion, hope as, 100–101
Contexts, contrasting, 122–123
Control
effort, 169–170
Grit, 161–162
necessity, 142
of thoughts and images, 169–170
Countertransference, 209–210, 211
Courtesy, 54–55
Cravings
influencing, 270
management, 269
mindfully riding, 269–270
techniques for reducing, 269–270
Creative narratives, 102–103
Crenshaw, K., 246
Crisis, 9, 91, 94, 121
defined, 37
Cross-purposes, 141–142

D

Decentering, 107
Decision making, 180
Defensiveness, 145, 230
Degree of change, 9–10
Degrees of conscious awareness, 106
Diagnostic and Statistical Manual of Mental Disorders, 39, 208, 266
Dialectical thinking, 31, 231
Difficult clients, 15–34, 52, 210–211
antisocial clients, 64
autonomy and freedom, respecting and encouraging, 62–63
back off, option of telling to, 57
at basis of techniques, 27–28
at basis of theory, 22–27
being courteous, 54–55
being persuasive, 55–56
boundaries setting, 65
change as skill, 28–29
chunks, addressing problems in, 64
client's meaning system, 60–61
compassion and caring, displaying, 63–64
concentric circle technique, 68
defense systems, bringing down, 62
defined, 20
emotional state matching, 65–66
empathy, increasing therapist capacity for, 59–60
framing techniques, 28
hesitancy, expecting, 64–65
"I Behind The Eye", 67
Joy case, 40, 205–206
Kathy case, 42–43
making, 35–48
matching therapist with, 67
metacognitive aspects of change, 29–31
metalogue, attending to, 57–58
perspicacity, as wisdom characteristic, 67
positive qualities, validating, 56–57
precursors of change, 17–19, 21–22, 33
problem, redefining, 61–62
realigning psychotherapy theories around change, 24–26
reflecting meanings before feelings, 58–59
relationship with, 54–68
requesting permission from, 54–55
Ricky case, 40–41, 206–207
as teachers, 236–237
theories for clients, 26–27
therapeutic change and client-specific factors, 16–17
therapeutic modalities, 68
therapeutic window, opening, 66
Tommy case, 39–40, 204–205
transtheoretical model and motivational interviewing, 31–33
uncooperativeness as protection of freedom, 63
value of model for therapy with, 20–22
"what's-in-it-for-me?" question, 63
Difficult therapists, 210–211
Difficulty. *See also* Grit, or the willingness to experience anxiety or difficulty
implicit learning and maladaptive beliefs, 41–43
Kurt case, 38, 217–218
and resistance as self-protection, 36–41
roots of, 41–43
therapist variables influencing therapeutic success, 43–47
training to be difficult by therapists, 37–39
training wisdom out of therapist, 46–47
wisdom and wise therapist, 44–46
Disappointment, 94
Discouragement, 94, 215, 235
Disequilibrium, 37
Disgusting, 235
Doubling, 265
Douglass, F., 249

Driving metaphor, 154
Drug hero technique, 263
Drug use, 95, 103, 129, 151. *See also* Addiction
admiring intention, 257–258
craving, influencing, 270
craving management, 269
drug hero technique, 263
existential-cognitive therapy of oppression for, 264–265
finding out reasons for, 256–257
first high, best high strategy, 259–260
Marco case, 267–269
pharmacotherapy for, 259–260
redirecting, 258–259
role-playing, 265–266
in service of family, 266–269
vs. therapy, 255–259
urge surfing technique, 269–270
value dissonance and realignment, 264
Dumping, 147

E

Education, 29
Effort, or will to change, 5, 18, 278, 167–168
absence of, 173–175
ambivalent client, 174–175
automatic thoughts, identifying and disputing, 180–181
clinical markers for, 172–173
conflicted client, 174–175
deficit in, 215–216
and efficient use of time, 170–171
empathic validation, 177–178
facilitating the decision, 180
falsely compliant client, 174
freedom of choice, 176
goal, clarification of, 176–177
graduated tasks, assigning, 177
leading to therapeutic change, 169–171
maladaptive beliefs about, 181–182
maneuvering, 170
mapping intentions, 179–180
metalogue, attending to, 181
metaphors for, 175–176
nature of, 168–169
negative self-talk, identifying and disputing, 180–181
passive–compliant client, 173–174
recognizing, 172–173
self-observation, 177
significant moments of change involving, 171–172
subpersonality approach to, 178–179
techniques for increasing, 175–182
will, effortful control of thoughts and images, 169–170
Ego strength, 94
Eidetic reduction, 71
Elliott, R., 3
Embodied cognition, 8
Emotional self-distancing, 151
Emotional state matching with difficult clients, 65–66
Empathic validation, 177–178
Empathy, 46, 51–52, 228, 233–234
as act of will, 231–232
conveying, 87
of difficult clients, strategies for increasing, 54. *See also* Difficult clients
as form of, 109
immunization, 124
increasing therapist capacity for, 59–60
inurement, 124
perceived, 186
prioritizing care over, 226–227
recognizing in others, 189–190
rehabilitating, 124
as remedy, 124–125
Empiricism, 24–26
Empowerment
oppressive clients, 252
strategies, 97
Empty chair technique, 27, 47
for addiction, 262–263
applications of, 262–263
for awareness, 120–121
with subpersonalities, 262
Enactivism, 8
Epictetus, 144
Epistemic injustices, 242
Equanimity, 234–235
Existential-cognitive therapy
for drug use, 264–265
stage 1, 250–251
stage 2, 252
stage 3, 253

F

Falsely compliant client, 174
False self, 62
Family therapy, 8, 68, 207–208
Farber, M., 94
Feeling good, as motivating force, 143–144
Feelings, identifying, 119–120
Finger technique, 175–176
First degree of awareness, 106
First-order change, 9
Focusing technique, 87, 119
FOM. *See* Freedom-from-oppression model
Frank, J., 90, 146
Frank, J. B., 26, 47, 55
Frank, J. D., 26, 47, 55, 57
Frankl, V. E., 90, 160, 280
Freedom, 161–162, 279–281
and CHANGES model, 285–286
freedom-for, 283, 285

freedom-from, 282, 284
freedom-to, 282, 284–285
freedom-with, 282–283, 285
modes of, 281–283, 284
respecting and encouraging, 62–63
as teleological approach to integration, 283–285
uncooperativeness as protection of, 63
Freedom-from-oppression model (FOM), 250
Freedom of choice, 176
Freud, S., 20, 107, 209
Freud, Sigmund, 3
Fromm-Reichmann, F., 210

G

Gendlin, E., 3, 6, 67, 87, 119
Genuineness, 46, 51, 53, 66
Gestalt therapy, 26, 28, 83, 87, 120, 262
Glass, G. V., 23
Glasser, W., 168
Goal, clarification of, 176–177
Goldfried, M. R., 22
Graduated tasks, assigning, 177
Graduate training, 43, 44, 46
Grand reframe of addictive behaviors, 255–259
Grit, or the willingness to experience anxiety or difficulty, 5, 18, 30, 278
absence of, 151–152
anxiety accompanying change, 146
clinical markers of, 150–151
control or freedom challenge, 161–162
deficit in, 215
defined, 145
internal dialogue avoiding anxiety, 155–157
leading to change, 147–149
maladaptive beliefs about, 164–165
metaphors, 152–155
ownership taking and blame stopping, 160–161
paradox, use of, 157–160
person owning problem, 147
philosophical roots of, 146–147
recognizing, 150–151
self-monitoring, 162–164
significant moments of change involving, 149
techniques for establishing, 152–165
Group therapy, 8, 68, 103, 125, 192–193, 207–208, 266
Guts metaphor, 154–155

H

Hadley, S. W., 43
Hanna, F. J., 4, 266
Harris, Sydney J., 5
Harvey, J., 241
Heidegger, M., 26, 72, 146–147, 250
Hermeneutical injustice, 242
Hesitance, 64–65
to confronting, 79
Hooking and unhooking approach, 227
Hope for change, 5, 18, 277
absence of, 93–94
clinical markers of, 92–93
clues to presence of, 92
as contagion, 100–101
converting threats into challenges, 98–100
creative narratives, 102–103
deficit in, 213
defined, 89
empowerment strategies, 97
hopelessness, reflecting, 97–98
and humor, 93
Jamesian Device, 96–97
leading to therapeutic change, 90–91
maladaptive beliefs, examining, 103–104
moments of change involving, 91–92
negative behaviors, reframing, 98
negative role models and, 95–96
recognizing, 92–93
relating and rewriting, 102–103
sense of worthiness and, 101–102
techniques for building, 94–104
telling stories of recovery, 103
Hopelessness, 91, 94, 97–98, 235
Horney, K., 216, 232
Hostility, with equanimity and humor, 234–235
House cleaning metaphor, 176
Human being, as active agent, 8–9
Human Change Processes (Mahoney), 37
Hume, D., 25, 169
Humor, 93, 234–235
Husserl, E., 26, 71–72, 83, 107

I

Idea of making changes, 79
Implicit learning, 41–42
Implicit oppression, 241–242
Individual therapy, 68
Inductive reasoning process, 58
Integrative existential-cognitive therapy. *See* Existential-cognitive therapy
Intelligence, 44, 46
Intentional unawareness, handling, 123–124
Intentions, mapping, 179–180
Interpersonal input, 125
Interpersonal savvy, role of, 7–8
In vivo confronting, 80
Marvin case, 80–81
Rusty case, 81–83

Involuntary client, 132
Involvement, 19

J

James, W., 9, 25, 26, 96–97, 135, 169, 280
Jamesian device technique, 96–97, 252
James–Lange theory of emotion, 26
Jung, C. G., 135, 237

K

Kaivalya, 281
Kant, I., 25, 197
Kiesler, D. J., 227
Kottler, J. A., 227

L

Lambert, M. J., 43
Lazarus, A. A., 15, 16, 46, 53, 68, 190, 216
Learned optimism, 89
Lewin, K., 74
Liberation, 253
  defined, 247
  oppressive clients, 253
  perspicacity, leveraging, 248–249
  and power of perception, 247–249
Life space, 74
Linehan, M. M., 53, 121
Loevinger, J., 119
Logical positivism, 24
Loneliness, 142–143
  Jeannie case, 143
Long-winded explanation, engaging in, 226
Low, A. A., 168

M

Mahoney, M. J., 37
Maladaptive beliefs, 41–43
  awareness, 125
  effort, 169, 181–182
  grit, 164–165
  hope, 103–104
  necessity, 144
  social support, 193–194
Mandela, N., 280
Maneuvering, 170
Matthew effect, 286
Meaning system, 41, 58, 62–63
  establishing, 60–61
  operating within, 61
Medawar, P., 72
Medical model, 284
Meichenbaum, D. B., 118
Merleau-Ponty, M., 72
Merton, R.K., 286
Metacognition
  within active agency, 9
  aspects of change, 29–31
  awareness, 106–107
  confronting as, 70
  defined, 29
  therapeutic change and, 6–7
Metalogue, 57–58
  attending to, 181, 232–233
Metaphors, 78–79
  awareness, 113–114
  effort, 175–176
  grit, 152–154
Metaphysics, 25
MI. *See* Motivational interviewing
Mind–body problem, 8–9
Mindfulness, 73, 87, 231, 269, 281
  craving, 269–270
Miracle question technique, 88
Mirroring technique, 85
Moksha, 281
Motivation, 19, 51
  feeling good as, 143–144
Motivational interviewing (MI), 32–33
Mountain overlook metaphor, 114
Mukti, 281
Multiculturalism, 239–240

N

Nardini, J. E., 90
Narratives, creative, 102–103
Necessity for change, 5, 18, 127, 277–278
  absence of, 130–131
  by aligning therapy, 133–135
  clinical markers of, 129–130
  cognitive dissonance, 128–129, 138–141
  deficit in, 214–215
  feeling good, as motivating force, 143–144
  involuntary client, 132
  leading to therapeutic change, 128–129
  loneliness, giving voice to, 142–143
  maladaptive beliefs about, 144
  nonverbal cues, 129–130
  rate of, 132
  recognizing, 129–130, 142
  secondary gains or cross-purposes, 141–142
  significant moments of change involving, 129
  subpersonalities, 135–137
  techniques for instilling, 131–144
  verbal communications, 129–130
Negative behaviors, reframing, 98, 115
Negative role models, impact of, 95–96
Negative self-belief, 251
Negative self-talk, 118, 172, 180–181, 232
Nirvana, 281

Nonattachment, cultivating, 232
"No pain, no gain" principle, 152–153
North, C., 59
Northup, S., 249
"Number one on the pyramid" technique, 61

O

Observer, establishing, 121–122
Observing self, 121
Old pipes and dirty water metaphor, 154
Onion peeler technique, 83–84
Opportunity cost models, 70
Oppression, 253
  Carlos case, 243–245
  civilized, 241
  defined, 240
  by deprivation, 240
  epistemic injustices, 242
  by force, coercion, or duress, 240
  implicit, 241–242
  integrative existential-cognitive therapy for, 250–253, 265–266
  nature of, 240–242
  systemic, 245–247
  trauma of, 242–247
Ownership taking, 160–161

P

Paradox, use of
  awareness, 122
  grit, 157–158
Passive–compliant client, 173–174
Patanjali's yoga, 72
Peer recognition, 95
Perceived potency scale, 4
Perception, power of, 247–249
Perls, F. S., 78
Permission asking, 54–55
Personality disorder, 29, 78, 123, 189
  antisocial, 61, 124, 176, 203
  borderline, 135
  defined, 208
  histrionic, 109
  narcissistic, 124
  schizoid, 188
Perspicacity, 67, 240, 253
  defined, 247
  leveraging, 248–249
  transcultural, 248
Persuasion, 55–56
Pharmacotherapy, 259–260
Phenomenology, 30, 71–73, 107, 287
Philosophical psychotherapy, 60
Popper, K., 24
Positive qualities, validating, 56–57
Positivism, logical, 24
Power struggle, 142, 228
Powers, W. T., 108
Pragmatism, 25
Precursors
  actual presence of, 200–202
  assessing, 198
  rating, 199–200
Precursors of change, 5–6, 17–19, 33. *See also specific precursors*
  defined, 5, 17
  in group and family therapies, 8
  increasing through relationship, 52–53
  interaction and interdependence of, 21–22
Problem-solving approach, 38
Prochaska, J. O., 19, 23, 31
Projective identification, 235
Psychoeducation, 56, 87
Psychological theories, 19, 24

Q

Quality of goodness, 56–57

R

Rank, O., 168
Rating, for necessity, 132
Reality testing, 73
Reciprocal inhibition, 71
Reframing awareness
  automatic thoughts, 116
  negative behaviors, 115
  as savoir faire, 115–116
  situation, 116
Refusal dialogue, 266
Refusal skills, 265–266
Rehabilitating empathy, 124
Relationship. *See* Therapeutic relationship
Representational confronting, 85–87
Resistance to change
  defined, 20
  difficulty and, 36–41
  therapist-initiated, 36–39
Resoluteness, 147
Resolute perception, 71, 80
Responsibility, 147, 161, 280
Retelling of life story, 103
Rhetoric, 55–56
Rigpa, 108
Ritchie, M. H., 4
Rogers, C. R., 8, 52, 72
Role-play, 27, 47, 59–60
  for awareness, 120
  drug use, 265–266
  social support, 192
  stages of, 265–266
"Rug-out-from-under" feeling, 64
Russell, B., 25

S

Samyama, 72
Sartre, J. P., 72
Schizophrenia, 59
Secondary gains, 141–142
Second degree of awareness, 106
Second-order change, 9
Self-awareness, 44, 60, 117, 215, 221, 231, 274
Self-compassion, 232
Self-control, 161–162
Self-defence, 230
Self-determination, 37, 240, 252
Self-disclosure, 58, 68, 143, 188, 214
Self-distancing, 107, 151
Self-forgiveness, 102
Self-monitoring, 27, 108, 121
  grit, 162–164
Self-observation, 108, 177
Self-protection, 36–41
Self-talk, 83, 119, 155, 177
  de-automated, 29
  negative, 118, 172, 180–181, 232
  verbal, audible, 265
Seligman, M. E. P., 89
Sense of worthiness, hope and, 101–102
Single principle imperialism, 21
Skinner, B. F., 280
Smith, M. L., 23
Social justice and advocacy models, 284, 285
Social support for change, 5, 18, 183, 278–279
  absence of, 187–189
  cautionary note, 189
  clinical markers for, 186–187
  concentric-circle technique, 190–191
  deficit in, 216
  empathy, recognizing, 189–190
  group therapy, 192–193
  leading to therapeutic change, 184–185
  lost sources, recovering, 192
  maladaptive beliefs about, 193–194
  opposing people, identifying, 191
  rating, 200
  recognizing, 186–187
  significant moments of change involving, 185–186
  sources of, 190
  standard social skills techniques, 192
  techniques for growing, 189–194
  trust, exploring, 192
Spitting in the client's soupis technique, 139, 141
Splitting, 112
Standard social skills techniques, 192
Stein, D. M., 43
Sternberg, R. J., 67, 247
Storytelling of recovery, 103
Strean, H. S., 26
Strupp, H. H., 43
Stuffed closet metaphor, 153
Subpersonalities, 178–179
  for addiction, 260–261
  addressing, 135
  empty chair with, 262
  Tracey case, 135–137
  using, 135
Substance use. *See* Addiction; Drug use
Suicide, 94
Supervision
  during clinical training, 225
  therapist interference. *See* Therapist interference
Systematic desensitization, 71
Systemic oppression, 245–247
Systems model, 284, 285

T

Talking cure, 19, 78, 80
Telling stories of recovery, 103
Testimonial injustice, 242
Therapeutic change
  awareness of problem leads to, 106–109
  confronting leading to, 70–75
  degree of change, 9–10
  first-order change, 9
  fundamental questions about, 4–5
  human being as active agent, 8–9
  as metacognitive skill, 6–7
  model of change for. *See* Difficult clients
  people opposing, 191
  precursors in group and family therapies, 8
  precursors of, 5–6
  processes, 31
  rate and magnitude of, 274, 275
  rating client potential for, 197–208
  rating therapist interference in, 209–222
  role of interpersonal savvy in facilitating, 7–8
  second-order change, 9
  stages of, 31
  techniques oriented around principles, 8
Therapeutic humor, 93
Therapeutic modalities, 68
Therapeutic relationship
  with difficult clients, 54–68
  increasing precursors through, 52–53
  orienting around change, 51–68
  purpose of, 53–54
Therapeutic window, 66
Therapist effectiveness, 42–43
Therapist-initiated resistance, 38
Therapist interference, 211
  accusing clients of defensiveness, 230

care, prioritizing, 226–227
challenging clients, 227
CHANGES assessment, 211–221
confrontation, avoiding, 229
confronting with impatience, 229
credentials or degrees, asserting, 228
disliking client, 230
long-winded explanation, engaging in, 226
losing focus in session, 227
markers of, 226–230
power struggles, 228
rating, 209–222
rigidity and flexibility, 229–230
self-defence, 230
wanting to be liked, 226
Therapy enhancing behavior, 53
Therapy veterans, 236
Third degree of awareness, 106
Threats, conversion to challenges, 98–100
Gwen case, 99–100
Time, efficient use of, 170–171
Training and supervision
therapist interference. *See* Therapist interference
Transference, 209
Transtheoretical model (TTM), 31–33
Trauma-focused somatic therapies, 41
Trauma of oppression, 242–243
Carlos case, 243–245
systemic oppression, anger related, 245–247
Trust, 186–187, 192
Truth theory, 96–97
TTM. *See* Transtheoretical model

U

Unaware person, example of, 114–115
Uncooperativeness, as protection of freedom, 63
Upanishads, 107
Urge surfing, 269–270

V

Vaccine effect, of confronting, 75
Value dissonance and realignment, 264
Vickery, P. J., 250
Visceral reactions, 233
Voodoo death, 90

W

Wachtel, P. L., 16, 23
Warmth, 46, 51, 64
Washing metaphor, 154
Weiss, P., 281
Wellness model, 284–285
"What's-in-it-for-me?" question, 63, 177
Whitaker, C., 8, 26, 59, 207, 266
Whitehead, A. N., 25
Will. *See also* Effort, or will to change
as effortful control of thoughts and images, 169–170
nature of, 168–169
Willingness. *See* Grit, or the willingness to experience anxiety or difficulty
Winnicott, D. W., 62
Wisdom, 46, 231
characteristics of, 45
confronting with compassion and empathy, 233–234
defined, 44
difficult clients, as teachers, 236–237
discouragement, 235
disgust, 235
empathy as act of will, 231
hostility with equanimity and humor, 234–235
metalogue, attending to, 232–233
nonattachment, cultivating, 232
perspicacity as, 67
projective identification, 235
self-compassion, 232
therapy veterans, asking, 236
tolerance and acceptance, 236
training out of therapist, 46–47
visceral reactions monitoring, 233
and wise therapist, 44–46
Wise mind, 121–122
Working through transference, 78, 148, 152
Workout machines metaphor, 153
Workout metaphor, 152–153

Y

Yalom, I. D., 103, 125, 183
Yoga psychology, 72

ABOUT THE AUTHORS

Brett D. Wilkinson, PhD, LMHC, is an associate professor in the Department of Counseling and Graduate Education at Purdue University Fort Wayne (PFW). He serves as editor-in-chief of the *Journal of Humanistic Counseling* and founding director of the PFW Institute for Counseling Research. His research examines therapeutic change mechanisms, experiential-humanistic interventions, and cognitive complexity development. He is also author of the textbook *Educational Psychology for Learners,* now in its third edition. He has 15 years of therapy experience working with adolescents, adults, couples, and families in community agencies and private practice and provides consulting services on the CHANGES model, reflective supervision practices, and experiential interventions.

Fred J. Hanna, PhD, is a professor and codesigner of the PhD program in Counseling at Adler University in Chicago. He is also a senior faculty associate at Johns Hopkins University, where he taught graduate counseling courses for 25 years, including 11 years full time, leaving as a full professor. Fred has authored or coauthored over 70 peer-reviewed and professional publications. An award-winning teacher, he has also delivered well over 500 presentations at conferences, seminars, trainings, and workshops across America. Fred was the recipient of the 2019 Humanistic Impact Award, a national award granted by the Association for Humanistic Counseling. He also received the Adler University Social Justice Award for 2020. Fred has served as a consultant and trainer to the medical, mental health, corrections, business, and education communities, including at such places as the Johns Hopkins School of Medicine Department of Psychiatry; the Fort Peck Sioux Reservation in Montana; the Department of Psychiatry at Yale University; and a wide variety of school systems, community agencies, prisons, and criminal justice settings from coast to coast. His research interests have focused on developing the CHANGES model as well as developing, publishing, and presenting many evidence-based, innovative psychotherapy techniques and strategies designed for application in the areas of client motivation, therapy resistance, addictions, diversity and multiculturalism, oppression, liberation, trauma, spirituality, criminality, defiant adolescents, personal development, and difficult personalities. He is also an accomplished world traveler, having explored many remote areas in Asia over a period of 2 years.